Solar and Astrophysical Magnetohydrodynamic Flows

NATO ASI Series

Advanced Science Institutes Series

A Series presenting the results of activities sponsored by the NATO Science Committee, which aims at the dissemination of advanced scientific and technological knowledge, with a view to strengthening links between scientific communities.

The Series is published by an international board of publishers in conjunction with the NATO Scientific Affairs Division

A **Life Sciences**	Plenum Publishing Corporation
B **Physics**	London and New York
C **Mathematical and Physical Sciences**	Kluwer Academic Publishers
D **Behavioural and Social Sciences**	Dordrecht, Boston and London
E **Applied Sciences**	
F **Computer and Systems Sciences**	Springer-Verlag
G **Ecological Sciences**	Berlin, Heidelberg, New York, London,
H **Cell Biology**	Paris and Tokyo
I **Global Environmental Change**	

PARTNERSHIP SUB-SERIES

1. **Disarmament Technologies**	Kluwer Academic Publishers
2. **Environment**	Springer-Verlag / Kluwer Academic Publishers
3. **High Technology**	Kluwer Academic Publishers
4. **Science and Technology Policy**	Kluwer Academic Publishers
5. **Computer Networking**	Kluwer Academic Publishers

The Partnership Sub-Series incorporates activities undertaken in collaboration with NATO's Cooperation Partners, the countries of the CIS and Central and Eastern Europe, in Priority Areas of concern to those countries.

NATO-PCO-DATA BASE

The electronic index to the NATO ASI Series provides full bibliographical references (with keywords and/or abstracts) to more than 50000 contributions from international scientists published in all sections of the NATO ASI Series.
Access to the NATO-PCO-DATA BASE is possible in two ways:

– via online FILE 128 (NATO-PCO-DATA BASE) hosted by ESRIN,
Via Galileo Galilei, I-00044 Frascati, Italy.

– via CD-ROM "NATO-PCO-DATA BASE" with user-friendly retrieval software in English, French and German (© WTV GmbH and DATAWARE Technologies Inc. 1989).

The CD-ROM can be ordered through any member of the Board of Publishers or through NATO-PCO, Overijse, Belgium.

Series C: Mathematical and Physical Sciences – Vol. 481

Solar and Astrophysical Magnetohydrodynamic Flows

edited by

Kanaris C. Tsinganos

**Department of Physics,
University of Crete,
Heraklion, Greece**

Kluwer Academic Publishers

Dordrecht / Boston / London

Published in cooperation with NATO Scientific Affairs Division

Proceedings of the NATO Advanced Study Institute on
Solar and Astrophysical Magnetohydrodynamic Flows
Heraklion, Crete, Greece
June 11–23, 1995

A C.I.P. Catalogue record for this book is available from the Library of Congress.

ISBN-13:978-94-010-6603-7   e-ISBN-13: 978-94-009-0265-7
DOI: 10.1007/978-94-009-0265-7

Published by Kluwer Academic Publishers,
P.O. Box 17, 3300 AA Dordrecht, The Netherlands.

Kluwer Academic Publishers incorporates the publishing programmes of
D. Reidel, Martinus Nijhoff, Dr W. Junk and MTP Press.

Sold and distributed in the U.S.A. and Canada
by Kluwer Academic Publishers,
101 Philip Drive, Norwell, MA 02061, U.S.A.

In all other countries, sold and distributed
by Kluwer Academic Publishers Group,
P.O. Box 322, 3300 AH Dordrecht, The Netherlands.

Printed on acid-free paper

Table of Contents

PART I:

SOLAR MAGNETOHYDRODYNAMIC FLOWS

1. Questions and Conjectures on the Origin of Stellar and Galactic Magnetic Fields

2. Magnetic Flux Tubes and the Solar Dynamo

6. The Spontaneous Formation of Current-Sheets in Astrophysical Magnetic Fields

7. Magnetohydrodynamic Processes in the Solar Corona : Flares, Coronal Mass Ejections and Magnetic Helicity

8. Reconnection of Magnetic Lines of Force

9. New Developments in Magnetic Reconnection Theory

10. Hydrodynamic and Magnetohydrodynamic Turbulence

11. Numerical Simulations of Solar and Astrophysical MHD Flows

12. Further Thoughts on the Solar Corona as a Minimum Energy System

13. ULYSSES Observations of the Solar Wind Out of the Ecliptic Plane

W.C. Feldman, J.L. Phillips, B.L. Barraclough and C.M. Hammond 265

14. Stellar Winds

Keith B. MacGregor 301

15. Some Implications of Solar and Stellar Activity

PART II:

ASTROPHYSICAL MAGNETOHYDRODYNAMIC FLOWS

18. Numerical Simulation of Plasma Flow in the Magnetosphere of an Axisymmetric Rotator

Sergei V. Bogovalov 411

19. Critical Points and Separatrix Characteristics in Solar and Astrophysical MHD Flows

Kanaris Tsinganos, Christophe Sauty, George Surlantzis, Edoardo Trussoni and John Contopoulos 427

20. Collimation of MHD Outflows

21. The Parker Instability in the Interstellar Medium

22. Multifluid Magnetohydrodynamics and Star Formation

23. Jets Associated with Young Stars

24. Molecular Outflows from Protostars

25. Jets and MHD Flows Associated with Symbiotic Stars

Menas Kafatos 585

26. Phenomenology and Modelling of Large-scale Jets

Attilio Ferrari, Silvano Massaglia, Gianluigi Bodo and Paola Rossi 607

27. MHD Accretion-ejection Flows in Active Galactic Nuclei

Guy Pelletier, J. Ferreira, G. Henri and A. Marcowith 643

28. Interaction of Turbulent Accretion Disks with Embedded Magnetic Fields

29. Magnetic Reconnection in Accretion Disc Coronae

30. Relativistic Outflows in the Galaxy

31. Stationary Relativistic MHD Flows

Prof. Eugene Parker

PREFACE

At the dawn of the Space Age in the late fifties, a conceptual revolution in astronomy was the prediction by Eugene Parker that an extremely hot solar corona cannot be confined by solar gravity, but expands supersonically into interplanetary space to fill the whole solar system. By the end of our century, modern observations including those recently with the Hubble Space Telescope, have revealed that the Universe is replete with analogous outflows from all kinds of objects, ranging from the rich varieties of stars to the galaxies. Parker initiated the use of the magnetohydrodynamic (MHD) approach to describe such cosmical phenomena. Since then, MHD is increasingly used in several key problems of solar plasmas, such as, magnetoconvection and magnetic field generation, sunspots and prominences, MHD equilibria in plasma loops and arcades, magnetic nonequilibrium and coronal heating, coronal mass ejections and the acceleration of the solar wind. On the other hand, the same MHD approach is also increasingly used in several central and puzzling aspects of more distant astrophysical plasmas, such as, stellar winds across the main sequence, the dynamics of the interstellar medium and the galactic center, collimated outflows from young stellar objects and accretion disks, molecular outflows and planetary nebulae, jets associated with enigmatic binaries and symbiotic stars, relativistic flows associated with superluminal microquasars in our own galaxy, astrophysical jets from nearby galaxies, or far away active galactic nuclei and quasars probably fueled by supermassive black holes. However, although the underlying similarities in the physics of these phenomena are rather striking, nevertheless quite often researchers in one field are unaware of the methods employed in the study of the other.

The present book has been developed from review lectures given by leading astrophysicists at the NATO Advanced Study Institute (ASI) on *Solar and Astrophysical Magnetohydrodynamic Flows* which was held June 11 – 23, 1995 in Heraklion, Crete, Greece. With Prof. Parker having been the inspiration for many to study the role of magnetism in the Cosmos, it was no surprise that a large number of experts in MHD and participants from 25 countries gathered to honour him on the occassion of his retirement from the University of Chicago, after almost 50 years in solar physics and astrophysics. What attracted them was also the ambitious scientific program naturally dedicated to Eugene Parker's interests. Thus, appropriately, special reviews were presented on topics Parker has immensely contributed, such as, new developments in the Parker instability in the interstellar medium, the spontaneous formation of electric current sheets in magnetized plasmas, magnetic reconnection, sunspot modelling, coronal heating, stellar winds and also a generalisation of its model of the solar wind via exact self-similar solutions of the MHD equations. Should we also guess that an extra reason that attracted so many participants was perhaps the tempting prospect to spend two weeks along the north beach of the island of Crete, inspired to discuss astrophysics by the spectacular view of the azur blue Greek Aegean sea, a few km from the historic Minoan palaces, near the birth place of the Cretan painter D. Theotokopoulos (El Greco) at the Fodele Beach Hotel and Conference Center ?

The book contains 31 chapters on solar (Part I) and astrophysical (Part II) MHD flows with emphasis on their common hydromagnetic physics. Part I starts with a discussion of the generation, storage and eruption of magnetic fields in our Sun deep in its convective layers, as prototypes of the fields of other stars, galaxies and accretion disks. The erupted through the photosphere magnetic field forms intense flux tubes and sunspots and thus siphon flows and Alfvén waves along such magnetic flux tubes are next addressed. The impressive high resolution pictures of the Sun in X-rays by the *Yokhoh* satellite are reviewed and indeed show a solar atmosphere constituted of many steady and transient magnetic loops like those expected to emerge from deep within the solar convective zone. The X-ray data also show clear evidence of intense heating due to magnetic reconnection. Thus, the spontaneous formation of current sheets in astrophysical magnetic fields and the various MHD processes in the solar corona (flares, coronal mass ejections, etc) are treated next. Thorough overviews on reconnection of the magnetic lines of force and HD/MHD turbulence are also given. These ideas are next tested by numerical simulations of solar and astrophysical MHD flows (magnetic loops, flares, solar X-ray jets as well as the nonlinear evolution of the Parker instability and magnetically accelerated astrophysical jets), together with some thoughts on the global shape of the solar corona as a minimum energy system. Very recent data from the south and north solar polar pass of the *Ulysses* spacecraft are showing, for example, that the solar wind is indeed a rather meridionally anisotropic outflow with the proton speed at large heliolatitudes almost twice its equatorial value. Stellar winds from solar- as well as non solar-type stars are reviewed and the first part ends with a discussion of some general implications of solar and stellar activity, such as the profound effects of the variation of solar brightness to the terrestrial climate.

Part II on astrophysical MHD flows starts with an amazing updated review of the very recent Hubble Space Telescope observations of galactic and extragalactic jets showing explicitly for the first time their association with accretion disks, as well as their intricate structure and properties. This review, together with all the experience gained by the previously discussed solar MHD flows, sets the stage and provides the challenge for a theoretical investigation of the ubiquitous astrophysical MHD flows. Self-similar analytical models of MHD flows with streamline collimation towards the rotation and magnetic axes are reviewed, together with some basic novel physical properties of the MHD equations and theorems on the general tendency of rotating magnetized outflows to focus around the system axis. A comprehensive overview is given of the Parker instability and subsequent stellar formation in the frame of multifluid MHD. Next, we return to a detailed observational review of jets associated with young stars, some questions posed by the related molecular outflows and an exposition of the detailed properties of symbiotic stars and their jets. An extensive review of the morphology and modelling of large-scale jets and accretion-ejection structures from active galactic nuclei is given in the following before we focus on physical properties in accretion disks with embedded magnetic fields and their coronae. Finally, observed superluminal outflows in the Milky Way and theoretical studies of relativistic MHD outflows complete the second part of the book on astrophysical MHD flows.

We are grateful to the NATO Scientific Affairs Division for its generous funding of this Advanced Study Institute, the University of Crete and Founation for Research and Technology Hellas for partial support, the MITOS company for helping us with the organisation of the ASI, the staff of the Fodele Beach and Conference Center for their philoxenia, and Georgia Vlastou for creating a detailed index to this book.

Of course we owe particular thanks to the authors of this volume for producing really excellent tutorial reviews which communicate the essential flavor and the latest ideas in one of the most interesting areas of modern astrophysical research.

Kanaris Tsinganos
University of Crete

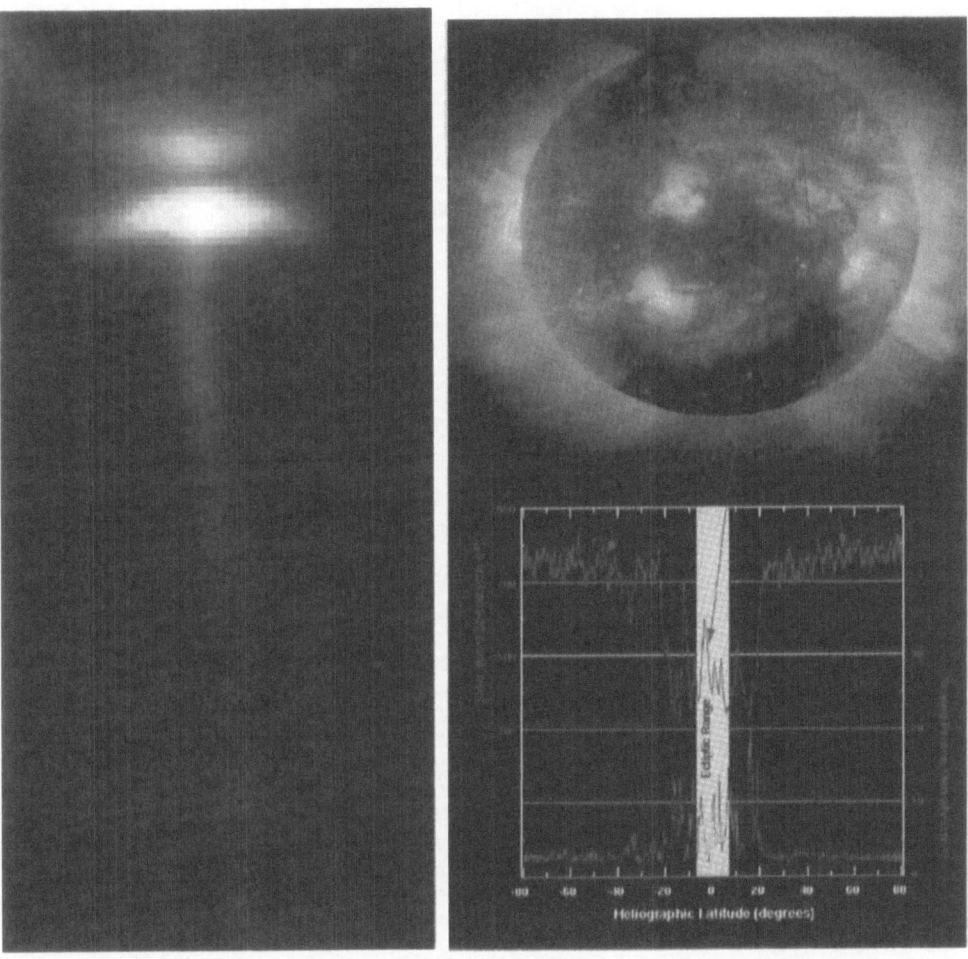

Figure 1. *Right:* Identification of the coronal origins of the two different flow states of the solar wind observed by *Ulysses*. The X-ray image of the corona was kindly provided by the Yohkoh mission of ISAS, Japan (see chapters 5, 13). *Left:* The HH 30 stellar jet and its associated accretion disk based on data by Burrows et al 1996 (see chapters 16, 23, 31).

PARTICIPANTS

AGAPITOU, Vassiliki
Queen Mary and Westfield College
Astronomy Unit
Mile End Road, London E1 4NS
V.Agapitou@qmw.ac.uk
U.K.

ANAGNOSTOPOULOS, Georgios
Democritos University of Thrace
67100 Xanthi
anagnosto@xanthi.cc.duth.gr
GREECE

ANASTASIADIS, Anastasios
Department of Physics
Aristotle University of Thessaloniki
54006 Thessalonki,
anastasi@helios.astro.auth.gr
GREECE

ANDREEV, Alexander
Crimean Astrophysical Observatory
334413 Nauchny, Crimea
andre@crao.crimea.ua
UKRAINE

APPL, Stefan
11, rue de l'Universite
Universite Louis Pasteur
F-67000 Strasbourg
appl@cdsxb9.u-strasbg.fr
FRANCE

BASU, Shantanu
Physics-Astronomy Department
Michigan State University
East Lansing, MI 48824
basu@msupa.pa.msu.edu
USA

BERGHMANS, David
Centre for Plasma Astrophysics
Celestijnenlaan 200 B
B-3001 Heverlee (Leuven)
David.Berghmans@wis.kuleuven.ac.be
BELGIUM

BERGER, Mitchell
Mathematics
University College London
Gower Street, London WC1E 6BT
mberger@math.ucl.ac.uk
U.K.

BESKROVNAYA, Nina
Pulkovo Observatory
196140 Pulkovo, St.Petersburg
conf@pulkovo.spb.su
RUSSIA

BIRETTA, John
Space Telescope Science Institute
3700 San Martin Dr
Baltimore MD 21218
biretta@stsci.edu
USA

BOGOVALOV, Sergei
Moscow Engineering Physics Institute
Kashirskoje Shosse 31
Moscow, 115409
bogoval@mx.iki.rssi.ru
RUSSIA

CALIGARI, Peter
Kiepenheuer Institute for Solar
Physics, Schoeneckstrasse 6
D-79104 Freiburg
cale@kis.uni-freiburg.de
GERMANY

CAMENZIND, Max
Landessternwarte Konigstuhl
69117, Heidelberg
mcamenzi@hp2.lsw.uni-heidelberg.de
GERMANY

CARTLEDGE, Nicholas
The North Haugh
St Andrews KY16 9SS, Scotland
nick@ac.uk.st-andrews.dcs
UK

CATTANEO, Fausto
The University of Chicago
Enrico Fermi Institute
Laboratory for Astrophysics
and Space Research
933 E 56th Str, Chicago, Il 60637
cattaneo@mhd2.uchicago.edu
USA

CHARBONNEAU, Paul
High Altitude Observatory
National Center for Atmospheric Research
P.O. Box 3000, Boulder, Co 80307–3000

paulchar@dante.hao.ucar.edu
USA

CONTOPOULOS, John
NASA/Goddard Space Flight Center
Postal Address: Code 665
Greenbelt, MD 20771
conto@rosserv.gsfc.nasa.gov
USA

DAVIDSON, Jacqueline
NASA / Ames Research Center
MS 245-6
Moffett Field, CA 94035-1000
davidson@cma.arc.nasa.gov
USA

De BRUYNE, Peter
Centre for Plasma Astrophysics
Celestijnenlaan 200 B, B-3001 Heverlee
Peter.DeBruyne@wis.kuleuven.ac.be
BELGIUM

DEL ZANNA, Luca
Mathematical Sciences Department
The University
St. Andrews, KY16 9SS, Scotland
luca@dcs.st-andrews.ac.uk
UK

DE PONTIEU, Bart
MPIf Extraterrestrische Physik (MPE)
Giessenbachstrasse
D-85740 Garching-bei-Muenchen
bdp@mpe.mpe-garching.mpg.de
GERMANY

DORCH, S. Bertil F.
Copenhagen University Observatorium

Oester Voldgade 3
DK-1350 Copenhagen K
dorch@astro.ku.dk
DENMARK

DOROTOVIC, Ivan
Sloval Central Observatory
P.O. Box 42, SK-94701 Hurbanovo
rybansky@auriga.ta3.sk
THE SLOVAK REPUBLIC

DOWNES, Turlough
School of Cosmic Physics,
Dublin Institute for Advanced Studies
5 Merrion Square, Dublin 2,
tpd@cp.dias.ie
IRELAND

DUMITRACHE, Cristiana
Rumanian Academy
Astronomical Institute
Str. Cutitul de Argint 5
75212 Bucharest
crisd@imar.ro
ROMANIA

ELSTNER, Detlef
Astrophysikalisches Institut Potsdam
An der Sternwarte 16
D-14482 Potsdam-Babelsberg
E-mail: delstner@aip.de
GERMANY

EMONET, Thierry
Istituto de Astrofisica de Canarias
Calle Via Lactea
32200 La Laguna, Tenerife
temonet@ll.iac.es
SPAIN

ERDELYI, Robert
Katholieke Universiteit Leuven,
Center of Plasma-Astrophysics
Celestijnenlaan 200 B
B-3001 Leuven-Heverlee
rerdelyi@solar.stanford.edu,
mwekb05@cc2.kuleuven.ac.be
BELGIUM

FENDT, Christian
Landessternwarte, Konigstuhl
D-69117 Heidelberg
GERMANY, and
Lund Observatory
Box 43, S-22100 Lund
chris@astro.lu.se
SWEDEN

FELDMAN, William C.
Ms D466, Los Alamos National Laboratory
Los Alamos, NM 87545
wfeldman@sst.lanl.gov
USA

FERRARI, Attilio
Osservatorio Astronomico di Torino
Strada dell' Osservatorio 20
I-10025, Pino Torinese
ferrari@astrto.ph.unito.it
ITALY

FERREIRA, Joao
The University of St Andrews
The North Haugh
St Andrews, Fife KY16 9SS, Scotland
jmtdsf@st-andrews.ac.uk
UK

FERREIRA, Jonathan
Landessternwarte, Konigstuhl
D-69117 Heidelberg
jferreir@hp2.lsw.uni-heidelberg.de
GERMANY

FERRIZ-MAS, Antonio
Instituto de Astrofisica de Canarias
Calle Via Lactea
38200, La Laguna, Tenerife,
afm@ll.iac.es
SPAIN

FLEISCHER, Juergen
Theoretische Physik IV
Ruhr-Universitaet
D-44780 Bochum
jf@tp4.ruhr-uni-bochum.de
GERMANY

FLETCHER, Lyndsay
Sterrekundig Instituut
Postbus 80.000, 3508 TA Utrecht
L.Fletcher@fys.ruu.nl
THE NETHERLANDS

FOGLIZZO, Thierry
MPI fur Astrophysik
Karl-Schwarzschild-Str. 1
Postfach 15 23
85740 Garching-bei-Munchen
foglizzo@mpa-garching.mpg.de
GERMANY

FRAIX-BURNET, Didier
Laboratoire d' Astrophysique
de l' Observatoire de Grenoble
BP 53, 414 rue de la piscine,
F-38041, Grenoble Cedex 9,
fraix@gag.obsrv-gr.fr
FRANCE

GALSGAARD, Klaus
The University of St Andrews
The North Haugh
St Andrews KY16 9SS, Scotland
klaus@dcs.st-and.ac.uk
UK

GEORGOULIS, Manolis
Department of Physics
Aristotle University of Thessaloniki
54006 Thessalonki
georgoul@astro.auth.gr
GREECE

GOEDBLOED, Hans
FOM-Institute for Plasma Physics
P.O. Box 1207, 3430 BE Nieuwegein
goedbloed@sara.nl
THE NETHERLANDS

GONTICAKIS, Constantin
Institut d' Astrophysique Spatiale (IAS)
Batiment 121, Universite Paris XI
91405 Orsay Cedex
costis@iaslab.ias.fr
FRANCE

GOOSSENS, Marcel
Katholieke Universiteit Leuven
Department of Mathematics
Celestijnenlaan 200 B
B-3001 Heverlee (Leuven)
FGAHA01@cc1.kuleuven.ac.be
BELGIUM

GOUVEIA DAL PINO, Elisabete

University of Sao Paulo,
Instituto Astronomico e Geofisico
Miguel Stefano, 4200,
Sao Paulo, SP, 04301-904
dalpino@astro1.iagusp.usp.br
BRAZIL

GVARAMADZE, Vasilii
Abastumani Astrophysical Observatory
Georgian Academy of Sciences
A. Kazbegi Ave., 2-A, Tbilisi, 380060
dzogin@esoc1.bitnet
GEORGIA

HADJIDIMITRIOU, Despina
University of Crete, Dept. of Physics
71409 Heraklion, Crete,
dh@iesl.forth.gr
GREECE

HAERENDEL, Gerhard
MPIf Extraterrestrische Physik (MPE)
Giessenbachstrasse
D-85740 Garching-bei-Muenchen
hae@mpe.mpe-garching.mpg.de
GERMANY

HANASZ, Michal
Inst. of Astronomy
Nicolaus Copernicus University
ul. Chopina 12/18, PL-87-100 Torun
mhanasz@astri.uni.torun.pl
POLAND

HENRIKSEN, Dick
Department of Physics
Queens University at Kingston

Ontario
henriksn@lola.phy.queensu.ca
CANADA

HEYVAERTS, Jean
Observatoire de Strasbourg,
11 rue de l' Universite
F-67000, Strasbourg,
heyvaert@cdsxb6.u-strasbg.fr
FRANCE

HORNIG, Gunnar
Theoretische Physik IV
Ruhr-Universitaet, D-44780 Bochum
gh@tp4.ruhr-uni-bochum.de
GERMANY

IIJIMA, Takashi
Osservatorio Astrofisico
I–36012 Asiago (Vi)
IIJIMA%ASBTRAS@ipdunivx.unipd.it
ITALY

INNES, Davina
Max Planck Institut fuer Aeronomie
Postfach 20, 37189 Katlenburg-Lindau
innes@linax1.mpae.gwdg.de
GERMANY

JAHN, Krzystof
Astronomical Observatory
Warsaw University
Al. Ujazdowskie 4, 00-478 Warszawa
crj@sirius.astrouw.edu.pl
POLAND

JARDINE, Moira
University of St. Andrews
Dept. of Physics and Astronomy

Fife, KY16 9SS, Scotland
mmj@st-and.ac.uk
UK

KAKOURIS, Alexandros
Physics Dept., University of Athens
Panepistimiopolis, GR 157 83 Zografos
akakour@atlas.uoa.ariadne-t.gr
GREECE

KASSOTAKIS, Gregorios
National Observatory of Athens
Institute of Ionospheric and Space
Research, Athens
greg@thission.iisr.noa.ariadne-t.gr
GREECE

KAFATOS, Menas
George Mason University
Space Science Program
CSI, Science & Technology I #116
Fairfax, VA, 22030
mkafatos@compton.gmu.edu
USA

KHANNA, Ramon
Landessternwarte
K"onigstuhl, 69117 Heidelberg
rkhanna@mail.lsw.uni-heidelberg.de
GERMANY

KYLAFIS, Nikolaos
University of Crete
Department of Physics
71409 Heraklion, Crete,
kylafis@iesl.forth.gr
GREECE

LAWRENCE, John K.

Department of Physics and Astronomy
California State University, Northridge
18111 Nordhoff Street
Northridge, CA 91330-8268
jkl@pinet.aip.org
USA

LERY, Thibaut
Observatoire Astronomique de Strasbourg
11,rue de l'Universite
67000 Strasbourg
LERY@astro.u-strasbg.fr
FRANCE

LESCH, Harald
Max-Planck-Institut for Radioastronomy
Auf dem Huegel 69, 53121 Bonn
V398cah@mpifr-bonn.mpg.de
GERMANY

LIFSCHITZ, Alexander
University of Illinois at Chicago
Department of Mathematics
851 S. Morgan Str., Chicago IL 60607
lifschitz@math.uic.edu
USA

LIMA, Joao
Centro de Astrofísica
Universidade do Porto
Rua do Campo Alegre, 823
4150 Porto
jalima@ncc.up.pt
PORTUGAL

LORENTE, Rosario
Universidad Complutense
Avda Complutense, 28040 Madrid
lor@ucmast.fis.ucm.es

SPAIN

LOW, Boon Chye
High Altitude Observatory
National Center for Atmospheric Research
P.O. Box 3000, Boulder, Co 80307–3000
low@solo.hao.ucar.edu
USA

MACGREGOR, Keith
High Altitude Observatory
National Center for Atmospheric Research
P.O. Box 3000, Boulder, Co 80307–3000
mac@hao.ucar.edu
USA

MADJARSKA, Maria Sotirova
Institute of Astronomy
72 Trakia Blvd., 1784 Sofia
madjarsk@bgearn.bitnet
BULGARIA

MANOLAKOU, Constantina
Laboratory of Astronomy
Department of Physics
Aristotle University of Thessaloniki
540 06 Thessaloniki
student4@astro.auth.gr
GREECE

MASSAGLIA, Silvano
Osservatorio Astronomico di Torino
Strada dell'Osservatorio, 20
10025 Pino Torinese (TO)
massaglia@to.astro.it
ITALY

MATHIOUDAKIS, Mihalis
2150 Kittredge Str.

Center for EUV Astrophysics
University of California at Berkeley
Berkeley, CA 94704
mihalis@cea.berkeley.edu
USA

MENDOZA, Cesar
University of St. Andrews
Dept. Maths and Compt. Sciences
St. Andrews KY16 9SS, Fife, Scotland
cesar@dcs.st-andrews.ac.uk
UK

MIRABEL, F.
Service d' Astrophysique
Centre de Etudes de Saclay
91191 Gif–sur–Yvette
mirabel@ariane.saclay.cea.fr
FRANCE

MONTAGNE, Marc
Observatoire Midi-Pyrenees
14, Av. E. Belin, 31400 Toulouse
montagne@obs-mip.fr
FRANCE

MORENO-INSERTIS, Fernando
Istituto de Astrofisica de Canarias
Calle Via Lactea
32200 La Laguna, Tenerife
fmi@kis.uni-freiburg.de
SPAIN

MOUSCHOVIAS, Telemachos Ch.
University of Illinois at
Urbana-Champaign
Departments of Physics and Astronomy
1002 West Green Street, Urbana, IL 61801
tchm@astro.uiuc.edu

USA

MOUSSAS, Xenophon
National University of Athens
Panepistimiopolis
Gr 15783 Zographos, Athens
xmoussas@atlas.uoa.ariadne-t.gr
GREECE

NINOS, George
University of Athens, Astronomy & Astrophysics,
Space Physics Group
Anaxandrou 3, 116 31 Athens
gninos@atlas.uoa.ariadne-t.gr
GREECE

OCANA, Gerardo
School of Mathematical Sciences
Queen Mary and Westfield College
Mile End Road, London E1 4NS
gof@qmw.ac.uk
UK

ODSTRCIL, Dusan
Solar Department
Astronomical Institute
25165 Ondrejov
odstrcil@asu.cas.cz
CZECH REPUBLIC

PAATZ, Gernot
Landessternwarte Koenigstuhl
D-69117, Heidelberg
gpaatz@hp2.lsw.uni-heidelberg.de
GERMANY

PALEOLOGOU, Efthimios
University of Crete
Department of Physics
71409 Heraklion, Crete,
papamast@iesl.forth.gr
GREECE

PAPAMASTORAKIS, John
University of Crete
Department of Physics
71409 Heraklion, Crete,
papamast@iesl.forth.gr
GREECE

PARKER, Eugene
The University of Chicago
Enrico Fermi Institute
Laboratory for Astrophysics
and Space Research
933 E 56th Str, Chicago, Il 60637
parker@odysseus.uchicago.edu
USA

PAVLOS, Georgios
Democritos University of Thrace
67100 Xanthi
pavlos@xanthi.cc.duth.gr
GREECE

PELLETIER, Guy
Laboratoire d' Astrophysique
de l' Observatoire de Grenoble
BP 53, 414 rue de la Piscine
F-38041, Grenoble Cedex 9
pelletie@gag.observ-gr.fr
FRANCE

PETER, Hardi
Max-Planck Institut fuer Aeronomie

P.O. Box 20
D-37189 Katlennurg-Lindau
hpeter@osf1.mpae.gwdr.de
GERMANY

PISANKO, Yuri V.
Institute of Applied Geophysics,
Rostokinskaya St.9
Moscow 129128
geophys@sovamsu.sovusa.com
RUSSIA

PISHKALO, Nikolai
Astronomical Observatory of Kiev University
Observatorna vul., 3
Kiev-53, 254053
aoku@gluk.apc.org
UKRAINE

POUQUET, Annick
Observatoire de la Cote d' Azur
Avenue Copernic F-06130, Nice Cedex
pouquet@obs-nice.fr
FRANCE

PRIEST, Eric
St Andrews University
Department of Mathematics
The University, St Andrews KY16 9SS
Scotland
eric@dcs.st-andrews.ac.uk
UK

PSALTIS, Demetrios
University of Illinois at
Urbana-Champaign
103 Astronomy Bldg.
1002 W. Green St., Urbana, IL 61801

demetris@astro.uiuc.edu
USA

RAFIOS, Xenophon
University of Athens
Nuclear and Particle Physics
Solonos 104, GR 106 80 Athens
GREECE

RASTAETTER, Lutz
Theoretische Physik IV
Ruhr-Universitaet, D-44780 Bochum
lr@tp4.ruhr-uni-bochum.de
GERMANY

RAY, Tom
Dublin Institute for Advanced Studies
School of Cosmic Physics
5 merrion Square, Dublin 2
tr@cp.dias.ie
IRELAND

REKOWSKI, Matthias
Astrophysikalisches Institut Potsdam
An der Sternwarte 16
mrekowski@aip.de
GERMANY

REZZOLLA, Luciano
SISSA, Via Beirut 2-4
34013 Trieste
rezzola@tsmi19.sissa.it
ITALY

ROSNER, Robert
The University of Chicago
Enrico Fermi Institute
Laboratory for Astrophysics
and Space Research

933 E 56th Str, Chicago, Il 60637
rrosner@mhd5.uchicago.edu
USA

RYUTOVA, Margarita
Lawrence Livermore National Laboratory
Institute of Geophysics and
Planetary Physics
L-413, Livermore, CA 94550,
ryutova@igpp.llnl.gov
USA

SARRIS, Emmanuel
National Observatory of Athens and
Democritos University of Thrace
67100 Xanthi
sarris@thission.iisr.noa.ariadne-t.gr
GREECE

SAUTY, Christophe
Observatoire de Meudon
Section d' Astrophysique
DAEC, 92195 Meudon
FRANCE

SCHILHAM, Arnold
Sterrekundig Instituut
Princetonplein 5, Postbus 80.000
NL-3508 TA Utrecht
schilham@fys.ruu.nl
THE NETHERLANDS

SCHLICHENMAIER, Rolf
MPIf Extraterrestrische Physik
Giesenbachstr. 1
D-85740 Garching bei M"unchen
ROS%MPEBV2@mpe.mpe-
garching.mpg.de
GERMANY

SCHMIDT, Hermann Ulrich
Max-Planck-Institute for Astrophysics
Karl Schwarzchild Str. 1
Postfach 15 23, 85740 Garching,
petra@MPA-Garching.MPG.DE
GERMANY

SCHUSSLER, Manfred
Kiepenheuer Institut fur Sonnenphysik
Schoneckstrasse 6, D-779104
Freiburg im Breisgau
msch@kis.uni-freiburg.de
GERMANY

SHIBATA, Kazunari
National Astronomical Observatory
Mitaka, Tokyo 181
shibata@spot.mtk.nao.ac.jp
JAPAN

SURLANTZIS, George
University of Crete
Department of Physics
71409 Heraklion, Crete,
sourl@iesl.forth.gr
GREECE

THOMAS, John
Department of Mechanical Engineering
University of Rochester
233 Hopeman Engineering Building
Rochester, N.Y., 14627-0132
jhth@db1.cc.rochester.edu
USA

TIRRY, Wouter
Centre for Plasma Astrophysics
Celestijnenlaan 200 B

B-3001 Heverlee (Leuven)
Wouter@cpa.wis.kuleuven.ac.be
BELGIUM

TORKELSSON, Ulf
Sterrekundig Instituut
Postbus 80000
3508 TA Utrecht
torkel@fys.ruu.nl
THE NETHERLANDS

TOTH, Gabor
Sterrekundig Instituut
Postbus 80000, 3508 TA Utrecht
toth@fys.ruu.nl
THE NETHERLANDS

TRUSSONI, Edo
Osservatorio Astronomico di Torino
Strada dell' Osservatorio 20, I-10025,
Pino Torinese
trussoni@astrto.ph.unito.it
ITALY

TSAGOURI, Ioanna
University of Athens
Nuclear and Particle Physics
Solonos 104, GR 106 80 Athens
GREECE

TSINGANOS, Kanaris
University of Crete
Department of Physics
71409 Heraklion, Crete
tsingan@iesl.forth.gr
GREECE

TSUNETA, Saku
University of Tokyo

Institute of Astronomy
Osawa, Mitaka, Tokyo 181
tsuneta@sxt2.ioa.s.u-tokyo.ac.jp
JAPAN

ULTCHIN, Yigal
Technion-Israel Institute of Technology
Haifa 32000
yigal@phastro.technion.ac.il
ISRAEL

VLAHAKIS, Nektarios
University of Crete
Department of Physics
71409 Heraklion, Crete,
vlahakis@iesl.forth.gr
GREECE

VARDAVAS, Ilias
University of Crete
Department of Physics
71409 Heraklion, Crete,
vardavas@iesl.forth.gr
GREECE

VENTURA, Joseph
University of Crete
Department of Physics
71409 Heraklion, Crete,
ventura@iesl.forth.gr
GREECE

WALSH, Robert
University of St. Andrews
Dept. Maths and Compt. Sciences
St. Andrews KY16 9SS, Fife, Scotland
robert@dcs.st-andrews.ac.uk
UK

xl

XILOURIS, Manolis
University of Crete
Department of Physics
71409 Heraklion, Crete,
vlahakis@iesl.forth.gr
GREECE

YOKOYAMA, Takaaki
National Astronomical
Observatory of Japan
Mitaka, Tokyo 181
yokoyama@spot.mtk.nao.ac.jp
JAPAN

ZIMMER, Frank
Radioastronomisches Institut
der Universität Bonn
Max-Planck Institut fuer Aeronomie
Auf dem Hügel 71, D-53121, Bonn
fzimmer@astro.uni-bonn.de
GERMANY

PART I:
SOLAR MAGNETOHYDRODYNAMIC FLOWS

QUESTIONS AND CONJECTURES ON THE ORIGIN OF STELLAR AND GALACTIC MAGNETIC FIELDS

EUGENE PARKER

The University of Chicago,
Enrico Fermi Institute and Depts. of Physics and Astronomy
933 East 56th Street, Chicago, Illinois 60637 USA
(parker@odysseus.uchicago.edu)

Abstract. The magnetic fields of the Sun and the Galaxy, as prototypes of the fields of other stars and galaxies, appear to originate in some form or variant of the well known $\alpha\omega$-dynamo. However, the $\alpha\omega$-dynamo can function only in the presence of irreversible diffusion and dissipation of the small-scale internal magnetic fields, requiring effective diffusion coefficients of the order of 10^{12} and 10^{25} cm^2/sec for the Sun and Galaxy, respectively. The internal motions of scale ℓ and velocity v provide the product ℓv with these magnitudes, but the effect is not one of irreversible diffusion. So the $\alpha\omega$-dynamo does not work in the manner conventionally associated with turbulent diffusion. It is hypothesized instead that the magnetic field of the Sun is in an intensely fibril state throughout the convective zone, with rapid reconnection wherever nonparallel fibrils are pressed together by the fluid motions. The result of this scenario would seem to be a sufficiently rapid irreversible interdiffusion of fields for successful operation of the solar dynamo. For the Galaxy it is hypothesized that rapid reconnection between the extended magnetic lobes or Ω-loops that form the halo accomplished the irreversibility necessary for the operation of the galactic dynamo. However, it must be emphasized that until these conjectures are firmly established, it cannot be said that we understand the origin of the magnetic fields of stars and galaxies. The fact that there is at present no known alternative to the $\alpha\omega$-dynamo is not to say that the $\alpha\omega$-dynamo operation is fully understood.

1. Introduction

Planets, stars, galaxies, and clusters of galaxies are composed of non magnetic material, and it is inferred from observations that they possess general magnetic fields of such strength as to dominate the surrounding space.

1

K. C. Tsinganos (ed.), Solar and Astrophysical Magnetohydrodynamic Flows, 1–16.
© *1996 Kluwer Academic Publishers.*

This surrounding magnetosphere, corona, or halo is invariably filled with suprathermal ($10^6 - 10^7$K) gas as well as fast particles ($> 1kev$ or 10^7K). Indeed, it is largely the polarized electromagnetic emission from the suprathermal particles that provides the evidence for the enclosing magnetic field.

The universal phenomenon of magnetic fields with enclosed suprathermal particle populations is the basis for the X-ray and non thermal radio emission from planets, stars and galaxies. It suggests that magnetic fields are generated within the central body, presumably induced by the internal motions of electrically conducting fluids. It also suggests that the magnetic field is significantly deformed by the fluid motions within, or strongly buffeted by winds from without, so as to provide the suprathermal particle population. That is to say, the maintenance of the observed suprathermal particle population implies an abundance of magnetic free energy whose dissipation provides the suprathermal population.

The purpose of this paper is to review present understanding of the generation of magnetic fields in astronomical objects. The purpose of the next paper is to review present understanding of the coronal and halo heating that produces the suprathermal gas. As we shall see, there are several fundamental questions that are not yet fully answered, popular ideas to the contrary not withstanding. The assertion that magnetic fields of celestial bodies are induced by fluid motions was first made by Elsasser (1946a) in the context of the magnetic dipole field of Earth. He showed that no other process known to physics was sufficient, and then set about to understand what fluid motions in the liquid metallic core of Earth might generate the observed magnetic field. For otherwise the field would decay in a characteristic time of the order to 3×10^4 years as a consequence of the resistivity of the liquid Fe - Ni core. He showed (Elsasser, 1946b, 1947, 1950a,b, 1954, 1956, 1957) that the nonuniform rotation of the convecting core shears the poloidal (dipole) field and produces a strong azimuthal field within the core without effect on the poloidal field. Cowling's (1932) theorem, that an axi-symmetric field cannot be sustained by fluid motions, made it clear that there must be another step in the induction process, necessarily involving fluid motions on a smaller scale in order to avoid axi-symmetry. It was then pointed out that the many individual convective cells within the liquid core are cyclonic as a consequence of the rapid spin of Earth and that the interaction of the individual cyclonic convective cell with the azimuthal magnetic field produces a local loop of field that is rotated by the cyclonic motion about a vertical axis providing loops with nonvanishing projection on the meridional planes. The resistivity plays a central role then in irreversibly merging these many loops into a single pattern of circulation of magnetic field in the meridional planes. That is to say, the resistivity destroys the small-scale Fourier components of the magnetic field (in a characteristic time of 10^3 years or less) leaving only the large-scale poloidal field (with a characteristic resistive decay time of about 3×10^9 years) (Parker, 1955, 1957, 1970, 1979; Backus, 1958; Moffatt, 1978). In the absence of resistivity the deformation of the magnetic field into many loops in the meridional plane would be (a) reversible and (b) the small-scale components would grow so rapidly that their Maxwell stress would soon overpower the cyclonic convection, striving to push the system back to the initial axis-symmetric azimuthal magnetic field. Thus the resistivity introduces the essential irreversibility for production of poloidal field. It goes without saying that the resistive diffusion serves also to smooth the meridional profile of the azimuthal field, although the large-scale azimuthal magnetic field remains largely reversible, i.e., reversing the nonuniform flow would destroy much of the azimuthal field even in the total absence of resistivity.

The combination of nonuniform rotation ω and cyclonic convection provides a regenerative dynamo within a simply connected volume of electrically conducting fluid. Designating the helicity or cyclonic portion as the α-effect, the result is the so called $\alpha\omega$ -dynamo. There are many variations of the $\alpha\omega$-dynamo and many interesting higher order effects (cf. Moffatt, 1978; Parker, 1979; Krause and Rädler, 1980; see also the formulation by Braginskii, 1964a,b, 1975; Soward, 1972; Soward and Roberts, 1976). But the basic process is the combination of overall nonuniform rotation and small-scale cyclonic convection in the presence of enough resistive diffusion to render the α-effect irreversible.

With the discovery (Babcock and Babcock, 1955) of the general oscillatory magnetic field of the Sun, it was pointed out (Parker, 1955, 1957) that the $\alpha\omega$-dynamo provides periodic migratory dynamo waves, suggestive of the observed equatorward migration of azimuthal field in the convective zone of the Sun. Preliminary modeling of the solar $\alpha\omega$-dynamo (Parker, 1957, 1971b, 1993; Leighton, 1969; Steenbeck and Krause, 1969a, b; Gilman, 1969a,b) showed that the effective resistive diffusion coefficient η must be of the general order of $10^{12} \mathrm{cm}^2/\mathrm{sec}$ in order to provide the necessary irreversibility over the observed dimensions ($\sim 10^{10}\mathrm{cm}$) in a matter of a few years. The electrical conductivity $\sigma \cong 2 \times 10^7 T^{\frac{3}{2}}/\mathrm{sec}$ in the convective zone ($10^4 < T < 2 \times 10^6$ K) provides η in the range 10^3 to $5 \times 10^6 \mathrm{cm}^2/\mathrm{sec}$. So it became clear that the turbulent diffusion must play the major role in providing η. With the characteristic mixing length expression $\eta_t = 0.1\ell v$ for the turbulent diffusion coefficient, the mixing length ℓ ($\sim$ one pressure scale height) and the associated convective velocity v (cf. Spruit, 1974) provide η_t in the general range $10^{11} - 10^{13} \mathrm{cm}^2/\mathrm{sec}$. So the $\alpha\omega$-dynamo fitted nicely into the picture.

The success of the $\alpha\omega$-dynamo with the magnetic fields of the Sun suggested that in one form or another it is the basis for the oscillating magnetic fields of other stars as well.

The magnetic field of the Galaxy was the next challenge. The galactic magnetic field of some $2 - 3 \times 10^{-6}$ gauss is anchored in the gaseous disk of the Galaxy and lies along the spiral arms with the direction of the field alternating from one arm to the next, as one might expect of a magnetic field initially lying along a diameter of a disk in which the angular velocity increases inward toward the center. Given that the Galaxy has rotated some 50 times at the radius of the Sun and 100 times in the central regions, the question naturally arises as to why the arms of the Galaxy, and the magnetic field embedded in the arms, are not tightly wound. In fact the four known arms (Scutum, Sagittarius, Orion, Perseus) lie at radial distances of about 5,6,8, and 10 kpc from the center of the Galaxy, respectively, with the Sun located on the edge of the Orion arm. This indicates a net winding of no more than two or three revolutions.

The resolution of the winding problem for the spiral arms is that they represent spiral density waves (Lin, Yuan, and Shu, 1968) propagating radially outward with a phase velocity of the order of 10 km/sec, as compared with the rotational velocity of 250 km/sec. The outward phase velocity of the individual spiral arms unwinds the spiral as fast as the nonuniform rotation winds it up. The net result is the spiral pattern rotating more or less rigidly, in spite of the substantial radial variation in the actual angular velocity of the interstellar gas.

The magnetic field of the Galaxy is tied to the gas, of course, so it cannot move radially with the phase velocity of the density wave. The explanation is, then, that the magnetic field is a migrating $\alpha\omega$-dynamo wave that is phase locked to the density wave (Fujimoto and Sawa, 1989; Tesa and Chiba, 1990; Rüdiger, Elstner, and Schultz, 1993; Denner, Brandenburg, and Thomason, 1993; Moss, et al, 1993). The creation of the magnetic field of the Galaxy by an $\alpha\omega$-dynamo operating in the nonuniformly rotating gaseous disk seems simple enough at first sight (Parker, 1971a, 1979; Ruzmaikin, Shukurov, and Sokoloff, 1988; Krause, Rädler, and Rüdiger, 1993). The motion of the gas has a characteristic scale of the order of 10^2 pc and an rms velocity of 5-10 km/sec. Combined with the angular velocity $1 - 2 \times 10^{-15}$ radians/sec the result is an $\alpha\omega$-dynamo that requires an effective diffusion coefficient of the order of $10^{25} - 10^{26}$ cm/sec. Obviously some form of turbulent diffusion is required to achieve this enormous effect, and it was gratifying to note that the usual mixing length formula 0.1 ℓv with the standard scale $\ell \sim 10^2 pc$ of the interstellar gas motions $v \sim 5 - 10$ km/sec yields something of the order of 3×10^{25} cm^2/sec.

Another question is the enhanced magnetic fields in galaxies undergoing epochs of star formation – the so called star burst galaxies. They exhibit magnetic fields of 10×10^{-6} gauss instead of the usual $2-3 \times 10^{-6}$ gauss. The only explanation seems to be a more vigorous dynamo activity brought on by the formation of large numbers of massive, and hence short lived, stars, producing supernovae and extensive interstellar regions of hot gas (Ko and Parker, 1988; Ko, 1990, 1993).

2. The Difficulty

Unfortunately there is important physics missing from the $\alpha\omega$-dynamo as applied to stars and galaxies, and that is the physics of the essential irreversible turbulent diffusion. The problem is that the mean azimuthal magnetic fields are sufficiently strong in the Sun and in the Galaxy as to have Maxwell stress densities comparable to the Reynolds stress density of the turbulence that is supposed to provide the turbulent diffusion, $\eta_t \sim 10^{12}$ cm^2/sec and 10^{25} cm^2/sec, respectively. That is to say, the estimated mean fields are comparable to the equipartition field B_{eq} for the turbulence,

$$\frac{B_{eq}^2}{8\pi} = \frac{1}{2}\rho < v^2 > .$$

This does not prevent substantial local displacement l of the magnetic field with the turbulent velocity $< v^2 >^{\frac{1}{2}}$. Indeed one expects magnetic fluctuations ΔB comparable to B_{eq} under the circumstances. The problem is that the displacements take on the form of Alfven waves. The waves interact nonlinearly with each other, of course, so that a hierarchy of shorter waves may be produced, but the result is not a turbulent cascade to dissipation at large wave numbers. This may be seen for a Kolmogoroff spectrum in which the characteristic velocity $v(\lambda)$ associated with a small scale λ varies as $\lambda^{\frac{1}{3}}$. Thus, if the dominant eddies have a scale l and are in approximate equipartition with the mean magnetic field, then on smaller scales $(\lambda < l)$, $\frac{1}{2}\rho v^2(\lambda) = (B_{eq}^2/8\pi)(\lambda/l)^{\frac{2}{3}}$, so that $\Delta B(\lambda) < B_{eq}$ for $\lambda < \ell$. It follows that these weak small-scale perturbations are linear Alfven waves propagating in the varying large scale field $B + \Delta B(\ell, t)$. It is interesting to note then (Godreich and Sridhar, 1994) that there is nonlinear interaction in the Fourier components with wave vectors nearly perpendicular to the mean field, providing a cascade of energy to small-scales. But the effect is only to dissipate the strong Alfven waves, with no significant enhancement of the diffusion of the large-scale mean field (Parker, 1996).

Another way to state it is that diffusion of the mean field implies displacement of magnetic flux bundles over dimensions of the order of l, with diffusion or small-scale reconnection cutting off the magnetic connections so that the flux does not return to its initial location. This is equivalent to the statement that the magnetic Reynolds number must be of the order of unity, when in fact it is very large compared to unity.

To put it another way, the dominant eddies provide the quantity $v(l)\ell$ with the dimensions of diffusivity, but there is nothing significantly irreversible about these oscillatory motions.

The conclusion is that we must search further for the necessary dissipation and irreversibility. Whatever the resolution of this dilemma, it is clear that the answers may be quite different for the Sun and for the Galaxy. We consider each in turn.

3. The Sun

The appearance of magnetic flux in a bipolar magnetic region on the surface of the Sun in amounts up to 10^{23} Maxwells suggests that the azimuthal flux in that hemisphere is not less. If the flux comes from the lower half (10^{10}cm) of the convective zone from a latitude band comparable in width to the 10^{10}cm width of the bipolar active regions at the surface ($\sim 8°$ of latitude), the mean azimuthal flux density is not less than 10^3 gauss. Thus the azimuthal field is at least of the same general order as the maximum equipartition field of about 3×10^3 gauss in the convective zone (cf. Spruit, 1974). The helioseismological evidence that the principal shear (rotational gradient $\Delta\omega$) lies immediately below the bottom of the convective zone (in the so called overshoot layer) suggests that the azimuthal field lies principally in that same relatively thin layer ($\sim 3 \times 10^9$cm), indicating a mean field of at least 3×10^3 gauss.

We have suggested (Parker, 1993b) that this confinement of the azimuthal field to the overshoot layer may greatly relieve the problem with the irreversibility of the α-effect in the convective zone above. For if there is no more than 10-100 gauss of azimuthal field and perhaps a 10 gauss poloidal field in the convective zone, then the mean field is small compared to the equipartition field, allowing some degree of turbulent mixing of the field.

It is not obvious, however, that this is adequate by itself to produce the full irreversible turbulent diffusion necessary for effective operation of the solar dynamo. For the fact is that the small-scales necessary for effective turbulent resistive diffusion are associated with magnetic filaments that have been stretched longitudinally to where the field intensity is larger than the mean field by the factor $R^{\frac{1}{2}}$, where R is the magnetic Reynolds number $v(l)\ell/\eta$. For the lower convective zone of the Sun, $l \sim 10^9$ cm, $v(l) \sim 3 \times 10^3$ cm/sec, and $\eta \sim 10^4$ cm^2/sec. It follows that $R \sim 3 \times 10^8$ so that $R^{\frac{1}{2}} \sim 2 \times 10^4$. But the mean field in the convective zone is presumably ~ 10 - 10^2 gauss, suggesting that the small-scale fields would be $2 \times 10^5 - 2 \times 10^6$

gauss. The turbulence cannot stretch so strong a mean field into sufficiently thin filaments as to provide significant resistive diffusion.

This suggests that the convection may be inhibited on all scales by the Maxwell stresses of the small-scale fields. But this is evidently not the case because there is no noticeable inhibition of the convection at the surface of the Sun and because it appears that the convection delivers heat to the surface of the Sun with an efficiency that requires free irreversible mixing of the fluid elements. We speculate that the free efficient convection is possible because the final asymptotic effect of the convection on the magnetic field is not so much to draw the field into progressively thinner filaments of increasingly large intensity as it is to isolate the field in intense magnetic fibrils with relatively field-free fluid in the wide spaces between the concentrated fibrils. This clearly is the condition at the visible surface. It can be shown that the concentration and isolation of the magnetic flux into widely space intense fibrils represents a minimum total (magnetic plus thermal) energy state (Parker, 1984a), suggesting that the field may be in a fibril state throughout the convective zone. The fact is that for a given total magnetic flux and total convective heat transport, the more concentrated the fibrils the larger the magnetic energy but the more volume available for free convective heat transport. The minimum total energy under solar conditions suggests magnetic fibrils of a few kilogauss, of the same order of magnitude as the 1-2 kilogauss observed at the surface. Zwaan (1985) among others has argued for the general fibril state throughout the convective zone on the grounds that the magnetic fields observed emerging through the surface of the Sun are clearly in a fibril state prior to emerging. Recent studies (D'Silva and Choudhuri, 1993; D'Silva, 1993; Fan, Fisher, and DeLuca, 1993) indicate fibrils of 10^5 gauss at the bottom of the convective zone. The point to be made here is that the postulated fibril state also appears to explain the operation of the solar dynamo, avoiding the prohibitive growth of small-scale fields in the turbulent convection.

The physical picture of a fibril field in the convective zone is entirely different from the traditional concept of a continuous field in the presence of turbulence. A fibril field can be understood as the tendency for any coordinated eddy or swirl to exclude the magnetic field (Parker, 1963, 1979, §16.5.3; Clark, 1965; Weiss, 1966). Once the field is partially excluded, it coordinates and concentrates the eddy in the reduced field region, causing further exclusion of field, etc. The end result is evidently the extreme fibril state, such as is observed at the surface of the Sun.

Fibril magnetic fields are carried along with the larger eddies in the fluid, with the Lorentz force exerted by the field transferred to the fluid by the aerodynamic drag of the fibril on the fluid. Fibrils with diameters of the order of $10^2 - 10^3$km are carried along in the larger eddies with relatively little slip. The mean field equations are easily formulated (Parker, 1984b,c). Vishniac (1995) has explored the dynamics and consequent magnetohydrodynamics of fibril fields in more detail, providing limits on the fibril field intensity, fibril diameter, and fibril curvature. The essential point for the present concern with effective irreversible diffusion is the rapid reconnection wherever nonparallel fibrils are pressed together. Thus, when one bundle of fibrils comes in contact with another bundle with a different orientation, they cut across each other rapidly and irreversibly. The minimum speed at which one fibril cuts across and reconnects with another is of the order of $V_A/R^{\frac{1}{2}}(r)$ where V_A is the Alfven speed in the fibril and R is the magnetic Reynolds number $V_A r/\eta$ for a fibril of radius r. With $r \sim 10^7$cm for a fibril of 10^4 gauss in the lower convective zone where $\rho = 0.1$ gm/cm^3, it follows that $V_A \sim 10^4$cm/sec. Then $R \sim 10^7$ and the minimum cutting speed is of the order of 3 cm/sec traversing 10^7 cm in 3×10^6 sec. In fact the cutting and reconnection may proceed much more rapidly. The maximum rate, suggest by the Petscheck mode, is of the order of $V_A/\ln R$ and may be 10^2 times larger. The time to cut across the fibril radius r may be as little as 3×10^4 sec. The point is simply that fibril fields in the convective zone of the Sun are subject to displacement and deformation by the convection, and the deformation becomes irreversible through rapid reconnection wherever nonparallel fibril fields come in contact. Thus the cyclonic α-effect, with the many twisted loops in the azimuthal field, is accomplished by the severance (reconnections) of the individual loops from the azimuthal field along with the amalgamation of individual loops into a general circulation of magnetic field in meridional planes to provide the poloidal field. Similarly the production of azimuthal fibrils through shearing of poloidal fibrils by the nonuniform rotation is quickly assimilated into the mean azimuthal field by rapid reconnection between anti-parallel fibrils. Vainshtein, Parker, and Rosner (1993) provide a crude model of the solar dynamo based on fibril magnetic fields. It is clear that much theoretical work remains to be done, and it is also clear that it will be difficult, if not impossible, to construct quantitative models of the unobservable fibril state of the magnetic field of the Sun in the deep convective zone without detailed knowledge of the precise structure of the magnetic fibrils – their field intensity, diameter, degree of twisting, etc.

4. The Galaxy

The dynamo that appears to be operating in the gaseous disk of the Galaxy provides quite a different problem from the Sun. In the first place the characteristic time scale for the galactic dynamo is of the order of 10^8 years, so that we cannot in the brief time span of the human race observe its operational procedure. Second, the magnetic field of the Galaxy is not in a fibril state, for the reason that the interstellar gas pressure does not generally exceed the magnetic pressure, of the order of 0.4×10^{-12} dynes/cm^2. On the other hand, the magnetic field in the spiral arm is unstable (Parker, 1966b,c; 1967, 1969, 1979 §13.7) with characteristic times of the order of 10^7 sec. We have suggested that the dissipation and irreversibility arises in the nonlinear state of the instability (Parker, 1990, 1992, 1993a). The essential point is that the magnetic field along the spiral arms in the gaseous disk of the Galaxy is confined by the weight of the interstellar gas. That is to say, the gas rests on the horizontal magnetic field. In fact, relativistic cosmic ray gas inflates the magnetic field with a pressure comparable to the magnetic pressure, and the combined magnetic and cosmic ray pressure supports the gas. Consider, then, the result of a slight local uplifting, causing the field lines to be elevated locally. The gas in the elevated portion of the magnetic field drains downward along the sloping field lines on each side. This unloads the elevated portion of the field, which then expands farther upward, forced by both the magnetic pressure and the cosmic ray pressure. The thermal gas accumulates in the low places along the field, increasing their burden of gas and pushing them downward toward the central plane of the gaseous disk. The local upward bulge of the magnetic field (with a characteristic length of 100 pc or more) soon finds itself in a runaway situation. Abandoned by the confining interstellar gas, it is inflated by the cosmic ray gas (Parker, 1966a, 1969; Lerche and Parker, 1966) which pushes it upward at speeds of the order of 40 km/sec. The result is illustrated schematically in Fig. 1a, showing the largely vertical alternating magnetic fields extending out in both directions from the gaseous disk. The magnetic field of the Galaxy forms a sequence of contiguous Ω-loops. The estimated rate of inflation of the individual magnetic lobe or Ω-loop comes from the cosmic ray life τ of about $10^{14}/N$ sec in the gaseous disk with mean gas density of N atoms/cm^3, (during which the spallation of the heavier nuclei among the cosmic ray particles indicates a penetration through about 5 gm/cm^2). If the half-thickness of the gaseous disk is h, the mean outflow at the surface of the disk is h/τ. For $h = 2 \times 10^{20}$cm the result is an outflow velocity of $20N$ km/sec. The estimated mean density in the disk (Schmidt, 1957) is $N \sim 2$ atoms/cm^3.

The outward inflation of these Ω-loops of magnetic field may be substantially augmented in some locations by fountains of hot interstellar gas (10^6K) from OB associations and perhaps by a galactic wind from the central regions of the Galaxy. But we suggest (Parker, 1965, 1966b,c; 1967, 1968, 1969; Jokipii and Parker, 1969; Hartquist and Morfill, 1986) that the cosmic ray inflation may be the major overall effect, particularly at great height above the disk. The net result is the creation of a galactic halo of

10

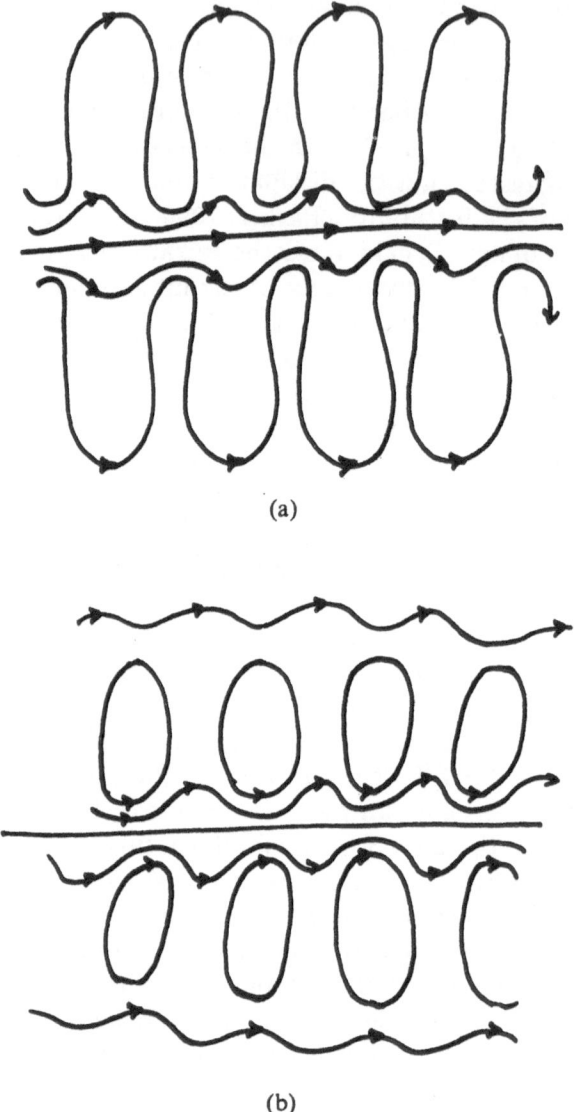

Figure 1. (a) A schematic drawing of the lobes of magnetic field bulging outward from the spiral arm. (b) A schematic drawing of the same fields after reconnection between adjacent lobes, with net loss of magnetic flux to the spiral arm. There remains behind a residue of local closed loops of field anchored in the spiral arm.

cosmic rays, hot gas, and more or less vertical magnetic field of $1 - 2 \times 10^{-6}$ gauss. Such halos are observed for spiral galaxies viewed edge on (see the various papers in Beck, Kronberg, and Wielebinski, 1990) and we expect that the Galaxy is similarly equipped.

The implication for the galactic dynamo is that the inflated lobes of field are twisted by the Coriolis force as they rise and expand upward from the rotating disk. Each vertical leg of an Ω-loop is surrounded by vertical legs of opposite sign of the neighboring Ω-loops, so the legs reconnect between neighboring lobes. The result is shown schematically in Fig. 1b, from which it can be seen that many loops of field remain behind with their lower ends embedded in the gaseous disk and presumably rotated somewhat from their original orientation along the spiral arm. The rotated loops form the poloidal field, probably with further reconnection and readjustment among themselves dissipating all field components except the largest scale. There is of course a net loss of azimuthal magnetic flux in the process, to be replaced through continued shearing of the poloidal field. We propose (Parker, 1990, 1992a,b) that the α-effect is carried out in this way, as an integral part of the dynamics of the galactic halo.

The galactic reconnection rates cannot be computed with any more certainty than for fibrils in the convective zone of the Sun, but the estimated order of magnitude seems about right. The Alfven speed in the outer halo where the number density may be 10^{-4} atoms/cm^3 and the field is $1 - 2 \times 10^{-6}$ gauss is 200-400 km/sec. A reconnection velocity of 1 km/sec cuts a distance of 50pc in 5×10^7 years in both directions from the reconnection site. Thus it completely reconnects flux regions of $10^2 pc$ in the time of 5×10^7 years in which the Ω-loop is extended outward a distance of 2 kpc by the cosmic ray inflation. This time is less than the characteristic dynamo regeneration time of 10^8 years and provides a halo with the extent observed in other spiral galaxies. The reconnection and associated irreversibility of the α-effect is rapid enough to provide the necessary $\alpha\omega$-dynamo.

In conclusion it seems not implausible that the galactic halo is the site of the α-effect, with the irreversibility arising from the reconnection between adjacent lobes or Ω-loops in the galactic magnetic field. The complicated dynamics of the gaseous disk of the Galaxy and the formation of the halo depends in detail on the precise state of the interstellar gas and cosmic ray production across the gaseous disk - something presently unknown and unknowable. The large Reynolds numbers and magnetic Reynolds numbers combined with rapid reconnection between opposite $10^2 pc (3 \times 10^{20}$cm) magnetic field components across scales of 10^{10}cm precludes theoretical dynamical studies. So it is not clear that the galactic dynamo can ever be raised above the level of conjecture. The existence of the galactic dynamo is an hypothesis to explain the coincidence of the winding of the spiral magnetic fields with the density waves represented by the spiral arms and to explain the enhanced magnetic flux in star burst galaxies. The existence of the dynamo implies a vigorous dissipation and irreversibility of magnetic loops, which can be understood only in the context of the rapid reconnection between alternating magnetic fields of the galactic halo.

It is interesting to look further into the cosmos, where observations find evidence for magnetic fields filling regions with dimensions of a Mpc or more. For instance Kim and Kronberg (1990) infer a rms magnetic field of at least 1.5×10^{-6} gauss with a characteristic internal coherence scale of 40 kpc throughout the coma cluster of galaxies, based on extensive radio observations. Schlickeiser and Rephaeli (1990) estimate magnetic fields of $0.04 - 0.15 \times 10^{-6}$ gauss in several clusters, based on radio and hard X-ray emissions. Giovannini, et al (1990) infer a magnetic field of about 0.3×10^{-6} gas over a region of $1.4 \, Mpc$ through the coma - A1367 supercluster. Vallee (1990) estimates 1.5×10^{-6} gauss across several Mpc in the Virgo cluster, assuming the electron density of $10^{-6}/cm^3$ inferred from the intensity of X-ray emission (see also Vallee, MacLeod, and Broten, 1986; Kronberg, 1987). Battaner, Florido, and Sanchez-Saavedra (1990) speculate that the observed correlated warping of the disks of local galaxies can be explained by a magnetic field of the order of $0.01 - 0.1 \times 10^{-6}$ gauss and a scale of 20 Mpc.

These inferences raise a number of questions. First of all, what confines the fields of $1 - 2 \times 10^6$ gauss over the dimensions of a cluster of galaxies? Second, there are no definitive polarization measures with which to establish the coherence length of the cluster fields. We are left with the fundamental question of whether the fields are tangled and intensified on scales of 10-30 kpc through the turbulent stretching and wrapping of an initial weak primordial field ($\sim 10^{-9}$ gauss), as suggested by Sokoloff, Ruzmaikin, and Shukurov (1990). The intercluster extension of magnetic regions would suggest that there is a component of the field extending coherently over the scale of one or more Mpc. The essential point is that if the intracluster and intercluster magnetic fields are coherent (as distinct from an accumulation of smaller scale loops escaping from individual galaxies) their creation does not follow from any dynamo mechanism of which we are aware. A primordial field of 10^{-9} gauss is as mysterious as a contemporary field of $10^{-7} - 10^{-6}$ gauss. So if the fields exist as conjectured, they imply physical processes which are presently unknown to us.[1] This possibility must be kept in mind when discussing the origin of the magnetic field in galaxies. The foregoing speculations on the nature of the galactic dynamo may be overlooking some important physical effect.

Acknowledgement. This work was supported in part by the National Aeronautics and Space Administration under NASA grant NAGW-2122.

---

[1] Ideas on the present magnetic field of the Galaxy in the absence of a galactic $\alpha\omega$-dynamo can be found in Kulsrud (1990), Tajima, et al. (1992), and Kulsrud and Anderson (1992).

References

Babcock, H.W. and Babcock, H.C. 1955, The Sun's magnetic field 1952 - 1954, *Astrophys. J.* **121**, 349.

Backus, G.E. 1958, A class of self-sustaining dissipative spherical dynamos, *Ann. Phys.* **4**, 372.

Battaner, E., Florido, E. and Sanchez-Saavedra, M.L. 1990, Intergalactic magnetic fields and the morphology of spiral galaxies, in Galactic and Intergalactic Magnetic Fields, IAU Symp. No. 140, Heidelberg. 19-23 June, 1989, Kluwer Academic Publishers, Dordrecht, ed. R. Beck, P.P. Kronberg, and R. Wielebinski, p. 504.

Beck, R., Kronberg, P.P., and Wielenbinski, R. eds. 1990, Galactic and Intergalctic Magnetic Fields, Kluwer Academic Publishers, Dordrecht.

Braginskii, S. I. 1964a, self-excitation of a magnetic field during the motion of a highly conducting fluid, *Sov. Phys. JETP* **20**, 726.

Braginskii, S.I. 1964b, Theory of the hydromagnetic dynamo, *Sov. Phys. JETP* **20**, 1462.

Braginskii, S.I. 1975, Nearly axis-symmetric model of the hydromagnetic dynamo of the Earth, *Geomag. Aeron.* **15**, 122.

Clark, A. 1965, Production and dissipation of magnetic fields by differential fluid motions, *Phys. Fluids* **7**, 1299.

Cowling, T.G. 1933, The magnetic field of sunspots, *Mon. Not. Roy. Astron. Soc.* **94**, 39.

Donner, K.J. Bradenburg, A., and Thomasson, M. 1993, Galactic dynamos and dynamics in The Cosmic Dynamo, IAU Symp. No. 157, Potsdam, 7-11, September, 1992, Kluwer Academic Publishers, Dordrecht, ed. F. Krause, K.W. Rädler, and G. Rüdiger, p. 333.

D'Silva, S. 1993, Can equipartition fields produce the titlts of bipolar magnetic regions? *Astrophys. J.* **407**, 395.

D'Silva, S. and Choudhuri, A.R. 1993, A theoretical model for tilts of bipolar magnetic regions, *Astron. Astrophys.* **272**, 621.

Elsasser, W.M. 1946a, Induction effects in terrestrial magnetism. Part I. Theory *Phys. Rev.* **69**, 106.

Elsasser, W.M. 1946b, Induction effects in terrestrial magnetism. Part II. The secular variation, *Phys. Rev.* **70**, 202.

Elsasser, W.M. 1947, Induction effects in terrestrial magnetism. Part III. Electric modes, *Phys. Rev.* **72**, 821.

Elsasser, W.M. 1950a, The hydromagnetic equations, *Phys. Rev.* **79**, 183.

Elsasser, W.M. 1950b, The Earth's interior and geomagnetism, *Rev. Mod. Phys.* **22**, 1.

Elsasser, W.M. 1954, Dimensional relations in magnetohydrodynamics, *Phys. Rev.* **95**, 1.

Elsasser, W.M. 1956, Hydromagnetic dynamo theory, *Rev. Mod. Phys.* **28**, 135.

Elsasser, W.M. 1957, The terrestrial dynamo, *Proc. Natl. Acad. Sci.* **43**, 14.

Fan, Y., Fisher, G.H., and DeLuca, E.E. 1993, The origin of morphological asymmetries in bipolar active regions, *Astrophys. J.* **405**, 390.

Fujimoto, M. and Sawa, T. 1989, Generation and maintenance of bisymmetric spiral magnetic fields and interaction with spiral density waves, *Geophys. Astrophys. Fluid Dyn* **50**, 159.

Gilman, P.A. 1969a, A Rossby wave dynamo for the Sun. I, *Solar Phys.* **8**, 316.

Gilman, P.A. 1969b, A Rossby wave dynamo for the Sun. II, *Solar Phys.* **9**, 3.

Giovannini, G., Kim, K.T., Kronberg, P.P. and Venturi, T. 1990, Evidence for a large-scale magnetic field in the Coma-A1367 super cluster, in Galactic and Intergalactic magnetic fields, IAU symp. No. 140, Heidelberg, 19-23 June 1989, Kluwer Academic Publishers, Dordrecht, ed. R. Beck, P.P. Kronberg, and R. Wielebinski, p. 492.

Goldreich and Sridhar

Hartquist, T.W. and Morfill, G.E. 1986, A model of static cosmic-ray supported galactic coronae, *Astrophys. J.* **311**, 518.

Jokipii, J.R. and Parker, E.N. 1969, Cosmic-ray life and the stochastic nature of the galactic magnetic field, *Astrophys. J.* **155**, 799.

Kim, K.T. and Kronberg, P.P. 1990, The strength and structure of the intercluster magnetic field in the coma-cluster of galaxies, in Galactic and Intergalactic Magnetic Fields. IAU Symp. No 140, Heidelberg, 19-23 April, 1989. Kluwer Academic Publishers, Dordrecht, ed. R. Beck, P.P. Kronberg, and R. Wielebinski, p. 483.

Ko, C.M. 1990, A study of the kinematic dynamo equation with time dependent coefficients, *Astrophys. J.* **360**, 151.

Ko, C.M. 1993, A similarity solution to the kimenatic $\alpha^2\omega$ dynamo equation with time dependent coefficients, *Astrophys. J.* **410**, 128.

Ko, C.M. and Parker, E.N. 1988, Intermittent behavior of galactic dynamo activity, *Astrophys. J.* **341**, 828.

Krause, F.K. and Rädler, K.H. 1980, Mean-Field Magnetohydrodynamics and Dynamo Theory, Pergamon Press, New York.

Krause, F., Rädler, K.H., and Rüdiger, G. editors. 1993, The Cosmic Dynamo, IAU Symp. No. 157, Potsdam 7-11 Sept. 1992, Kluwer Academic Publishers, Dordrecht.

Kronberg, P.P. 1987, In Interstellar Magnetic Fields, Springer, Berlin, ed. R. Beck and R. Graves, p. 86.

Kulsrud, R.M. 1990, The present state of a primordial magnetic field, in Galactic and Intergalactic Magnetic Fields. IAU Symp. No. 140, Heidelburg, 19-23 June, 1989, Kluwer Academic Publishers, Dordrecht, ed. R. Beck, P.P. Kronberg, and R. Wielebinski, p. 527.

Kulsrud, R.M. and Anderson, S.W. 1992, The spectrum of random magnetic fields in the mean field dynamo theory of the galactic magnetic field, *Astrophys. J.* **396**, 606.

Leighton, R.B. 1969, A magneto-kinematic model of the solar cycle, *Astrophys. J.* **156**, 1.

Lerche, I. and Parker, E.N. 1966, Nonrelativistic equations of bulk motion of a relativistic gas, *Astrophys. J.* **145**, 106.

Lin, C.C., Yuan, C. and Shu, F.H. 1968, On the spiral structure of disk galaxies. III Comparison with observations, *Astrophys. J.* **155**, 721, erratum **156**, 797.

Moffatt, H.K. 1978, magnetic Field Generation in Electrically Conducting Fluids, Cambridge University Press, Cambridge.

Moss, D., Brandenburg, A., Donner, K.J., and Thomasson, M. 1993, in the Cosmic Dynamo, IAU Symp. No. 157, Potsdam, 7-11 Sept., 1992. Kluwer Academic Publishers, Dordrecht, ed. F. Krause, K.H. Rädler, and G. Rüdiger, p. 339.

Parker, E.N. 1955, Hydromagnetic dynamo models, *Astrophysic. J.* **122**, 293.

Parker, E.N. 1957, The solar hydromagnetic dynamo, *Proc. natl. Acad. Sci.* **43**, 8.

Parker, E.N. 1963, Kinematical hydromagnetic theory and its application to the low solar photosphere, *Astrophys. J.* **138**, 552.

Parker, E.N. 1965, Cosmic rays and their formation of a galactic halo, *Astrophys. J.* **142**, 584.

Parker, E.N. 1966a, The kinetic properties of the cosmic ray gas, *Astrophys. J.* **144**, 916.

Parker, E.N. 1966b, The dynamical state of the interstellar gas and field, *Astrophys. J.* **145**, 811.

Parker, E.N. 1966c, General dynamical effects of cosmic rays in the Galaxy, in Proc. of the Ninth International Conference on Cosmic Rays, Vol 1, 126, The Physical Society, London.

Parker, E.N. 1967, Dynamical state of the interstellar gas and field. II. Nonlinear growth of clouds and forces in three dimensions, *Astrophys. J.* **149**, 517.

Parker, E.N. 1968, Dynamical properties of cosmic rays, Chap. 14 in Nebulae and Interstellar Matter, Vol. VII Stars and Stellars Systems. University of Chicago Press, Chicago, ed. B.M. Middlehurst and L.H. Aller.

Parker, E.N. 1969, the galactic effects of the cosmic-ray gas, *Space Sci. Rev.* **9**, 651.

Parker, E.N. 1970, The generation of magnetic fields in astrophysical bodies: I. The dynamo equations, *Astrophys. J.* **162**, 665.

Parker, E.N. 1971a, The generation of magnetic fields in astrophysical bodies: II. The galactic field, *Astrophys. J.* **163**, 255.

Parker, E.N. 1971b, The generation of magnetic fields in astrophysical bodies: IV. The solar and terrestrial dynamos, *Astrophysic. J.* **164**, 491.

Parker, E.N. 1979, Cosmical Magnetic Fields, Oxford, Clarendon Press.

Parker, E.N. 1984a, Stellar fibril magnetic systems. I. Reduced energy state, *Astrophys. J.* **283**, 343.

Parker, E.N. 1984b, Stellar fibril magnetic fields. II. Two dimensional magnetohydrodynamic equations, *Astrophys. J.* **294**, 47.

Parker, E.N. 1984c, Stellar fibril magnetic fields. III. Convective counterflow, *Astrophys. J.* **293**, 57.

Parker, E.N. 1990, Spontaneous discontinuties in galactic magnetic fields and the creation of galactic radio halos, in Galactic and Integalactic Magnetic Fields. Proc. IAU Symp. No. 140, Heidelberg, June 19-23, 1989, Kluwer Academic Publishers, Dordrecht, ed. R. Beck, P.P. Kronberg, and R. Wielebinski, P. 169.

Parker, E.N. 1992, Fast dynamos cosmic rays and the galactic magnetic field, *Astrophys. J.* **401**, 137.

Parker, E.N. 1993a, Galactic cosmic rays and galactic halo X-ray emission, in Currents in Astrophysics and Cosmology, Cambridge University Press, Cambridge, ed. G.G. Fazio and R. Silberberg, p. 3.

Parker, E.N. 1993b, A solar dynamo surface wave at the interface between convection and nonuniform rotation, *Astrophys. J.* **408**, 707.

Parker, E.N. 1996

Rüdiger, G., Elstner, D. and Schultz, M. 1993, The galactic dynamo: modes and models in The Cosmic Dynamo, IAU Symp. No. 157, Potsdam, 7-11 Spet. 1992, Kluwer Academic Publishers, Dordrecht, ed. F. Krause, K.H., Rädler, and G.Rüdiger, p. 321.

Ruzmaikin, A.A. Shukurov, A.M., and Sokoloff, D.D. 1988, Magnetic Fields of Galaxies, Kluwer Academic Publishers, Dordrecht.

Schlickeiser, R. and Rephaeli, Y. 1990, Intracluster magnetic fields from X-ray and radio measurements, in Galactic and Intergalactic Magnetic Fields, IAU Symp. No. 140, Heidelberg, 19-23 June 1989, Kluwer Academic publishers, Dordrecht, ed. R. Beck, P.P. Kronberg, and R. Wielebinski, p. 487.

Schmidt, M. 1957, Spiral structure in the inner parts of the galactic system derived from the hydrogen emission at 21 cm wave length, *Bull. Astron. Inst. Netherlands* **13**, 247.

Sokoloff, D.D., Ruzmaikin, A.A. and Shukurov, A. 1990, Intermittent magnetic fields generated by turbulence in galaxies and galaxy cluster, in Galactic and Intergalactic Magnetic Fields, IAU Symp. No. 140, Heidelberg, 19-23 June 1989, Kluwer Academic Publishers, Dordrecht, ed. R. Beck, P.P. Kronberg, and R. Wielebinski, p. 499.

Soward, A.M. 1972, A kinematic theory of large magnetic Reynolds number dynamos, *Phil. Trans. Roy. Soc. London* **272**, 431.

Soward, A.M. and Roberts, P.H. 1976, *Magnitnaya Gidrodinanika* 1,3.

Spruit, W.C. 1974, A model of the solar convection zone, *solar Phys.* **34**, 277.

Steenbeck, M. and Krause, F. 1969a, Zur dynamotheorie stellarer and planeterar magnet felder I. Berechnumg sonnenähnlicher wechsel feldgeneratoren, *Astron. Nachr.* **291**, 49.

Steenbeck, M. and Krause, F. 1969b, Zur dynamotheorie stellarer and planetarer magnetfelder II. Berechnung planeten ähnlicher gleich feldgeneratoren, *Astron. Nachr.* **291**, 271.

Tajima, J., Cable, S., Shibata, K. and Kulsrud, R.M. 1992, On the origin of cosmical magnetic fields, *Astrophys. J.* **390**, 309.

Tosa, M. and Chiba, M. 1990, in Galactic and Intergalactic Magnetic Fieldss, IAU Symp. No. 140, Heideberg, 19-23 June 1989, ed. Kluwer Academic Publishers, Dordrecht, ed. R. Beck, P.P. Kronberg, and R. Wielebinski, p.127.

Vainshtein, S.I., Parker, E.N. and Rosner, R. 1993, On the generation of "strong" magnetic fields, *Astrophys. J.* **404**, 773.

Vallee, J.P. 1990, A possible excess rotation measure and large-scale magnetic field in the Virgo supercluster of galaxies, *Astron. J.* **99**, 459.

Vallee, J.P., Macleod, J.M. and Broten, N.W. 1986, A large-scale magnetic field in the galaxy cluster A2319, *Astron. Astrophys.* **156**, 386.

Vishniac, E.T. 1995, The dynamics of flux tubes in a high β plasma, *Astrophys. J.* (in press).

Walker, M.R. and Barenghi, C.F. 1994, High resolution numerical dynamos in the limit of a thin disk galaxy, *Geophys. Astrophys. Fluid Dyn.* **76**, 265.

Weiss, N.O. 1966, The expulsion of magnetic flux by eddies, *Proc. Roy. Soc. A.* **293**, 310.

Zwaan, C. 1985, the emergence of magnetic flux, *Solar Phys.***100**, 397.

MAGNETIC FLUX TUBES AND THE SOLAR DYNAMO

Storage, instability and eruption of magnetic flux

MANFRED SCHÜSSLER

Kiepenheuer-Institut
Schöneckstr. 6, D-79104 Freiburg
GERMANY, msch@kis.uni-freiburg.de

Abstract. The observed properties of sunspot groups on the surface of the Sun are consistent with the concept of magnetic flux tubes emerging from deep within the solar convection zone. In order to maintain coherence and orientation during their rise in the turbulent convective flows, the magnetic field strength in these flux tubes should always exceed the equipartition value with respect to the kinetic energy density of the convective motions.

Through linear stability analysis and nonlinear numerical simulations a consistent picture of the storage, instability, and rise of magnetic flux tubes has emerged in recent years, which is in accordance with the observed basic properties of sunspot groups: magnetic flux tubes are generated and stored in mechanical equilibrium in a subadiabatically stratified overshoot layer below the convection zone; they become unstable with respect to an undulatory (Parker-type) instability once the field strength exceeds a critical value; flux loops form, move through the convection zone and give rise to bipolar sunspot regions when they emerge at the surface. Both the stability criteria and the constraints set by the observed properties of sunspot groups (orientation, tilt angle, asymmetry, proper motions) require that the field strength at the bottom of the convection zone should be of the order of 10^5 G, an order of magnitude larger than the equipartition field strength with respect to the convective flows.

We present the line of arguments which leads to this view and discuss a number of open questions in connection with the implied super-equipartition fields. These include the problem of how such strong fields are generated, the rôle played by the 'explosion' process, and the need for non-conventional dynamo models.

K. C. Tsinganos (ed.), Solar and Astrophysical Magnetohydrodynamic Flows, 17–37.
© *1996 Kluwer Academic Publishers.*

1. The dynamo problem

The solar activity cycle with its various manifestations and regularities (quasi-periodic variation of the number of sunspots, 22-year magnetic cycle, equatorward drift of the sunspot zone, reversal of the polar magnetic fields, etc.) is believed to be the manifestation of a hydromagnetic dynamo mechanism operating in the solar convection zone (see the monographs by Moffat, 1978, and Parker, 1979). The 'classical' picture of the solar dynamo is based on two basic effects: differential rotation generates toroidal magnetic field by shearing a poloidal field, which itself is produced from the toroidal field through (turbulent) convective flows under the influence of rotationally induced Coriolis forces. The concept of field regeneration by 'cyclonic convection' (Parker, 1955a) found a mathematical basis in the theory of *mean-field magnetohydrodynamics* (Krause & Rädler, 1980). The resulting dynamo equations for the mean magnetic field permit periodic wave solutions, which are in accordance with basic phenomena of the solar cycle like the temporally periodic mean field and its migration in heliographic latitude – if the parameters are suitably chosen.

A dynamo process as sketched above is basically 'kinematic': it is required that the magnetic field lines are passively carried and twisted by the convective flows. As the field strengths grows in the course of the dynamo process the back-reaction of the magnetic field on the generating velocity fields through the Lorenz force becomes important. This limits the field strength that can be reached in the course of the dynamo process (at least to order of magnitude) to the 'equipartition field' B_{eq}, i.e., the field strength for which the magnetic energy density equals the kinetic energy density of the generating motions:

$$\frac{B_{eq}}{8\pi} \simeq \frac{1}{2}\rho v_c^2 \,, \tag{1}$$

where ρ is the density and v_c the velocity amplitude. For the lower half of the solar convection zone we find $B_{eq} \simeq 10^4$ G if we use a mixing-length estimate of the convective velocity.

2. Storage of magnetic flux

2.1. THE BUOYANCY DILEMMA

The connection between the dynamo working in the interior of the convection zone and the observed surface phenomena of solar activity is commonly made by invoking *magnetic buoyancy* (Parker, 1955b; Jensen 1955). Differential rotation is thought to create a strong toroidal magnetic field, which rises towards the surface since the magnetic pressure leads to a deficit of gas density in the magnetized plasma. Sunspot groups and magnetic active regions form as the rising field breaks through the surface of the Sun.

In fact, the observed properties of sunspot groups indicate the emergence of a coherent magnetic structure, which is not passively carried by convective flows and originates from a source region of well-ordered magnetic flux in the solar interior:

- The magnetic flux of an active region erupts rapidly (within a few days) while the lifetime of the region can be much longer (up to months).

- Sunspots form out of small, 'active' fragments, which follows paths determined by subsurface dynamics (e.g. Strous, 1994). The initial polarity mix in an emerging active region rapidly disentangles to form a well-defined bipolar structure.

- Hale's polarity rules are nearly strictly obeyed (Howard, 1989) and the polarities are oriented basically in east-west direction with a systematic *tilt angle*, which depends on the latitude (e.g. Wang & Sheeley, 1991).

These properties are in accordance with the 'rising tree' picture (Zwaan 1978, 1992; Zwaan & Harvey, 1994) of a partially fragmented magnetic structure that rises towards the surface and emerges in a dynamically active way. Only later, after the initial stage of flux emergence, the surface fields come progressively under the influence of convective flow patterns.

The dynamics of such concentrated magnetic structures can be conveniently described using the concept of *isolated magnetic flux tubes.* In ideal MHD we define them as bundles of magnetic field lines (constant magnetic flux), which are separated from the non-magnetic environment by a tangential discontinuity (surface current). As a consequence, the coupling between the tube and its environment becomes purely hydrodynamic and mediated by pressure forces, so that the flux tube can move through a perfectly conducting surrounding plasma, similar to solid body in a fluid. This is different from a diffuse field, which has to follow all plasma motions because of the flux freezing condition.

If the diameter of the flux tube is small compared to all other relevant length scales (scale heights, wavelengths, radius of curvature, etc.) the *thin flux tube approximation* can be employed, a quasi-1D description that greatly simplifies the mathematical treatment (Spruit 1981, Ferriz-Mas & Schüssler 1993). The forces which are most important for the dynamics of a magnetic flux tube are the *buoyancy force*, the *magnetic curvature force* (if the tube is non-straight), the *Coriolis force* (due to rotation), and the *aerodynamic drag force* (for motion relative to the surrounding plasma). Parker (1975) first pointed out that the buoyancy force has important consequences for dynamo models: on the one hand, it is the obvious mechanism to drive the rise and eruption of magnetic flux generated by the dynamo;

on the other hand, however, it is much too efficient in doing so since fields of equipartition strength are lost from the convection zone within about a month. This is much shorter than the 11-year time scale for field generation and amplification by the dynamo. As a consequence, magnetic flux is lost from the convection zone much faster than it can be regenerated by the dynamo process.

The buoyancy problem is actually made even more severe by the superadiabatic stratification of the convection zone. If the field strength is large enough to suppress the convective energy exchange between the flux tube and the surrounding medium the evolution of the tube in the deep convection zone is nearly adiabatic since the radiative thermal exchange time is very large (unless the tube diameter is extremely small). If a flux tube in initial temperature equilibrium with its environment rises from a starting depth z_0 to depth z, the density difference between the plasma within a flux tube and the surrounding medium becomes,

$$\frac{\Delta\rho}{\rho} = -\frac{1}{\beta}\left(\frac{1}{\gamma} + \nabla_{\rm ad}\frac{\beta}{\beta_0}\right) + \frac{\Delta S_e}{c_p}, \tag{2}$$

where $\beta = 8\pi p/B^2$ and β_0 is the corresponding value at z_0; $\Delta S_e = S_e(z) - S_e(z_0)$ is the entropy difference, which is related to the superadiabaticity of the stratification, $\delta = \nabla - \nabla_{\rm ad}$, and the pressure scale height, H_p, by $dS_e/dz = -c_p\delta/H_p$. The first term in the equation describes magnetic buoyancy while the second term represents the convective buoyancy. For $\delta > 0$ (i.e., in the convection zone) the convective buoyancy increases the density deficit and therefore leads to an even faster flux loss from the convection zone. On the other hand, a stable stratification ($\delta < 0$) opens the possibility to compensate the magnetic buoyancy by the second term in Eq. (2) and to store magnetic flux without buoyant loss. A subadiabatically stratified layer of *convective overshoot* below the convection zone combines the possibility for flux storage with the existence of a velocity field (overshooting convection) for the operation of the dynamo; it has therefore been proposed by Spiegel & Weiss (1980) and others as the site of the solar dynamo and the storage location of the generated toroidal magnetic flux.

2.2. FLUX TUBES IN MECHANICAL EQUILIBRIUM

It is not known whether magnetic flux is stored within the overshoot region in the form of a more or less homogeneous layer of diffuse field or in the form of isolated magnetic flux tubes. Most of the observed magnetic flux at the solar surface forms a hierarchy of concentrated magnetic structures (Solanki 1993) and we may conjecture that a similar organization might prevail throughout the convection zone and overshoot region. Moreover,

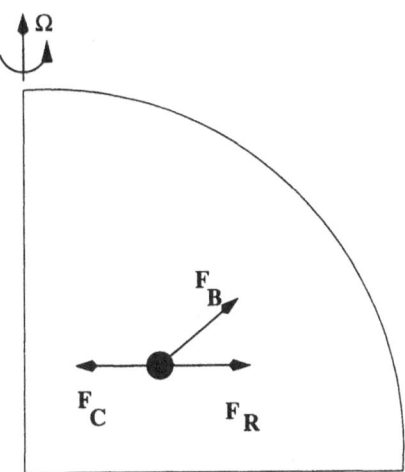

Figure 1. Forces on an axisymmetric flux tube. The buoyancy force (F_B), rotationally induced force (F_R), and magnetic curvature force (F_C) are shown in a meridional plane. The cross section of the flux tube is indicated by the black disk. While F_R and F_C are perpendicular to the axis of rotation, the buoyancy force is (anti)parallel to the radial direction of gravity.

layers of magnetic flux are notoriously unstable with respect to the magnetic Rayleigh-Taylor instability (e.g., Cattaneo & Hughes, 1988), which leads to tubular structures. Therefore, we assume in what follows that the magnetic flux is stored within the convective overshoot layer in the form of isolated toroidal flux tubes, i.e., flux rings in planes parallel to the equatorial plane.

The equilibrium of such a flux ring is determined by the balance between the buoyancy force, the magnetic curvature force, and the (Coriolis and centrifugal) forces due to rotation. The geometry of these forces is sketched in Fig. 1. Since the curvature force and the rotational force are perpendicular to the axis of rotation, the component of the buoyancy force parallel to the axis of rotation cannot be balanced unless the flux tube is exactly within the equatorial plane. Consequently, the buoyancy force must vanish in mechanical equilibrium and the rotational and curvature forces have to balance each other. This leads to the following two conditions (Moreno-Insertis et al., 1992):

$$\Delta\rho = 0, \tag{3}$$

$$\Omega_i^2 = \Omega_e^2 + v_A^2/\varpi^2. \tag{4}$$

The rotation rate Ω_i of the plasma within the flux tube has to be somewhat larger than that of the surrounding medium, Ω_e. v_A is the Alfvén velocity

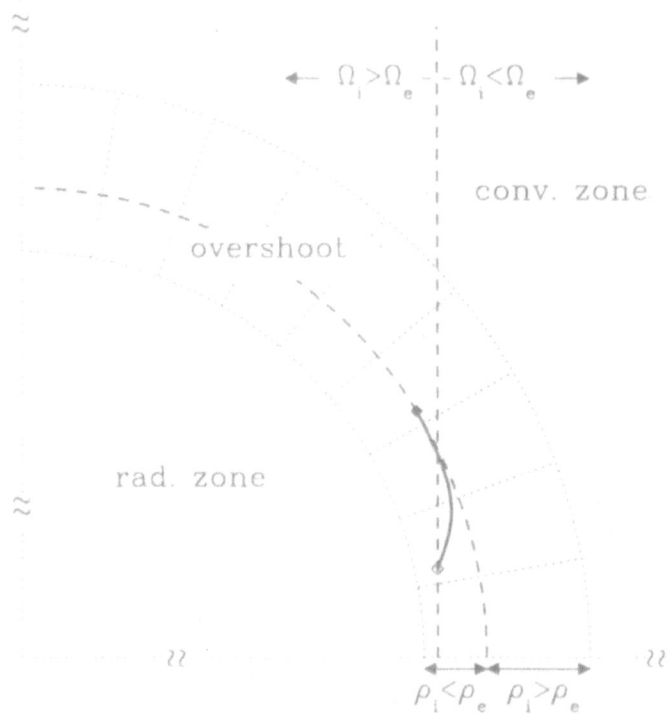

Figure 2. Approach of an axisymmetric flux tube towards its position of mechanical equilibrium given by Eqs. 3 and 4, sketched in a meridional plane. The flux tube is initially in temperature equilibrium with its environment and rotates at the same rate. The resulting unbalanced buoyancy and curvature forces lead to a rising and poleward motion until the equilibrium position is reached.

inside the flux tube and ϖ its distance from the axis of rotation. How can the flux tube actually reach such an equilibrium state, which is characterized by faster rotation and lower temperature (for $\Delta\rho = 0$) than its environment ? To answer this question assume a flux tube in the overshoot layer which is in temperature equilibrium with its environment and rotates with the same angular velocity. The tube is buoyant and rises; in a subadiabatically stratified layer, however, the tube becomes cooler than its environment and the buoyancy force diminishes until the equilibrium condition with vanishing density deficit is reached. Similarly, the unbalanced curvature force leads to a poleward meridional motion of the tube, which leads to an increase of its rotation speed because of angular momentum conservation. The poleward motion proceeds until the resulting rotational force balances the curvature force. The approach of the flux tube towards its equilibrium position is sketched in Fig. 2.

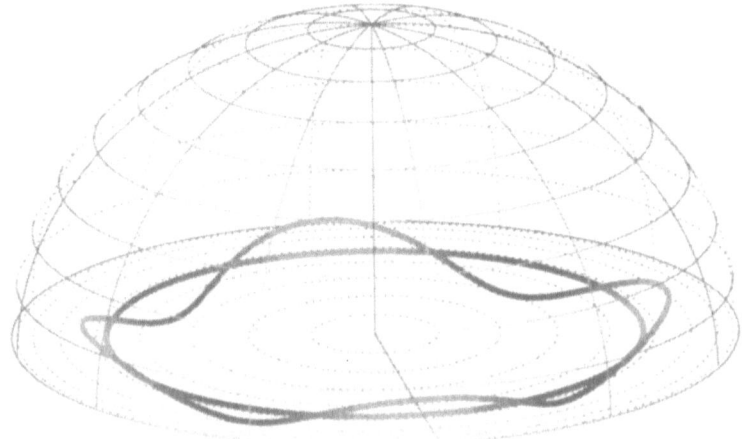

Figure 3. Undulatory instability of a toroidal flux tube. In the case sketched here, the initial flux ring is perturbed by a displacement of azimuthal wavenumber $m = 4$. A downflow of plasmas from the crests into the troughs lets the summits rise while the valleys sink.

3. Instability and eruption

3.1. UNDULATORY INSTABILITY

Since the buoyancy force is compensated by the stable stratification of the overshoot layer, the rise of flux tubes towards the surface and their final eruption must have another cause. As the field strength of a flux tube is intensified by differential rotation, at some stage it exceeds the threshold for the onset of the *undulatory instability* (Spruit & van Ballegooijen, 1982). This instability, which is a flux-tube analog to the Parker instability, in most cases sets in for non-axisymmetyric perturbations of an equilibrium flux ring as sketched in Fig. 3. A downflow of plasma along the field lines within the flux tube leads to an upward buoyancy force acting on the outward displaced parts and a downward force on the troughs, so that the perturbation grows. As a consequence, flux loops form, rise through the convection zone, and finally emerge at the surface to form sunspot groups and active regions. The sinking parts of the tube reach a new equilibrium at the bottom of the overshoot region and effectively 'anchor' the erupted loop.

A rather general linear analysis of the undulatory instability of flux rings at arbitrary latitude has been given by Ferriz-Mas & Schüssler (1993, 1994, 1995), including the effects of (differential) rotation, stratification,

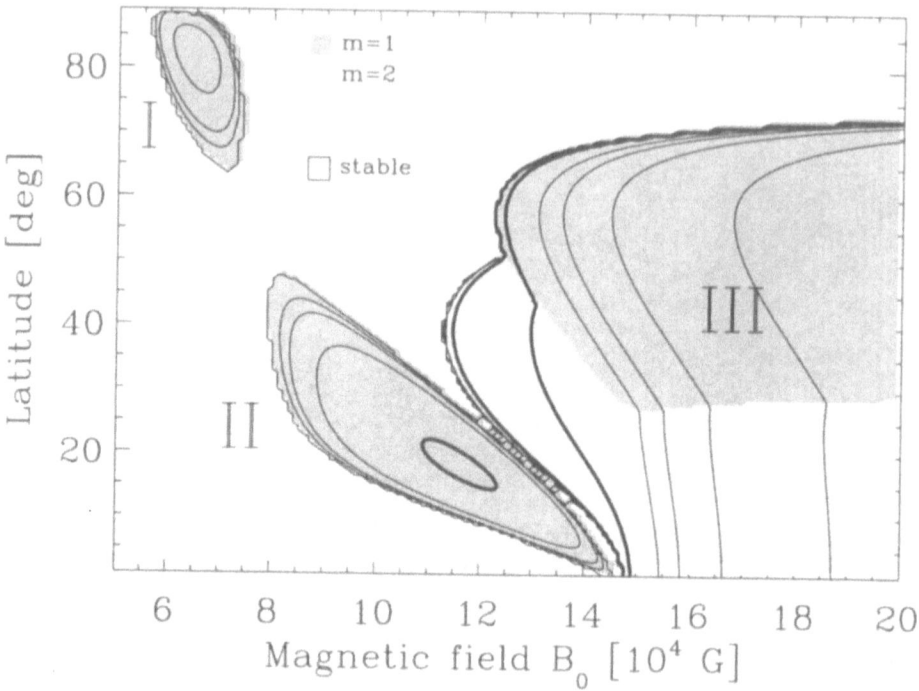

Figure 4. Stability diagram for the undulatory instability of toroidal flux rings in a model of the solar overshoot layer. Shaded regions indicate unstable flux tubes in the plane (initial) field strength vs. equilibrium latitude of the flux tubes. The degree of shading denotes the azimuthal wave number of the fastest growing unstable mode. The contours give lines of constant growth time.

and spherical geometry. Fig. 4 shows the result for flux rings stored in the overshoot region of a solar model calculated with a non-local mixing-length formalism (Skaley & Stix, 1991). The figure gives regions of undulatory instability in the plane of field strength vs. latitude. The degree of shading indicates the value of the azimuthal wavenumber for the mode with the largest growth rate. The contours correspond to lines of constant growth time of the instability.

The growth rates corresponding to the stability diagram shown in Figure 4 are given in Figure 5 in a perspective display. The values are small (corresponding to slow growth of the instability with e-folding times of more than a year) for the island-like regions in Fig. 4. The growth rates increase strongly for growing field strength in the third region of instability which, therefore, might be considered most important concerning the eruption of magnetic flux towards the solar surface. The slowly growing instabilities

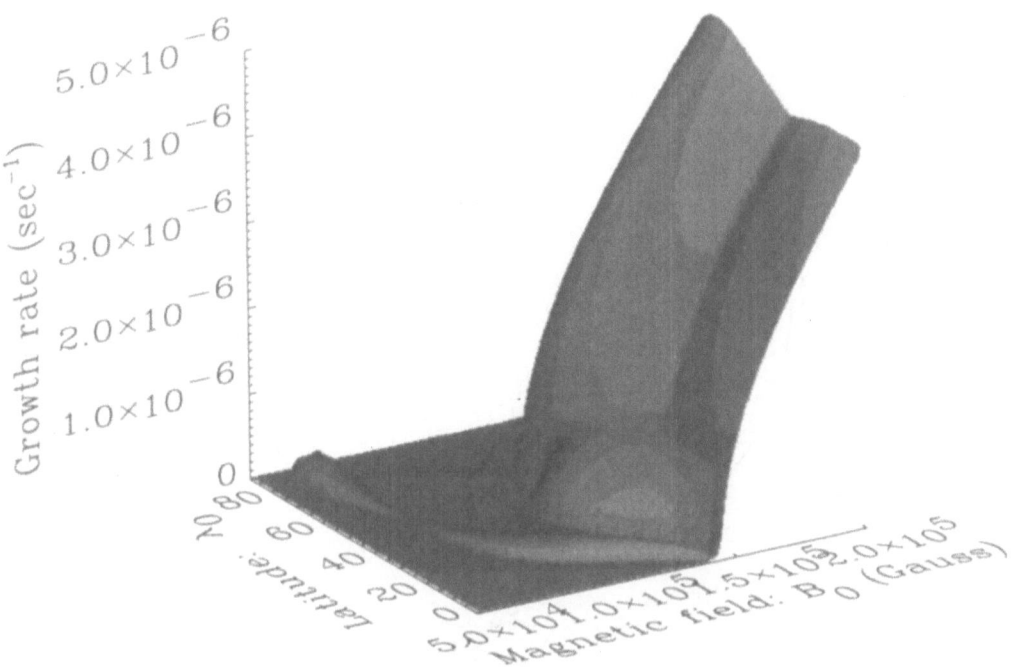

Figure 5. Growth rate of the undular instability (in s^{-1}) for the stability diagram in the (B_0, λ_0) plane given in Figure 4.

in the 'islands' are of potential relevance for the dynamo process (Ferriz-Mas et al., 1994). Fig. 5 demonstrates that significant growth rates of the instability (growth time below one year) are first reached in low latitudes and for field strength of the order of 10^5 G. The precise value is somewhat dependent on the value of the superadiabaticity ($\delta = -2.6 \cdot 10^{-6}$ in the case shown), but the order of magnitude is the same for various models of the solar overshoot layer (e.g., van Ballegooijen, 1982; Schmitt et al., 1984).

3.2. NONLINEAR EVOLUTION OF THE INSTABILITY

The threshold value of $\simeq 10^5$ G is about an order of magnitude larger than the equipartition field strength, which was previously thought to be the limit of field amplification by a turbulent dynamo. The results of numerical simulations for the nonlinear development of the undulatory instability support the requirement for such large initial field strengths.

Choudhuri & Gilman (1987) realized first that rotation has a strong

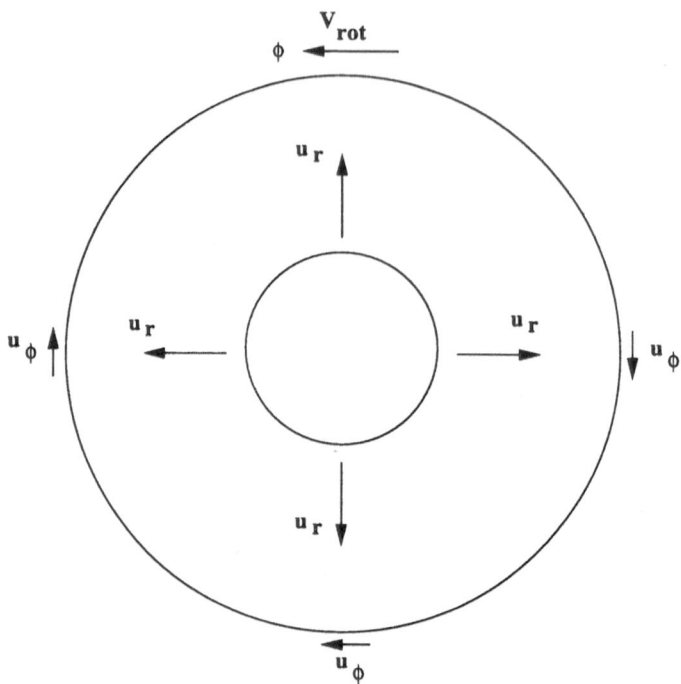

Figure 6. Effect of rotation on an expanding flux ring, which is seen from above along the axis of rotation. The expansion motion u_r causes a flow u_ϕ against the direction of rotation due to angular momentum conservation. This leads to an *inward* directed Coriolis force, which acts against the expansion of the flux ring.

influence on axisymmetric flux tubes rising due to buoyancy: the Coriolis force deflects the flux rings to high latitudes unless their initial field strength is larger than about 10^5 G. The following simple example illustrates the effect (see Fig. 6). Assume a flux ring in the equatorial plane, which has been set into an expanding motion with velocity u_r (by the action of a buoyancy force, for example). In a rotating system, this motion causes a Coriolis force which drives a flow u_ϕ along the tube and against the direction of rotation, reflecting the conservation of angular momentum. This velocity, in turn, causes an *inward* directed Coriolis force, which acts as a restoring force with respect to the expansion of the flux ring. The Coriolis acceleration in cylindrical coordinates (r, ϕ, z) is given by

$$A_C = 2\Omega \left(u_\phi, -u_r, 0 \right), \tag{5}$$

where Ω is the angular velocity. The linearized equation of motion in the absence of all other forces can be written as

$$\dot{u}_\phi = -2\Omega u_r, \tag{6}$$

$$\dot{u}_r = 2\Omega u_\phi, \tag{7}$$

which has the elementary solution

$$u_\phi, u_r \propto \sin(2\Omega t). \tag{8}$$

Consequently, the rise of the flux ring is transformed into an *inertial oscillation* with a period 2Ω, whose amplitude depends on the strength of the initial buoyancy force. If we are outside the equatorial plane, however, the Coriolis force affects only the component of the buoyancy force perpendicular to the axis of rotation (see Fig. 1); the trajectory of the flux ring then becomes the superposition of an inertial oscillation in the plane perpendicular to the axis of rotation and a rise parallel to the axis. If the Coriolis force dominates (i.e., if the rise time is longer than the period of the inertial oscillation) this leads to the eruption of magnetic flux in high latitudes, which is in striking conflict with the observed concentration of solar active regions in low-latitude bands.

To order of magnitude, the ratio of buoyancy force (F_B) to Coriolis force (F_R) can be estimated as

$$\frac{|\mathbf{F}_B|}{|\mathbf{F}_R|} \simeq \frac{B^2/(8\pi H_p)}{2\rho v_{\mathrm{rise}}\Omega} = \left(\frac{B}{B_{\mathrm{eq}}}\right)\left(\frac{\mathrm{Ro}}{2}\right), \tag{9}$$

where $\mathrm{Ro} = v_c/(2H_p\Omega)$ is the *Rossby number*, v_c is the convective velocity, and the equipartition field strength, B_{eq}, has been defined in Eq. (1). The rise velocity, v_{rise}, has been assumed here to be of the order of the Alfvén velocity (Parker, 1975). For the lower convection zone of the Sun we have $\mathrm{Ro} \simeq 0.2$. Consequently, magnetic fields must have field strengths of at least

$$B \simeq 10\, B_{\mathrm{eq}} \simeq 10^5\,\mathrm{G}$$

in order to avoid being dominated by the Coriolis force when erupting to form sunspots and active regions.

The simple estimates given above are confirmed by numerical simulations of the rise of magnetic flux tubes. Choudhuri (1989) showed that emerging flux loops (as they may result from the undulatory instability) are deflected poleward in the same way as rising flux rings unless the initial field strength at the bottom of the convection zone is of the order of 10^5 G. D'Silva & Choudhuri (1993) found that the *tilt angle* of sunspot groups with

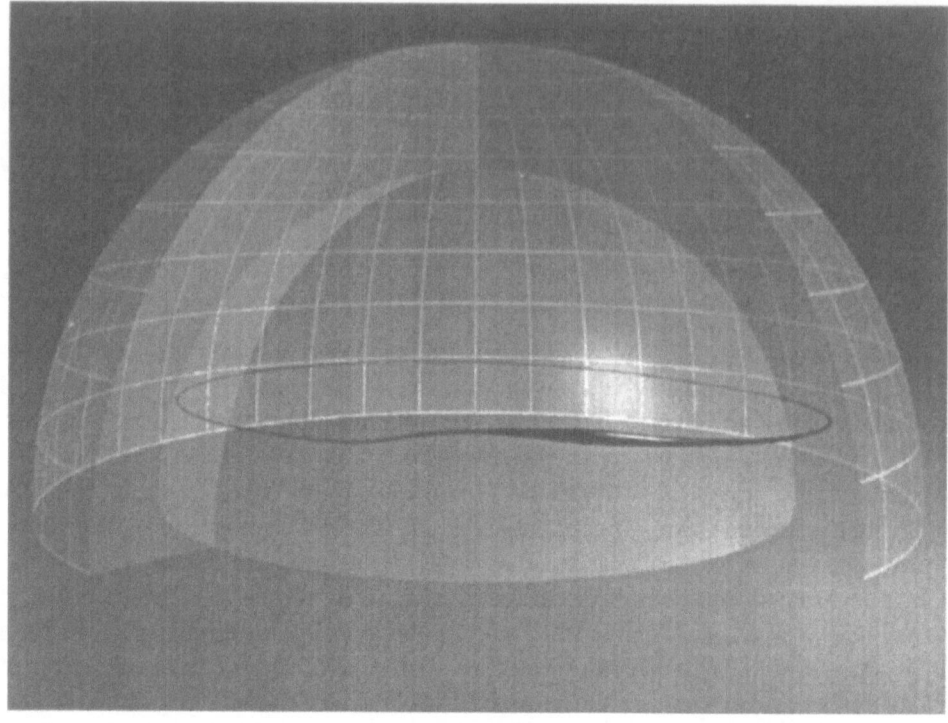

Figure 7. Three-dimensional view of an emerging flux loop resulting from the undulatory instability. The northern hemisphere is represented. The two transparent half-spheres correspond to the solar surface and the bottom of the convection zone, respectively..

respect to the East-West direction can be understood as a consequence of the Coriolis force acting on the expanding motion along a rising flux loop. The observed sizes of the tilt angles and their dependence on latitude again requires initial field strengths of the order of 10^5 G. Similar results were found by Fan et al. (1993), who performed a simulations of the same kind, but with a different numerical code.

While the simulations mentioned above start from a strongly buoyant flux loop in temperature equilibrium, which has already entered the convection zone, the approach of Schüssler et al. (1994) and Caligari et al. (1995) starts from flux tubes stored in force equilibrium within the overshoot layer. As we have seen, such flux tubes become unstable and form growing and rising flux loops precisely within the range of field strengths that is required for the observed emergence latitudes and tilt angles. The resulting loops are less buoyant than assumed in the simulations of the other groups; this has the consequence that the tendency for angular momentum conservation re-

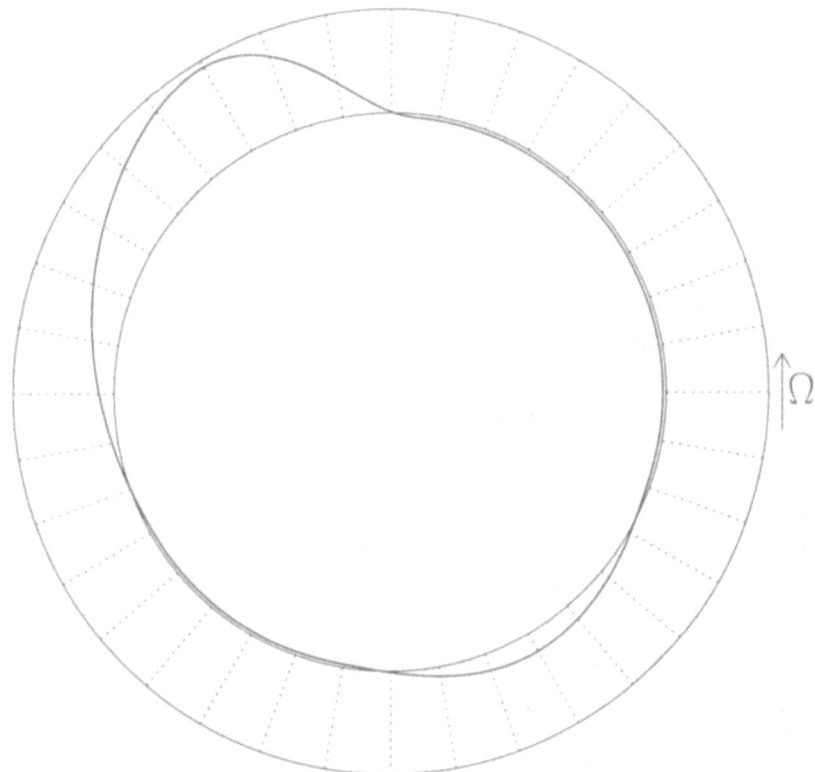

Figure 8. Polar view corresponding to the flux tube configuration of Fig. 7. Shown is a projection of the tube on the equatorial plane. Note the distinct geometric asymmetry between the two legs of the emerging loop. Since the unstable mode has an azimuthal wave number $m = 2$, a second loop can be seen emerging at about 180° longitude distance from the first.

tards the upper parts of the rising loop and leads to a distintive *asymmetry:* the part preceding in the direction of rotation is more inclined with respect to the vertical than the following part (Moreno-Insertis et al., 1994). This has observable consequences, namely the well-known difference in proper motion between preceding and following spots of young sunspots groups (e.g., Gilman & Howard, 1985; Strous 1994), and the geometrical asymmetry of the opposite-polarity parts of active regions and sunspot groups (van Driel-Gesztelyi & Petrovay, 1990).

Figs. 7 and 8 show an example of the simulation results of Caligari et al. (1995). The two transparent half-spheres in Fig. 7 correspond to the solar surface and the bottom of the convection zone, respectively. The observer looks face-on onto the erupting part of the flux tube, which can be recognized by its expansion. The tilt with respect to the direction of rotation is clearly visible. A slight poleward deflection of the erupted loop is visible, the effect being moderate since the initial field strength is large enough for buoyancy to dominate over the Coriolis force. At this stage, when the top of the rising loop is only a few thousand kilometers below the surface, the simulation has to be stopped since the tube diameter exceeds the local scale height and thin flux tube approximation breaks down. Fig. 8 gives a polar view on the same tube, i.e., a projection on the equatorial plane seen from above the North pole. The asymmetry between the preceding and the following part of the loop (with respect to the direction of rotation) is clearly visible: the following part is more vertical than the preceding part. The rise of such an asymmetric structure leads to the characteristic proper motions and geometrical asymmetries of young sunspot groups.

In summary we may say that the combined approach of linear stability analysis and numerical simulation leads to a consistent model for the storage, instability and eruption of magnetic flux, which is in accordance with the observational results for sunspot groups and indicates the existence of a layer of strong, super-equipartition field in a stably stratified overshoot layer below the main body of the solar convection zone.

4. Generation of super-equipartition fields

4.1. STRETCHING BY DIFFERENTIAL ROTATION

Which process generates the strong fields that are indicated by the studies of magnetic flux tube dynamics ? Since the field strength should be at least ten times larger than the equipartition field, the magnetic energy density is two orders of magntitude larger than the mean kinetic energy density of the convective motions. This largely excludes flux expulsion by convection as a mechanism. It is conceivable that convective flows could *locally* be much stronger (for instance, in concentrated downflows) and compress the field; however, such local concentrations correspond to large azimuthal wave numbers ($m > 10$, say) for which the undulatory instability requires field strength well in excess of 10^6 G.

An obvious possibility for field intensification is stretching by differential rotation in a layer of radial shear. Since the magnetic pressure is much smaller than the gas pressure for all conceivable field strengths in the lower convection zone, the change in field strength of a stretched flux tube is simply proportional to its elongation, $\Delta B/B \simeq \Delta l/l \simeq \Delta v \cdot \tau/d$, where Δv is the difference in rotation speed over the thickness of the shear layer, d, and τ is the elapsed time. Assuming $\Delta v = 100$ m/s (from helioseismology) , $d = 10^4$ km (estimated thickness of the overshoot layer) and a required amplification factor $\Delta B/B = 1000$ (from 100 G to 10^5 G) we find $\tau \simeq 3$ years, which appears to be a reasonable value for the amplification time.

However, the dynamical aspect of the problem has to be taken into account: as the field strength grows, the tension forces lead to an increasing resistance of the tube against further stretching. Assume a loop with a radius of curvature equal to the thickness of the shear layer, $R_c \simeq 10^4$ km, being stretched by the velocity difference $\Delta v = 100$ m·s^{-1} over the layer. Further stretching is inhibited when the resisting tension force, $B^2/4\pi R_c$, balances the aerodynamic drag force, $C_D \rho (\Delta v)^2/a$, which provides the stretching. C_D is the drag coefficient (of order unity) and a the radius of the tube. Equating both forces and inserting values yields a relation between the field strength that can be reached by stretching and the tube radius, viz.

$$B \simeq 3 \cdot 10^4 \, a_8^{-1/2} \text{ G} , \qquad (10)$$

where a_8 is the radius in units of 10^8 cm. The smaller the radius the stronger is the coupling of the tube to the shear flow by means of the drag force. Eq. (10) shows that, in order to reach field strengths in excess of 10^5 G, the tube radius must be smaller than 100 km, i.e., only very thin tubes, which contain much less magnetic flux than a large active region, can be sufficiently stretched. It is conceivable, however, that a larger tube may fray into smaller tubes near the apex of the stretched loop due to the action of the interchange and Kelvin-Helmholtz-like instabilities and could thus continue to be stretched.

Another problem with differential rotation as a generator for super-equipartition fields arises from energy considerations. The kinetic energy of *differential rotation* in the shear layer, $E_{\Delta\Omega}$, in a latitude belt of $\pm 30°$ can easily be estimated as

$$E_{\Delta\Omega} \simeq 5 \cdot 10^{37} \, v_4^2 \, d_9 \text{ erg} , \qquad (11)$$

where v_4 is the velocity difference in units of 10^4 cm/s and d_9 the thickness of the shear layer in units of 10^9 cm. On the other hand, the magnetic energy of a toroidal field with a magnetic flux of 10^{24} Mx (rough estimate of flux erupting during one activity cycle of the Sun, see Howard 1974) and a field strength of 10^5 G is

$$E_M \simeq 10^{39} \text{ erg} . \qquad (12)$$

We see from these estimates that the kinetic energy of differential rotation at any given time is about an order of magnitude smaller than the required magnetic energy for one (half) cycle. Consequently, a very efficient mechanism is required that continuously restores the shear layer and feeds energy into the differential rotation. No such mechanism has been proposed to date.

It is tempting to draw the attention of the reader to the following coincidence (?). The magnetic energy has to come, in one way or the other, from the energy of the convection providing the solar energy output. Dividing $E_M \simeq 10^{39}$ erg by 11 years, we find that the result is about 10^{-3} of the solar luminosity. If the magnetic energy is accumulated mainly during activity minimum and released during activity maximum, this would correspond to a luminosity variation between minimum and maximum of about 0.2%, the Sun being fainter during minimum and brighter during maximum. This is indeed the sign and the amplitude of the observed variations (e.g., Fröhlich, 1994).

4.2. 'EXPLOSION' OF MAGNETIC FLUX TUBES

A quite different mechanism for the intensification of magnetic fields stored in the overshoot region is possibly related to the 'explosion' phenomenon (Moreno-Insertis et al. 1995). This has to do with the *superadiabatic* stratification of the convection zone and the *adiabatic* evolution of a flux tube due to the very long thermal exchange time with its environment. We illustrate the process with a simplified model. Assume that the dynamical evolution of a flux tube in the convection zone is adiabatic and quasi-static, so that hydrostatic equilibrium (along the field lines) is maintained. If the plasma in the flux tube was isentropic initially, the internal gas pressure, p_i, as a function of height, z, is given by a polytropic stratification,

$$p_i(z) = p_{i,0} \left(1 - \frac{\nabla_{\mathrm{ad}} z}{H_{i,0}} \right)^{1/\nabla_{\mathrm{ad}}}, \tag{13}$$

where $H_{i,0}$ is the internal pressure scale height at $z = 0$ and ∇_{ad} is the adiabatic logarithmic temperature gradient. In the external medium we assume a polytropic stratification with (constant) temperature gradient ∇, which is given by

$$p_e(z) = p_{e,0} \left(1 - \frac{\nabla z}{H_{e,0}} \right)^{1/\nabla}, \tag{14}$$

where $H_{e,0}$ is the external pressure scale height at $z = 0$. The temperature difference between inside and outside as a function of height is

$$\Delta T \equiv T_i - T_e = T_{i,0} - T_{e,0} + \frac{\mu g}{\mathcal{R}} (\nabla - \nabla_{\mathrm{ad}}) z \tag{15}$$

(μ: mean molecular weight, g: gravitational acceleration, $\mathcal{R}$: gas constant). Since the external stratification is superadiabatic ($\nabla - \nabla_{\mathrm{ad}} > 0$), ΔT grows and the internal scale height becomes larger than the external scale height. Consequently, $p_i(z)$ decreases less rapidly with height than $p_e(z)$ and there is a critical height, z_e, at which we have $p_i(z_e) = p_e(z_e)$. Since the balance of total pressure between the flux tube and its surroundings is maintained, this

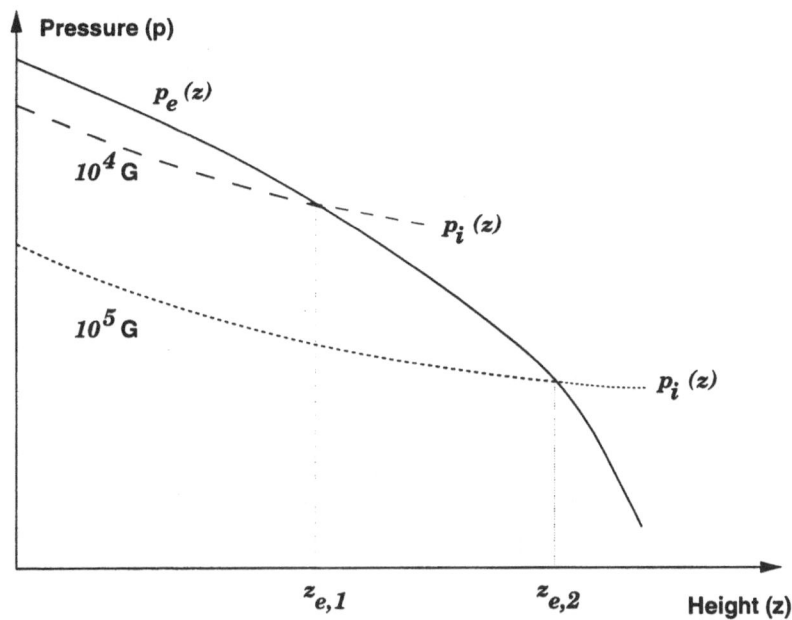

Figure 9.   Schematic representation of the external (p_e) and internal (p_i) pressure profiles as a function of height. Due to the superadiabaticity of the external stratification the internal pressure decreases slower with height, so that the curves intersect at the 'explosion height', z_e, where $B \to 0$. The value of z_e depends on the initial field strength: increasing field strength entails a larger initial pressure difference and a larger explosion height.

means that $B^2/8\pi = p_e - p_i \to 0$ for $z \to z_e$ and conservation of magnetic flux formally demands that the radius of the tube becomes infinite at that point: the flux tube 'explodes' when the top of the rising loop approaches the height z_e.

Fig. 9 schematically illustrates the origin of the explosion phenomenon by showing the gas pressure as a function of height for the external medium (thick curve) and the internal gas pressure for two flux tubes of different initial field strength. Since the initial pressure difference increases for growing field strength, the explosion height increases with B: flux tubes with stronger fields explode nearer to the surface.

In the numerical simulations the explosion manifests itself by a sudden dramatic weakening of the field when the apex of a rising loop reaches the explosion height; the simulations cannot be continued after that event since the thin flux tube approximation breaks down and the flux tube looses its identity as a coherent object. The explosion only occurs, however, if the evolution is slow enough to allow hydrostatic equilibrium to be approximately maintained. This is the case if the tube radius is sufficiently small to allow the aerodynamic drag force to slow down the rising motion; for a

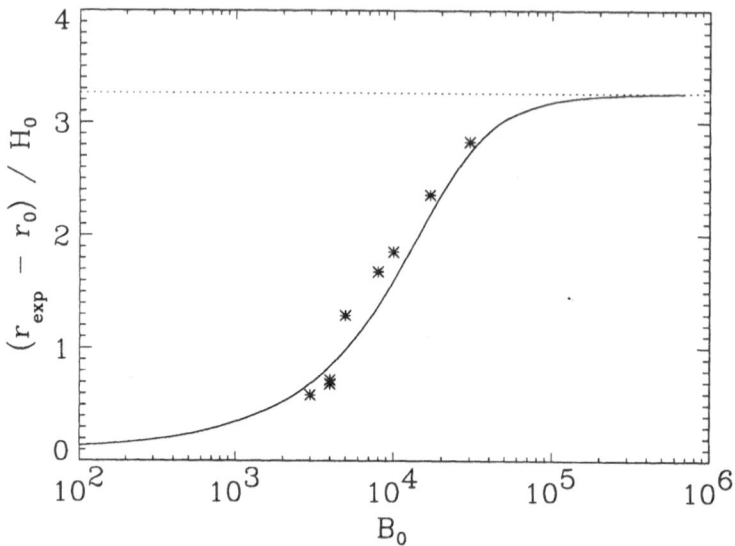

Figure 10. Explosion height (height above the overshoot layer, in scale heights) as a function of initial magnetic field strength (in G). The full curve gives the estimate based on the assumption of hydrostatic equilibrium. The asterisks denote the numerical results for flux tube evolution ending in an explosion. The dashed horizontal line indicates the solar surface.

field of 10^4 G this requires a magnetic flux below $\simeq 10^{20}$ Mx.

A comparison between the prediction on the basis of hydrostatic equilibrium (Eqs. 13 and 14) and the actual explosion height in the numerical simulations is given in Fig. 10. The predicted explosion height (height above the overshoot layer in units of the scale height there) is given as a function of the initial field strength (full line). The asterisks indicate the results of the numerical calculation for exploding flux tubes. The good agreement supports the validity of our interpretation of the explosion phenomenon. The horizontal dashed line in Fig. 10 gives the height of the solar surface; we see from the figure that flux tubes with field strengths above 10^5 G can reach the surface without exploding, even if they contain so little flux that they practically evolve near hydrostatic equilibrium. Flux tube with equipartition field strength of 10^4 G, on the other hand, explode about one scale height above the overshoot region, i.e., in the lower half of the convection zone.

The abrupt weakening of the magnetic field and the concomitant inflation of the upper parts of a flux loop when it approaches the explosion radius could provide a mechanism for the intensification of the field strength in the deepest reaches of a flux tube: matter is 'sucked' up from below into the inflated summit region, the gas pressure in the submerged part of the tube decreases, and the magnetic field strength increases correspondingly. This is related to the mechanism proposed by Parker (1994): a section of a stable (weak-field or equipartition) flux tube is lifted from the overshoot region by a convective upflow and carried into the superadiabatic convection zone to form an 'Omega-loop'. The establishment of hydrostatic equilibrium along the tube drains gas from the submerged parts and so increases their field strength. Moreno-Insertis et al. (1995) have shown that a sizeable field intensification through this process requires that the flux loop approaches its explosion height very closely. Only then can the enormous amount of matter which must be 'sucked up' into the loop be accomodated by its drastically inflated apex. In practice, however, this close approach cannot be distinguished from the explosion itself: the field strength at the top of the loop decreases dramatically, the flux tube looses its coherence there and becomes passive with respect to the convective motions: the remaining tube will consist of two more or less inclined tubular 'stumps' connected by a large web of tangled weak field. We may speculate that the further evolution following the explosion may be driven by the buoyancy of the high-entropy material within the flux tubes and lead to a continuous out-flow of plasma from the stumps. If this process occurs in the middle of the convection zone, then a noticeable field concentration in the deepest part of the flux tube could perhaps ensue.

5. Strong fields and dynamo theory

If the toroidal flux system in the overshoot layer indeed has a field strength of the oder of 10^5 G, this creates a problem for conventional theory of turbulent dynamo action. The toroidal field is much stronger than the equipartition value so that convection cannot efficiently twist it to create a poloidal field component via the 'α-effect'. Parker (1993) has proposed a spatial separation of the regions where the toroidal field is created (in the overshoot layer by differential rotation) and where the poloidal field is generated (in the convection zone proper by the conventional α-effect). Both regions are connected by diffusion so that the resulting dynamo wave takes on the character of a *surface wave* propagating along the interface between the convection zone and the overshoot layer.

Instead of invoking diffusion, Durney (1995) and Choudhuri et al. (1995) suggest that the poloidal field generated in the convection zone is transported into the overshoot layer by a *meridional flow* (poleward at the top, equatorward at the bottom). The latter authors find that their model shows equatorward propagation of the dynamo wave (in accordance with the observed latitude drift of the sunspot zone) even if the sign combination of α-effect and angular velocity gradient would lead to poleward propagation in the absence of a meridional flow (Yoshimura, 1975).

For both approaches, the explosion phenomenon discussed in the preceding section may provide a source of weak field in the convection zone. The turbulent flows can act upon such fields and give rise to the α-effect. This is also relevant for a 'dynamic dynamo' operating on the basis of intrinsically strong fields (e.g. Ferriz-Mas et al., 1994), which requires a weak-field dynamo as a starter and backup system. This approach is based on the influence of rotation on the non-axisymmetric instability of toroidal flux tubes: the Coriolis force leads to growing helical waves, which give rise to a non-vanishing α-effect. A scenario for a 'flux tube dynamo' based upon this effect together with differential rotation has been described by Schüssler (1993).

In contrast to the conventional kinematic dynamo models, in the framework of such a model the velocity field is not prescribed, but consistently determined from the (linearized) MHD equations. Strong fields are no problem but actually required for the mechanism to operate, since the undulatory instability must be excited. On the other hand, such a dynamo is not truly self-excited since the instability requires the field strength to exceed a certain threshold value; a turbulent dynamo as a 'starter' is required to work in the background.

6. Conclusions

A consistent picture for the storage, instability and eruption of magnetic flux tubes in the solar overshoot layer and convection zone has emerged during the last years. Its basic ingredients are:

- Toroidal magnetic flux in a mechanical equilibrium configuration is stored within the stably stratified overshoot region.
- Undular instability of flux tubes leads to loop formation when the field strength exceeds the threshold value of $\simeq 10^5$ G.
- Simulation results for the nonlinear development are in accordance with observed basic properties of active regions.

The generation of the required super-equipartition fields seems only marginally feasible with differential rotation. It is unclear at present whether the formation of 'Omega-loops' or the explosion process provide an alternative mechanism for the intensification of toroidal fields.

In any case, super-equipartition fields require a modified dynamo model. This could be a dynamo with spatial separation of the generation processes for toroidal and poloidal field or a 'dynamic dynamo' based upon the instability of strong magnetic flux tubes.

References

Caligari, P., Moreno-Insertis, F., Schüssler, M. 1995, ApJ, 441, 886
Cattaneo, F., Hughes, D.W. 1988, J. Fluid Mech., 196, 323
Choudhuri, A.R. 1989, Sol. Phys., 123, 21
Choudhuri, A.R., Gilman, P.A. 1987, ApJ, 316, 788
Choudhuri, A.R., Schüssler, M., Dikpati, M. 1995, A&A, in press
D'Silva, S., Choudhuri, A.R. 1993, A&A, 272, 621
Durney, B.R. 1995, Sol. Phys., in press
Fan, Y., Fisher, G.H., DeLuca, E.E. 1993, ApJ, 405, 390
Ferriz-Mas, A., Schmitt, D., Schüssler, M. 1994, A&A, 289, 949
Ferriz-Mas, A., Schüssler, M. 1993, Geophys. Astrophys. Fluid Dyn., 72, 209
Ferriz-Mas, A., Schüssler, M. 1994, ApJ, 433, 852
Ferriz-Mas, A., Schüssler, M. 1995, Geophys. Astrophys. Fluid Dyn., in press
Fröhlich, C. 1994, in: The Sun as a Variable Star, J.M. Pap, C. Fröhlich, H.S. Hudson, S.K. Solanki (eds.), Cambridge University Press, Cambridge, 28
Gilman, P.A., Howard, R.F. 1985, ApJ, 295, 233
Howard, R.F. 1974, Sol. Phys., 38, 283
Howard, R.F. 1989, Sol. Phys., 123, 271
Jensen, E. 1955, Ann. d'Astrophys., 18, 127
Krause, F., Rädler, K.-H. 1980, Mean-Field Magnetohydrodynamics and Dynamo Theory, Pergamon, Oxford
Moffat, H.K. 1978, Magnetic Field Generation in Electrically Conducting Fluids, Cambridge University Press, Cambridge
Moreno-Insertis, F., Caligari, P., Schüssler, M. 1995, ApJ, 452, 894
Moreno-Insertis, F., Schüssler, M., Caligari, P. 1994, Sol. Phys., 153, 449
Moreno-Insertis, F., Schüssler, M., Ferriz-Mas, A. 1992, A&A, 264, 686
Parker, E.N. 1955a, ApJ, 122, 293
Parker, E.N. 1955b, ApJ, 121, 491
Parker, E.N. 1975, ApJ, 198, 205
Parker, E.N. 1979, Cosmical Magnetic Fields, Clarendon, Oxford
Parker, E.N. 1993, ApJ, 408, 707
Parker, E.N. 1994b, ApJ, 433, 867
Schmitt, J.H.M.M., Rosner, R., Bohn, H.-U. 1984, ApJ, 282, 316
Schüssler M. 1993, in: The Cosmic Dynamo, F. Krause, K.-H. Rädler, G. Rüdiger (eds.), IAU-Symp. No. 157, Kluwer, Dordrecht, 27
Schüssler, M., Caligari, P., Ferriz Mas, A., Moreno-Insertis, F. 1994, A&A, 281, L69
Skaley, D., Stix, M. 1991, A&A, 241, 227
Solanki S.K. 1993, Space Sci. Rev., 61, 1
Spiegel E.A., Weiss N.O. 1980, Nature, 287, 616
Spruit H.C. 1981, A&A, 102, 129
Spruit, H.C., van Ballegooijen, A.A. 1982, A&A, 106, 58
Strous, L.K. 1995, PhD thesis, University of Utrecht
van Ballegooijen A.A. 1982, A&A, 106, 43
Van Driel-Gesztelyi, L., Petrovay, K. 1990, Sol. Phys., 126, 285
Wang, Y., Sheely, N. 1991, ApJ, 375, 761
Yoshimura H. 1975, ApJ, 201, 740
Zwaan, C. 1978, Sol. Phys., 60, 213
Zwaan, C. 1992, in: Sunspots–Theory and Observations, eds. J. H. Thomas & N. O. Weiss (Dordrecht:Kluwer), 75
Zwaan, C., Harvey, K.L. 1994, in: Solar magnetic fields, eds. M. Schüssler & W. Schmidt, (Cambridge University Press), 27

SIPHON FLOWS IN SOLAR MAGNETIC FLUX TUBES AND SUNSPOTS

JOHN H. THOMAS
Department of Mechanical Engineering, Department of Physics
and Astronomy, and C.E.K. Mees Observatory
University of Rochester
Rochester, New York 14627
USA

Abstract. The behavior of steady siphon flows along thin, arched magnetic flux tubes in the solar atmosphere is discussed, with particluar attention to the case of "flexible" flux tubes in the solar photosphere, where the plasma beta is of order unity. Qualitative features of subcritical and supercritical siphon flows are illustrated by the simple case of isothermal flow in an isothermal external atmosphere. More realistic flows, including representations of the temperature-stratified external solar atmosphere, the radiative transfer of energy between the flux tube and its surroundings, and variable ionization fraction along the flow are also discussed. The jump conditions for an adiabatic tube shock in a supercritical siphon flow are analyzed in some detail. Siphon flows have been detected in intense photospheric flux tubes arching across a magnetic neutral line on the Sun. Siphon flows also offer the most plausible explanation of the Evershed effect in sunspots. Recent observational and theoretical evidence concerning the siphon-flow model of the Evershed effect is discussed.

1. Introduction

Our subject here is the flow that is driven along an arched magnetic flux tube in the solar (or stellar) atmosphere by a pressure difference between the two footpoints of the arch. These siphon flows were first studied by Meyer and Schmidt (1968) in the context of the Evershed flow in sunspots. Some subsequent studies examined siphon flows in coronal magnetic loops (Pikel'ner 1971; Glencross 1980; Cargill and Priest 1980, 1982; Noci 1981;

39

K. C. Tsinganos (ed.), Solar and Astrophysical Magnetohydrodynamic Flows, 39–60.
© *1996 Kluwer Academic Publishers.*

Priest 1981). All of these studies assumed an effectively rigid flux tube whose magnetic configuration is unaffected by the flow, a valid assumption in the case of the corona where the plasma beta is low and where the magnetic field is in a space-filling, force-free state.

A different approach to calculating steady siphon flows in coronal loops (low plasma beta), which leads to two-dimensional solutions in a space-filling magnetic field, is based on the method of an ignorable coordinate developed by Tsinganos (1982). This method finds exact solutions of the equations of magnetohydrostatic equilibrium and the Bernoulli equation for steady, polytropic flow. Tsinganos, Surlantzis, and Priest (1993) found a class of solutions whose topology is representative of an arcade of coronal loops. More recently, Surlantzis *et al.* (1994) have extended this approach to include the steady shock front in a supersonic flow and the effect of the flow on the magnetic field geometry (as a small perturbation).

For flux tubes in the solar photosphere, where the plasma beta is of order unity, and to some extent in the low chromosphere, where the plasma beta is still not small, it is inappropriate to assume that the flux tube is rigid and that its geometry can be specified independent of the flow. In this case one may consider an isolated flux tube, surrounded by field-free gas (or gas permeated by some ambient magnetic field), and free to expand or contract in response to pressure changes associated with the siphon flow inside the tube in order to maintain lateral pressure balance with the surrounding atmosphere (Thomas 1984, 1988). Moreover, because the siphon flow affects the magnetic forces and creates its own non-negligible inertial force in this case, the equilibrium path of the flux tube through the atmosphere cannot be specified independently, but instead must be computed along with the siphon flow. Thus, in the photosphere we must deal with an isolated, flexible flux tube rather than an embedded, rigid flux tube appropriate for the corona (low-beta limit). Siphon flows in isolated, flexible flux tubes have been studied extensively in recent years, with theoretical models of increasing realism. These flows will be the main topic of this paper.

2. Steady Siphon Flows in Isolated Magnetic Flux Tubes

2.1. BASIC EQUATIONS

We shall assume that the flux tube is thin, in the sense that its radius is much smaller than both its radius of curvature and the scale height of the surrounding atmosphere. Then we can consider the flow inside the tube to be one-dimensional, *i.e.*, locally parallel and uniform across the cross-section of the tube, so that all flow variables are functions only of the tangential coordinate s along the tube.

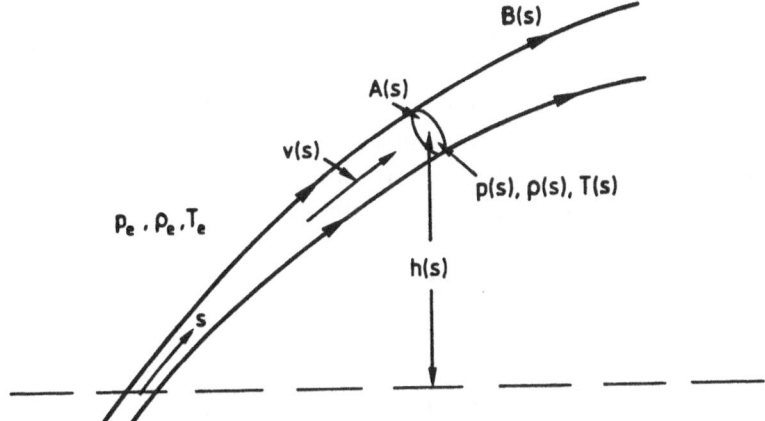

Figure 1. Basic geometry of a thin, isolated magnetic flux tube

The basic equations for steady, one-dimensional siphon flow along an isolated flux tube are the following (Thomas 1988):

$$\text{Continuity}: \quad \frac{d}{ds}(\rho v A) = 0 \quad \Rightarrow \quad \rho v A = Q = \text{constant}, \qquad (1)$$

$$\text{Momentum}: \quad v\frac{dv}{ds} = -\frac{1}{\rho}\frac{dp}{ds} - g\frac{dh}{ds}, \qquad (2)$$

$$\text{Equation of state}: \quad p = \rho R T, \qquad (3)$$

$$\text{Energy}: \text{ e.g.,} \quad T = T_e \text{ (isothermal)} \quad \text{or} \quad \frac{p}{p_0} = \left(\frac{\rho}{\rho_0}\right)^{\gamma} \text{(adiabatic)}, \quad (4)$$

$$\text{Magnetic flux conservation}: \quad \frac{d}{ds}(AB) = 0 \quad \Rightarrow \quad AB = \Phi = \text{constant}, \quad (5)$$

$$\text{Lateral pressure balance}: \quad p + \frac{B^2}{8\pi} = p_e. \qquad (6)$$

Here s is the tangential coordinate along the flux tube (see Fig. 1) and the other symbols have their usual meaning. Note that in the case of a rigid, embedded flux tube (in the limit of low plasma beta), where the tube geometry $h(s)$ and $A(s)$ are specified independent of the flow, equations (1)–(4) constitute a complete set of equations for the four variables $v(s)$, $p(s)$, $\rho(s)$, and $T(s)$.

For our present case of a flexible flux tube, where $h(s)$ and $A(s)$ depend on the flow, we must supplement equations (1)–(6) with the following equations describing the equilibrium path $h(x)$ of the flux tube through the surrounding atmosphere (Degenhardt 1989; Thomas and Montesinos 1990):

$$\frac{dh}{dx} = \tan\theta, \tag{7}$$

$$\left(\frac{B^2}{4\pi} - \rho v^2\right)\frac{d\theta}{dx} = -(\rho_e - \rho)g. \tag{8}$$

Here h is the altitude of the flux-tube axis, x is horizontal distance, and θ is the inclination of the flux-tube axis with respect to the horizontal (see Fig. 1). Equation (8) describes a balance among the net force due to magnetic tension, the inertial force due to the flow along curved streamlines, and the magnetic buoyancy force. Equations (7) and (8) follow from the normal component of the vector momentum equation, while equation (1) follows from the tangential component of the vector momentum equation. (Note that equations (1), (2), and (5) can also be written with x as the independent variable instead of s.)

2.2. ISOTHERMAL SIPHON FLOWS

Most of the important qualitative features of siphon flows can be illustrated by means of the simple case of isothermal flow and an isothermal external atmosphere. In this case, the basic flow equations lead to the following relations between velocity and height and between area and height (Thomas 1988):

$$\left(1 - \frac{v^2}{c_t^2}\right)\frac{dv}{v} = \frac{dh}{2H}, \tag{9}$$

$$\left(1 - \frac{v^2}{c_t^2}\right)\frac{dA}{A} = \left(1 - \frac{v^2}{c_1^2}\right)\frac{dh}{2H}, \tag{10}$$

where $H = RT_e/g = c^2/g$ is the constant density scale height of the external atmosphere, $c = (RT)^{1/2} = (RT_e)^{1/2}$ is the isothermal sound speed, $a = (B^2/4\pi\rho)^{1/2}$ is the Alfvén speed, $c_t = [c^2a^2/(c^2 + a^2)^{1/2}]$ is the tube speed, and $c_1 = [c^2a^2/(2c^2 + a^2)]^{1/2}$ is another characteristic speed. Note that in this case of isothermal flow the sound speed c is constant along the flux tube but the other characteristic speeds a, c_t, and c_1 vary along the tube.

The tube speed c_t is the speed of propagation of an axisymmetric pulse along an isolated thin flux tube (Defouw 1976), and as such it plays the role of the critical speed for siphon flows in a flexible flux tube, rather than the sound speed c, as in the case of a rigid flux tube (see Tsinganos *et al.* 1996 for further discussion of critical speeds for MHD flows). In general, we have the ordering

$$c_1 < c_t < \min(c, a). \tag{11}$$

There are two qualitatively different kinds of flow, *subcritical* flow ($v < c_t$) and *supercritical* flow ($v > c_t$), playing roles equivalent to subsonic and supersonic flow in the case of a rigid flux tube (Thomas 1984).

We can illustrate the qualitative behavior of siphon flows just by considering the velocity-height relation (9) and area-height relation (10). Consider a thin, isolated flux tube in the form of a symmetric arch, with a siphon flow going from left to right along the tube, as shown in Figure 2a. If the arch is low enough and the initial velocity (at the upstream footpoint $h = 0$) is low enough, the flow will remain subcritical throughout the arch. Assume for the moment that $v < c_1$ throughout the flow also. Then, the three terms in parentheses in equations (9) and (10) are positive everywhere along the tube and hence dv and dA have the same sign as dh, so the flow accelerates and the tube expands up to the top of the arch and then the flow decelerates and the tube contracts along the downstream leg of the arch, as illustrated in Figure 2a. The flow speed and the cross-sectional area both have maximum values at the top of the arch (where $dh = 0$), and from the continuity equation (1) we see that the density reaches a minimum at the top of the arch. The flow is symmetric about the top of the arch and the gas pressure is the same at the two footpoints. (This is a consequence of our neglect of viscosity; a steady viscous subcritical siphon flow will require a pressure difference between the footpoints to drive the flow.)

If the initial velocity or arch height are adjusted to produce a somewhat faster, but still subcritical flow, such that $c_1 < v_{max} < c_t$, then the term in parentheses on the right-hand side of the area-height relation (10) changes sign at some point on the way up to the top of the arch and the tube contracts beyond that point, producing a local maximum of cross-sectional area, or *bulge point*, at the point where $v = c_1$, as illustrated in Figure 2b. There is a symmetrically placed bulge point in the downstream leg as well. A static flux tube in a stratified atmosphere expands monotonically with height. The addition of a siphon flow in the tube creates a Bernoulli effect which reduces the internal gas pressure and thus reduces the rate of expansion with height. A bulge point is produced when the flow reaches speeds that are high enough that the Bernoulli effect completely counteracts the expansion due to stratification and causes the flux tube to contract with height.

44

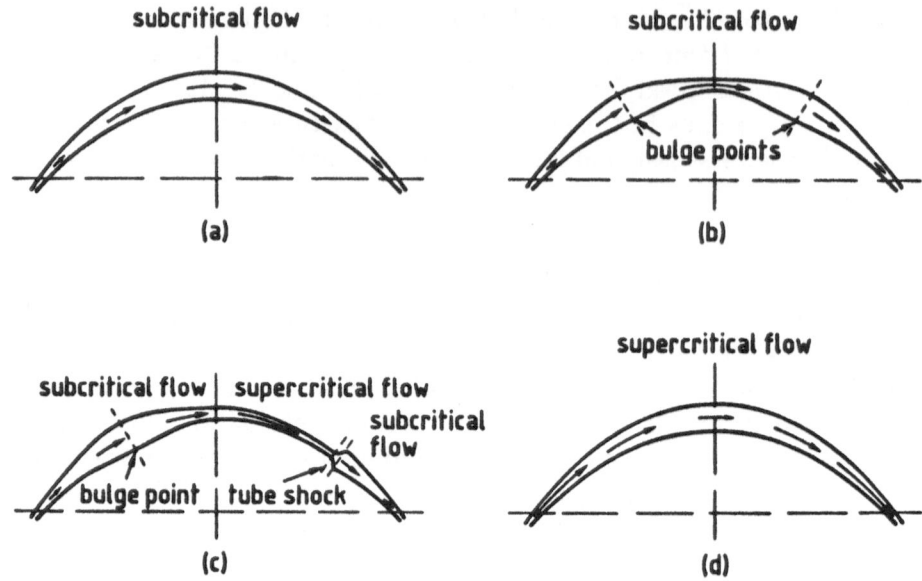

Figure 2. Schematic diagram of various types of steady, isothermal siphon flows in an arched, isolated magnetic flux tube in an isothermal external atmosphere. (a) Purely subcritical flow with $v < c_1 < c_t$ everywhere. (b) Purely subcritical flow with $c_1 < v_{max} < c_t$ and upstream and downstream bulge points. (c) Critical flow undergoing a smooth transition from $v < c_t$ to $v > c_t$ at the top of the arch and a discontinuous transition (*tube shock*) back to subcritical flow somewhere along the downstram leg of the arch. (d) Purely supercritical flow with $v > c_t$ everywhere. (From Thomas 1988.)

The velocity-height relation (9) shows that the flow can undergo a smooth transition ($dv > 0$) from subcritical ($v < c_t$) to supercritical ($v > c_t$) flow only at the top of the arch where $dh = 0$. A flow in which the velocity reaches the value $v = c_t$ somewhere is called a *critical flow*. For isothermal flow we can only have $v = c_t$ at the top of the arch. (The same is true for adiabatic flow; see Montesinos and Thomas 1989.) For a critical flow there are two possible solutions for the downstream leg of the flow, depending on the gas pressure at the downstream footpoint. If that pressure is the same as at the upstream footpoint, then the flow decelerates and is purely subcritical in the downstream leg. This is the limiting case of the symmetrical subcritical flows discussed above, with $v_{max} = c_t$ at the top of the arch. If the pressure is lower at the downstream footpoint than at the upstream footpoint, then the flow continues to accelerate and is super-critical in the downstream leg of the arch. In this case of supercritical flow, all three terms in parentheses in equations (9) and (10) are negative, so the velocity increases and the area decreases with decreasing height in the downstream leg. However, the supercritical flow solution corresponds to a very low internal gas pressure at the downstream footpoint. Generally, the actual pressure at the downstream footpoint will be higher than this value; in this case the flow must decelerate abruptly to subcritical speed across a standing *tube shock* somewhere along the downstream leg of the arch (see Fig. 2c). The velocity and magnetic field strength decrease abruptly across the tube shock, while the cross-sectional area increases abruptly. The location and strength of the standing tube shock depend on the gas pressure at the downstream footpoint; the shock is stronger and further downstream for lower downstream pressures. However, the supercritical flow between the top of the arch and the tube shock is the same in all these cases. A critical flow is "choked" in the sense that the total mass flux Q along the tube can not be increased by lowering the pressure at the downstream footpoint; this only serves to delay the subcritical-supercritical transition, *i.e.*, to move the standing tube shock further downstream. We will discuss the standing tube shocks further in section 3.

The flow equations also allow solutions that are supercritical everywhere along the flux tube, as illustrated in Figure 2d. However, these solutions are clearly unrealistic. There are interesting analogies between the siphon-flow equations and the equations for compressible flow through a Laval nozzle or the equations for a spherically symmetric solar wind, which the reader is encouraged to consider.

2.3. MORE REALISTIC SIPHON FLOWS

The simple example of isothermal flow in an isothermal external atmosphere discussed above is useful for demonstrating the qualitative behavior of siphon flows, but is inadequate for detailed modeling of actual siphon flows in the solar atmosphere. More realistic models have been developed which include realistic representations of the temperature-stratified solar atmosphere (Degenhardt 1989; Thomas and Montesinos 1991), the radiative transfer between the flux tube and its surroundings (Degenhardt 1991; Montesinos and Thomas 1993), and the varying ionization fraction of the flowing gas (Montesinos and Thomas 1993).

Degenhardt (1991) studied the effect of radiative transfer on purely subcritical siphon flows using the relaxation-time approach of Spiegel (1957; see also Unno and Spiegel 1966). He showed that the addition of radiative transfer between the flux tube and its surroundings, with finite, nonzero relaxation time, produces siphon flows that are asymmetric about the top of the arch and in which the critical point can occur somewhere other than the top of the arch, in contrast to the simpler cases of isothermal or adiabatic flow.

Montesinos and Thomas (1993) extended the treatment of radiative siphon flows to include critical siphon flows with standing tube shocks, variable ionization fraction of the flowing gas, and a more accurate representation of the radiative transfer. A few details of our treatment of the radiative transfer, which is based largely on earlier work of others, may be of interest to the reader. The steady-state energy equation for one-dimensional flow along the flux tube can be written in terms of the specific entropy S of the flowing gas in the form

$$v\frac{dS}{ds} = \left(\frac{dS}{dt}\right)_{rad}, \tag{12}$$

where $(dS/dt)_{rad}$ is the radiative exchange. In the layers beneath the solar surface we used the Spiegel formulation

$$\left(\frac{dS}{dt}\right)_{rad} = \frac{c_p}{\tau_{rad}T}(T_e - T), \tag{13}$$

where c_p is the specific heat at constant pressure, T and T_e are the temperatures inside and outside the flux tube, and τ_{rad} is the variable radiative cooling time given by

$$\tau_{rad} = \frac{c_p}{16\kappa\sigma T^3}[1 - \kappa\rho r \operatorname{arccot}(\kappa\rho r)]^{-1}, \tag{14}$$

where κ is the gray Rosseland opacity per unit mass, ρ is the mass density, and σ is the Stefan-Boltzmann constant. The term $\kappa\rho r$ in equation (14) is the optical thickness of the flux tube, and the formulation (13), (14) goes smoothly from radiative diffusion in the optically thick limit ($\kappa\rho r \gg 1$) in deeper layers to the optically thin limit ($\kappa\rho r \ll 1$) of Newton's law of cooling near the visible surface. For values of the opacity κ, we interpolated in the tables of Rogers and Iglesias (1992).

For the visible layers above the solar surface we used a method developed by Kalkofen and Ulmschneider (1977), Ulmschneider et al. (1978), and Herbold et al. (1985), which is more accurate than Newton's law of cooling for the larger temperature differences between the tube and its surroundings that occur at these higher levels. Using the gray LTE approximation, we write

$$\left(\frac{dS}{dt}\right)_{rad} = \frac{4\sigma\kappa}{T}(T_e^4 - T^4). \tag{15}$$

This formulation is more general than Newton's law of cooling but is equivalent to it for small temperature differences between the tube and its surroundings. For values of the opacity κ in equation (15) we used the empirical formula of Ulmschneider et al. (1978),

$$\kappa = 1.376 \times 10^{-23}p^{0.738}T^5 \tag{16}$$

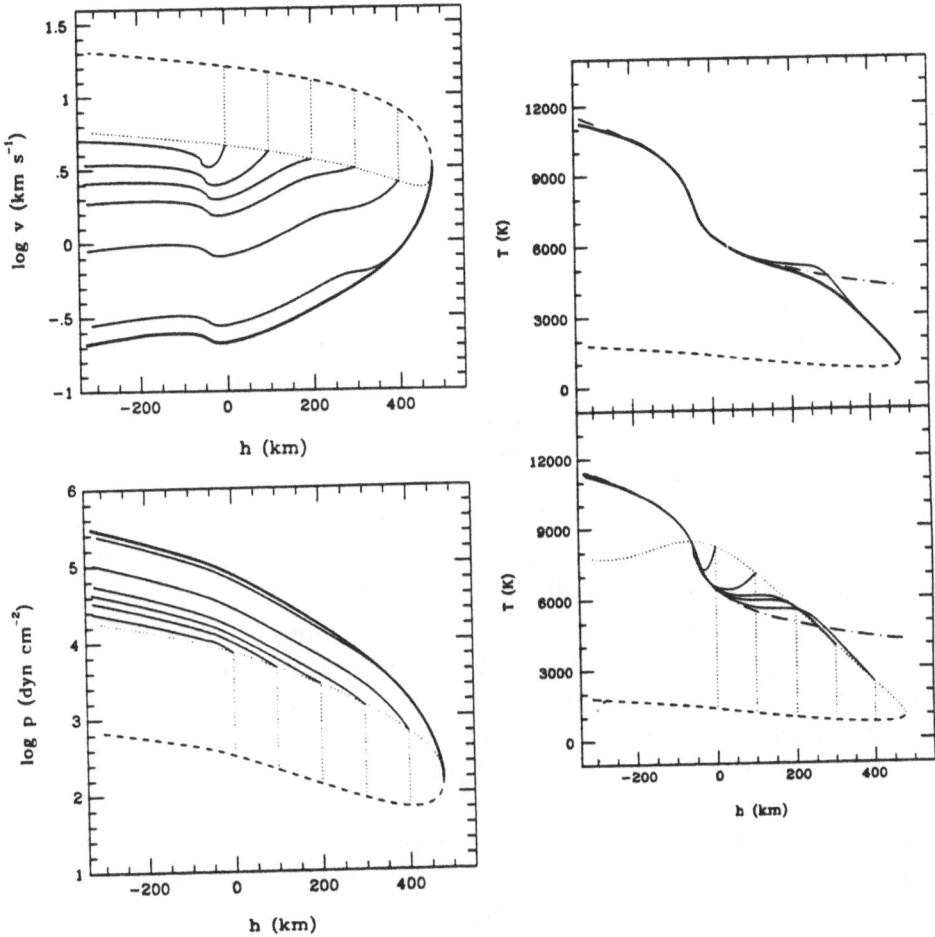

Figure 3. Examples of radiative critical siphon flows reaching up to the temperature minimum in the solar atmosphere, with standing tube shocks at several different heights ($h = 0$, 100, 200, 300, and 400 km). Shown are the velocity v, gas pressure p, and temperature T as functions of height h. Each plot shows the upstream subcritical branch (*heavy solid curve*), the supercritical downstream branch (*dashed curve*), the locus of possible values just downstream of the shock (*dotted curve*), the jumps across the shock at selected heights (*vertical dotted lines*), and the subcritical flow downstream of the shock (*solid curves*). The temperature plots show the purely subcritical case (upper panel) and the supercritical downstream branches with shocks (lower panel) separately and also show the temperature in the external atmosphere (*dash-dot curve*). (From Montesinos and Thomas 1993.)

(in cgs units), obtained by fitting a plane to the relation $\kappa(p, T)$ given by the opacity tables of Kurucz (1979).

Figure 3 shows an example of a critical radiative siphon flow reaching up to the height of the temperature minimum, with standing tube shocks at several different heights. To trace a particular flow along the arched tube, beginning at the upstream end, follow the thick solid curve (subcritical flow) up to the top of the arch (maximum h), follow the dashed curve (supercritical flow) down to one of the shocks (upstream side), follow the vertical dotted line to the downstream side of the shock discontinuity, and then follow the corresponding thin solid line (subcritical flow) to the downstream end. The "choked" flow in the upstream leg of the arch is the same in each case, but successively lower gas pressures at the downstream footpoint correspond to tube shocks further downstream. The internal temperature T increases sharply across each of the tube shocks and then relaxes radiatively to the external temperature as we go downstream of the shock. For shocks near the top of the arch (at $h = 400$ km, for example), where opacity is low, the flow is nearly adiabatic for some distance after the shock and significant radiative exchange does not begin until around $h = 200$ km, where the opacity is sufficiently large. On the other hand, for a low shock (at $h = 100$ km, for example) where the opacity is already high, radiative relaxation occurs over a short distance along the tube.

For siphon flows in arches reaching up to heights of about 300 km or more, such as that shown in Figure 3, the temperature of the flowing gas near the top of the arch is much lower than that of the surroundings because the radiative exchange is weak at those heights and the flow is nearly adiabatic. The temperature in actual solar cases will probably not be as low as that shown in Figure 3 (~ 1000 K) because the adiabatic cooling will be offset to some extent by mechanical heating. Nevertheless, we have suggested (Montesinos and Thomas 1993) that siphon flows along arched flux tubes may contribute to the cool component of the solar atmosphere near the temperature minimum inferred from observations (see Ayres 1981).

3. Tube Shocks in Supercritical Siphon Flows

In a critical siphon flow, the flow undergoes a smooth transition from subcritical to supercritical speed somewhere near the top of the arch and continues to accelerate along the downstream leg of the arch. At some point, however, it must decelerate suddenly across a shock front in order for the flow to conform to the required pressure at the downstream footpoint of the arch. This standing "tube shock" in a flexible flux tube is different from an ordinary hydrodynamic shock in a rigid flux tube in that the cross-sectional area of the flux tube increases abruptly across the shock as a consequence of the required lateral pressure balance with the external atmosphere.

Because of the rapid change in cross-sectional area across a tube shock, magnetic curvature forces are significant and the flow and magnetic field are necessarily two-dimensional within the shock. Thus, the detailed structure of a tube shock cannot be studied within the thin flux tube approximation. However, provided that the shock is thin, in the sense that its thickness is much less than the scale height of the surrounding atmosphere, then for purposes of computing a siphon flow we can treat it as a discontinuous transition and apply the appropriate jump conditions.

The jump conditions for an adiabatic tube shock were first derived by Herbold *et al.* (1985). Letting subscripts 1 and 2 denote the upstream and downstream sides of the shock, respectively, the jump conditions (assuming no change in the ratio of specific heats γ across the shock) are the following:

$$\rho_1 v_1 A_1 = \rho_2 v_2 A_2, \tag{17}$$

$$A_1[\rho_1 v_1{}^2 + 2(p_e - p_1)] = A_2[\rho_2 v_2{}^2 + 2(p_e - p_2)], \tag{18}$$

$$\frac{1}{2}v_1{}^2 + \frac{\gamma}{\gamma - 1}\frac{p_1}{\rho_1} = \frac{1}{2}v_2{}^2 + \frac{\gamma}{\gamma - 1}\frac{p_2}{\rho_2}, \tag{19}$$

$$B_1 A_1 = B_2 A_2, \tag{20}$$

$$p_1 + \frac{B_1{}^2}{8\pi} = p_2 + \frac{B_2{}^2}{8\pi} = p_e, \tag{21}$$

$$s_1 \leq s_2. \tag{22}$$

These conditions correspond, in the order given, to conservation of mass, momentum, energy, and magnetic flux, lateral pressure balance, and the second law of thermodynamics.

In order to elucidate the properties of tube shocks it is helpful to introduce three "Mach numbers" based on the three characteristic speeds c, a, and c_t. (Here c is the adiabatic sound speed.) Thus, we define the *Mach number* $M_c \equiv v/c$, the *Alfvén Mach number* $M_a \equiv v/a$, and the *tube Mach number* $M_t \equiv v/c_t$. From the definition of the tube speed, $c_t{}^2 \equiv c^2 a^2/(c^2 + a^2)$, we see that these three Mach numbers are related according to

$$M_t{}^2 = M_c{}^2 + M_a{}^2, \tag{23}$$

so that a Cartesian plot of M_a versus M_c also corresponds to a polar plot with coordinates (r, ϕ) given by

$$r = M_t, \quad \tan\phi = \frac{c}{a} = \left(\frac{\gamma\beta}{2}\right)^{1/2}, \tag{24}$$

i.e., the radial distance r equals the tube Mach number and the polar angle ϕ is a measure of the plasma beta. Such a plot, shown here in Figure 4, is helpful in understanding the implications of the jump conditions for a tube shock. The jump conditions (17)–(22) imply the following inequalities involving the three Mach numbers:

$$M_{t2} < 1 < M_{t1}. \quad M_{a2} < M_{a1} < 1, \quad M_{c2} < M_{c1}, \quad M_{c2} < 1. \qquad (25)$$

The first inequality, which follows from the entropy condition (22) (second law of thermodynamics), requires that the flow must go from supercritical to subcritical across the shock, which means that the points 1 and 2 in Figure 4, corresponding to the upstream and downstream sides of the shock, must lie outside and inside the unit quarter circle, respectively. The condition $M_{a2} < M_{a1} < 1$ means that the flow speed is sub-Alfvénic on both sides of the shock and the Alfvén Mach number decreases across the shock. In fact, Cheng (1992) has shown that in any steady flow along an isolated magnetic flux tube, the quantity $(1 - M_a{}^2)\rho v \cos \theta$ is constant along the tube, where θ is the slope of the tube axis, as in equation (7). Thus, the term $(1 - M_a{}^2)$ must have the same sign everywhere along the tube, and so a siphon flow that is sub-Alfvénic anywhere is sub-Alfvénic everywhere. This same result is true for the two-dimensional siphon flows studied by Tsinganos *et al.* (1993).

The flow must be subsonic downstream of the shock ($M_{c2} < 1$), but it need not be be supersonic upstream of the shock; there are allowed, "weak shock" solutions with $M_{c1} < 1$ (but $M_{t1} > 1$) with point 1 lying in the region bounded by the unit circle and the lines $M_c = 1$ and $M_a = 1$ in Figure 4. However, for strong shocks ($M_{t1} \gg 1$) the angle ϕ_1 is small, the upstream plasma beta β_1 is small, and $M_{c1} \simeq M_{t1} \gg 1$.

The jump conditions (17)–(22) can be solved analytically. It is convenient to introduce the following dimensionless ratios of downstream to upstream quantities:

$$X_v \equiv \frac{v_2}{v_1}, \ X_p \equiv \frac{p_2}{p_1}, \ X_\rho \equiv \frac{\rho_2}{\rho_1}, \ X_A \equiv \frac{A_2}{A_1} = \frac{B_1}{B_2}, \qquad (26)$$

where the last equality follows from the condition of flux conservation (20). The jump conditions (17)–(22) can then be written in terms of these dimensionless quantities, and through elimination we can derive a cubic equation for the area ratio X_A. After factoring out the trivial solution $X_A = 1$ (the case of no shock), we obtain the following quadratic equation for the area ratio (Ferriz-Mas and Moreno-Insertis 1987):

$$[\gamma(1 + \beta_1)(\alpha_1 - \beta_1)]X_A{}^2 - [(\gamma - 1)\alpha_1 + (2 - \gamma)\beta_1]X_A - \alpha_1 = 0, \qquad (27)$$

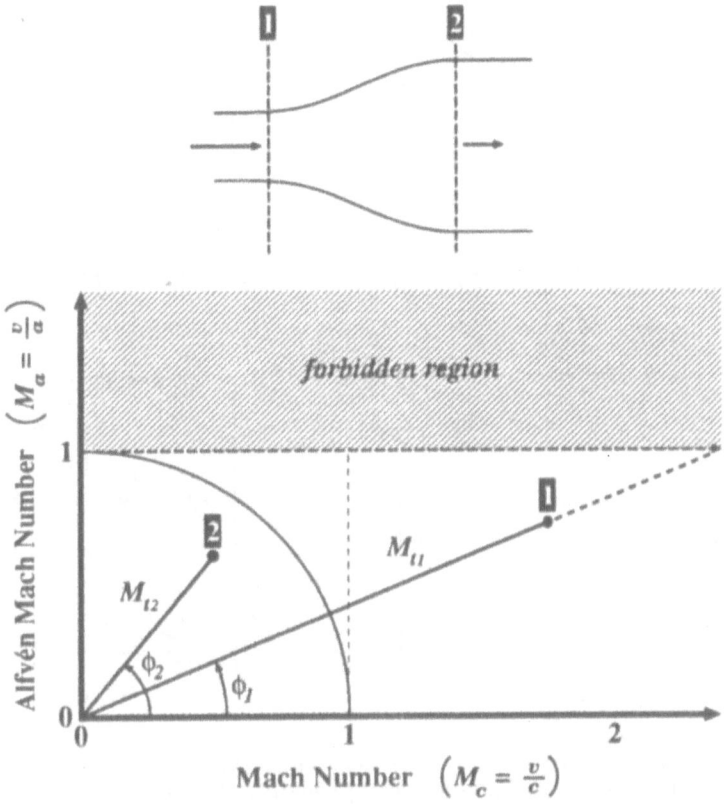

Figure 4. Graphical representation of the upstream (1) and downstream (2) states of a standing tube shock on the M_c, M_a plane. See text for discussion.

where $\beta_1 = 8\pi p_1/B_1{}^2$ is the upstream plasma beta and $\alpha_1 = 2p_1/\rho_1 v_1{}^2 = 2/\gamma M_{c1}{}^2$. Only one of the two solutions of this quadratic equation is physically meaningful. Once the ratio X_A is determined, the other ratios can be determined using the following expressions (which follow from the jump conditions):

$$X_v = 1 - \frac{\alpha_1}{\beta_1}\left(\frac{X_A - 1}{X_A}\right), \quad X_p = 1 + \frac{1}{\beta_1}\left(\frac{X_A{}^2 - 1}{X_A{}^2}\right), \quad X_\rho = \frac{1}{X_v X_A}. \quad (28)$$

Figure 5 shows plots of the pressure, density, velocity, and area ratios as functions of the square of the upstream tube Mach number, $M_{t1}{}^2$ for various values of the upstream plasma beta, β_1. The pressure ratio p_2/p_1 increases monotonically with increasing $M_{t1}{}^2$, whereas the density ratio ρ_2/ρ_1 reaches a local maximum with increasing $M_{t1}{}^2$, as noted by Ferriz-Mas and Moreno-Insertis (1987). The ratio of tube radii, $(A_2/A_1)^{1/2}$, increases with

52

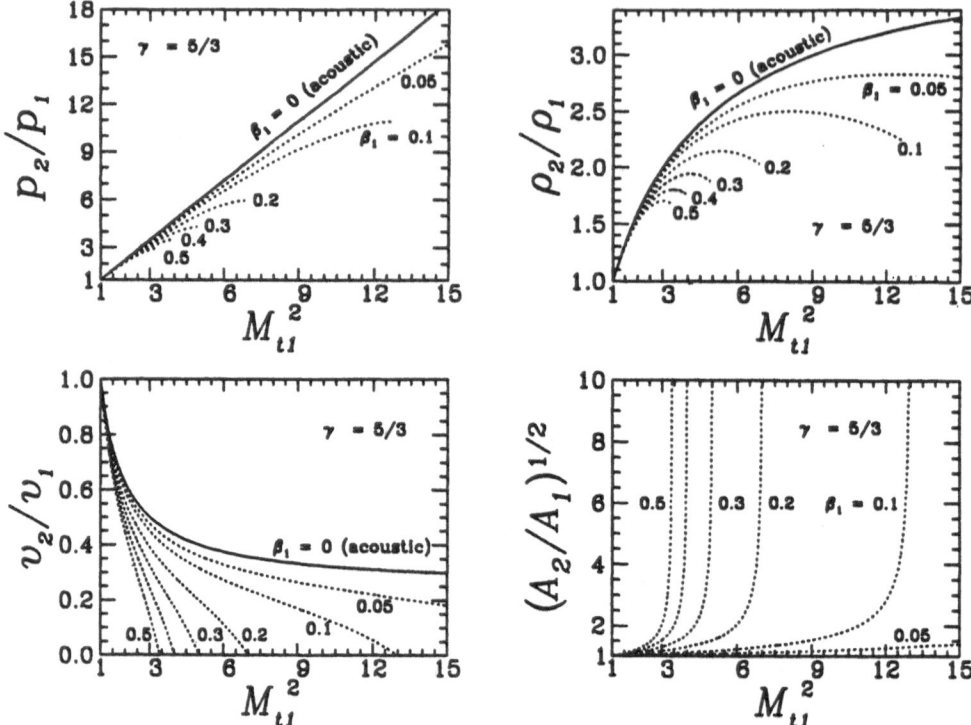

Figure 5. Plots, for an adiabatic tube shock, of ratios of upstream and downstream pressure, density, velocity, and area versus the square of the upstream tube Mach number, M_{t1}^2, for several values of the upstream plasma beta, β_1

increasing M_{t1}^2, and the rate of increase becomes large and the ratio blows up at a value of M_{t1}^2 which corresponds to $M_{a1} = 1$, *i.e.*, to the case where point 1 lies on the boundary of the "forbidden region" in Figure 5. The velocity ratio goes to zero at this value of M_{t1}^2 too, and indeed the thin flux tube approximation breaks down as the area ratio A_2/A_1 becomes large for these very strong shocks. In the limit $\beta_1 \to 0$ the tube shock relations go over smoothly to the relations for an ordinary hydrodynamic shock in a rigid flux tube; this "acoustic" limit for $\beta_1 = 0$ is also shown in each panel of Figure 5.

If we allow variations in the ionization fraction of the flowing gas, in response to temperature and pressure changes, then jump condition (19) must be rewritten in terms of the more general adiabatic exponent $\Gamma_3 \equiv 1 + (\partial \ln T / \partial \ln \rho)_s$. Since the ionization fraction jumps across the shock, the value of Γ_3 and the value of the mean molecular weight of the gas both jump across the shock. In this case the jump conditions must be solved iteratively. (See Montesinos and Thomas 1993 for details.)

4. Siphon Flows in Intense Photospheric Flux Tubes

Although application to the Evershed effect (discussed in the next section) is what motivated our early work on siphon flows in isolated flux tubes, we also suggested that siphon flows might be found in some of the intense magnetic elements that constitute most of the magnetic flux in the photosphere outside of sunspots (Thomas 1988, 1990; Thomas and Montesinos 1991). A siphon flow might occur in an arched magnetic flux tube connecting neighboring magnetic elements (footpoints) of opposite polarity, and this siphon flow might contribute to the intensification of the magnetic field strength through the Bernoulli effect.

In Thomas and Montesinos (1991) we described the expected observational signature of such a siphon flow: essentially, a pair of magnetic elements of opposite polarity, separated by a distance of a few arcseconds, with the magnetic field in each element slightly inclined to the vertical and toward the other element (consistent with the two elements being connected by an arched flux tube), and with an upflow in one element (the upstream footpoint) and a downflow and somewhat greater field strength in the other element (the downstream footpoint).

This predicted signature was then indeed discovered by Rüedi, Solanki, and Rabin (1992), in observations of the infrared spectral lines Fe I 15648.5 Å and Fe I 15652.9 Å. Their results showed a magnetic element of strength 1200 G and an upflow of order 1 km s^{-1} on one side of a magnetic neutral line and magnetic element of strength 1500 G and a downflow of order 1 km s^{-1} on the other side. Subsequently, we modeled these observations by computing synthetic spectral line profiles for computed siphon flows (Degenhardt *et al.* 1993) and found a good match with the observed profiles for a supercritical siphon flow in an arch reaching up to the temperature minimum ($h \approx 500$ km) with the tube shock near the top of the arch, very similar to the siphon flows shown in Figure 3. Further observations of such flows in photospheric flux tubes would be of considerable interest.

5. Siphon Flows and the Evershed Effect in Sunspots

The *Evershed effect*, consisting of a shift of a spectral line core and an asymmetry of the line profile in the penumbra, is a feature of all fully developed sunspots. For a photospheric spectral line the line shift is invariably toward the red in the limb-side penumbra and toward the blue in the center-side penumbra, and the shift is maximal when the slit lies along a line connecting the sunspot with disk center and disappears when the slit lies perpendicular to this line. The effect weakens as the sunspot approaches disk center. All of this suggests a nearly horizontal, radial outflow of gas in the penumbra, which indeed was Evershed's (1909) original interpretation of his observational results. The effect decreases with increasing line strength, *i.e.*, with increasing height, and reverses in the strongest lines, forming the *reverse* Evershed effect, or radial inflow, in the penumbral chromosphere. The normal (photospheric) Evershed effect generally increases outward across the penumbra, reaching a maximum near the outer penumbral boundary, and usually stops rather abruptly at or near the penumbral boundary.

High-resolution observations have shown that the Evershed effect is concentrated in the dark penumbral filaments, where the magnetic field is nearly horizontal (Beckers and Schröter 1969; Title *et al.* 1992, 1993; Rimmele 1993). Also, recent high-resolution observations have disclosed that the Evershed effect is time dependent (Shine *et al.* 1990, 1994; Rimmele 1994). The flow in individual filaments waxes and wanes and tends to repeat in the same location with a time scale of 10-15 minutes. The time behavior in one filament seems to be uncorrelated with that in neighboring filaments (Rimmele 1994).

An illustration of the Evershed effect seen at high spatial resolution is provided in Figure 6, based on observations made by Shine *et al.* (1994) at the Swedish Solar Observatory on La Palma in the Canary Islands . This figure shows a continuum image of a sunspot away from disk center (in NOAA AR 5612, 31 July 1989) and a corresponding doppler image in the photospheric line Fe I 5576 Å. The filamentary nature of the photospheric Evershed effect, its increase outward across the penumbra, and its abrupt disappearance near the penumbral boundary are readily apparent in this figure. The maximum Evershed flow speeds seen in this sunspot in Fe I 5576 Å are 2–3 km s^{-1}. Flow speeds as high as 4–6 km s^{-1}, approaching the sound speed, are sometimes seen in photospheric lines in sunspot penumbrae, and the speed of the reverse Evershed flow in the chromosphere is often much higher, 10–20 km s^{-1}.

Meyer and Schmidt (1968) were the first to propose that the Evershed effect is caused by siphon flows along magnetic flux tubes in the penumbra. They explained the Evershed effect and its reversal in the chromosphere as follows (see also Spruit 1981). A siphon flow in an arched flux tube will be directed toward the footpoint with the lower gas pressure, which, for footpoints at equal gravitational potential, is the footpoint with higher magnetic field strength (due to the balance of total pressure along the equipotential surface). A low-lying flux tube with one footpoint in the penumbra, with field strength of say 1000 G, connecting to a magnetic element in the surrounding photosphere with field strength 1500 G (or another sunspot, with even higher field strength) will have a siphon flow directed radially outward, producing the normal Evershed effect. A higher-arching flux tube originating in the umbra, where the field strength is say 2000 G, and connecting to a similar magnetic element outside the sunspot will

55

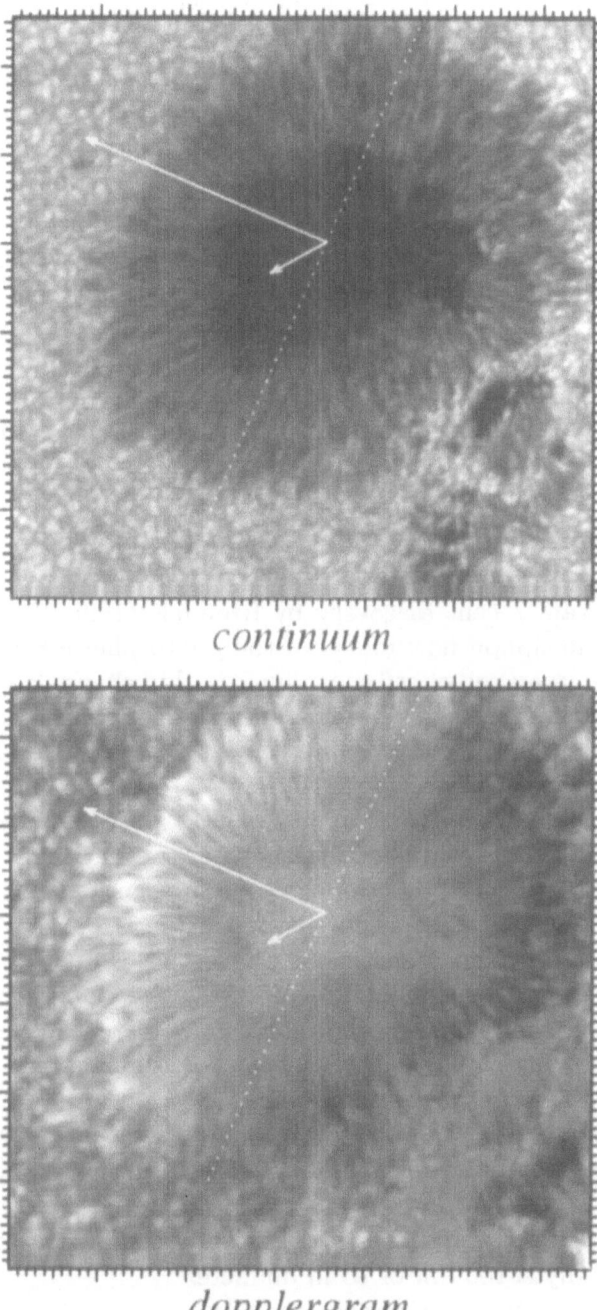
continuum

dopplergram

Figure 6. The Evershed flow in a sunspot, illustrated by a continuum image and a dopplergram in Fe I 5576. The long arrow points to the center of the solar disk and the short arrow points to solar north. Doppler velocities are shown according to a gray scale, with white corresponding to a blue shift and black corresponding to a red shift. Tick marks are spaced by 1". (From Shine *et al.* 1994.)

have a siphon flow directed radially inward toward the spot, producing the reverse Evershed effect in the chromosphere. One can imagine, then, a spatial array of flux tubes carrying siphon flows with a gradation between outward and inward flow with increasing height in the penumbra. This natural explanation of the normal and reverse Evershed effect is a particularly attractive feature of the siphon-flow model.

We have modeled the photospheric Evershed effect in some detail using our formulation for flexible flux tubes appropriate for the penumbral photosphere (Thomas and Montesinos 1993). Our model is based on the recently developed picture of a sunspot penumbra as a deep structure with a two-component (nearly horizontal and inclined), fluted magnetic field (see the review by Thomas and Weiss 1992). We consider siphon flows along arched flux tubes embedded in an atmosphere with a uniform, inclined magnetic field, representing the bright component of the penumbral atmosphere with its more nearly vertical magnetic field. The flow speeds and the heights and spans of the flux-tube arches that we find are consistent with observed Evershed flows in the penumbra.

An important recent discovery by Rimmele (1995a) adds considerable support for the siphon-flow model of the photospheric Evershed effect. He finds that the Evershed effect is confined to thin channels that are elevated above the continuum height over most of the penumbra, consistent with a siphon flow along a thin, arched flux tube. Figure 7, taken from Rimmele's paper, shows time-averaged (over 40 minutes) doppler velocity maps of the center-side penumbra in three photospheric spectral lines. The extended dark (blue-shifted) velocity features seen in the Fe I 5691 Å line, formed at a height of 160 km above the continuum, are not seen in the C I 5380 Å line formed at a height of only 40 km; instead, we see in each case only a small, dot-like velocity feature at a positon corresponding to the inner end of the velocity feature in Fe I 5691 Å. We can interpret these dot-like features in the C line as the footpoints of arched, elevated magnetic flux tubes carrying siphon flows, which appear elongated in the higher Fe line because the tube has become nearly horizontal at that height. In Fe I 5576 Å, formed at a height of 310 km, we see these flow channels only near the outer penumbral boundary, indicating that at this height we are already above the flux tube over much of the penumbra. From this, Rimmele concludes that the Evershed flow is generally confined to narrow channels (flux tubes) only a 100 km or so in diameter.

More recently Rimmele (1995b) has observed the Evershed effect in a sunspot located precisely at disk center, so that only the relatively weak vertical component of the Evershed flow is measured. By computing a two-hour time average of the velocity, he is able to extract the weak, quasi-steady Evershed motion from the background of oscillatory and convective

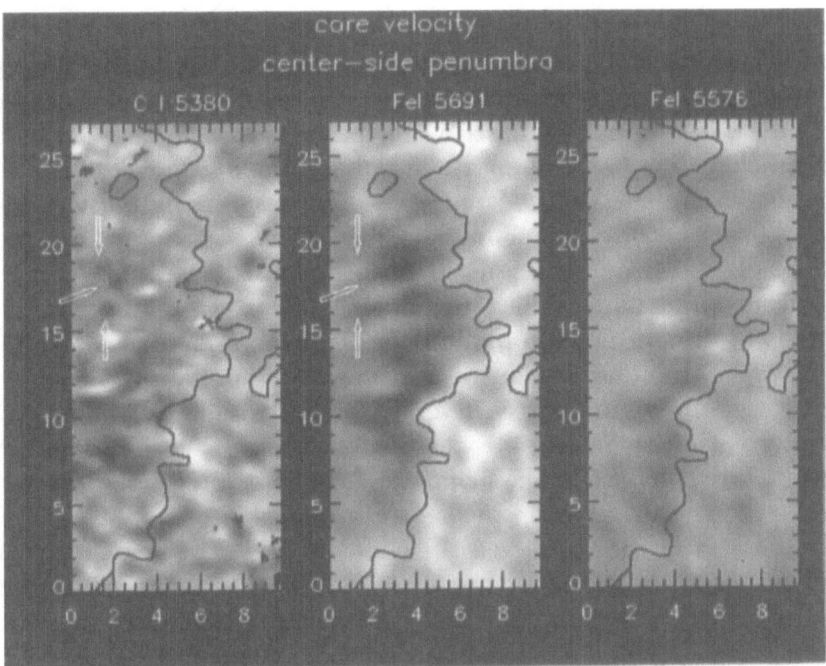

Figure 7. Time-averaged doppler velocity maps of the center-side penumbra of a sunspot measured in the cores of three spectral lines: C I 5380 Å (height = 40 km), Fe I 5691 Å (height = 160 km), and Fe I 5576 Å (height = 310 km). Line-of-sight velocity is displayed on a gray scale with dark corresponding to a blue shift. The black countour line indicates the outer boundary of the penumbra as seen in white light. These three velocity maps show that the Evershed effect is confined to narrow, elevated channels (flux tubes). (From Rimmele 1995a.)

motions. The resulting averaged velocity maps show distinct channels which extend beyond the outer boundary of the penumbra by as much as 10,000 km. The velocity distribution along these channels is consistent with a low, arched flux tube carrying the Evershed flow. Within the penumbra the channels coincide with the dark penumbral filaments seen in white light and in the core of the temperature-sensitive Mn I 5394.7 Å line; however, outside the penumbra the velocity channels appear relatively bright in the core of the Mn line, indicating a temperature excess of some 200 K. Rimmele suggested that this temperature excess may be due to heating across a tube shock in the siphon flow. Another possibility is that the reduced gas density associated with the siphon flow increases the transparency of the flux tube and thus allows us to see deeper in the atmosphere through the flux tube, to hotter layers. We are currently investigating both of these possibilities with actual calculations of siphon flows and synthetic line profiles.

Some important questions still need to be answered before the siphon-flow model provides a completely satisfactory explanation of the Evershed effect. One question has to do with the observed spatial distribution of the Evershed effect in the penumbra and outside the sunspot. Sometimes the photospheric Evershed effect is observed to extend along individual channels beyond the white-light boundary of the penumbra and into the surrounding photosphere, ending at an intense photospheric magnetic feature (e.g., Börner and Kneer 1992; Rimmele 1995b). This is consistent with the kinds of siphon flows computed by Thomas and Montesinos (1993). However, from the observations it appears that there are not enough such cases to account for all of the Evershed flow in the penumbra. Another possibility is that some of the Evershed flux tubes reach up to and extend outward along the base of the magnetic "canopy" outside the sunspot and return to the surface far from the sunspot. Indeed, Solanki, Montavon, and Livingston (1994) have observed the outward Evershed flow in the canopy outside a sunspot, with flow speeds of 0.5–2 km s^{-1}, but they point out that these flows along the canopy are not sufficient to carry all of the Evershed mass flux seen in the penumbra. Their estimate of the flow speeds and mass flux may be too low, however, if the flow channels remain narrow and only occupy part of the range of height of formation of their strong spectral lines (Thomas 1994; Rimmele 1995a). A third possibility, strongly suggested by recent observations of Title et al. (1995), is that many of the flux tubes carrying the Evershed flow submerge below the surface again, at a very shallow angle, just within the outer penumbral boundary. In time sequences of white-light images these flux tubes can be seen as dark streaks extending outward beyond the penumbral boundary and appearing and disappearing, as if they were moving up and down very near the visible surface. Because of the shallow re-entry angle, these features need not show magnetic polarity opposite that of the sunspot; indeed, because of the decreasing Wilson depression across the penumbra, even a thin, horizontal flux tube would submerge below the visible surface within the visible penumbra. We can imagine, then, that the flux tubes carrying the Evershed flow are divided among the three types of configurations mentioned here, in a way that is consistent with the observations, but this has yet to be demonstrated in a specific model.

Another important question has to do with the observed time dependence of the Evershed flow along individual penumbral filaments. Thus far we have studied only steady siphon flows in connection with the Evershed effect, but it is clear that real solar siphon flows will be transient in nature because of changing pressures at the footponts of the flux tubes. We have speculated (Thomas 1994) that the time scale of the siphon flow is controlled by the lifetime of the intense magnetic element forming the downstream footpoint of the flux tube. Theory and observations suggest that the lifetime of such a magnetic element is comparable to the lifetime of individual solar granules, which is roughly 10 minutes, in agreement with the observed time scale of the Evershed effect. We intend to extend our siphon-flow calculations to include time dependence in order to address this issue.

Acknowledgements

I am grateful to Kanaris Tsinganos for his extraordinary hospitality during this NATO ASI in Crete. I thank Dick Shine and Thomas Rimmele for providing me with Figures 6 and 7. I also thank Benjamin Montesinos for his collaboration and Donald Stanchfield for his help in preparing Figures 4 and 5. My research on siphon flows has been supported by NASA grants NAGW-2123 and NAGW-2444 and by a fellowship from the John Simon Guggenheim Memorial Foundation.

References

Ayres, T. R. 1981, *Astrophys. J.*, **244**, 1064.
Beckers, J. M., and Schröter, E. H. 1969, *Solar Phys.*, **10**, 384.
Börner, P., and Kneer, F. 1992, *Astron. Astrophys.*, **259**, 307.
Cargill, P. J., and Priest, E. R. 1980, *Solar Phys.*, **65**, 251.
Cargill, P. J., and Priest, E. R. 1982, *Geophys. Astrophys. Fluid Dyn.*, **20**, 227.
Cheng, J. 1992, *Astron. Astrophys.*, **259**, 296.
Defouw, R. J. 1976, *Astrophys. J.*, **209**, 266.
Degenhardt, D. 1989, *Astron. Astrophys.*, **222**, 297.
Degenhardt, D. 1991, *Astron. Astrophys.*, **248**, 637.
Degenhardt, D., Solanki, S., Montesinos, B., and Thomas, J. H. 1993, *Astron. Astrophys.*, **279**, L29.
Evershed, J. 1909, *Mon. Not. Roy. Astron. Soc.*, **69**, 454.
Ferriz-Mas, A., and Moreno-Insertis, F. 1987, *Astron. Astrophys.*, **179**, 268.
Glencross, W. M. 1980, *Astron. Astrophys.*, **83**, 65.
Herbold, G., Ulmschneider, P., Spruit, H. C., and Rosner, R. 1985, *Astron. Astrophys.*, **145**, 157.
Kalkofen, W., and Ulmschneider, P. 1977, *Astron. Astrophys.*, **57**, 193.
Kurucz, R. 1979, *Astrophys. J. Suppl.*, **40**, 1.
Meyer, F., and Schmidt, H. U. 1968, *Zeits. Angew. Math. Mech.*, **48**, 218.
Montesinos, B., and Thomas, J. H. 1989, *Astrophys. J.*, **337**, 977.
Montesinos, B., and Thomas, J. H. 1993, *Astrophys. J.*, **402**, 314.
Noci, G. 1981, *Solar Phys.*, **69**, 63.
Pikel'ner, S. B. 1971, *Solar Phys.*, **17**, 44.
Priest, E. R. 1981, in *Solar Active Regions*, ed. F. Q. Orrall (Boulder, CO: Colorado Associated University Press), p. 213.
Rimmele, T. R. 1993, Ph.D. dissertation, University of Freiburg.
Rimmele, T. R. 1994, *Astron. Astrophys.*, **290**, 972.
Rimmele, T. R. 1995a, *Astron. Astrophys.*, **298**, 260.
Rimmele, T. R. 1995b, *Astrophys. J.*, **445**, 511.

60

Rogers, F. J., and Iglesias, C. A. 1992, *Astrophys. J. Suppl.*, **79**, 507.

Rüedi, I., Solanki, S. K., and Rabin, D. 1992, *Astron. Astrophys.*, **261**, L21.

Shine, R., Smith, K., Tarbell, T., Title, A., and Scharmer, G. 1990, *Bul. Amer. Astron. Soc.*, **22**, 878.

Shine, R. A., Title, A. M., Tarbell, T. D., Smith, K., Frank, Z. A., and Scharmer, G. 1994, *Astrophys. J.*, **430**, 413.

Solanki, S. K., Montavon, C. A. P., and Livingston, W. 1994, *Astron. Astrophys.*, **283**, 221.

Spiegel, E. A. 1957. *Astrophys. J.*, **126**, 202.

Spruit, H. C. 1981, in *The Physics of Sunspots*, ed. L. E. Cram and J. H. Thomas (Sunspot, NM: Sacramento Peak Observatory), p. 359.

Surlantzis, G., Démoulin, P., Heyvaerts, J., and Sauty, C. 1994, *Astron. Astrophys.*, **284**, 985.

Thomas, J. H. 1984, in *Small-Scale Dynamical Processes in Quiet Stellar Atmospheres*, ed. S. L. Keil (Sunspot, NM: National Solar Observatory), p. 276.

Thomas, J. H. 1988, *Astrophys. J.*, **333**, 407.

Thomas, J. H. 1990, in *Physics of Magnetic Flux Ropes*, ed. C. T. Russell, E. R. Priest, and L. C. Lee (Washington: American Geophysical Union), p. 133.

Thomas, J. H. 1994, in *Solar Surface Magnetism*, ed. R. J. Rutten and C. J. Schrijver (Dordrecht: Kluwer), p. 219.

Thomas, J. H., and Montesinos, B. 1990, *Astrophys. J.*, **359**, 550.

Thomas, J. H., and Montesinos, B. 1991, *Astrophys. J.*, **375**, 404.

Thomas, J. H., and Montesinos, B. 1993, *Astrophys. J.*, **407**, 398.

Thomas, J. H., and Weiss, N. O. 1992, in *Sunspots: Theory and Observations*, ed. J. H. Thomas and N. O. Weiss (Dordrecht: Kluwer), p. 3.

Title, A. M., Frank, Z. A., Shine, R. A., Tarbell, T. D., Topka, K. P., Scharmer, G. B., and Schmidt, W. 1992, in *Sunspots: Theory and Observations*, ed. J. H. Thomas and N. O. Weiss Dordrecht: Kluwer, p. 195.

Title, A. M., Frank, Z. A., Shine, R. A., Tarbell, T. D., Topka, K. P., Scharmer, G. B., and Schmidt, W. 1993, *Astrophys. J.*, **403**, 780.

Title, A. M., Frank, Z. A., Shine, R. A., Tarbell, T. D., Simon, G. W., and Brandt, P. N. 1995, *Bul. Amer. Astron. Soc.*, **27**, 978.

Tsinganos, K. 1982, *Astrophys. J.*, **252**, 775.

Tsinganos, K., Surlantzis, G., and Priest, E. R. 1993, *Astron. Astrophys.*, **275**, 613.

Tsinganos, K., Sauty, C., Surlantzis, G., Trussoni, E., and Contopoulos, J. 1996, this volume.

Ulmschneider, P., Schmitz, F., Kalkofen, W., and Bohn, H. U. 1978, *Astron. Astrophys.*, **70**, 487.

Unno, W., and Speigel, E. A. 1966, *Publ. Astron. Soc. Japan*, **18**, 85.

MHD WAVES IN MAGNETIC FLUX TUBES

Resonant Alfvén waves

MARCEL GOOSSENS AND MICHAEL RUDERMAN
Centre for Plasma Astrophysics K.U.Leuven
Celestijnenlaan 200B B-3001 Heverlee BELGIUM
Marcel.Goossens@wis.kuleuven.ac.be

Abstract. This review is concerned with a recent development in the analytic theory of resonant Alfvén waves in nonuniform magnetic flux tubes in the solar atmosphere. It discusses how an analytic study of simplified versions of the dissipative MHD equations for plasmas with high Reynolds numbers leads to the fundamental conservation law and simple analytic solutions for resonant Alfvén waves. Asymptotic analysis of the analytic solutions gives jump conditions that connect the solutions across the dissipative layer. The analytic results enable us to understand the basic dynamics and energetics of resonant Alfvén waves and help us with the interpretation of results of of numerical simulations in dissipative MHD. In addition, they make an accurate computation of resonant Alfvén waves possible without having to solve numerically the dissipative MHD equations.

1. Introduction

Well, you've learned the first rule of conferences, kid. Never go to lectures. Unless you're giving one yourself, of course. Or *I'm giving one.*
Prof. Morris Zapp to Perse McGarrigle in "Small World" by D. Lodge.

Observations give evidence of the almost ubiquitous presence of MHD waves in the solar atmosphere. These waves are far from being completely understood because of their intrinsic physical complexity due to the structuring, the non-uniformity and the anisotropy of the magnetic plasmas in the solar atmosphere. The solar atmosphere is a gigantic plasma, which from the photosphere to the corona, is strongly structured by the magnetic field and exhibits inhomogeneities in density, pressure and magnetic fields on small scales. Each region in the solar atmosphere has its own magnetic plasma structures and observations show that many of these magnetic structures support MHD waves. A summary of the different magnetic structures in the solar atmosphere and of observational information and theory on the

61

K. C. Tsinganos (ed.), Solar and Astrophysical Magnetohydrodynamic Flows, 61–83.
© *1996 Kluwer Academic Publishers.*

MHD waves that these structures support can be found in e.g. Edwin and Roberts (1987), Roberts (1988), and Roberts (1991).

The present review focusses on resonant MHD waves. Resonant MHD waves are directly related to the non-uniformity and the anisotropy of the magnetic plasma and are the natural wave modes of the individual magnetic surfaces. They are characterized by a resonant condition in the sense that at a certain magnetic surface the frequency of the wave becomes equal to the local Alfvén frequency or the local slow frequency which are the natural oscillation frequencies of the individual magnetic surfaces. This resonance causes the amplitude of the wave to become infinite in the absence of dissipation; dissipation however limits the amplitude to large but finite values. All MHD waves have the ability to transport and dissipate energy and to heat plasmas. However, in plasmas with high viscous and magnetic Reynolds numbers as in the sun and stars, resonant MHD waves are by far more efficient for dissipating energy of the wave motions and for heating plasmas than the discrete wave modes.

Resonant waves play an important role for explaining the heating of the solar corona. The heating of the solar corona is a fundamental problem for solar physics and astrophysics since most stars have a hot corona just like the sun (see e.g. Narain and Ulmshneider, 1995). Resonant absorption of MHD waves is an efficient means for dissipating energy of MHD waves in nonuniform plasmas. It was first studied as a means for the supplementary heating of fusion plasmas (see e.g. Tataronis and Grossmann, 1973; Grossmann and Tataronis, 1973; Chen and Hasegawa, 1974; Hasegawa and Chen, 1974) and subsequently proposed as a mechanism for heating magnetic flux tubes in the solar corona by Ionson (1978). Since the original suggestion, resonant absorption has remained a popular mechanism for explaining the heating of the solar corona (see e.g. Kuperus et al., 1981; Ionson, 1985; Davila, 1987; Hollweg 1990, 1991; Goossens, 1991). Recently, resonant absorption has been considered in relation to the observed loss of power of acoustic oscillations in sunspots. The observed loss of acoustic power in the vicinity of sunspots offers the unique possibility to obtain indirect information on the subphotospheric structure of the solar magnetic field which is a necessary condition for understanding the magnetic activity of the sun. This requires a good theoretical understanding of the waves in magnetic flux tubes. Coupling of acoustic waves to resonant Alfvén waves and absorption of these Alfvén waves inside sunspots has been proposed as a possible explanation of this phenomenon (see e.g. Hollweg, 1988; Sakurai et al., 1991b; Goossens and Poedts 1992; Stenuit et al., 1993).

Resonant MHD waves can be studied from three related viewpoints. They can be studied as a driven problem, an eigenvalue problem and an initial value problem. The driven problem involves an external source which excites driven oscillations in the plasma. Here the frequency of the incoming wave is prescribed. The eigenvalue problem is concerned with finding the free eigenmodes of an MHD equilibrium in the absence of any external forcing. The intial value problem is concerned with the free evolution in time of the system after an initial perturbation has been applied to it. The driven problem is important when studying the heating of magnetic flux tubes and the loss of acoustic power in sunspots. The relation between the eigenvalue problem and the driven problem was analysed by Goossens and Hollweg (1993). The study of driven Alfvén waves involves forced oscillations in dissipative MHD. This means that the time dependent non-linear equations of dissipative MHD have to be integrated in the presence of a time varying force term. Most studies have used linear theory of wave motions superimposed on an ideal equilibrium state. Even in the context of linear MHD, it is common practice to circumvent the time integration by considering the asymptotic or steady state of the Alfvén waves. In the asymptotic state all the perturbed quantities oscillate with the same frequency as the incoming wave so that the time dependency can be removed out of the mathematical formulation. In what follows we shall exclusively focus on the asymptotic state in linear MHD.

The physical basis of resonant MHD waves can be understood from spectral theory of linear ideal MHD. The linear ideal MHD spectrum of a nonuniform plasma contains an Alfvén continuum and a slow continuum in addition to discrete eigenfrequencies. The spatial solutions that correspond to these continuum frequencies are non-square integrable (see e.g. Goedbloed, 1983). Consider now an MHD wave with given wave numbers and a frequency within the Alfvén continuum of the flux tube. If this wave impinges on the flux tube it excites an Alfvén wave on the magnetic surface where the local Alfvén frequency equals the frequency of the incoming wave. Part of the energy of the wave will be absorbed and the energy fluxes of the incoming and outgoing waves will be different. In ideal MHD this driven problem is governed by a set of differential equations that are singular at the point where the local Alfvén frequency equals the frequency of the wave. The spatial solutions are non-square integrable, and ideal MHD cannot be used to describe the resonant behaviour. Non-ideal effects such as viscosity and electrical resistivity will remove the singularity but at the same time the order of the system of differential equations is raised. The steep gradients of the spatial solutions cause efficient damping in non-ideal MHD.

The large values of the viscous and magnetic Reynolds numbers in the solar atmosphere imply that the dissipative terms in the MHD equations are unimportant except in narrow layers. In the case of resonant absorption of MHD waves the dissipative terms are only important in a narrow layer around the position where the frequency of the wave equals the local Alfvén frequency. Outside this narrow layer the MHD waves are accurately described by the equations of ideal MHD. This observation led Sakurai et al. (1991a) to design a method for obtaining the solutions of resonant Alfvén waves which does not require solution of the full set of dissipative MHD equations in the whole of the flux tube. It suffices to solve the equations of ideal MHD in the regions to the left and the right of the dissipative layer. The solutions of the dissipative MHD equations are used to connect the ideal MHD solutions over the dissipative layer. The cumbersome part is to solve the equations of dissipative MHD in the dissipative layer and in regions to the left and the right of the dissipative layer that overlap with the regions where ideal MHD is valid. In general the combination of the dissipative layer and the overlap regions is only a tiny fraction of the flux tube so that it is possible to use simplified dissipative MHD equations which allow analytical solutions. If one is only interested in the global behaviour of the solution, the dissipative solution can be seen as a means to connect the ideal solutions across the disipative layer. As a consequence, once these connection formulae are found, it is not any longer necessary to solve the dissipative MHD equations.

In an early version of this approach, Hollweg (1987a,b) and Hollweg and Yang (1988) noted that in planar geometry the Eulerian perturbation of total pressure, P', could be taken to be approximately constant across the dissipative layer. Thus $P' =$ constant represented the desired connection formula, and it allowed for a straightforward description of the structure of the velocity amplitude in the dissipative layer. However, $P' =$ constant was assumed ad hoc, and has to be replaced with a different conservation law in cylindrical geometry. Sakurai et al.(1991a) obtained the fundamental conservation law for cylindrical geometry at the Alfvén resonance point in ideal MHD. They assumed that this ideal conservation law remains valid in dissipative MHD and obtained the solutions to the dissipative MHD equations for the radial component of the Lagrangian displacement ξ_r and the Eulerian perturbation of total pressure P' in terms of double integrals of Hankel functions of order 1/3 of a complex argument. An asymptotic expansion of these solutions enabled Sakurai et al. to obtain the connection formulae for ξ_r and P'. These connection formulae in combination with solutions of ideal MHD were then used by Sakurai et al. (1991b) to study absorption of acoustic waves in sunspots and by Goossens and Hollweg (1993) to obtain conditions for maximal and also total absorption.

The dissipative MHD solutions obtained by Sakurai et al. (1991a) are too complicated to give us any information about the solutions in the dissipative layer. Goossens et al. (1995) extended the analysis of Sakurai et al. (1991a). They showed that the fundamental conservation law found in ideal MHD remains indeed a conservation law in dissipative MHD. They also obtained simple analytical solutions to the dissipative MHD equations for the three components of the Lagrangian displacement and for the Eulerian perturbation of total pressure. These solutions allow a straightforward physical interpretation of the dynamics and energetics of resonant Alfvén waves in the dissipative layer.

This review paper closely follows Sakurai et al. (1991a) and Goossens et al. (1995) (see also the recent review paper by Goossens and Ruderman, 1995). It is organized as follows. In Section 2 we list the basic set of linear differential equations that govern linear motions superimposed on a static equilibrium state in linear resistive MHD. In Section 3 we collect results obtained by Sakurai et al. (1991a) on resonant Alfvén waves in ideal MHD which will return in a modified form in resistive MHD. In Section 4 the linear resistive MHD equations of Section 2 are simplified by restricting the analysis to the subclass of linear displacements that correspond to resonant Alfvén waves in ideal MHD. In Section 5 the solutions to these linear resistive MHD equations are given in an interval that embraces the dissipative layer and two overlap regions where ideal MHD is also valid. Section 6 summarizes the main results of the present paper.

2. Asymptotic state of resonant Alfvén waves

I am afraid that I rather give myself away when I explain. Results without causes are much more impressive.
Sherlok Holmes in "The Stock-Broker's Clerk" by Sir Arthur Conan Doyle.

In what follows we shall concentrate on the asymptotic state of Alfvén waves in linear MHD. All perturbed quantities are put proportional to

$$\exp(-i\omega t) \tag{1}$$

with $\omega > 0$ the frequency of the incoming wave or the external forcing term. The asymptotic state is in principle only valid for $t \to +\infty$, but in practice it gives an accurate description for $t \gg t_{crit}$ where $t_{crit} \sim (R_m)^{1/3}$ and R_m is the magnetic Reynolds number (see Kappraff and Tataronis,1977; Poedts et al. 1990b).

Detailed results based on large-scale numerical simulations of the asymptotic state of Alfvén wave heating were obtained by Grossmann and Smith (1988) in ideal MHD, by Poedts et al. (1989ab, 1990a) in resistive MHD and by Ofman et al. (1994a) and Erdelyi and Goossens (1995) in visco-resistive MHD. Time dependent computations of Alfvén wave heating in linear dissipative MHD have been carried out by Poedts et al. (1990b), and Poedts and Kerner (1992) and in non-linear dissipative MHD by Ofman et al. (1994bc, 1995a). In what follows we shall exclusively focus on the asymptotic state in linear MHD.

The present review considers Alfvén waves in 1-dimensional magnetic flux tubes. It is obvious that 1-dimensional magnetic flux tubes are a simplification of reality. Sunspots and coronal loops are complicated structures with spatial variations of the physical quantities in two or even three directions. Macroscopic magnetic structures that exist over relatively long time spans in the solar atmosphere have to satisfy the equilibrium and stability limits set by ideal MHD. As a consequence an equilibrium state with well defined magnetic surfaces is very probably a good approximation of the global magnetic equilibrium of a sunspot and also of a coronal loop. Spatial variations of the equilibrium quantities in different directions affect the waves in different ways. For example, radial stratification causes resonant absorption of acoustic oscillations in sunspots (see e.g. Hollweg, 1988; Lou, 1990; Sakurai et al., 1991b; Goossens and Poedts, 1992), while axial stratification causes conversion of acoustic waves into slow waves producing a downward drainage of wave energy (Spruit and Bogdan, 1992; Cally and Bogdan, 1993). It is evident that these two effects should be combined and that there is a need for studying waves in 2-dimensional equilibrium states. The continuous part of the spectrum of linear ideal MHD has been studied by e.g. Pao (1975) and Goedbloed (1975) for 2-dimensional fusion plasma equilibrium states and by Goossens et al. (1985), Poedts et al. (1985), Poedts and Goossens (1987, 1988) for 2-dimensional magnetic structures in the solar corona. Resonant waves have been studied in 2-dimensional equilibrium states by Thompson and Wtight (1993), Wright and Thompson (1994), and Tirry and Goossens (1995). Effects of finite lenght in the axial direction or line tying and driving at the footpoints of the magnetic field lines have been studied numerically by Halberstadt (1994), Halberstadt and Goedbloed (1993, 1995ab) in linear MHD, and by Poedts and Boyton (1995) in non-linear MHD. This review aims to provide a clear understanding of the basic properties of driven resonant Alfvén waves in simple 1-D equilibrium configurations which are believed to play an essential role in more complicated situations.

The static equilibrium state of the flux tube is idealized as a cylindrically symmetric column of plasma. We use a system of cylindrical coordinates r, φ, z with the z-axis coinciding with the axis of symmetry of the cylinder. The components of the equilibrium magnetic field $\mathbf{B}(0, B_\varphi, B_z)$ as well as the pressure p and density ρ are functions of the radial coordinate r only. They satisfy the radial force balance equation

$$\frac{d}{dr}\left(p + \frac{B^2}{2\mu}\right) = -\frac{B_\varphi^2}{\mu r} \, . \tag{2}$$

The linear displacements superimposed on this back-ground state are governed by the linearized versions of the MHD equations. The perturbations are fully 3-dimensional in space, all perturbed quantities depend on the three spatial coordintes r, φ, z and on time t and velocity $\mathbf{V}'$ and Lagrangian displacement ξ have three components. Since the aim is to study the resonant behaviour of Alfvén waves in dissipative MHD, we have to include a dissipative term that removes the ideal singularity. Electrical resistivity is a prime candidate in this respect (very similar results are obtained if viscosity is taken to be the dissipation mechanism). The equations of linear resistive MHD are:

$$\frac{\partial \rho'}{\partial t} + \nabla \cdot (\rho \mathbf{V}') = 0 , \tag{3}$$

$$\frac{\partial p'}{\partial t} + V_r' \frac{dp}{dr} = \frac{\gamma p}{\rho} \left(\frac{\partial \rho'}{\partial t} + V_r' \frac{d\rho}{dr} \right) , \tag{4}$$

$$\rho \frac{\partial \mathbf{V}'}{\partial t} = -\nabla p' + \frac{1}{\mu}[(\nabla \times \mathbf{B}) \times \mathbf{B}' + (\nabla \times \mathbf{B}') \times \mathbf{B}] , \tag{5}$$

$$\frac{\partial \mathbf{B}'}{\partial t} = \nabla \times (\mathbf{V}' \times \mathbf{B}) + \eta \nabla^2 \mathbf{B}' , \tag{6}$$

In these equations η is the coefficient of magnetic diffusivity and γ is the ratio of specific heats. These two quantities are assumed to be constant. The other notations are the same as in Sakurai et al. (1991a). Equation (4) is the equation for the variation of internal energy not including non-ideal effects. In particular the term due to finite electrical conductivity is ignored in this equation (the adiabatic equation). The inclusion of finite electrical conductivity in the generalized form of Ohm's law (equation (6)) but not in the energy equation, is an approximation often made in MHD and based on the fact that finite electrical conductivity has its most important effect in equation (6). Poedts et al. (1990b) have included the resistive term in the equation for the variation of internal energy in their numerical study of the temporal evolution of resonant absorption. They found that inclusion of the resistive term in the energy equation has virtually no effect. The adiabatic equation (4) is a very good approximation of the energy equation.

Since the equilibrium quantities depend on r only we can Fourier-analyze the perturbed quantities with respect to φ and z and put them proportional to

$$\exp[i(m\varphi + kz)]. \tag{7}$$

Here m (an integer) and k are the azimuthal and axial wave numbers. As the time-dependence $\exp(-i\omega t)$ has already been factored out the perturbed quantities are functions of r only and the differential equations are reduced to ordinary differential equations.

3. Resonant Alfvén waves in linear ideal MHD

Singularity is almost invariably a clue.
Sherlok Holmes in "The Boscombe Valley Mystery" by Sir Arthur Cannon Doyle.

Dissipative MHD is only necessary to treat the dissipative layer. Part of the basic physics of resonant Alfvén waves can be understood in the context of linear ideal MHD. The aim of the present section is to collect results obtained by Sakurai et al. (1991a) in ideal MHD in such a way that expressions which return in a modified form in resistive MHD are both well-defined and well-distinguished.

All but two of the perturbed variables may be eliminated from the linear ideal MHD equations leading to a set of two first-order differential equations for the radial component of the Lagrangian displacement, ξ_r and the perturbed total pressure P' ($P' = p' + \vec{B}.\vec{B}'/\mu$):

$$\begin{cases} D\dfrac{d(r\xi_r)}{dr} = C_1 r\xi_r - C_2 r P', \\[2mm] D\dfrac{dP'}{dr} = C_3\xi_r - C_1 P'. \end{cases} \tag{8}$$

Equations (8) govern the linear motions of a compressible cylindrical plasma. They were first obtained in this form by Appert et al. (1974). The coefficient functions D, C_1, C_2, and C_3 depend on the equilibrium quantities and on the frequency ω. They take the form

$$D = \rho(c^2 + v_A^2)(\omega^2 - \omega_A^2)(\omega^2 - \omega_C^2), \tag{9}$$

$$C_1 = \frac{2}{\mu r}B_\varphi^2\omega^4 - (c^2 + V_A^2)(\omega^2 - \omega_C^2)\frac{2mf_B}{\mu r^2}B_\varphi, \tag{10}$$

$$C_2 = \omega^4 - (c^2 + V_A^2)(\omega^2 - \omega_C^2)\left(\frac{m^2}{r^2} + k^2\right), \tag{11}$$

$$\begin{aligned} C_3 = {}& D\left[\rho(\omega^2 - \omega_A^2) + \frac{2B_\varphi}{\mu}\frac{d}{dr}\left(\frac{B_\varphi}{r}\right)\right] \\ & + \frac{4\omega^4 B_\varphi^4}{\mu^2 r^2} - 4\rho(c^2 + v_A^2)(\omega^2 - \omega_C^2)\omega_A^2\frac{B_\varphi^2}{\mu r^2} . \end{aligned} \tag{12}$$

The symbols have their usual meaning: c is the adiabatic velocity of sound, v_A is the Alfvén velocity, ω_A is the local Alfvén frequency, and ω_C is the local cusp frequency. The squares of these quantities are defined as

$$c^2 = \gamma p/\rho, \quad v_A^2 = B^2/(\mu\rho), \quad \omega_A^2 = f_B^2/\mu\rho,$$

$$\omega_C^2 = c^2 \omega_A^2/(c^2 + v_A^2), \quad f_B = \vec{k}.\vec{B}, \quad \vec{k} = (0, m/r, k).$$

In a nonuniform plasma c^2, v_A^2, ω_A^2, and ω_C^2 are functions of position. The variation of ω_A with position plays a fundamental role in the present discussion of resonant Alfvén waves.

The other perturbed quantities (p', ρ', etc.) can be computed once ξ_r and P' are known. For later use we note that the component of the displacement vector in the magnetic surfaces and perpendicular to the magnetic field lines ($\xi_\perp = (\xi_\varphi B_z - \xi_z B_\varphi)/B$) is related to ξ_r and P' as

$$(\omega^2 - \omega_A^2)\xi_\perp = \frac{i}{\rho B}(g_B P' - 2f_B B_\varphi B_z \xi_r/\mu r), \tag{13}$$

where

$$B = (B_\varphi^2 + B_z^2)^{1/2}, \quad g_B = \frac{m B_z}{r} - k B_\varphi. \tag{14}$$

Equations (8) define an eigenvalue problem with ω^2 as eigenvalue parameter when they are supplemented with boundary conditions. In a driven problem ω is prescribed. Equations (8) have regular singular points at the zeroes of the coefficient function D. As a consequence we have mobile regular singular points at the positions r where

$$\omega^2 = \omega_A^2(r), \quad \omega^2 = \omega_C^2(r). \tag{15}$$

Equation (15a) defines the Alfvén resonance point, while equation (15b) defines the slow or cusp resonance point. Since $\omega_A^2(r)$ and $\omega_C^2(r)$ are functions of position, these two equations define two continuous ranges in the spectrum which are classically referred to as the Alfvén continuum and the slow or cusp continuum. In what follows we shall concentrate on the Alfvén resonance.

Let us now focus on a frequency in the Alfvén continuum and determine the spatial behaviour of the corresponding perturbation close to the singular point where the condition $\omega^2 = \omega_A^2(r_A)$ is satisfied. It is convenient to introduce the new radial variable s defined as

$$s = r - r_A \tag{16}$$

and use series expansions of the coefficient functions around $s = 0$ to obtain simplified versions of the relevant differential equations. These simplified versions are valid in the interval $[-s_A, s_A]$ around the point of resonance where the linear Taylor polynomial is a valid approximation of $\omega^2 - \omega_A^2(r)$; hence s_A has to satisfy

$$s_A \ll \left| \frac{2(\omega_A^2)'}{(\omega_A^2)''} \right|. \tag{17}$$

The simplified versions of equations (8) close to the Alfvén resonance point are (Sakurai et al. 1991a):

$$\begin{cases} s\Delta \dfrac{d\xi_r}{ds} = \dfrac{g_B}{\rho B^2} C_A(s), \\[3mm] s\Delta \dfrac{dP'}{ds} = \dfrac{2f_B B_\varphi B_z}{\mu r_A \rho B^2} C_A(s), \end{cases} \tag{18}$$

where

$$C_A(s) = g_B P' - \frac{2f_B B_\varphi B_z \xi_r}{\mu r_A}. \tag{19}$$

In equations (18) all equilibrium quantities are evaluated at $s = 0$ $(r = r_A)$, and

$$\Delta = \frac{d}{dr}(\omega^2 - \omega_A^2). \tag{20}$$

The fact that the right hand members of equation (18) have the common factor $C_A(s)$ which is a linear combination of ξ_r and P' is a key point in the present analysis. It suggests that we should obtain a differential equation for $C_A(s)$. This is readily done by taking the appropriate linear combination of equations (18):

$$s \frac{dC_A(s)}{ds} = 0. \tag{21}$$

Since the large solutions of ξ_r and P' (i.e. solutions containing logarithmic terms) have to be continuous, equation (21) implies that

$$C_A(s) \equiv g_B P' - \frac{2f_B B_\varphi B_z \xi_r}{\mu r_A} = \text{constant}. \tag{22}$$

Condition (22) is the fundamental conservation law at the Alfvén resonance point (Sakurai et al.,1991a). An extensive review of the theory of conservation laws and its importance for studying resonant MHD waves is given by Goossens and Ruderman (1995).

The solutions for ξ_r and P' take the form

$$\begin{cases} \xi_r(s) = \dfrac{g_B}{\rho B^2 \Delta} C_A \ln(|\,s\,|) + \begin{cases} \xi_- & s < 0 \\ \xi_+ & s > 0, \end{cases} \\[5mm] P'(s) = \dfrac{2f_B B_\varphi B_z}{\mu r_A \rho B^2 \Delta} C_A \ln(|\,s\,|) + \begin{cases} P'_- & s < 0 \\ P'_+ & s > 0. \end{cases} \end{cases} \tag{23}$$

The jumps in ξ_r and P' are due to dissipative effects and will be specified later.

Let us now return to the conservation law at the Alfvén resonance point. In an equilibrium with a straight magnetic field ($B_\varphi = 0$), the conservation law reduces to

$$[P'] = 0. \tag{24}$$

The solution(23) for ξ_r remains unchanged, while P' can be considered to be constant across the point of resonance. The approximate constancy of P' was the key ingredient in the physical discussions of resonance absorption given by Hollweg (1987a,b;1988) and Hollweg and Yang (1988). Thus the discovery of the conservation law $C_A =$ constant puts the approach by Hollweg and Yang on a rigourous mathematical footing, and extends the approach to cylindrical geometry.

For an equilibrium with a curved magnetic field we can use equation (13) to rewrite the conservation law in terms of $\xi_\perp$:

$$s\xi_\perp = i\frac{C_A}{\rho B \Delta}. \tag{25}$$

Equation (25) expresses that $s\xi_\perp$ is constant across the resonant layer, or that $\xi_\perp$ has a $1/s$-singularity and a $\delta(s)$ contribution which dominate the $\ln|s|$ singularity and the jump found for ξ_r and P'. The dominant singularities in the solution reside in the components in the magnetic surfaces and perpendicular to the magnetic field lines (see also Goedbloed, 1983). The dominant dynamics of the wave is contained in $\xi_\perp$ and the wave is polarized in the magnetic surfaces perpendicular to the magnetic field lines. The nonuniform plasma supports an Alfvén wave that is confined to the magnetic surface where the dispersion relation for Alfvén waves in a uniform plasma is locally satisfied. The confinement of the Alfvén wave is not absolute. The Alfvén wave is linked to the outside world as it is coupled to a wave with components normal to the magnetic surfaces. From a dynamic point of view we can consider the wave as an almost pure Alfvén wave confined to its resonant magnetic surface and polarized perpendicular to the magnetic field lines. From the point of view of the energetics ξ_r, the component normal to the magnetic surfaces, is essential since it is this quantity which provides the unidirectional transfer of energy to the resonant surface.

4. Resistive MHD equations for resonant Alfvén waves.

I can open your eyes,
take you wonder by wonder
over sideways and under
on a magic car.

Awhole new world,
a new fantastic point of view.
*A Whole New World from "Aladdin", S.Menken/T. Rice-Wonderland/
Walt Disney.*

The aim is to find out how the singular solutions for the resonant Alfvén waves found in ideal MHD are modified by dissipation. For this purpose it suffices to consider non-zero electrical resistivity since this removes the singularity present in the ideal equations. We are not interested in the equations that govern arbitrary linear displacements of a cylindrical plasma in resistive MHD. Our interest goes exclusively to the subclass of linear displacements that correspond to resonant Alfvén waves in ideal MHD. The restriction to this subclass of perturbations makes a significant simplification of the equations of resistive MHD possible. The effects of dissipation are generally small and only important in the vicinity of the ideal resonance. In this dissipative layer the derivatives of the perturbed quantities with respect to r are much larger than those with respect to z and φ. In addition the derivatives of equilibrium quantities can be neglected in comparison with the derivatives of the perturbed quantities with respect to r. Therefore we retain in the dissipative terms only the r-derivatives of the perturbed quantities while we neglect those of the equilibrium quantities. We write $\partial^2 \mathbf{b}/\partial r^2$ instead of $\nabla^2 \mathbf{b}$.

Just as in ideal MHD all but two of the perturbed variables may be eliminated from the linear resistive MHD equations leading to a set of two differential equations of third order for ξ_r and P'. The details of the derivation can be found in Goossens et al. (1995).

The equations are (Sakurai et al. (1991a); Goossens et al. (1995))

$$\begin{cases} D_\eta \dfrac{d(r\xi_r)}{dr} = C_1 r\xi_r - C_2 r P', \\[2ex] D_\eta \dfrac{dP'}{dr} = C_3 \xi_r - C_1 P', \end{cases} \tag{26}$$

where D_η is the differential operator

$$D_\eta = \rho(c^2 + V_A^2)(\omega_\eta^2 - \omega_A^2)(\omega^2 - \omega_C^2), \tag{27}$$

and ω_η^2 denotes the differential operator

$$\omega_\eta^2 = \omega^2 \left(1 - i\frac{\eta}{\omega}\frac{d^2}{dr^2} \right). \tag{28}$$

The notation ω_η^2 leads to resistive equations which are formally as similar as possible to the ideal equations.

Equations (26) are the equations that govern the resonantly driven linear displacements of a cylindrical plasma in resistive MHD. They are a set of two differential equations of third order. They are formally the same as equations (8), but the coefficient function D is replaced by the differential operator D_η. The singularities are removed from the equations, but the order of the set of differential equations is raised from 2 (in ideal MHD) to 6 (in non-ideal MHD), and in addition the coefficient of the derivative of highest order is proportional to η

The equation for $\xi_\perp$ is (Goossens et al.(1995))

$$\left(\omega_\eta^2 - \omega_A^2\right)\xi_\perp = \frac{i}{\rho B}\left(g_B P' - \frac{2}{\mu r}f_B B_\varphi B_z \xi_r\right) , \qquad (29)$$

Equation (29) is the resistive generalization of equation (13). Equation (29) is formally the same as equation (13), but ω^2 is now replaced by the second order differential operator ω_η^2. As a consequence the resistive equation for $\xi_\perp$ is a differential equation of second order for which $s = 0$ is not a singular point.

5. Resistive MHD solutions for resonant Alfvén waves close to the ideal resonance position

Whheeeeeeeee! Europe, here we come! Or, Asia, or America, or wherever. It's June, and the conference season is well and truly open. $\cdots$ The whole academic world seems to be on the move. Half the passengers on transatlantic flights these days are university teachers. $\cdots$ Write a paper and see the world.
"Small World" by D. Lodge.

The aim is to determine the solutions to the set of dissipative MHD equations (26) and (29) in the vicinity of the critical point r_A defined by the condition $\omega_A^2(r_A) = \omega^2$. The radial variable s defined by equation (16) and series expansions of the coefficient functions around $s = 0$ are used to obtain simplified versions of the dissipative MHD differential equations. Just as in ideal MHD these simplified versions are valid in the interval $[-s_A, s_A]$ around the point of resonance where the linear Taylor polynomial is a valid approximation of $\omega^2 - \omega_A^2(r)$; hence s_A has to satisfy again the inequality (17). All the remainig equilibrium quantities are replaced by their values at $s = 0$. The simplified versions of equations (27) are

$$\begin{cases} \left(s\Delta - i\omega\eta\dfrac{d^2}{ds^2}\right)\dfrac{d\xi_r}{ds} = \dfrac{g_B}{\rho B^2}C_A(s), \\[4mm] \left(s\Delta - i\omega\eta\dfrac{d^2}{ds^2}\right)\dfrac{dP'}{ds} = \dfrac{2f_B B_\varphi B_z}{\mu r_A \rho B^2}C_A(s), \end{cases} \qquad (30)$$

where $C_A(s)$ is again given by equation (19). Equations (30) are the resistive generalizations of the ideal equations (18). The ideal factor $s\Delta$ is now replaced by the second order differential operator $s\Delta - i\omega\eta\frac{d^2}{ds^2}$. The set of two differential equations (18) is replaced by a set of two differential equations of third order. The ideal singularity at $s = 0$ is obviously absent from the resisitive equations (30). As in equations (18) all equilibrium quantities are evaluated at $s = 0$ $(r = r_A)$.

Again there is the common factor $C_A(s)$ in the right hand sides of equations (30). It is straightforward to obtain the differential equation for $C_A(s)$ in dissipative MHD by taking a linear combination of equations (30):

$$\left(s\Delta - i\omega\eta\frac{d^2}{ds^2}\right)\frac{dC_A(s)}{ds} = 0. \tag{31}$$

Equation (31) has equation (21) as ideal counterpart. Equation (29) reduces to

$$\left(s\Delta - i\omega\eta\frac{d^2}{ds^2}\right)\xi_\perp = i\frac{C_A(s)}{\rho B}. \tag{32}$$

This is the resistive generalization of equation (25). The equations (30-32) have no singularities at $s = 0$ in contrast to their ideal counterparts (18), (21), (25).

Dissipation is important when the terms $s\Delta$ and $\omega\eta\frac{d^2}{ds^2}$ in the left hand sides of equations (30-32) are comparable. This results in a dissipative layer with a thickness measured by the quantity δ_A

$$\delta_A = \left(\frac{\omega\eta}{|\Delta|}\right)^{1/3}. \tag{33}$$

The thickness of the dissipative layer therefore scales as $(\eta/ |d\omega_A/dr|)^{1/3}$, a result already obtained by Kappraff and Tataronis (1977) and Hollweg and Yang (1988) and numerically verified by Poedts et al. (1990a). In view of the very large values of the magnetic Reynolds number in the solar corona we have that

$$\frac{s_A}{\delta_A} \gg 1. \tag{34}$$

Inequality (34) is important for the present discussion. It implies that the interval of validity of the simplified versions of the dissipative MHD equations embraces the dissipative layer and in addition contains two overlap regions to the left and the right of the dissipative layer where ideal MHD is valid. Sakurai et al. (1991a) introduce a new scaled variable

$$\tau = \frac{s}{\delta_A} \tag{35}$$

which is of order 1 in the dissipative layer, but in view of inequality (34) $s \to \pm s_A$ corresponds to $\tau \to \pm\infty$.

With this new variable the equations (30-32) take the form

$$
\begin{cases}
\left\{ \dfrac{d^2}{d\tau^2} + i\,\mathrm{sign}(\Delta)\tau \right\} \dfrac{d\xi_r}{d\tau} = i\dfrac{g_B}{\rho B^2 \,|\,\Delta\,|}C_A, \\[3mm]
\left\{ \dfrac{d^2}{d\tau^2} + i\,\mathrm{sign}(\Delta)\tau \right\} \dfrac{dP'}{d\tau} = i\dfrac{2f_B B_\varphi B_z}{\rho B^2 \mu r_A \,|\,\Delta\,|}C_A, \\[3mm]
\left\{ \dfrac{d^2}{d\tau^2} + i\,\mathrm{sign}(\Delta)\tau \right\} \dfrac{dC_A}{d\tau} = 0, \\[3mm]
\left\{ \dfrac{d^2}{d\tau^2} + i\,\mathrm{sign}(\Delta)\tau \right\} \xi_\perp = \dfrac{-C_A}{\delta_A \,|\,\Delta\,|\,\rho B}.
\end{cases}
\qquad (36)
$$

Sakurai et al. (1991a) did not obtain the differential equations (36c) and (36d) for C_A and $\xi_\perp$ in dissipative MHD. They focussed on the dissipative equations for ξ_r and P' in an attempt to find the jumps in these quantities across the dissipative layer. They assumed that the ideal conservation law (22) remains valid in dissipative MHD. This assumption implies that the right hand sides of equations (36ab) are constant. Sakurai et al. (1991a) then obtained solutions of the dissipative equations (36ab) in terms of a double integral of Hankel functions of order 1/3 of a complex argument. Asymptotic expansions of these solutions enabled them to obtain the jumps in ξ_r and P'.

A related approach was used by Hollweg (1987b) and Hollweg and Yang (1988) in a discussion of the viscous dissipative layer for planar geometry. In essence, their approach was to take P' to be constant across the dissipative layer as the outset. This was equivalent to neglecting the inertia in the dissipative layer. Since $C_A = g_B P'$ for planar geometry, they thus started with the constancy of C_A, and then derived an equation equivalent to (36d), which was solved in terms of Airy functions with imaginary argument. The advantage of this approach was the reduction of the 6th-order dissipative problem to the 2nd-order equation (36d). They were also able to explicitly show that the total viscous heating (equivalent to the r-integral of the volumetric heating rate) in the dissipative layer was independent of the viscosity coefficient, and that this heating was equal to the corresponding energy lost from the waves by resonant absorption. The heating, and the energy lost from the waves by resonant absorption, was found to be adequate to heat active region loops. But the difficulty is that the coronal viscosity is so small that all the heating takes place in a very thin layer. If the same amout of energy could be distributed over the diameter of an active region loop, then one would have a reasonable model of coronal heating.

Goossens et al. (1995) use a more compact and straightforward method for obtaining the solutions to equations (36). They first show that C_A is also a constant in dissipative MHD for $\mid s \mid \leq s_A$ so that the ideal conservation law continues to be a conservation law in dissipative MHD. This proves the assumption by Sakurai et al. (1991a) and guarantees that the jump conditions found by Sakurai et al. (1991a) are correct. Then they obtain compact analytical solutions for ξ_r, P', and $\xi_\perp$ in the dissipative and the overlap regions which allow a very simple mathematical and physical interpretation.

Inspired by the techniques used by Boris (1968) and Mok and Einaudi (1985) for incompressible perturbations they looked for solutions in integral form. In order to obtain the solutions to equations (36) it is instructive to consider the following two differential equations of second order:

$$\begin{cases} \left\{ \dfrac{d^2}{d\tau^2} + i\,\mathrm{sign}(\Delta)\tau \right\} \Psi(\tau) = 0, \\[2mm] \left\{ \dfrac{d^2}{d\tau^2} + i\,\mathrm{sign}(\Delta)\tau \right\} F(\tau) = -1. \end{cases} \tag{37}$$

It is straightforward to show that

$$F_{1,2,3}(\tau) = \int_{\Gamma_{1,2,3}}^{\infty} \exp(iu\tau\,\mathrm{sign}(\Delta) - u^3/3)du, \tag{38}$$

are three solutions of equation (37b). Here Γ_1 is the postive real axis and Γ_2 and Γ_3 are the two rays in the complex plane that start at the origin and go to $\infty \exp(2\pi i/3)$ and to $\infty \exp(-2\pi i/3)$ respectively.

It immediately follows that

$$\Psi_1(\tau) = F_3(\tau) - F_1(\tau), \quad \wedge \quad \Psi_2(\tau) = F_3(\tau) - F_2(\tau) \tag{39}$$

are two linearly independent solutions of equation (37a). It is easily shown that $\Psi_1(\tau) = -i\pi(Ai(\zeta) - iBi(\zeta))$, while $\Psi_2(\tau) = -2\pi i Ai(\zeta)$ where the argument of the Airy functions is $\zeta = i\tau\,\mathrm{sign}(\Delta)$. The general solutions of the second order differential equations (37a) and (37b) then take the form

$$\begin{aligned} \Psi(\tau) &= A_1(F_3(\tau) - F_1(\tau)) + A_2(F_3(\tau) - F_2(\tau)), \\ F(\tau) &= F_1(\tau) + A_1(F_3(\tau) - F_1(\tau)) + A_2(F_3(\tau) - F_2(\tau)), \end{aligned} \tag{40}$$

where A_1, and A_2 are two yet undetermined constants.

It is obvious that we are only interested in the solutions of equations (37) that are bounded, in particular they have to remain bounded for $\tau \to$

$\pm\infty$. The asymptotic behaviour of $F_1(\tau)$, $F_2(\tau)$, and $F_3(\tau)$ was determined by Goossens et al. (1995). They showd that out of the three functions $F_1(\tau)$, $F_2(\tau)$, and $F_3(\tau)$ only the function $F_1(\tau)$ is bounded everywhere. The functions $F_2(\tau)$ is unbounded at $\tau \to -\infty \, \mathrm{sign}(\Delta)$, and the function $F_3(\tau)$ at $\tau \to +\infty \, \mathrm{sign}(\Delta)$. This implies that both $\Psi_1(\tau)$ and $\Psi_2(\tau)$ become unbounded, and that we have to take

$$A_1 = A_2 = 0 \tag{41}$$

to obtain bounded solutions for $\Psi(\tau)$ and $F(\tau)$ so that

$$
\begin{aligned}
\Psi(\tau) &\equiv 0, \\
F(\tau) &\equiv F_1(\tau).
\end{aligned}
\tag{42}
$$

Equation (42a) implies that the bounded solution of equation (36c) satisfies

$$
\begin{aligned}
\frac{dC_A(\tau)}{d\tau} &= 0, \\
C_A(\tau) &= constant.
\end{aligned}
\tag{43}
$$

The ideal conservation law obtained by Sakurai et al. (1991a) continues to hold in dissipative MHD. Equation (42b) also implies that

$$
\begin{aligned}
\frac{d\xi_r}{d\tau} &= -i\frac{g_B}{\rho B^2 |\Delta|} C_A F(\tau), \\
\frac{dP'}{d\tau} &= -i\frac{2 f_B B_\varphi B_z}{\mu r_A \rho B^2 |\Delta|} C_A F(\tau),
\end{aligned}
\tag{44}
$$

and

$$\xi_\perp = \frac{C_A F(\tau)}{\rho B \delta_A |\Delta|}. \tag{45}$$

Integration of equations (44a-b) gives the solutions in dissipative MHD for ξ_r, P', and $\xi_\perp$ that remain finite for $\xi_\perp$ and only logariphmically grow for ξ_r and P' when $|\tau| \to \infty$ (and thus can be matched with the outer ideal solutions) as:

$$
\left\{
\begin{aligned}
\xi_r &= -\frac{g_B C_A}{\rho B^2 \Delta} G(\tau) + C_\xi, \\
\\
P' &= -\frac{2 f_B B_\varphi B_z C_A}{\rho B^2 \mu r \Delta} G(\tau) + C_P, \\
\\
\xi_\perp &= \frac{C_A}{\delta_A |\Delta| \rho B} F(\tau),
\end{aligned}
\right.
\tag{46}
$$

where C_ξ and C_P are constants of integration and

$$G(\tau) = \int_0^\infty \frac{e^{-u^3/3}}{u} \{\exp(iu\tau\,\mathrm{sign}(\Delta)) - 1\}du. \tag{47}$$

This function was also used by Davila (1987) for constructing the solution for the normal component of velocity in planar geometry. ¿From the definition of C_A, equation (19), it follows that the constants C_ξ and C_P cannot be independently chosen, but are related by $C_A = g_B C_P - 2f_B B_\varphi B_z C_\xi / \mu r_A$.

Straightforward Mac Laurin expansions give absolute convergent power series for all τ for $G(\tau)$ and $F(\tau)$:

$$\begin{cases} F(\tau) = \displaystyle\sum_{n=0}^\infty a_n \tau^n , \\[2mm] G(\tau) = i\tau\,\mathrm{sign}(\Delta) \displaystyle\sum_{n=0}^\infty \frac{a_n}{n+1}\tau^n , \end{cases} \tag{48}$$

where

$$a_n = \frac{3^{n/3}}{3^{2/3}} \Gamma\left(\frac{n+1}{3}\right) \frac{[i\,\mathrm{sign}(\Delta)]^n}{n!} \tag{49}$$

and Γ is the gamma-function. These power series can be used to obtain solutions for ξ_r, P', and $\xi_\perp$. In particular for $|\tau| \le 1$, they show that in resistive MHD all the physical quantities take finite values in the dissipative layer and at the ideal resonance position where the ideal MHD solutions diverge. ¿From equation (49) it immediately follows that the coefficients a_{2k} with even indices are purely real, while the coefficients a_{2k+1} with uneven indices are purely imaginary. This implies that $Re F(\tau)$ and $Re G(\tau)$ are even functions of τ, while $Im F(\tau)$ and $Im G(\tau)$ turn out to be uneven functions of τ.

The power series given in equation (48) are not well suited to find out how the resisitive MHD solutions have to be connected to the ideal MHD solutions. To this end we determine the asymptotic behaviour of the resistive MHD solutions for $\tau \to \pm\infty$. The asymptotic behaviour of $G(\tau)$ for $\tau \to \pm\infty$ was calculated by Goossens et al. (1995). With the use of this result they obtained the following asymptotic expansions for ξ_r and P' for $\tau \to \pm\infty$

$$\begin{cases} \xi_r \simeq \dfrac{g_B C_A}{\rho B^2 \Delta}\left(\ln|\tau| + \dfrac{2\nu}{3} + \dfrac{1}{3}\ln 3 - \dfrac{i\pi}{2}\mathrm{sign}(\Delta\tau)\right) + C_\xi , \\[4mm] P' \simeq \dfrac{2f_B B_\varphi B_z C_A}{\mu r_A \rho B^2 \Delta}\left(\ln|\tau| + \dfrac{2\nu}{3} + \dfrac{1}{3}\ln 3 - \dfrac{i\pi}{2}\mathrm{sign}(\Delta\tau)\right) + C_P . \end{cases} \tag{50}$$

where ν is the Euler constant. These results provide us with the asymptotic behaviour of ξ_r, and P' in the overlap regions where ideal MHD is also valid. In the overlap regions the asymptotic versions of the dissipative solutions (50) and the ideal solutions (23) represent the same solutions. The asymptotic versions recover the logarithmic behaviour of $Re(\xi_r)$, and $Re(P')$ already found in ideal MHD in equation (23), but show that this logarithmic behaviour is only valid away from the ideal resonance position for large $|\tau|$.

Comparison of the asymptotic versions of the dissipative MHD solutions with the ideal MHD solutions enables us to obtain expressions for $\xi_\pm$, and $P'_\pm$ in equations (23) (see Goossens et al. 1995) and in particular they allow us to obtain the jumps in ξ_r and P' :

$$\begin{cases} [\xi_r] = -i\pi \dfrac{g_B C_A}{\rho B^2 \mid \Delta \mid} \\[4mm] [P'] = -i\pi \dfrac{2 f_B B_\varphi B_z C_A}{\rho B^2 \mu r_A \mid \Delta \mid} \end{cases} \tag{51}$$

These jumps and the conservation law were first derived by Sakurai et al. (1991a). An important property of resonant Alfvén wave heating is that the jumps are independent of η. This implies that the amount of absorbed wave energy and the total amount of resistive heating in the dissipative layer are also independent of η.

The asymptotic behaviour of $F(\tau)$ for $\tau \to \pm\infty$ is calculated in Goossens et al. (1995). They obtain the following asymptotic expansion for $\xi_\perp$

$$\xi_\perp \simeq \frac{i C_A}{\tau \rho B \delta_A \Delta} \tag{52}$$

This asymptotic version recovers the $1/\tau$ behaviour of $Im(\xi_\perp)$, already found in ideal MHD in equation (25), but it shows that this behaviour is only valid away from the ideal resonance position for $\tau \to \pm\infty$. In order to understand fully the relation between the ideal and resistive solution for $\xi_\perp$ it is necessary to find out what has happened to the ideal $\delta(s)$ contribution to $\xi_\perp$. Goossens et al. (1995) determined the limit of the resistive solution of $\xi_\perp$ for $\delta_A \to 0$, as a function of s and found that

$$\lim_{\delta_A \to 0} \xi_\perp = \frac{C_A}{\rho B} \left[\frac{\pi}{|\Delta|} \delta(s) + \frac{i}{\Delta} \mathcal{P} \left(\frac{1}{s} \right) \right] . \tag{53}$$

where $\mathcal{P}$ denotes the principal Cauchy part. Thus the $\delta(s)$ contribution to $\xi_\perp$ can be thought of as arising from $Re F(\tau)$, which is an even function of τ. The amplitude of $Re\xi_\perp$ at $s = 0$ is proportional to $1/\delta_A$, while $Re\xi_\perp$

becomes small if $\mid s \mid \geq \delta_A$. The area under $Re\xi_\perp(s)$ is thus independent of δ_A, leading to the δ-function as $\delta_A \to 0$.

The real and imaginary parts of $F(\tau)$ and $G(\tau)$ are shown in Goossens et al. (1995). Except for the normalization, the curves for $F(\tau)$ appear to be identical to results obtained previously by Hollweg (1987b) and Hollweg and Yang (1988). Hollweg considered the structure of the viscous dissipative layer in an incompressible plasma, while Hollweg and Yang considered the viscous dissipative layer around the Alfvén resonance in a compressible plasma. In both cases the geometry was planar. Our present work shows that the same behaviour occurs also in cylindrical geometry for the resistive layer.

In ideal MHD the dominant dynamics of resonant Alfvén waves resides in the perpendicular component of the displacement $\xi_\perp$. The logarithmic singularity and the jump contribution to the radial component of the displacement ξ_r are overruled by the s^{-1} singularity and the δ-function contribution to $\xi_\perp$. In resistive MHD all these singularities disappear and all physical variables take finite values. To determine the dominant dynamics in resistive MHD Goossens et al. (1995) estimated the relative importance of ξ_r and $\xi_\perp$ close to the ideal resonance position. They found that

$$\frac{\mid \xi_\perp \mid}{\mid \xi_r \mid} \sim R_m^{1/3} \tag{54}$$

where $R_m = \omega L^2/\eta$, is the magnetic Reynolds number. Since R_m is very large, of the order of $10^{10} - 10^{12}$ in the solar atmosphere, equation (54) implies that the dominant dynamics continues to reside in the perpendicular components as was found by Poedts et al (1989ab,1990a).

6. Conclusions

He surveyed the scene with complacency. He felt sure he was going to enjoy his stay here. $\cdots$ All you had to do, to come and stay in this idyllic retreat, $\cdots$, given every facility for reflection and creation, was to apply.
Observation by Prof. Morris Zapp in "Small World" by D. Lodge.

This review has shown how analytic theory gives a full understanding of the complicated physics of driven resonant Alfvén waves in 1-dimensional nonuniform untied magnetic flux tubes. The equations of linear resistive MHD for linear displacements superimposed on a static equilibrium state have been simplified for displacements that correspond to resonant Alfvén waves in ideal MHD. Series expansions of the coefficient functions around the ideal Alfvén resonance are used to obtain versions of the linear resistive equations that govern the Alfvén waves in a region which contains the dissipative layer and two overlap regions where ideal MHD is also valid. This set of linear resistive MHD equations is solved in a consistent manner by the use of integral solutions. The fundamental conservation in ideal MHD found by Sakurai et al. (1991a) remains valid in dissipative MHD. This implies that the total resistive heating in the dissipative layer, and the amount of absorbed wave energy, are independent of the resistivity. Compact analytical solutions have been obtained for ξ_r, P', and $\xi_\perp$ which allow a straightforward mathematical and physical interpretation of resonant Alfvén waves in the dissipative layer. As such they enable us to understand the basic physics of resonant Alfvén waves. They also help us with the interpretation of the results of large-scale numerical simulations. Asymptotic expansion of these solutions lead to jump conditions. These jump conditions and the conservation law make it possible to determine the absorption of Alfvén waves without having to solve the dissipative MHD equations. This procedure was used by Sakurai et al. (1991b) for studying the absorption of acoustic oscillations in sunspots. It was generalized to stationary equilibrium states by Goossens et al. (1992) and Erdélyi et al. (1995). Goossens and Hollweg (1993) used this scheme to obtain conditions for maximal and total absorption and to explain the variation of the spatial solutions with frequency. Keppens et al. (1994) used this scheme to determine the scattering and resonant absorption of acoustic oscillations by an ensemble of magnetic flux tubes (see also Keppens, 1995ab). Stenuit et al. (1995) designed a simple numerical scheme for the computation of resonant Alfvén waves based on the conservation law and connection formulae. They compared absorption coefficients calculated on the basis of the ideal MHD equations in combination with the connection formulae with those obtained by numerical integration of the dissipative equations. They found excellent agreement providing numerical support and verification of the consistency of the method using connection formulae. Keppens (1995c) applied the connection formulae to slow resonant waves in hot magnetic fibrils and reconsidered the impedance matching conditions obtained by Goossens and Hollweg (1993) for optimal driving. Tirry and Goossens (1995) extended the analysis on conservation laws and connection formulae to 2-dimensional equilibrium states. Ruderman et al. (1995) extended the analysis to nonlinear theory for resonant slow waves.

References

Appert, K., Gruber, R. and Vaclavik, J. (1974) *Phys. Fluids* **17**, 1471.

Boris, J.P. (1968) Resistively modified normal modes of an inhomogeneous incompressible plasma. *Ph.D. Thesis*, Princeton University, UMI Dissertation Services, Ann Arbor Michigan, p. 172.

Cally, P.S. and Bogdan, T.J. (1993) *Astrophys. J.* **402**, 721.

Chen, L. and Hasegawa, A. (1974) *Phys. Fluids* **17**, 1399.

Davila, J.M. (1987) *Astrophys. J.* **317**, 514.
Erdélyi, R. and Goossens, M. (1995) *Astron. Astrophys.* **294**, 575.
Erdélyi, R., Goossens, M. and Ruderman, M.S. (1995) *Solar Phys.*, **161**, 123.
Edwin, P. and Roberts, B. (1987) MHD Waves in the Solar Atmosphere, in: *Small-scale Plasma Processes, Proceedings of the 21st ESLAB Symposium ESA SP-275*, p. 169.
Goedbloed, J.P. (1975) *Phys. Fluids* **15**, 1090.
Goedbloed, J.P. (1983) Lecture Notes on Ideal Magnetohydrodynamics. *Rijnhuizen Report* 83-145, 113.
Goossens, M. (1991) MHD waves and wave heating in non-uniform plasmas, in: eds. Priest, E.R. and Hood A.W., *Advances in Solar System Magnetohydrodynamics*, Cambridge University Press, Cambridge, p. 135.
Goossens, M., Hollweg, J.V. and Sakurai, T. (1992) *Solar Phys.* **138**, 233.
Goossens, M. and Hollweg, J.V. (1993) *Solar Phys.* **145**, 19.
Goossens, M., Poedts, S. and Hermans, D. (1985) *Solar Phys.* **102**, 51.
Goossens, M. and Poedts, S. (1992) *Astrophys. J.* **384**, 348.
Goossens, M. and Ruderman, M.S. (1995) *Physica Scripta* **51**, 1.
Goossens, M., Ruderman, M.S. and Hollweg, J.V. (1995) *Solar Phys.* **157**, 75.
Grossman, W. and Smith, R.A. (1988) *Astrophys. J.* **332**, 476.
Grossman, W. and Tataronis, J. (1973) *Z. Phys.* **261**, 217.
Halberstadt, G. (1994) Magnetic heating of the solar corona. Photospheric excitation and coronal dissipation of Alfvén waves. *Ph.D. thesis*, Vrije Universiteit Amsterdam, Amsterdam.
Halberstadt, G. and Goedbloed, J. (1993) *Astron. Astrophys.*, **280**, 647.
Halberstadt, G. and Goedbloed, J. (1995a) *Astron. Astrophys.*, **301**, 559.
Halberstadt, G. and Goedbloed, J. (1995b) *Astron. Astrophys.*, **301**, 577.
Hasegawa, A. and Chen, L. (1974) *Phys. Fluids* **17**, 1924.
Hollweg, J.V. (1987a) *Astrophys. J.* **312**, 880.
Hollweg, J.V. (1987b) *Astrophys. J.* **320**, 875.
Hollweg, J.V. (1988) *Astrophys. J.*, **335**, 1005.
Hollweg, J.V. (1990) *Computer Phys. Rep.* **12**, 205.
Hollweg, J.V. (1991) Alfvén waves, in: eds. Ulmschneider, P., Priest, E.R. and Rosner, R., *Mechanisms of Chromospheric and Coronal Heating*, Springer-Verlag, Berlin, p. 423.
Hollweg, J.V. and Yang, G. (1988) *J. Geophys. Res.* **93**, 5423.
Ionson, J.A. (1978) *Astrophys. J.* **226**, 650.
Ionson, J.A. (1985) *Solar Phys.* **100**, 289.
Kappraff, J.M. and Tataronis, J. (1977) *J. Plasma Physics* **18**, 209.
Keppens, R., Bogdan, T.J. and Goossens, M. (1994) *Astrophys. J.* **436**, 372.
Keppens, R. (1995a) On the interaction of acoustic oscillations with magnetic flux tubes in the solar photosphere, *Ph.D. Thesis*, K.U.Leuven.
Keppens, R. (1995b) *Solar Phys.*, **161**, 251.
Keppens, R. (1995c) *Astrophys. J.*, submitted.
Kuperus, M., Ionson, J.A., and Spicer, D. (1981) *Ann. Rev. Astron. Astrophys.* **19**, 7.
Lou, Y.-Q. (1990) *Astrophys. J.* bf 350, 452.
Mok, Y. and Einaudi, G. (1985) *J. Plasma Phys.* **33**, 199.
Narain, U. and Ulmschneider, P. (1995) *Space Sci. Rev.*, accepted for publication.
Nayfeh, A.H. (1981) Introduction to Perturbation Techniques, Wiley-interscience, New York.
Ofman, L., Davila, J.M. and Steinolfson, R.S. (1994a) *Astrophys. J.* **421**, 360.
Ofman, L., Davila, J.M. and Steinolfson, R.S. (1994b) *Geophys. Res. Let.* **21**, 2259.
Ofman, L., Davila, J.M. and Steinolfson, R.S. (1994c) Nonlinear studies of coronal heating by the resonant absorption of Alfvén waves, in: eds. Rusin, V., Heinzel, P. and Vial, J.-C., *IAU Colloquium 144, Solar Coronal Structures*, VEDA, Bratislava, p. 473.
Ofman, L. and Davila, J.M. (1995) *J. Geophys. Res.*, submitted.
Pao, Y.P. (1975), *Nuclear Fusion* **18**, 1629.

Poedts, S., Hermans, D. and Goossens, M. (1985) *Astron. Astrophys.* **151**, 16.

Poedts, S. and Goossens, M. (1987) *Solar Phys.* **109**, 265.

Poedts, S. and Goossens, M. (1988) *Astron. Astrophys.* **198**, 331.

Poedts, S., Goossens, M. and Kerner, W. (1989a) *Solar Phys.* **123**, 83.

Poedts, S., Kerner, W. and Goossens, M. (1989b) *J. Plasma Physics* **42**, 27.

Poedts, S., Goossens, M. and Kerner, W. (1990a) *Astrophys. J.* **360**, 279.

Poedts, S., Goossens, M. and Kerner, W. (1990b) *Computer Phys. Comm.* **59**, 75.

Poedts, S. and Kerner, W. (1992) *J. Plasma Phys.* **47**, 139.

Poedts, S. and Boynton, G.C. (1995) *Astron. Astrophys.*, accepted.

Roberts, B. (1988) Solar Magnetohydrodynamics, in: ed. B. Buti *Cometary and Solar Plasma Physics*, World Scientific, Singapore.

Roberts, B. (1991) MHD waves in the Sun, in: eds. Priest, E.R. and Hood A.W., *Advances in Solar System Magnetohydrodynamics*, Cambridge University Press, Cambridge, p. 105.

Ruderman, M.S., Hollweg, J.V. and Goossens, M. (1995), in preparation.

Sakurai, T., Goossens, M. and Hollweg, J.V. (1991a) *Solar Phys.* **133**, 227.

Sakurai, T., Goossens, M. and Hollweg, J.V. (1991b) *Solar Phys.* **133**, 247.

Spruit, H.C. and Bogdan, T.J. (1992) *Astrophys.J.* **391**, L109.

Stenuit, H., Poedts, S. and Goossens, M. (1993) *Solar Phys.* **147**, 13.

Stenuit, H., Erdélyi, R. and Goossens, M. (1995) *Solar Phys.*, **161**, 139.

Tataronis, J. and Grossmann, W. (1973) *Z. Phys.* **261**, 203.

Thompson, M.J. and Wright, A.N. (1993) *J. Geophys. Res.* **98**, 15541.

Tirry, W. and Goossens, M. (1995) *J. Geophys. Res.*, accepted.

Wright, A.N. and Thompson, M.J. (1994) *Phys. Plasmas* **1**, 691.

THE DYNAMIC SOLAR CORONA IN X-RAYS WITH YOHKOH

SAKU TSUNETA
Institute of Astronomy, University of Tokyo
Osawa 2-21-1, Mitaka city, Tokyo 181, JAPAN
tsuneta@sxt2.ioa.s.u-tokyo.ac.jp

Abstract. *Yohkoh* is revolutionizing our understanding of the solar corona and the behavior of magnetized plasmas in general. It appears that all the transient heating, including solar flares, which have time scales of 10–100 Alfvén transit times, is due to magnetic reconnection. This transient heating is sometimes associated with global structural changes in the coronal magnetic fields. Magnetic reconnection with its associated slow shocks is a powerful engine to convert magnetic energy to plasma kinetic and thermal energies. The *Yohkoh* observations also show the existence of steadily heated plasmas with temperature of 2–4 MK, both in active regions and in the quiet Sun. The mechanism of the steady heating has not yet been understood.

1. Introduction

The *Yohkoh* satellite was launched on 1991, August 30 for observations of solar flares and the solar corona in X-ray and gamma-ray wavelengths. Excellent observations have been done over the last four years. The satellite has experienced the entire transition from the maximum to the minimum of the last solar cycle and is now ready to proceed to observations of the beginning of the next solar cycle. *Yohkoh* (Ogawara *et al.* 1991) carries two X-ray telescopes: the hard X-ray telescope has imaging capability above 40 keV with high spatial and time resolution. These hard X-ray images allow us to obtain new information on where and how energetic electrons are accelerated in solar flares. The soft X-ray telescope (SXT) is a grazing incidence X-ray (5-50 Å) telescope to observe the solar corona and solar flares with high spatial (3 arcsec) and time (2 sec) resolution (Tsuneta *et al.* 1991, Acton *et al.* 1992). SXT is sensitive to plasmas with temperatures

85

K. C. Tsinganos (ed.), Solar and Astrophysical Magnetohydrodynamic Flows, 85–108.
© *1996 Kluwer Academic Publishers.*

from 2 MK through 20 MK. Two other instruments onboard are the Bragg crystal spectrometer to observe the iron, calcium, and sulfur lines from flare plasmas, and the wide-band spectrometers to observe flare spectra from 5 keV to 10 MeV. For details on the instrumentation onboard *Yohkoh* the interested reader is referred to Volume **136** of *Solar Physics* (1991). The discoveries made so far by *Yohkoh* cover a wide area of solar physics. In the following we give a representative sample (Tsuneta and Lemen 1993, Uchida 1993, Tsuneta 1994).

(1) *Solar flares as an on-going magnetic reconnection* (Tsuneta *et al.* 1992a, Tsuneta 1993, 1996a, Masuda *et al.* 1994, Shibata *et al.* 1995, Yokoyama 1995). The *Yohkoh* data, for the first time, show unambiguously that magnetic reconnection (Priest 1996) is responsible for significant energy release in solar flares; magnetic reconnection serves as a highly efficient engine to convert magnetic energy into plasma kinetic and thermal energies with standing slow shocks.

(2) *Micro flares and coronal heating* (Shimizu *et al.* 1992, 1994, Shimizu 1995): Small transient brightenings are frequently seen in active regions. Reconnection of multiple loops is involved in these brightenings. The observed number of the brightenings is, however, not enough to explain the active region heating.

(3) *Global structural changes of the quiet solar corona* (Tsuneta *et al.* 1992b, Tsuneta 1996b). The X-ray movies show frequent structural changes of coronal magnetic fields on global and local scales with X-ray brightening (heating). Magnetic reconnection is again a primary agent for the large-scale restructuring of the quiet Sun.

(4) *Temperature structure of active regions* (Yoshida and Tsuneta 1996, Yoshida *et al.* 1995, Ichimoto *et al.* 1995). The temperatures of active region structures range from 3 MK to 10 MK. The loop structures with shorter lifetime ($<$ a few hours) generally have higher temperatures (5–8 MK) than loops with longer lifetime (2–4 MK). The 6–7 MK plasma is transiently heated by magnetic reconnection, whereas the 2–4 MK plasma is more steadily and uniformly heated. Two different mechanisms are apparently involved in the coronal heating.

(5) *The temperatures of the quiet Sun and coronal holes* (Hara *et al.* 1992, 1994). The temperature of coronal holes obtained with the *Yohkoh* SXT is about 2 MK, which is higher than those obtained by the *Skylab* observations (Doschek and Feldman 1977). The discrepancy may indicate the multi-temperature nature of coronal hole plasmas. The temperature of the quiet Sun (regions outside active regions and coronal holes) is also about 2 MK. There is no systematic difference in temperatures between coronal holes and the quiet Sun.

(6) *Strong X-ray emission from buoyant emerging fluxes* (Shibata *et al.*

(1992, 1994). Magnetic flux emerging from below the photosphere is bright in X-rays. Strong heating occurs through reconnection between the emerging flux and the existing coronal fields (Yokoyama and Shibata 1995). Shibata *et al.* (1992, 1994) also discovered X-ray jets, which appear to be driven by the slingshot effect of the reconnected field lines.

(7) *Expansion of active region magnetic fields* (Uchida *et al.* 1992). Active region magnetic fields sometimes show an almost continuous expansion. It appears that mass and magnetic fluxes from these active regions are injected to the interplanetary space by the expansion.

(8) *Hard X-ray observations.* The *Yohkoh* hard X-ray telescope showed that hard X-ray flares essentially have double source structures, which are located at the footpoints of the soft X-ray loop. Furthermore, the intensities of the double sources correlate in time within 100 ms (Sakao 1994). This clearly indicates that a rich population of non-thermal electrons (upto $\sim$ 100 keV) are created in flare loops in association with magnetic reconnection. It is, however, not known whether non-thermal acceleration is directly related to reconnection process or not (Tsuneta 1995). In addition to the footpoint hard X-ray sources, Masuda *et al* (1994) discovered a hard X-ray source above the soft X-ray loop. Although the location is consistent with the reconnection picture, the nature of the loop-top hard X-ray source is not fully understood. The interested reader is referred to Kosugi (1994) for a review of the *Yohkoh* hard X-ray observations.

This paper is concerned with some of these new findings with SXT as well as with the theoretical problems raised by the new observations.

2. The X-ray Corona

2.1. THE OVERALL X-RAY CORONA

Figure 1 is a *Yohkoh* soft X-ray image of the quiet Sun. We can see complex structures, which are completely different from optical images of the Sun. The structures seen in Figure 1 include active regions, large scale magnetic structures, coronal holes on both polar regions, and X-ray jets (Shibata *et al.* 1992) from a bright region located on the north-west sector. X-ray bright points are seen in the coronal holes. These complex structures represent magnetic fields in the corona, because the gyro-frequency ω_B is higher than the electron-ion collision frequency ν_c (Spitzer 1962) in the solar corona;

$$\omega_B = 1.8 \times 10^4 B >> \nu_c = 4.7 \times 10^{-11} \frac{n \ln \Lambda(T)}{T^{1.5}}, \qquad (1)$$

where B is the magnetic field strength $[10^{2-3}$ G], n is the density $[10^{8-10}$ cm^{-3}] and T the temperature [0.1-2 keV]. $\ln \Lambda(T)$ is the Coulomb logarithmic term, which equals usually about 20. Thus, coronal magnetic fields

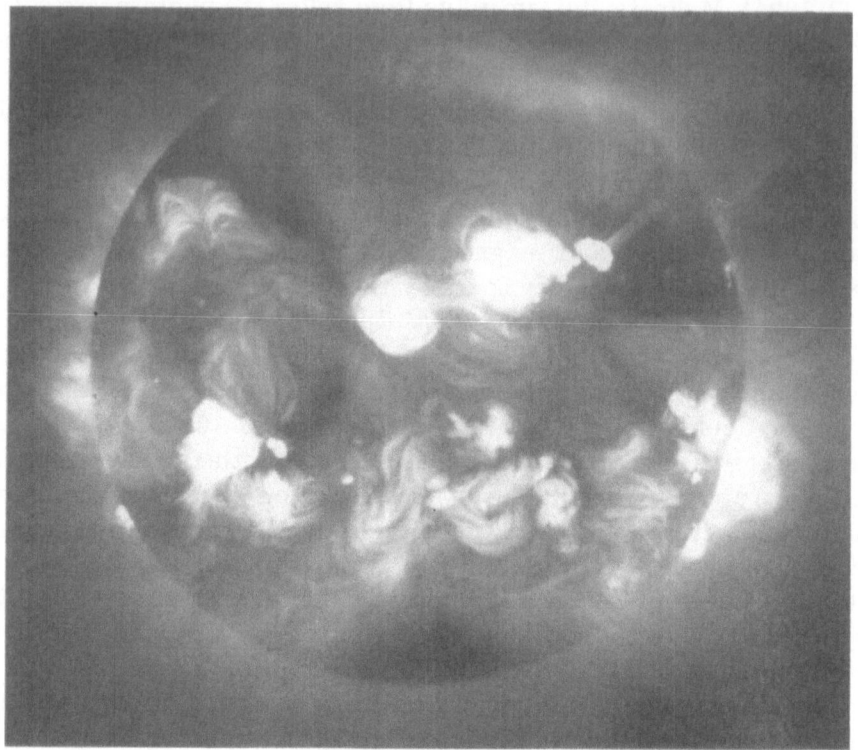

Figure 1. A *Yohkoh* soft X-ray image of the Sun

suppress the classical perpendicular transport of mass and heat. Although high resolution soft X-ray imaging provides us with the only means to "see" the coronal magnetic structures, we need to be careful that not all the magnetic structures are seen in X-ray images; they are recognized in the X-ray images only when the temperatures and/or densities of the plasmas loaded on the particular field lines are enhanced. The temperatures and densities are in turn sensitive to the heat deposition in the plasma. This point is important and should be kept in mind when analyzing X-ray images.

Figure 2 shows an X-ray image with the corresponding optical image also taken by *Yohkoh*, Hα and a magnetograph image. The bright regions in X-rays (active regions) correspond to strong magnetic pairs on the photosphere. Furthermore, most of the enhancements seen in X-rays correspond to smaller and weaker magnetic pairs. This one-to-one correspondence shows that the energy input to the corona is generally related to the magnetic field strength on the photosphere.

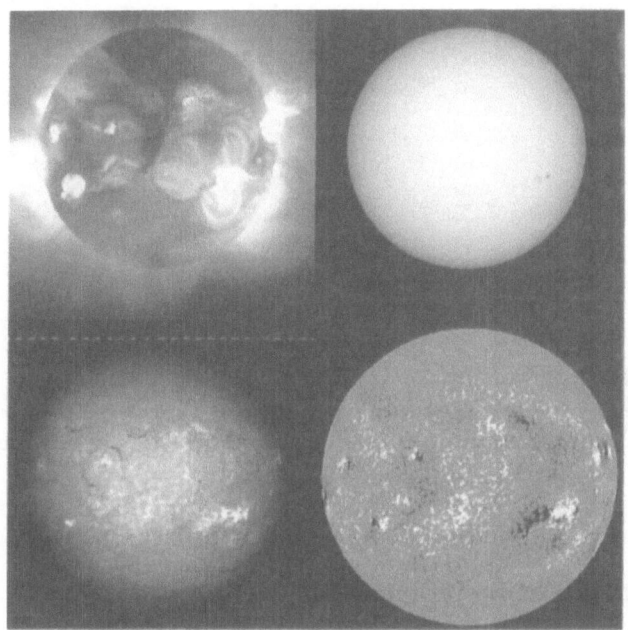

Figure 2. X-ray (top left) and white light (top right) images both taken by the *Yohkoh* SXT together with the Hα (bottom left) and magnetograph (bottom right) images. The gray scale of the magnetograph image indicates the polarity and magnitude of the longitudinal component of the photospheric magnetic fields.

2.2. THE ACTIVE REGION CORONA

Figure 3 is the X-ray and optical images of an active region both taken by *Yohkoh*. The X-ray image shows two kinds of magnetic loops in the active region; diffuse dark region and thin small loops, which are brighter than the diffuse loops.

An important question here is that despite the fact that the active region corona is filled with intense magnetic fields, *why are these particular small loops bright?* In other words, *why are most of the magnetic structures dark in X-rays?* The structures seen in the X-ray images must be closely related to the mechanism of the preferential heat injection to the particular magnetic structures. This question will be partially answered in the subsequent sections.

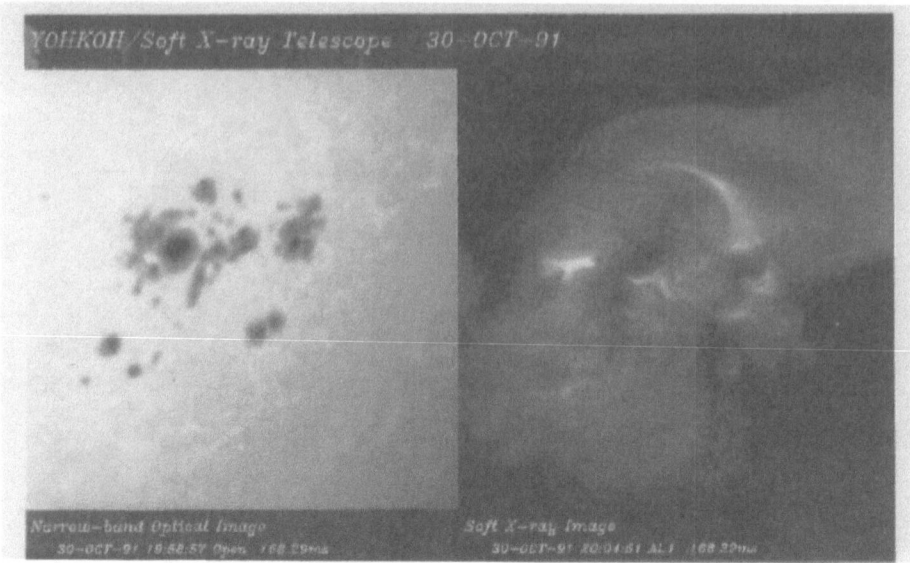

Figure 3. X-ray and optical images of an active region both taken by *Yohkoh*.

3. Transient and Steady Loops in the Solar Corona

3.1. TRANSIENT LOOPS IN THE SOLAR CORONA

The X-ray images of the solar active regions show ubiquitous small brightenings (Figure 4): the X-ray intensities of the brightenings range from those of small flares to the detection limit of the SXT. The physical parameters derived from *Yohkoh* observations are summarized in Table 1.

TABLE 1. Physical parameters of transient brightenings (Shimizu 1995)

Physical parameter	Value
Temperature	4 – 8 MK
Volume emission measure	$10^{44.5} - 10^{47.5}$ cm^{-3}
Electron density	$2 \times 10^9 - 2 \times 10^{10}$ cm^{-3}
Gas pressure	5 – 20 dyne cm^{-2}
Loop length	$5 \times 10^3 - 4 \times 10^4$ km
Loop width	$2 \times 10^3 - 7 \times 10^3$ km
Duration	2 – 7 min
Total energy loss (radiation+conduction)	$5 \times 10^{26} - 5 \times 10^{29}$ erg
Scaled total energy loss (Hudson 1991)	$10^{25} - 10^{29}$ erg
Thermal energy content	$10^{26} - 10^{29}$ erg

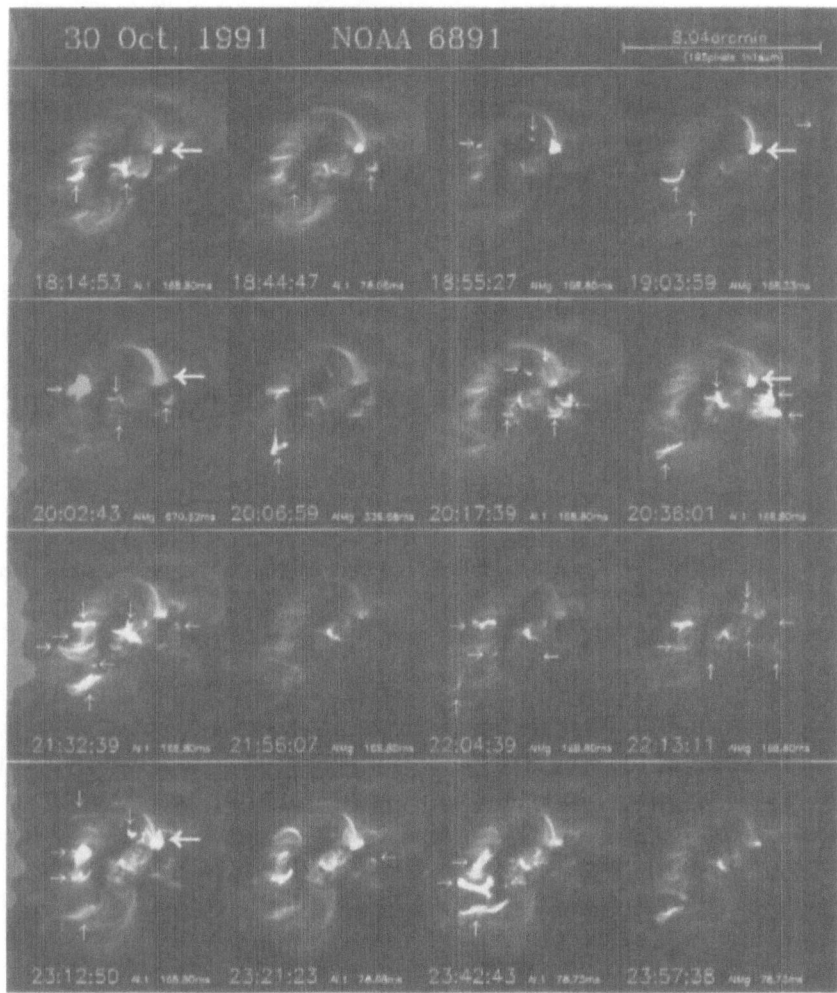

Figure 4. X-ray images of the active region NOAA 6891. Frequent microflaring can be seen, as indicated by arrows (Shimizu *et al.* 1992). The FOV is 8 arcmin, and the pixel size is 2.5 arcsec.

The transient brightenings often appear to be of multiple loops (Figure 5); two or more loops, that are almost parallel each other, or in contact near one of the footpoints, simultaneously brighten (Shimizu *et al.*, 1992, 1994). This morphological signature implies that magnetic reconnection of separate loop structures is involved in the transient heatings.

3.2. CAN TRANSIENT BRIGHTENINGS EXPLAIN CORONAL HEATING?

Parker (1988) suggested that numerous nano-flares due to magnetic reconnection of loops tangled by photospheric motions can heat the corona. Are

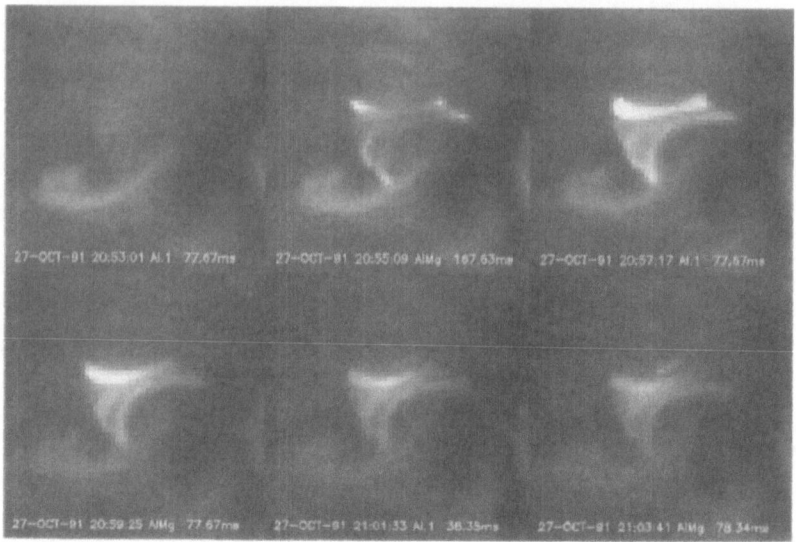

Figure 5. X-ray evolution of a microflare that occurred on 1991, October 27. Multiple footpoint brightenings are followed by brightening of the entire loops.

the transient brightenings discovered with *Yohkoh* the tip of the iceberg of those nano flares? To investigate the micro-nano flare heating hypothesis, Shimizu (1995) investigated the intensity size distribution of about 300 transient brightenings. He found a power-law occurrence distribution with power index of 1.4–1.7 (Figure 6) over 3 orders of magnitude. The distribution deviates from the power law in the fainter end, probably due to the detection limit of SXT. The other end overlaps with small flares. Dennis (1985) and Crosby, Aschwanden, Dennis (1993) found the power-law size distribution of the hard X-ray flares obtained by the *Solar Maximum Mission*. The power index is close to the one obtained here. The combined data set implies that the power law extends over many orders of magnitude from major flares to the transient brightenings. This indicates that flares and the transient brightenings discovered with *Yohkoh* are caused by the same physical mechanism.

Theses observations do not appear to directly support the nano-flare heating hypothesis: (1) Shimizu (1995) concluded that the energy supplied to active regions by the observed transient brightenings is only 10–20 % at most, and that fainter events with occurrence rate higher than the extrapolated power law are required to explain the coronal heating. (The occurrence rate has to have sharp increase below the *Yohkoh* detection limit.) (2) They appear to be associated with reconnection of different *macroscopic* magnetic structures identifiable with SXT. If the microflares are due to warping and interweaving of the field lines driven by the photospheric motion within the

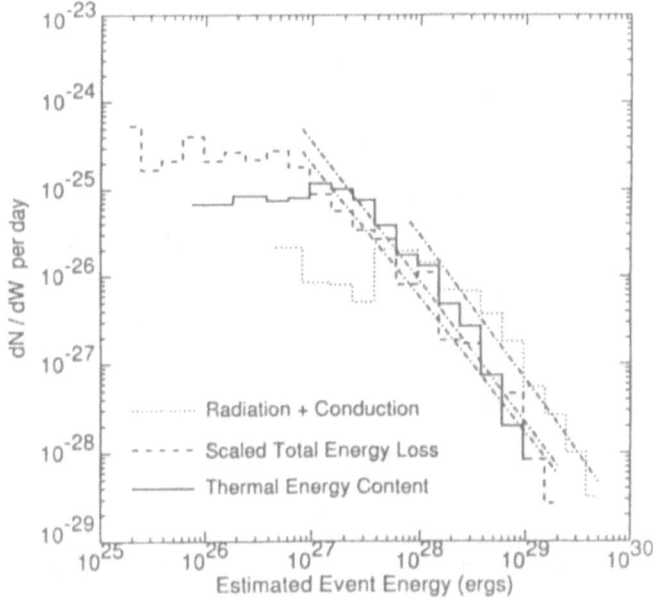

Figure 6. Frequency distribution of the transient brightenings as a function of the total energy. The three lines indicate the three different distributions; radiation + conduction loss (dotted line), scaled total energy loss (following Hudson 1991, dashed line), and the thermal energy content (solid line). Each distribution can be represented by a single power-law with index of 1.5–1.6 (dashed-dotted line), although the distribution deviates from the power-law at the lower end of energy range (Shimizu 1995).

single loop structures (Parker 1988), we should have observed brightenings within single clean loop structures.

It is, however, too early to rule out the nano-flare heating hypothesis. There is still a possibility that ensemble of unseen nano-flare contributes to the heating of the active region corona. Individual nano-flares may be too weak to be detected in the relatively strong background coronal emissions. The temperature enhancement due to individual nano-flares may be too low to be detected by the SXT, which has virtually no sensitivity below 2 MK. We need an X-ray telescope which has higher sensitivity over wider temperature range, especially in a temperature range below 2–3 MK as well as high time cadence to observationally verify the nanoflare hypothesis.

In addition to the transient brightenings, there are definitely loops steadier than these transient brightenings. These steady structures will be discussed in the next section.

94

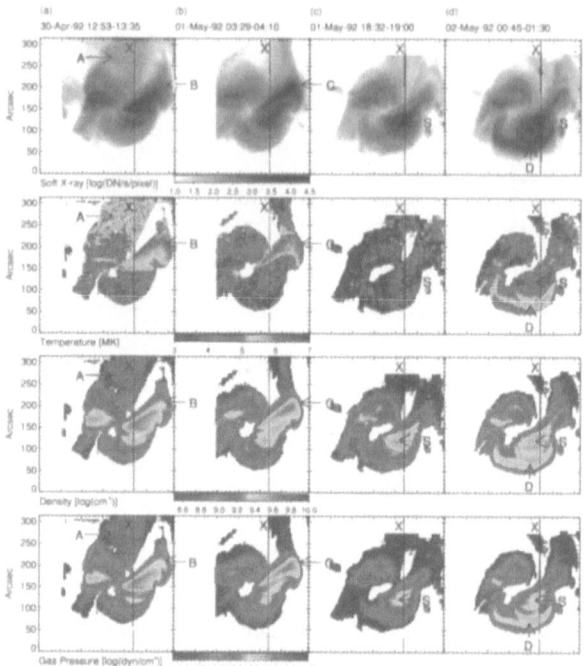

Figure 7. Soft X-ray (thin aluminum filter), temperature, electron density, and gas pressure maps of an active region on the center of the solar disk from 1992 April 30 through May 2. The soft X-ray maps are produced by summing frames over 60 min. The times shown at the top of each panel are the start and end times of the summation. The temperatures were derived from the two summed images with different broadband filters. Color pictures can be found in Yoshida and Tsuneta (1996)

4. Temperature Structure of Active Regions

All the analyses made so far on the temperature structure of coronal loops have been concentrated on individual X-ray loops (Kano and Tsuneta 1995, 1996; Yoshida *et al.* 1995; Tsuneta *et al.* 1992; Tsuneta 1996). We, however, need to remember that the high temperature corona apparently exists outside distinct loops and in the diffuse region. In this section, we discuss the temperature structure of an entire active region.

Figure 7–(d) shows a few loops with similar lengths (marked by S) around the center of the active region. The temperature is rather uniform, and it is hard to identify these loop structures in the temperature map: in other words, the loops appear to lose their identity in temperature. This indicates that the bundle of loops may be heated quasi-uniformly rather than heated selectively. The temperature of this region appears almost constant over a day [Figure 7 (b)–(d)]. On the other hand, loops D located south of the active region are transiently heated to around 5.5 MK, while those of loops S in the center of the active region stay at 4 MK.

The region A in Figure 7–(a) is low X-ray intensity diffuse corona, but has temperatures of about 6 MK. This source is transiently heated, and decays in about an hour. We also notice the remarkable cusps B and C with temperature of 6.5 MK in the West of Figure 7–(a) and (b). (The cusp C appears again after the disappearance of the cusp B.) The temperatures of the cusp structures are generally higher than 6 MK, and the outer edge of the cusp has higher temperature.

The temperatures of all the loop structures in panel (c) are as low as 3–4 MK. We can not identify any loop structure corresponding to the soft X-ray loops in the temperature map.

4.1. CUSP RECONNECTION AND LOOP–LOOP RECONNECTION

The brightening marked D in Figure 7–(d) is a transient brightening discussed in section 3.1. At least, two different loops are heated to 5 MK in this particular case. Such heating of multiple loops is a general tendency associated with transient brightenings. This strongly supports the speculation by Shimizu *et al.* (1992, 1994) that the heating is due to magnetic reconnection of multiple loop structures. The fact that two different macroscopic loop structures are significantly heated simultaneously can be explained only by reconnection of those two separate loop structures. Without reconnection, these loops are robustly insulated by magnetic fields from the surrounding corona. These observations show that magnetic reconnection of different loop systems can be a strong transient heat source.

Cusp-structures seen in the Figure 7–(a) and (b) are often seen in these temperature maps. Although these "mini-cusps" are small in energy and spatial size, and are not recognized as flares, the similarity in the morphology and the temperature structure suggests that the physical mechanism involved in the heating of the mini-cusps would be essentially the same as the large cusp-type flares (see section 6). The magnetic reconnection at a neutral sheet formed above the loop top is responsible for the energy release of these mini-cusps. It appears that the formation of the cusp structures is universal phenomena ranging from large flares to these mini-cusps.

The cusp-reconnection and loop-loop reconnection are the two primary configurations for transient heatings. Large flares are of cusp-type configuration, whereas loop-loop reconnection appears to be seen only in small transients.

4.2. TEMPERATURE STRUCTURE AND TIME VARIABILITY

Figure 8 shows the temperature distribution along the X axis indicated in Figure 7 as a function of time over 50 hours. The Figure clearly shows several transient heatings; A, B, C, and D in the "sea" of the steady temperature component. The two 6 MK peaks located at 13 UT on 1992 April 30, and 2 UT on May 1 correspond to the two cusp brightenings B and C in Figure 7–(a), (b). The high temperature region (5–6 MK) extended over to the north at 13 UT on April 30 corresponds to the region A in Figure 7–(a). The compact high temperature plasma at 1 UT on May 2 is due to the loops D in Figure 7–(d)

Figure 8 shows all the high temperature regions (> 6 MK) have lifetime of a few hours or so. The observed decay time scale is compared with the estimated radiative and conductive time scales. (The rise time is smaller than the time resolution of the temperature maps.) The conductive time scale of these transiently heated plasmas in the corona is given by

$$t_{\mathrm{cond}} = \frac{3nk_{\mathrm{B}}Tl}{\kappa_0 T^{5/2} \cdot (T/l)}, \tag{2}$$

where $\kappa_0 T^{5/2}$ is the Spitzer thermal conductivity ($\kappa_0 = 10^{-6}$ [erg/sec/cm/K$^{3.5}$], Spitzer 1962), and l is the assumed scale length of the loop. The conduction time scale t_{cond} is $1000 \sim 1500$ sec for $T = 6 \sim 7$ MK, $l = 10^5$ km, and $n = 10^{9.5}$ cm^{-3}. The radiative time scale is

$$t_{\mathrm{rad}} = \frac{3nk_{\mathrm{B}}T}{\Lambda(T)n^2}, \tag{3}$$

where $\Lambda(T)$ is the radiative loss function. Since the temperature is $T \sim 6\text{–}7$ MK, and the radiative loss function is $\Lambda(T) = 6 \times 10^{-20}T^{-0.5} \sim 2.3\text{-}2.4 \times 10^{-23}$ erg/sec/cm^{-3}, the radiative cooling time t_{rad} is $3\text{-}4 \times 10^4$ sec. Thus, we conclude that these high temperature plasmas are transiently heated, and then cool down essentially with the conductive time scale.

In addition to the transient high temperature plasmas, there is definite steady plasma component. The temperature of this component is lower (3–5 MK), and has longer lifetime ($\sim$ a day or longer). The radiative and conductive cooling time scales are 350 and 50 minutes, respectively for $n = 10^{9.5}$ cm^{-3}, $T = 4.5$ MK, and $l = 10^{10}$ cm. The observed time scale is much longer than the conductive and radiative time scales. This indicates that the steady plasmas are continuously heated.

The properties of these transient plasmas with lifetime shorter than several hours are summarized as follows. (1) The temperatures are higher than 5 MK, and do not depend on the soft X-ray intensities. For instance, the faint region A in Figure 7–(a) has temperature around 6 MK [see Yoshida and Tsuneta (1996) for interpretation]. (2) The loops often have "mini-cusp" or interacting loop structures as shown in Figure 7. (3) The cusp-like loops have higher temperatures on the outer edges than those on the inner edges.

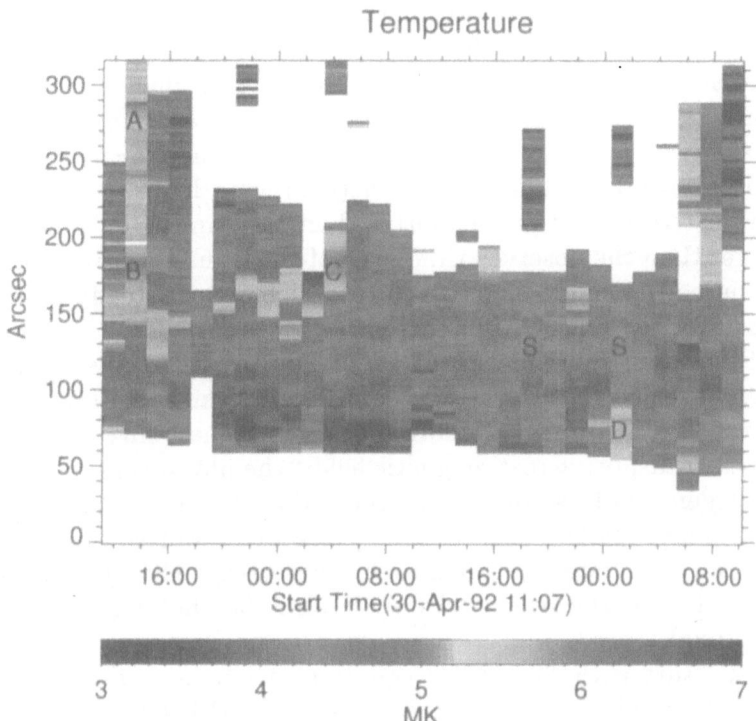

Figure 8. The temperature distribution along the line X shown in Figure 7 as a function of time. The transient heatings as well as steady corona are clearly seen. The regions marked by A, B, C, and D correspond to the region with the same marks in Figure 7 (Yoshida and Tsuneta 1996).

On the other hand, steady loops with life time longer than several hours appear to have the following features: (1) The temperatures of the loop tops are from 4 to 5 MK, and those of the footpoints are around 3 MK. (2) The difference in temperature between nearby loops is too small to recognize individual loop structures in the temperature maps.

This indicates that the coronal heating consists of the basal heating and the transient heating. The basal heating mechanism maintains the plasma temperature around 3–4 MK, and the occasional transient heating due to magnetic reconnection produces higher temperature plasmas with temperatures upto 10 MK.

5. The Scaling Laws

There has been a long standing issue of whether there exist simple relations between properties (temperature, length, pressure, and heating rate) of the quasi-steady coronal loops (Craig, McClymont, and Underwood 1978). In this section, we discuss these "scaling laws", following Rosner, Tucker and Vaiana (1978) and Kano and Tsuneta (1995, 1996).

"Quasi-steady" here means that no appreciable change in the soft X-ray brightness or structure occur over a critical time scale, defined as the faster of the time scales for either radiative or conductive energy loss. This critical time scale is 30 min–1 hr for active region loops. (The sound and Alfvén transit times are much shorter than this time.) This time scale is much longer than the observed time scale of the transient brightenings, and there are numerous single-loop structures which appear steady in shape and brightness over the critical time scale.

The *Skylab* and subsequent *Yohkoh* imaging observations of the Sun in soft X-rays revealed that the solar corona is not a single entity, but consists of ubiquitous isolated magnetic flux tubes. Since these flux tubes suppress the classical transport across magnetic fields, the physical properties of single loops depend on their *in situ* present and past heating rate, and do not depend on the properties of nearby loops. In other words, the neighboring loops do not "communicate" with each other except through magnetic reconnection. This is the reason why the scaling law holds for a wide range of solar coronal loops.

Here we start with the energy equation of magnetic flux tubes under the assumptions of (a) steady state $\partial/\partial t = 0$, (b) velocity field $v = 0$ and (c) homogeneous pressure along the loop. We consider the situation that the coronal loop anchored on the cool photosphere is steadily heated. The input energy is conductively dumped toward the photosphere, while partially losing energy by radiation. The total input of energy is balanced by the total radiation and conduction energies. The energy equation is

$$E_{\mathrm{H}} + E_{\mathrm{R}} - \boldsymbol{\nabla} \cdot \boldsymbol{F}_{\mathrm{C}} = 0, \tag{4}$$

where E_{H} is the heat input rate, $-E_{\mathrm{R}}$ is the radiative loss and $\boldsymbol{F}_{\mathrm{C}}$ is the conductive heat flux along the magnetic field. These quantities change only along magnetic field lines. The radiative loss and the heat conduction are

$$-E_{\mathrm{R}} = n_{\mathrm{e}}^2 \, \Lambda(T) = \frac{p^2}{4k_{\mathrm{B}}^2} \frac{\Lambda(T)}{T^2}, \tag{5}$$

$$\boldsymbol{\nabla} \cdot \boldsymbol{F}_{\mathrm{C}} = -\frac{\mathrm{d}}{\mathrm{d}s}\left(\kappa_{\|} \frac{\mathrm{d}T}{\mathrm{d}s}\right), \tag{6}$$

where p is the constant pressure, $\Lambda(T)$ is the radiative loss function, s is the coordinate along the loop from the footpoint and $\kappa_{\|}$ is the thermal conductivity along magnetic field lines; $\kappa_{\|} = 10^{-6}T^{5/2}$ [erg/K/cm/sec] $\equiv \kappa_0 T^{5/2}$ (Spitzer 1962).

We review the scaling law derived from equation (4), following Kano and Tsuneta (1995). We assume that the radiative loss function is a power of the temperature T and that the heating function E_H is the sum of a power of the temperature and a δ function at an arbitrary point s_0 :

$$\Lambda(T) = \chi T^\alpha \ , \quad E_H = \mathcal{E}_1 T^\beta + \mathcal{E}_2 \delta(s - s_0), \tag{7}$$

where $\mathcal{E}_1$, $\mathcal{E}_2$, and β are arbitrary constants, and s is the distance from the footpoint. We assume $\chi = 1.5 \times 10^{-19}$ and $\alpha = -0.5$ (Rosner, Tucker and Vaiana 1978) for solar conditions. The scaling law is obtained by integrating equation (4):

$$T[\text{K}] = 1.4 \times 10^3 (p[\text{dyn cm}^{-2}] \, L \, [\text{cm}])^{0.33}, \tag{8}$$

where T is the peak temperature of the loop, p the pressure of the loop, and L the length of the loop, if the heat is injected at one position of the loop top ($\mathcal{E}_1 = 0$). If the loop is uniformly heated along the loop ($\mathcal{E}_2 = 0$), the scaling law is then

$$T = 1.1 \times 10^3 (pL)^{0.33}. \tag{9}$$

The numerical factor is slightly different between the two extreme cases. The power index of the right hand side of the equations does not depend on the heat input profile along the loop. The total energy input F_T [erg/cm^{2}sec^{-1}] is

$$F_T = \int E_H ds = \mathcal{E}_2 = 1.5 \times 10^5 p^{1.16} L^{0.17}, \tag{10}$$

for the case of the delta-function heat input ($\mathcal{E}_1 = 0$), and

$$F_T = \mathcal{E}_1 L = 1.7 \times 10^5 p^{1.16} L^{0.17}, \tag{11}$$

for the case of the uniform input ($\mathcal{E}_2 = 0$). Again only the numerical factor slightly depends on how the heat is applied along the magnetic fields. The two scaling laws are thus robust for a wide range of heat deposition profiles along the magnetic fields. This is one of the reasons that the scaling laws are useful.

5.1. SCALING LAW OBTAINED WITH YOHKOH

Kano and Tsuneta (1995) obtained the maximum temperature $T_{\max}$ [K], the pressure p [dyn/cm^2] and the length L [cm] of 43 steady solar coronal loop structures observed with SXT, and found a clear correction as shown in Figure 9. The best fit equation is

$$T_{\max} = 6.3 \times 10^4 (p \cdot L)^{1/5.5}. \tag{12}$$

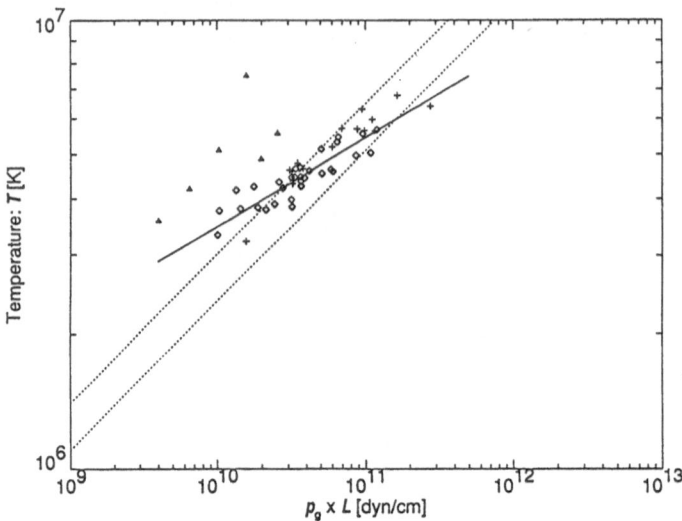

Figure 9. The correlation between the maximum temperature and the product of the gas pressure and the loop length (the temperature scaling law). The crosses indicate the steady loops. The diamonds and the filled triangles indicate the steady loops and the transient brightenings in Kano and Tsuneta (1995). The solid line is the fitted line for the steady loops (diamonds), and the two dotted lines are the theoretical temperature scaling laws (see text). The transient brightenings systematically deviate from the line.

The best-fit equation, however, systematically deviates from the theoretical scaling law $T_{\max} = 1.4 \times 10^3 (p \cdot L)^{1/3}$ derived by Rosner, Tucker and Vaiana (1978). Kano and Tsuneta (1995) argue that the discrepancy can be due to either the assumption of isothermal plasma in the temperature analysis or the uncertainty on the loop-length estimation. Though there is a possibility that there is a fundamental physical reason for the discrepancy, the data may be essentially consistent with the scaling law derived theoretically.

5.2. TEMPERATURE DISTRIBUTION ALONG LOOPS

Kano and Tsuneta (1996) also obtained the temperature distributions along the steady loops, and discovered the correlation between the energy input and the gas pressure ("energy scaling law" hereafter in this paper). Figures 10 show the two extreme types of the temperature distributions along the loops; "trapezoidal" type and "triangular" type. Figure 10c shows a 3.5 power of the temperature distribution, which is proportional to the conductive flux. The nominal conductive flux is almost constant along the loop except around the loop top, and suggests that the energy input is concentrated at the loop top, especially for the loops with the triangular temperature distribution.

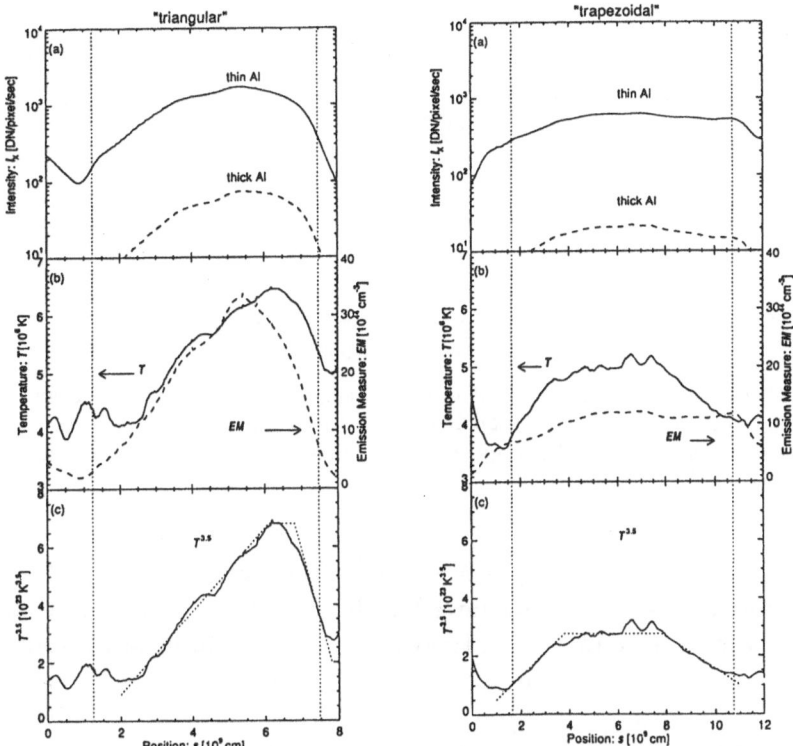

Figure 10. Left: "Trapezoidal" temperature distribution along the loop. (a) The X-ray intensities taken with the thin (solid line) and the thick aluminum (dashed line) filters. (b) The temperature (solid line) and the emission measure (dashed line) distributions. (c) The 3.5 power of the temperature distribution, whose gradient gives the thermal conductive flux. The dotted lines in these plots show the positions of the apparent footpoints. Right: Same as left figures, but for "triangular" temperature distribution along the loop (Kano and Tsuneta 1996).

They also find that the emission measure EM peaks around the loop top (dashed lines in Figure 10b). The nominal pressure at the loop top is a few times higher than the footpoint pressure. This fact appears to be inconsistent with the dynamical stability criterion. Possible change in the loop filling factor along the loops is discussed in Kano and Tsuneta (1996).

5.3. THE ENERGY SCALING LAW

Kano and Tsuneta (1996) discovered that the total energy loss $\mathcal{L}_T$, which is the sum of the conductive loss near the footpoints and the radiative loss from the entire loop, is well correlated with the peak gas pressure (Figure 11);

$$\mathcal{L}_T = 1.1 \times 10^7 p_g^{0.99}. \tag{13}$$

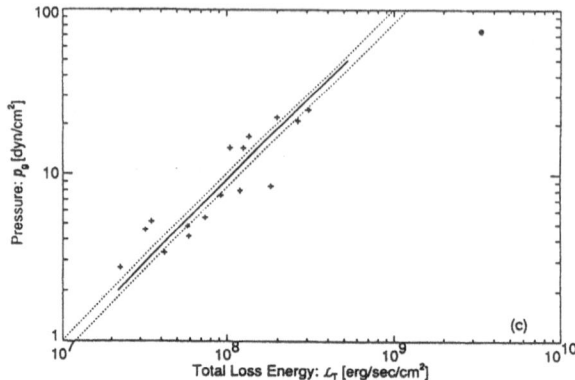

Figure 11.   The scatter diagrams of the total energy loss with the peak gas pressure of the steady coronal loops. The solid line is the fitted line $\mathcal{L}_T = 1.1 \times 10^7 p_g^{0.99}$, and the dotted lines are the theoretical energy scaling laws (see text) for a 6×10^6 K plasma (Kano and Tsuneta 1996).

The temperature of the 16 loops is almost the same; $T \sim 6 \times 10^6$ [K], and the theoretical energy scaling laws for the 6×10^6 K plasma are shown in Figure 11 with dotted lines. The theoretical energy scaling law is remarkably consistent with the observed one.

6. Solar Flares and Magnetic Reconnection

There are strong pieces of evidence from *Yohkoh* observations that magnetic reconnection (Petschek 1964) is essentially responsible for the flare energy release. Evidence so far accumulated is summarized as follows (Tsuneta *et al* 1992a, Tsuneta 1996): (1) The height and the foot-point separation of the soft X-ray loop increase as a function of time. This is due to the rise of the X-point location along the neutral sheet as reconnection goes on. (2) The loop top region has a cusp-like structure, implying the location of a reconnection site at the top of the flare loop. It would be hard to create a sharp cusp-like structure at the loop top without invoking the singular X-point structure. (3) The outer loops have higher temperatures in the decay phase. This is consistent with a flare energy supplied from the reconnection process near the top of the loop.

6.1. PHYSICAL PARAMETERS OF THE RECONNECTION SITE

Among those pieces of evidence to support magnetic reconnection as the flare energy source, the temperature structure gives key quantitative information to study the structure of the reconnection site. The observations of the 1992 February 21 flare (Tsuneta *et al* 1992a, Tsuneta 1996a) show that the outer loops systematically have higher temperatures, reaching the peak (12 MK) far outside the apparent bright X-ray loop where the X-ray intensity is only 1−5 % of the peak. This is strong evidence that a reconnection point is located higher in the corona, and that the attached standing isothermal slow shocks (Petschek 1964, Forbes and Malherbe 1991, Ugai 1992, Yokoyama 1996) heat the downstream plasma. The bright soft X-ray loops are the reconnected flux tubes subsequently filled with evaporated plasma. The physical parameters of the upstream and the downstream of the slow shocks and the separatrices are determined from the X-ray observations. We estimate that the outflow speed is ~ 800 km sec^{-1} (Mach speed ~ 1), the inflow speed ~ 56 km sec^{-1} (Alfvén Mach number ~ 0.07), and the reconnection height $\sim 8 \sim 18 \times 10^4$ km above the apparent top of the flare loop (6×10^4 km). The estimated energy produced by the slow shocks is close to the estimated magnetic energy in the upstream that is fed to the reconnection site. Detailed analysis can be found in Tsuneta (1996a).

Yohkoh observations show that magnetic reconnection serves as a highly efficient engine to convert magnetic energy into plasma kinetic and thermal energies with standing slow shocks. This implies an important role of magnetic reconnection for energetic phenomena in other astrophysical as well as magnetospheric contexts.

6.2. HOW IS THE NEUTRAL SHEET STRUCTURE CREATED?

We now have evidence that the neutral sheet structure is formed around the start of flares, and that magnetic reconnection with slow shocks is responsible for the heating of the flare loops. An important question here is how the system dynamically reaches such structure from an initial configuration. We have found many examples of erupting structures, under which flare loops are later created (Tsuneta 1993, Shibata *et al.* 1995). It appears that these eruption phenomena create the neutral sheet structure, where reconnection for the primary energy release occurs. What is the driving force of the eruption? The open field structure would have a magnetic energy higher than the closed potential loop structure with the same photospheric boundary condition. The formation of the neutral sheet structure apparently needs to overcome this potential difference. The energy released in the magnetic reconnection consumes this potential difference. The eruption and the reconnection are thus two key mechanisms for solar flares. The reconnection process begins to be understood, whereas the dynamical formation of the neutral sheet is poorly understood.

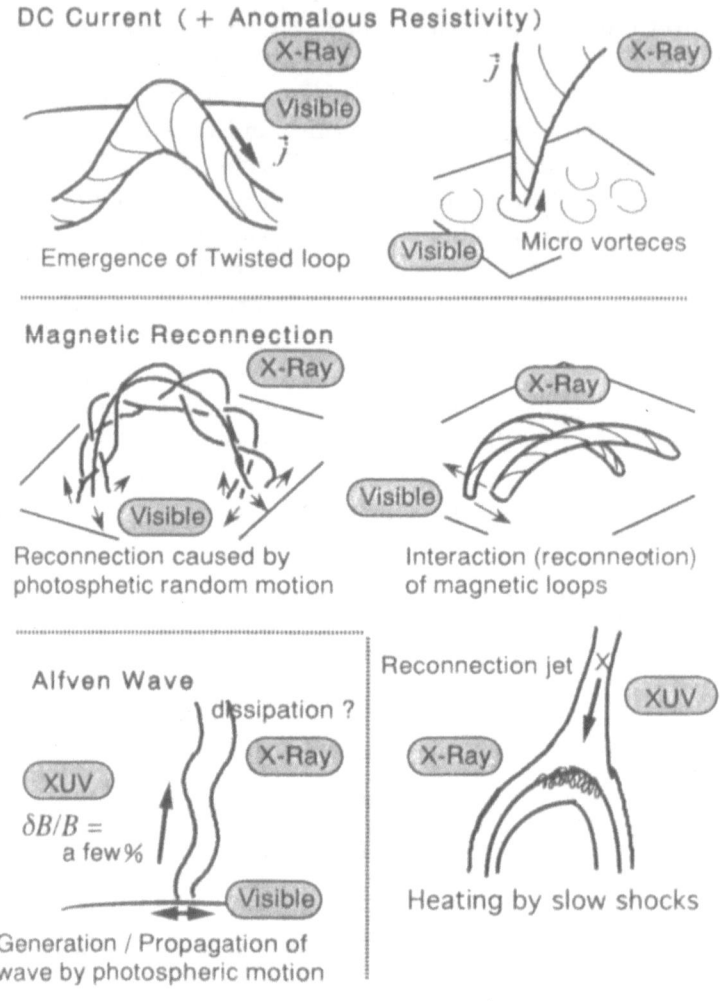

DC Current (+ Anomalous Resistivity)

Emergence of Twisted loop

Micro vorteces

Magnetic Reconnection

Reconnection caused by
photosphetic random motion

Interaction (reconnection)
of magnetic loops

Alfven Wave

dissipation ?

$\delta B/B =$ a few %

Generation / Propagation of
wave by photospheric motion

Reconnection jet

Heating by slow shocks

Figure 12. Candidates of Coronal Heating Mechanisms and SOLAR-B observations

7. Magnetic Reconnection and Transient Phenomena

7.1. TIME SCALE OF RECONNECTION

Table 2 indicates the physical parameters of the transient phenomena introduced in this paper. Although the time scales and the magnitudes of the energy involved differ over many orders of magnitude, the observed time scale τ appears to have a common value some 10 to 100 times the Alfvén transit times τ_A, and the released energy is roughly proportional to $B^2/8\pi$. A recent laboratory experiment on magnetic reconnection (also listed in Table 2) shows a similar time scale (Yamada *et al.* 1990). On the other hand, steady loops have much longer life times in terms of the Alfvén

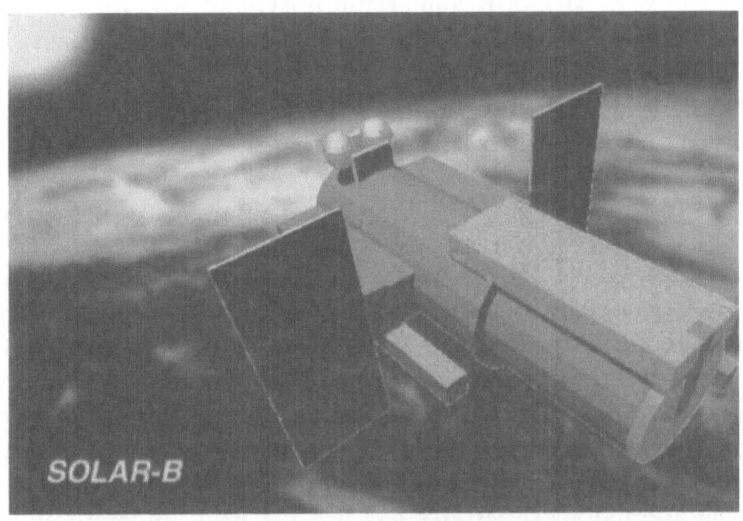

Figure 13. The Solar-B Satellite

transit times.

TABLE 2. Time Scales of Transient Phenomena Observed by *Yohkoh* (Tsuneta 1994)

	B(gauss)	Length	n(cm^{-3})	Time scale τ
Quiet Sun Restructuring	10	5 x 10^4 km	10^9	10 hr $\sim 50\tau_A$
Transient Brightenings	100	5 x 10^3 km	10^{10}	1 min $\sim 30\tau_A$
Flares	100-500	10^4 km	10^{11}	1 min $\sim 20\tau_A$
Steady loops	100	5 x 10^3 km	10^{10}	1 hr $\sim 10^4\tau_A$
Laboratory	500	15 cm	3 $\times$ 10^{14}	30μsec$\sim 10 - 60\tau_A$

Note: τ_A is the Alfvén transit time obtained from the tabulated parameters. The time scales should be regarded as order-of-magnitude estimates to demonstrate the gross similarity in the time scales of the various transient phenomena (except steady coronal loops).

This time scale is related to the speed of the reconnection, ie the inflow speed of reconnection. The time scale of the reconnection τ is given by

$$\tau \sim \frac{L}{v_{\text{in}}}, \tag{14}$$

where L is the scale size of reconnection upstream site, and v_{in} the inflow speed. The fact that $\tau \sim (10-50)\tau_A$ for all types of the transients indicates that the inflow speed is commonly $v_{in} \sim (0.1-0.02)V_A$. Detailed analysis of the 1992 February 21 flare (Tsuneta 1996) shows that the inflow speed is 0.07 Alfvén velocity, which is consistent with the above rough estimate.

7.2. WHY DOES RECONNECTION OCCUR IN THE SOLAR CORONA?

The *Yohkoh* data have shown unambiguous evidence of magnetic reconnection. Since the electric conductivity of the solar corona is large (high magnetic Reynolds number), the small scale size (Petschek 1964) of the diffusion region ($\sim$ ion-Larmor radius) combined with an anomalous resistivity (Yokoyama and Shibata 1994) are the conditions for fast reconnection. Though we do need the anomalous resistivity with resistivity $\sim 10^6$ as high as the Spitzer resistivity to explain the observed time scales, the origin of the anomalous resistivity has not yet been understood. The high magnetic Reynolds number of the system makes reconnection more difficult to occur. At the same time, this may be the reason why we observe intense impulsive transient heatings in the solar corona: if the magnetic Reynolds number of the system is lower, we would have observed a more continuous dissipation rather than an impulsive dissipation of magnetic energies.

8. From Yohkoh to Solar-B

Figure 12 shows various candidate mechanisms to heat the corona. We now know that loop-loop reconnection and the reconnection at the neutral sheet (cusp) are responsible for the transient heatings. What mechanism is responsible for the steady heating? The coronal magnetic fields are constantly agitated through their footpoints by random velocity fields due to granulation and super-granulation motions as well as some systematic velocity fields on the photosphere. The solar corona may be regarded as a statistical self-organized critical system (Bak, Tang and Wiesenfeld 1988) against such disturbances (noise) applied from the photosphere, where new magnetic fluxes are also constantly emerging. The solar corona may continuously adjusts (relaxes) itself toward the lower energy state through reconnection process with heat output as the result. Otherwise, the coronal magnetic fields should have had much more complex configuration with larger energy.

To understand this process, it is vital to observe both the magnetic activity at the photospheric and chromospheric levels and the coronal X-activity with high spatial and time resolution. The Solar-B satellite (Figure 13) to be launched in early 2000s is under feasibility study. The satellite carries high resolution X-ray as well as optical telescopes with the possible addition of an XUV spectrograph. The optical telescope is designed such that it can take precision vector magnetic images with high spatial and time resolution. Figure 12 also shows how Solar-B tackles, for instance, the coronal heating problem with this set of instruments.

Acknowledgement. The author would like to thank Kanaris Tsinganos for his comments on the paper.

References

Acton, L., **Tsuneta, S.**, Ogawara, Y., Bentley, R., Bruner, M., Canfield, R., Culhane, L., Doschek, G., Hiei, E., Hirayama, T., Hudson, H., Kosugi, T., Lang, J., Lemen, J., Nishimura, J., Makishima, K., Uchida, Y., Watanabe, T. (1992) *Science* **258**, 618.
Bak, P., Tang, C., and Wiesenfeld, K. (1988) *Phys. Rev. A* **38**, 364.
Craig, I., McClymont, A., and Underwood, J. (1978) *Astron. and Astrophys.* **70**, 1.
Crosby, N., Aschwanden, M., and Dennis, B. (1993) *Solar Phys.* **143**, 275.
Dennis, B. (1985)*Solar Phys.* **100**, 465.
Doschek, G. A. and Feldman, U. (1976) *Astrophys. J.* **212**, L143.
Forbes, T. G. and Malherbe, J. M. (1991) *Solar Phys.* **135**, 361.
Hara, H., Tsuneta, S., Acton, L. W., Lemen, J. R., and McTiernan, J. M. (1992) *Publ. Astron. Soc. Japan* **44**, L135.
Hara, H., Tsuneta, S., Acton, L. W., Bruner, M. E., Lemen, J. R., and Ogawara, Y. (1994) *Publ. Astron. Soc. Japan* **46**, 493.
Hudson, H. S. (1991) *Solar Phys.* **133**, 357.
Ichimoto, K., Hara, H., Takeda, A., et al. (1995) *Ap. J.* **445**, 978.
Kano, R. and Tsuneta, S. (1995) *Astrophys. J.* in press.
Kano, R. and Tsuneta, S. (1996) *Publ. Astron. Soc. Japan* submitted.
Kosugi, T. (1994) *Proc. Kofu Symposium* eds. Enome, S. and Hirayama, T. Nobeyama Radio Observatory, NRO Report No. 360, p11.
Masuda, S., Kosugi, T., Hara, H., and Tsuneta, S. (1994) *Nature* 371, 495.
Ogawara, Y., Takano, T., Kato, T., Kosugi, T., Tsuneta, S., Watanabe, T., Kondo, I., and Uchida, Y. (1991) *Solar Phys.* **136**, 1.
Parker, E. N. (1988) *Ap. J.* **330**, 474.
Petschek, H. E. (1964) *AAS-NASA Symposium on Solar Flares*, ed. W. N. Hess (NASA SP-50) p 425.
Priest, E. R. (1996) in these proceedings.
Rosner, R., Tucker, W., and Vaiana, G. (1978) *Astrophys. J.* **220**,643
Sakao, T. (1994) *Ph. D. Thesis*, University of Tokyo.
Shibata, K., Ishido, Y., Acton, L. W., Strong, K. T., Hirayama, T., Uchida, Y., McAllister, A., Matsumoto, R., Tsuneta, S., Shimizu, T., Hara, H., Sakurai, T., Ichimoto, K., Nishino, Y., and Ogawara, Y. (1992) *Publ. Astron. Soc. Japan* **44**, L173.
Shibata, K., Nitta, N., Strong, K. T., Matsumoto, R., Yokoyama, T., Hirayama, T., Hudson, H. S., and Ogawara, Y. (1994) *Astrophys. J. (Letters)* **431**, L51.
Shibata, K., Masuda, S., Shimojo, M., Hara, H., Yokoyama, T., Tsuneta, S., Kosugi, T., and Ogawara, Y., (1995) *Astrophys. J. (Letters)* **451**, L83.
Shimizu, T., Tsuneta, S., Acton, L. W., Lemen, J. R., and Uchida, Y. (1992) *Publ. Astron. Soc. Japan* **44**, L147.
Shimizu, T., Tsuneta, S., Acton, L. W., Lemen, J. R., Ogawara, Y., and Uchida, Y. (1994) *Astrophys. J.* **422**, 906.
Shimizu, T. (1995) *Publ. Astron. Soc. Japan* **38**, 251.
Spitzer, L. (1962) *Physics of Fully Ionized Gases* Interscience, New York.
Tsuneta, S., Acton, L. W., Bruner, M. E., Lemen, J. R., Brown, W., Caravalho, R., Catura, R., Freeland, S., Jurcevich, B., Morrison, M., Ogawara, Y., Hirayama, T.,

108

and Owens, J. (1991) *Solar Phys.* **136**, 37.

Tsuneta, S., Hara, H., Shimizu, T., Acton, L. W., Strong, K. T., Hudson, H. S., and Ogawara, Y. (1992a) *Publ. Astron. Soc. Japan* **44**, L63.

Tsuneta, S., Takahashi, T., Acton, L. W., Bruner, M. E., Harvey, K., and Ogawara, Y. (1992b) *Publ. Astron. Soc. Japan* **44**, L211.

Tsuneta, S. and Lemen, J. R. (1993) *Physics of Solar and Stellar Coronae* eds., Linsky, J. and Serio, S., Kluwer Academic Publishers, Dordrecht, p113.

Tsuneta, S. (1993) *The Magnetic and Velocity Fields of Solar Active Regions* eds. Zirin, H., Ai, G., and Wang H., Astronomical Society of the Pacific, San Francisco, p239.

Tsuneta, S. (1994) *Solar Active Region Evolution* eds. Balasubramanian, K. and Simon, G., Astronomical Society of the Pacific, San Francisco, p338.

Tsuneta, S. (1995) *Publ. Astron. Soc. Japan* **47**, 691.

Tsuneta, S. (1996a) *Astrophys. J.* in press.

Tsuneta, S. (1996b) *Astrophys. J. (Letters)* in press.

Uchida, Y. (1993) *Physics of Solar and Stellar Coronae* eds., Linsky, J. and Serio, S., Kluwer Academic Publishers, Dordrecht, p97.

Uchida, Y., McAllister, A., Strong, K. T., Ogawara, Y., Shimizu, T., Matsumoto, R., and Hudson, H. S. (1992) *Publ. Astron. Soc. Japan* **44**, L155.

Ugai, M. (1992) *Phys. Fluids B* **4**, 2953.

Yamada, M., Ono, Y., Hayakawa, A., and Katsurai, M. (1990) *Physical Review Letters* **65**, 721.

Yokoyama, T. and Shibata, K. (1994) *Astrophys. J. (Letters)* **436**, L197.

Yokoyama, T. and Shibata, K. (1995) *Nature* **375**, 42.

Yokoyama, T. (1995) *Ph. D. Thesis* National Astronomical Observatory, Tokyo.

Yoshida, T., Tsuneta, S., Golub, L., and Strong, K. T. (1995) *Publ. Astron. Soc. Japan* **47**, L15.

Yoshida, T. and Tsuneta, S. (1996) *Astrophys. J.* in press.

THE SPONTANEOUS FORMATION OF CURRENT-SHEETS IN ASTROPHYSICAL MAGNETIC FIELDS

B.C. LOW
High Altitude Obervatory
National Center for Atmospheric Research
P.O. Box 3000
Boulder, CO 80307-3000, USA
low@solo.hao.ucar.edu

Abstract. This article is an introduction to Parker's idea that electric current sheets form spontaneously in astrophysical magnetic fields under the condition of high electrical conductivity. Upon formation, the current sheets will collapse to such small widths as to result in resistive reconnection of magnetic fields and heating, despite the very large but finite electrical conductivity. This mechanism is an attractive explanation of the ubiquitous association between magnetic fields and heated plasmas in many astrophysical situations. The hydromagnetic process of this mechanism is illustrated, using a well-studied two-dimensional Cartesian model involving a quadrupolar magnetic field with or without a magnetic null point. The purpose of this illustration is to acquaint the reader with the basic physics in terms of elementary mathematical results and familiar properties which are possible to obtain for this simple model. The general and more complicated processes in three-dimensional magnetic fields is treated in Parker's latest (1994) monograph on this subject.

1. Introduction

Gene Parker proposed in 1972 that under the typical astrophysical conditions of high electrical conductivity, a magnetic field of complex topology will, in the course of its evolution, spontaneously form electric current-sheets, or magnetic tangential discontinuities, (Parker 1972, 1979, 1994). When volumes of plasma are dynamically forced into each other, they may squeeze out of the way the layers of plasma separating them. These ex-

K. C. Tsinganos (ed.), Solar and Astrophysical Magnetohydrodynamic Flows, 109–131.
© 1996 *Kluwer Academic Publishers.*

pelled layers tend to have magnetic flux surfaces as boundaries since they must carry their frozen-in magnetic flux along with them. As such a layer is being squeezed out and two volumes of plasma converge, a magnetic tangential discontinuity forms, simply as the consequence of two topologically unrelated magnetic fields coming to meet at a contact surface. In the complete absence of electrical resistivity, the entire layer may be expelled to produce a current sheet of zero thickness. The current density in the sheet becomes integrably infinite in that limit. In a real plasma with a small but finite resistivity, the collapse of the current sheet is overtaken by resistive reconnection when the sheet thickness gets below a small but finite critical size. This results in plasma heating.

If the plasma is resistive, the expulsion of a layer of plasma from between two converging volumes of plasma does not take place cleanly between magnetic flux surfaces; the material boundaries of plasma volumes in this case can slip resistively across magnetic flux surfaces. The coming into contact of two flux surfaces to form a current-sheet therefore depends crucially on a high degree of flux-freezing. The important point made here is that the spontaneous formation of electric current sheets is a consequence of high electrical conductivity. In other words, high conductivity not only allows for ready flow of electric currents but also the development of extreme current densities leading to heating by resistive dissipation. This property may explain the observation that a large variety of astrophysical plasmas, e.g., the solar corona, are both hot and magnetized under the condition of high conductivity.

Parker's idea is a fundamental theoretical discovery. It was in the eighties when his 1972 paper received its overdue attention by the community, whereupon a lively debate set in with some critics attempting to demonstrate that the idea could not work (e.g., Antiochos 1987, van Ballegooijen 1985). The physics of the idea is new and subtle so that the answers to various criticisms turned out to be very instructive and useful in the later works by Parker and others to develop the theory. This short review is intended to be an introduction to the theory. The current state of the theory is described in Parker's new monograph "Spontaneous Current Sheets in Magnetic Fields" published by Oxford University Press in 1994. The interested reader is referred to this book as well as the references given in this article for a more complete account of the theory.

2. The Parker Problem

Consider a fluid medium of perfect electrical conductivity. The question at hand is to demonstrate that in response to dynamical stress, the boundaries of two spatially separate volumes of plasma, identified by magnetic flux surfaces, may press into each other to form a current sheet. The two surfaces would first touch at a point and then spread the point into a contact surface across which the magnetic field is tangential and discontinuous. One way to study this process is to numerically simulate it with the full set of magnetohydrodynamic equations:

$$\rho \left(\frac{\partial \mathbf{v}}{\partial t} + \mathbf{v} \cdot \nabla \mathbf{v} \right) = -\nabla p + \frac{1}{4\pi} \left(\nabla \times \mathbf{B} \right) \times \mathbf{B} - \rho \nabla \Phi + \mathbf{F}, \qquad (1)$$

$$\frac{\partial \rho}{\partial t} + \nabla \cdot (\rho \mathbf{v}) = 0, \qquad (2)$$

$$\frac{\partial \mathbf{B}}{\partial t} = \nabla \times (\mathbf{v} \times \mathbf{B}), \qquad (3)$$

$$\left(\frac{\partial}{\partial t} + \mathbf{v} \cdot \nabla \right) (p\rho^{-\gamma}) = S, \qquad (4)$$

where ρ, $\mathbf{v}$, p, $\mathbf{B}$, and Φ are, respectively, the density, velocity, pressure, magnetic field, and gravitational potential. We have also inserted the source terms $\mathbf{F}$ and S to allow for viscous forces and entropy loss or gain. There are two nontrivial challenges to a direct numerical approach although some important simulations have been reported in the literature (e.g., Mikic, Schnack, & Van Hoven 1989, Otani & Strauss 1988, Strauss & Otani 1988).

The first challenge is that current-sheet formation is the consequence of extremely high electrical conductivity, as represented by an enormous value of the magnetic Reynolds number, for example, of the order of $10^{12} - 10^{18}$ for the solar corona. The values of the magnetic Reynolds number attained in numerical simulations are severely limited by computer hardware and available numerical techniques. The second challenge has to do with our relative ignorance of the physical properties of this process. To formulate clearly articulated questions for quantitative exploration with the powerful tools of numerical simulation, we need to start with an anlytical and physical understanding.

What Parker did in his 1972 paper was to pose a magnetostatic problem which brings out the basic physical issues about current-sheet formation without the complications of time-dependent magnetohydrodynamics. Figure 1 is Parker's well-known sketch describing the problem he posed (Parker 1979, 1994). Consider a uniform magnetic field sandwiched and anchored between a pair of plates perpendicular to the field. The medium is perfectly conducting. By moving the magnetic footpoints in a random manner on one of the boundary plates, the field can be given a complex field-line topology, as sketched in Figure 1. Let us restrict ourselves to continuous footpoint

112

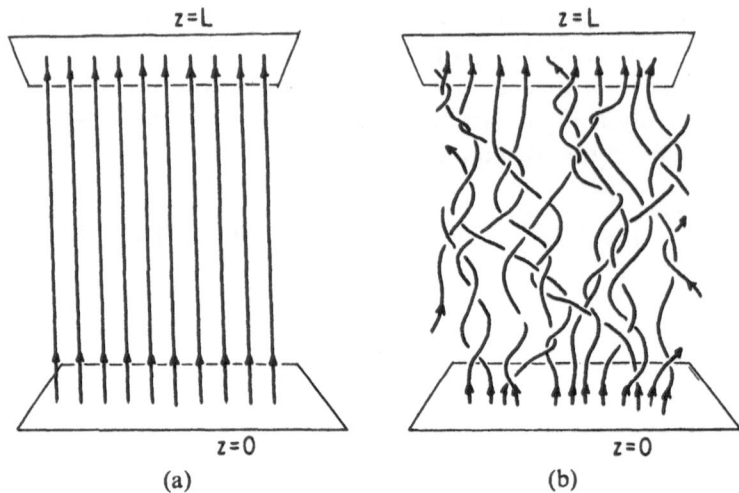

Figure 1. A sketch taken from Parker (1994) of the arbitrary winding of the lines of force of a continuous magnetic field anchored at two rigid plates, at $z = 0$ and $z = L$, beginning with (a) the initially uniform field and becoming twisted and interlaced (b) as the result of footpoint displacements imposed at one of the boundary plates.

displacements so that the distorted field is also continuous. Anchor the displaced magnetic footpoints rigidly at the two boundary plates in order to lock in the complex field topology. Suppose the field, which is not (in general) in equilibrium, is then allowed to relax to an equilibrium state under the condition of perfect conductivity. The actual dynamical relaxation is not important. What is crucial is that the magnetic flux is frozen into the plasma during the relaxation. For example, we may take the medium to be kinematically viscous. Then, as motion is generated dynamically in the ensuing evolution, it is damped away by viscous forces at the expense of the total energy in the closed system. Eventually, the system is drained of its free energy to arrive at a static state.

The interesting question is whether the static state at the end of such an evolution is a continuous magnetic field. Parker concluded that under the constraint of preserving the complex field topology locked into the field, the end state is, in general, one with current sheets. In the relaxation process, discontinuous plasma displacements are unavoidable as magnetic volumes press into each other to evolve to an equilibrium state. Although the field at the beginning of the relaxation process is continuous, these discontinuous plasma displacements produce discontinuities in the final relaxed magnetic field. If the discontinuities are unavoidable, it follows that the static equilibrium problem for the end state will not admit any solution with a continuous magnetic field. Demonstration of the absence of continuous solutions is thus the central task of the Parker problem.

Magnetostatic equilibrium is described by

$$\frac{1}{4\pi} \left(\nabla \times \mathbf{B}\right) \times \mathbf{B} = \nabla p + \rho \nabla \Phi. \tag{5}$$

The end state of relaxation in the Parker problem is a solution to this equation subject to two conditions, namely, (i) the boundary flux distributions at the two rigid walls are given, and (ii) the field topology is prescribed to be the same as that of the magnetic field at the start of the relaxation process. In a fully developed time-dependent MHD flow, many dynamical effects may conceivably result in the formation of current sheets. Those are not dealt with in the Parker problem whose essential point is the following. Even if the dynamical effects incidental to an actual physical MHD evolution are suppressed, and the magnetic field is artificially adjusted to attain an equilibrium state, the imposition of a fixed magnetic field topology is in general such that the equilibrium state requires the presence of embedded current sheets.

Traditionally when we solve the equilibrium equation (5), we look for a mathematical solution describing an everywhere smooth magnetic field, since we are dealing with partial differential equations. The solution is subject to boundary and other suplementary conditions to determine it in some unique manner. For the Parker problem, the imposition of the field topology combined with the traditional boundary conditions may overdetermine the problem so that in the strict mathematical sense, the problem has no solutions. In this case, we need to admit discontinuity to construct generalized solutions which meet the full set of conditions of the problem dictated by physical consideration.

The first attempt by Parker to establish his result, in his 1972 paper, was criticized by van Ballegooijen (1985). This led van Ballegooijen to reject Parker's result and to take the opposite position that no electric current sheets would form in general in the system depicted in Figure 1. In van Ballegooijen's conclusion, small scale magnetic structures of decreasing sizes will form with the progressive deformation of the field by the footpoint motions at the boundary, in the manner of a cascade in wave number from small to large wave numbers, analogous to classical fluid turbulence (van Ballegooijen 1986). The distinction between the two physical pictures is that in the Parker picture, current sheets will collapse to zero thicknesses when they form, whereas in the van Ballegooijen picture, current sheets form with progressively small scales but not with zero thickness; taking the electrical conductivity to be infinite in both pictures.

In the assessment of this author, Parker's picture has since been established by the later works of Parker and others. In the rest of this article, the basis for that assessment is given. To begin, let us review what the two early papers of Parker and van Ballegooijen accomplished.

Parker (1972) treated the case where the footpoint motions at the boundary are confined to eddy scales of some dimension l which is small compared to the separation L between the two boundary plates. Then using $\epsilon = \frac{l}{L}$ as an expansion parameter, the desired magnetostatic solution may be obtained as a small departure from the initial state of a uniform magnetic field, assuming that the field is everywhere continuous and mathematically well behaved. Without having to deal directly with the imposed field topology, Parker's expansion shows that force-balance in the relaxed state forbids any variation in the direction z perpendicular to the two boundary plates. This absence of z-variation is a (spatially) local condition of equilibrium, not related directly to the field topology which is a global property. Clearly this constraint cannot be met by all possible complex field topologies obtained by footpoint motions. The interweaving of lines of forces can create a variation in z which is intrinsic to the topology and cannot go away during relaxation. It therefore follows that fields with such topologies cannot satisfy local force balance. The conclusion then follows that the equilibrium magnetic field cannot be continuous except for those fields topologically compatible with no variation in z.

A crucial question about Parker's analysis is whether his expansion scheme is capable of exhaustively generating all possible solutions in the small ϵ regime. Van Ballegooijen showed that it is not by directly calculating for a new set of solutions based on a reordering of the terms in the small-parameter expansion of the governing equation (5). Take the case of the total twist of a flux tube of many turns to be distributed along the entire length of the flux tube. Consider the situation of the twist of a single rotation ending up along a small portion of length l_* of the long flux tube. The length l_* may still be large compared to the eddy length l. Then redefining the expansion parameter to be $\epsilon = l/l_*$, van Ballegooijen obtained a different set of equations governing the different orders of expansion. The lowest order equation takes a form in strict analogy with the two dimensional time-dependent vorticity equation:

$$\frac{\partial \psi}{\partial t} = \frac{\partial \left(\psi, \nabla^2 \psi \right)}{(x, y)}, \tag{6}$$

where ψ is the classical velocity potential of an incompressible flow. Van Ballegooijen's equation is precisely this equation with t replaced by z; x and y being the other two Cartesian coordinates, and ψ being related to the lowest order perturbation in the magnetic-field problem.

In van Ballegooijen's formulation of the problem, invariance in z is no longer a required local condition for equilibrium, and it was concluded that smooth-equilibrium solutions exist for all imposed field topologies. This conclusion is partly based on observing that in two-dimensional incompressible hydrodynamics, governed by the same equation (6), an initially smooth state will continue to be smooth at all subsequent times. Still, the conclusion is too strong because it is saying something about global field topology without dealing with this global property in its treatment.

This is not a trivial point. Equation (6) is a spatially local extrapolation of the magnetic field from one base plate to the other, identifying z with t. It has not built into itself any requirement of an imposed global field topology. Starting with a set of data at one plate, one needs to carry out the extrapolation from it to the opposite plate to obtain the global field and its topology. Consider the supposition that the family of fields generated in this manner from all possible smooth data at one plate have the same set of topologies as the one produced by all possible footpoint motions in the Parker problem. If that supposition is true, van Ballegooijen's conclusion is valid. In the next section, we will provide a basis to see that the supposition is not true. Hence, van Ballegooijen's formulation does not demonstrate that current-sheets will not form in a topologically complex magnetic field (Low 1990). It is an important method for generating those solutions which are smooth even if they have complex three-dimensional topologies. It does not say anything about other equilibrium magnetic fields, if they exist, which are necessarily discontinuous by virtue of their complex topologies, evidently of a different kind from that of the van Ballegooijen set.

3. Current-Sheet Formation by Flux Expulsion

The convergence of two magnetic flux surfaces to form a current sheet is, in general, a three-dimensional process, with an interplay between variations in the two independent directions in each of the flux surfaces and variations in the direction out of these surfaces (Low 1989). Theoretical demonstration of this process is hindered by the inherent difficulty of treating three-dimensional problems. If we simplify the problem by imposing a geometric symmetry, the motion of the plasma may be so restricted that the expulsion of plasma from between flux surfaces is suppressed. Nevertheless, it is useful to consider special circumstances under which symmetric systems do exhibit current-sheet formation, as a way of demonstrating the basic physics and developing our intuition for the same process in more complex situations.

With such a motivation, we review current-sheet formation in a two-dimensional Cartesian system treated by many authors, beginning with sheet formation at a magnetic null point (Aly 1987, Aly & Amari 1989, Hu & Low 1982, Hu et al. 1995, Low 1982, 1987, Low & Hu 1983, Moffatt

1987, Priest & Raadu 1975, Sneyd 1993, Sweet 1969, Syrovatskii 1981, Titov 1992, Vekstein & Priest 1992, 1993, Vekstein, Priest, & Amari 1991, Wolfson 1989, Zweibel & Proctor 1990). Although the process involving a null point is well understood, we will give it a rigorous mathematical treatment (Low 1982) in order to identify specific physical effects in terms of familiar properties and then go on to show that the same effects can produce current-sheets in the absence of null points.

3.1. A TWO-STEP EVOLUTIONARY PROCESS

First some mathematical preliminaries. Consider a *continuous* magnetic field independent of x in the infinite half space $z > 0$ taken to be a perfectly conducting atmosphere which is so tenuous that its pressure is negligible compared to the magnetic pressure. Let us for the present also ignore gravity to keep the system physically simple. Then, allowing for the possible presence of electric currents in the atmosphere, the equilibrium state of the magnetic field is one in which the Lorentz force vanishes:

$$\left(\nabla \times \mathbf{B}\right) \times \mathbf{B} = 0. \tag{7}$$

The magnetic field independent of x can be written in the form

$$\mathbf{B} = \left(B_x, \frac{\partial A}{\partial z}, -\frac{\partial A}{\partial y}\right), \tag{8}$$

in terms of the magnetic stream function A and its x-component B_x. In this form, the magnetic field is automatically solenoidal with lines of constant A being the projections of the lines of force on the $y - z$ plane. Applying the curl operator across equation (8),

$$\nabla \times \mathbf{B} = \left(-\nabla^2 A, \frac{\partial B_x}{\partial z}, -\frac{\partial B_x}{\partial y}\right) \tag{9}$$

gives the electric current density apart from constant factors.

The force-free equation (7) dictates that B_x is a strict function of A and the two are related by

$$\nabla^2 A + B_x(A)\frac{dB_x}{dA} = 0. \tag{10}$$

To construct a solution, one approach is to prescribe the form $B_x(A)$ explicitly in some manner so that equation (10) poses a problem for A as the unknown, subject to prescribed boundary values of A at $z = 0$, and the vanishing of $\mathbf{B}$ at infinity in the domain $z > 0$.

We are going to consider three examples of the following two-step evolution process. Take a given magnetic field $\mathbf{B}_{initial}$ which is force-free in $z > 0$ with boundary flux distribution

$$A|_{z=0} = F_{initial}(y), \qquad (11)$$

where $F_{initial}(y)$ is calculated from $\mathbf{B}_{initial}$. Note that prescribing A at $z = 0$ gives the normal field component $B_z = \frac{\partial A}{\partial y}$ at that boundary. This initial field is subject to a footpoint displacement on $z = 0$ without introducing any magnetic flux across this lower boundary. Continue the boundary displacement into a plasma displacement in the domain $z > 0$ in some manner, deforming the initial field into some state $\mathbf{B}_{deformed}$ according to induction equation (3). The deformed field is everywhere continuous in $z > 0$ but is, in general, not force-free in $z > 0$. The footpoints are then rigidly anchored in their displaced positions while the deformed field in $z > 0$ is allowed to adjust to a new force-free state $\mathbf{B}_{end}$, taking the atmosphere to have perfect electrical conductivity. In the first step of this process, the boundary footpoint displacement changes the flux distribution from that given by equation (11) to

$$A|_{z=0} = F_{end}(y), \qquad (12)$$

where $F_{end}(y)$ is related uniquely by the induction equation to $F_{initial}(y)$ and the specific footpoint displacement employed at $z = 0$. The end-state $\mathbf{B}_{end}$ of the second step of the process, provided it is a continuous magnetic field, is a solution of equation (7) subject to boundary condition (12), and subject to the requirement that it has exactly the same field topology under the assumption of perfect conductivity. The interesting question we pursue is whether the solution for $\mathbf{B}_{end}$ is everywhere continuous in $z > 0$ or is required to have magnetic discontuities.

3.2. CURRENT-SHEETS AT MAGNETIC NULL POINTS

Consider a force-free magnetic field $\mathbf{B}$ lying in the $y - z$ plane, i.e., $B_x = 0$. If this magnetic field is everywhere continuous in $z > 0$, then it is potential in $z > 0$, with A satifying the Laplace equation:

$$\nabla^2 A = 0. \qquad (13)$$

This follows from the force-free equation (7) applied to the planar magnetic field. This equation requires the electric current density to be either zero or else parallel to the magnetic field, in order that no Lorentz force is exerted on the tenuous atmosphere. For the planar field, the current density is strictly in the x direction, everywhere perpendicular to the field; set

$B_x = 0$ in equation (9) to see this. Therefore, the current density of a planar force-free magnetic field must vanish everywhere in the domain, leading to equation (13).

Consider a quadrupolar, potential magnetic field lying in the $y - z$ plane, of the form shown in Figure 2a. This magnetic field has an X-type null point in the domain. Take this field to be the initial field $\mathbf{B}_{initial}$ in the two-step evolutionary process described above and subject it to a deformation associated with a prescribed footpoint displacement at $z = 0$. This footpoint displacement is taken strictly in the y direction so that $\mathbf{B}_{deformed}$ remains in the $y - z$ plane and has the same field topology as $\mathbf{B}_{initial}$. It follows that the end-state $\mathbf{B}_{end}$ must also remain in the $y - z$ plane and has the same field topology as $\mathbf{B}_{initial}$.

If $\mathbf{B}_{end}$ is a continuous field, this planar equilibrium field must be a potential field. Let us then consider the potential field $\mathbf{B}_{pot}$ satisfying equation (13) subject to boundary condition (12) and the condition that $|\mathbf{B}|$ vanishes at infinity in $z > 0$. We have given this potential field a different notation to allow for the possiblity that $\mathbf{B}_{end}$ may not be identified with $\mathbf{B}_{pot}$. These boundary conditions alone are sufficient to determine $\mathbf{B}_{pot}$ unqiuely with no freedom to additionally demand that this field has a prescribed field topology. The question then arise whether the unique field $\mathbf{B}_{pot}$ may have just the right field topology for the deformed field to relax into it under the condition of perfect conductivity.

For the arbitrarily prescribed footpoint displacement, the potential field $\mathbf{B}_{pot}$ has a topology different from the common topology of $\mathbf{B}_{initial}$ and $\mathbf{B}_{deformed}$. In this general case, then, $\mathbf{B}_{pot}$ is not available to the relaxing magnetic field under the condition of perfect conductivity. Since the potential field $\mathbf{B}_{pot}$ is the unique force-free field, everywhere continuous in $z > 0$ and matching the boundary condition (12), it follows that the force-free field $\mathbf{B}_{end}$ cannot be continuous. Figure 2 shows an example of this general case.

There are four topologically distinct flux bundles in the four sectors around the X-type null point, characterized with a certain amount of flux in each of the four sectors. For the field $\mathbf{B}_{initial}$ in Figure 2a, its associated $\mathbf{B}_{pot}$ in Figure 2c has a different distribution of fluxes in the four sectors. The difference in the topologies of these two fields can also be seen from the change in magnetic connections of the 4 footpoints marked $\alpha, \beta, \gamma, \delta$ going from Figure 2a to 2c.

The potential field $\mathbf{B}_{pot}$ is not available as an end-state and the true end-state $\mathbf{B}_{end}$ is discontinuous as shown in Figure 2b. This discontinuous state is described by the equation

$$\nabla^2 A = \delta(S),$$
(14)

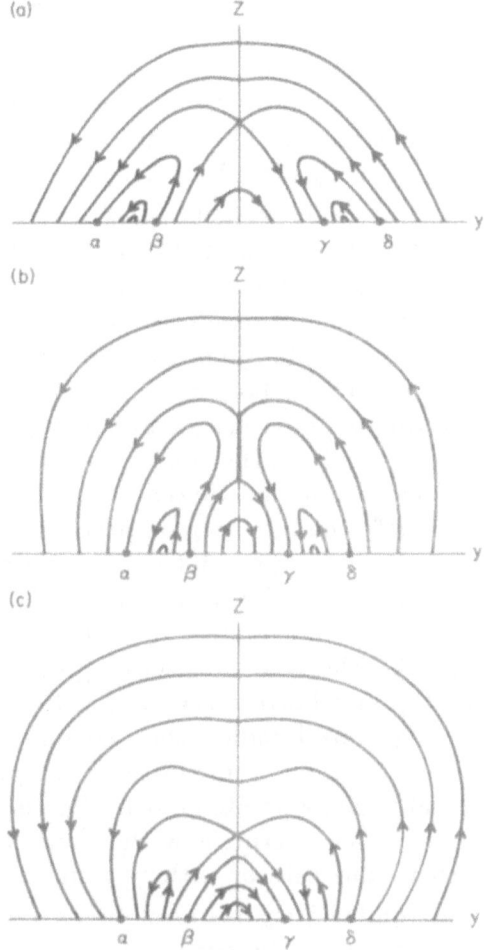

Figure 2. Formation of an electric current sheet in an X-type magnetic null point, taken from Low & Wolfson (1988). (a) Quadrupolar potential magnetic field with an X-type null point; $\mathbf{B}_{initial}$ in the text. Two lines of force with footpoints labeled $\alpha, \beta, \gamma, \delta$ are identified. Under the frozen-in condition, the relaxation of the magnetic field initially in the state shown in (a), in response to smoothly conveying footpoint displacement in the y-direction must result in a *nearly everywhere* potential state $\mathbf{B}_{end}$ shown in (b), with a current sheet represented by the vertical thick line. The footpoints $\alpha, \beta, \gamma, \delta$ are identified in their displaced positions, showing the same field-line connectivities among these footpoints as obtaining in $\mathbf{B}_{initial}$. (c) The *everywhere smooth* potential field $\mathbf{B}_{pot}$ having the same normal flux distribution at $z = 0$ as $\mathbf{B}_{end}$. The footpoints $\alpha, \beta, \gamma, \delta$ are identified, showing that this potential field requires these footpoints to have connectivities different from those of $\mathbf{B}_{initial}$ and $\mathbf{B}_{end}$.

subject to the same boundary condition (13). This equation replaces equation (13) by introducing an integrable current density at some current sheet surface S. The surface S is an unknown in equation (14), to be constructed such that the relaxed state $\mathbf{B}_{end}$ has the correct distribution of fluxes in the four sectors of the quadrupolar field, and such that S as a material surface is in force balance. The latter condition requires that the total magnetic pressure is continuous across S. The reader is referred to Low (1989) for a variational formulation of this free-boundary problem.

The relaxation of the field from $\mathbf{B}_{deformed}$ to $\mathbf{B}_{end}$ may be visualized to take place in a highly viscous fluid in the following manner. In the example in Figure 2, the field $\mathbf{B}_{deformed}$ is produced from $\mathbf{B}_{initial}$ by a boundary footpoint displacement compressing towards the origin from both sides, followed by freezing the footpoints in their new positions. Around the original X-type null point, the Lorentz force generated in the field would push the plasma in the two side sectors horizontally into each other as the top and bottom sectors yield to the converging side sectors (Charbonneau & Low 1991). The sectors may be viewed as flexible fluid volumes defined by the magnetic fluxes frozen into their respective volumes. The state $\mathbf{B}_{end}$ is created when the side sectors press into each other and spread the original X-type null point into a line to form the current sheet. In this process, the top and bottom sectors are completely squeezed out of the way to become physically detached. Another way of describing this process is to note that the parting of the top and bottom sectors produces a gap in the plasma within which two other flux bundles, the two side sectors, come together to meet, not at point but along a line in the $y - z$ plane. This process has an interesting optical analogy pointed out by Parker (1989a, 1989b, 1991), which shows in quite general terms how the same process may take place in more complex situations with or without a magnetic null point.

If the electrical conductivity is taken to be truly infinite, there can be no field-line reconnection. The current sheet forming in $\mathbf{B}_{end}$ will then collapse to a zero thickness. In a real plasma with a large but finite electrical conductivity, resistive reconnect will set in during the relaxation when the current sheet width has narrowed below a small but finite critical size. With the complete dissipation of the currents in $z > 0$, the field topology would have changed to one compatible with $\mathbf{B}_{pot}$ which then becomes available as an end-state. Non-resisitive evolution can take the deformed state only towards $\mathbf{B}_{end}$, creating the condition of an extremely large current density. Resistivity sets in to dissipate the current in the sheet despite the large but finite condutivity, whereupon making $\mathbf{B}_{pot}$ available as a truly final relaxed equilibrium state. During the current dissipation, the plasma is heated.

It should be pointed out that not every boundary footpoint displacement will produce a current-sheet. It is easy to devise special magnetic footpoint displacements on $z = 0$ such that the changed boundary flux distribution (12) gives a potential field $\mathbf{B}_{pot}$ which has exactly the sectorial flux distribution of $\mathbf{B}_{deformed}$. In that case, the field $\mathbf{B}_{end}$ and $\mathbf{B}_{pot}$ are the same and the magnetic field will relax by non-resistive evolution to an equilibrium state without forming any current-sheet. This result obtains only for boundary footpoint displacements specially tailored to satisfy the requirement on the sectorial flux distribution. For an *arbitrary* footpoint displacement, this requirement is not satisfied. In this sense, the deformation of a planar quadrupolar field with an X-type null point, *in general*, produces current sheets.

3.3. CURRENT-SHEET FORMATION WITHOUT A NULL POINT

The formation of the current sheet in the previous example has its cause in the interaction of distinct flux bundles, and does not require the presence of the X-type null point. To see this consider the case where $\mathbf{B}_{initial}$ takes the form of the quadrupolar potential field in Figure 3a, which does not have a null point in the domain $z > 0$ (Low 1987). We can easily construct such a field by taking a global quadrupolar potential field and set the photosphere $z = 0$ such that the X-type null point is located below $z = 0$. The field in $z > 0$ is composed of three flux bundles separated by the separatrix line which we shall refer to as Γ. This separatrix line is tangential to the base $z = 0$ at the origin. Nowhere in $z > 0$ is the field zero.

Now repeat the two-step evolution of the previous example. Subject the initial field to a footpoint displacement on the boundary $z = 0$, horizontally-directed in the y-direction, to change the flux distribution from an initial form given by equation (11) to some final form given by equation (12). The deformed magnetic field $\mathbf{B}_{deformed}$ remains in the $y - z$ plane and has the same field topology as $\mathbf{B}_{initial}$. In particular, the deformed field has no null points in $z > 0$. Next let the deformed field relax to a new force-free state $\mathbf{B}_{end}$.

For boundary footpoint displacements which compress towards the origin, and if the compression is large enough, the potential field $\mathbf{B}_{pot}$ matching the final flux distribution (12) can be shown to be a quadrupolar magnetic field with an X-type null point in $z > 0$; having a topology of the type shown in Figure 3c. The field $\mathbf{B}_{pot}$ is therefore not accessible via a non-resistive evolution, and once again, the proper end-state of relaxation $\mathbf{B}_{end}$ must be discontinuous with a current sheet. Figures 3b and 3c show sketches of $\mathbf{B}_{end}$ and $\mathbf{B}_{pot}$ having these properties.

Again it should be emphasized that special footpoint displacements can be prescribed such that $\mathbf{B}_{pot}$ has the same topology as that of $\mathbf{B}_{deformed}$, and the end-state $\mathbf{B}_{end}$ may be identified with $\mathbf{B}_{pot}$. For the arbitrary foot-

122

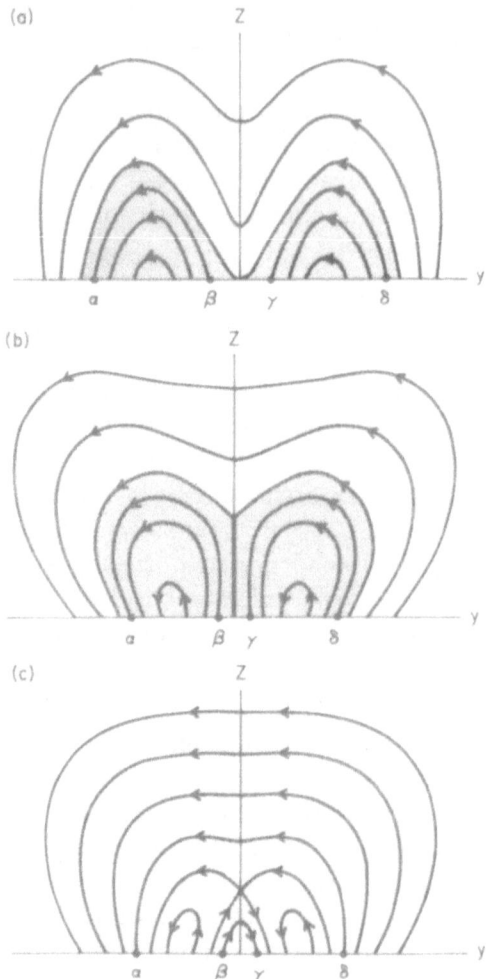

Figure 3. Formation of an electric current sheet in a quadrupolar magnetic field initially without a null point by a converging footpoint displacement in the y-direction, taken from Low & Wolfson (1988). (a) The initial potential state $\mathbf{B}_{initial}$ where two lines of force are identified by their footpoints $\alpha, \beta, \gamma, \delta$. The separatrix line named Γ in the text separates the magnetic flux into the two shaded bipolar fields and the overlying third bipolar field. The magnetic field deformed by the imposed footpoint displacement must settle into the nearly everywhere potential state $\mathbf{B}_{end}$ shown in (b) with the vertical current sheet. This field has the same field topology as $\mathbf{B}_{initial}$, as indicated by the same connectivities among footpoints $\alpha, \beta, \gamma, \delta$ in (a) and (b). The everywhere potential field $\mathbf{B}_{pot}$ with the same normal flux distribution at $z = 0$ as $\mathbf{B}_{end}$ is shown in (c), which displays connectivities among footpoints $\alpha, \beta, \gamma, \delta$ different from those in (a) and (b).

point displacement, the two fields $\mathbf{B}_{end}$ and $\mathbf{B}_{pot}$ have different topologies.

In the relaxation, the Lorentz force of $\mathbf{B}_{deformed}$ pushes the two shaded bipolar fields under the separatrix line Γ towards each other. This compression expels the flux located above Γ holding the two bipolar fields apart. The current sheet in Figure 3b forms when this flux is completely expelled to let the two shaded bipolar fields come into a line-contact, across which the tangential field jumps discontinuously. In the final state, the field is everywhere potential in $z > 0$ except at the current sheet whose equilibrium is assured with a continuous magnetic pressure across it. If resistivity is admitted to dissipate away the electric current sheet, the magnetic field may then *change topology* to relax to the potential field $\mathbf{B}_{pot}$. The physics of this process is really the same as in the previous example, namely, flux expulsion to form a current sheet. The previous example was well known for many years and may have been the origin of a common (mistaken) notion that a null point is necessary for the formation of current sheets.

A cautionary remark is in order about the interpretation of the process in Figure 3. If we take $\mathbf{B}_{pot}$ to vary parametrically with the form of F_{end} in equation (12), we will find that with a compressive footpoint displacement which is continuously increased in magnitude, the null point of $\mathbf{B}_{pot}$ first appears right on $z = 0$ and then rises (parametrically) into $z > 0$. This should not be construed to be the physical emergence of a null point from below $z = 0$ because the force-free assumption is valid only in $z > 0$. The force-free or potential state in $z > 0$ may not be physically continued into the region $z < 0$ where a very different, non-force-free state obtains. Moreover, the parametric sequence of $\mathbf{B}_{pot}$ so constructed is not a physical evolution. In a succession of the two-step evolutionary process, the field is not constantly force-free, but is subject to repeated deformation out of and relaxation to the force-free state. The potential states in the above parametric sequence are accessible to the relaxing field in $z > 0$, provided they are topologically equivalent to the relaxing field $\mathbf{B}_{deformed}$. At the parametric point when the potential field $\mathbf{B}_{pot}$ develops a null point on $z = 0$, the relaxed state is no longer found in this parametric sequence but instead bifurcates into the distinctly different sequence made up of the set $\mathbf{B}_{end}$ with current sheets.

3.4. CURRENT-SHEET FORMATION IN A SHEARED MAGNETIC FIELD

Current-sheet formation in the quadrupolar field takes on truly novel properties if the field is sheared with field-aligned electric current density, that is, in the case where $B_x \neq 0$ in equation (8) (Low & Wolfson 1988).

It is useful to relate the x-component B_x of the field to the footpoint separation Δx, in the x direction, of a bipolar line of force. Integrate along a given line of force to determine the x coordinates of the two footpoints on $z = 0$. Taking the difference, we obtain

$$\Delta x(A) = \int_{A=constant} \frac{B_x}{|\nabla A|} ds, \tag{15}$$

where the subscript of the integral indicates that the integration is along a line of constant A with path length s, in the y-z plane, from one footpoint to the other. The dependence of Δx on A generates the footpoint separations for the different lines of force.

The mathematical problem for a force-free magnetic field can now be posed in terms of footpoint separation $\Delta x(A)$. Supposed we are given a field in $z > 0$ with the normal flux distribution on $z = 0$ given by equation (12) in terms of stream function A. If we further prescribe $\Delta x(A)$, the field topology is fixed, and we have sufficient conditions for determining the force-free field described by equation (10). We need to construct $B_x(A)$ such that it is related to $\Delta x(A)$ by equation (15). Under the force-free condition, B_x is a strict function of A so that equation (15) simplifies to

$$\Delta x(A) = B_x(A) \int_{A=constant} \frac{1}{|\nabla A|} ds. \tag{16}$$

and, substituting away B_x in equation (10), we obtain

$$\nabla^2 A + \frac{1}{2}\frac{d}{dA}\left[\Delta x(A) \int_{A=constant} \frac{1}{|\nabla A|} ds\right]^2 = 0, \tag{17}$$

to determine A as the unknown, subject to boundary condition (12). In general this boundary value problem has solutions which necessarily have discontinuous derivatives in A. These solutions describe equilibrium states with discontinuous magnetic fields.

For the purpose of making a physical point, consider the situation in which the planar magnetic field in Figure 3a is subject to a footpoint displacement directed strictly in the x direction, zero in $y < 0$, increasing from zero at $y = 0$ to some non-zero value in $y > 0$. This footpoint displacement is continued into the domain $z > 0$ in order that the deformed magnetic field is everywhere continuous in $z > 0$. Next freeze the footpoints on $z = 0$ and allow the deformed magnetic field to relax to a new equilibrium state with the imposed footpoint separation locked into the field. In contrast to the previous two examples, the field $\mathbf{B}_{deformed}$ has a topology different from $\mathbf{B}_{initial}$; the former is sheared whereas the latter is planar. In the subsequent relaxation, the sheared topology of $\mathbf{B}_{deformed}$ is to be preserved. The end-state $\mathbf{B}_{end}$ clearly is not a potential field but a solution to equation (17) with a prescribed $\Delta x(A)$. This integro-partial differential equation is formidable. Fortunately, it is possible to show that the end-state $\mathbf{B}_{end}$ associated with the above special footpoint displacement is necessarily discontinuous without having to construct the solution explicitly.

Consider the relaxation of the different parts of the quadrupolar field in Figure 3a after subjecting it to the above footpoint displacements. Begin with the left shaded biopolar field beneath the separatrix line Γ. The lines of force of this bipolar field remains in the y-z plane after its deformation by the imposed footpoint displacement, because the imposed footpoint displacement is zero in $y \leq 0$. Therefore this bipolar field will relax to a configuration within which $\Delta x(A) = 0$ and $B_x = 0$. Outside of this region, every line of force has $\Delta x(A) \neq 0$. The bipolar region under Γ in $y \geq 0$ has both footpoints of each line of force differentially displaced. For the field above Γ, each line of force has one undisplaced footpoint in $y < 0$ and the other footpoint in $y > 0$ displaced by a finite amount $\Delta x(A)$. During the relaxation, lines of force with non-zero $\Delta x(A)$ must distribute the acquired $B_x \neq 0$ uniformly along each line so that $B_x = B_x(A)$. This happens in the bipolar field lying above the Γ and the bipolar region in $y > 0$ beneath Γ. It follows that in the relaxed state, B_x must be discontinuous across Γ, going from the left shaded bipolar field, where $B_x = 0$, to the bipolar field just above Γ with footpoints on the two sides of the origin.

The x-component of the magnetic field in this two-dimensional Cartesian system manifests as a magnetic pressure term; see equation (10). In the relaxation, the parts of the deformed magnetic field which acquired $B_x \neq 0$ will therefore expand outward into regions where B_x is weak or zero. Hence, the relaxed equilibrium state $\mathbf{B}_{end}$ will take on the form as sketched in Figure 4 shown in field-line projections on the $y-z$ plane. During the relaxation to attain $\mathbf{B}_{end}$, as the parts of the field with non-vanishing B_x expand into the left shaded bipolar field with $B_x = 0$, the overlying flux above Γ, which initially keeps the two underlying bipolar fields apart, may be expelled out of the way in the same manner as encountered in the examples treated in sections 3.2 and 3.3. Thus, as shown in Figure 4, the separatrix line Γ in the relaxed state would be folded into a triple-point with a line extension connecting with the origin. Along the line extension the two underlying bipolar fields come into contact. Therefore, the separatrix line will become quite complicated in the relaxed state. Across Γ, B_x is discontinuous from the overlying bipolar field into the left shaded bipolar region in $y < 0$. We can also expect B_x to be discontinuous across Γ from the bipolar field above Γ into the bipolar field below in $y > 0$. This is because, the lines of force on the two sides of Γ in this region of space have different footpoints and have B_x distributed along different path lengths along the respective lines of force. The line extension from the triple point to the origin is also a current sheet since the bipolar field in $y > 0$ is sheared with $B_x \neq 0$ and

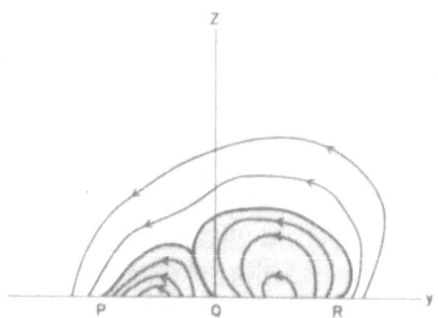

Figure 4. The end-state $\mathbf{B}_{end}$, taken from Low (1990), resulting from subjecting the quadrupolar initial potential field in Figure 3a to a continuous footpoint displacement chosen to be directed in the x direction, of zero magnitude in $y \leq 0$ and nonzero in $y > 0$. Shown are the lines of force projected on the y-z plane with the separatrix line, named Γ in the text, separating the magnetic flux into the two shaded bipolar fields and the third overlying bipolar field. The field everywhere except the left shaded bipolar field has a non-vanishing x-component. The separatrix line has been folded into a triple point connected by a line extension to the origin. The footpoint of this line extension is identified by Q; P and R are the original two footpoints of the separatrix line.

the bipolar field in $y < 0$ is planar with $B_x = 0$.

The above analysis suffices as a rigorous demonstration of current-sheet formation although it is an extremely difficult task to obtain the discontinuous solution for $\mathbf{B}_{end}$. There is an additional property of this solution which can be deduced without obtaining the solution explicitly. This property concerns the geometry of Γ in the neighborhood of where it threads the lower boundary $z = 0$ in the relaxed state $\mathbf{B}_{end}$. This property has implications for the problem treated by Parker and van Ballegooijen.

In the solution $\mathbf{B}_{end}$, the separatrix line must satisfy force balance by having the magnetic pressure continuous across it. Let $B_\perp$ be the magnitude of the component of $\mathbf{B}_{end}$ in the y-z plane. Across Γ, the jump in partial magnetic pressure due to a discontinuous B_x must be exactly compensated by the jump in partial pressure due to $B_\perp$, as expressed by

$$< B_\perp^2 >_\Gamma = - < B_x^2 >_\Gamma, \qquad (18)$$

where the angular brackets indicate the jump of a quantity across the curve indicated in their subscript. Let θ be the angle the separatrix line makes with the horizontal at a given point of the separatrix. Then, by definition

$$B_y = B_\perp \cos\theta, \quad B_z = B_\perp \sin\theta. \qquad (19)$$

Now take the jump relationship along the separatrix line down to where it intersects the boundary $z = 0$. Note that B_x is constant along a line of force $A = constant$. Therefore, on the separatrix line, the right hand side of equation (18) is a constant and $B_\perp^2$ is required to have a fixed discontinuity all the way down to $z = 0$. This requirement cannot be met if $\theta = \theta_0 \neq 0$ on $z = 0$. By definition, on $z = 0$, $B_z = B_\perp \sin\theta_0$ is continuous, so that if $\theta_0 \neq 0$, $B_\perp$ must be continuous across Γ on $z = 0$. A careful analysis of this requirement will show that it can be met only if $\theta \to \theta_0 = 0$ as $z \to 0$ along Γ, with $B_\perp \to \infty$ in such a way that the product $B_\perp \sin\theta$ is finite and equals B_z on $z = 0$. Moreover $B_\perp^2$ must tend to infinity along the two sides of Γ as $z \to 0$ in such a way that it has a finite jump across Γ at $z = 0$ to ensure that jump condition (18) is satisfied.

With $\theta_0 = 0$, the separatrix line in $\mathbf{B}_{end}$ must thread the boundary $z = 0$ tangentially as shown in Figure 4. The infinity in $B_\perp$ mathematically arises from the requirement that Γ intersects the boundary horizontally and yet maintains a finite (normal component) B_z at the point of intersection. This is a singularity approached only in the sense of a limit during an actual evolutionary relaxation. In such a relaxation, the jump in $B_x(A)$ cannot be sustained by the field near the footpoint of Γ and this part of the separatrix line would yield to the side with the larger value of B_x. Progressively the separatrix line is pressed onto the boundary $z = 0$. If resistivity is completely neglected both in the plasma and at the boundary, an infinity in $B_\perp$ will develop at the intersection of Γ with $z = 0$. The magnetic energy associated with this infinity of the field at a point is integrable so that the issue of infinite energy is avoided. The point infinity in magnetic field strength at this point also implies an infinite Lorentz force density which is also integrable. This local stress is taken up by the assumed rigid boundary $z = 0$ taken in the strict but extreme mathematical sense. ¿From the physical point of view, resistivity is only small and not zero and the rigid boundary approximation must eventually breakdown as this extreme limit is approached in an actual relaxation evolution. In the presence of a very small but finite resistivity, the current density would build up to such enormous values that resistive dissipation would cause the magnetic field to diffuse at the boundary, changing the normal flux distribution at $z = 0$. The yield of the boundary to the build up of Maxwell stress will also result in a re-distribution of the the boundary flux, in a more realistic model. In such a model, the infinite growth of $B_\perp$ at the point of intersection of Γ with $z = 0$ is physically avoided (Low 1990).

The important point to make here is that when a current sheet forms spontaneously in the plasma, the sheet current may extend all the way along the magnetic field to the boundary to produce integrably infinite values in the tangential field components at the line of intersection between the current-sheet surface and the boundary.

3.5. INFLUENCE OF PRESSURE AND GRAVITY

One of the issues encountered in the study of the quadrupolar two-dimensional magnetic field is whether current-sheet formation may be suppressed altogether in the presence of pressure and gravity. It was suggested that the weight of the plasma may prevent magnetic flux from being expelled to form a current sheet (Karpen, Antiochos & DeVores 1990). For completeness of our discussion, we present a simple analysis of this issue. The reader is referred to the original works for quantitative demonstrations (Low 1991, 1992).

Consider the second example in Figure 3a. In this example, the compressive horizontal displacements of magnetic footpoints results in the expulsion of upward-concaved magnetic lines of force above the origin, so that the two shaded bipolar fields beneath the separatrix line Γ could come to meet in a current sheet. The question we wish to analyse is whether the weight of the plasma located at the the bottom of each of these upward-concave magnetic lines of force might serve to suppress the expulsion. The first thing to note is that the compressive boundary displacement of the magnetic footpoint produces an increase of curvature in these upward-concave magnetic lines of force, which enhances the upward-directed, local magnetic tension force. This tension force is the driver of the expulsion. If the atmosphere is hydrostatically stratified under the influence of gravity, the enhanced tension force not only has to drive the motion of expulsion but also overcome the weight of the plasma. On the other hand, the deformation of the system brought about by footpoint displacements produces pressure inbalance along the magnetic field. Since the Lorentz force is perpendicular to the field, the subsequent dynamical evolution will also need to establish a purely hydrostatic equilibrium along each line of force. In Figure 3a, this process for the lines of force above Γ would imply siphon flows which generally tend to drain plasma out of the upward lifted part of the field in the neighborhood of the z axis. The drained plasma will flow along the field out of the up-lifted region and down to the footpoints located away from the origin. This lightens the load originally weighing on the upward-concave part of the field. In the hydrostatically stratified atmosphere, the lightening of the load also leads to a buoyancy effect which aids the expulsion of the local field. Therefore, in the presence of pressure and gravity, siphon flows and buoyancy would promote rather than impede the flux expulsion process to produce current sheets.

To be sure, circumstances exist under which siphon flows could be adding mass to a field to increase its weight loading and keep a flux bundle from being expelled out of the way. In that case, the formation of current sheets may be suppressed, but, as pointed above, there are other circumstances under which gravity and pressure have the opposite effect of promoting current-sheet formation. The key point about the sheet-formation process is that under conditions of a very high electrical conductivity, the magnetic flux is so tightly frozen into the plasma that individual flux bundles behave like integral entities with the possibility of one bundle pushing its way to make contact with a spatially distant bundle. That happens whenever the forces cannot be balanced geometrically, a situation likely to arise readily in a natural system without a geometric symmetry (Parker 1989a, Tsinganos, Distler & Rosner 1984). This effect will take place irrespective of whether the Lorentz force acts alone, balancing the magnetic pressure force against the magnetic tension force, in a low-β plasma, or, in the presence of other forces.

4. Discussion and Conclusion

This article is intended as an introduction to Parker's idea of the spontaneous formation of electric current sheets. The quadrupolar Cartesian magnetic field is a well-studied simple model which allows for simple mathematical treatment of this process in terms of properties which are physically intuitively easy to see. In particular, this model shows that the same process creating current sheets in the familiar setting of a magnetic null point works much the same way in the absence of a null point. The key topological factor is the interaction between distinct magnetic flux bundles such that two bundles may expel a third out of the way to come into contact. In the two dimensional Cartesian plane, separatrix lines are essential to separate the magnetic flux into distinct bundles. If the symmetry of the two-dimensional Cartesian plane is broken, or more generally, in a fully three-dimensional system without a special symmetry, a geometrically richer variety of plasma motion becomes available to naturally separate the magnetic flux into bundles (Low 1989, Parker 1989a, 1989b, 1990b). In this setting, separatrix lines or surfaces no longer have the special role as seen in the two-dimensional Cartesian plane. Current sheets simply form whenever a layer of magnetized plasma is squeezed out of the way of a pair of magnetic flux surfaces. The high conductivity dictates that the expelled plasma be bounded by flux surfaces as the least restrained way for a layer of plasma to be expelled. In a given situation, the dynamics and the global field topology dictate which of the infinite number of flux surfaces will suffer the discontinuous slipping of plasma along them.

In Parker's recent works, the last point is well illustrated by several models ingeniously constructed to consider the consequences of imposing symmetry-breaking deformation on an initially symmetric equilibrium state. In the subsequent adjustment of the deformed system, the lack of symmetry leads to the flux-expulsion phenomenon. In one case, a bundle of flux is singled out of an otherwise orderly array of parallel bundles to be displaced into mis-alignment with the array (Parker 1983, 1987). In another, symmetry is broken by changes in the geometric shape of the boundary (Harm & Kulsrud 1985, Parker 1990a). Even more interesting is the idea that more than one equilibrium state may be available to a magnetic field, of which one has a continuous magnetic field but is not the lowest-energy state. A perturbation of this smooth equilibrium state could result in the field evolving towards a lower energy state which is inherently discontinuous. This example demonstrated with the close packing of parallel bundles of magnetic flux is very instructive (Parker 1990b, Parker and Vainshtein 1986).

To close our article, we return to van Ballegoijen's treatment of the Parker problem. In this treatment, a rich family of force-free field solutions may be generated by extrapolating from known quantities at one boundary plate to the other using equation (6). A basic assumption is that these known quantities are all well behaved and regular at the starting boundary plate. We have demonstrated in the previous section that if a current sheet does form in the plasma interior, it will extend along the magnetic field to intersect the boundaries where it would create singular tangential components of the field at the boundaries. From this property, it follows that the van Ballegooijen extrapolation will automatically exclude those force-free field solutions which are necessarily discontinuous. An important point to note in section 3.4 is that an infinitesimal amplitude footpoint displacement may still result in boundary fields which are unbounded at localized regions. This suggests that methods other than perturbational analysis are needed to construct the discontinuous force-free fields in the Parker problem (Rosner & Knobloch 1982). This and many other interesting unsolved problems in the current-sheet formation are worthy of our attention in the near future.

Acknowledgement

With this article I wish to express my deep gratitude to Gene Parker for much that I have learned from him about science and life. Kanaris Tsinganos kindly read and commented on this paper. The National Center for Atmospheric Research is sponsored by the U.S. National Science Foundation.

References

Aly, J.J. (1987) in *Proc. Workshop Interstellar Magnetic Fields*, ed. R. Beck, R. Gräves, p. 240, Springer-Verlag, New York.
Aly, J.J. and Amari, T. (1989) *Astron. Astrophys.* **221**, 287.
Antiochos, S.K. (1987) *Astrophys. J.* **312**, 886.
Charbonneau, P. and Low, B.C. (1991) in *Cool Stars, Stellar Systems, and the Sun*, ed. M.S. Giampapa and J.A. Bookbinder, p. 531, Astron. Soc. Pacific, San Francisco.
Hahm, T.S. and Kulsrud, R.M. (1985) *Phys. Fluids* **28**, 2412.
Hu, Y.Q. and Low, B.C. (1982) *Sol. Phys.* **81**, 107.
Hu, Y.Q., Wang, J.X., Ai, G. and Nie, Y.P. (1995) *Sol. Phys.* **159**, 251.
Karpen, J.T., Antiochos, S.K. and DeVores, C.R. (1990) *Astrophys. J.* **356**, L67.
Low, B.C. (1982) *Rev. Geophys. Space Sci.* **20**, 145.
Low, B.C. (1987) *Astrophys. J.* **323**, 358.
Low, B.C. (1989) *Astrophys. J.* **340**, 558.
Low, B.C. (1990) *Ann. Rev. Astron. Astrophys.* **28**, 491.
Low, B.C. (1991) *Astrophys. J.* **381**, 295.
Low, B.C. (1992) *Astron. Astrophys.* **253**, 311.
Low, B.C. and Hu, Y.Q. (1983) *Sol. Phys.* **84**, 83.
Low, B.C. and Wolfson, R. (1988) *Astrophys. J.* **324**, 574.
Mikic, Z., Schnack, D.D., and Van Hoven, G. (1989) *Astrophys. J.* **338**, 1148.
Moffatt, H.K. (1987) in *Advances in Turbulence*, ed. G. Comte-Bellot, J. Meathieu, p 228, Springer-Verlag, New York.
Otani, N.F. and Strauss, H.R. (1988), *Astrophys. J.* **325**, 468.
Parker, E.N. (1972) *Astrophys. J.* **174**, 499.
Parker, E.N. (1979) *Cosmical Magnetic Fields*, Oxford University Press, Oxford.
Parker, E.N. (1983) *Astrophys. J.* **264**, 635.
Parker, E.N. (1987) *Astrophys. J.* **318**, 876.
Parker, E.N. (1989a) *Geophys. Astrophys. Fluid Dyn.* **45**, 169.
Parker, E.N. (1989b) *Geophys. Astrophys. Fluid Dyn.* **46**, 105.
Parker, E.N. (1990a) *Geophys. Astrophys. Fluid Dyn.* **52**, 183.
Parker, E.N. (1990b) *Geophys. Astrophys. Fluid Dyn.* **53**, 43.
Parker, E.N. (1991) *Phys. Fluids* **133**, 2652.
Parker, E.N. (1994) *Spontaneous Current Sheets in Magnetic Fields*, Oxford University Press, Oxford.
Priest, E.R. and Raadu, M.A. (1975) *Sol. Phys.* **43**, 177.
Rosner, R. and Knoblock, E. (1982) *Astrophys. J.* **262**, 349.
Sneyd, A.D. (1993) *Geophys. Astrophys. Fluid Dyn.* **70**, 195.
Strauss, H.R. and Otani, N.F. (1988) *Astrophys. J.* **326**, 418.
Sweet, P.A. (1969) *Ann. Rev. Astron. Astrophys.* **7**, 149.
Syrovatskii, S.I. (1981) *Ann. Rev. Astron. Astrophys.* **19**, 163.
Titov, V.S. (1992) *Sol. Phys.* **139**, 401.
Tsinganos, K.C., Distler, J., and Rosner, R. (1984) *Astrophys. J.* **278**, 409.
Vainshtein, S.I. and Parker, E.N. (1986) *Astrophys. J.* **304**, 821.
van Ballegooijen, A.A. (1985) *Astrophys. J.* **298**, 421.
van Ballegooijen, A.A. (1986) *Astrophys. J.* **311**, 1001.
Vekstein, G.E. and Priest, E.R. (1992) *Astrophys. J.* **384**, 333.
Vekstein, G.E. and Priest, E.R. (1993) *Sol. Phys.* **146**, 119.
Vekstein, G., Priest, E.R. and Amari, T. (1991) *Astron. Astrophys.* **243**, 492.
Wolfson, R. (1989) *Astrophys. J.* **344**, 471.
Zweibel, E.G. and Proctor, M.R.E. (1990) in *Topological Fluid Mechanics*, ed. H.K. Moffat, p. 187, Cambridge University Press, Cambridge.

MAGNETOHYDRODYNAMIC PROCESSES IN THE SOLAR CORONA: FLARES, CORONAL MASS EJECTIONS AND MAGNETIC HELICITY

B.C. LOW
High Altitude Obervatory
National Center for Atmospheric Research
P.O. Box 3000
Boulder, CO 80307-3000, USA
low@solo.hao.ucar.edu

Abstract. The magnetized, million-degree solar corona evolves in cycles of about eleven years, in dynamical response to newly generated magnetic fluxes emerging from below to eventually reverse the global magnetic polarity. Over the larger scales, the corona does not erupt violently all the time. Violent events like the flares and episodic ejections of material into interplanetary space occur frequently, several times a day, but they often originate in long-lived magnetic structures which form continually throughout the solar cycle. In this paper, the creation, stability, and eventual eruption of these structures are discussed from basic principles, drawing on recent advances in observation and theory. A global view is offered in which different pieces of observation relate physically, with distinct roles for the conservation of magnetic helicity and the releases of magnetic energy in dissipated and ordered forms.

1. Introduction

The solar atmosphere is a unique natural laboratory for the direct observation of a rich variety of fluid and magnetohydrodynamic (MHD) phenomena involving astronomical length and time scales. In the past two decades, our observational knowledge of these complex phenomena has expanded significantly, through the deployment of several generations of groundbased and space borne instruments. There has also been progress in the theoretical understanding of the fundamental processes of these phenomena. This is

133

K. C. Tsinganos (ed.), Solar and Astrophysical Magnetohydrodynamic Flows, 133–149.
© *1996 American Institute of Physics.*

especially true of the phenomena in the outer, extended part of the so-lar atmosphere called the corona (Athay 1988, Parker 1979, Low 1990). I will review briefly the basic theoretical questions and their possible an-swers concerning the large-scale structure, stability, gradual evolution, and eruption of the corona in the MHD description. A plausible global picture (Low 1993a) will emerge in which otherwise disjoint phenomena begin to relate meaningfully, and MHD processes in turbulent and ordered forms play separate but complementary roles.

2. The Solar Corona

In approximately eleven-year cycles, dynamo action in the interior of the Sun generates fresh magnetic fluxes which continually make their way into its atmosphere. Although the precise dynamo mechanism is still being de-bated, its consequences are well observed (Parker 1979, Gilman 1986). The new magnetic fluxes appear on the visible solar surface - the photosphere, as belts of sunspots in the mid-latitudes of the two hemispheres. These sunspot belts migrate to the equator in the course of the solar cycle to eventually decay away, culminating in the reversal of the global magnetic polarity at the end of the cycle.

The consequences of the solar dynamo are especially dramatic in the corona. This part of the atmosphere is mainly fully ionized hydrogen main-tained with stability in a temperature range of several million degrees, in contrast to the photosphere at $5000K$. It has a hydrostatic scale height of the order of a solar radius ($R_\odot \sim 7 \times 10^{10}$ cm) which explains its extended form made familiar by (scattered) white-light images of the total eclipse (Sime, Fisher & Mickey 1985). The thermal radiation of the corona is prin-cipally in the Extreme Ultra-Violet and X-rays (Athay 1988, Tsuneta et al. 1992). Direct imaging in these radiations in space is another means of observing the corona, which does not require occulting the bright photo-sphere. An important implication of its high temperature is that over scale lengths comparable to $R_\odot$, both thermal and electrical conductivities are sufficiently large for the corona to be treated in most considerations as a perfect conductor of heat and electric currents (Parker 1979).

The magnetic field is fundamental to the dynamics and structure of the corona. The tenuity ($n \sim 10^8$ $protons$ cm^{-3} at the base) and the ionized state of the corona do not allow direct measurement of its magnetic fields. Simple extrapolations from measured photospheric magnetic fields give an average intensity of 10 $gauss$ at the coronal base. In this region, the plasma β is typically 10^{-1} or smaller, again in contrast to the photosphere ($n \sim 10^{17}$ $protons$ cm^{-3}) where dense plasma is able to confine the kilo-gauss magnetic fields of sunspots.

Fresh magnetic fluxes are continually injected into the low corona from below by the solar dynamo. Flaring activities are obviously the hydromagnetic and plasma responses of the corona to this flux injection, although the nature of these complex processes are not understood well (de Jager & Svestka 1986). Observation suggests that a significant part of the emergent magnetic flux is in the form of lines of force with the two ends anchored in the dense photosphere. Hence, the magnetic field in the corona continues to be energized by the photosphere through the turbulent transport ($\sim \frac{1}{2}$ km s^{-1}) of the magnetic footpoints. This is the basis for the popular notion that magnetic energy can build up on a time scale long compared to the typical coronal Alfvén transit time of the order of minutes, to be released in spurts of 10^{31-32} erg as sporadic flares (Hagyard & Rabin 1986, Gold 1964, Sakurai 1989). On the larger scales, a magnetic structure observed in some plasma signature, say, H_α emissions, may evolve quasi-statically and then erupts. On the small scales, of the order of the photospheric scale height ($\sim 2 \times 10^7$ cm), which is below attainable observational spatial resolutions, the build up of magnetic energy is believed to proceed in a turbulent manner with the spontaneous formation of magnetic tangential discontinuities under the condition of high electrical conductivity (Parker 1979, 1988, 1989, Strauss & Otani 1988). The steepening to a tangential discontinuity leads to its resistive dissipation, despite the high conductivity. Thus a small part of the built-up energy goes to heat the corona. This is a remarkable process in which high conductivity itself promotes resistive dissipation (Parker 1979, 1989, Low 1990). Coronal heating is a central problem which is far from resolved, observationally and theoretically (Parker 1988, Kuperus, Ionson, and Spicer 1981).

While flares are frequent, with a rate as high as a few a day at activity maximum, it is interesting that the corona is not in a violent turbulent state everywhere all the time. Large scale structures do persist for periods of time long compared to the characteristic rise times of various MHD instabilities. To begin with, for magnetic-energy storage to work, it must be possible to store energy in stable field-aligned currents in the low-β corona, until the amount required for a flare has accumulated (Low 1985, 1990, Hagyard & Rabin 1986). This possibility was demonstrated by recent proofs of linear stability of force-free magnetic fields $\mathbf{B}$ satisfying, for a suitable function α,

$$\nabla \times \mathbf{B} = \alpha \mathbf{B}, \tag{1}$$

$$\nabla \cdot \mathbf{B} = 0, \tag{2}$$

in the ideal MHD description (Low 1988a, b, Hood 1992). Among the models are fields with realistic three-dimensional geometries. More impressive

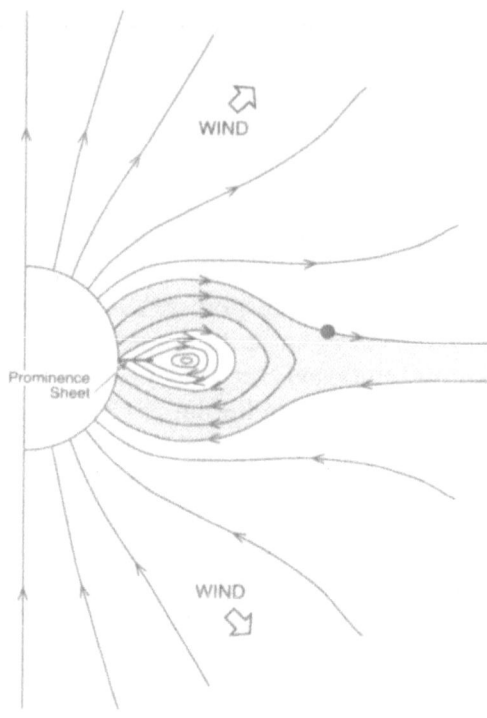

Figure 1. Sketch of the helmet-streamer in an axisymmetric corona. Shown are the arrowed, magnetic lines of force in a meridian plane; the high density part of the helmet represented by the shaded region; the cavity with a set of nested closed lines of force threading a prominence sheet; and steady solar winds flowing in the open fields as indicated. On lines of force which are closed or anchored at two ends to the Sun represented by the half circle, a magnetic component in the azimuthal direction may exist. The boundaries between the high density part of the helmet with the open-field regions and with the cavity may carry magnetic tangential discontinuities. The conspicuous dot on one of these boundaries is for the purpose of discussion in the text.

are the helmet-shaped coronal structures in white-light eclipse photographs sketched in Figure 1. Continuous observations of these so-called helmet-streamers using a coronagraph, which artificially occults the bright photosphere, have revealed that they may persist with stability for as long as a half rotation of the Sun, about two weeks.

3. Helmet-Streamers, Mass Ejections and the Solar Wind

3.1. LARGE-SCALE EQUILIBRIUM

The equilibrium of the large-scale corona is controlled by two natural effects, the tendency of the heated coronal plasma to expand into interplanetary space as the solar wind (Parker 1963, Hundhausen 1977) and the opposing tendency of a bipolar magnetic field frozen into the plasma to resist being opened up. A million-degree corona cannot be everywhere gravitationally bound. The point is simple in the absence of the magnetic field. Thermal conduction would carry the high temperature in a static atmosphere out to very many solar radii. Given the inverse-square law of Newtonian gravity, the hydrostatic pressure tends to a constant at large distances which, in the case of the Sun, is orders of magnitude too large to be confined by reasonable interstellar gas and magnetic pressures (Parker 1963). Therefore the corona is in a state of expansion converting its thermal energy into the kinetic energy of the wind flow. This wind reaches supersonic speeds a few solar radii away from the Sun and connects with the interstellar medium via a hydrodynamic shock.

The magnetic field complicates the physics with the following basic effects. The Lorentz force exerted by the magnetic field is composed of an isotropic pressure force and a tension force. The magnetic pressure enhances the tendency of the coronal plasma to expand. The magnetic tension force can resist expansion if the magnetic curvature is towards the Sun. Consequently, the quiescent corona is in a dichotomy of two distinct dynamical states (Low 1990, Parker 1963, Hundhausen 1977, 1988, Pneuman & Kopp 1971). Where the wind dominates, the magnetic field is flow-aligned and open into interplanetary space, neglecting the weak effect of solar rotation in the corona. Where the magnetic field has a strong sunward tension force, pockets of plasma can be trapped in static equilibrium. This occurs in the low corona over long (inversion) lines separating photospheric regions of opposite magnetic polarities, to give rise to the helmet-streamers.

Subject to mass conservation in the stratified atmosphere, the flow speed is everywhere negligible within one or two solar radii under any reasonable quiescent condition (Parker 1963, Hundhausen 1977). Hence, the low corona is approximately static as described in the one-fluid approximation by:

$$\frac{1}{4\pi}\left(\nabla \times \mathbf{B}\right) \times \mathbf{B} - \nabla p - \rho \nabla \Phi = 0, \tag{3}$$

where p, ρ and Φ are the pressure, density and solar gravitational potential, respectively. Since the Lorentz force is perpendicular to the magnetic field, Eq. (3) describes hydrostatic columns of plasma confined in individual magnetic flux tubes kept in lateral equilibrium by the Lorentz force. The same basic conclusion about the need for a wind flow is reached for those flux tubes extending too far from the Sun (Hundhausen, Hundhausen & Zweibel 1981). Thus the static solution of Eq. (3) must blend into a wind-dominated solution a few solar radii away. Generally, the hydrostatic pressure of the corona under the influence of efficient thermal conduction declines with height less rapidly than the pressure of a diverging magnetic field subject to flux conservation. Hence, although the plasma β is a small number in the nearly static lower corona, it increases with height to unity beyond which the plasma and its flow dominate, and the magnetic field is everywhere open.

3.2. INTERNAL STRUCTURES OF THE CORONAL HELMET

The closed field region of the helmet-streamer usually has a characteristic three-part structure, a high-density dome, a low-density cavity at the base, and a quiescent prominence in the cavity (Sime, Fisher & Mickey 1985, Saito & Tandberg-Hanssen 1973, Serio et al. 1978, Sturrock & Smith 1968); see Fig. 1. The dome is a shell of plasma draped over an elongated cavity overlying a long polarity inversion line on the photosphere. The prominence is a plasma condensation thermally shielded and suspended by the magnetic field, with densities two orders higher and temperatures two orders lower than the corresponding quantities in the surrounding corona. The prominence tends to be in the form of a vertical sheet, so that, as a first approximation, the sheet thickness may be neglected compared to other length scales as sketched in Fig. 1.

Prominences present challenging questions about their formation, mechanical and thermal equilibrium, and their role in solar activity (Tandberg-Hanssen 1974, Priest 1989). Observation has revealed that the magnetic fields in the prominence (beneath a coronal helmet) and in the photosphere below generally point in opposite directions (Leroy 1989). This constraint implies the magnetic topology in the prominence-cavity region as sketched in Fig. 1 (Low 1993b, Hundhausen & Low 1994, Low & Hundhausen 1995). The significant point made in (Low 1993b) is that the cavity contains a rope of twisted magnetic field running in the azimuthal direction around the axis of symmetry. In the idealized geometry of Fig. 1, this rope projects onto the closed lines in the cavity which carry a significant azimuthal magnetic component. The low pressure of the cavity is compensated by an enhancement of its magnetic intensity relative to that permeating the helmet dome. The flux rope is force-free, exerting no force, except at the prominence sheet. The right hand rule shows that the prominence sheet has a positive, discrete current flowing in the azimuthal direction, of the same sign as the azimuthal current of the force-free flux rope centered above it. The attraction between the two like-signed currents dominates locally, and is a principal means of support for the weight of the prominence. The force relationship is mutual, and it can also be said that the prominence serves as an anchor for the cavity magnetic flux rope which would otherwise rise buoyantly in the stratified atmosphere (Parker 1979, Low 1981).

The reality of the cavity magnetic flux rope is not yet confirmed although its possible manifestations as filament channels (Martin et al. 1993, Straka, Papagiannis & Kogut 1975) in the lower atmosphere and as ejected plasmoids, the interplanetary magnetic clouds (Gosling 1990, Rust 1994) are persuasive. In any case, its existence has very interesting implications, one of which we will follow up in the next section. A point about geometry is worth noting. In the real corona, the helmet arcade is of a finite length. The cavity flux rope and its embedded prominence bend downward at their respective two ends to blend into the magnetic structures at the base of the corona. Under the assumption of axisymmetry in Fig. 1 to keep the model simple, both the cavity flux rope and prominence sheet are completely detached from the solar surface. What is captured of the real situation by the idealized model is that the prominence and cavity are elongated magnetic structures with a main part of their lengths elevated in the corona.

The equilibrium of the coronal helmet as a whole may be described in terms of the cavity flux rope being anchored by the prominence beneath it and weighted down by the heavy helmet dome over it. No mature theory exists to explain how the force balance among the different constituents of this global structure is inherently stable as it must be from its observed long life. Recent calculations have given some insight (Hundhausen & Low 1994, Low & Hundhausen 1995). The boundary between the cavity and the helmet dome is usually sharp (Saito & Tandberg-Hanssen 1973), suggesting that it is a magnetic tangential discontinuity across which the total pressure is continuous. Crossing this boundary from the dome into the cavity, plasma pressure decreases and magnetic intensity increases abruptly. As shown in Fig. 1, the enhanced magnetic field of the cavity is convex to the high density plasma of the helmet dome. This magnetic boundary is thus stable by the magnetic-curvature force. On the other hand, this boundary is Rayleigh-Taylor unstable since heavy plasma sits over light plasma along it. For a magnetically dominated atmosphere, the former stabilizing influence can dominate.

The helmet dome and its long streamer has a distinctly shaped boundary with the low-density, high field-strength, external (open-field) region. This boundary has a certain stabilizing feature which can be identified without calculation. Take a typical point on the upper part of the boundary shown in Fig. 1. The stronger magnetic field in the exterior is convex to the high density region of the helmet-streamer. Hence, along this part of the boundary, *both* stratification and magnetic curvature are stabilizing. It will be interesting for future work to demonstrate that these two stabilizing agents could overcome the Kelvin-Helmholtz instability expected of the velocity shear separating the static from the wind-flow region. Future work should also address the observed stability of field-aligned and cross-field currents in the continuous part of the plasma.

3.3. DISRUPTIONS INTO CORONAL MASS EJECTIONS

The large scale corona is not absolutely steady but evolves quasi-steadily in response to the continual injection of fresh magnetic fluxes from below and to slowly changing physical conditions at the base. A major discovery of the last two decades is that the corona undergoes major reconfiguration as often as one or two times a day with the ejection of some 10^{15} g of plasma per event into interplanetary space (Low 1990, 1993c, Hundhausen 1988, 1995, MacQueen 1980, Kahler 1987, 1992). A majority of these ejections originate from the disruption of a helmet-streamer (Illing & Hundhausen 1986), displaying a characteristic three-part structure, a leading bright shell accounting for much of the ejected mass, a trailing low-density cavity, and often, an erupted prominence in the cavity (Hundhausen 1988, 1995). The erupted prominence may carry an additional mass of the same order as that in the leading shell. A reasonable interpretation is that as a helmet-streamer persists with stability over time, days to weeks, it eventually transits (Low 1981, 1984, Wolfson 1982, Low, Munro & Fisher 1982, Sime 1989) eruptively via a runaway instability or a loss of equilibrium into a mass ejection with its three-part structure readily identified with the pre-eruption helmet dome, cavity and quiescent prominence (Low 1990, 1984, Hundhausen 1995). Hence, the helmet-streamer has a dual role in coronal dynamics: as a structure which determines the large-scale quiescent corona in terms of magnetically open and closed regions, and as the agent of coronal reconfiguration.

The mass ejections account for a mass loss rate which is less than a percent of that due to the more or less steady solar wind (MacQueen 1980). The significance of the mass ejections lies in the opening up of initially closed magnetic fields, setting the stage for a greater amount of mass to subsequently escape as a part of the solar wind. Another deeper significance will be proposed in section IV. What is especially interesting is that the ejected mass has speeds (Gosling et al. 1976, Hundhausen, Holzer & Low 1987) lying in a broad range, 10 to 10^3 km s^{-1}, with a median of 350 km s^{-1}, which is above the sound speed of about 120 km s^{-1} for the corona and below the estimated coronal Alfvén speed of typically 700 km s^{-1}. The gravitational escape speed near the base of the corona is about 500 km s^{-1}, and it is interesting that many mass ejections show constant Lagrangian speeds well below this escape speed. Hence, gravity and the magnetic field are both important in the motion of the mass ejection (Low 1984). The kinetic energy of the expelled mass and the work done to overcome gravity are of comparable magnitudes and add up to 10^{31-32} erg which is of the

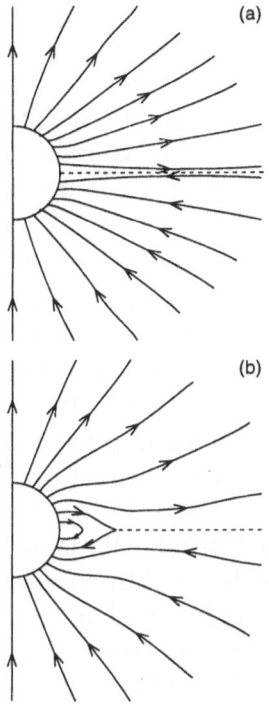

Figure 2. The relaxation by magnetic reconnection of a fully open, axisymmetric magnetic field (2a) to a partially open state (2b). The dashed lines indicate magnetic tangential discontinuities. Both magnetic fields are poloidal.

same order as the energy of a typical flare. Note that the former are ordered energies, in contrast to the dissipated energies liberated in a flare.

The association between mass ejections and flares (Low 1981, Hundhausen 1995, Kahler 1992) gives a crucial clue to the physical nature of the former. The observed association is statistically strong. More important, whenever a flare occurs in close proximity of the site from which the mass ejection is launched, the flare erupts some 10 minutes after the mass ejection is in full motion into the upper corona (Hundhausen 1995, Harrison 1986). This definite ordering of the two events can be understood in terms of the mass ejection opening up an initially closed bipolar magnetic field (in a background of open fields) into a fully open configuration with a field-reversal layer which now extends to the base of the corona as shown in Fig. 2a. In Fig. 1, the field-reversal layer lies in the high-β outer corona whose flow keeps the upper field open. In Fig. 2a, this field-reversal layer extends into the low-β lower corona where, upon formation, it rapidly collapses to reconnect. Magnetic energy is dissipated as a flare while a part of the open field returns to a closed geometry (Kopp & Pneuman 1976), see Fig. 2b. (If not for the high-β condition of the outer corona, all the open field would have closed back on the Sun.) This is a distinct kind of flare (Tsuneta et al. 1992, deJager & Svestka 1986, Hundhausen 1995, Kahler 1992, Hiei, Hundhausen & Sime 1993, Hiei, Sime & Hundhausen 1993). The reconnected, closed field is initially hot, energizing two rows of magnetic footpoints by thermal conduction and particle beams, and the helmet may reform (Hiei, Hundhausen & Sime 1993, Hiei, Sime & Hundhausen 1993).

The mass ejection and its associated flare are quite distinct MHD processes, the former involving mainly ordered, large-scale motions whereas the latter is highly turbulent and dissipative in nature. This point is salient in theoretical considerations (Low 1981, 1984, Wolfson 1982) and interpretation of observations (Hundhausen 1988, 1995, Kahler 1992, Sime 1989, Harrison 1986) It was recently confirmed by observing the two associated phenomena simultaneously, for the first time, in scattered white-light and soft X-rays, using the High Altitude Observatory coronagraph and the soft X-ray imager in the YohKoh satellite launched in 1990. A mass ejection was found visible in white light but had no significant detections in the high-temperature (3 to $5 \times 10^6 K$) window of the X-ray instrument (Hiei, Hundhausen & Sime 1993, Hiei, Sime & Hundhausen 1993). The latter instrument observed only the highly-heated flare plasma created after the mass ejection had moved out of the low corona.

It is believed that the mass ejection is driven with energy stored in the pre-eruption, coronal magnetic field. No agents in the photosphere could drive the mass ejection at a speed high enough to keep pace with the rapid motions, typically at the Alfvenic speeds, in the corona. Accounting for the mass-ejection energy is not a trivial problem (Low 1993c). A first approach is to take an axisymmetric magnetostatic structure in the low corona as described by Eq. (3) and ignore the solar wind except to demand that the magnetic field above a few solar radii is everywhere open (Hundhausen, Hundhausen & Zweibel 1981). In this approximation, force-free magnetic fields with all lines of force anchored to the coronal base cannot have an energy in excess of that in any fully open field having the same flux distribution at the coronal base (Aly 1991, Sturrock 1991). This severe constraint means that the shearing of an anchored magnetic field through quasi-static force-free states cannot accumulate enough energy to spontaneously open up, let alone supply the energy of the expelled plasma. It points to the necessity of storing energy in cross-field currents, associated with plasma pressure gradients and the gravitational force (Low & Smith 1993). In particular, a detached magnetic flux rope held down by anchored magnetic fields and plasma weight - the helmet-streamer, may have the needed energy.

The energy in the detached magnetic field is fully available by its expansion to do work on the plasma. In contrast, a magnetic field anchored at both ends to the coronal base in any expulsion of plasma would be stretched out so that it takes up rather than gives up energy. The energetics of the mass ejection therefore brings us back to the idea of a flux rope in the helmet cavity.

4. Magnetic Flux Ropes and Magnetic Helicity

Assuming that the helmet cavity is a magnetic flux rope, what is its origin? We have pointed out some insights into how the helmet-streamers could be stable for long periods of time as observed, but why do such structures form in the course of a solar cycle? In fact, observation suggests that high-latitude helmet-streamers may form, persist for a time, erupt away, and reform over photospheric regions which survive from one solar cycle into the next (Martin et al. 1993). Physical context and possible answers to the above two questions can be based on a recent proposal (Low 1993a) described below.

Let us first recall the theory of Taylor relaxation for the spontaneous reversal of magnetic fields in laboratory pinch devices (Taylor 1974, 1986). In a perfect electrical conductor, the magnetic helicity $h(V_M) = \int_{V_M} \mathbf{A} \cdot \mathbf{B} dV$ is conserved in any volume V_M bounded by a magnetic flux surface; $\mathbf{A}$ is the magnetic vector potential. In the turbulent state of a highly conducting low-β plasma, as characterized by a large magnetic Reynold's number, energy can be released resistively by rapid magnetic reconnection, which does not conserve the helicity $h(V_M)$. The hypothesis was proposed that for a contained plasma with no magnetic flux threading across it, the total helicity H in the container is approximately conserved. In other words, during magnetic reconnection, the decay rate of the total helicity H is small compared to that of the energy (Berger 1984). The end state into which the plasma finally settles is by definition stable at a minimum of magnetic energy which conserves H. This end state is a force-free magnetic field with a constant value for α chosen to preserve H (Woltjer 1958). The basic structure of these equations is that of the Helmholtz equation, with solutions which successfully predicted the observed field reversals in the laboratory pinch (Taylor 1986).

Attempts (Norman & Heyvaerts 1983, Heyvaerts & Priest 1984) to apply Taylor's theory to the solar atmosphere overlooked some obstacles to a simple carry over (Low 1985, Aly 1993). Solar magnetic fields generally thread across their natural boundaries, so that magnetic helicity is not gauge invariant. This problem can be addressed by a suitable formulation of a gauge-invariant relative helicity (Jensen & Chu 1984, Berger & Field 1986,

Finn 1986). Another problem is that solar magnetic fields are not isolated in a finite volume by rigid containers. The infinite space of an atmosphere implies that a constant-α force-free magnetic field matching a given arbitrary distribution of the boundary normal flux has an infinite amount of magnetic energy (Seehafer 1978, Berger 1985, Aly 1992, Laurence & Avellaneda 1993). The exception to this result is the $\alpha = 0$ potential magnetic field. If Taylor's theory is carried to its logical conclusion for the infinite domain, the $\alpha = 0$ potential field is, in fact, the end state of the relaxation to a minimum-energy, force-free field with a conserved net helicity. It would take an infinitely long time to spread the finite, net helicity over the infinite domain. In such a process, the helicity density approaches zero everywhere so that the magnetic field approaches the potential state. The force-free Taylor state for the infinite domain is thus physically not interesting or relevant. It is probable that reconnection in a flaring coronal magnetic field never gets to the point of spreading the net helicity over the entire atmosphere. The spread of helicity slows down as reconnection proceeds, and the magnetic field soon attains a slowly evolving, intermediate or quasi-relaxed state. Since the plasma β is not small everywhere in the corona, the relevant quasi-relaxed state is not a force-free magnetic field, but one subject to gravitational stratification and pressure gradients. The extension of Taylor's theory to define this physically more relevant quasi-relaxed state has not yet been developed.

The following intuitive essence of the Taylor theory suffices for our purpose here. In a nearly perfectly conducting plasma undergoing magnetic reconnection, resistive dissipation of currents is enhanced not necessarily by an increase of resistivity but by the steepening of magnetic gradients basically as an ideal MHD effect. Resistive dissipation takes place at almost randomly distributed sites around which currents are channeled in the otherwise perfect conductor. The essential point is that, unlike linear circuits, the entire current cannot be shorted out and a certain amount of electric currents always survives in the end, as estimated quantitatively by the conservation of total helicity.

Returning to the corona, it seems reasonable to identify the initial flaring of a twisted magnetic field (Kurokawa 1987, Leka et al. 1993) emerging into the corona with the above turbulent phase of rapid magnetic energy release. If we accept that not all the electric currents can be shorted away by magnetic reconnection, this magnetic field will relax to a stable state which, in general, still carries a significant amount of current associated with a net total helicity. A slow evolution of the magnetic field driven by the photospheric displacement of magnetic footpoints as well as the emergence of more magnetic flux from below may lead to additional flaring. Each flaring leaves the magnetic field in a relaxed, stable state trapping locally a net helicity and its associated, ordered electric current. This would explain recent observations of stable, highly-sheared magnetic structures (Hagyard & Rabin 1986, Athay, Jones & Zirin 1985). If this hypothesis is valid, the traditional attempt to identify the end state after a flare with a potential field is a mistaken undertaking; see the review by Hagyard & Rabin (1986). In other words, not all the free energy relative to the potential state in a given preflare magnetic field is available for liberation in a flare (Berger 1984).

As an active region (of sunspots) decays, the distribution of magnetic polarity on the photosphere becomes more orderly with relatively long and straight polarity inversion lines. In this late stage of an active-region evolution, the helmet-streamers tend to be especially long-lived before giving in to disruptions as mass ejections.

If we accept that flaring generally leaves behind a significant net helicity, with the post-mass-ejection flare being a notable exception as we shall see; it follows that the helmet-streamer is where magnetic helicity locally accumulates. Although the accumulated helicity implies a significant free magnetic energy, that energy is permanently locked since helicity cannot be shed locally except over coronal resistive time scales of the order of years (Berger 1984). It is the ever changing physical condition at the coronal base and the continual injection of magnetic flux that ultimately create conditions for the corona to get rid of the pockets of accumulated helicity and trapped energies. This is by the bodily ejection of the helmet-streamer out of the corona as an *ideal* MHD process, a follow-up scenario which has no counter part in the laboratory device. The mass ejection leaves behind a highly stressed state in the form of an open, unsheared magnetic field, with little or no magnetic helicity. The low level of helicity allows a relaxation to a nearly potential state; this is reflected by the fact that both the stressed and relaxed fields in Fig. 2 are poloidal. Hence, the flare which follows the mass ejection can liberate practically all the stress energy in a final throe.

The obvious place where the accumulated magnetic helicity may be lodged prior to eruption is the twisted magnetic flux rope in the helmet cavity; see Rust (1994) for an alternative suggestion. This flux rope probably originates in the solar interior (van Ballegooijen & Martens 1990). In solar physics, we are accustomed to the idea of a magnetic field emerging into the corona in the form of bipolar lines of force with the two ends remaining anchored to the photosphere. The magnetic lines of force above the photosphere cannot be directly observed and one must infer their geometry from the emissions of the plasma embedding it. Perhaps as the result of this observational constraint, it has seldom been discussed whether a twisted magnetic flux rope with a large-scale coherence could make its way, in its entirety or a great length of it, through the photosphere into the corona;

but see Rust (1994). If this takes place, its passage at the photospheric level is hard to discern since the high-β turbulent plasma there and below would have churned the flux rope into small scale structures. However, the large-scale coherence would remain as some net magnetic helicity in the tangled field under the high-conductivity condition. As bits and pieces of the field make their way by a systematic buoyancy into the corona, they would expand in the low-β environment and press into each other to undergo reconnection. The corona is thus heated to maintain its high temperature, and flares erupt, all in a complicated course of events. Eventually, the small scale structures dissipate away to leave behind a stable field with its endowed large-scale coherence or net helicity. If this attractive dynamo origin of the cavity flux rope is accepted, it becomes interesting as to whether the magnetic flux and helicity carried away by the mass ejections are significant for the operation of the solar dynamo.

5. Summary and Conclusion

Coronal activity is a collection of plasma processes manifesting from the passage of magnetic fields through it from below, generated by the solar dynamo in cycles of approximately eleven years. This global process culminating in the reversal of the solar magnetic dipole at the end of each cycle involves the turbulent disspation of magnetic energy - the flares and heating of the corona. Not so well recognized is the fact that the process also involves the spawning of stable large-scale magnetic structures, the helmet-streamers, to be eventually expelled out of the corona by ordered motions carrying away an energy comparable to that of a typical flare. The turbulent, highly dissipative, as well as largely ideal MHD processes play their distinct roles, each liberating a comparable amount of energy stored in the magnetic fields.

The formation of stable structures containing considerable magnetic energy can be understood in terms of the conservation of magnetic helicity, drawing upon the analogy with the Taylor relaxation of plasmas in the laboratory setting. We have argued that these structures would form naturally from the flaring magnetic fields entering the corona which are endowed with a net helicity not destroyed by magnetic reconnection. The stable, post-flare magnetic field may thus contain a considerable amount of energy associated with its helicity. If we accept that magnetic fields enter the corona highly twisted, their interaction and ultimate reorganization during their coronal relaxation may produce a magnetic flux rope, which, by a variety of observational and theoretical considerations, has a ready identification with the helmet cavity. Since magnetic helicity is generally difficult to remove locally in a highly conducting plasma, it follows that bodily expulsion of the entire helmet-streamer, as an ideal MHD process, is the inevitable means of its removal if the corona is not to accumulate pockets of magnetic helicity over a solar cycle. Thus, the mass ejection finds its natural role in coronal activity.

We have synthesized what seems to be an attractive global picture of the evolution of the corona in the course of an eleven-year solar cycle. This picture provides a physical understanding as a whole of certain hitherto disjoint observations. It also provides context for various basic questions concerning coronal evolution and discussion of their possible answers. It will be interesting to see if the proposed picture will ultimately be confirmed, modified, or rejected by future observations and theoretical work to pin down the underlying physical ideas.

Acknowledgement This article, based on a talk given to the American Physical Society - Plasma Physics Division, was published in the *Journal of Physics of Plasmas*, Volume 1, 1689 (1994) and is reprinted here, unchanged except for an updating of references, with copyright permission granted by the Americal Institute of Physics. The National Center for Atmospheric Research is sponsored by the National Science Foundation. I thank Paul Charbonneau for comments on the original manuscript.

148

References

Aly, J.J. (1991) *Astrophys. J.* **375**, L61.
Aly, J.J. (1992) *Sol. Phys.* **138**, 133.
Aly, J.J. (1993) *Phys. Fluids* **85**, 151.
Athay, R.G. (1988) in *Multiwavelength Astrophysics*, edited by F. A. Cordova, p. 7, Cambridge University Press, New York.
Athay, R.G., Jones, H.P. and Zirin, H. (1985) *Astrophys. J.* **288**, 363.
Berger, M.A. (1984) *GAFD* **30**, 79.
Berger, M.A. (1985) *Astrophys. J. Suppl.* **59**, 433.
Berger, M. and Field, G. (1986), *J. Fluid Mech.* **147**, 133.
de Jager, C. and Z. Svestka (editors) (1986), *The Physics of Solar Flares, Adv. Space Res.* **6**.
Finn, J.M. (1986) *Phys. Fluids* **29**, 2630.
Gilman, P.A. (1986), in *Physics of the Sun*, edited by P. A. Sturrock, T. E. Holzer, D. M. Mihalas, and R. K. Ulrich, Vol. 1, Chap. 5, Reidel, Dordrecht.
Gold, T. (1964) in *The Physics of Solar Flares*, edited by W. N. Hess, p. 389, NASA, Washington, D.C..
Gosling, J.T. (1990) in *Physics of Magnetic Flux Ropes*, edited by C. T. Russell, E. R. Priest, and L. C. Lee, p. 343, AGU Publ., Washington, D.C..
Gosling, J.T., Hildner, E., MacQueen, R.M., Munro, R.H., Poland, A.I. and Ross, C.L. (1976) *Sol. Phys.* **48**, 389.
Hagyard, M.J. and Rabin, D.M. (1986) *The Physics of Solar Flares, Adv. Space Res.* **6**, 7.
Harrison, R.A. (1986) *Astron. Astrophys.* **162**, 283.
Heyvaerts, J. and E. R. Priest, Astron. Astrophys. **137**, 63 (1984).
Hiei, E., Hundhausen, A.J. and Sime, D.G. (1993) *Geophys. Res. Lett.*, **20**, 2785.
Hiei, E., Sime, D.G. and Hundhausen, A.J. (1993) *Geophys. Res. Lett.*, submitted.
Hood, A.W. (1992) *Plasma Phys. and Con. Fusion* **34**, 411.
Hundhausen, A.J. (1977) in *Coronal Holes and High Speed Wind Streams*, edited by J. B. Zirker, Chap. VII, Colorado Associated University Press, Boulder.
Hundhausen, A.J., (1988) in *Proc. Sixth International Solar Wind Conf.*, Vol. 1, edited by V. Pizzo, D. G. Sime and T. E. Holzer, p. 181, National Center for Atmospheric Research, Boulder.
Hundhausen, A.J. (1995) in *The Many Faces of the Sun*, edited by K. Strong, J. Saba and B. Haisch, Springer-Verlag, New York (in press).
Hundhausen, A.J., Holzer, T.E. and Low, B.C. (1987) *J. Geophys. Res.* **92**, 11,173.
Hundhausen, J.R. and Low, B.C. (1994) *Astrophys. J.*, **429**, 876.
Hundhausen, J.R., Hundhausen, A.J. and Zweibel, E.G. (1981) *J. Geophys. Res.* **86**, 11,117.
Illing, R.M.E. and Hundhausen, A.J. (1986) *J. Geophys. Res.* **91**, 10,951.
Jensen, T. and Chu, M. (1984) *Phys. Fluids* **27**, 2881.
Kahler, S. (1987) *Rev. Geophys.* **25**, 663.
Kahler, S. (1992) *ARAA* **30**, 113.
Kopp, R. and Pneuman, G.W. (1976) *Sol. Phys.* **50**, 85.
Kuperus, M., Ionson, J.A. and Spicer, D.S. (1981) *ARAA* **19**, 7.
Kurokawa, H. (1987) *Sol. Phys.* **113**, 259 (1987).
Laurence, P. and Avellaneda, M. (1993) *GAFD* **69**, 201.
Leka, K.D., van Driel-Gesztelyi, L., Canfield, R.C., Anwar, B., Metcalf, T.R., Mickey, D.L. and Nitta, N. (1993) *Bull. AAS* **25**, 1187.
LeRoy, J.L. (1989) in *Dynamics and Structure of Solar Prominences* edited by E.R. Priest, Chap. 4, Kluwer Academic, Utrecht.
Low, B.C. (1981) *Astrophys. J.* **251**, 352.
Low, B.C. (1984) *Astrophys. J.* **281**, 392.
Low, B.C. (1985) *Sol. Phys.* **100**, 309.

Low, B.C. (1988a) *Sol. Phys.* **115**, 269.

Low, B.C. (1988b) *Astrophys. J.* **330**, 992.

Low, B.C. (1990) *ARAA* **28**, 491.

Low, B.C. (1993a), *Bull. AAS* **25**, 1218.

Low, B.C. (1993b) *Astrophys. J.* **409**, 798.

Low, B.C. (1993c) *Adv. Space Res.* **13**, 63.

Low, B.C. and Hundhausen, J.R. (1995) *Astrophys. J.*, **442**, 818.

Low, B.C. and Smith, D.F. (1993) *Astrophys. J.* **410**, 412.

Low, B.C., Munro, R.H. and Fisher, R.R. (1982) *Astrophys. J.* **254**, 335.

MacQueen, R.M. (1980), *Phil. Trans. R. Soc. Lon.* **A297**, 605.

Martin, S.F., Bilimoria, R. and Tracadas, P.W. (1993) in *Solar Surface Magnetism*, edited by R.J. Rutten and C.J. Shrijver, p. 303, Dordrecht: Kluwer.

Norman, C.A. and Heyvaerts, J. (1983) *Astron. Astrophys.* **124**, L1.

Parker, E.N. (1963) *Interplanetary Dynamical Processes*, Interscience, New York.

Parker, E.N. (1979) *Cosmical Magnetic Fields*, Oxford University Press, Oxford.

Parker, E.N. (1988) *Astrophys. J.* **330**, 474 (1988).

Parker, E.N. (1989) *GAFD* **45**, 159.

Pneuman, G.W. and Kopp, R.A. (1971) *Sol. Phys.* **18**, 258.

Priest, E.R., editor (1989) *Dynamics and Structure of Solar Prominences*, Kluwer Academic, Utrecht.

Rust, D.M. (1994) *Geophys. Res. Lett.* **21**, 241.

Saito, K. and Tandberg-Hanssen, E. (1973) *Sol. Phys.* **31**, 105.

Sakurai, T. (1989) *Sol. Phys.* **121**, 347.

Seehafer, N. (1978) *Sol. Phys.* **58**, 215.

Serio, S., Vaiana, G.S., Godoli, G., Motta, S., Pirronello, V. and Zappala, R.A. (1978) *Sol. Phys.* **59**, 65.

Sime, D.G. (1989) *J. Geophys. Res.* **94**, 151.

Sime, D.G., Fisher, R.R., Mickey, D.L. (1985) *Astrophys. J.* **333**, L103.

Straka, R.M., Papagiannis, M.D. and Kogut, J.A. (1975) *Sol. Phys.* **45**, 131.

Strauss, H.R. and Otani, N.F. (1988) *Astrophys. J.* **326**, 418.

Sturrock, P.A. (1991) *Astrophys. J.* **380**, 655.

Sturrock, P.A. and Smith, S.M. (1968) *Sol. Phys.* **5**, 87.

Tandberg-Hanssen, E. (1974) *Solar Prominences*, Reidel, Dordrecht.

Taylor, J.B. (1974) *Phys. Rev. Lett.* **33**, 1139.

Taylor, J.B. (1986) *Rev. Mod. Phys.* **58**, 741.

Tsuneta, S., Hara, H., Shimizu, T., Acton, L.W., Strong, K.T., Hudson, H. and Ogawara, Y. (1992) *PASJ* **44**, L63.

van Ballegooijen, A.A. and Martens, P.C.H. (1990) *Astrophys. J.* **361**, 283.

Wolfson, R.L.T. (1982) *Astrophys. J.* **255**, 774.

Woltjer, L. (1958) *Proc. Nat. Acad. Sci.* **44**, 489.

RECONNECTION OF MAGNETIC LINES OF FORCE

E. R. PRIEST

Mathematical and Computational Sciences Department
St Andrews University
ST ANDREWS, KY16 9SS
Scotland

Abstract. Parker laid the foundation for the subject of magnetic reconnection in his fundamental early papers. We first of all summarise his contributions and give a new generalisation of the Sweet-Parker relations for a current sheet in which the outflow pressure is an extra parameter. Then we review the models for fast reconnection that have since been proposed, beginning with the Petschek mechanism and continuing to the more general Almost-Uniform and Nonuniform families. A comparison with numerical experiments is also made and the conditions under which fast reconnection exists are elucidated.

1. Introduction

It is indeed a great pleasure to honour Gene Parker, who has always been my greatest MHD hero ! The bulky box of his publications in my office is one of my most treasured possessions. The three qualities which I most admire are: his great independence of thought, which has inspired so many topics; the combination of deep physical intuition and cunning but simple mathematical modelling (the anonymous referee of his solar wind paper who said that "Parker is incompetent and knows nothing of hydrodynamics or solar physics" didn't quite assess him accurately!); and finally his fluent style of presentation - it is a real enjoyment to be carried along by his persuasive style like a conjurer with no notes and aided only by a piece of chalk and blackboard. One of his favourite themes is that "the Sun is cleverer than we are"; I can remember him in reaction to a particular theory commenting "I can imagine mother nature laughing up her sleeve at that one". We

151

K. C. Tsinganos (ed.), Solar and Astrophysical Magnetohydrodynamic Flows, 151–170.
© 1996 *Kluwer Academic Publishers.*

are fortunate that these skills have been applied to magnetohydrodynamics rather than politics or commerce !

Let me take you back to the 1950's. Giovanelli [6] had suggested that solar flares may occur near an X-type neutral point, and then Cowling [2] pointed out that, if a solar flare is due to ohmic dissipation, you need a current sheet only a few metres thick to power it. Next Dungey [4] suggested that a current sheet could form due to an instability of the magnetic field near an X-type neutral point and that "lines of force could be broken and rejoined"; he coined the word "reconnection" at the 1956 IAU symposium. Sweet [20] stressed that conditions far from a neutral point and plasma pressure may play important roles in forming a current sheet or "colliding layer" and was the first to model such a sheet. He considered the potential field with an X-point due to four collinear sources in the photosphere shown in Figure 1: if the sources approach one another and the magnetic field remains frozen to the plasma, a narrow layer of strong current forms around the null point. The field flattens and hydrostatic equilibrium implies that the plasma pressure inside the current sheet is of order the external magnetic pressure. Plasma is squeezed out of the ends of the sheet at the internal sound speed, just as a fluid would be squeezed between two approaching plates. The time-scale for ohmic dissipation is about 10^4 secs, very much smaller than the local diffusion time of 10^{10} secs.

Later Parker [10] clarified Sweet's presentation and went to the core of the problem by deriving scaling laws for reconnection. Even though much of Gene's work *is* prophetic and 1957 seems to be before the 1958 of Sweet's paper, in this case it is in fact after, since 1958 refers to the 1956 IAU Symposium in Stockholm ! Indeed, it was when listening to Sweet's talk that he realised how to model the process in terms of MHD and so eagerly went back to his room that evening and worked out the details. In a massive 34-page paper [11], he then gave an in-depth development of the mechanism and its application to solar flares. He modelled the internal structure across the current sheet and, in passing, made the challenging comment that "it would be instructive if the exact equations could be integrated on a machine", a challenge that took nearly thirty years to be accomplished properly. In the paper he included compressibility and other effects for enhancing reconnection, such as ambipolar diffusion, plasma turbulence and fluid instabilities, but he concluded that the mechanism is too slow by a factor of a hundred to explain energy release in a solar flare (at least in the initial stages).

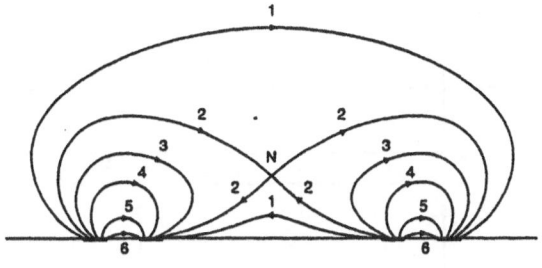

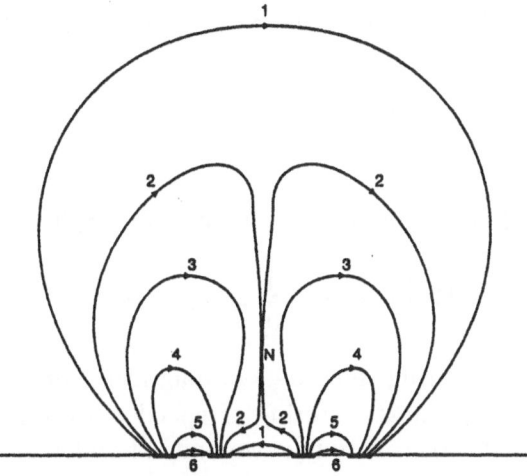

Figure 1. The creation of a current sheet in the solar corona in response to the motion of two bipolar photospheric sources. (After Sweet [20]).

2. Sweet-Parker Reconnection

The key questions in this field are: what is the nature of energy conversion and what is the rate at which it occurs? In other words, what is the speed (v_i) at which magnetic field lines (of strength B_i, say) are carried in towards the reconnection furnace? The main emphasis has been on the steady-state process, partly because it is simpler and partly because the solar flare is essentially a steady-state energy release for many thousands of Alfvén times (though of course modulated in a time-dependent way).

Consider a simple diffusion region of dimensions $2l$ and $2L$ (Figure 2). In a steady state the field lines are carried in at the same speed as they are

154

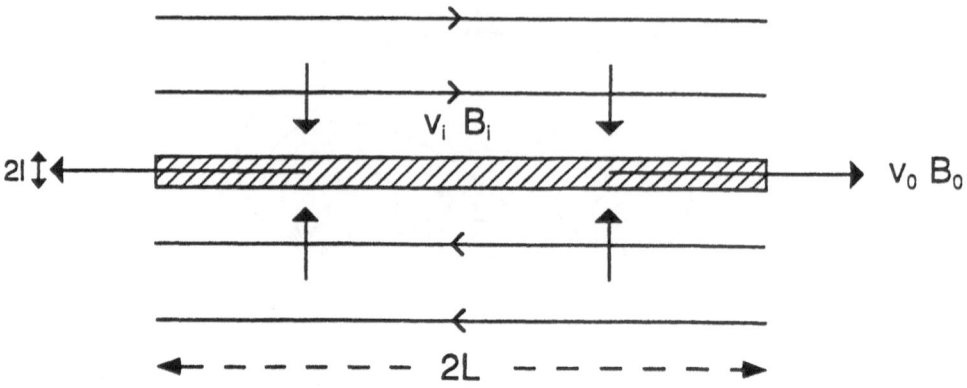

Figure 2.   The notation for a Sweet-Parker current sheet.

trying to diffuse out, so that

$$v_i = \frac{\eta}{l},$$

(1)

where η is the magnetic diffusivity. This is a consequence of Ohm's Law

$$\mathbf{E} + \mathbf{v} \times \mathbf{B} = \frac{\mathbf{j}}{\sigma} = \eta \nabla \times \mathbf{B}$$

and is obtained by equating the orders of magnitude of $\mathbf{v} \times \mathbf{B}$ and $\eta \nabla \times \mathbf{B}$.

The conservation of mass entering at speed v_i over two sides of length $2L$ and leaving at speed v_0 through two ends of width $2l$ gives

$$L v_i = l v_0$$

(2)

when the density is uniform.

Eliminating l between these two Sweet-Parker relations (1) and (2) gives the reconnection rate as

$$v_i = \left(\frac{\eta v_o}{L}\right)^{\frac{1}{2}},$$

(3)

and the sheet half-width (l) and outflow field strength (B_0) follow from (2) and flux conservation

$$v_i B_i = v_o B_o,$$

(4)

respectively, as

$$l = L\frac{v_i}{v_o}, \quad B_0 = B_i\frac{v_i}{v_o}.$$

However, what is the outflow velocity (v_0) that appears in (3)? The equation of motion is

$$\rho(\mathbf{v}.\nabla)\mathbf{v} = \mathbf{j} \times \mathbf{B} - \nabla p, \tag{5}$$

where the current density in the current sheet (from Ampère's Law $\mathbf{j} = \nabla \times \mathbf{B}/\mu$) is approximately

$$j \approx \frac{B_i}{\mu l}.$$

Thus, if the pressure gradient is neglected, the x-component (i.e. along the sheet) of (5) gives in order of magnitude

$$\rho\frac{v_0}{L}v_0 \approx \frac{B_i}{\mu l}B_0.$$

But $\nabla.\mathbf{B} = 0$ implies that

$$\frac{B_0}{l} \approx \frac{B_i}{L}, \tag{6}$$

and so this gives the outflow speed as

$$v_0 = \frac{B_i}{\sqrt{(\mu\rho)}} \equiv v_{Ai}, \tag{7}$$

namely the Alfvén speed.

Then, in terms of the magnetic Reynolds number

$$R_m = \frac{Lv_A}{\eta}$$

and Alfvén mach number

$$M = \frac{v}{v_A},$$

Equation (3) gives in dimensionless form the *Sweet-Parker reconnection rate* as

$$M_i = \frac{1}{R_{mi}^{\frac{1}{2}}}. \tag{8}$$

When the sheet length (L) is identified with the global (external) length-scale (L_e) and appropriate values ($10^6 - 10^{12}$) of the corresponding global magnetic Reynolds number are inserted in (8), we obtain reconnection rates of between 10^{-3} and 10^{-6} of the Alfvén speed.

What is now the effect of a pressure gradient? It too may accelerate the plasma along the sheet from the stagnation point at the centre of the current sheet, and in Figure 2 both the inflow pressure (p_i) and outflow pressure (p_0) may be imposed. When the inflow speed is small enough $(v \ll v_A)$ that the inertial term is negligible and the sheet is thin enough $(l \ll L)$ that the magnetic tension is negligible, the y-component (across the sheet) of the equation of motion (5) becomes

$$0 \approx -\frac{\partial}{\partial y}\left(p + \frac{B_x^2}{2\mu}\right),$$

and so integrating across the sheet gives the pressure at the neutral point as

$$p_N = p_i + \frac{B_i^2}{2\mu}. \tag{9}$$

By comparison, the x-component (along the sheet) is

$$\rho v_x \frac{\partial v_x}{\partial x} = j B_y - \frac{\partial p}{\partial x},$$

and, when evaluated half-way along the sheet, it gives

$$\rho \frac{v_0}{2} \frac{v_0}{L} = \frac{B_i}{\mu l} \frac{B_0}{2} - \frac{p_0 - p_N}{L},$$

or, after substituting for B_0 from (6),

$$v_0^2 = v_{Ai}^2 + \frac{2(p_N - p_0)}{\rho}. \tag{10}$$

The dimensionless version of the resulting reconnection rate (3) after using (8) for p_N is

$$M_i = \frac{[2 + 2\beta_i(1 - p_0/p_i)]^{\frac{1}{4}}}{R_{mi}^{\frac{1}{2}}}. \tag{11}$$

This shows how the values of the inflow plasma beta $(\beta_i = 2\mu p_i/B_i^2)$ and pressure ratio (p_0/p_i) produce departures from the Sweet-Parker rate (8). In particular, when there is no pressure gradient along the sheet $(p_0 = p_N)$ we recover $v_0 = v_{Ai}$ from (10), and when the ambient pressure is uniform $(p_0 = p_i)$ we find $v_0 = \sqrt{2}v_{Ai}$.

Next, let us consider the energetics. For example, the steady-state incompressible MHD equations are

$$\rho(\mathbf{v}.\nabla)\mathbf{v} = -\nabla p + \mathbf{j} \times \mathbf{B}, \qquad \nabla.\mathbf{v} = 0, \tag{12}$$

$$\mathbf{E} + \mathbf{v} \times \mathbf{B} = \frac{j}{\sigma}, \qquad \nabla.\mathbf{B} = 0, \tag{13}$$

$$\mathbf{j} = \frac{\nabla \times \mathbf{B}}{\mu}. \tag{14}$$

Here equations (12) essentially determine p and $\mathbf{v}$, (13) determine $\mathbf{B}$, (14) gives $\mathbf{j}$ and then a heat energy equation determines the temperature. The equations of electromagnetic energy and mechanical energy are derived from the above set and so give no extra information about the physical variables (although they do determine the partition of the energy). Thus, for example, $\nabla \times \mathbf{E} = 0$ together with (13a) and (14) imply

$$-\int \mathbf{E} \times \mathbf{H}.d\mathbf{S} = \int \frac{j^2}{\sigma} + \mathbf{v} \cdot \mathbf{j} \times \mathbf{B}\, dV,$$

which expresses the fact that an inflow of electromagnetic energy produces ohmic heating and work done by the Lorentz force. Since $E \approx v_i B_i$ and $j \approx B_i/(\mu l)$, in order of magnitude it becomes

$$v_i B_i \cdot \frac{B_i}{\mu} \cdot 4L = \frac{B_i^2}{2\mu^2 l^2 \sigma} \cdot 4Ll + \frac{v_0}{2} \cdot \frac{B_i}{\mu l} \cdot B_0 \cdot 4Ll.$$

Using (4) for B_0, this reduces to

$$v_i = \frac{\eta}{2l} + \frac{v_i}{2}$$

or

$$v_i = \frac{\eta}{l}.$$

This is of course the standard result (1), which is not surprising since it too was a consequence of Ohm's Law. However, from two lines above we can deduce that the inflowing electromagnetic energy is converted half into ohmic heat and half into the work done by $\mathbf{j} \times \mathbf{B}$ (which in turn gives purely kinetic energy when there is no pressure gradient). In other words, a Sweet-Parker layer converts magnetic energy into the kinetic energy and heat of hot fast streams of plasma (with equipartition when no work is done by a pressure gradient).

Again, if we take the dot product of $\mathbf{v}$ with (12a) and use (12b), we find

$$\int_S (\tfrac{1}{2}\rho v^2)\mathbf{v}.d\mathbf{S} = -\int_S p\mathbf{v}.d\mathbf{S} + \int_V \mathbf{v}.\mathbf{j} \times \mathbf{B}\, dV,$$

which states that the kinetic energy changes due to the work done by pressure gradients and magnetic forces. In order of magnitude it just gives the previous equations (7) or (10) for v_0.

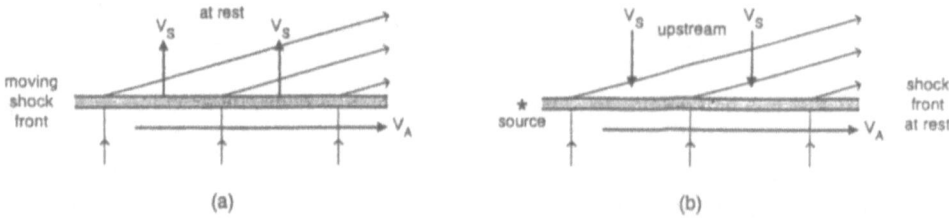

Figure 3. Slow-mode shock wave that is (a) propagating and (b) standing

3. Almost-Uniform Potential Reconnection [Petschek, 1964]

The rate of Sweet-Parker reconnection was recognised to be too slow for solar flares and so the hunt was on for a faster mechanism. At a conference on solar flares, Petschek [13] came up with a solution which was a stroke of genius. Gene Parker, one of the attendees, immediately acclaimed it as the solution they were seeking - at least, he did so after having stayed up late the previous evening probing the details with Petschek privately! Petschek realised that a slow magnetoacoustic shock wave is another way of converting magnetic energy. In the switch-off limit, it propagates into a medium at rest (Figure 3a) with speed

$$v_s = \frac{B_N}{\sqrt{(\mu\rho)}},$$

namely the Alfvén speed based on the normal field component (B_N) to the shock front. It turns the magnetic field towards the normal and so decreases the field strength; at the same time it heats the plasma and accelerates it to the Alfvén speed (v_A), two-fifths of the inflowing magnetic energy going into heat and three-fifths into kinetic energy.

If, on the other hand, the upstream plasma moves (down) at speed v_s, then the shock will be stationary (Figure 3b). If you imagine a film with field lines coming down and passing through the shock, the kink in an individual field line will move to the right - in other words, the shock is

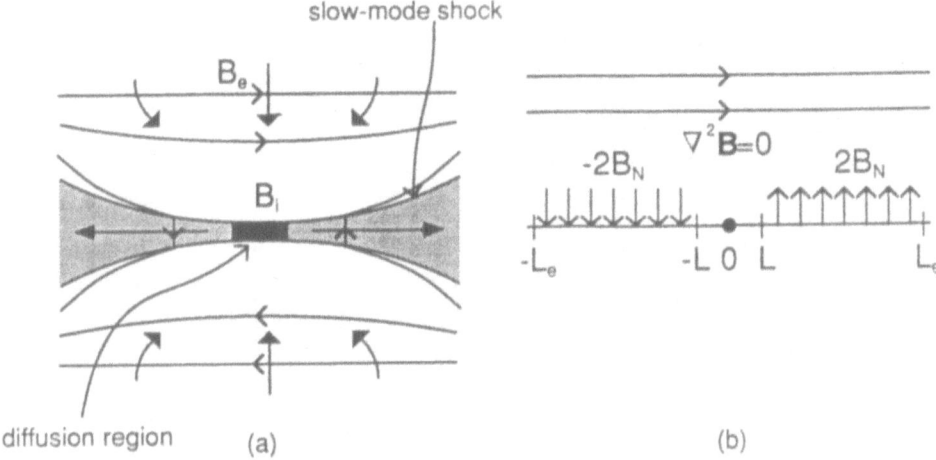

Figure 4. (a) Petschek's mechanism. (b) The domain for calculating the inflow solution

generated by a source at the left-hand end of the shock front. Furthermore, just as a hydrodynamic shock is generated by an airplane when it moves supersonically, so a slow magnetoacoustic shock is generated when plasma moves faster than the slow magnetoacoustic wave speed.

Petschek made the key suggestion that the Sweet-Parker diffusion region could be limited to a small section of the boundary between opposing fields and that such a region would act as a source for four slow magnetoacoustic shock waves which propagate in different directions and stand in the flow when a steady state is reached (Figure 4). "Fast" reconnection has a reconnection rate, M_e, the "external" value of the Alfvén Mach number (M) far upstream, which is much larger than the Sweet-Parker rate (8). Petschek's model is "Almost-Uniform Potential Reconnection" in the sense that the inflow region has a current-free magnetic field which is only slightly curved and is a small perturbation to a uniform field $B_e \hat{x}$, with $M_e (\ll 1)$ being the small parameter. It has a small diffusion region and standing slow-mode shocks which accelerate and heat the plasma and provide most of the energy conversion. As the plasma approaches the diffusion region the magnetic field falls from a value B_e to

$$B_i = B_e(1 - 4(M_e/\pi)\log(L_e/L)). \qquad (15)$$

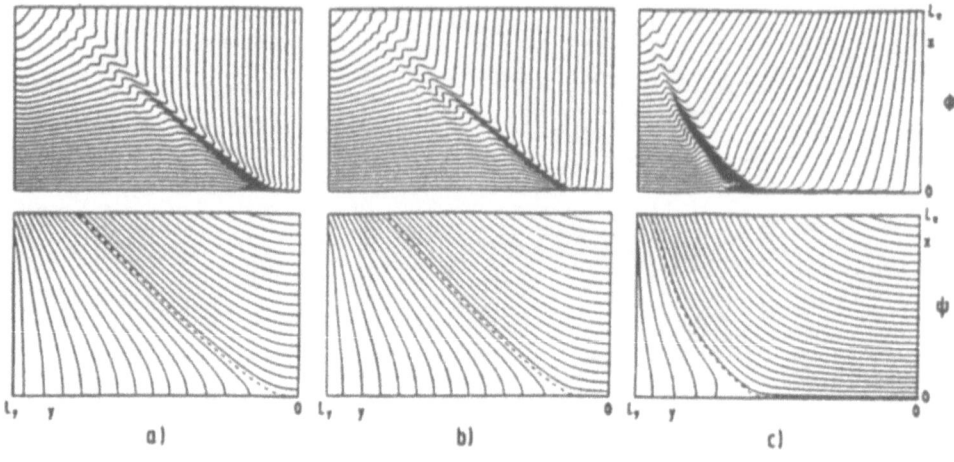

Figure 5. Biskamp's [1] experiment showing streamlines (above) and magnetic field lines (below) in one quadrant for $M_e = 0.042$ and $R_{me} = $ (a) 1746, (b) 3492, (c) 6984.

Flux conservation ($v_i B_i = v_e B_e$) may be written

$$\frac{M_e}{M_i} = \left(\frac{B_i}{B_e}\right)^2,$$

(16)

and the Sweet-Parker relations (1) and (2) with $v_0 = v_{Ai}$ determine the diffusion region dimensions as

$$\frac{L}{L_e} = \frac{1}{R_{me} M_e^{1/2} M_i^{3/2}}, \quad \frac{l}{L_e} = \frac{1}{R_{me} M_e^{1/2} M_i^{1/2}}$$

(17)

in terms of R_{me}, M_e, M_i.

After substituting for B_i and L from (15) and (17) into Equation (16), it determines $M_i(M_e, R_{me})$, and so the scalings (17) reduce to $L/L_e \approx 1/(R_{me} M_e^2), l/L \approx 1/(R_{me} M_e)$, so that L and l decrease as either R_{me} or M_e increases. Also, the maximum reconnection rate is $M_e^* \approx \pi/(8 \log R_{me})$, which follows by putting $B_i = \frac{1}{2} B_e$ in (15). It should be noted that, for the other reconnection regimes that we shall discuss, the procedure is very similar: Equations (16) and (17) still hold, but (15) is replaced by different expressions for the appropriate inflow solution.

Around this time, Gene Parker [12] did something without realising what he was doing - perhaps for the only time in his life! He discovered an exact nonlinear solution of the MHD equations - and very few such

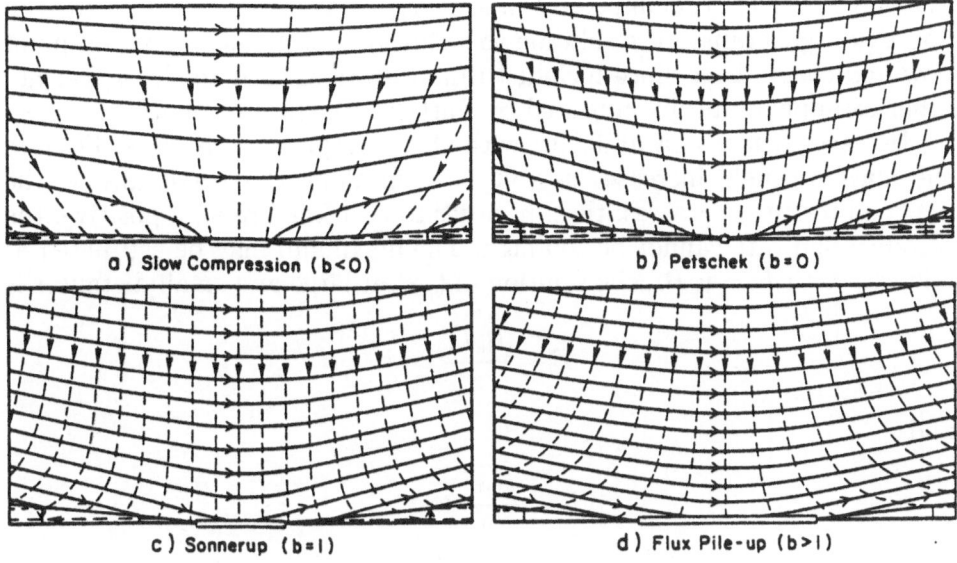

a) Slow Compression (b<0)

b) Petschek (b=0)

c) Sonnerup (b=1)

d) Flux Pile-up (b>1)

Figure 6. Almost-Uniform reconnection regimes

solutions exist. It was for magnetic annihilation of straight field lines under a balance between outwards diffusion and inwards advection. He considered a stagnation-point flow with components

$$v_x = -\frac{V_e x}{L_e} \quad , \quad v_y = \frac{V_e y}{L_e}$$

satisfying $\nabla.\mathbf{v} = 0$. The magnetic field is unidirectional

$$\mathbf{B} = B(x)\hat{\mathbf{y}}$$

and so Ohm's Law becomes

$$E - \frac{V_e x}{L_e}B = \eta \frac{dB}{dx} \tag{18}$$

with E constant for a steady state. This determines the magnetic field $B(x)$ in a kinematic sense for the given flow. However, Sonnerup and Priest [19] later pointed out that this is in fact an exact MHD solution, since it automatically satisfies the equation of motion (since the velocity is irrotational), which simply determines the pressure as

$$p = p_e + \frac{1}{2}\rho(v_e^2 - v^2) + \frac{1}{2\mu}(B_e^2 - B^2)$$

in terms of the external imposed values p_e, v_e, B_e at $x = L_e, y = 0$. Sonnerup and Priest generalised the solution to a three-dimensional stagnation-point flow and others have since included time-variations, and so this is yet another example of a topic initiated by Parker's fruitful imagination.

However, there is a difficulty with this solution, which arises because the magnetic field increases as you come in along the x-axis and so the plasma pressure decreases: thus, in order to keep the pressure positive. the reconnection rate cannot be too high. The maximum may be estimated as follows. With $E = v_e B_e$, the solution of (18) at large and small distances is

$$B = \frac{B_e L_e}{x}$$

and

$$B = \frac{B_e v_e}{\eta} x,$$

respectively. At the location

$$x = \frac{L_e}{R_{me}^{1/2} M_e^{1/2}}$$

where these two expressions are equal, we have the edge of the diffusion layer and the magnetic field reaches a maximum of

$$B_{\max} = B_e R_{me}^{1/2} M_e^{1/2}.$$

Thus, in order to avoid negative plasma pressures, we need the corresponding magnetic pressure to be smaller than the external plasma pressure (p_e), assuming a negligible effect from the inertial term ($\frac{1}{2}\rho v_e^2$). This may be written as a condition for the reconnection rate (M_e) as

$$M_e < \frac{\beta_e}{R_{me}}$$

in terms of the plasma beta (β_e) and magnetic Reynolds number (R_{me}). In other words

$$v_e < \frac{\eta}{L_e}\beta_e,$$

and to attain rates larger than the Sweet-Parker rate ($R_{me}^{-1/2}$) one needs an enormous plasma beta, namely,

$$\beta_e > R_{me}^{1/2}.$$

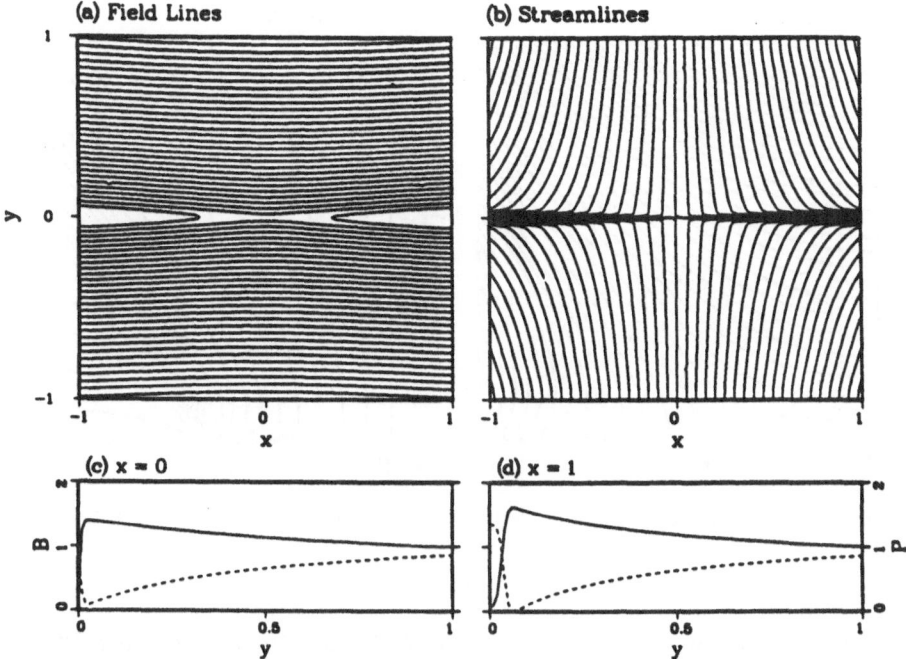

Figure 7. (a) The magnetic field lines and (b) streamlines for Almost-Uniform Reconnection with $b = 2, M_e = 0.1, R_{me} = 2000$. The variation of magnetic field (continuous) and pressure (dashed) along (c) $x = 0$ and (d) $x = 1$ ([21]).

Recent diffusive reconnective solutions of Craig and Henton [3] possess the same disadvantage.

For ten years all settled down and seemed well understood - until, that is, Biskamp [1] and Lee and Fu [9] came along and responded to Parker's wish by conducting some beautiful numerical experiments (Figure 5). In Biskamp's experiment L and l increase as either M_e or R_{me} increase, and so he concluded that Petschek's mechanism (and, therefore, fast reconnection) does not exist for large magnetic Reynolds numbers (R_{me}). By contrast, with different boundary conditions Lee and Fu [9] found L to decrease with increasing R_{me}. Both experiments have several interesting features:

(i) the nature of the inflow varies from weakly converging to strongly diverging streamlines;

(ii) the inflow field lines are highly curved and the shock angle large;

(iii) there are strong jets of plasma flowing out along the separatrices;

(iv) spikes of reversed current are present at the ends of the diffusion region creating reverse curvature in the outflowing field lines.

However, there is now a new generation of theoretical models which generalise Petschek's mechanism in several ways and give at least a par-

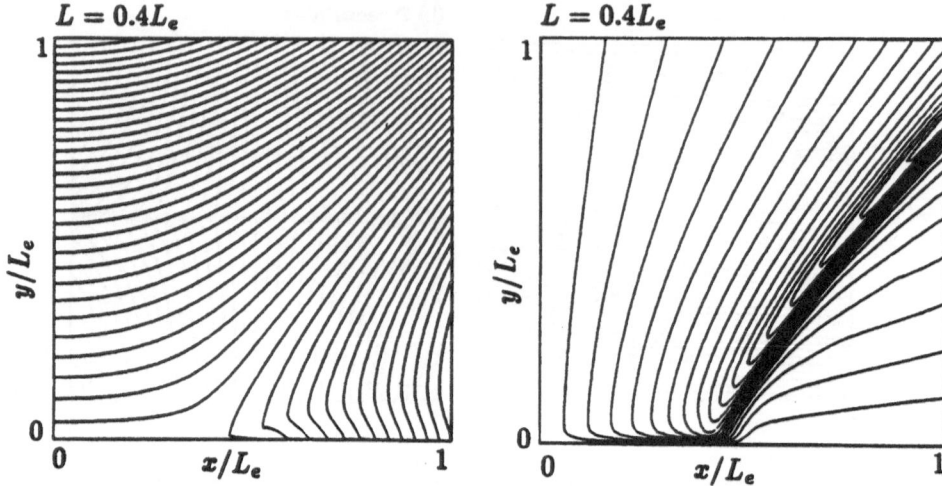

Figure 8. Field lines and streamlines in one quadrant for Non-Uniform Reconnection ([16])

tial explanation for many of these properties. A family of *Almost-Uniform* models is nonpotential, include pressure gradients and is able to reproduce feature (i). Also another family of *Non-Uniform* models reproduces the remaining features (ii) - (iv).

4. Almost-Uniform Non-Potential Reconnection (Priest and Forbes,1986,1992)

Priest and Forbes [15] solve the two-dimensional, incompressible, ideal steady MHD equations in the inflow region by linearising about a uniform field B_e

$$\mathbf{B} = B_e\hat{\mathbf{x}} + \mathbf{B}_1 + ..., \qquad \mathbf{v} = v_e\hat{\mathbf{y}} + ...$$

At first order they find

$$\nabla^2 A_1 = \frac{\mu}{B_e}\frac{dp_1}{dy}$$

for the flux function A_1 such that $(B_{1x}, B_{1y}) = (\partial A_1/\partial y, -\partial A_1/\partial x)$. The solution gives the required expression for the inflow field as

$$B_i = B_e(1 + 16M_e(b-1)\tanh \tfrac{1}{2}\pi(L_e/L)\pi^{-2}\sin(\tfrac{1}{2}\pi L/L_e) + ...) \qquad (19)$$

There are different regimes, depending on the value of the parameter (b), which is such that $v_x(L_e, L_e)$ is proportional to $(b-2/\pi)v_e^2$. When $b < 0$ the inflow is converging and produces a slow-mode compression regime (Fig 2a). When $b = 0$ and $b = 1$ one finds Petschek and Sonnerup-like reconnection, respectively. When $b > 1$, the inflow is diverging, and we have a flux pile-up regime in which $B_i > B_e$ and the diffusion region tends to be long. When $b > 0$ the reconnection rate is faster than Petschek and when $b > 1$ there is no maximum in the reconnection rate within the limitations of the theory. The main result is that the type of regime, the rate of reconnection and the scaling of L all depend on b and therefore on the inflow boundary conditions. If you keep b fixed you find one scaling, but if, say, you keep the value of v_x or the angle of the flow at the corner (L_e, L_e) fixed, then b varies with M_e and the scaling is quite different.

Yan et al [21] have conducted a numerical experiment on Almost-Uniform Reconnection. On the inflow boundary they set $B_x = B_e$, $\partial j/\partial y = 0$, $v_y = -v_e$, $\partial w/\partial y = 0$ and on the outflow boundary they put $\partial B_y/\partial x = 0$, $\partial w/\partial x = 0$ and impose from the Almost-Uniform theory the value of v_y, which depends on b. When $b = 0$ a Petschek flow results; when $b = 1$ Sonnerup-like flow is found; when $b = 2$ they find an example of flux pile-up reconnection (Figure 7) in which it can be seen that, as one approaches the diffusion region along the axis $x = 0$, the magnetic field increases and the plasma pressure decreases in a strong slow-mode expansion. When $b = -3$ they find an example of slow-mode compression reconnection with converging streamlines; the field decreases and the pressure increases as the diffusion region is approached. The configurations and scalings agree with Almost-Uniform Theory, provided the magnetic diffusivity (η) is enhanced in the diffusion region. However, when η is uniform the states are no longer sustained, although the cause has not been identified; one possibility is that the shocks become so thick numerically that they are no longer resolved; another is that the boundary conditions imposed in the numerical experiment were too loose: in particular, it would be interesting to impose the pressure on the outflow boundary in future.

5. Non-Uniform Reconnection

5.1. POTENTIAL RECONNECTION (PRIEST AND LEE, 1992)

In this family of solutions the inflow magnetic field is potential ($j = 0$). Priest and Lee [17] write $(v_x, v_y) = (\partial\psi/\partial y, -\partial\psi/\partial x)$, $(B_x, B_y) = (\partial A/\partial y,$

$-\partial A/\partial x$) and have three parts to their analysis. First, in the upstream region they assume $v \ll v_A$, $c_S \ll v_A$ so that $j = 0$ and choose the magnetic field components as

$$B_y + iB_x = B_i \left(\frac{z^2}{L^2} - 1 \right)^{\frac{1}{2}} \tag{20}$$

in terms of the complex variable $z = x + iy$. This has a current sheet, a cut in the complex plane, stretching between $z = -L$ and $z = L$. The equations $\mathbf{E} + \mathbf{v} \times \mathbf{B} = 0$ and $\nabla \cdot \mathbf{v} = 0$ are then used to deduce $\mathbf{v}$ and ψ. In the second stage the position of the "shock" passing through the end-point $z = L$ is deduced from $\psi + A = \text{constant}$, and the "shock" relations across it are applied. Finally, the downstream flow and field are deduced from the ideal MHD equations together with boundary conditions at the shock and along the outflow boundary. An example when $L = 0.4L_e$ is shown in Figure 8, which exhibits the properties (ii) - (iv) of Biskamp's experiment. One can see the highly curved inflow magnetic field lines, the separatrix jet and the inverse curvature caused by the reversed current spike at the end of the diffusion region.

5.2 NON-POTENTIAL RECONNECTION (PRIEST AND FORBES, 1992)

However, Priest and Lee did not find the same scaling for L as Biskamp. The problem is that, like Petschek's potential solution, there is no extra degree of freedom which allows us to impose the extra particular boundary condition of Biskamp. Nevertheless, we have realised that it is possible to generalise the potential solution to include pressure gradients and so produce a Nonpotential Non-Uniform solution with such a freedom. In the inflow region, the equation of magnetostatic balance ($\mathbf{j} \times \mathbf{B} = -\nabla p$) with $(B_x, B_y) = (\partial A/\partial y, -\partial A/\partial x)$ reduces to

$$\nabla^2 A = -j(A) \tag{21}$$

If the flux function is written as $A = A_p + A_{np}$, where A_p is the potential solution that Priest and Lee used, and A_{np} is an extra nonpotential part, (21) becomes

$$\nabla^2 A_{np} = -j(A_p + A_{np}), \tag{22}$$

which in general is an extremely hard equation to solve analytically because of the non-linear function of $A_p + A_{np}$ on the right-hand side. However, there does exist a useful simple solution with uniform current, so that $j(A) = -cB_e/L$, say, and the field becomes

$$B_y + iB_x = B_i \left(\frac{z^2}{L^2} - 1 \right)^{\frac{1}{2}} - ic\frac{B_e}{L}y. \tag{23}$$

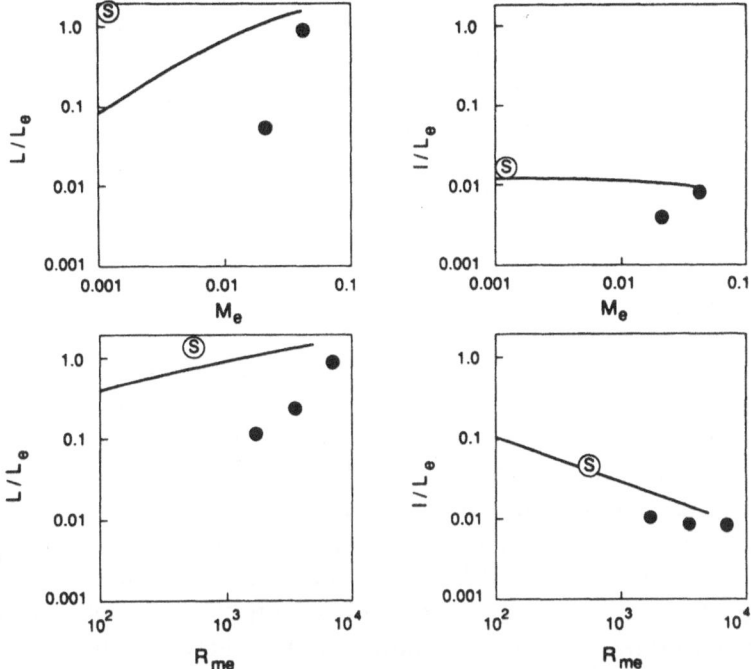

Figure 9. Comparison of scalings of the length (L) and width (l) of the diffusion region with reconnection rate (M_e) and magnetic Reynolds number (R_m) between Non-Uniform Nonpotential theory with the extra Biskamp boundary condition (solid curves) and the numerical experiment (dots).

Here c is a parameter analogous to the parameter b in Almost-Uniform theory: $c < 0$ tends to produce converging flow and $c > 0$ diverging flow. Also, c can be determined by imposing some extra boundary condition. Biskamp imposed $v_y = 0$, $B_x = 0$ on the outflow boundary and $v_y(x)$ on the inflow. His key condition, however, was to prescribe the normal component of the magnetic field, namely $B_y = B_e x/L$, on the inflow boundary. Although we do not have the freedom in our analytical theory to impose the functional form of $B_y(x)$ along this boundary, we can impose the same value as Biskamp at the corner $x = L_e$ where $B_y = B_e$ and use this to determine c. The resulting scalings of L and l with M_e and R_{me} are shown in Figure 9. Since the analytical theory does not have precisely the same boundary conditions as the experiments we do not expect to be able to reproduce the experimental scalings exactly, but they do have very similar qualitative characteristics. Unlike the original Petschek scaling, we now have L increasing with both M_e and R_{me}, albeit not as rapidly as in the experiment. Furthermore, the maximum reconnection rates are in very close agreement with the theoretical and experimental values scaling as $3.4 R_{me}^{-1/2}$

and $3.5R_{me}^{-1/2}$, respectively. Thus we conclude that it is a combination of the X-type background and the fixed value of the normal magnetic field on the inflow boundary which produces these scalings and prevents fast reconnection in Biskamp's experiment. Jin and Ip [8] and Lee and Fu [9] have also presented results on numerical reconnection experiments. Their scalings are different from Biskamp [1] and have also been compared with the theory by Priest and Forbes [16].

6. Conclusions

Gene Parker started us on a voyage of discovery of discovery about the nature of two-dimensional reconnection, for which we are most grateful. The extensive differences in scaling laws of the sheet length and maximum reconnection rate that are found both in theories and in experiments are clearly a direct consequence of the wide variety of boundary conditions. The nature and value of the imposed boundary conditions are crucial, with much physics being hidden within them. The "correct" boundary conditions depend on the particular application of reconnection; spontaneous reconnection due to some localised instability and unaffected by distant magnetic fields would require free boundary conditions and would tend to produce potential reconnection, either Almost-Uniform or Non-Uniform, depending on the initial state; driven reconnection depends on the details of the driving but in general tends to give rise to non-potential reconnection - for instance, diverging flow in an Almost-Uniform state gives fast flux pile-up reconnection.

We have seen that analytical theory and numerical experiment go hand in hand, complementing and stimulating one another. Conditions on both the inflow and outflow boundaries play an important role in determining the nature of the inflow, the length of the diffusion region and the strength of separatrix jets and reversed current spikes. The Almost-Uniform and Non-Uniform regimes are equally valid - they just have different boundary conditions.

Although we have been discussing mainly incompressible models, the inclusion of compressibility is not expected to alter the basic features and conclusions. It has already been incorporated in the Almost-Uniform models, for instance, by Jardine and Priest [7]. Another point is that, when the diffusion region becomes too long, it goes unstable to secondary tearing and an impulsive bursty regime of reconnection ensues (Priest [14]; Lee and Fu [9] and Forbes and Priest [5]). This is particularly likely for the flux pile-up regime.

In conclusion, by adopting boundary conditions similar to Biskamp [1], we have been able to reproduce approximately his scalings. These results therefore suggest that fast reconnection can indeed exist, given appropriate boundary conditions and possibly a locally enhanced magnetic diffusivity, produced for example by current-induced micro-instabilities in the diffusion region. Fast reconnection, either Almost-Uniform or Non-Uniform is then a prime candidate for the rapid energy conversion that is often seen in solar, space and astrophysical plasmas.

Acknowledgement

I am most grateful to Tom Bogdan, Art Hundhausen and Boon Chye Low for their hospitality at High Altitude Observatory where this paper was written.

References

1. Biskamp, D (1986) Magnetic reconnection via current sheets *Phys. Fluids* **29**, 1520–1531.
2. Cowling, T.G. (1953) Solar electrodynamics, in *The Sun* (ed. G. Kuiper), Chapter 8, University of Chicago Press.
3. Craig, I.J.D. and Henton, S.M. (1995) "Exact solutions for steady-state incompressible magnetic reconnection", preprint.
4. Dungey, J.W. (1953) Conditions for the occurrence of electrical discharges in astrophysical systems, *Phil. Mag.* **44**, 725–738.
5. Forbes, T G and Priest, E R (1987) A comparison of analytical and numerical models for steadily driven reconnection *Rev. Geophys* **25**, 1583–1607.
6. Giovanelli, R.G. (1947) Magnetic and electric phenomena in the Sun's atmosphere associated with sunspots, *Mon. Not. Roy. Astron. Soc.* **107**, 338–355.
7. Jardine, M and Priest, E R (1988) Weakly nonlinear theory of fast steady-state magnetic reconnection *J Plasma Phys.* **40**, 143–161.
8. Jin, S P and Ip, W H (1991) Two-dimensional compressible MHD simulation of a driven reconnection process *Phys. Fluids* **B3**, 1927–1936.
9. Lee, L C and Fu, Z F (1986) Multiple x-line reconnection *J Geophys Res.* **91**, 6807–6815.
10. Parker, E.N. (1957) Sweet's mechanism for merging magnetic fields in conducting fluids, *J. Geophys. Res.* **62**, 509–520.
11. Parker, E N (1963) The solar flare phenomenon and theory of reconnection and annihilation of magnetic fields *Ap. J. Supp* **8**, 177–211.
12. Parker, E.N. (1973) Comments on the reconnection rate of magnetic fields, *J. Plasma Phys.* **9**, 49–63.
13. Petschek, H E (1964) Magnetic field annihilation *AAS-NASA Symp. on Phys. of Solar Flares*, NASA SP-50, 425–439.
14. Priest, E R (1986) Magnetic reconnection on the Sun *Mit. Astron. Ges.* **65**, 41–51.
15. Priest, E R and Forbes, T G (1986) New models for fast steady-state magnetic reconnection *J Geophys Res.* **91** 5579–5588.
16. Priest, E R and Forbes T G (1992) Does fast reconnection exist? *J Geophys Res.*, **97**, 16757–16772.
17. Priest, E R and Lee, L C (1990) Nonlinear magnetic reconnection models with separatrix jets *J Plasma Phys.* **44**, 337–360.
18. Scholer, M (1989) Undriven magnetic reconnection in an isolated current sheet *J Geophys Res.*, **94**, 8805–8812.
19. Sonnerup, B.U.O. and Priest, E.R. (1975) Resistive MHD stagnation point flows at

a current sheet *J. Plasma Phys.* **14**, 417–431

20.   Sweet, P A (1958) The neutral point theory of solar flares *IAU Symp.* **6**, 123–134.
21.   Yan, M, Lee, L C and Priest, E R (1992) Fast magnetic reconnection with small shock angles *J Geophys Res.* **97**, 8277–8293.

NEW DEVELOPMENTS IN MAGNETIC RECONNECTION THEORY

E. R. PRIEST
Mathematical and Computational Sciences Department
St Andrews University
ST ANDREWS, KY16 9SS
Scotland

Abstract.

Although two-dimensional reconnection is now fairly well understood, there have recently been some interesting new developments. Four distinct types of reconnection are possible, namely: viscous (or kinematic) reconnection; extra-slow (or linear) reconnection; slow (or Sweet-Parker) reconnection; and fast (Almost-uniform or Nonuniform) reconnection. An antireconnection theorem has been discovered which states that steady two-dimensional reconnection (with flow across the separatrices) is impossible for slow inviscid flow (and it has also been generalised to three dimensions). Solutions for linear reconnection have been presented and a new theory for the self-consistent time-dependent collapse of an X-point to form a reconnecting current sheet has been developed.

The new field of three-dimensional reconnection is only just beginning. Schindler et al have suggested a concept of general magnetic reconnection and Priest et al have proposed a model for "magnetic flipping". A magnetic null in three dimensions has a skeleton consisting of a spine and a fan, which are respectively an isolated field line and a surface of field lines that pass through the null point. At such a null, reconnection may occur either by "spine reconnection" or by "fan reconnection", in which singular resistive behaviour occurs either at the spine or the fan, respectively. When no null points are present, reconnection can still occur at so-called "quasi-separatrix surfaces", which are regions where the mapping of field lines has extremely steep gradients and the field lines become disconnected from the plasma.

K. C. Tsinganos (ed.), Solar and Astrophysical Magnetohydrodynamic Flows, 171–194.
© *1996 Kluwer Academic Publishers.*

1. Introduction

The theory of reconnection, which Gene Parker played a large part in initiating, is now highly developed and so I can only touch on a few aspects
here. In two dimensions, it is fairly well understood. The classical regimes of
Sweet-Parker [33] and Petschek [17] have been replaced by a new generation
of fast regimes. These include Almost-Uniform Reconnection (Priest and
Forbes [20]; Jardine and Priest [13]), with weakly curving inflow fields and
a variety of sheet lengths and reconnection rates, and Nonuniform Reconnection (Priest and Lee [24]) with highly curved field lines and separatrix
jets. Fast reconnection has been shown convincingly to exist and there is
now good agreement (Priest and Forbes [23]) between the theories and
numerical experiments (Biskamp [2]; Yan et al [35, 36]).

The first of the two main equations of MHD is the equation of motion

$$\rho \frac{d\mathbf{v}}{dt} = -\nabla p + \mathbf{j} \times \mathbf{B}, \tag{1}$$

which expresses the fact that plasma is accelerated by pressure gradients
and magnetic forces ($\mathbf{j} \times \mathbf{B}$), where the current ($\mathbf{j}$) is given by Ampère's
law

$$\mathbf{j} = \frac{\nabla \times B}{\mu}. \tag{2}$$

The second main equation is the *induction equation*,

$$\frac{\partial \mathbf{B}}{\partial t} = \nabla \times (\mathbf{v} \times \mathbf{B}) + \eta \nabla^2 \mathbf{B}, \tag{3}$$

where $\eta = (\mu\sigma)^{-1}$ is the *magnetic diffusion coefficient*. It states that magnetic fields change in time due to two effects (on the right) namely transport
of the magnetic field with the plasma and diffusion through it. The ratio
of these two effects is the magnetic Reynolds number VL/η, which is enormous in most of the universe (typically $10^6 - 10^{12}$ in the Sun), so that
the magnetic field is frozen to the plasma to a very high approximation.
The exception is in singularities, where magnetic field lines may break and
reconnect and magnetic energy is released.

For steady two-dimensional reconnection, the integral of (3) is Ohm's
Law

$$\mathbf{E} + \mathbf{v} \times \mathbf{B} = \eta \nabla \times \mathbf{B}, \tag{4}$$

where the electric field ($\mathbf{E}$) is uniform and constant. In addition, if viscous
forces are also included, (1) becomes

$$\rho(\mathbf{v}.\nabla)\mathbf{v} = -\nabla p + \mathbf{j} + \mathbf{B} + \rho\nu\nabla^2\mathbf{v} \tag{5}$$

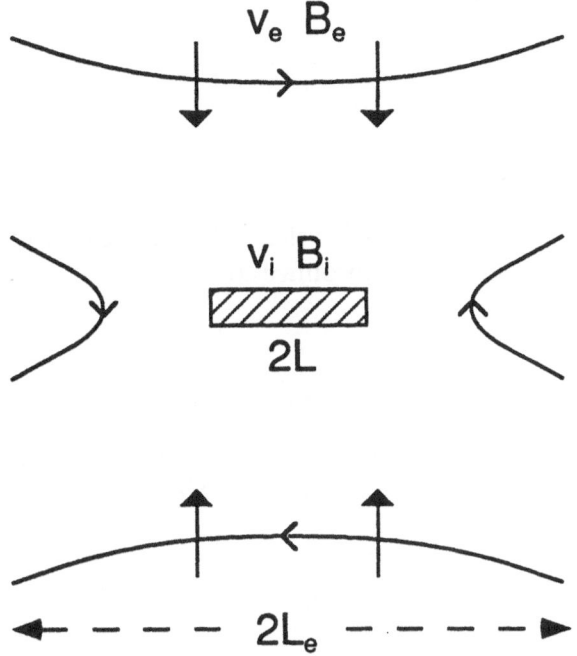

Figure 1. The global nomenclature for reconnection

when the flow is incompressible (with ρ uniform). Furthermore, if we consider a finite region of size L_e and suppose the flow speed and magnetic field at the boundary of the region are v_e and B_e, respectively, then there are four typical speeds

$$v_e, \qquad v_{Ae} = \frac{B_e}{\sqrt{(\mu\rho)}}, \qquad v_{de} = \frac{\eta}{L_e}, \qquad v_{\nu e} = \frac{\nu}{L_e},$$

namely the flow speed, Alfvén speed, global magnetic diffusion speed and global viscous diffusion speed, respectively. Three dimensionless speeds may therefore be formed such as

$$M_e = \frac{v_e}{v_{Ae}}, \qquad R_{me} = \frac{v_{Ae}}{v_{de}}, \qquad Re = \frac{v_{Ae}}{v_{\nu e}},$$

namely the Alfvén Mach number, magnetic Reynolds number and Reynolds numbers (based on the Alfvén speed). Then the basic equations (4) and (5) may be nondimensionalised by writing

$$\mathbf{v}^* = \frac{\mathbf{v}}{v_e}, \mathbf{B}^* = \frac{\mathbf{B}}{B_e}, E^* = \frac{E}{v_e B_e}, p^* = \frac{p}{B_e^2/\mu}, \mathbf{j}^* = \frac{\mathbf{j}}{B_e/(\mu L_e)}, \nabla^* = L_e \nabla,$$

so that (4) and (5) become

$$\mathbf{E}^* + \mathbf{v}^* \times \mathbf{B}^* = \frac{1}{R_{me}M_e}\nabla^* \times \mathbf{B}^*, \tag{6}$$

$$(\mathbf{v}^*.\nabla^*)\,\mathbf{v}^* = -\frac{1}{M_e^2}\nabla^* p^* + \frac{1}{M_e^2}\mathbf{j}^* \times \mathbf{B}^* + \frac{1}{R_e M_e}\nabla^{*2}\mathbf{v}^*. \tag{7}$$

Four different types of reconnection [27] may be considered as follows:
(i) Viscous (or kinematic) reconnection
when

$$M_e \ll \frac{1}{R_e}, \quad 1,$$

so that viscosity dominates with (7) determining $\mathbf{v}$ and (6) giving $\mathbf{B}$;
(ii) Extra-slow (or linear) reconnection [4]
when

$$M_e \ll \frac{1}{R_{me}}, \quad 1,$$

so that force balance (7) determines $\mathbf{B}$, while (6) gives $\mathbf{v}$;
(iii) Slow (Sweet-Parker) reconnection
when

$$\frac{1}{R_{me}} < M_e < \frac{1}{R_{me}^{\frac{1}{2}}};$$

(iv) Fast (almost-uniform or nonuniform) reconnection
when

$$M_e > \frac{1}{R_{me}^{\frac{1}{2}}}.$$

2. Two-Dimensional Magnetic Reconnection Theory

Here I want to describe two recent developments in two-dimensional theory

2.1. LINEAR RECONNECTION

In two dimensions, the simplest *null point* (where the magnetic field vanishes) has magnetic field components

$$B_{0x} = y, \quad B_{0y} = x, \tag{8}$$

and magnetic field lines which are rectangular hyperbolae and asymptote to the separatrices $y = \pm x$. Now the simplest procedure is to linearise about such a field and ask what is the nature of the resulting slow reconnection (taking the Alfven Mach number $M_A = v/v_A \ll 1$)?

The linearised Ohm's law and equation of motion are then

$$\mathbf{E}_1 + \mathbf{v}_1 \times \mathbf{B}_0 = \frac{\mathbf{j}_1}{\sigma}, \tag{9}$$

$$0 = \mathbf{j}_1 \times \mathbf{B}_0 - \nabla p_1. \tag{10}$$

where $\mathbf{j}_1 = \nabla \times \mathbf{B}_1/\mu$ and $\nabla.\mathbf{B}_1 = 0$. There was considerable difficulty finding reconnection solutions to these equations numerically and then we discovered an *Anti-reconnection Theorem* [27], which states that steady two-dimensional reconnection (with flow across the separatrices) is impossible for slow, inviscid flow. The proof is simple: in two dimensions $\nabla \times \mathbf{E}_1 = 0$ implies that $\mathbf{E}_1$ is constant; then the curl of (10) implies that

$$(\mathbf{B}_0.\nabla)\mathbf{j}_1 = 0,$$

so that $\mathbf{j}_1$ is constant along field lines; but $\mathbf{E}_1 = \mathbf{j}_1/\sigma$ at the X-point from (9) and so $\mathbf{E}_1 = \mathbf{j}_1/\sigma$ all along a separatrix (a field line through the X-point); therefore, by (9), $\mathbf{v}_1 \times \mathbf{B}_0 = 0$ along the separatrix and so no field lines are carried across it - i.e. there is no reconnection (with trans-separatrix flow).

Consider the possible incompressible solutions of (9) and (10) in more detail by writing

$$\mathbf{v} = \nabla \times (\psi \hat{\mathbf{z}}), \quad \mathbf{B} = \nabla \times (A\hat{\mathbf{z}}),$$

in terms of the stream function (ψ) and flux function (A), so that (4) and (1) become

$$E + (\mathbf{B}.\nabla)\psi = -\eta\nabla^2 A, \tag{11}$$
$$\rho(\mathbf{v}.\nabla)\psi = (\mathbf{B}.\nabla)j, \tag{12}$$

where the electric current and vorticity are $j = -\nabla^2 A/\mu$ and $\omega = -\nabla^2\psi$. Now a trivial solution is a potential field with no flow

$$\mathbf{v}_0 = 0 \quad, \quad \mathbf{j}_0 = 0,$$

such as the field of a simple null point

$$A_0 = \tfrac{1}{2}r^2 \sin 2\phi,$$

for which

$$B_{0x} = y \quad , \quad B_{0y} = x,$$

when r is measured from the origin and ϕ from the separatrix $y = x$. Now linearise for $M_e = v_e/v_{Ae} \ll 1$ about this solution by writing

$$A = A_0 + M_e A_1 + \ldots, \quad \psi = M_e \psi_1 + \ldots$$

so that (11) and (12) become

$$
\begin{aligned}
E_1 + (\mathbf{B_0} . \nabla)\psi_1 &= -\eta \nabla^2 A_1, \\
0 &= \mathbf{B_0} . \nabla j_1.
\end{aligned}
\tag{13}
$$

When $\eta = 0$, the appropriate solution of (13) is

$$\psi_1 = \tfrac{1}{2} \log | \tan \phi |,$$

so that

$$v_{1r} = \frac{1}{r \sin 2\phi}, \quad v_{1\phi} = 0,$$

which represents inflow for $\pi/4 < \phi < \pi/2$ and outflow for $0 < \phi < \pi/4$, but it has a singularity on the separatrix at $\phi = 0$. Craig and Rickard [5] therefore resolved the singularity by including a nonzero magnetic diffusivity, for which the solution of (13) is

$$\psi_1 = \tfrac{1}{2}(1 - j_1(A_0)/R_{me}) \log | \tan \phi |$$

and the current $j_1(A_0)$ is chosen to be peaked at the separatrix.

However, the difficulty with this solution is that, in agreement with the antireconnection theorem, the magnetic field lines diffuse across the separatrices rather than being carried by the flow as required by classical reconnection.

One way to try and overcome the theorem is to include viscosity. This is technically very tough but we have managed to find self-similar solutions (Priest et al [27]) , as shown in Figure 2 for several values of the small parameter $\epsilon = \nu\eta/(L_e{}^2 v_A{}^2)$. As ϵ decreases, so the width of the current peak at the separatrix reduces like $\epsilon^{1/4}$.

Another way to overcome the theorem is to introduce nonlinearity by allowing Alfvenic flow. This is what happens in the classical regimes of Sweet and Parker and Petschek and the newer Almost-Uniform and Nonuniform regimes.

2.2. X-POINT COLLAPSE

The collapse of an X-point due to the motions of the distant sources of magnetic field was first proposed by Dungey [7] and, more recently, Craig and McClymont [3] have included resistivity for small oscillations about an X-type equilibrium. Green [10] and Somov and Syrovatsky [31] used complex variable theory to deduce the resulting equilibrium state with a current sheet that may develop from an X-type field. Thus, if one starts with an X-point field

$$B_y + iB_x = z \equiv x + iy, \qquad (14)$$

it may evolve into a field

$$B_y + iB_x = \left(z^2 + L^2\right)^{1/2} \qquad (15)$$

with a cut in the complex plane (a current sheet) stretching from $z = -iL$ to $z = iL$ and Y-points at the ends (middle of Figure 2). It may, however, instead evolve to the field

$$B_y + iB_x = \frac{z^2 + a^2}{(z^2 + L^2)^{1/2}} \qquad (16)$$

with reverse currents and singularities near the ends of the sheet where the field becomes infinte (right of Figure 2). Thus the question arises: does the X-type field (14) collapse to the form (15), as has commonly been assumed, or to (16), or to some other state?

The above complex-variable solutions model a slow evolution through a series of equilibria. Recently, however, we have made some progress on this question (Priest, Titov and Rickard, [26]). We have discovered some nonlinear, self-similar, compressible solutions for dynamic time-dependent formation of a current sheet from an X-point. When the plasma velocity is much greater than the sound speed and much less than the Alfvén speed the dimensionless equation of motion is of the form

$$\epsilon\rho\frac{D\mathbf{v}}{Dt} = \mathbf{j} \times \mathbf{B}. \qquad (17)$$

Expanding $\mathbf{v}$ and $\mathbf{B}$ in powers of ϵ we then have in two dimensions a series of states with

$$\mathbf{j} = \mathbf{0} \qquad (18)$$

around the sheet to lowest order. The plasma velocity ($\mathbf{v}_\perp$) perpendicular to the magnetic field is given from the motion of the field lines, i.e. from the induction equation

$$\frac{\partial \mathbf{B}}{\partial t} = \nabla \times (\mathbf{v} \times \mathbf{B}). \qquad (19)$$

178

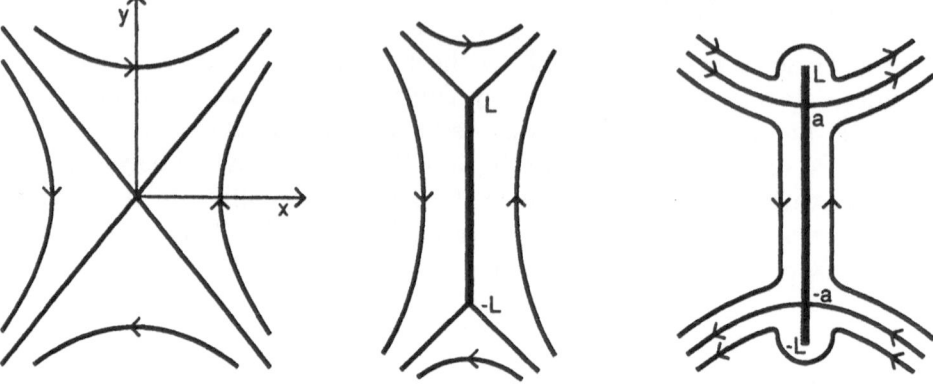

Figure 2. X-point Collapse

Also the velocity parallel to the magnetic field is determined by the condition that $D\mathbf{v}/Dt$ be perpendicular to the zeroth-order magnetic field; this is a consequence of the first-order part of (17). What this means physically is that, as the magnetic field lines move, they rotate and the flow along them is given by a balance between the Coriolis and centrifugal forces associated with the rotation of the field lines. The potential magnetic field sequence may not, however, be imposed at will: it has to be determined in a self-consistent way from (18), (19) and

$$\frac{D\mathbf{v}}{Dt} \cdot \mathbf{B} = 0.$$

The simplest solution we have found of this type has magnetic and velocity components

$$B_y + iB_x = \frac{\frac{1}{2}z + \sqrt{\frac{1}{4}z^2 - t}}{2(\frac{1}{4}z^2 - t)},$$

$$v_x + iv_y = \frac{1}{\frac{1}{2}z + \sqrt{\frac{1}{4}z^2 - t}},$$

where $z = x + iy$ is the usual complex variable. Individual plasma elements converge on the x-axis along straight lines (Figure 3a) and they force a current sheet of length $4\sqrt{t}$ to form (Figure 3b). As the magnetic field collapses, the current sheet grows in length and the magnetic dissipation increases. The ends of the sheet move with speed $1/\sqrt{t}$. They swallow up

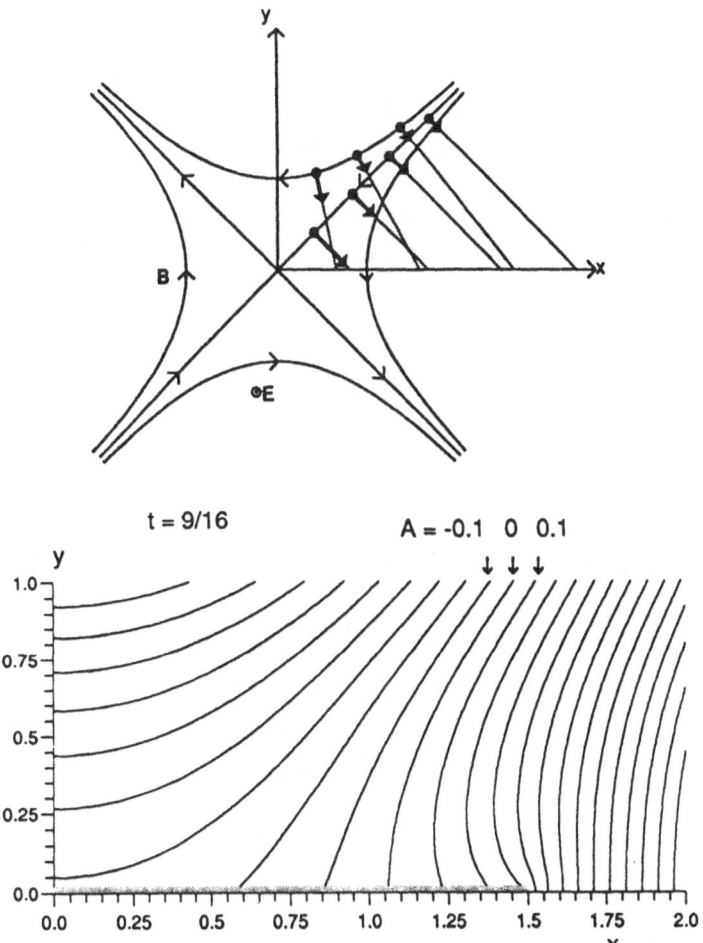

Figure 3. Magnetic field lines at $t = 0$ and 9/16 (Priest, Titov and Rickard, 1995)

part of the magnetic flux and cause the remainder of it to pile up around a region of reversed current.

3. Three-Dimensional Reconnection

A general magnetic configuration in two dimensions (Figure 4) contains *separatrix curves*, which separate the plane into topologically distinct regions, in the sense that all the field lines in one region start at a particular source and end at a particular sink. The separatrices intersect in an *X-point* where the field vanishes and the field is locally hyperbolic. Reconnection then occurs by the breaking and rejoining of field lines at the X-point and a transfer of flux across the separatrices from one topological region to another. In three dimensions, some configurations have similar properties, with *separatrix surfaces* separating the volume into topologically different regions. They intersect each other in a *separator*, a field line which ends at null points or on the boundary. A start has recently been made on trying to understand 3D reconnection: for example, Schindler et al [29] have proposed a general theory, while Priest and Forbes [21], Lau and Finn [14] have focussed on null points, and Priest and Forbes [22] have set up a theory for magnetic flipping in the absence of null points. Here we outline some recent progress that Priest and Titov [28] and Priest and Demoulin [19] have been making on reconnection at nulls and in the absence of nulls, respectively.

3.1. RECONNECTION AT NULL POINTS

The simplest null point has field components

$$(B_x, B_y, B_z) = (x, y, -2z) \tag{20}$$

or, in cylindrical polars

$$(B_r, B_\phi, B_z) = (R, 0, -2z),$$

which satisfy $\nabla . \mathbf{B} = 0$ and whose field lines from

$$\frac{dx}{B_x} = \frac{dy}{B_y} = \frac{dz}{B_z}$$

are

$$y = cx, \quad z = \frac{k}{x^2}.$$

Two quite distinct families of field lines pass through the null point. The *null spine* is an isolated field line which approaches the null from above and below along the z-axis: neighbouring field lines form two bundles which spread out as they approach the xy-plane. The *null fan* is a surface of field lines (the xy-plane) which spread out from the null. Thus the spine and the fan form the skeleton of the field lines near the null.

The kinematics of steady reconnection may be studied by solving

$$\nabla \times \mathbf{E} = 0 \tag{21}$$

and

$$\mathbf{E} + \mathbf{v} \times \mathbf{B} = 0. \tag{22}$$

Equation (21) implies that the electric field can be written as $\mathbf{E} = -\nabla \Phi$ in terms of a potential (Φ) and then (22) implies that $(\mathbf{B} \cdot \nabla)\Phi = 0$ or

$$\Phi = \Phi(c, k), \tag{23}$$

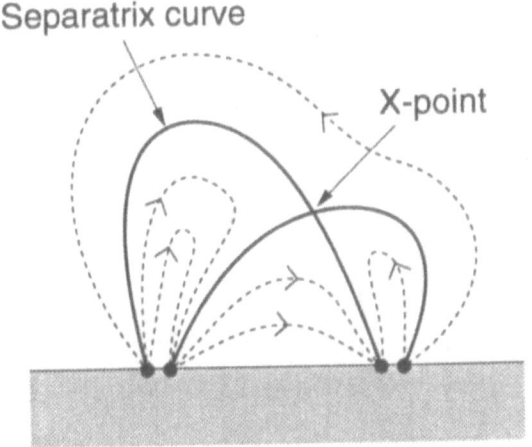

(a)

(b)

Figure 4.   (a) Separatrix curves in 2D. (b) Separatrix surfaces in 3D.

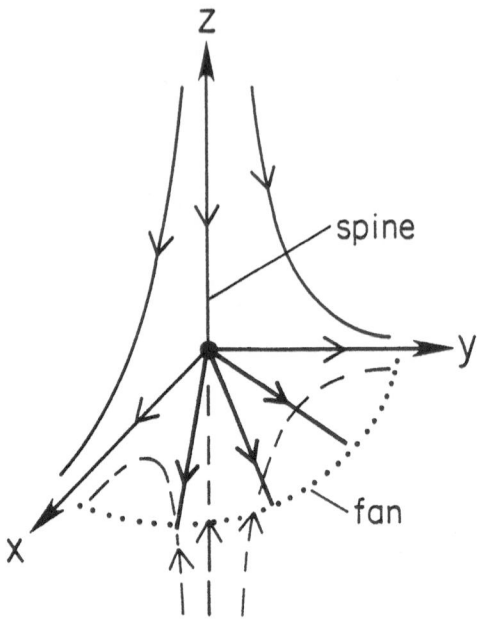

Figure 5. The structure of a 3D null point

where c and k are constants describing a field line. Also (22) implies that the velocity normal to the magnetic field is

$$\mathbf{v}_\perp = \frac{\mathbf{E} \times \mathbf{B}}{B^2}. \qquad (24)$$

Thus the boundary conditions determine $\Phi(c, k)$ and so $\mathbf{E}$ and $\mathbf{v}_\perp$ follow. This approach has been applied to a single null, a pair of nulls and fields without nulls, as follows.

First of all, let us consider a 3D null point and ask what is the nature of reconnection near such a point? How do magnetic flux surfaces reconnect? If you impose continuous footpoint motions on any cylindrical surface surrounding the spine and crossing the fan such as $R = 1$, then a singular motion is driven along the spine axis, so we refer to such a process as *spine reconnection*. Consider field lines in any vertical plane and suppose the footpoints on $R = 1$ move down on the right and up on the left. Then the field lines approach the null, break and reconnect, while the other ends move across the top and bottom of the cylinder through the spine. Suppose the same process occurs in all other vertical planes but modulated in ϕ: then

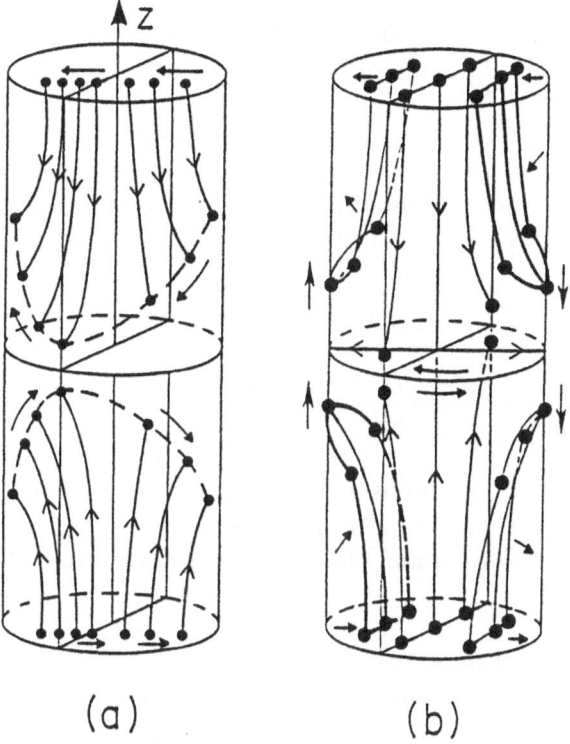

(a)   (b)

Figure 6. Motion of field lines and flux surfaces in fan reconnection

what are the implications for flux surfaces formed from footpoints that lie initially on the top edge and then move down? The other ends of the field lines (on the top or bottom surface) move in and through the spine. The flux surface therefore moves in and touches the null, forming a fold along the spine. It then reconnects with an oppositely placed flux surface and unfurls from the spine like a cylindrical bubble. A simple solution for the equations of the flux surfaces is

$$R^2 z = \pm 1 - t \sin \phi \qquad (25)$$

and a simple solution for the field line velocity is

$$(v_{\perp R}, v_{\perp \phi}, v_{\perp z}) = \frac{E_0(\phi)}{R(R^2 + 4z^2)}(2z, 0, R). \qquad (26)$$

It can be seen that a singularity exists all along the spine ($R = 0$) and so a key question is whether it can be resolved by diffusion.

If instead you impose continuous motions on surfaces (such as $z = \pm 1$) that cross the spine, then singular behaviour is driven at the fan surface.

Suppose the footpoints on the top move in a straight line from right to left (Figure 6). Then the other ends twirl around the z-axis like a swirling skirt. Consider a flux surface made of field lines whose footpoints march across the top in a straight line: the flux surface distorts and becomes a vertical surface plus a semicircle. It then breaks and reconnects with a similar flux surface on the opposite side of the null point. During this process, magnetic field lines rotate rapidly in one direction above the fan and in the opposite direction below it. A simple solution has field line velocity

$$(v_{\perp x}, v_{\perp y}, v_{\perp z}) = \frac{1}{(x^2 + y^2 + z^2)(4 + y^2 z)^{3/2}} \tag{27}$$
$$\left(\frac{2xy(z^3 - 1)}{z^{1/2}}, \frac{2(x^2 + 4z^2 + y^2 z^3)}{z^{1/2}}, (4 + y^2 z + x^2 z) y z^{1/2} \right),$$

from which we note that there is a singularity at the fan ($z = 0$), and so again the question is: can it be resolved by diffusion?

In the simplest case, namely linear reconnection, the answer is no, and it leads to an

Antireconnection Theorem:
Steady MHD reconnection in 3D with convective plasma flow across the spine or fan of a radial null point is impossible in an inviscid plasma with a highly subAlvénic flow and a uniform magnetic diffusivity.

The implication of the theorem is that probably nonlinearity is required for such reconnection.

4. Reconnection without Null Points

When there is a null point, two-dimensional reconnection is associated with the fact that the mapping of field lines from one footpoint to another is discontinuous. For example, with the simple X-point field

$$B_x = x, \quad B_y = -y \tag{28}$$

the point (x_0, y_0) on a boundary will map to (x_1, y_1) in such a way that, when (x_0, y_0) crosses a separatrix, the point (x_1, y_1) suddenly jumps in location (Figure 7). In a classic paper Schindler et al [29] realised that, when no nulls (or bald patches, [34]) are present, the mapping is continuous, so that separatrices do not exist and the 2D concept of reconnection based on flux transfer across separatrices is no longer applicable. They proposed instead a concept of "general reconnection" to include all effects of local nonidealness that produce a component ($E_{\parallel}$) of electric field parallel to

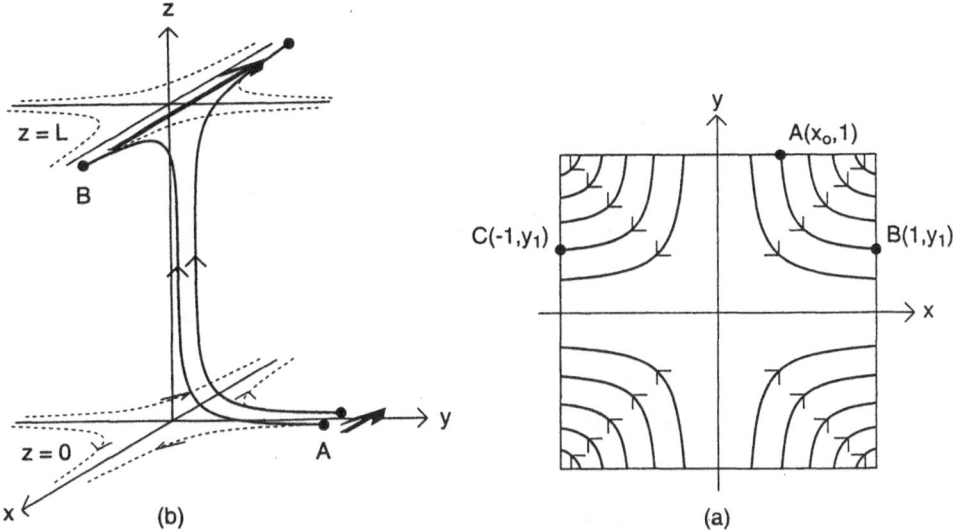

Figure 7. The mapping of footpoints for (a) a 2D X-field and (b) a 3D sheared X-field

the magnetic field. Furthermore, Priest and Forbes [21] suggested a local definition of reconnection in terms of potential singular lines, while Priest and Forbes [22] proposed a concept of *magnetic flipping* and Lau and Finn [14] analysed kinematic reconnection.

Here I shall report on a new theory for reconnection without nulls (Priest and Demoulin, [19]) which builds on the above ideas. It has four steps for investigating reconnection in a given 3D configuration:

(i) first, the volume of consideration is surrounded by a closed surface S;

(ii) then the mapping of footpoints of field lines from one part of S to another is calculated; for instance, if a small component ($B_z = l \leq 1$) is added to (28) to create a sheared X-field the mapping becomes continuous, so that, as the point (x_0, y_0) crosses the y-axis in the plane $z = 0$, say, the other end (x_1, y_1) in the plane $z = 1$ moves continuously (Figure 7);

(iii) next, so-called *quasi-separatrix layers* (QSL) are identified where the gradients of the mapping are very large;

(iv) finally, reconnection occurs when there is a breakdown of ideal MHD and a change of connectivity of plasma elements: this takes place in quasi-separatrix layers where the field line velocity greatly exceeds the plasma velocity - it may be driven by slow regular boundary motions of footpoints (across a QSL).

It may be noted that these definitions of a QSL and magnetic reconnection involve a mapping to a boundary and therefore refer to global properties of a configuration. The concept of a quasi-separatrix layer may be defined formally as follows. Split the surface into parts S_0 and S_1 where the field lines enter and leave the volume, respectively, and set up orthogonal coordinates (u, v) in S and w normal to S. Then field lines map (u_0, v_0) in S_0 to (u_1, v_1) in S_1. Next, form the *displacement gradient tensor*

$$F = \begin{pmatrix} s_1 \partial u_1/\partial u_0 & s_2 \partial u_1/\partial v_0 \\ s_3 \partial v_1/\partial u_0 & s_4 \partial v_1/\partial v_0 \end{pmatrix} \tag{29}$$

from the gradients of the mapping functions $u_1(u_0, v_0)$ and $v_1(u_0, v_0)$ and the scaling factors s_i and evaluate the norm

$$N = \sqrt{\left[\left(s_1 \frac{\partial u_1}{\partial u_0}\right)^2 + \left(s_2 \frac{\partial u_1}{\partial v_0}\right)^2 + \left(s_3 \frac{\partial v_1}{\partial u_0}\right)^2 + \left(s_4 \frac{\partial v_1}{\partial v_0}\right)^2\right]}. \tag{30}$$

Finally, define a quasi-separatrix layer as the region where

$$N \gg 1.$$

The properties of F are as follows. First of all, a difference δu_0 and δv_0 in footpoint positions maps to

$$\begin{pmatrix} \delta u_1 \\ \delta v_1 \end{pmatrix} = F \begin{pmatrix} \delta u_0 \\ \delta v_0 \end{pmatrix}.$$

Secondly, a surface element dS_0 transforms to

$$dS_1 = J dS_0,$$

where $J = s_1 s_4 (\partial u_1/\partial u_0)(\partial v_1/\partial v_0) - s_2 s_3 (\partial u_1/\partial v_0)(\partial v_1/\partial u_0)$ is the Jacobian. Thus flux conservation $(B_1 dS_1 = B_0 dS_0)$ implies that

$$B_1 = \frac{B_0}{J},$$

where J is finite and nonzero if the field has no nulls or singularities. Thirdly, the displacement gradient tensor may be written as the product

$$F = RU$$

of one matrix (R) representing a rotation through ϕ and another (U) representing a stretching by λ_+ (an eigenvalue) along $\mathbf{e}_+$ (an eigenvector) together with a compression by λ_- along $\mathbf{e}_-$. Thus a quasi-separatrix layer (where $N \gg 1$) is associated with a large expansion along one direction and a large compression along the other, such that N is approximately equal to the largest eigenvalue

$$N \approx \lambda_{\max}.$$

Consider as an example the sheared X-field

$$(B_x, B_y, B_z) = (x, -y, l) \tag{31}$$

inside a cube with $l \ll 1$. The mapping from the base S_0 to the top and sides (S_1) is given by

$$x_1 = x_0 e^{z_1/l} \quad , \quad y_1 = y_0 e^{-z_1/l}. \tag{32}$$

Thus, when the point $A(x_0, y_0, 0)$ on S_0 is so close to the y-axis that $2x_0 < \epsilon$, A maps to a point B on the top ($z_1 = 1$) and

$$F = \begin{pmatrix} \epsilon^{-1} & 0 \\ 0 & \epsilon \end{pmatrix},$$

while

$$N \approx \frac{1}{\epsilon},$$

where

$$\epsilon = e^{-1/l} \ll 1.$$

On the other hand, when $\epsilon < 2x_0 < 1$, A maps to C on the side ($x_1 = \frac{1}{2}$) and the elements of F and the value of N are of order unity. The resulting variations of x_1, y_1, z_1, N with x_0 are shown in Figure 8, which reveals the quasi-separatrix layer as a very narrow region of width ϵ where $N \gg 1$. When $l = 0.1$ the value of N in the QSL is 10^4, and even when l is as large as 0.3, N is about 28 in the QSL. If the cube is replaced by a hemisphere

188

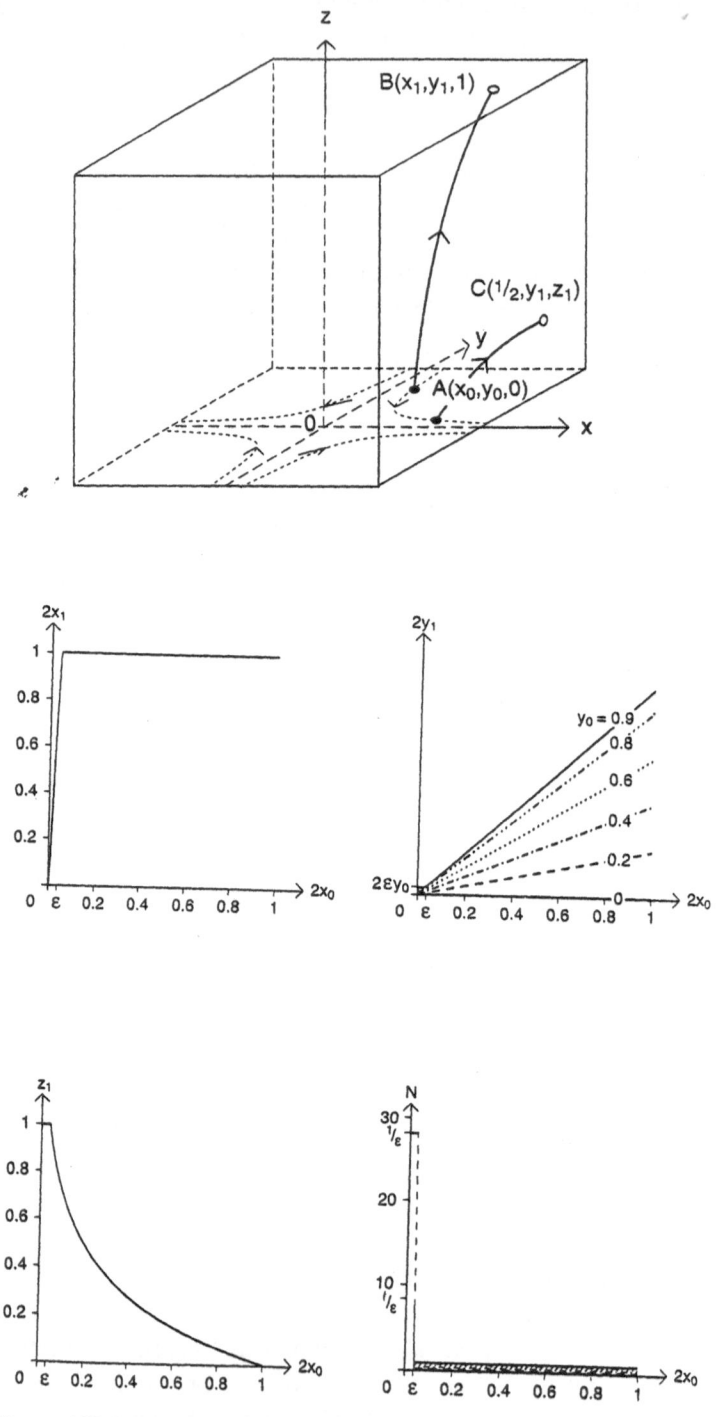

Figure 8. Sheared X-field in a cube together with the variations of x_1, y_1, z_1 and N with x_0 and y_0

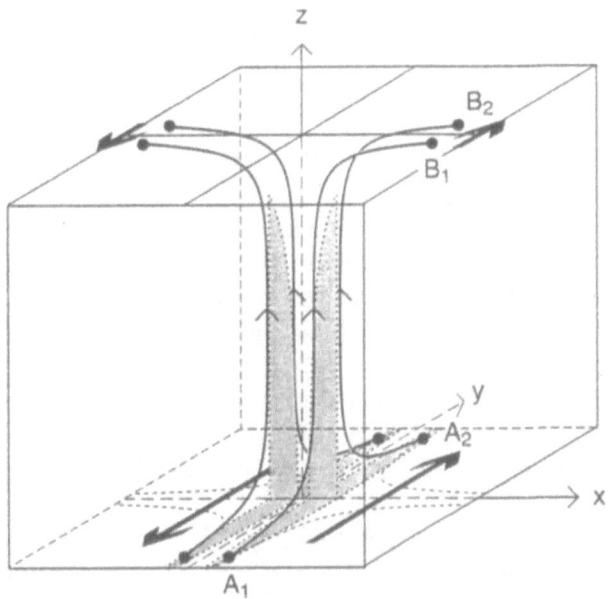

Figure 9. Quasi-separatrix layers (shaded) produced by footpoint motion on the top side of a cube.

or sphere similar forms are produced but the functions become continuous and differentiable.

Having located a quasi-separatrix layer, we may now consider kinematic reconnection satisfying (21) and (22) and producing a potential (23) and field line velocity (24). Thus suppose we impose the field line velocity components $v_{\perp 1x}$ and $v_{\perp 1y}$ on the top side ($z = 1$) of the cube and deduce the function $\Phi(x_1, y_1)$ together with $\mathbf{E}$ and $\mathbf{v}_\perp$ throughout the cube. The resulting electric field on the base ($z = 0$) of the cube has components

$$E_{x0} = \frac{\partial \Phi}{\partial x_1}\frac{\partial x_1}{\partial x_0} + \frac{\partial \Phi}{\partial y_1}\frac{\partial y_1}{\partial x_0},$$

$$E_{y0} = \frac{\partial \Phi}{\partial x_1}\frac{\partial x_1}{\partial y_0} + \frac{\partial \Phi}{\partial y_1}\frac{\partial y_1}{\partial y_0},$$

which depend partly on the electric field components on the top ($E_{x1} = \partial\Phi/\partial x_1$, $E_{y1} = \partial\Phi/\partial y_1$) and partly on the gradients of the mapping functions ($x_1(x_0, y_0)$) and $y_1(x_0, y_0)$). Thus $\mathbf{E}_0$ is large where the gradients of the mapping are large, namely in the quasi-separatrix layer. For example, if on the top ($z = 1$) and side ($x = \frac{1}{2}$) of the cube we impose

$$v_{\perp 1x} = 0, \quad v_{\perp 1y} = v_0 x_1$$

and

$$v_{\perp 1x} = 0, \quad v_{\perp 1y} = \tfrac{1}{2}v_0,$$

respectively, then the resulting velocity on the base $(z = 0)$ along the x-axis $(y = 0)$ is

$$
v_{\perp y 0} = \begin{cases} \frac{v_0 x_0}{\epsilon^2} & \text{if} \quad |x_0| < \frac{1}{2}\epsilon \\ \frac{v_0}{4 x_0} & \text{if} \quad x_0 > \frac{1}{2}\epsilon \end{cases} .
$$

This peaks at $x_0 = \frac{1}{2}\epsilon$ with a value of $v_0/(2\epsilon)$ and so, if this peak value exceeds the Alfvén speed, there will exist two diffusive layers centred on $x_0 = \pm\frac{1}{2}\epsilon$ where the field lines are unfrozen and flip rapidly through the plasma. In other words the field lines move quicker than the plasma and become disconnected from it.

The above approach has also been applied to models of three-dimensional twisted flux tubes, such as are thought to exist for many solar flares (Demoulin, Priest and Lonie [6]). They take the form

$$
\begin{aligned}
B_x &= -(z - a)^2 + b^2(1 - y^2/c^2) \\
B_y &= d \\
B_z &= x
\end{aligned}
$$

where a, b, c, d are constants. The configuration is shown diagrammatically in Figure 10 with three types of field line, namely below the flux tube, within the tube and above it. Also shown are examples of calculated quasi-separatrix layers, with ends that curl up like an umbrella handle and become increasingly complex as the twist increases. The connectivity of points in the layers is also indicated in the bottom right-hand panel.

5. Reconnection in Solar Coronal Heating

One important example of reconnection in the Sun is a large solar flare (e.g. Priest, [18]; Forbes, [8]); but another is the coronal heating problem - indeed, it has recently been demonstrated that part of the corona is highly likely to be heated directly by reconnection.

The corona has a three-fold structure of coronal holes, coronal loops and x-ray bright points, which was revealed by soft x-ray images from Skylab. In these pictures, the active regions above sunspot groups are rather fuzzy and the bright points are mainly just unresolved points of emission. However, recently, the remarkable NIXT photographs from rocket flights (at a temperature of 2-3 x 10^6K and with a five-times better spatial resolution of 500 km) have shown that active regions consist of many fine-scale loops and that bright points appear to include several interacting loops.

X-ray bright points (XBP) were discovered in 1970 from rocket images and were studied with Skylab by Golub et al [9]. They are situated above

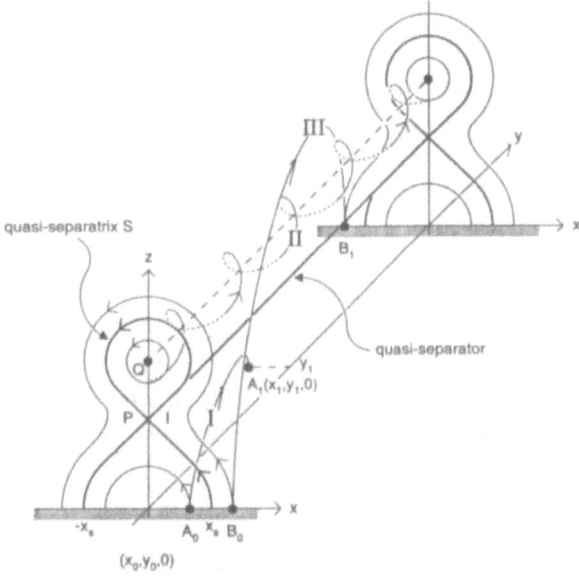

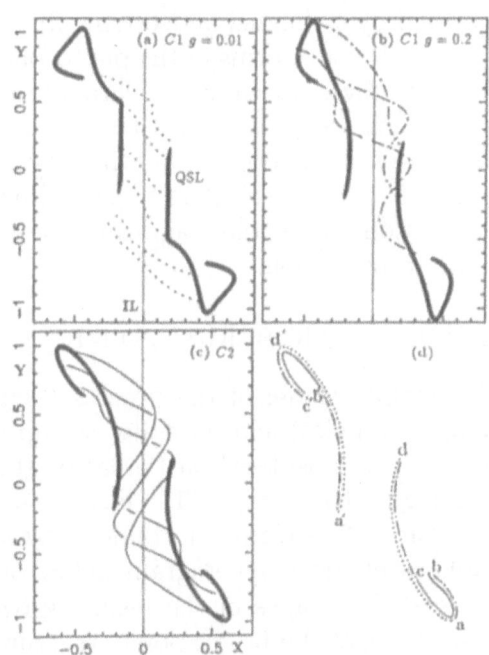

Figure 10. Twisted flux tube showing three types of field lines I, II, and III. Quasi-separatrix layers viewed from above together with sample field lines of type I (dotted), II (solid) and III (dash-dotted) and the connectivity of points on the quasi-separatrix layers.

pairs of opposite polarity magnetic fragments in the photosphere. It was natural to assume that the magnetic fragments represent emerging flux and this became the standard explanation for bright points (Heyvaerts et al [12]). However, Harvey [11] showed that two-thirds of the bright points instead lie above so-called "cancelling magnetic features" (CMF), where pairs of photospheric magnetic fragments are approaching and cancelling (Martin [15]).

So what is happening in an XBP/CMF event? It cannot be simple submergence of magnetic flux since this would not explain the brightening (which starts well before cancellation). Also, no chromospheric fibrils join the fragments and the fragments are initially widely separated (and therefore unconnected). We need to include the effect of the ambient magnetic field to which the fragments are initially connected.

5.1. CONVERGING FLUX MODEL

We have proposed a Converging Flux Model (Priest et al [25]), which has three phases. In the Pre-interaction Phase a pair of oppositely directed magnetic fragments in the photosphere are unconnected and approach one another. They are separated by a channel of overlying flux, which is squeezed by the approach until a null point forms in the photosphere. In the Interaction Phase the null point moves upwards and coronal reconnection creates an x-ray bright point, whose structure consists of two newly reconnected and heated flux tubes, one a small loop linking the fragments and the other a large loop (as seen in NIXT and Yohkoh images) linking to distant locations. In the Cancellation Phase the fragments come into contact and cancel by photospheric reconnection.

5.2. APPLICATION TO SPECIFIC BRIGHT POINTS

We have developed a simple model of the process (regarding the photospheric sources as magnetic poles) and have also set up a numerical MHD experiment. Finally, we have considered particular bright points that were observed on the NIXT flight of July 11, 1991 in Fe XVI at 2-3 x 10^6K. In the full disc image there is an active region surrounded by a collection of five bright points. The photospheric magnetogram below one of them shows four discrete sources and the shapes of the resulting overlying field lines compare well with the shape of the bright point structures which resemble the wings and body of a beautiful sea bird. When the central source moves, flux is transferred between the regions and the field lines that have reconnected brighten to give shapes that compare well with the bright point. Indeed, coronal reconnection driven by footpoint motion may produce the restructuring events seen by Yohkoh. It may also represent an elementary heating event that could heat not just bright points but also other coronal loops and even coronal holes (Axford [1]) such as the nanoflares of Parker, Sturrock and Poletto and Kopp.

6. Conclusion

Magnetic reconnection theory is at a very lively stage at present. In two dimensions new theories for the collapse of an X-point to form a current sheet and for linear reconnection have recently been set up. However, the three-dimensional nature of reconnection promises to be a topic for much exploration in future. At null points reconnection may take place by spine reconnection or fan reconnection. Also in the absence of null points reconnection may occur within quasi-separatrix layers where the norm of the displacement gradient tensor is large. Finally, the three-dimensional applications to the solar corona are beginning to be explored, both in solar flares and coronal heating.

Acknowledgements

I am delighted to acknowledge fruitful discussions with Pascal Demoulin, Terry Forbes and Slava Titov and financial support from the United Kingdom Particle Physics and Astronomy Research Council. I am also most grateful to Tom Bogdan, Art Hundhausen and Boon Chye Low for hospitality at the High Altitude Observatory where this paper was written.

References

1. Axford, W. I. (1993) The origin of the solar wind *Proc. 2nd SOHO Workshop*, Elba, ESA, in press.
2. Biskamp, D. (1986) Magnetic reconnection via current sheets *Phys. Fluids* **29** 1520–1531
3. Craig, I.J.D. and McClymont, A.N. (1991) Dynamic magnetic reconnection at an X-type neutral point *Astrophys. J.* **371**, L41–L44
4. Craig, I.J.D. and McClymont, A.N. (1993) Linear theory of fast reconnection at an X-type neutral point *Astrophys. J.* **405**, 207–215
5. Craig, I.J.D. and Rickard, G.J. (1993) Fast magnetic reconnection and the coalescence instability *Phys. Fluids* **B5**, 956–964
6. Demoulin, P., Priest, E.R. and Lonie, D.P. (1995) 3D magnetic reconnection without null points. 2 Application to twisted flux tubes, *Astron. Astrophys.*, in press.
7. Dungey, J.W. (1953) Conditions for the occurrence of electrical discharges in astrophysical systems *Phil. Mag.* **44**, 725–738
8. Forbes, T. G. (1991) Magnetic reconnection in solar flares *Geophys. Astrophys. Fluid Dyn.* **62**, 15–36
9. Golub, L., Krieger, A.S., Silk, J., Timothy, A. and Vaiana G. (1974) Solar x-ray bright points *Astrophys. J.* **189**, L93–L97
10. Green, R.M. (1965) Models of annihilation and reconnection of magnetic fields *IAU Symp.* **2**, 398–404

194

11.   Harvey, K.L. (1984) Solar cycle variation of ephemeral active regions *Proc. 4th European Meeting on Solar Phys.* ESA SP 220, 235–236

12.   Heyvaerts, J., Priest, E.R. and Rust, D.M. (1977) An emerging flux model for the solar flare phenomenon *Astrophys. J.* **216**, 123–137

13.   Jardine, M. and Priest, E.R. (1988) Weakly nonlinear theory of fast steady-state magnetic reconnection *J Plasma Phys.* **40**, 143–161

14.   Lau,Y.T. and Finn, J.M. (1990) 3D kinematic reconnection in the presence of field nulls and closed field lines *Astrophys. J.* **350** 672–691

15.   Martin, S.F. (1984) in *Small-Scale Dynamical Processes in Stellar Atmospheres* (ed S. Keil) Sac. Peak Observatory, p30.

16.   Parnell, C., Golub, L. and Priest, E.R. (1993) The 3D structure of X-ray bright points *Solar Phys.* **151**, 57–74

17.   Petschek, H.E. (1964) Magnetic field annihilation *AAS-NASA Symp. on Phys. of Solar Flares* NASA SP-50, 425–439

18.   Priest, E.R. (1991) The magnetohydrodynamics of energy release in solar flares *Phil. Trans. Roy. Soc. Lond. A.* **336**, 363–380

19.   Priest, E.R. and Demoulin, P. (1995) 3D magnetic reconnection without null points. 1. Basic theory of magnetic flipping, *J. Geophys. Res.*, in press.

20.   Priest, E.R. and Forbes, T.G. (1986) New models for fast steady state reconnection *J. Geophys. Res.* **91** 5579–5588

21.   Priest, E.R. and Forbes, T.G. (1989) Steady reconnection in three dimensions *Solar Phys.* **119** 211–214

22.   Priest, E.R. and Forbes, T.G. (1992) Magnetic flipping – reconnection in three dimensions without null points *J. Geophys. Res.* **97** 1521–1531

23.   Priest, E.R. and Forbes, T.G. (1992b) Does fast magnetic reconnection exist? *J. Geophys. Res.* **97**, 16757–16772

24.   Priest, E.R. and Lee, L.C. (1990) Nonlinear magnetic reconnection models with separatrix jets *J Plasma Phys.* **44** 337–360

25.   Priest, E.R., Parnell, C.E. and Martin, S.F. (1994) A converging flux model of an X-ray bright point and an associated cancelling magnetic feature *Astrophys. J.* **427** 459–474

26.   Priest, E.R., Titov, V.S. and Rickard, G.K. (1995) The formation of magnetic singularities by nonlinear time-dependent collapse of an X-type magnetic field *Phil. Trans. Roy. Soc. Lond.*, **351**, 1–37.

27.   Priest, E.R., Titov, V.S., Vekstein, G.E., and Rickard, G.J. (1994) Steady linear X-point magnetic reconnection *J. Geophys. Res.* **99**, 21467–21480.

28.   Priest, E.R. and Titov, V.S. (1995) Magnetic reconnection at 3D null points, *Phil. Trans. Roy. Soc.*, in press.

29.   Schindler, K., Hesse, M. and Birn, J. (1988) General magnetic reconnection, parallel electric fields and helicity *J. Geophys. Res.* **93** 5547–5557

30.   Shibata, K., Ishido, Y., Acton, L., Strong, K., Hirayama, T., Uchida, Y., McAllister, A., Matsumoto, R., Tsuneta, S., Shimizu, T., Hara, H., Sakurai, T., Ichimoto, K., Nishino, Y. and Ogawara, Y. (1992) Observations of X-ray jets with the Yohkoh soft X-ray telescope *Pub. Astron. Soc. Japan* **44**, L173–L180

31.   Somov, B.V. and Syrovatsky, S.I. (1976) Hydrodynamic plasma flow in a strong magnetic field *Proc. Lebedev. Phys. Inst.* **74**, 13–72

32.   Strong, K., Harvey, K., Hirayama, T., Nitta, N., Shimizu, T. and Tsuneta, S. (1992) Observations of the variability of coronal bright points by the soft x-ray telescope on Yohkoh *Pub. Astron. Soc. Jap.* **44**, L161–L166

33.   Sweet, P.A. (1958) The neutral point theory of solar flares *IAU Symp.* **6**, 123–134

34.   Titov, V.S., Priest, E.R. and Demoulin, P. (1993) Conditions for the appearance of 'bald patches' at the solar surface *Astron. Astrophys.*, **276**, 564–570

35.   Yan, M., Lee, L.C. and Priest, E.R. (1992) Fast magnetic reconnection with small shock angles *J. Geophys. Res.* **97** 8277–8293

36.   Yan, M., Lee, L.C. and Priest, E.R. (1993) Magnetic reconnection with large separatrix angles *J. Geophys. Res.* **98**, 7593–7602.

HYDRODYNAMIC AND MAGNETOHYDRODYNAMIC TURBULENCE: A RAPID OVERVIEW

ANNICK POUQUET
OCA–CNRS 1362
B.P. 229 F–06304
NICE Cedex 04
FRANCE
pouquet@obs–nice.fr

Abstract. This text aims at being a soft introduction to hard turbulence. A rapid trip through some of the key concepts in turbulence today is given, with their possible consequences for applications to a few selected topics, in particular the solar environment (intermittency in the solar wind, and heating of the solar corona), and the role of turbulence in the interstellar medium. Several one–dimensional models are discussed, and the MHD case is particularly emphasized.

1. Introduction

Some of the key concepts relevant to turbulence, together with a few examples taken from MHD as can be encountered in astrophysical flows are illustrated in this short review on fluid and MHD turbulence. What is turbulence made of? Since the times of Kolmogorov and in fact before, we see it as being a superposition of three kinds of structures: (i) large–scale (L_0) energy–containing eddies which are not universal but reflect both the instability mechanisms that give rise to them and the boundary conditions that contain them (although nonlinearities can build excitation at scales larger than L_0, see §5.4); (ii) inertial–range structures that transport the energy in a loss–less fashion; and (iii) small–scale dissipative eddies. The actual mechanisms whereby a large–scale blob of vorticity transforms into a dissipative scale are largely unknown, except for the initial exponential phase, although some ideas are emerging (more on that later).

This constant flux engine (to within intermittency corrections, see §3.1 and §3.2) leads to a power–law distribution of energy among modes $E(k) \sim$

K. C. Tsinganos (ed.), Solar and Astrophysical Magnetohydrodynamic Flows, 195–216.
© *1996 Kluwer Academic Publishers.*

k^{-n}, with n depending presumably on the problem at hand; for example $n = 5/3$ for the classical three–dimensional Navier–Stokes (NS) incompressible turbulence (Kolmogorov, 1941), and $n = 3/2$ for MHD turbulence and waves (Iroshnikov, 1963; Kraichnan, 1965; hereafter referred to as the IK phenomenology; see *e.g.* Pouquet (1996) for a recent review). This engine to produce small scales stems from the nonlinearity of the primitive equations: Fourier transforming them leads to a convolution term, *i.e.* mode coupling. The 5/3 law for incompressible flows can be recovered from simple dimensional analysis by stating that the mean flux of energy to small scale $\bar{\epsilon} = dE^V/dt \sim v^3/l$ is constant (here, the sensity is constant and can be taken equal to unity), and evaluating v^3 through the relationship $v^2 \sim kE^V(k)$. The boundary between weak and strong turbulence on the one hand, and between cascading and shock formation on the other hand is somewhat fuzzy, in part because power–law solutions for the energy spectra obtain *i.e.* because in all cases a multitude of modes are actually excited, and also because of the presence of coherent structures in both cases (either shocks or vorticity filaments).

One dimensional models of nonlinear behavior are treated in Section 2 (with an application to the problem of the mono–sidedness of proto–stellar jets), the two–dimensional case in Section 4 (with an application to the heating of the solar corona), and the three–dimensional case in Section 5, where the dynamo problem (generation of *both* small–scale and large–scale magnetic fields) is also briefly mentioned. Section 3 deals mostly with intermittency (*i.e.* the scarcity of strong small–scale structures, and the departures from Gaussian probability distribution functions, as observed in the solar wind, in solar magnetic fields and in the interstellar medium), and Section 6 describes succintly a plausible model for the interstellar medium at the scale of the kiloparsec. Section 7 is a brief conclusion.

2. The one–dimensional case

2.1. A FLUID MODEL

There are basically three cartoons that can be drawn for nonlinear mode coupling: a velocity discontinuity due to nonlinear steepening; a shock smoothed by dissipation; and a balance between steepening and dispersive effects, leading to soliton–like behavior. In order to see this, let us write the following *ad hoc* one–dimensional equation:

$$\frac{\partial u}{\partial t} + A_1 u \frac{\partial u}{\partial x} = A_2 \nu \frac{\partial^2 u}{\partial x^2} + A_3 \chi \frac{\partial^3 u}{\partial x^3} + A_4 \mu \frac{\partial^4 u}{\partial x^4} , \qquad (1)$$

and let us consider various cases according to what the A_i parameters are. Of course, the left–hand side of the above equation, with $A_1 = 1$, is the Lagrangian derivative (following the motion of the fluid particle), which is thus the 'natural" choice for this parameter. However, we shall see below that for one particular case (the Korteweg de Vries equation), another choice may be convenient for setting the equation in its "classical" form. As for the other A_i ($i \neq 1$) parameters, there are either 1 or 0, depending on what equation we want to examine in the following.

When $A_i \equiv 0 \ \forall i \neq 1$, and $A_1 = 1$, the temporal evolution of the velocity starting from smooth initial conditions leads to the development of a shock (velocity discontinuity) forming in a time L_0/U_0 where L_0 and U_0 are respectively the characteristic scale and velocity difference in the system. Of course this discontinuity will be prevented either by viscosity ($A_2\nu \neq 0$) or by dispersion ($A_3\chi \neq 0$). For example with $A_3 = A_4 = 0$ and $A_1 = A_2 = 1$, one has Burgers' equation. Its solution can be found by changing it to a linear equation, using the Hopf–Cole transformation; writing that the velocity derives from a *scalar* potential:

$$u(x,t) = \partial_x s(x,t)$$

leads to $s_t + u^2/2 = \nu u_x + f(t)$. Choosing now for the potential $s(x,t)$:

$$s(x,t) = -2\nu ln\Psi(x,t) \ ,$$

we obtain:

$$\Psi_t = \nu\Psi_{xx} - \Psi f(t)/(2\nu) \ .$$

In order to arrive at the desired heat equation, we have to eliminate the last term in the above equation; to that effect, we use the gauge freedom of the transformation between $u(x,t)$ and its potential. Indeed, the fields $\tilde{s}(x,t) = -2\nu\tilde{\Psi}(x,t)$ with $\tilde{\Psi}(x,t) = \Psi(x,t)G(t)$, and $s(x,t)$ yield the *same* physical velocity field $u(x,t)$. We now have

$$\Psi_t = \nu\Psi_{xx} - \Psi[\dot{G}(t)/G(t) + f(t)/(2\nu)] \ .$$

The heat equation obtains with the choice of gauge $\dot{G}(t)/G(t) = -f(t)/(2\nu)$.

We thus see that in the dissipative case, an internal boundary layer forms of thickness $\sqrt{\nu}$ (the solution in the vicinity of the smoothed shock is locally proportional to $\tanh(x/2\nu)$). We conclude that the actual ratio of excited length scales in the problem varies as $R_V^{1/2}$ (and not R_V) where

$$R_V = U_0 L_0/\nu$$

is the (kinetic) Reynolds number, with ν the kinematic viscosity. Other power laws in this scaling can arise. For example, Kolmogorov scaling gives

$R_V^{+3/4}$, obtained by equating at the dissipative scale $\ell = \ell_D$ the viscous time ℓ^2/ν and the eddy turn–over time

$$t_\ell \sim \ell/v_\ell \ ,$$

hence the Kolmogorov length $\ell_D^{(K)} \sim R_V^{-3/4}$ (in units of L_0). In MHD, the length analogous to the Kolmogorov scale within the framework of the IK phenomenology (that is, with an energy spectrum $E(k) \sim k^{-3/2}$) leads to a $R_V^{+2/3}$ scaling.

On the other hand, returning to equation (1) and retaining only nonlinearity and dispersion ($A_1 = 6$, $A_2\nu = A_4\mu = 0$, $A_3\chi = -1$), i.e. writing the Korteweg de Vries equation or in short KdV, a different solution emerges, namely $u = sech^2(\xi/2)$. Here, the change of variable into a traveling wave solution

$$u(x,t) = \frac{1}{2}c_0 f(\xi)$$

with

$$\xi = c_0^{1/2}(x - c_0 t)$$

is useful (we have anticipated that the amplitude of the solitary wave is a function of its speed). You can check that this is a solution by integrating once, multiplying by f', and integrating again; by supposing that for $\xi \to \infty$, the function and its derivative vanish, you arrive at the relationship $[f']^2 = f^2(-f+1)$, hence the $1/cosh^2$ solution. This corresponds to a soliton – propagating the faster the stronger – in which there is exact balance between dispersion and nonlinear steepening (see e.g. Drazin and Johnson (1989) for an introduction). We can simply remark that soliton equations have an infinite number of invariants. Following Kruskal (1974), we write $I_n = \int T_n(x)dx$ where I_n is the nth invariant of flux X_n, viz.:

$$\partial_t T_n + \partial_x X_n = 0 \ .$$

The first invariant $T_1 =< u >$ is the momentum and the second invariant $T_2 =< u^2/2 >$ the energy; $T_3 =< u^3/3 - 6\partial_x u^2 >$. These invariants can be generated by the Miura transformation for even powers of ϵ (the odd powers do not lead to independent invariants):

$$\partial_t w + \partial_x \left(w^2/2 + (\epsilon^2/18)w^3 + 6(\partial^2 w/\partial x^2) \right) = 0$$

with $u = w + i\sqrt{6}\epsilon\partial_x w + (\epsilon^2/6)w^2$. Finally, note that the solitonic solutions to the KdV equation can be built by searching the extremum of I_n subject to the constraints that all I_p are constant $\forall p < n$ (recall that, from the theorem of Noether (1918), conservation equations can be deduced from the invariance of a Lagrangian system). Note that, for the KdV equation with a friction term $\nu'u$, self–organization at large scale of solitons take place, a phenomenon that does not occur when the usual Laplacian term is used (Hasegawa, 1985).

With $A_3\chi = 0, A_1 = 1$ and $A_2 = A_4 = -1$, one has the Kuramoto–Sivashinsky equation where the ν term is a forcing term (involving the velocity field itself) and the highest derivation term is dissipative.

All such one–dimensional equations have been studied in their own right, in part because they emerge naturally in a variety of physical (and other) problems – using *e.g.* asymptotic techniques. They can be viewed as well as prototypes for competing effects in non–linear phenomena; see for example the large body of literature in the context of magnetospheric physics (*e.g.* Spangler, 1990; 1994 and references therein) for the Derivative NonLinear Schrodinger equation or DNLS, as applied *e.g.* to the Earth's bow shock; in the context of the interstellar medium, see Adams, Fatuzzo & Watkins (1994); and for the thermal instability and the dynamics of fronts, see Elphick, Regev, and Shaviv (1992).

In MHD, the following equations were proposed by Thomas (1968, 1970) with $\delta = 1$ as a model for MHD interactions in the spirit of the Burgers' model, and by Burgers (with $\delta = -1$) as an extension of its first equation to compressible gas:

$$\frac{\partial u}{\partial t} + u\frac{\partial u}{\partial x} = \delta b\frac{\partial b}{\partial x} + \nu\frac{\partial^2 u}{\partial x^2} \tag{2}$$

$$\frac{\partial b}{\partial t} + u\frac{\partial b}{\partial x} = \delta b\frac{\partial u}{\partial x} + \eta\frac{\partial^2 b}{\partial x^2} \tag{3}$$

With $\nu \equiv 0, \eta \equiv 0$, these equations conserve for $\delta = 1$ a "pseudo" energy (namely $< u^2 + b^2/3 >$), and for $\delta = -1$, the cross–correlation between the velocity and the magnetic field $< ub >$. They show a dynamo effect, they lead to shock–like as well as cusp–like solutions (Passot, 1987; Ponty, 1990) and follow a Painlevé integrability property in some cases (Passot & Pouquet, 1986). When $\delta = 1$ (*resp.* -1), one in fact deals with an isentropic γ–law, with $\gamma = -1$ (*resp.* 3). In the former case, it may correspond to complex behavior of astrophysical flows when heating and cooling are added to the energy equation, as occurs in the interstellar medium (see §6.3).

2.2. A NUMERICAL MODEL

Because, for the Burgers equation, the thickness of the viscous layer scales as $\sqrt{\nu}$, this means that a large fraction of the interacting modes are in fact in the dissipation range. In three dimensions, using the ansatz $L_0/(2\Delta x) \sim R_V^a$ (where Δx is the grid spacing) and with $a = 1$ to simplify the argument, one can show that the fraction of modes pertaining to the dissipative range varies like $(1 - R_V^{-3/2})$ which tends to unity for $R_V \to \infty$. Already for $R_V \sim 10^3$ (*i.e.* on a grid of $N = 10^9$ points, with the scaling of a chosen here), *only* of the order of 10^6 modes are in the inertial (and energy–containing) ranges. Clearly some sort of grid refining or adaptation is needed here. But for a turbulent flow, where many shock–like or filament–like structures develop, such a grid may become too cumbersome to handle. On the other hand, in order to obtain a gain by a factor of two in resolution (*i.e.* taking a grid with $\Delta x' = \Delta x/2$), the CPU cost is 16 times heavier (for an explicit temporal scheme), and the memory requirement 8 times more.

One can rely on other methods, for example by trying to reduce the number of modes in the dissipative range. Many codes dealing with compressible flows use the Euler (conservative) fluid – and MHD – equations, with some sort of numerical viscosity in shocks. Another such method consists in setting in equation (1) $A_2 = 0$ and $A_4 = -1$, *i.e.* using for the multi–dimensional problem a bi–Laplacian (in such "hyperviscosity" methods as they are called, powers of k^2 or ∂^2 higher than 2 can be used). One may also decimate modes in the dissipation range, retaining only one mode out of two in some interval at the onset of the dissipation range, one out of four a bit further in wavenumber, and so on. Such methods have been tested and reproduce well the global energetics of the flow, but *not* necessarily the detailed structures at small scale (see *e.g.* Meneguzzi, Politano, Pouquet and Zolver (1996) and references therein).

A method vastly different in its conception has been developed recently in the context of ocean modeling (Eby and Holloway, 1994). It consists in introducing a small tendency (through a Laplacian term with a negative viscosity) for the computed oceanic circulation to evolve towards its preferred state evaluated from statistical mechanics (although the flow is both driven and damped), taking into account the invariants of the inviscid equations, a state which depends on the bottom topography of the oceanic floor because of the induced vorticity due to the rotation of the Earth. The reason why it apparently works so well may be due to the fact that the ocean has a slow response time (typically of the order of $1,000$ days) to the quick meteorological (say, Eolian) excitation that drives it. It is tempting to remark here that, similarly, the response time of the interstellar medium at large scale is at least a factor 100 longer than the small–scale thermal driving (heating and cooling linked with star formation, see §6).

2.3. THE HADA EQUATION AND THE ASYMMETRIC STABILITY OF PROTO–STELLAR JETS

The compressible MHD equations, when including the Hall term in a generalized Ohm's law, can be written as:

$$\frac{\partial \rho}{\partial t} + \nabla \cdot (\rho \mathbf{u}) = 0$$

$$\rho \left(\frac{\partial \mathbf{u}}{\partial t} + \mathbf{u} \cdot \nabla \mathbf{u} \right) = -\nabla \left(P + \frac{|B|^2}{2} \right) + \mathbf{B} \cdot \nabla \mathbf{B}$$

$$\frac{\partial \mathbf{B}}{\partial t} = \nabla \times (\mathbf{u} \times \mathbf{B}) - \frac{1}{\xi} \nabla \times \left(\frac{1}{\rho} (\nabla \times \mathbf{B}) \times \mathbf{B} \right)$$

$$\nabla \cdot \mathbf{B} = 0 \ , \tag{4}$$

where $\mathbf{B}$ has been normalized to a velocity. The ξ factor, in front of the dispersive Hall term is the ionization fraction of the gas (provided it is not too low, see Bacciotti, Chiuderi and Pouquet (1996), since in that case it would be dominated by the ambipolar drift term.

In the context of magnetospheric physics, Hada (1993) has recently proposed a set of coupled equations stemming from the one–dimensional version of the MHD equations above, but with several simplifying assumptions; the Hada system is in fact very close to the system of Thomas (see equations (3)) and thus can provide an *a posteriori* justification for them. Now, $b = b_y + i b_z$ is a complex divergence–free magnetic field (transverse to a uniform $B_0 \equiv 1$ field in the x–direction):

$$\frac{\partial u}{\partial T} = -\frac{\partial}{\partial X} (\Gamma u^2 + \Delta u + |b|^2/2) \tag{5}$$

$$\frac{\partial b}{\partial T} = -\frac{\partial}{\partial X} (ub) - \frac{i}{\xi} \frac{\partial^2 b}{\partial X^2} \tag{6}$$

$$\frac{d}{dt} (\mathcal{P}/\rho^\gamma) = 0 \tag{7}$$

with $\Gamma = \frac{\gamma+1}{2}$. Note that in Hada (1993), the medium is assumed to be fully ionized (corresponding to $\xi \equiv 1$). In this complex–variable formulation, one has $b = i \partial_x a$ where $a = a_y + i a_z$.

These equations are conservative (although dissipative terms could be added, *e.g.* by making ξ complex) with invariants $< (u^2 + b^2)/2 >$ (energy) and $< (ab^\star + ba^\star)/2 > = < (a_z \partial_x a_y - a_y \partial_x a_z) >$ the magnetic helicity. These equations stem from an asymptotic expansion around a uniform magnetic field in the x direction, uniform density and zero velocity as the basic state. A scaling on time and space is taken so as to conserve the linear dispersion relation $\omega \sim -k^2$ stemming from the Hall term and the adiabatic equation $d_t(\mathcal{P}/\rho^\gamma)$ is transformed using the further constraint that within the plasma, the ratio of gas to magnetic pressure is very close to unity (without that extra assumption, one gets the DNLS equation). The Hada equations can be taken as an extension of the equations of Burgers and of Thomas to a case where dispersive effects are included.

Hada showed (using a modulational instability) that there is stability (*resp.* instability) according to the polarization of the propagating wave, a result that extends to the case $\xi \neq 1$ (Bacciotti *et al.*, 1996). In fact this asymmetry in the instability diagram leading to the formation of weak shocks can be used to explain the asymmetry in the emission knots of protostellar jets (Bacciotti *et al.*, *op. cit.*), which are observed to be mono-sided in many instances, as for example HH34 (Reipurth & Heathcote, 1992) and HH111 (Reipurth, Raga and Heathcote, 1992; see also T. Wray, this Volume, as well as several other Lectures on the topic of jets).

These Herbig–Haro knots can also be interpreted as fragments – or bullets (Stone and Norman, 1994; Stone, Xu & Mundy, 1995) emanating from stellar winds subject to hydrodynamical (or MHD) instabilities (*i.e.* due to the interaction of the shell produced by the stellar wind, and the jet itself), as may also occur in supernovae remnants (Franco, Ferrara, Rozcyczka, Tenorio–Tagle and Cox, 1993), and in fact in planetary nebulae as well (MacLow, 1995).

3. Diagnostics

3.1. STRUCTURE FUNCTIONS

In the absence of a systematic theory of turbulent flows, one relies on models, on phenomenology, on experiments (including numerical experiments) and observations. Astrophysical flows have high Reynolds numbers and are often magnetized. The diagnostics one can perform on such flows are two-fold. On the one hand, because of the strong noise of the signal, one deals with statistics. On the other hand, one analyzes the physical structures that develop (such as intense localized vortices, as those that develop in the atmosphere of the Earth). The energy spectrum (*e.g.* the two–point one time Fourier transform of the velocity correlation function) is the most common statistical diagnostic, but one also can consider higher moments. Defining the structure functions as usual as

$$\delta v_\ell = u(x + \ell) - u(x) \ ,$$

where $u(x)$ is the longitudinal component of the velocity (or magnetic field) and assuming self–similar behavior in a range of (inertial) scales, one writes:

$$< \delta v_\ell^p >\sim \ell^{\zeta_p} \; , \tag{8}$$

where ζ_p are the unknown scaling exponents to be determined.

Energy is being transferred at an average rate $\bar{\epsilon} = U_0^3/L_0$. With the further assumption that the energy dissipation in a sphere of radius ℓ scales as well as $\epsilon_\ell = u_\ell^3/\ell$ (Kolmogorov refined similarity hypothesis), and defining τ_p as the anomalous scaling exponent of transfer – and dissipation – (viz. $< \epsilon_\ell^p >\sim \ell^{\tau_p}$), one obtains:

$$\zeta_p = p/3 + \tau_{p/3} \; . \tag{9}$$

When no fluctuations of ϵ are allowed, the Kolmogorov (1941) scaling is recovered, namely $\zeta_p = p/3$. When fluctuations in the energy transfer rate are allowed leading to intermittency, a different scaling obtains. Several models have been proposed to explain the observed departure of scaling exponents of high–order structure functions from a linear $p/3$ law, some of which have also been extended to MHD (see Carbone (1994), and for an introduction, Biskamp (1994)). Such models are restricted by exact relations for some ζ_p exponents; for example, for incompressible homogeneous isotropic and stationary fluids, $\zeta_3 \equiv 1$, stemming from the conservation of energy (see for example in Frisch (1995) for details). Similarly, for a passive scalar θ such as temperature, one has $< \delta\theta_\ell^2 \delta v_\ell >\sim \ell$ (Yaglom, 1949); this result can be extended to two–dimensional MHD, replacing the scalar θ by the (scalar) magnetic potential (Caillol, 1995; Caillol, Politano and Pouquet, 1995), although it is an active scalar – it acts on the velocity field through the Lorentz force. This relationship is well verified numerically (although the inverse cascade of magnetic potential leads to a non–stationary behavior).

3.2. THE SHE–LEVÊQUE MODEL

Recently, a model has been proposed (She and Levêque, 1993, referred hereafter as SL) to determine the ζ_p exponents, and which agrees to better than 1% with experimental data (using the Extended Self–Similarity, or ESS, of Benzi, Ciliberto, Tripicciona, Baudet, Massaioli and Succi, 1993). The SL model relies on equation (9) and furthermore supposes that (i) the characteristic time of the most intermittent structure scales as $t_\ell \sim \ell^{2/3}$; (ii) the co–dimension of dissipative structures is two (i.e. they are filaments, as indicated –clearly for vorticity, not so clearly for dissipation – by several numerical simulations for three–dimensional incompressible flows, see e.g. Vincent and Meneguzzi, 1991) and (iii) there is an underlying log–Poisson statistics of energy transfer (a multiplicative process leading to a log–distribution, and Poisson statistics because of the rarity of events); note that richer statistics can obtain for such flows (Dubrulle, 1994; Schertzer, Lovejoy and Schmidt, 1995). The latter condition specifically reads, following Dubrulle (1994):

$$< \Pi_l^p >= B_p < \Pi_l >^{\frac{1-\beta^P}{1-\beta}} \, , \tag{10}$$

with $\beta \neq 1$ and

$$< \Pi_\ell >= \epsilon_\ell / \epsilon^{(\infty)} \tag{11}$$

where $\epsilon^{(\infty)}$ is the maximum energy available to dissipate in the flow. This relationship has been verified in the laboratory (Ruiz–Chavarria, Baudet and Ciliberto, 1995), at Taylor Reynolds numbers R_T ranging from 100 to 800 (where R_T is based on the Taylor scale λ_T constructed from the energy spectrum $E(k)$ as $\lambda_T \sim [\int k^2 E(k)dk/ \int E(k)dk]^{-1/2}$, in other words weighing the small scales that develop in the process of forming turbulent flows). Note that the assumption that $\epsilon \sim (\partial U/\partial x)^2$ is made (which holds for strictly homogeneous and isotropic turbulence) in this experimental verification.

In the SL model, $\epsilon^{(\infty)}$ is evaluated as

$$\epsilon^{(\infty)} = U_0^2/t_\ell$$

and is assumed to be finite (which seems to be verified experimentally up to $R_\lambda \sim 2,000$, in so far as the probability distribution functions of the velocity gradients are not power laws in the wings, but remain stretched exponentials. Finally, β^{SL} is set at $\beta = 2/3$ (corresponding to the Kolmogorov scaling for t_ℓ); this leads to $\zeta_p^{SL} = \frac{p}{9} + 2(1 - (2/3)^{p/3})$.

The model has been extended to MHD in the context of the IK phenomenology with the assumption that dissipative structures are now sheets instead of filaments (Grauer, Krug and Marliani, 1994); this leads (with $t_\ell \sim \ell^x$) to $x = 1/2$ and $\beta = 1/2$ (see below, equation (13)); it implies $\zeta_4^B \equiv 1$, a relationship for which there is no proof as yet.

These models can be generalized further to a variable scaling time and variable co–dimension, both for fluids (Dubrulle, 1994) and in MHD (Politano and Pouquet, 1995) with in the latter case, using the IK phenomenology:

$$\zeta_p^B = \frac{p}{4}(1 - x_B) + C_0^B(1 - \beta_B^{p/4}) \tag{12}$$

together with

$$C_0^B = \frac{x_B}{1 - \beta_B} \, . \tag{13}$$

Comparing with the observational data of Burlaga (1990) obtained with the Voyager spacecraft, the law fits well the data with $x_B \sim 1/2$ and $\beta_B \sim 1/2$, although the error bars are too large in order to permit to discriminate among various fluid (Kolmogorov like) and MHD (IK like) models, although a recent analysis (Carbone, Veltri and Bruno, 1995) tends to show a different behavior at high order for fluid and for MHD flows. Temporally well–resolved data sets will be very useful in this instance (as for example with the Ulysses spacecraft).

Although the linear scaling laws for the anomalous exponents of structure functions may differ (*i.e.* $p/3$ for Kolmogorov fluids, and $p/4$ for IK magnetofluids) due to different characteristic times for the basic interactions, the intermittency corrections may fall into one (or a few) universality classes. Indeed, using normalized exponents $\tilde{\zeta}_p^U = \zeta_p^U/\zeta_3^U$ for the U variable, the same approximate values obtain experimentally (Ciliberto, private communication) for Navier–Stokes turbulence, convection and MHD numerical simulations in two space dimensions. In that light, it should be remarked that when computing the predicted theoretical values for the She-Leveque model of fluid turbulence and for the SL–MHD model, normalizing the latter by ζ_3^B, one finds values that are very close; for example, $\zeta_2^{V,SL} \approx 0.696$ vs. $\tilde{\zeta}_2^B \approx 0.692$, $\zeta_6^{V,SL} \approx 1.78$ vs. $\tilde{\zeta}_6^B \approx 1.79$ and $\zeta_{10}^{V,SL} \approx 2.59$ vs. $\tilde{\zeta}_{10}^B \approx 2.66$. Only accurate data will allow to compute high–order structure functions, and thus to discriminate between models; this means ideally data sets of $\sim 10^{10}$ or more points. Hence the need, in solar wind data, to take the longest records possible with a good signal–to–noise ratio.

4. The two–dimensional case

4.1. ARE NONLINEAR INTERACTIONS SELF–DEFEATING IN MHD?

We saw in §2.1 that nonlinear interactions between modes can lead, in simple cases at least, to singularities corresponding in the language of structures to the formation of shocks (for compressible flows) and presumably strong velocity gradients. But if there was a way to damp these nonlinear interactions, they would not be so efficient and the flow could evolve in a quasi–linear manner. For example, writing first the MHD equations in the incompressible case as:

$$\frac{\partial \mathbf{u}}{\partial t} + \mathbf{u} \cdot \nabla \mathbf{u} = -\frac{1}{\rho_0} \nabla \mathcal{P} + \nu \nabla^2 \mathbf{u} + \mathbf{j} \times \mathbf{b} \ ,$$

$$\frac{\partial \mathbf{b}}{\partial t} = \nabla \times (\mathbf{u} \times \mathbf{b}) + \eta \nabla^2 \mathbf{b} \ ,$$

$$\nabla \cdot \mathbf{u} = 0 \ ; \ \nabla \cdot \mathbf{b} = 0 \ ,$$

with $\mathbf{b} = \mathbf{B}/\sqrt{\mu_0 \rho_0}$ the Alfvén velocity, μ_0 is the permeability, η is the magnetic diffusivity and $\mathcal{P}$ the pressure, we see immediately that $\mathbf{v} = \pm \mathbf{b}$ is an exact solution of the conservative equations ($\nu \equiv 0$, $\eta \equiv 0$). If such a flow is attractive in the space of solutions, no further evolution will take place on the dynamical time–scale (of course some dissipation will take place on the slow diffusive time scale when small amounts of viscosity and magnetic diffusivity are reintroduced).

It has been shown that the correlation coefficient

$$\rho_C = 2 < \mathbf{v} \cdot \mathbf{b} > / < v^2 + b^2 >$$

grows with time (see Pouquet (1993) and references therein). So the flow does evolve towards a $\mathbf{v} = \pm \mathbf{b}$ state ! In fact, even when $\rho_C \sim 0$ globally (recall that it is a volume integral of a non positive definite quantity), there may be (and in fact, there are, as observed for example in numerical simulations, see Passot, Politano, Pouquet and Sulem, 1990) within the flow regions where locally $| \rho_C | \sim 1$ *i.e.* rather static regions at the border of which all action takes place in thin dissipative layers. Closure models of MHD turbulence do indicate that ρ_C is close to zero only around the dissipative scale. This image of MHD turbulence could be backed up by new numerical data at high Reynolds number, at least in two dimensions. In three–dimensional incompressible fluid turbulence, a depletion of non-linearities (from what is expected for a purely random Gaussian velocity field) has been observed on numerical data (Kraichnan and Panda, 1988). It could be attributed to a local quasi bi–dimensionalization of the swirling flows that are induced (through the Biot–Savart law) by the strong vortex filaments that develop in such 3D fluids (see §5.2). The equivalent analysis has not been done for MHD flows.

4.2. DOES RECONNECTION OCCUR AT A FINITE RATE FOR $R \to \infty$?

Because the Reynolds numbers for the sun are so large, it is often assumed that dissipation mechanisms are negligible. But there is in fact the *a priori* possibility that dissipation can occur at a finite rate for $R \to \infty$, provided strong enough velocity gradients develop. This may be of importance for astrophysical flows and in particular in the framework of heating the solar corona, a problem for which recent observations (those of YOHKOH in particular, see Tsuneta, this Volume) clearly indicate that magnetic reconnection plays a fundamental role in shaping the dynamical behavior of the corona (*i.e.*, here we do not choose the static framework considered elsewhere in this Volume, see *e.g.* Low; and Parker).

This question, for a flow that develops small scales in time, amounts to: is there a finite limit, as the Reynolds number $R \to \infty$, for the dissipation $\mathcal{D}$? For an incompressible MHD fluid, the dissipation is given by

$$\mathcal{D} \sim \nu < \omega^2 > + \eta < j^2 > \ ,$$

where $< \omega^2 >$ and $< j^2 >$ are the kinetic and magnetic enstrophies, or in other words the total squared vorticity and current density. The numerical or analytical determination of a limit for $\mathcal{D}$ remains an open challenge in three dimensions (and for incompressible neutral fluids as well) unless a mean magnetic field B_0 is present, strong enough to suppress neutral points of the random field.

In two dimensions, there is clear numerical evidence that at early time the flow does not develop any singularity: the growth of small scales is only exponential. However, for later times, it has been shown using both the primitive equations (Politano, Pouquet and Sulem, 1989) and hyper-viscosity codes (Biskamp and Welter, 1989; Passot, Politano, Pouquet and Sulem, 1990) that $\mathcal{D} \rightarrow \mathcal{D}_0$, a finite value. This phenomenon is probably linked to the tearing mode instability which leads to the breaking of long thin current sheets into small islands, thus producing more dissipation (catastrophically). Is it a numerical artefact, due to either too long a time of computation, or (more probable) too low a Reynolds number ? Or is it the case that indeed $\mathcal{D}_0 \neq 0$?

4.3. A PLAUSIBLE MODEL FOR HEATING THE SOLAR CORONA

As just discussed, the formation and disruption of current layers may be responsible for the heating of the solar corona. It is well known that there is a whole spectrum (both in intensity and in duration) of solar flares, from the nano–flare to the major eruption; this spectrum in fact follows a power-law and the model of Lu, Hamilton, McTierman and Bromund (1993) is very successful in advocating a connection between such observations and the phenomenon of avalanches (and sand piles), linked to self–organized criticality. This model has been extended (see for example Vlahos (1993)) to more complex cases but the basic mechanism remains that, when a gradient becomes too large, the system relaxes to a lower gradient configuration by re–organizing the neighboring points, a reorganization which may propagate in some cases to the whole volume, leading then to major disruptions. The question remains of what is the physical mechanism responsible for the MHD equivalent to the "avalanches". A recent calculation of two–dimensional MHD incompressible fluids with a forcing term that mimics the random motions of foot points of solar flux tubes shows that the distribution of peaks in the dissipation is random, with numerous small events and rare big events (Einaudi, Velli, Politano and Pouquet, 1996). Similar computations in three–dimensional MHD are performed by Galsgaard and Nordlund (this Volume). This phenomenon is linked to the intermittency of current layers, both spatial and temporal. However, it is not clear whether the major flares can also be produced in such a fashion.

5. The three–dimensional case

5.1. THE BETCHOV RELATION

A simple relation derived by Betchov (1956) links the development of small scales in turbulence (and hence the growth of enstrophy) to the formation of vortex sheets. Starting from the Navier–Stokes equations (4) with $\mathbf{B} \equiv 0$, and omitting for brevity to write pressure and viscous terms, one can derive the equation for the time derivative of the velocity gradient tensor; dotting it with $\partial_i u_j$ and averaging over the whole flow using the hypothesis of isotropy, homogeneity and incompressibility, Betchov showed that:

$$\frac{D < \omega^2 >}{Dt} \sim -s < a_s b_s c_s > , \qquad (14)$$

where s is positive and where a_s, b_s and c_s are the three eigenvalues of the symmetrized $\partial_i u_j$ matrix. Since $a_s + b_s + c_s = 0$ because of incompressibility, the sign in the equation above is linked to that of b_s, *i.e.* the intermediate eigenvalue. In other words, small scales are produced with on average two positive eigenvalues of the rate of strain tensor: an initial roundish blob of vorticity is being stretched in two directions and squeezed in the third direction. Hence the production of vorticity sheets.

5.2. THE STRUCTURING OF VORTICITY INTO FILAMENTS

Numerical data – and now experiments in the laboratory – both indicate that the vorticity, when strong, is structured into filaments, except at early time where numerical data clearly indicates that sheets are formed (see *e.g.* Brachet, Meneguzzi, Vincent, Politano and Sulem, 1991). The reason is simple: these sheets are unstable (either to Kelvin–Helmholtz instability (Neu, 1984), or through a self–focusing mechanism (Passot, Politano, Sulem, Angillela and Meneguzzi, 1995) and filaments form that are both very strong and long–lived, although in fact they contain little vorticity (roughly 1 %): the background vorticity may still be organized into sheets. Similar structures are found in the compressible case (Porter, Pouquet and Woodward, 1995). Finally, let us remark that because of the analogy between the vorticity equation in the incompressible case (and neglecting diffusion coefficients)

$$D_t \omega = \omega \cdot \nabla \mathbf{v}$$

and the induction equation $D_t \mathbf{b} = \mathbf{b} \cdot \nabla \mathbf{v}$, it has been conjectured that the magnetic field will develop in a similar fashion to vorticity, and indeed the small–scale dynamo favors flux tubes (Galloway and Frisch, 1986), at least in the kinematic phase. The nonlinear problem is poorly explored till now, because of the high cost of computing in three dimensions (but see Brandenburg, Jennings, Nordlund, Rieutord, Stein & Tuominen, 1995).

5.3. TRANSPORT COEFFICIENTS AND THE EXCESS OF MAGNETIC ENERGY IN THE SMALL SCALES

Because we do not know how to solve nonlinear equations in general, we cheat and simplify the problem. The concept of eddy viscosity arises from the idea that, by coupling modes nonlinearly and transferring energy to the small scales where it is ultimately dissipated, such an effect can be modeled by an enhanced dissipation. A formalization of such a concept can be achieved through the development of statistical theories – the pioneer of which is the DIA (Kraichnan, 1959) – which lead to a set of integro-differential equations for the energy spectrum.

Many such transport coefficients may be thus computed in a variety of physical contexts. Both small–scale velocity and magnetic fields will contribute to the dissipation of the velocity field; but because the induction equation is *linear* in **B**, in fact *only* small kinetic scale – to first order – participate to its dissipation. This has been shown in the context of closures of turbulence (see *e.g.* Pouquet, Frisch and Léorat, 1976) in three dimensions. In other words, we can model in the framework of that approximation the temporal evolution of the kinetic and magnetic energy spectra as diffusion equations with an extra term for the kinetic energy. On that basis, we can postulate that slightly more magnetic energy will be present in the small scales, a phenomenon observed in many instances. A global excess of magnetic energy can in fact be derived from statistical ensemble of MHD turbulence (Stribling and Matthaeus, 1990). Note that if either ambipolar drift or the Hall term are added to a generalized Ohm's law, leading to terms nonlinear in **B** in the induction equation, this result may not hold any longer (the existence or not of inverse magnetic cascades in these cases would be interesting to study). Different expressions emerge in two dimensions (see Pouquet, 1978) because of the inverse cascade of magnetic potential linked to a negative magnetic diffusivity, involving small magnetic scales as well.

5.4. TRANSPORT COEFFICIENTS AND THE DYNAMO PROBLEM

In fact in three dimensions, small–scale (kinetic and magnetic) when helical, lead to the destabilization of a large–scale magnetic field through the well–known α–effect (Steenbeck, Krause and Rädler, 1966) in the kinematic regime (*i.e.* with a given velocity field since, at small amplitude, the Lorenz force is negligible). The efficiency of this mechanism has been put to doubt recently because even at low levels (well below equipartition between kinetic and magnetic energy) a large–scale magnetic field seems to have a strong effect on the underlying Lagrangian chaos of the velocity, presumably quenching the α–effect to quasi non–existence.

However, in the turbulent context, the nonlinear dynamo is viewed simply as the manifestation – exemplified with the help of transport coefficients – of the inverse cascade of magnetic helicity postulated on the basis of statistical equilibria of a truncated system of Fourier modes for which the quadratic invariants are conserved (cross correlation, magnetic helicity and total energy). Such inverse cascades are rather well documented (Hasegawa, 1985; see also Pouquet, 1993) in different contexts, *e.g.* two–dimensional Navier–Stokes and MHD incompressible turbulence, three–dimensional MHD, the Strauss equations (*i.e.* MHD in the presence of a strong uniform magnetic field), as well as for the Hasegawa–Mima equations for drift–wave turbulence – and Rossby wave turbulence (although the two may differ, see Hasegawa, 1985; and Kukharin, Orszag and Yakhot, 1995).

The important point is that such a mechanism leads to a non–stationary evolution: there is no real saturation of the growth of magnetic energy. It can in fact be shown in the framework of a simple model (Kraichnan, 1979) confirmed by numerical experiments – albeit at low resolution (Meneguzzi, Frisch and Pouquet, 1981) – that magnetic helicity H^M grows *linearly* with time, and because of the Schwartz inequality $E^M(k) > H^M(k)/k$, magnetic energy continues to grow as well. Of course, when the magnetic excitation reaches the size of the overall system, boundary effects come into play and this mechanism will be altered.

5.5. THE DYNAMICAL DEVELOPMENT OF SINGULARITIES

When assuming that the skewness of the velocity field

$$S_k \sim < \partial^3_{xxx} u > / [< \partial^2_{xx} u >^{3/2}]$$

is time–independent, Betchov in fact showed that $\frac{D<\omega^2>}{Dt} \sim < \omega^2 >$, leading to a singularity in a finite time. Experimentally and numerically, it is known that the skewness becomes constant after the enstrophy has reached its maximum, *i.e.* when dissipation sets in. Other models have been devised over the years which lead to singularity when (arbitrarily) neglecting in the primitive equations some (important ?) term, or when setting for all times that some specific symmetry be enforced.

Following a theorem by Beale, Kato and Majda (1989), a good criterion for singularity is to follow in time the development of the peak vorticity ω_p (or the peak strain), because the theorem states that *if* a singularity occurs, *then* the supremum of vorticity must blow–up. There is numerical evidence that such singularities might occur (Kerr, 1985; 1993; 1995; Pelz and Boratav, 1995) when using specific initial conditions. In the analysis of Bhattacharjee, Ng & Wang (1995), a singularity might occur depending on the sign of $\partial_x^4 \mathcal{P}$ (where $\mathcal{P}$ is the pressure) for an initial velocity field satisfying $u_x = f(y)$, $u_y = f(z)$ and $u_z = f(x)$ (see also Bhattacharjee & Wang, 1992); furthermore, this model agrees with the analysis of Lundgren (1982) as to yielding a $E^V(k) \sim k^{-5/3}$ spectrum.

6. Turbulence in the interstellar medium

6.1. EVIDENCE OF TURBULENCE THROUGH LINE WIDTHS

The interstellar medium (or ISM) is a complex superposition of structures (as for example seen with IRAS and now ISO) encompassing a wide variety of scales, densities and temperatures. Several authors have advocated that it is turbulent (see for example the reviews by Scalo (1985) and by Falgarone (1995)). Evidence for such turbulence is to be seen in the power–law behavior when scaling velocity dispersion against scale of cloud (as first emphasized by Larson in 1981), in the presence of self–similar structures over four order of magnitude in scales, in the computation of fractal dimensions for the boundaries of clouds, and in the non–gaussian wings of velocity spectra, as for example studied in the High Latitude Cloud Ursa–HLC in the $^{12}CO(2-1)$ transition (Falgarone, Lis, Phillips, Porter, Pouquet and Woodward, 1994). The fact that with a superposition of modes with a given (Kolmogorov) spectrum but with random phases, one can reproduce almost as well (Dubinskii, Narayan and Philipps, 1995) the observational data (except for large excursions in the vorticity) shows that numerical simulations in three dimensions are still largely under–resolved as far as high Reynolds number turbulence is concerned ... Intermittency may also affect the chemistry balance in the medium, allowing for local high temperatures which may give rise to emission lines unexplained otherwise (Falgarone and Puget, 1995).

6.2. A MODEL OF THE ISM AT THE KILOPARSEC SCALE

Given the wide range of scales (from the kiloparsec down to the dense cores of 0.01 pc), densities (from roughly 1 particle per cm^3, up to 10^5), and temperatures (from $10^6 K$ in the coronal phase, down to $10 K$ in dense cores), a direct numerical model of the medium is difficult, even if one resorts to adaptative grids. One can take a different approach, namely resolving scales down to a certain size below which (or above, or both) , the physics will not be computed exactly but modeled. For example, radiative processes leading to heating because of ionization winds can be introduced by hand in the energy budget (this stellar heating is due to the eventual collapse of condensations as they form); cooling can be introduced as well following the classical temperature–dependent model of Dalgarno and McCray (1972). Several models have been developed in the literature (see *e.g.* Rosen, Bregman and Norman, 1993; Rosen and Bregman, 1994; Stone and Norman, 1994). In one such model (Vazquez, Passot and Pouquet, 1995a), the resulting computation on a two–dimensional grid, and for fiducial values of the parameters, leads to a realistic description of the medium. In particular, all ingredients (self–gravitation, heating and cooling, and hydrodynamical phenomena) are necessary to obtain an ISM that resembles our own, organized in three phases (hot tenuous pervading gas, cold dense clouds and expanding shells within which self–propagating enhanced stellar formation occurs), and for which there is an energetic balance between all actors. Furthermore, the flow is subsonic with little vorticity in most of the medium in the hot phase, and is mostly supersonic with strong local vorticity in the regions of star formation. Note that these expanding shells are linked to the ionization winds of newly–formed OB stars (see for a discussion of winds of massive young stars, MacLow, 1995).

6.3. AN EFFECTIVE THERMODYNAMICAL LAW

The radiative (heating and cooling) processes occur on a typical time–scale significantly shorter than the hydrodynamical time scale, by a factor of roughly one hundred. Hence, a pressure and temperature equilibrium obtains rapidly, and the observed condensations (associated with Giant Molecular Clouds) must be held by turbulent (*ram*) pressure. Assuming a constant heating law Γ, and a cooling law Λ that is both density– and temperature–dependent (namely $\Lambda = \rho T^n$, with n varying in different temperature intervals, as in the Dalgarno–McCray (1972) model), one can easily show, following Elmegreen (1991), that the ensuing equilibrium pressure writes $\mathcal{P}_{equi} \sim \rho^{\gamma_{eff}}$ with $\gamma_{eff} = 1 - 1/n$ (in fact in Elmegreen (1991), a more general law is written with a more general assumed behavior of heating and cooling). For most of the medium which is in the temperature range $2,000 < T < 8,000$, this leads to an effective gamma–law gas with $\gamma_{eff} = 1/3$; indeed, $\mathcal{P} - \rho$ scatter plots obtained from numerical simulations in two space dimensions agree with such an estimate (Vazquez–Semadeni *et al.*, *op. cit.*) with, in the MHD case, a shallower effective law.

A more realistic model can be built along several lines. On the one hand, one can take into account the self–shielding of clouds to UV radiation, leading to a density–dependent diffuse heating; this leads to more realistic values for the temperature of the cold clouds, and to an enhanced density contrast ($\rho_{max}/\rho_{min} \sim 1,000$) (Vazquez, Passot and Pouquet, 1995b). On the other hand, both rotation and magnetic fields should be included.

6.4. THE MHD CASE

In the presence of a strong uniform magnetic field, Alfvén waves propagate in the ISM (see *e.g.* Falgarone and Puget, 1986; Pudritz and Gómez de Castro, 1991). But such waves may steepen, and the approach encompassing turbulence as described in the preceding Section may then be followed again. Furthermore, both rotation and magnetic fields are present in the ISM, with a dynamical role (Mouschovias, 1976ab; and this Volume). That role may not be the same at the kiloparsec scale and at the dense core scale where it is thought that proto–stellar collapse occurs quasi–statically at first, because the tension in magnetic field lines is an efficient braking mechanism. A linear analysis by Elmegreen (1994) – in the context of the galactic spiral arms – shows that rotation and magnetic fields have opposite effects on the gravitational instability (similar in fact to the convective case). In the linear context, in the absence of rotation, the magnetic field simply produces an enhanced pressure (stabilizing the Jeans' length) for the radial modes. Rotation without magnetic field is stabilizing; but a magnetic field reduces the stabilizing of rotation for the azimuthal modes. In the nonlinear context that can be studied with numerical simulations, it is also shown that the magnetic field can have a "pressure –cooker" effect (also envisioned, in the somewhat different context of (one dimensional) spherical symmetry, in Slavin and Cox, 1992, 1993): a closed loop produces an inward Lorenz force that confines the plasma (as in tokamaks), preventing shell expansion and thus sweeping of new material, hence a global reduction in the star formation rate, but within the loop leading to strong turbulence.

In order to reach a more realistic description of the ISM, other effects must be included and other scales (the molecular cloud, the dense core) must be studied. One natural expansion of the previous calculations is to investigate whether the inclusion of supernovae in the heating mechanism will lead to a coronal phase (at a temperature of $10^6 K$), and how pervasive is that phase (what is its filling factor). Furthermore, the evolution of proto–stellar outflows and the formation of density shells within them may be somewhat similar, as already mentioned, to the evolution of supernovae shells (note that rapid cooling is essential in the process, see Chevalier, Blondin, and Emmering (1992), and Stone *et al.* (1995) and references therein). At smaller scales, chemistry and ambipolar drift should be included as well.

7. Conclusion

It is of course impossible to introduce key concepts in turbulence research today in so brief a Lecture. The domain is very active, in part (a personal bias) because new experiments and observations are better at pinpointing the insufficiencies of models. Clearly in Astrophysics this is the case, as observers reach a greater range of scales (such as with the solar THEMIS instrument, or with the radio–wavelengths array at Bures or the millimeter BIMA array for studying the interstellar medium for ground–based instruments, as well as numerous spacecrafts Ulysses, CLUSTER, ISO). Our taking into account of a multitude of scales in complex interactions already proves useful, as this text attempted to show through but a few examples.

8. Acknowledgements

Kanaris Tsinganos has provided us with a stimulating environment in a wonderful setting for the ASI Meeting. This Lecture draws in some places from collaborative works, not all of which are published as yet (with F. Bacciotti, C. Chiuderi and H. Politano). I also wish to thank J.D. Fournier and G. Holloway for interesting discussions.

References

Adams F., Fatuzzo M. & R. Watkins (1994) *Astrophys. J.* **426**, 629.

Bacciotti, F., C. Chiuderi & Pouquet, A. (1996) *Astrophys. J.*, to appear.

Benzi, R., Ciliberto, S., Tripicciona, R., Baudet, C., Massaioli, F. & Succi, S. (1993) *Phys. Rev. E* **48**, R29.

Bhattacharjee, A. & Wang X. (1992) *Phys. Rev. Lett.* **69**, 2196.

Bhattacharjee, A., Ng C.S. & Wang X. (1995) *Phys. Rev. E*, to appear.

Beale J., Kato T. & Majda A. (1989) *Comm. Math. Phys.* **94**, 61.

Biskamp, D. (1994) **Nonlinear Magnetohydrodynamics**, Cambridge University Press.

Biskamp, D. & Welter H. (1989) *Phys. Fluids B* **1**, 1964.

Betchov, R. (1956) *J. Fluid Mech.* **1**, 467.

Brandenburg, A., Jennings, R., Nordlund, A., Rieutord, M., Stein R. & Tuominen, I. (1996) *J. Fluid Mech.*, to appear.

Brachet, M.E., Meneguzzi, M., Vincent, A., Politano, H. & Sulem, P.L. (1992) *Phys. Fluids A* **4**, 2845.

Burlaga, L. (1991) *J. Geophys. Res.* **96**, 5847.

Caillol, P. (1995), Stage de DEA, Université de Paris VI.

Caillol, P., Politano, H. & Pouquet, A. (1995) Preprint Observatoire de Nice.

Carbone, V. (1994) *Phys. Rev. E* **50**, R671.

Carbone, V., Veltri, P.L. & Bruno, R. (1995) *Phys. Rev. Lett.* **75**, 3110.

Chevalier, R., Blondin, J. & Emmering, R. (1992) *Astrophys. J.* **392**, 118.

Dalgarno, A. & McCray (1972) *ARA&A* **10**, 375.

Dubinskii, J., Narayan, R. & Philipps T. (1995) *Astrophys. J.* **448**, 226.

Dubrulle, B. (1994) *Phys. Rev. Lett.* **73**, 959.

Drazin, P. & Johnson R. (1989) **Solitons: an introduction**, Cambridge University Press, Texts in Applied Mathematics.

Eby, M. & Holloway, G. (1994) *J. Phys. Oceanogr.* **24**, 2577.

Einaudi, G., Velli, M., Politano, H. & A. Pouquet (1996) *Astrophys. J. Lett.*, January 10 issue.

Elmegreen, B. G. (1991) *Astrophys. J.* **378**, 139.

Elmegreen, B. G. (1994) *Astrophys. J.* **433**, 39.

Elphick, C., Regev, O. & Shaviv, N. (1992) *Astrophys. J.* **392**, 106.

215

Falgarone, E. (1995) in "Small–scale structures in fluids and MHD", M. Meneguzzi, A. Pouquet & P.L. Sulem Editors, Notes in Physics **462**, Springer–Verlag.

Falgarone, E., D.C. Lis, T.G. Philips, D. Porter, Pouquet, A. & P. Woodward (1994) *Astrophys. J.*, **436**, 728.

Falgarone, E. & Puget, J. L. (1986) *Astron. Astrophys.*, **162**, 235.

Falgarone, E. & J. Puget (1995) *Astron. Astrophys.*, **293**, 840.

Franco, J., Ferrara, A., Rozcyczka, M., Tenorio–Tagle, G. & Cox, D. (1993) *Astrophys. J.* **407**, 100.

Frisch, U. (1995) **Turbulence: The legacy of Kolmogorov**, Cambridge University Press.

Galloway, D.J. & Frisch U. (1986) *Geophys. Astrophys. Fluid Dyn.* **36**, 53.

Grauer R., J. Krug & C. Marliani (1994) *Phys. Letters A*, **195**, 335.

Hada, T. (1993) *Geophys. Res. Lett.* **20**, 2415.

Hasegawa, A. (1985) *Advances in Physics* **34**, 1.

Iroshnikov, P. (1963) *Sov. Astron.* **7**, 566.

Kerr, R. (1985) *J. Fluid Mech.* **153**, 31.

Kerr, R. (1993) *Phys. Fluids* **A5**, 1725.

Kerr, R. (1995) in "Small–scale structures in fluids and MHD", M. Meneguzzi, A. Pouquet & P.L. Sulem Editors, Lecture Notes in Physics **462**, Springer–Verlag.

Kolmogorov, A. (1941) *Dokl. Akad. Nauk SSSR* **30**, 299.

Kraichnan, R.H. (1959) *J. Fluid Mech.* **5**, 497.

Kraichnan, R.H. (1965) *Phys. Fluids* **8**, 1385.

Kraichnan, R.H. (1979) *Phys. Rev. Lett.* **42**, 1677.

Kraichnan, R.H. & Panda, R. (1988) *Phys. Fluids Lett.* **31**, 2395.

Kruskal, M. (1974) *Lectures in Applied Mathematics* **15**, 61.

Kukharin, N., Orszag, S. & Yakhot, V. (1995) *Phys. Rev. Lett.* **75**, 2486.

Lu, E., Hamilton R., McTiernan M. & Bromund K. (1993) *Astrophys. J.* **412**, 841.

Lundgren, T. (1982) *Phys. Fluids* **25**, 2193.

MacLow, M–M (1995) *Nature* **377**, 287.

Meneguzzi, M., Politano, H., Pouquet, A. & M. Zolver (1996) *J. Comp. Phys.*, **123**, Nb. 1.

Mouschovias, T. (1976a) *Astrophys. J.* **206**, 753.

Mouschovias, T. (1976b) *Astrophys. J.* **207**, 141.

Neu, J.C. (1984) *J. Fluid Mech.*, **143**, 253.

Noether, E. (1918) *Nachr. Ges. Göttingen Math. Phys. Kl.*, 235.

Passot, T. (1987) Thèse, Université de Paris VII.

Passot, T. & Pouquet, A. (1986) *Phys. Lett. A* **118**, 121.

Passot, T., Politano, H., Pouquet A. & Sulem, P.L. (1990) *Theor. Comput. Fluid Dyn.*, 1, 47.

Passot, T., Politano, H., Sulem, P.L., Angillela J. R. & Meneguzzi, M. (1995) *J. Fluid Mech.* **282**, 313.

Pelz, R. & Boratav O. (1995) in "Small–scale structures in fluids and MHD", M. Meneguzzi, A. Pouquet & P.L. Sulem Editors, Lecture Notes in Physics **462**, Springer–Verlag.

Politano, H., Pouquet, A. & Sulem, P.L. (1989) *Phys. Fluids B* **1**, 2230.

Politano, H. & Pouquet, A. (1995) *Phys. Rev. E* **52**, 6361.

Ponty, Y. (1992) Rapport de stage de Maîtrise, Université de Nice.

Porter, D., Pouquet, A. & Woodward, P. (1995) in "Small–scale structures in fluids and MHD", M. Meneguzzi, A. Pouquet & P.L. Sulem Editors, Notes in Physics **462**, Springer–Verlag.

Pouquet, A., U. Frisch, & J. Léorat (1976) *J. Fluid Mech.*, **77**, 321.

Pouquet, A. (1978) *J. Fluid Mech.* **88**, 1.

Pouquet, A. (1993) in **Les Houches Summer School on Astrophysical Fluid Dynamics**, July 1987 (also Preprint High Altitude Observatory, December 1987). Les Houches Session **XLVII**, p. 139; Eds. J. P. Zahn & J. Zinn–Justin, Elsevier.

216

Pouquet, A. (1996) in Proceedings European School of Astrophysics, C. Chiuderi & G. Einaudi Editors, Springer–Verlag, to appear.

Pudritz, R. E. & Gómez de Castro, A. I. (1991) in **Fragmentation of Molecular Clouds and Star Formation**, Eds. E. Falgarone, F. Boulanger & G. Duvert (Dordrecht, Kluwer Academic Press), 317.

Reipurth, B. & Heathcote, S. (1992) *Astron. Astrophys.* **257**, 693.

Reipurth, B., Raga, A. & Heathcote, S. (1992) *Astrophys. J.* **392**, 145.

Rosen, A., Bregman, J. N. & Norman, M. L. (1993) *Astrophys. J.* **413**, 137.

Rosen, A. & Bregman, J. N. (1994) *BAAS* **25**, 1394.

Ruiz–Chavarria, G., Baudet, C. & Ciliberto, S. (1995) *Phys. Rev. Lett.* **74**, 1986.

Scalo, J. (1985) in **Protostars and Planets** Vol. II, p. 201, D. Black & M. Matthews Eds, Tucson, University of Arizona Press, Tucson.

Schertzer, D., Lovejoy, S. & Schmidt, F. (1995) in "Small–scale structures in fluids and MHD", M. Meneguzzi, A. Pouquet & P.L. Sulem Editors, Notes in Physics **462**, Springer–Verlag.

She, Z.S. & E. Lévêque (1994) *Phys. Rev. Lett.* **72**, 336.

Slavin, J. & Cox, D. (1992) *Astrophys. J.* **392**, 131.

Slavin, J. & Cox, D. (1993) *Astrophys. J.* **417**, 187.

Spangler, S. (1990) *Phys. Fluids* **2**, 407.

Spangler, S. (1994) in Proc. Kyoto Conference on *Chaos and Nonlinearity in Space Plasmas* T. Hada Ed., World Scientific, Singapore.

Steenbeck, M., Krause, F. & Rädler, K.H. (1966) *Z. Naturforsch* **21a**, 369. See also the translation in P.H. Roberts & M. Stix, Technical Notes NCAR–TN/IA–60, p.29, Boulder, Colorado (1971).

Stone, J. & Norman, M. (1994) *Astrophys. J.* **420**, 237.

Stone, J., Xu, J. & Mundy, L. (1995) *Nature* **377**, 315.

Stribling, T. & Matthaeus, W. (1990) *Phys. Fluids B* **2**, 1979.

Thomas, J. (1968) *Phys. Fluids* **11**, 1245.

Thomas, J. (1970) *Phys. Fluids* **13**, 1877.

Vazquez, E., Passot, T. & Pouquet, A. (1995a) **441**, 702.

Vazquez, E., Passot, T. & Pouquet, A. (1995b) *Astrophys. J.*, **455**, 447.

Vincent, A. & Meneguzzi, M. (1991) *J. Fluid Mech.* **225**, 1.

Vlahos, L. (1993) *Adv. Space Res.* **13**, 122.

Yaglom,A. (1949) *Dokl. Aced. Nauk SSSR* **69**, 743.

NUMERICAL SIMULATION OF SOLAR AND ASTROPHYSICAL MHD FLOWS

Jets, Loops, and Flares

KAZUNARI SHIBATA
National Astronomical Observatory
Mitaka, Tokyo 181, JAPAN
shibata@spot.mtk.nao.ac.jp

Abstract. The recent development of supercomputers enabled us to perform two or three dimensional MHD numerical simulations of dynamic phenomena in the solar atmosphere and in astrophysical gas layers. We are now able to compare simulation movies with real observational movies to clarify the origin of various dynamic phenomena such as jets, emerging magnetic loops, and flares. In this article, I will first summarize the difficulties and richness intrinsic to solar and similar astrophysical MHD simulations, and then show several examples of simulation results which answer various questions presented by old and new observations of solar and astrophysical MHD flows, in particular, in relation to jets, loops, and flares. The subjects treated in this article are solar jets (spicules and surges) accelerated by nonlinear MHD waves, astrophysical jets ejected from accretion disks via magnetic forces, the nonlinear evolution of the Parker instability in galactic and accretion disks, emerging magnetic loops in the solar atmosphere, magnetic reconnection driven by the Parker instability as a model of solar X-ray jets and compact flares, and finally magnetic reconnection in protostellar magnetospheres.

1. Introduction

In the latter half of this century, it has been revealed that our universe is full of active phenomena, such as jets ejected from active galactic nuclei (AGN), bipolar jets from young stellar objects (YSO) (e.g., Burgarella et al. 1993), various bursts in close binary systems, flares and coronal loops in the Sun and stars (e.g., Haisch et al. 1991), and so on. In the case of the solar phenomena, it has been established that magnetic fields play a dominant role in generating enormous activity in the solar atmosphere and

217

K. C. Tsinganos (ed.), Solar and Astrophysical Magnetohydrodynamic Flows, 217–247.
© 1996 *Kluwer Academic Publishers.*

218

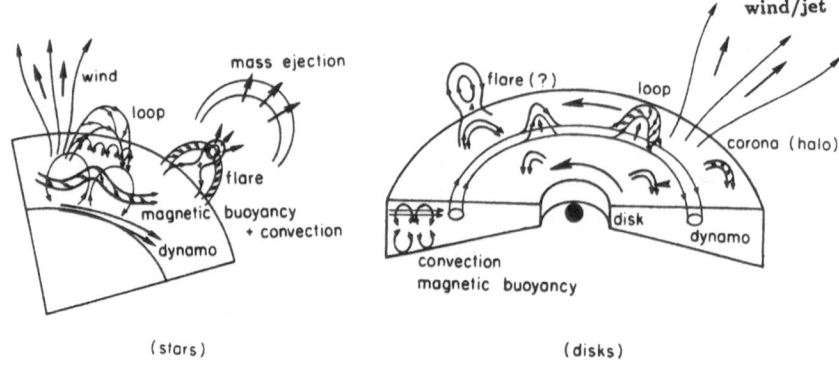

Figure 1. Schematic illustration of magnetic activity in stars and disks (accretion disks and galactic disks).

magnetohydrodynamics (MHD) is the basis for understanding macroscopic processes in these active phenomena (e.g., Parker 1979, Priest 1982). In the case of non-solar active phenomena, it has not yet been observationally established that magnetic fields are playing fundamental role, but recently many people began to recognize the importance of magnetic fields in astrophysical activity, such as in AGN and YSO's (e.g., Blandford 1993, Shu et al. 1987). In fact, recent developments in observations of radio and X-ray astronomy have revealed many active phenomena which are similar to solar magnetic phenomena as sketched in Fig. 1. Hence it is clear that MHD is the basis for understanding astrophysical magnetic activity, and for this reason, we have to solve the MHD equations.

As is well known, the MHD equations are a very complex set of nonlinear equations (e.g., Tsinganos et al. 1996). It is not easy to analytically solve them especially if we want to solve them in their full form (i.e., nonsteady, nonlinear, three dimensional, including non-ideal effects such as resistivity, viscosity, heat conduction, radiative cooling, and etc.). Even if spatial dimension is reduced to two or one, the equations are still complex enough to prevent analytical treatment, and only self-similar analytic solutions exist for axisymmetric systems (Blandford and Payne 1982, Tsinganos and Trussoni 1991, Tsinganos et al. 1993).

This is the reason why we have to perform numerical simulations using computers. Here, the word *numerical simulation* is defined as *to obtain numerical solutions of the time-dependent, nonlinear, MHD partial differential equations.*

It has been considered that theory and experiment are the two basic methods used in the physical sciences. The recent rapid development of computers has changed this situation dramatically, and now we can say that *theory, experiment, and computer simulations* are the three basic methods used in the physical sciences.

This is also true in our field, *solar and astrophysical MHD*. Owing to recent development of supercomuters, it is now possible to perform two or three dimensional MHD numerical simulations of dynamic phenomena in the solar atmosphere and astrophysical gas layers, such as acceretion disks and galactic disks, in realistic situations. We are now able to compare simulation movies with real observational movies (in the case of the Sun) to explore the origin of various dynamic phenomena such as jets, emerging magnetic loops, and flares. If we succeed to reproduce the observed phenomena in computers, it would be possible to find some hints (and even physical conditions of enigmatic phenomena) which are not easy to be found by observations or experiments. When the observational data are so scarce, we can study basic physics by performing numerical experiment of basic physical processes by changing parameters extensively. It is easy to compare AGN jets with solar jets according to numerical experiment. Even an experiment (with unrealistic physical parameters) to understand basic physics is possible by using numerical experiment.

Consequently, the purposes of numerical simulations in solar and astrophysical MHD are two-fold :

(1) realistic modeling of observed phenomena,

(2) numerical experiment of basic physical processes.

In this article, starting from the discussion on the difficulties and richness intrinsic to solar and astrophysical MHD simulations, I will discuss many examples of simulation results which answer various questions presented by old and new observations of solar and astrophysical MHD flows, in particular, in relation to jets, loops, and flares, keeping above two purposes in mind.

2. Difficulties in Simulating MHD Phenomena in the Solar Atmosphere and Astrophysical Gas Layers

Since there are many excellent papers and books on the method of numerical simulations (e.g., Tajima 1989), I will not discuss details of numerical methods. Instead, I will discuss here some basic points in numerical solar and astrophysical MHD which have not been discussed in other papers and books.

The solar atmosphere and astrophysical gas layers (galactic and accretion disks) are *gravitationally stratified gas layers*. This is a very basic point of numerical solar and astrophysical MHD. Many difficulties (and richness) in simulating solar and astrophysical MHD come from this point. There are essentially three basic properties of gravitationally stratified gas layer.

(1) *Large dynamic range in density (ρ), gas pressure (p_g), and β(= p_g/p_{mag})*. For example, the solar photosphere and chromosphere extends to more than $(10 - 15)H$, where

$$H = \frac{C_s{}^2}{\gamma g} = \frac{R_g T}{\mu g} \simeq 150 \left(\frac{\mu}{1.2}\right)^{-1} \left(\frac{T}{6000\text{K}}\right) \left(\frac{g}{g_\odot}\right)^{-1} \quad \text{km}$$

is the pressure scale height in these (nearly isothermal) layers. Namely, the density decreases by more than $5 - 7$ orders of magnitude from the photosphere to the corona. Since the grid size in the vertical (z) direction should be less than $0.15 - 0.3H$, this means that we require many grid points ($> 50 - 100$) in the vertical direction. Large dynamic range in $\beta \sim C_s{}^2/V_A{}^2$ is a result of large dynamic range of sound speed C_s and/or Alfven speed V_A, which limits the time step Δt significantly through the CFL condition, $\Delta t < \Delta x/(C_s{}^2 + V_A{}^2 + V^2)^{1/2}$, when we use explicit scheme.

(2) *Wave amplification through vertical propagation.* Since the density decreases rapidly with height, the amplitude of an MHD wave increases with height when the wave propagates from the bottom of the atmosphere to the top. For example, the energy flux carried by an acoustic wave (or a slow mode MHD wave) propagating along the vertical magnetic flux tube with cross section A becomes $\rho V_\parallel{}^2 C_s A = $ constant, if the WKB approximation holds. From this, we find

$$V_\parallel \propto \rho^{-1/2} A^{-1/2}$$

since the temperature is nearly constant in the photosphere and chromosphere within a factor of 2. Hence if $A_{cor}/A_{ph} \sim 100$ (typical value for the ratio of the cross-section of a flux tube in the photosphere, A_{ph}, to that in the corona, A_{cor}, in the quiet region of the Sun) and $\rho_{tr}/\rho_{ph} = 10^{-6}$, the amplitude of the acoustic wave (or slow mode) is enormously amplified by 2 orders of magnitude, say from 1 km/s in the photosphere to 100 km/s in the transition region if the wave remains a linear wave, where ρ_{tr} and ρ_{ph} are the densities of the transition region and photosphere, respectively. (Actually, nonlinear effects and the associated dissipation become important in the upper layer. See next section.) Note that in the case of MHD waves, approximate one dimensional propagation along the flux tube is realized (for slow mode and Alfven mode), so that the wave amplification is stronger than in the case of non-magnetic acoustic waves (e.g., Shibata 1983).

(3) *Various instabilities driven by gravitational energy.* In a gravitationally stratified gas layer, various instabilities occur owing to gravitational

free energy, such as convective instability, Rayleigh-Taylor instability, magnetic buoyancy instability, Parker instability, etc. These instabilities and resulting mass motion generate various MHD waves and electric currents, which could be the source of energy to heat the chromosphere and corona. Even a more vigorous activity such as flares could be a result of release of magnetic free energy which is stored by dynamo action and/or interaction of a flux tube with turbulent motion in the convection zone. Note that the magnetic field significantly enhances the activity of gravitationally stratified gas layers through *coupling with convection and rotation* and *magnetic buoyancy* (Parker 1979).

All these three factors introduce various difficulties in actual numerical simulations. For example, even a small numerical error at the bottom of the atmosphere can be amplified enormously in the upper atmosphere, and can affect the dynamics seriously. Hence the treatment of the lower boundary is the most important in this kind of numerical simulations. Various physical instabilities should be discriminated from numerical instabilities, and thus we must be well aware of the physics of these instabilities.

On the other hand, these factors introduce richness (and interesting physics) to the problem. Enormously amplified waves can accelerate a jet along the tube and can heat the atmosphere. Magnetic flux loops rapidly emerge from below the photosphere to the corona, and interact with ambient fields, leading to violent reconnection and jets.

These vigorous activity intrinsic in a gravitationally stratified magnetized gas layers, in turn, introduce another kind of numerical difficulties; i.e., strong MHD waves are generated by these active phenomena. These waves are reflected at the numerical boundary, which often produces physical/numerical resonances near the upper boundary, leading to numerical instability. Hence, the treatment of the upper boundary is also one of the most important (and difficult) problems in numerical solar and astrophysical MHD. There are a number of methods to overcome this difficulty, though there is no almighty one. The simplest and the safest "method" is to move the upper boundary as far as possible using non-uniform grids, which physically eliminates the chance of wave reflection.

3. Jets – Plasma Acceleration due to Nonlinear MHD Waves and Centrifugal Force

3.1. SOLAR JETS: SPICULES AND SURGES

3.1.1. *Slow Shock Acceleration*

As we discussed in the previous section, a small amplitude slow mode MHD wave (acoustic wave along a flux tube) grows enormously during the prop-

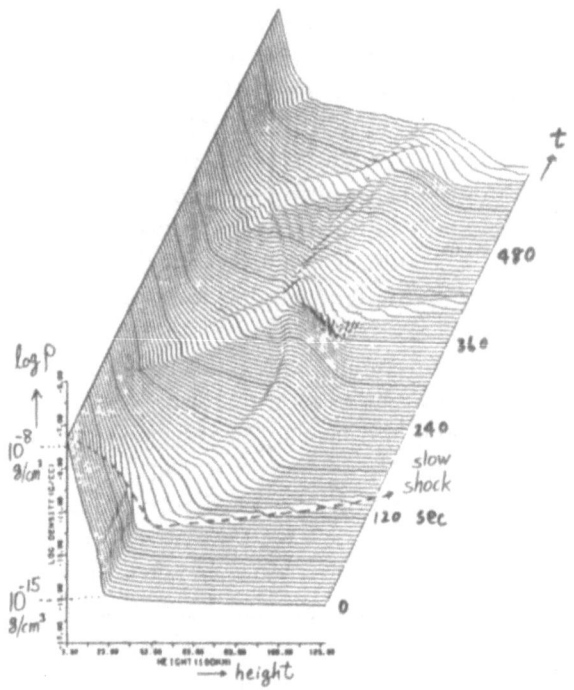

Figure 2. Spicule model by Suematsu et al. (1982). The density distribution is shown as a function of height and time. Note that the slow mode wave (acoustic wave in a flux tube) grows as it propagates upward and finally collides with the transition region, creating a spicule-like ejection.

agation along a vertical flux tube because of steep density gradients in these atmospheres. Even if the original velocity amplitude is well below the sound speed at the photospheric level, the velocity amplitude can easily exceed the sound speed in the chromosphere, and the wave can evolve into a shock. Once the shock is created, it can decay and heat the chromosphere. This is essentially the same as the scenario of classical acoustic shock coronal/chromospheric heating theory (e.g., Ulmschneider et al. 1991 for a review). The slow shock finally collides with the transition region (a kind of contact discontinuity) to accelerate the transition region as well as chromospheric plasmas in the upward direction. Such a chromospheric ejection might correspond to a spicule.

This scenario was first proposed by Osterbrock (1961), and was confirmed later by Suematsu et al. (1982) by performing full nonlinear 1D simulations (Fig. 2). In the simulation model of Suematsu et al. (1982), it was assumed that a sudden pressure enhancement due to a small flare-like bright point at the footpoint of spicules generates slow mode waves which eventually become shocks and accelerate spicules.

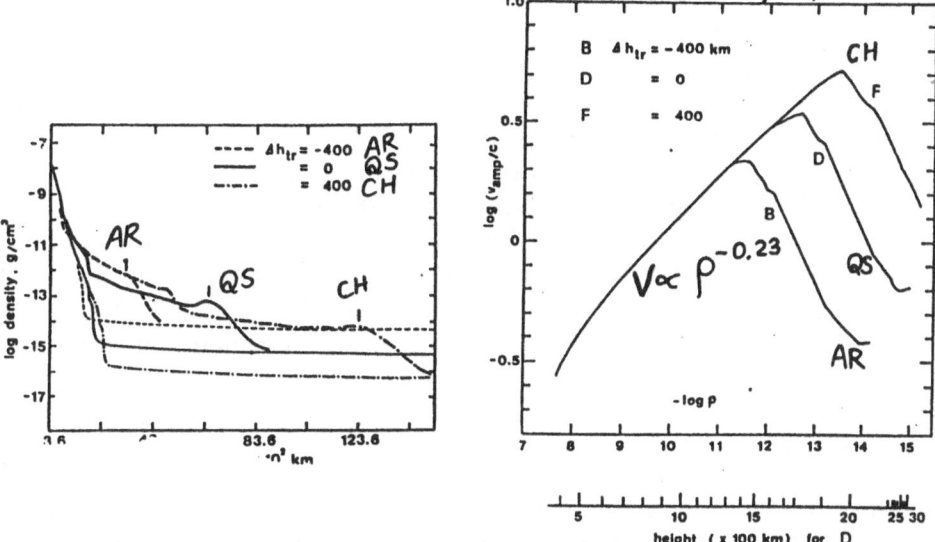

Figure 3. Right: Growth of a slow shock for various atmospheres with different transition region heights (Shibata and Suematsu 1982). Left: Density distribution in height for three "spicule" models with different initial transition region heights.

Shibata et al. (1982) showed that if the sudden pressure enhancement occurs above the middle chromosphere, the jets are directly accelerated by the *pressure gradient force* in the pressure enhancement region, not by the shock (see also Sterling et al. 1993 for the effect of radiation and conduction). In other words, if the sudden pressure increase (or energy release) occurs *below the middle chromosphere*, the jets are accelerated by the *slow shock*, not by the pressure gradient force in the energy release region. This can be also applied to other acceleration mechanisms, such as magnetic acceleration, if it occurs in the photosphere. That is, if the sudden plasma acceleration occurs *along the global flux tube* in the photosphere, the final jets ejected into the corona are not the jets accelerated in the photosphere, but those accelerated in the upper chromosphere by the *slow shock* which is generated by the plasma flow in the photosphere and has propagated to the upper chromosphere.

Shibata and Suematsu (1982) extended this model (*bright point - slow shock* model) to the cases of different initial transition region heights, and showed that spicules are taller in coronal holes and shorter in active regions; in the latter case, the height is so small that the ejection cannot be regarded as spicules. These theoretical predictions nicely fit with the observations that spicules are tall in polar coronal holes and absent in active region plages (Shibata and Suematsu 1982). The physical reason of this is as follows; (1) the rate of increase in velocity amplitude of strong shocks is $V_{\parallel} \propto$

$\rho^{-0.23}$ (Shibata and Suematsu 1982), and (2) the density ratio between the photosphere and the corona (ρ_{ph}/ρ_{co}) is larger in coronal holes and smaller in active regions than in quiet region (Fig. 3).

Hollweg (1982) proposed a similar but slightly different idea, i.e., the *rebound slow shock model* in which spicules are repeatedly accelerated by multiple shocks originating from the wake oscillating at the acoustic cut off frequency (see also Sterling and Hollweg 1988).

It is interesting to note that the physics of slow shock propagation discussed above is basically the same as that of shock propagation in type II supernova explosion (e.g., Ohno et al. 1961, Colgate and White 1967). The only difference is that the former is confined in a magnetic flux tube, but the latter is a spherical shock propagation.

3.1.2. *Nonlinear Alfven Wave (or Magnetic Twist) Acceleration*

Similarly to the slow mode, the amplitude of the Alfven mode also grows during the propagation along the vertical flux tube. In this case, from the conservation of energy flux $\rho V_\perp^2 V_A A = $ constant (WKB approximation) and $A \propto 1/B$, the amplitude $V_\perp$ becomes

$$V_\perp \propto \rho^{-1/4}.$$

Nonlinearity $V_\perp/V_A$ is then

$$V_\perp/V_A \sim B_\perp/B \propto \rho^{1/4} B^{-1}.$$

For the quiet solar atmosphere $(\rho_{tr}/\rho_{ph} \sim 10^{-5}, B_{tr}/B_{ph} \sim 10^{-2})$, we find that $V_{\perp,tr}/V_{\perp,ph} \sim 20$, $(V_\perp/V_A)_{tr}/(V_\perp/V_A)_{ph} \sim 5$. Thus a linear Alfven wave can evolve into a nonlinear wave if the flux tube expands significantly like a solar flux tube in a quiet region.

If the nonlinearity becomes important, the magnetic pressure force $\nabla(B_\perp^2/8\pi)$ can push the plasma parallel to the unperturbed flux tube, so that the plasma can be accelerated along the global flux tube. In other words, the nonlinear magnetic pressure force generates longitudinal plasma motion or a slow mode MHD wave. In the case of the nonlinear torsional Alfven wave (Hollweg et al. 1982, Shibata and Uchida 1985), not only the magnetic pressure force but also the centrifugal force $\rho V_\perp^2/r$ can accelerate the plasma along the global flux tube, where r is the radius of the flux tube.

Hollweg et al. (1982) studied the nonlinear propagation of torsional Alfven waves along the vertical (expanding) magnetic flux tube from the photosphere to the corona, by performing 1.5D MHD nonlinear simulations. They found that a spicule-like chromospheric ejection is generated when a large amplitude nonlinear torsional Alfven wave (equivalent to switch-on fast shock in this geometry) collides with the transition region. This process is similar to that in the case of the collision of a slow shock with the transition region (Fig. 2 and Shibata et al. 1982). (See also later related works by Mariska and Hollweg 1985, Hollweg 1992).

Shibata and Uchida (1985) studied the plasma acceleration associated with the sudden relaxation and propagation of the nonlinear magnetic twist, and applied the results to solar surges. They proposed that such sudden relaxation of nonlinear magnetic twist would be possible if the reconnection occurs between an untwisted flux tube and a twisted flux tube (Shibata and Uchida 1986b). On the other hand, if the magnetic twist is generated continuously by the interaction of the accretion disk with global poloidal magnetic fields, a similar magnetic twist jet would be produced and might explain astrophysical jets ejected from accretion disks. This *sweeping-magnetic-twist* jet model for astrophysical jets has been developed by Shibata and Uchida (1986a, 1987, 1990) incorporating the effects of accretion disk, and is applied to bipolar molecular flows in star forming regions by Uchida and Shibata (1985), which will be discussed in detail in the next subsection.

3.2. ASTROPHYSICAL JETS EJECTED FROM ACCRETION DISKS

3.2.1. *Nonsteady MHD Jets from Thin Disks – The Sweeping Magnetic Twist Mechanism*

If a strong poloidal magnetic field penetrates an accretion disks, what would occur ? Shibata and Uchida (1986a, 1987, 1990) studied this problem in detail as an initial value problem (see Figs. 4 and 5), and the results have been applied to bipolar molecular flows in star forming regions (Uchida and Shibata 1985). The initial condition of their model is as follows. A geometrically thin disk (with $(C_s/V_k)^2 \sim 10^{-2} - 10^{-3}$) is rotating around a point mass at the center. A uniform magnetic field (with $(V_A/V_k)^2 \sim 10^{-2} - 10^{-3}$) penetrates the disk vertically, and there is a non-rotating hot corona outside the disk. In the disk, the ratio of rotational velocity to the Keplerian velocity is taken to be a free parameter including both Keplerian and sub-Keplerian cases ($V_\varphi/V_k = 0.6 - 1.0$). In the Keplerian case, the rotating disk pulls the poloidal field lines toward the azimuthal direction, generating toroidal fields which propagate along the poloidal field lines as torsional Alfven waves. This process extracts angular momentum from the disk (i.e., *magnetic braking*), and hence the disk begins to contract, eventually producing bipolar jets when the toroidal fields get strong enough. In the sub-Keplerian case, these processes become more vigorous because the generation of toroidal fields is stronger and faster. Figures 4 and 5 show one typical example of such sub-Keplerian case. Note that the essential point of this process is independent of whether the rotaional velocity of the disk is Keplerian or not.

226

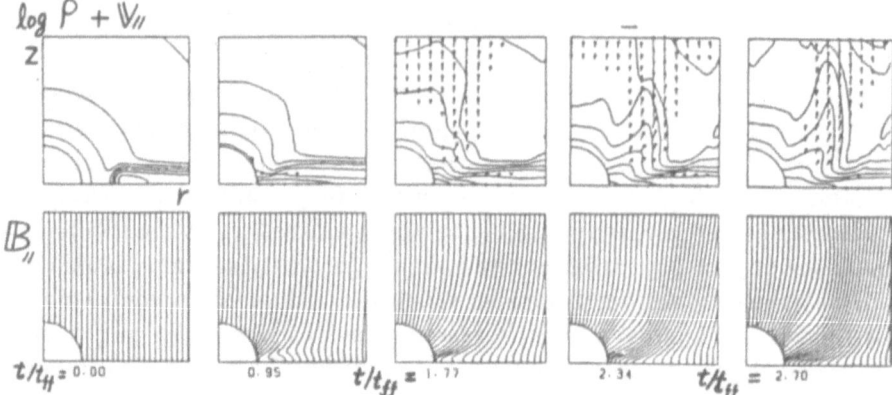

Figure 4. *Sweeping magnetic twist* jet model for astrophysical jets ejected from an accretion disk (Shibata and Uchida 1986a). (a) Density and velocity, (b) poloidal magnetic field lines. The times are in units of the free fall time at the inner edge of the initial disk.

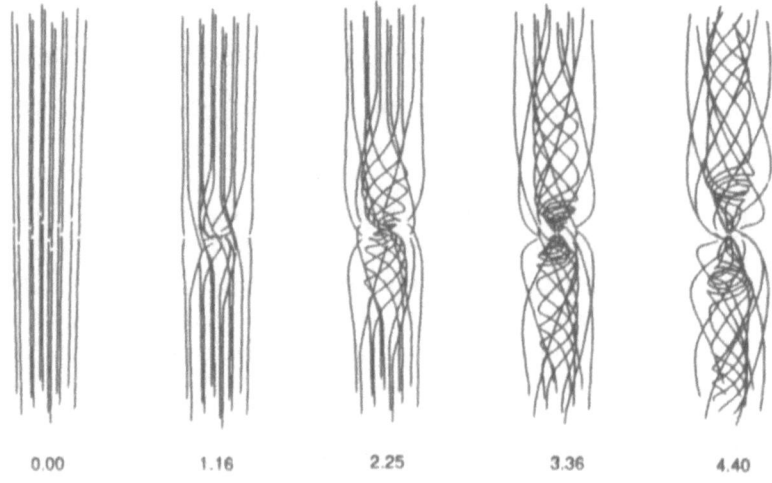

0.00   1.16   2.25   3.36   4.40

Figure 5. 3D representaion of magnetic field lines in the *sweeping-magnetic-twist* jet model (Shibata and Uchida 1990). The times are in units of the free fall time at the inner edge of the initial disk.

The acceleration of the jet is due to the $\mathbf{J} \times \mathbf{B}$ force (magnetic pressure gradient force $\nabla B_{\varphi}^{2}/8\pi$) and centrifugal force. This jet is a nonsteady version of a magneto-centrifugally driven wind/jet (e.g., Blandford and Payne 1982, Pudritz and Norman 1986). The jet has a hollow cylindrical shell structure with a helical motion in it, and these characteristics were actually found in the L1551 bipolar molecular flows (Uchida et al. 1987; though see Moriaty-Schieven and Wannier 1991 for a different interpretation) and

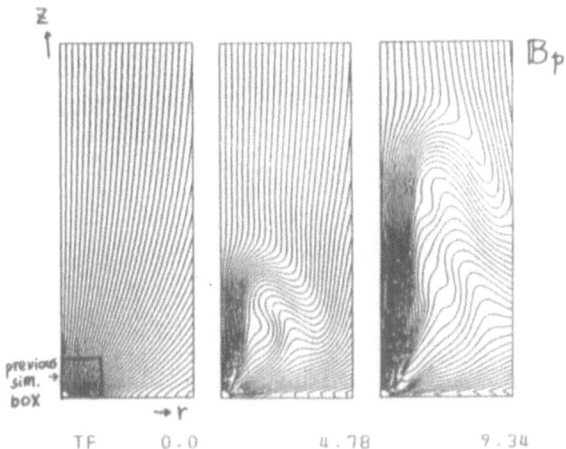

Figure 6. Large scale behavior of a *sweeping-magnetic-twist* jet when a jet propagates into diverging magnetic fields. Note the vigorous collimation by the toroidal fields. The times are in unit of the free fall time at the inner edge of the initial disk. A small square box in the left bottom corner of the $t = 0$ figure is the simulation box adopted by Shibata and Uchida (1986a).

in the Galactic center radio lobes (Sofue and Handa 1984, Uchida, Shibata, Sofue 1985, Shibata and Uchida 1987). The velocity of the jets produced by this mechanism is comparable to the Keplerian velocity at the footpoints of the jet, $V_{jet} \sim (1 - 2)V_k$, and is in proportion to $B^{0.5-0.7}$. [1] The latter is similar to Michel's minimum energy solution (Michel 1969)

$$V_\infty \sim (\Omega^2 B^2 R^2 / \dot{M})^{1/3} \propto B^{2/3},$$

where R is the radius of the disk at the footpoints of the jet.

Near the disk, the collimation of the jet is due to the poloidal field. If the initial poloidal field diverges at a larger distance, however, the effect of the toroidal field increases as the jet propagates farther, and eventually collimates the jet (Fig. 6). This is the same as in the collimation of a centrifugally driven jet/wind (Sakurai 1987, Heyvaerts and Norman 1989, Sauty and Tsinganos 1994).

3.2.2. *Relation to the Magneto-Rotational (Balbus-Hawley) Instability*

Stone and Norman (1994) studied the same problem as that of Shibata and Uchida (1986a), and confirmed the basic points on the production of jets.

---

[1]Sauty and Tsinganos (1994) also found $V_{jet} \sim (1 - 2)V_k$ in somewhat different situations.

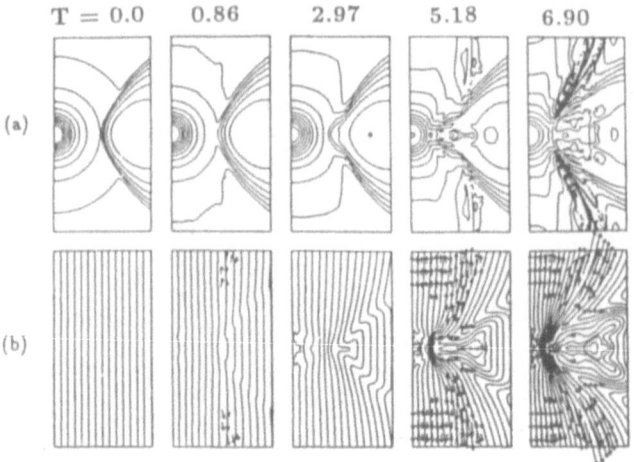

T = 0.0   0.86   2.97   5.18   6.90

(a)

(b)

Figure 7.   Sweeping magnetic twist jet ejected from a thick disk (Matsumoto et al. 1996a). The times are in units of the free fall time at the inner edge of the initial disk.

The new feature they found is the magneto-rotational instability (Balbus and Hawley 1991) when the initial magnetic field is weak. [2]

It is interesting to note that some of the key features of this magneto-rotational instability, such as the avalanche (accretion) flow at the surface of the disk, have already been seen in some results of Shibata and Uchida's simulations (1986, 1987, 1990) as well as in the early results of Matsumoto et al.'s simulations of magnetized thick disks before 1990 (see below). Although this feature was physically understood at that time, it has unfortunately never been explicitly emphasized nor been studied in detail. This is because the spatial resolution of these simulations was not good enough. ¿From this history, we can learn that we have to consider numerical results more seriously even if the spatial resolution is not so good. There are a lot of hints in numerical simulation results from which we can develop important theories.

3.2.3. *Nonsteady MHD Jets from Thick Disks*

In order to develop an AGN jet model, Matsumoto et al. (1996a) studied the case of thick disks penetrated by the poloidal field, by performing nonsteady 2.5D MHD simulations similar to Shibata and Uchida (1986a). Their results (see Fig. 7) show; (1) Avalanche (accretion) flow occurs along the surface of thick disks. (2) Magneto-rotational instability (Balbus and Hawley 1991)

---

[2]It may be said that even the basic physics of nonsteady MHD jet-accretion (Shibata and Uchida 1986a, Stone and Norman 1994) is the same as that of the Balbus-Hawley (1991) instability.

occurs inside thick disks. (3) The velocity of jets is again comparable to the Keplerian velocity, $V_{jet} \sim 0.6 - 2.3V_k \propto B^{0.15-0.25}$ if initially $(V_A/V_\varphi)^2 < 10^{-3}$. (4) The mass loss rate by the jet $\dot{M}$ is in proportion to the magnetic field strength; $\dot{M} \propto B$.

3.2.4. *Relation between Nonsteady and Steady Jets*

What is the relation between these nonsteady MHD jets and steady magnetically driven jets (such as in Blandford and Payne 1982) ? In order to answer this question, Kudoh and Shibata (1995) studied one dimensional (1.5D) steady magnetically driven jets along a fixed poloidal field line for a wide range of parameters, assuming the shape of the poloidal magnetic field line. There are two free parameters in their problem;

$$E_{mg} = \left(\frac{V_A}{V_k}\right)^2 \simeq 3.8 \times 10^{-4} \left(\frac{B}{1G}\right)^2 \left(\frac{M}{M_\odot}\right)^{-1} \left(\frac{n}{10^{12}\text{cm}^{-3}}\right)^{-1} \left(\frac{R}{15R_\odot}\right),$$

$$E_{th} = \frac{1}{\gamma}\left(\frac{C_s}{V_k}\right)^2 \simeq 6.5 \times 10^{-3} \left(\frac{T_d}{10^4\text{K}}\right) \left(\frac{M}{M_\odot}\right)^{-1} \left(\frac{R}{15R_\odot}\right),$$

where V_A is the Alfven speed based on the poloidal field, and the physical values in these equations are suitable for protostellar jets. They found that the inclination angle of poloidal magnetic field lines at the surface of accretion disks is very important to determine the properties of the jet as first noted by Blandford and Payne (1982) and later stressed by Cao and Spruit (1994). As the angle between the poloidal field and the disk surface decreases, the mass flux of the jets increases. If the angle becomes less than 60 degrees, a high mass flux jet with strong toroidal fields can arise, depending on the poloidal field strength. Namely, in such low angle case, the solution can be generally classified into two branches,

(1) *centrifugally driven jet*,
(2) *magnetic pressure driven jet*.

The former arises when the poloidal field is strong, i.e., the poloidal field is dominant near the surface of the disk, whereas the latter arises when the poloidal field is weak, i.e., the toroidal field is dominant near the disk (Fig. 8). In the former case, the main acceleration occurs below the Alfven point by the centrifugal force, whereas in the latter the acceleration occurs above the Alfven point by the magnetic pressure force of the toroidal field. The latter branch corresponds to the steady MHD jet model developed by Lovelace et al. (1991; see also related 2.5D work by Ustyugova et al. 1995). In the case of a *magnetic pressure driven jet*, (i.e., $E_{mg} < 0.01$), Kudoh and Shibata (1995) found;

(1) the mass loss rate is in proportion to the magnetic field strength, i.e.,

$$\dot{M} \propto E_{mg}^{0.5} \propto B,$$

. 230

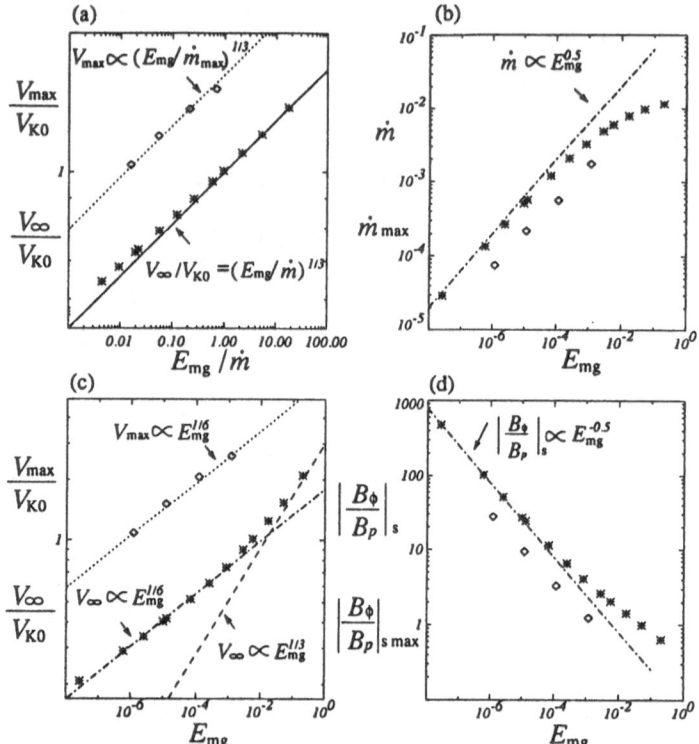

Figure 8. 1.5D steady and nonsteady magnetically driven jet models developed by Kudoh and Shibata (1995, 1996a,b). (a) Terminal velocity for steady case (asterisk) and maximum velocity for nonsteady case (diamond) as a function of $E_{mg}/\dot{m}$, where $\dot{m}$ is normalized by $\rho_d V_k R^2$. The solid line shows Michel's minimum energy solution. (b) The normalized mass flux versus E_{mg}. (c) Terminal (or maximum) velocity as a function of E_{mg}. (d) The ratio of the toroidal field to the poloidal field at the slow point, B_φ/B_p, versus E_{mg}. In these figures, the steady solution is denoted by asterisks, and the nonsteady solution is shown by diamonds.

(2) the terminal velocity V_∞ of the jet becomes Michel's minimum energy solution, but the dependence of V_∞ on B is weaker than had been thought for a centrifugally driven jet because of $\dot{M} \propto B$;

$$V_\infty \propto (E_{mg}/\dot{M})^{1/3} \propto B^{1/3},$$

(3) the terminal velocity is of the order of Keplerian velocity at the footpoint of the jet for a wide range of parameters; $V_\infty \propto 0.5 - 4.0 V_k$ for $10^{-5} < E_{mg} < 0.1$.

These results explain the simulation results of Matsumoto et al. (1996a) very well, and explain also previous 2.5D nonsteady MHD simulations (Shibata and Uchida 1986a, 1987, 1990, Stone and Norman 1994) which showed that the velocity of the nonsteady jet is of the order of the Keplerian velocity and the toroidal magnetic field is dominant near the disk. [3]

On the basis of these results, Kudoh and Shibata (1995) predicted the physical condition of the disk from which YSO optical jets and/or high velocity neutral winds are ejected. Namely, the jet production radius must be less than 20 $R_\odot$ (for one solar mass protostar) to explain the velocity of optical jets and/or high velocity neutral wind, and magnetic field strength > 3 G, and the temperature must be larger than 3000 K at $r \simeq 20R_\odot$.

Using the same poloidal field line configuration, Kudoh and Shibata (1996a,b) further studied nonsteady 1.5D MHD simulations assuming the initial and boundary conditions similar to those of Shibata and Uchida (1986a) to see the relation between nonsteady jets and steady jets. Note that the boundary condition of this problem is not suitable to get steady solutions. Nevertheless, such boundary conditions are adopted to see the long term behavior of the Shibata-Uchida simulation model. Note also that still at present it is not easy to perform long term 2.5D MHD simulations of this model. Although the boundary conditions are not suitable to get steady solution, Kudoh and Shibata found that the solution show some characteristics of the steady solution in the early stages of evolution, around 10 orbital periods. When the magnetic field is weak, the positions of Alfven and slow points agree well with those of the steady solution. Even in the cases where these positions do not agree with those of the steady solution, the dependences of maximum velocity and mass loss rate of jets upon E_{mg} agree with those found in the steady solution. Since the dynamics of 1.5D MHD model are common with that of 2.5D MHD model except for the accretion, we can say that characteristics of nonsteady jets found in 2.5D MHD models are essentially similar to those of the steady models. For example, the maximum velocity of jets found in 2.5D MHD model ($V_{jet} \sim 1 - 2V_k$) is explained by the steady models very well.

3.2.5. *Three Dimensional Propagation of MHD Jets*

Todo et al. (1993) have studied the stability of MHD jets by performing three dimensional MHD simulations. In their model, a jet was assumed initially to be propagating along helically twisted flux tube. The initial helical field is force free and stable to the kink instability. A dense region is located somewhere ahead of the jet, and the subsequent interaction of the helical jet with the dense region was studied in detail. (See Todo et al. 1992 for the 2.5D version, and Uchida et al. 1992 for the 1.5D version.) It was found that

[3]Exactly speaking, the results of Shibata and Uchida (1986a) are an intermediate case between *centrifugally driven jets* and *magnetic pressure driven jets*, and hence V_{jet} showed stronger dependence on B than in a pure magnetic pressure driven jet.

the collision of the *magnetic-twist-jet* with the dense region produce strong magnetic twists $(B_\varphi \gg B_z)$ (see also Shibata and Uchida 1990) which eventually become unstable to the kink instability in three dimensional space. The resulting magnetic field and velocity field configurations are similar to those observed around the filamentary sructures seen in some HH objects in star forming regions (Todo et al. 1993).

4. Loops – Nonlinear Evolution of the Parker Instability

4.1. MAGNETIC LOOPS IN GALACTIC AND ACCRETION DISKS

4.1.1. *Typical Nonlinear Evolution of the Parker Instability: The Most Unstable Mode*

The *Parker instability* is the undular mode of *magnetic buoyancy instability* (Hughes and Proctor 1988). It occurs when the gas layer, which is supported by horizontal magnetic field against gravity, is perturbed to undulate (Parker 1966, 1979; see Mouschovias 1996 for an elementary explanation). It occurs also for the horizontal isolated flux tube embedded in the non-magnetized plasmas (e.g., Moreno-Insertis 1986, Schüssler 1995).

Matsumoto et al. (1988) have first peformed full nonlinear 2D simulations of the Parker instability. They considered a local part of the gas disk rotating around a point mass M. Taking a local cartesian coordinate (x, z), with x in the azimuthal direction and z in the vertical direction, they assumed that the magnetic field is initially parallel to the equatorial plane (in the x direction). It is further assumed that the magnetized gas layer is isothermal and is in magneto-hydrostatic equilibrium with uniform Alfven and sound speeds. The pure 2D perturbations ($k_y = 0$, where k_y is the wavenumber perpendicular to the magentic field lines) are considered and the effects of rotation, Coriolis force, and cosmic rays are all neglected. Ideal MHD with an isothermal equation of state is assumed.

Figure 9 shows a typical example of the simulation results in the case of the most unstable mode with parameters $\epsilon = (GM/R)/[(1+1/\beta)C_s^2] = 6$, $\beta = 8\pi p/B^2 = 1$. In this case, the periodic boundaries are assumed at $x = 0$ and $x = X_{max}$ and the small periodic undular perturbation was given. As the instability develops, the gas slides down the rising magnetic loop, forming a dense structure in the valley. Note that the perturbed magnetic field lines cross the equatorial plane of the disk as predicted by the linear stability theory (Horiuchi et al. 1988, Giz and Shu 1993). The growth of the perturbation is saturated when the maximum velocity of downflow becomes comparable to the initial Alfven speed V_A. The maximum velocity of the rising motion of the magnetic loop is $0.3 - 0.5V_A$. In this case, since the Alfven speed is larger than the sound speed (because initial $\beta = 2(C_s/V_A)^2 = 1$),

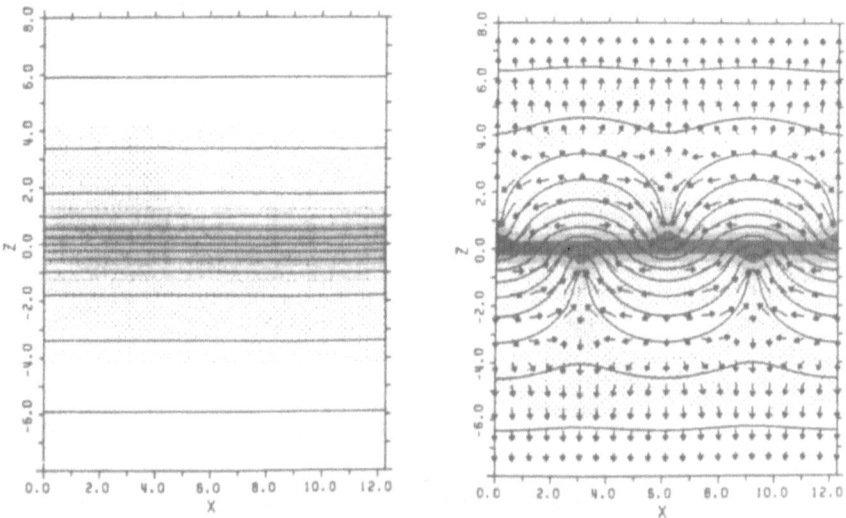

Figure 9. Left: Initial condition. Right: The nonlinear evolution of the Parker instability at $t/\tau = 5.99$, where τ is the linear growth time (Matsumoto et al. 1988). Distribution of density (grey scale), mangnetic field lines (solid curves) and velocity fields (arrows).

the downflow becomes supersonic (i.e., $V_{downflow} \sim V_A > C_s$), forming shock waves in the valley of the undulating field lines. This shock wave gradually propagates upward along the magnetic field line. After passing the shock, the downflow gas settles into a quasi-magneto-hydrostatic equilibrium state (Mouschovias 1974), forming a giant moledular cloud. (Similar results are obtained by Basu, Mouschovias, and Paleologou 1996.) On the other hand, if initial Alfven speed is less than the sound speed (e.g., $\beta = 10$), the downflow does not become supersonic so that the system does not reach the quasi static equilbrium state but undergoes nonlinear oscillation.

Whether shock waves occur or not is determined by whether the initial Alfven speed is larger than the sound speed or not, since the maximum speed of the downflow is comparable to the initial Alfven speed. Why is the downflow speed comparable to the initial Alfven speed ? The reason is as follows. The energy of the downflow comes from the gravitational potential energy stored initially in the magnetically supported gas layer, i.e., $gH \sim C_s^2 + V_A^2/2$. Since the plasmas settle into quasi-hydrostatic equilibrium along the curved magnetic loop, the final gravitational energy is written as $gH' \sim C_s^2$. Hence

$$V_{downflow}^2/2 \sim g(H - H') \sim V_A^2/2,$$

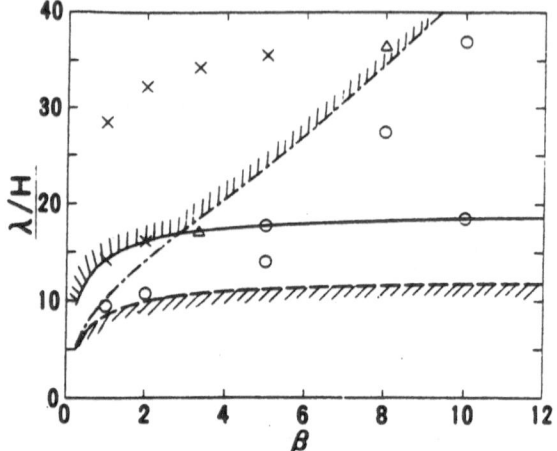

Figure 10. The parameter space for the *shock wave formation* (crosses) and the *nonlinear oscillation* (circles) (Matsumoto et al. 1990). This figure shows the wavelength λ vs. $\beta = p_g/p_{mag}$. The solid curve shows the most unstable wavelength, and the dash-dotted curve denotes λ_s (see the text). The dashed curve shows the critical wavelength of the linear Parker instability.

which leads to $V_{downflow} \sim V_A$. Note that this result is applicable only to the most unstable modes. If the instability occurs for longer wavelength modes, the downflow speed becomes larger (see Matsumoto et al. 1990 and next subsection).

4.1.2. *Condition of Shock Wave Formation and Nonlinear Oscillation*

In order to clarify the condition of shock wave formation and nonlinear oscillation in general situations, Matsumoto et al. (1990) have studied the effect of initial horizontal wavelength, λ. They adopted $\epsilon = 6$ model which corresponds to $p(z = \infty)/p_{disk} \simeq 2 \times 10^{-3}$, and studied various cases with $\lambda \neq \lambda_{max}$, where λ is the most unstable wavelength. They found (see Fig. 10) that

(1) If $\lambda > \lambda_s = (3.5\beta + 6)\Lambda$, shock waves are formed, and the system evolves into the quasi-static state, whereas if $\lambda < \lambda_s$, the nonlinear oscillation occurs. [4] Here $\Lambda = C_s^2(1 + 1/\beta)/g_{max}$.

(2) Once the nonlinear oscillation occurs, the nonlinear mode coupling occurs between different wavelength modes, increasing the wavelength of the perturbation with time up to λ_s. Hence eventually shock waves are formed, and the system evolves into the quasi-static equilibrium state. Note

---

[4]Note that this critical wavelength λ_s is different from the crtical wavelength λ_c for the linear Parker instability. This wavelength, λ_s, is determined by the nonlinear effect.

that this wavelength is longer than the linearly determined most unstable wavelength.

Summarizing these results, the length of magnetic loops in final equilibrium state is rougly given by $\max(\lambda_s, \lambda_{max})$, i.e.,

$$\lambda_{loop} \simeq \begin{cases} 6\pi[\beta/(\beta+2)]^{1/2}\Lambda = 6\pi(\beta+1)\beta^{-1/2}(\beta+2)^{-1/2}H, & (\text{ for } \beta < 3) \\ (3.5\beta+6)\Lambda = (3.5\beta+6)(1+1/\beta)H, & (\text{ for } \beta > 3) \end{cases}$$

where $H = C_s^2/g_{max}$ and $g_{max} = 0.385GM/R^2$. When applying this result to the actual accretion disk and galaxies, we can assume that H approximately corresponds to the thickness of the disk when $\beta > 1$. This result would be important to estimate the actual length of magnetic loops produced by the Parker instability in accretion disks and galactic disks.

4.1.3. Effect of Corona (Halo)

There is now increasing observational evidence that there is a hot corona around various astrophysical objects, such as stars, accretion disks, galactic disks, and so on. What is the effect of the corona upon the linear and nonlinear evolution of the Parker instability ? In the corona, the temperature is high so that the pressure scale height, H, is larger than that in the lower layer (disk or photosphere/chromosphere). Since the critical wavelength and the most unstable wavelength of the Parker instability are in proportion to H (Parker 1966, 1979), this means that the most unstable wavelength in the low temperature layer becomes *stable* in the corona. Consider the disk (either accretion disk or galactic disk) with a hot corona above and below it, and assume that the base height of the corona is $|z_{corona}|(> H)$ above/below the equatorial plane $z = 0$. Since the corona is stable for the most unstable wavelength in the disk, the instability is confined inside the disk, i.e., the effective wavelength in the vertical direction λ_z becomes comparable to z_{corona}. When λ_z decreases, the growth rate of the instability decreases and also the critical and most unstable wavelengths increase. Kamaya et al. (1996b) studied these points in detail.

It should be emphasized here that a similar stabilizing effect is found also in the case of non-uniform gravity in which the gravitational acceleration, g, decreases with height, such as in a point mass potential adopted by Matsumoto et al. (1988). This is because small g is equivalent to high temperature, T, since the instability wavelength is in proportion to the pressure scale height, $H \propto T/g$. In the simulation model of Matsumoto et al. (1988), the parameter $\epsilon = (GM/R)/[(1+1/\beta)C_s^2]$ is assumed to be 6. This value is much smaller than that expected for thin accretion disks and actual galactic disks, $\epsilon \sim 1000$. For this reason, this model was claimed to be irrelevant to real phenomena (e.g., Mouschovias 1996 private communication). However, considering the equivalence of small g and high T, the Matsumoto et al. model is effectively the same as that of a disk with a hot corona. This has been confirmed later by more realistic simulations (e.g., Shibata et al. 1990b). Hence the reader can safely apply the Matsumoto et al. (1988, 1990) results to real phenomena if he/she normalizes the length scale by the local pressure scale height in the disk (not by the radius of the disk).

Finally, it should be emphasized that the stabilizing effect of the corona works not only in the linear regime but also in the nonlinear regime. In fact, the ratio of the coronal pressure to the disk (or photospheric) pressure, p_{cor}/p_{ph} (or equivalely p_∞/p_{disk}), is very important for the nonlinear evolution of the Parker instability. If p_{cor}/p_{ph} (or p_∞/p_{disk}) is much smaller than unity, the upper part of the expanding magnetic loop cannot stop as long as p_{cor} is smaller than the magnetic pressure of the expanding loop. In this case, the expanding loop shows a self-similar evolution (Shibata et al. 1989a,b, 1990, and also section 4.2.1).

4.1.4. *Effect of Rotation and Shearing Motion*

Chou et al. (1995) performed 3D MHD simulations to study the effect of the Coriolis force on the nonlinear evolution of the Parker instability of an isolated magnetic flux tube in the isothermal gas layer which is rotating rigidly at angular speed Ω. They found that the rising magnetic flux tube is twisted by the Coriolis force to form a globally twisted flux tube (Fig. 11). This mechanism explains the observed tilt of bipolar sunspot groups.

Shibata and Matsumoto (1991), on the other hand, noted that the local twist may be generated in the downflow along the rising magnetic loop in addition to the global twist, since the downflowing blob or cloud contracts during the course of the downflow and suffer from the Coriolis force, similar to the usual contracting clouds in our Galaxy (Fig. 12). The sense of the rotation of the contracting cloud is the same as that of the rotation of the Galaxy. The twist accumulates in the valley of the undulating magnetic field lines, where the giant molecular cloud is formed. They applied this mechanism to explain the helical magnetic twist observed in a giant molecular cloud in the Galaxy (Uchida et al. 1990). Turbulent magnetic fields often observed in the valley of undulating magnetic fields (e.g., M31, see Beck et al. 1989) may also be explained by this mechanism. The twisted magentic field thus created is, in turn, favorable for the formation of the large scale cloud complex since the twist (magnetic shear) stabilizes the small scale interchange mode (Hanawa et al. 1992). It should be noted here that *the direction of this local twist in the rising magnetic loop is opposite to that of the global twist.* Hence, this mechanism may explain the origin

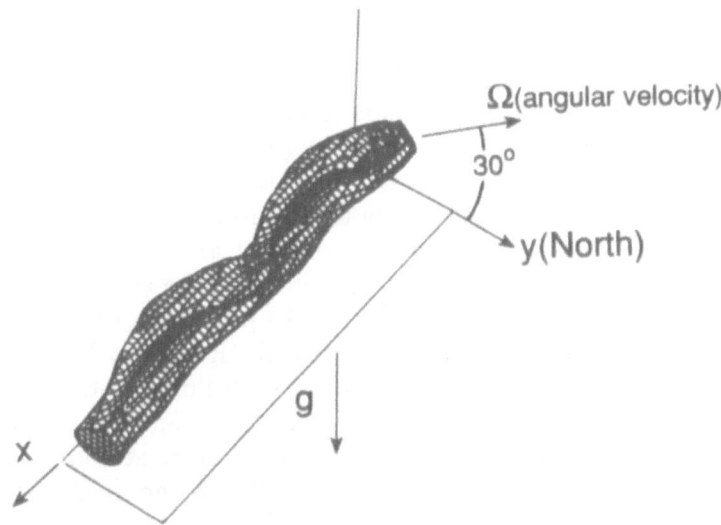

Figure 11. Twisted flux tube formed by the nonlinear evolution of the Parker instability in the presence of the Coriolis force (Chou et al. 1995).

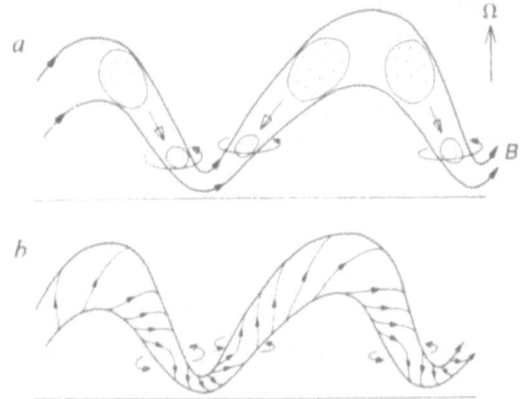

Figure 12. The formation of local twist by the effect of the Coriolis force on a contracting downflowing blob (or cloud) (Shibata and Matsumoto 1991).

of the magnetic twist observed in solar active regions (Pevtsov et al. 1994) and filaments (Rust 1994).

In actual rotating systems, there is a shear flow, i.e., differential rotation. What is the effect of a shearing flow on the Parker instability ? Shibata, Tajima, and Matsumoto (1990) studied this effect in a situation such that the horizontal magnetic flux (which is unstable to the Parker instability if there is no shear flow) is suffering from an amplification by the shear flow in an accretion disk. Usually the field amplification by the shear flow leads to instability or loss of equilibrium (as is well known in theoretical research of solar flares). However, against expectations, this case leads to an effective *stabilization* of the Parker instability, since the shearing time scale is shorter than the Parker time scale. As a result of this dynamic stabilization, the magnetic field cannot escape from the disk, and hence the disk evolves into the *low β* disk in which various violent phenomena (such as flares) can oocur *inside* the disk. On the other hand, if the shearing effect is weak, the magnetic flux can rapidly escape from the disk to form a high β disk, similarly to the conventional picture of an accretion disk. In this case, the disk-corona structure may be similar to the solar interior-corona structure. Shibata et al. (1990b) discussed the possibility of these two types of accretion disks, *low β* disk and *high β* disk. Recently, Mineshige et al. (1995) applied this idea to the observed two states of accretion disks, a low state and a high state (e.g., Miyamoto et al. 1995), and proposed that low states correspond to the *low β* disk. (As for a linear stability analysis of the Parker instability in a shearing flow parallel to the magnetic field, see Fogglizo and Tagger (1994).)

The shearing motion, on the other hand, can have a destabilizing effect [5] on a magnetized gas layer, i.e. a magneto-rotational instability (or Velikhov-Chandrasekhar instability or Balbus-Hawley instability). This instability gives one answer on the origin of viscosity in the accretion disk (e.g., Hawley, Gammie, and Balbus 1995, Matsumoto and Tajima 1995, Brandenberg et al. 1995). Since this instability occurs in wavelengths shorter than the Parker unstable wavelength, many interesting questions arise. Once this instability develops, does the disk magnetic field become turbulent and have only short scales ? In that case, is the Parker instability completely suppressed ? If the initial magnetic field is not so weak as in the case of star forming regions, what will happen ? The nonlinear coupling between the magneto-rotational instability and the Parker instability in realistic 3D situations will be an interesting and important subject for the future.

4.2. EMERGING SOLAR MAGNETIC LOOPS

4.2.1. *Self-similar Expansion of Magnetic Loops*

In the last subsection, we have considered the evolution of the periodic perturbation in the Parker instability. What is the result of non-periodic perturbations ?   In other words, what is the effect of side boundaries ? Shibata et al. (1989a) studied this effect for the Parker instability occurring in the isolated magnetic flux sheet embedded in the solar photosphere in order to make a model of emerging magnetic loop. (Note that emerging

[5] Even the Parker instability can be nonlinearly excited by the shearing motion when the gas layer is linearly stable (Kaisig et al. 1990).

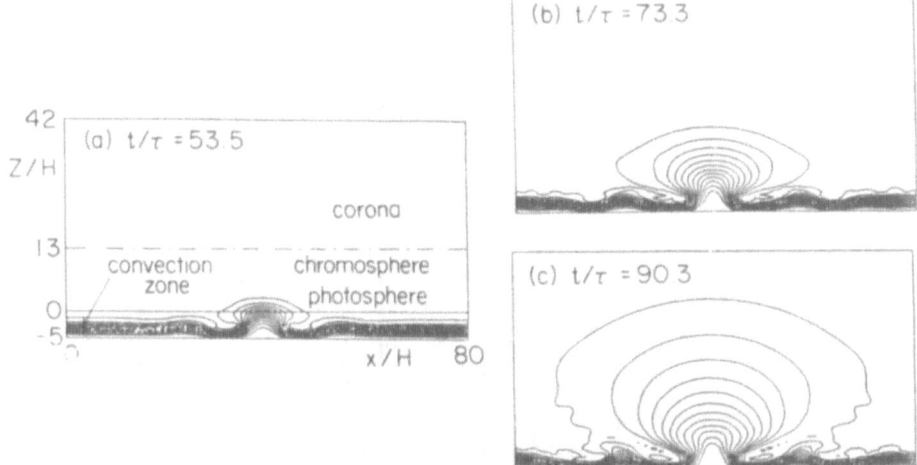

Figure 13. The self-similar evolution of the expanding magnetic loop as a result of the nonlinear evolution of the Parker instability (Shibata et al. 1990).

loops in the Sun are not periodic, at least, in the scale of active regions.) One important difference from the previous Matsumoto et al. model (1988, 1990) is that in the solar case the coronal pressure is much smaller than the photospheric pressure, $p_{cor}/p_{ph} \sim 10^{-7}$, so that the coronal pressure cannot work as a nonlinear stabilizing agent. Assuming a localized perturbation with the most unstable wavelength λ_{max}, Shibata et al. (1989a) compared the nonlinear evolution of the following three different cases; (1) $X_{max} = \lambda_{max}$, (2) $X_{max} = 2\lambda_{max}$, and (3) $X_{max} = 4\lambda_{max}$, where X_{max} is the horizontal box size.

They found that as X_{max} increases, the magnetic loop expands more horizontally. As a result, the magnetic tension force becomes smaller, so that the loop can continue to expand to upward direction and finally the expansion becomes approximately self-similar in the case of $X_{max} \geq 4\lambda_{max}$ (Fig. 13). Such self-similar behaviors are more clearly seen in a one dimensional distribution of the physical quantities along the height at the midpoint of the magnetic loop (Fig. 14). From this plot, we find

$$V_z \simeq \omega_n z,$$

$$\rho \propto z^{-4}, \qquad B_x \propto z^{-1}.$$

Here ω_n is the *nonlinear growth rate* in the Lagrangian frame in which the velocity increases with time as $V_z \propto \exp(\omega_n t)$. Interestingly, this *nonlinear growth rate* is found to be related to the linear growth rate ω of the Parker instability;

$$\omega_n \simeq \omega/2.$$

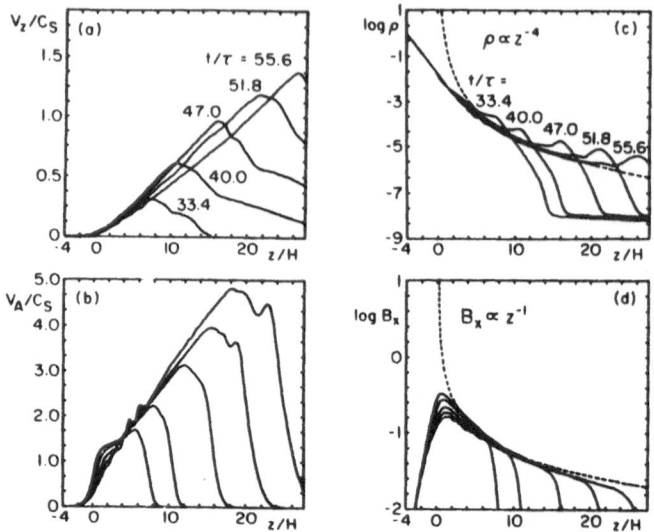

Figure 14. 1D distribution (in z) of physical quantities at the midpoint of the magnetic loop which is expanding self-similary as a result of the nonlinear evolution of the Parker instability (Shibata et al. 1989a). The times are in unit of $\tau = C_s/H$.

This relation holds in various situations, such as in the cases of the Parker-convective instability (Nozawa et al. 1992) and the Parker instability coupled with interchange modes (Matsumoto et al. 1993). We also find that the Alfven speed in the rising loop increases with height. This means that the plasma β decreases with height and explains why we have a low β corona in magnetic loops. As a magnetic loop expands, it tends to be current free since both thermal and gravitational forces become smaller than the magnetic force as the loop rises. Although Shibata et al. (1989a) found these results assuming an isolated magnetic flux initially, almost the same results are found by assuming continuous magnetic field distribution as an intitial condition (Kamaya et al. 1996a).

Shibata et al. (1989a) found an approximate self-similar solution for this problem, which explains these nonlinear simulation results very well (see also Shibata, Tajima, Matsumoto (1990a) for a detailed self-similar analysis and physical interpretation).

These results explain many observational characteristics of the emerging magnetic flux in the solar atmosphre (Shibata et al. 1989b). For example, the observations show that the rise velocity of the magetic loop in the photosphre ($z < 400$ km) is of order of $1 - 2$ km/s, while the loop expands at $10 - 20$ km/s in the upper chromosphere ($z \sim 4000 - 6000$ km). This is nicely explained by the equation found from the simulations and theory,

$$V_z \simeq \omega_n z \sim 0.06\, C_s(z/H) \sim 12\, (z/4000\ \mathrm{km})\ \mathrm{km/s}.$$

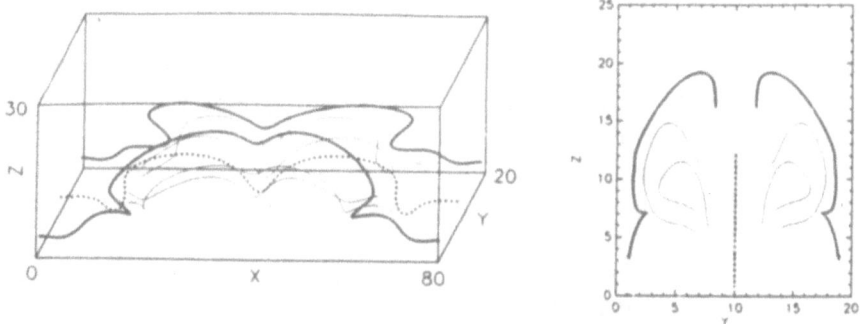

Figure 15. The 3D Parker instability coupled with interchnge mode (Matsumoto et al. 1993). Left: Magnetic field lines in 3D space. Right: Field lines in $y - z$ plane.

Observations also show that the downflow occurs at 30 – 50 km/s along the rising loops in the upper chromosphere, which is again explained by these simulation results.

4.2.2. *Three Dimensional Effect: Coupling with the Interchange Modes*

What will happen to the nonlinear evolution of the Parker instability in real 3D space ? In 3D space, the interchange mode can couple with the undular (Parker) mode. Matsumoto and Shibata (1992) and Matsumoto et al. (1993) studied this effect with full 3D simulations, and found that the resulting magnetic structure is a *interleaved structure* for both the galactic and the solar cases. It is also found that the *vortex motion* occurs around the rising magnetic loops, which produce *magnetic twist* or *torsional Alfven waves* (Fig. 15). Interestingly, even with such interleaved structure, the overall dynamics such as the self-similar expansion, is similar to that found from 2D simulations.

Matsumoto et al. (1993) also studied a 3D evolution of the emergence of the isolated magnetic flux tube, and found that the tube expands not only to the upward direction but also to the horizontal direction perpendicular to the tube. Such horizontal expansion decreases internal magnetic pressure significantly, and so the rise motion of the tube tends to be suppressed when the tube size (or the total magnetic flux of the tube) is smaller. It is found that there is a threshold magnetic flux for the tube to rise into the corona, which is about 0.3×10^{20}Mx. Interestingly, this agrees with observationally known threshold flux for the formation of arch filament system (e.g., Zwaan 1987).

4.2.3. *Emergence of Twisted Flux Tube*

Observations (Kurokawa 1989, Leka 1994) show that in some active regions the emerging flux is already twisted before emergenece. Matsumoto et al. (1996b) studied the nonlinear evolution of the coupling of the Parker instability with the kink instability as a model of emergence of a twisted flux tube. They found that the resulting magnetic flux tube show the double helix pattern and that their results eplain various observations such as the peculiar proper motion of sunspots and the apparent sheared S structure of coronal loops found by *Yohkoh*.

5. Flares – Magnetic Reconnection

5.1. SOLAR COMPACT FLARES AND X-RAY JETS

As we have seen above, a magnetic loop rapidly emerges from below the photosphere to the corona due to the Parker instability. Since the corona is filled with strong magnetic fields (i.e., low β plasma), such emerging magnetic loop would interact with pre-existing magnetic fields. What is the consequence of such interaction ? As an extention of previous emerging flux models (Shibata et al. 1989a,b, 1990c), Shibata, Nozawa, and Matsumoto (1992) first studied this process, i.e., the interaction between emerging flux and overlying anti-parallel horizontal coronal field, by performing high resolution 2D MHD simulations. They assumed a simplified anomalous resistivity model such that $\eta \propto (v_d - v_c)^2$ where $v_d = j/\rho$ is the drift velocity of the current j, and v_c is the critical velocity above which the anomalous resistivity sets in (Sato and Hayashi 1979, Ugai 1986). Their findings are as follows. Multiple magnetic islands are first created by the tearing instability in the current sheet between the emerging flux and the coronal field. These islands coalesce with each other by the coalescence instability, and are ejected out of the current sheet at high speed (of order of Alfven speed). The total energy released in this process is $\sim 10^{28} - 10^{29}$ ergs, comparable to that of compact solar flares. (This is an extension of the Heyvaerts et al. (1977)'s emerging flux flare model, and also of numerical simulations by Forbes and Priest (1984).) [6]

Yokoyama and Shibata (1994, 1995, 1996a,b; Yokoyama 1995) further developed this model, by performing comprehensive parameter survey, and extended the model to various situations.

(1) They succefully modeled the X-ray jets which are recently found by *Yohkoh* (Shibata et al. 1992b, 1994, Strong et al. 1992, Shimojo et al.

---

[6] As for a model of larger flares (e.g., Tsuneta et al. 1992, Shibata 1995), different approaches may be necessary; one such kind of modeling has been done by Magara et al. (1995).

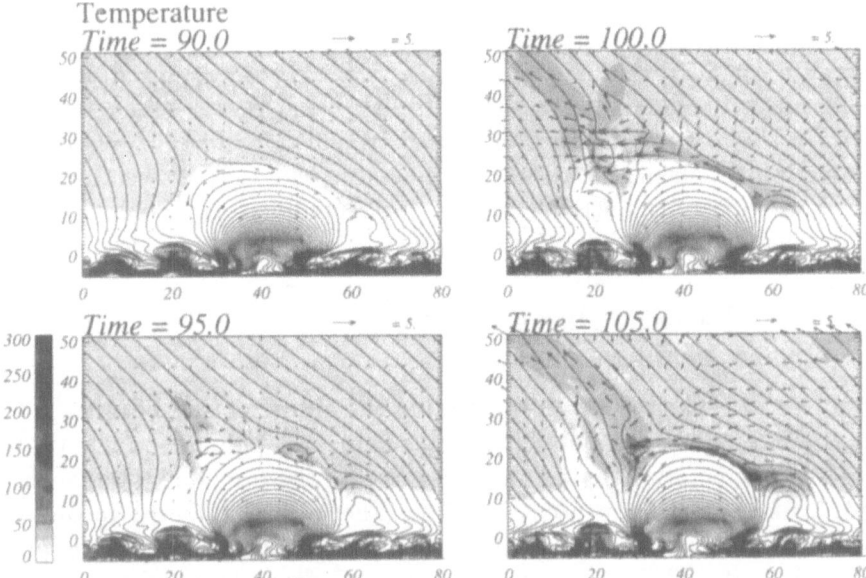

Figure 16. Magnetic reconnection driven by the emerging flux as a result of the Parker instability in the case of an oblique coronal field (Yokoyama and Shibata 1995a,b).

1995) by the emerging flux reconnection model (Yokoyama and Shibata 1995, 1996a,b). Figure 16 illustrates one typical example in which the initial coronal field is assumed to be oblique. Although the physical processes around the current sheet are similar to those found by Shibata et al. (1992a), it is found in this case that the magnetic islands (plasmoids) containing cool dense plasma are ejected sideway by the magnetic tension force, showing a whip-like motion. This is similar to the observed motion of Hα surges (e.g., Kurokawa and Kawai 1993, Canfield et al. 1995). Both plasmoids and the reconnection jets collide with the ambient oblique field (in the left side) to excite the fast mode MHD shock at the colliding point. The plasmoids disappear due to secondary reconnection with the oblique field, and the reconnection jets are decelerated significantly. However, plasmas behind the shock are accelerated again along the oblique field lines by the enhanced gas pressure behind the shock, and form a large scale hot jet which may correspond to observed X-ray jets. This model predicts coexistence of the hot and cool jets, and thus could explain observed coexistence of X-ray jets and Hα surges in some cases (Shibata et al. 1992b, Canfield et al. 1995, Okubo et al. 1995).

(2) Yokoyama and Shibata (1994), on the other hand, studied the basic physics of reconnection. They compared a *uniform resistivity* model with an *anomalous resistivity* model (similar to above Shibata et al. (1992a)'s

model) in reconnection driven by the Parker instability, and found that in the *uniform resistivity* model, the reconnection is slow and tends to be steady Sweet-Parker type, while in the *anomalous resistivity* model, it is fast and becomes nonsteady Petschek type, in agreement with Ugai (1995). In the *anomalous resistivity* case, it is also found that the formation and ejection of plasmoids are key processes, and that if the threshold of anomalous resistivity becomes higher, the reconnection becomes faster and more vilolent once reconnection occurs.

5.2. PROTOSTELLAR MAGNETOSPHERE

Hirose et al. (1995, Hirose 1994) studied the reconnection between a (non-rotating) protostellar bipolar magnetic field and poloidal magnetic fields carried by an accretion disk, as an extension of the previous Uchida-Shibata (1984, 1985) model for protostellar jets. They found that the reconnection itself accelerates plasmas directly from the disk, in addition to the general acceleration of disk plasmas via the $\mathbf{J} \times \mathbf{B}$ force (centrifugal force and magnetic pressure gradient force of toroidal fields). They suggested that optical jets may be generated by this reconnection acceleration. This reconneciton process is important to understand the evolution of protostellar rotation, and hence in the future the interaction between a *rotating* protostellar magnetosphere and a magnetized accretion disk should be studied in more detail.

6. Future Directions

In the following, I will summarize some future important subjects of *numerical solar and astrophysical MHD* in relation to jets, loops, and flares.

(1) Three dimensional MHD numerical simulations of magnetically driven jets from accretion disks should be studied from various points of view. If the initial poloidal field is weak, the Balbus-Hawley instability will modify the initial field significantly in the disk. Near the surface of the disk, the magnetic field lines are highly wound up, producing strong toroidal fields, which would be unstable to a non-axisymmetric instability such as the Parker instability. The magnetic reconnection in association with these instabilities will be very important. The 3D stability as well as long term behavior of the magnetically driven jet should also be studied in detail.

(2) In the case of the Parker instability in galactic disks, the effect of cosmic rays should be incorporated as emphasized by Parker (1966, 1979). Cosmic rays enable the nonlinear evolution of the Parker instability to proceed more vigorously. The effect of the Coriolis force and shearing motion on the nonlinear Parker instability should also be studied in detail with

high resolution 3D simulations, in relation to the formation of magnetic twist in various scales. The magnetic reconnection and resulting plasma heating will also be important in galactic disks and halos.

(3) In the case of the Sun, the generation and emergence of twisted flux tubes and resulting 3D reconnection would be one of the central problems in relation to the origin of flares. Since there are a lot of observational data for the solar case (such as *Yohkoh* data), it would be important to develop more realistic models incorporating various physical processes, such as heat conduction, evaporation, and radiative cooling (Hori et al. 1995), into the reconection problem (Yokoyama 1995).

References

Basu, S., Mouschovias, T. Ch., and Paleologou, E, V., 1996, *Astrophys. Lett. and Comm.*, in press.

Beck, R., et al., 1989, *A. Ap.*, **222**, 58.

Balbus, S. A., and Hawley, J. F., 1991, *Ap. J.*, **376**, 214.

Burgarella, D., Livio, M., and O'Dea, C. P. (eds), 1993, *Astrophysical Jets*, Cambridge Univ. Press.

Blandford, R. D., and Payne, D. G., 1982, *MNRAS*, **199**, 883.

Blandford, R. D., 1993, in *Astrophysical Jets*, eds. Burlarella, D. et al., Cambridge Univ. Press, p. 15.

Brandenburg, A., Nordlund, A., Stein, R. F., and Torkelsson, U., 1995, *Ap. J.*, **446**, 741.

Canfield, R. C., Reardon, K. P., Leka, K. D., Shibata, K., Yokoyama, T., and Shimojo, M., 1995, *Ap. J.*, submitted.

Cao, X., and Spruit, H. C., 1994, *A. Ap.*, **287**, 80.

Chou, W., Tajima, T., Matsumoto, R., and Shibata, K., 1996, *Proc. IAU Colloq. No. 153*, in press.

Colgate, S. A., and White, R. H. 1967, *Ap. J.*, **143**, 626.

Fogglizo, T., and Tagger, M., 1994, *A. Ap.*, **287**, 297.

Forbes, T. G., and Priest, E. R., 1984, *Solar Phys.*, **94**, 315.

Giz, A. T., and Shu, F. H. 1993, *Ap. J.*, **404**, 185.

Haisch, B., Strong, K. T., and Rodono, M., 1991, Ann. Rev. A. Ap., **29**, 275.

Hanawa, T., Matsumoto, R., and Shibata, K., 1992, *Ap. J. Lett.*, **393**, L71.

Hawley, J. F., Gammie, C. F., and Balbus, S. A., 1995, *Ap. J.*, **440**, 742.

Heyvaerts, J., Priest, E. R., and Rust, D. M.: 1977, *Ap. J.*, **216**, 123.

Heyvaerts, J., and Norman, C., 1989, *Ap. J.*, **347**, 1055.

Hirose, S., 1994, Ph. D. Thesis, University of Tokyo.

Hirose, S., et al., 1995, to be submitted.

Hollweg, J., 1982, *Ap. J.*, **257**, 345.

Hollweg, J., 1992, *Ap. J.*, **389**, 731.

Hollweg, J., Jackson, S., and Galloway, D., 1982, *Solar Phys.*, **75**, 35.

Hori, K., Yokoyama, T., Kosugi, T., and Shibata, K., 1996, *Proc. IAU Colloq. 153*, in press.

Horiuchi, T., Matsumoto, R., Hanawa, T. and Shibata, K., 1988, *Publ. Astron. Soc. Japan*, **40**, 147.

Hughes, D. W., and Procter, M. R. E. 1988, *Ann. Rev. Fluid Mech.*, **20**, 187.

Kaisig, M., Tajima, T., Shibata, K., Nozawa, S., and Matsumoto, R., 1990, *Ap. J.*, **358**, 698.

Kamaya, T., Mineshige, S., Matsumoto, R., Shibata, K., 1996a, *Ap. J. Lett.*, in press.

Kamaya, T., Mineshige, S., Matsumoto, R., Hanawa, T., Shibata, K., 1996b, (in prepa-

246

ration).

Kudoh, T:, and Shibata, K., 1995, *Ap. J. Lett.*, **452**, L41.

Kudoh, T., and Shibata, K., 1996a, *Astrophys. Lett. and Comm.*, in press.

Kudoh, T., and Shibata, K., 1996b, *Ap. J.*, (to be submitted).

Kurokawa, H., 1989, *Space Sci Rev.*, **51**, 49.

Kurokawa, H., and Kawai, G. 1993, *Proc. IAU Colloq. No. 141, ASP Conference Series 46*, ed. H. Zirin, G. Ai, and H. Wang, pp. 507-510.

Leka, K. D., 1994, Ph. D. Thesis, Hawaii Univ.

Lovelace, R. V. E., Berk, H. L., and Contopoulos, J., 1991, *Ap. J.*, **379**, 696.

Magara, T., Mineshige, S., Yokoyama, T., and Shibata, K., 1995, *Ap. J.*, submitted.

Matsumoto, R., Horiuchi, T., Shibata, K., and Hanawa, T., 1988, *Publ. Astron. Soc. Japan*, **40**, 171.

Matsumoto, R., Horiuchi, T., Hanawa, T., and Shibata, K., 1990, *Ap. J.*, **356**, 259.

Matsumoto, R., and Shibata, K., 1992, *Publ. Astr. Soc. Japan*, **44**, 167.

Matsumoto, R., Tajima, T., Shibata, K., and Kaisig, M., 1993, *Ap. J.*, **414**, 357.

Matsumoto, R., and Tajima, T., 1995, *Ap. J.*, **445**, 767.

Matsumoto, R., Uchida, Y., Hirose, S., Shibata, K., Hayashi, M. R., Ferrari, A, Bodo, G., and Norman, C., 1996a, *Ap. J.*, in press.

Matsumoto, R., Tajima, T., Chou, W., and Shibata, K., 1996b, *Proc. IAU Colloq. No. 153*, in press.

Michel, F. C., 1969, *Ap. J.*, **158**, 727.

Mineshige, S., Kusunose, M., Matsumoto, R., 1995, *Ap. J. Lett.*, **445**, L43.

Miyamoto, M., Kitamoto, S., Hayashida, K., and Egoshi, W., 1995, *Ap. J. Lett.*, **442**, L13.

Moreno-Insertis, F., 1986, *A. Ap.*, **166**, 291.

Moriarty-Schieven, G. H., and Wannier, P. G., 1991. *Ap. J.*, **373**, L23.

Mouschovias, T. Ch., 1974, *Ap. J.*, **192**, 37.

Mouschovias, T. Ch., 1996, in *Solar and Astrophysical MHD Flows*, K. Tsinganos (ed.), Kluwer, in press.

Nozawa, S., Shibata, K., Matsumoto, R., Tajima, T., Sterling, A. C., Uchida, Y., Ferrari, A., and Rosner, R., 1992, *Ap. J. Suppl.*, **78**, 267.

Ohno, Y., Sakashita, S., and Ohyama, N., 1961, *Prog. Theor. Phys. Suppl.*, **20**, 85.

Okubo, A., et al., 1996, *Proc. IAU Colloq. 153*, in press.

Osterblock, D. E., 1961, *Ap. J.*, **134**, 347.

Parker, E. N., 1966, *Ap. J.*, **145**, 811.

Parker, E. N., 1979, *Cosmical Magnetic Field* (Clarendon Press, Oxford), p. 314.

Pevtsov, A. A., Canfield, R. C., and Metclaf, T. R., 1994, *Ap. J. Lett.*, **425**, L117.

Pudritz, R. E. and Norman, C., 1986, *Ap. J.*, **301**, 571.

Priest, E. R.: 1982, *Solar Magnetohydrodynamics*, Reidel Pub., Dordrecht.

Rust, D. M., 1994, *Geophys. Res. Lett.*, **21**, 241.

Sakurai, T., 1987, *Publ. Astron. Soc. Japan*, **39**, 821.

Sato, T., and Hayashi, T., 1979, *Phys. Fluids*, **22**, 1189.

Sauty, C., and Tsinganos, K., 1994, *Astron. Ap.*, **287**, 893.

Schüssler, M., 1996, in *Solar and Astrophysical MHD Flows*, K. Tsinganos (ed.), Kluwer, in press.

Shibata, K., 1983, *Publ. Astron. Soc. Japan*, **35**, 263

Shibata, K., 1996, *Adv. Space Res.*, **17**, (4/5)9.

Shibata, K., Nishikawa, T., Kitai, R. and Suematsu, Y., 1982, *Solar Phys.*, **77**, 121.

Shibata, K., and Suematsu, Y., 1982, *Solar Phys.*, **78**, 333.

Shibata, K., and Uchida, Y., 1985, *Publ. Astron. Soc. Japan*, **37**, 31.

Shibata, K., and Uchida, Y., 1986a, *Publ. Astron. Soc. Japan*, **38**, 631.

Shibata, K., and Uchida, Y., 1986b, *Solar Phys.*, **103**, 299.

Shibata, K., and Uchida, Y., 1987, *Publ. Astron. Soc. Japan*, **39**, 559.

Shibata, K., et al., 1989a, *Ap. J.*, **338**, 471.

Shibata, K., Tajima, T., Steinolfson, R. and Matsumoto, R., 1989b, *Ap. J.*, **345**, 584.

Shibata, K., and Uchida, Y., 1990, *Publ. Astron. Soc. Japan*, **42**, 39.

Shibata, K., Tajima, T., and Matsumoto, R., 1990a, *Phys. Fluids*, **B2**, 1989.

Shibata, K., Tajima, T., and Matsumoto, R., 1990b, *Ap. J.*, **350**, 295.

Shibata, K., Nozawa, S., Matsumoto, R., Sterling, A. C., and Tajima, T., 1990c, *Ap. J. Lett.*, **351**, L25.

Shibata, K., and Matsumoto, R. 1991, *Nature*, **353**, 633.

Shibata, K., Nozawa, S., and Matsumoto, R., 1992a, *Publ. Astr. Soc. Japan*, **44**, 265.

Shibata, K., et al., 1992b, *Publ. Astr. Soc. Japan Letters*, **44**, L173.

Shibata, K., et al., 1994, *Ap. J. Lett.*, **431**, L51.

Shimojo, M., Hashimoto, S., Shibata, K., Hirayama, T., Hudson, H., and Acton, L., 1995, *Publ. Astr. Soc. Japan*, in press.

Shu, F. H., Adams, F. C., and Lizano, S., 1987, *Ann. Rev. A. Ap.*, **25**, 23.

Sofue, Y., and Handa, T., 1984, *Nature*, **310**, 568.

Sterling, A. C., and Hollweg, J. V., 1988, *Ap. J.*, **327**, 950.

Sterling, A. C., Shibata, K., Mariska, J. T., 1993, *Ap. J.*, **407**, 778.

Stone, J. M., and Norman, M., 1994, *Ap. J.*, **433**, 746.

Strong, K. T. et al., 1992, *Publ. Astr. Soc. Japan*, **44**, L161.

Suematsu, Y., Shibata, K., Nishikawa, T. and Kitai, R., 1982, *Solar Phys.*, **75**, 99.

Tajima, T., 1989, *Computational Plasma Physics*, Addison Wesley.

Todo, Y., Uchida, Y., Sato, T., and Rosner, R., 1992, *Publ. Astron. Soc. Japan*, **44**, 245.

Todo, Y., Uchida, Y., Sato, T., and Rosner, R., 1993, *Ap. J.*, **403**, 164.

Tsinganos, K., and Trussoni, A. A., 1991, *Astron. Ap.*, **249**, 156.

Tsinganos, K., Surlantzis, G., and Priest, E. R., 1993, *Astron. Ap.*, **275**, 613.

Tsinganos, K. et al., 1996, in *Solar and Astrophysical MHD Flows*, K. Tsinganos (ed.), Kluwer, in press.

Tsuneta, S., et al., 1992, *Publ. Astr. Soc. Japan*, **44**, L63.

Uchida, Y., and Shibata, K., 1984, *Publ. Astron. Soc. Japan*, **36**, 105.

Uchida, Y., and Shibata, K., 1985, *Publ. Astron. Soc. Japan*, **37**, 515.

Uchida, Y., Shibata, K. and Sofue, Y., 1985, *Nature*, **317**, 699.

Uchida, Y., et al., 1987, *Publ. Astron. Soc. Japan*, **39**, 907.

Uchida, Y., Mizuno, A., Nozawa, S., and Fukui, Y., 1990, *Nature*, **42**, 69.

Uchida, Y., Todo, Y., Rosner, R. and Shibata, K., 1992, *Publ. Astr. Soc. Japan*, **44**, 227.

Ugai, M., 1986, *Phys. Fluids*, **29**, 3659.

Ugai, M., 1995, *Phys. Plasmas*, **2**, 388.

Ulmschneider, P., Priest, E. R., and Rosner, R. (eds.), 1991, *Mechanisms of Chromospheric and Coronal Heating*, Springer Verlag.

Ustyugova, G. V., et al., 1995, *Ap. J. Lett.*, **439**, L39.

Yokoyama, T., 1995, Ph. D. Thesis, National Astronomical Observatory.

Yokoyama, T., and Shibata, K., 1994, *Ap. J. Lett.*, **436**, L197.

Yokoyama, T., and Shibata, K., 1995, *Nature*, **375**, 42.

Yokoyama, T., and Shibata, K., 1996a, *Astrophys. Lett. and Comm.*, in press.

Yokoyama, T., and Shibata, K., 1996b, *Publ. Astron. Soc. Japan*, in press.

FURTHER THOUGHTS ON THE SOLAR CORONA AS A MINIMUM ENERGY SYSTEM

P. CHARBONNEAU AND A.J. HUNDHAUSEN
High Altitude Observatory, NCAR[†]
P.O. Box 3000, Boulder, CO 80307, USA
paulchar@dante.hao.ucar.edu

Abstract. We conjecture that the global, large-scale structure of the solar corona represents a form of minimum energy state. We illustrate this conjecture with the help of an approximate model applicable to quiet solar minimum conditions. Possible implications and applications of the conjecture are discussed in the context of coronal mass ejections and of empirical modeling of the solar corona.

1. Introduction

The bright coronal features known as helmet streamers, often so striking on eclipse photographs and coronagraph images, are believed to be regions of closed magnetic fields where the hot coronal gas is effectively trapped. In contrast, in coronal holes the magnetic field is assumed to be open and extending more or less radially outward, permitting the expansion of the plasma to form the steady or quiet component of the solar wind. At epochs of low solar activity, the corona assumes a relatively simple and nearly axisymmetric configuration, characterized by a belt of helmet streamers straddling the equator and well-defined polar coronal holes of opposite magnetic polarities in the northern and southern hemispheres. (e.g., Hundhausen *et al.* 1981; Figure 1 herein). The "tops" of coronal helmet streamers generally suggest that closed magnetic fieldlines do not extend beyond heliocentric distances of 1.75 to 2.5 solar radii; quantitative reconstruction of the plasma and magnetic field structures (Gibson & Bagenal,

---

[†]The National Center for Atmospheric Research is sponsored by the National Science Foundation.

K. C. Tsinganos (ed.), Solar and Astrophysical Magnetohydrodynamic Flows, 249–264.
© *1996 Kluwer Academic Publishers.*

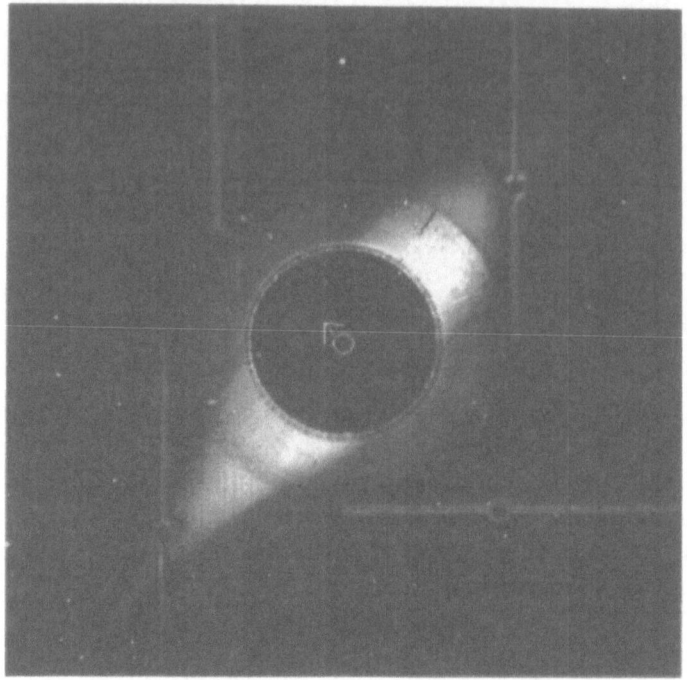

Figure 1. The solar minimum corona in white light, as seen on March 31 1986. Data from the HAO MkIII coronameter ($1.15\,R_\odot$ to $1.5\,R_\odot$) and Solar Maximum Mission Coronagraph ($1.5\,R_\odot$ to $4.0\,R_\odot$, four quadrant composite). The arrow indicates heliospheric North, and the dotted circle marks the radial extend of the solar limb.

1995) place the "cusp" at which open fieldlines of opposite polarity meet over the streamer at a similar heliocentric distance.

Helmet streamers appear to be surprisingly stable structures, often surviving many rotation periods. "Surprisingly", because even the quiet sun corona is probably not all that quiet. Figure 2 shows two composites of Yohkoh Soft X-ray and white light coronameter images. Both set of images are separated by one solar rotation period. The larger scales of the corona, as mapped in white light, show relatively little evolution in this time period, despite the fact the distribution and brightness of active regions, as seen in X-rays, have changed considerably. While streamers are long-lived objects, their demise often takes place through one of the more spectacular manifestation of solar activity: coronal mass ejections (hereafter CME). CMEs are believed to involve the opening of a large fraction of the originally closed helmet streamer magnetic field (e.g., Hundhausen 1994). Yet the solar corona is observed to readjust in a day or two to its quiet state after a CME event.

The observed (and remarkable) stability of the global, solar minimum

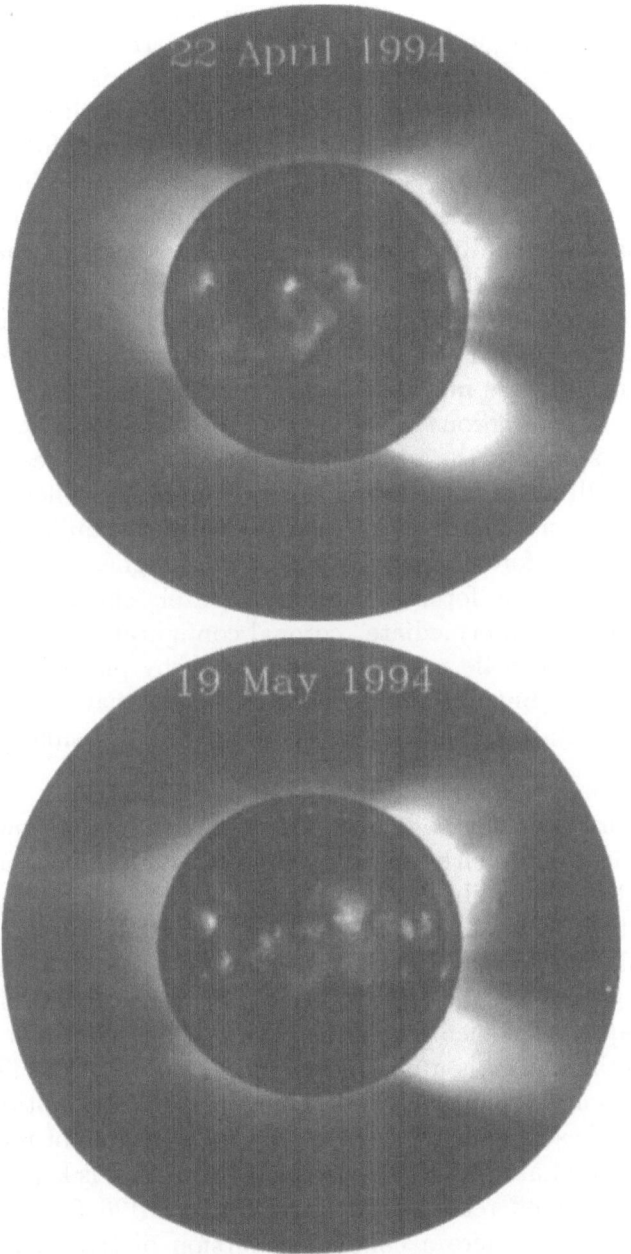

Figure 2.   The solar corona seen in white light and Soft X-rays. X-Ray data is from Yohkoh, white light from the HAO MkIII coronameter. Despite important changes in the distribution and brightness of active regions, the larger scales of the corona have evolved very little here, over timescales of the order of the solar rotational period, ~ 28 days as seen from Earth.

corona suggests that its large-scale configuration, i.e., the spatial distribution of plasma and magnetic fields, represent a form of minimum energy state. In this context, a particularly important result is due to Aly (1984, 1991), who showed that for a given boundary flux distribution, the minimum energy state for a force-free magnetic field in semi-infinite space with all fieldlines having at least one footpoint anchored at the boundary corresponds to a completely closed magnetic configuration, i.e., one where *both* ends of each magnetic fieldline are anchored on the lower boundary (see also Smith & Low, 1993). On the other hand, in the absence of a magnetic field the minimum energy state for a hot, tenuous plasma in a spherically symmetric gravitational field is one of transonic expansion to infinity. Clearly, these two constraints are mutually incompatible if one considers the *total* energy content of the corona, i.e., the sum of the magnetic and plasma contributions. As long as the frozen-in condition applies, steady-state flow along closed fieldlines is not possible. Either there is no flow at all, forcing the plasma into a higher energy, quasi-hydrostatic equilibrium state, or the magnetic field is forced open, leaving it in a higher energy state than in the absence of plasma forcing. This immediately suggests the following question: is there an "intermediate" coronal configuration characterized by a *partially open* large-scale magnetic field, whereby plasma flow proceeds along open fieldlines but not along closed fieldlines, that represent a minimum total energy state for the system as a whole, i.e., plasma and magnetic field?

We conjecture that this is indeed the case. In a paper now finally accepted for publication (Charbonneau & Hundhausen 1996; hereafter CH96), we illustrated the plausibility of this minimum energy conjecture. We did so by constructing sequences of axisymmetric, partially open coronal models including plasma and magnetic field, where the degree of magnetic field opening is controlled by a single adjustable parameter. We then verified that along a sequence of such coronal models with increasing degree of field opening, the total energy is a non-monotonic function of the field opening parameter. Section 2 below outlines the steps and assumptions involved in constructing these approximate models, and a sample of representative results are presented and briefly discussed. This section is kept deliberately succinct, the interested reader being referred to CH96 for further detail; rather than presenting here a condensed version of that paper, in what follows we prefer to further explore some aspects of the minimum energy conjecture that are either not addressed or only mentioned in passing in CH96. Accordingly, section 3 explores some implications for the onset of coronal mass ejections. Results of some recent and not-so-recent numerical simulations of the solar corona are also discussed, in the context of the minimum energy conjecture. Section 4 discusses a possible application of the conjecture as a constraint on empirical modelling of the corona.

2. An illustrative model

2.1. OVERVIEW OF MODEL CONSTRUCTION

The starting point of the modelling is the magnetic field configuration, for which we use a class of axisymmetric force-free solutions due to Low (1986). These solutions are expressed in terms of a 2-D generating function

$$\mathbf{B} \equiv (B_r, B_\theta, 0) = \frac{(f_B B_0)}{r \sin \theta} \left[\frac{1}{r} \frac{\partial Z}{\partial \theta} \hat{\mathbf{e}}_r - \frac{\partial Z}{\partial r} \hat{\mathbf{e}}_\theta \right], \tag{1}$$

in spherical polar coordinates (r, θ, ϕ) with r normalized to some reference radius r_0, and with $Z(r, \theta)$ given in closed form (See Low 1986, §3; CH96, §2.1). The numerical factor f_B is introduced so that $B_r(r_0) = \pm 2B_0$ at the poles, in analogy with the classical dipole. Figure 3 shows a few solutions from a one-parameter family of dipole-like solutions that are potential everywhere except in the equatorial plane, where stress-free current sheets are present. Two regions can be identified in these magnetic configurations; fieldlines anchored onto the photosphere at high latitudes extend to infinity, while at lower latitudes they cross the equatorial plane to anchor (anti)symmetrically in the lower hemisphere, thus defining a magnetically closed region (shaded on Fig. 3). The parameter a is directly proportional to the heliocentric extent of the closed region in the equatorial plane. A small (large) value of a corresponds to a small (large) closed region, $a = 1$ to a fully open field, and the limit $a \to \infty$ to a classical dipole.

The distinction between open and closed regions forms the basis of the strategy used in CH96 to construct approximate 2-D coronal models. With the frozen-in condition assumed to hold and given the high coronal temperature, any steady flow is ruled out in the closed region. This region is then modeled as being in strict hydrostatic equilibrium. For known base conditions and under the isothermality assumption, this is a trivial task. The high coronal temperatures also mean that in the absence of outside confinement, the coronal plasma cannot be in equilibrium and will expand outward in the form of a wind. Assuming that the plasma is effectively channeled along magnetic fieldlines and with the magnetic configuration considered given, the task is then basically the construction of 1-D thermally-driven single-fluid wind models for specified, non-radial flow geometries. This is a well defined and well understood problem in wind theory (see, e.g., Kopp & Holzer 1976). Going back to Fig. 3, consider then a "sector" in the open region as being defined by two neighbouring magnetic fieldlines. With the magnetic field known in closed form, the variation of the sector's cross section as a function of distance along the sector can be easily constructed.

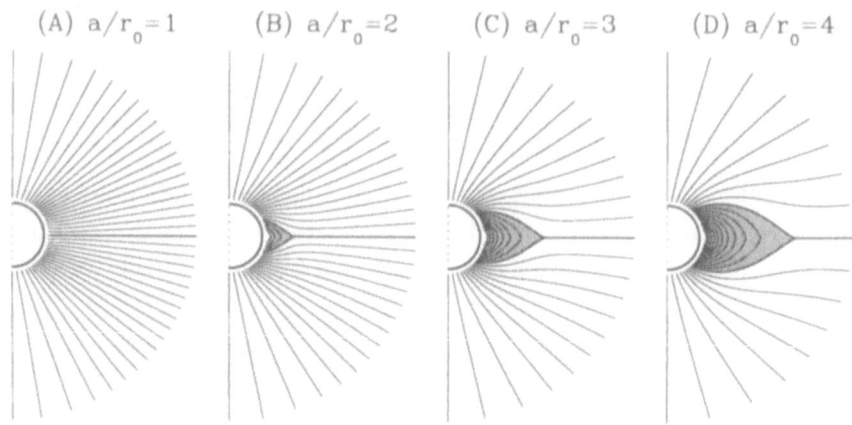

(A) $a/r_0=1$ (B) $a/r_0=2$ (C) $a/r_0=3$ (D) $a/r_0=4$

Figure 3. A few members of a sequence of partially open axisymmetric coronal configuration constructed using the Low (1986) force-free solutions. The parameter a is proportional to the equatorial extent of the magnetically closed region (shaded).

It is then a simply matter to determine the location of the sonic point, after which the full wind solution is obtained by requiring the solution to pass through this critical point. Such wind solutions can be defined for $j = 1, 2, ...J$ fieldline-based sector, the ensemble of J sectors thus defining the pseudo-2-D wind solution (see CH96, §2.3).

There are a number of physical inconsistencies associated with this procedure, the most obvious being that lateral force balance is in general *not* satisfied, due to the presence of unbalanced gas pressure gradients across neighbouring sectors. In general, unbalanced pressure gradients also exist across the boundary between the closed and open regions. While such inconsistencies obviously precludes using these solutions for detailed, quantitative studies of solar wind dynamics, they most likely remain adequate for illustrating aspects of global energetics, as discussed at some length in CH96 (§4). In particular, these models catch an essential feature of partially open coronal models, namely that fact that the density decreases outward more rapidly in the open, wind region than it does in the closed, hydrostatic region.

2.2. THE TOTAL ENERGY

The next step is to compute the total energy content of the resulting coronal configurations. This is done by considering separately the contributions from the magnetic field and from the plasma. The energy content of the magnetic field is in general given by

$$\mathcal{E}_B = \frac{1}{8\pi} \int\int\int_V |\mathbf{B}|^2 dV, \tag{2}$$

where the volume integral extends radially from r_0 to infinity. Using Gauss theorem and making use of the fact that our magnetic configurations are force-free and axisymmetric, this integral can be converted to the surface integral

$$\mathcal{E}_B = \frac{r_0^3}{4} \int_0^\pi \left[B_r^2 - B_\theta^2 \right]_{r=r_0} \sin\theta\, d\theta, \tag{3}$$

provided that the magnetic field strength falls off asymptotically faster than $1/r^2$ (see, e.g., Smith & Low 1993). With Z known analytically, this last integral is readily computed. For a classical dipole ($a \to \infty$ here), $\mathcal{E}_B = B_0^2 r_0^3/3$.

Consider now the energy content of the plasma. Three distinct contributions can be identified, namely thermal, gravitational and kinetic; in a volume element dV,

$$d\mathcal{E}_{\mathrm{th}} = \frac{3}{2}\rho c_s^2 dV, \tag{4a}$$

$$d\mathcal{E}_{\mathrm{g}} = \rho G M_\odot \left(\frac{1}{r_0} - \frac{1}{r} \right) dV, \tag{4b}$$

$$d\mathcal{E}_{\mathrm{u}} = \frac{1}{2}\rho u^2 dV. \tag{4c}$$

where c_s is the isothermal sound speed. The closed region is first divided into K radial sector extending from the reference radius r_0 out to the boundary of the closed region. The gravitational ($\mathcal{E}_{\mathrm{g},k}$) and thermal ($\mathcal{E}_{\mathrm{th},k}$) contributions to the total energy from the kth sector are obtained by integrating over the volume of each sector, which can be done analytically (see CH96, §3.3). Of course $\mathcal{E}_{\mathrm{u},k} = 0$ in the closed region since $u = 0$ there by virtue of the hydrostatic assumption. In the open region, the energy is also computed sector by sector, with the integrals being carried out numerically for each of the J sector defining the wind solution. A difficulty arise from the wind solutions being isothermal, which causes the kinetic energy integral to diverge in the limit $r \to \infty$, requiring truncation at some finite distance. We have chosen to integrate out to the sonic surface, but have verified (cf. CH96, §4) that the exact criterion adopted to truncate the integrals has little influence on the energy budget. This can be understood by noting that in both the wind and hydrostatic regions, all integrands in the plasma energy integrals are proportional to the local density. Consequently, all plasma contributions to the total energy are largely dominated by the first few scale heights above r_0.

The total energy $\mathcal{E}$ associated with a given coronal configuration is then obtained by summing contributions from all sectors in the open and closed regions, to which is added the magnetic energy:

$$\mathcal{E}(a, \rho_0, T, B_0) = \mathcal{E}_B + \sum_{k=1}^{K} [\mathcal{E}_{g,k} + \mathcal{E}_{th,k}] + \sum_{j=1}^{J} [\mathcal{E}_{g,j} + \mathcal{E}_{th,j} + \mathcal{E}_{u,j}]. \quad (5)$$

The total energy is in general a function of the field opening parameter a, of the assumed base magnetic field strength (B_0), and of the assumed base density (ρ_0) and temperature (T). These latter two quantities may in general be latitude-dependent.

2.3. SAMPLE RESULTS

Total energy curves are presented in CH96 for a number of combinations of base parameter values, including cases where ρ_0 and T assume different values at the base of the hydrostatic and wind regions. Sequences of models spanning $1.1 \leq a/r_0 \leq 6$ were constructed for base temperatures and densities $1.5 \leq T/10^6 K \leq 2.0$, $10^7 \leq n_0 \leq 10^8$ ($n_0 = \rho_0/\mu m_H \equiv$ protons/electrons per cubic centimeter), and base magnetic field strengths in the range $0.5 \leq B_0 \leq 5\,\mathrm{G}$. We only present here a small, illustrative subset of these models, defined by the following parameters values:

TABLE 1. Coronal base parameters

Case	T_H [10^6 K]	T_W [10^6 K]	$n_{0,H}$	$n_{0,W}$	B_0 [G]
I	1.5	1.5	5×10^7	5×10^7	2
II	1.5	1.5	10^8	5×10^7	2
III	2.0	1.5	5×10^7	5×10^7	1

The subscripts "H" and "W" refer to the hydrostatic (closed) and wind (open) regions, respectively. Figure 4 details the various contributions to the plasma energy in the open and closed region for case I. Figure 4(A) shows the variations as a function of a of the thermal and gravitational energies in the hydrostatic region, and Figure 4(B) the variations of thermal, gravitational, and kinetic energies in the wind region. In the closed region, the plasma energy content increases rapidly with a, directly reflecting the growth in size of the closed region, until the total energy content of the plasma is dominated by the contributions from the closed region, occurring for $a/R_\odot \gtrsim 3$. It is important to realize that the decrease of the plasma energy in the wind region with increasing a does *not* balance the

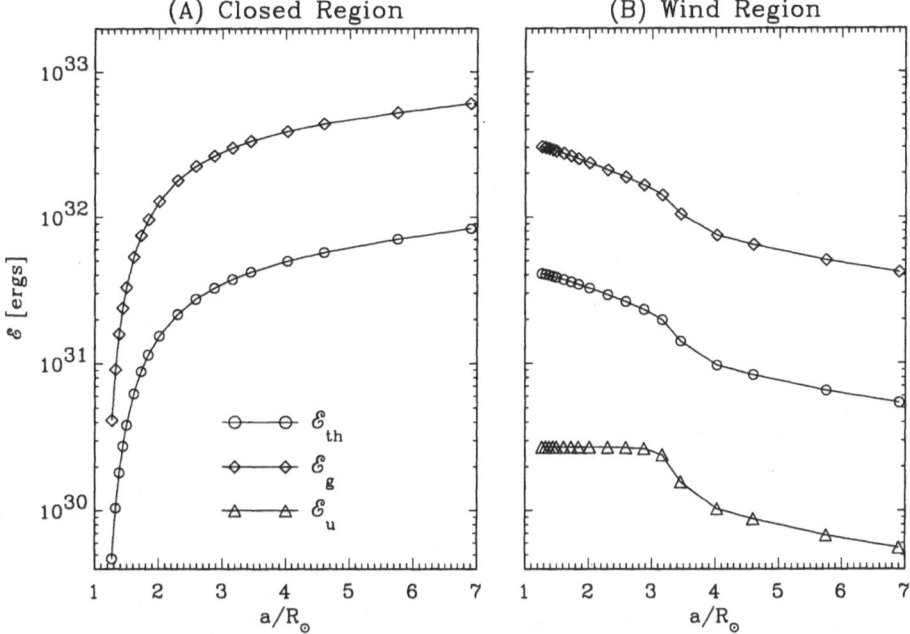

Figure 4. Contributions to the total plasma energy for a sequence of solutions corresponding to case I of Table 1. The various contributions in the closed (hydrostatic) and open (wind) regions are illustrated separately. The gravitational potential energy is defined in terms of the height above the photosphere. Curve coding is identical in both panels.

increase of the plasma energy in the hydrostatic region, and that this occurs primarily because of the differing scale heights in the two regions. For latitude-independent base conditions (cf. case I) the net variation in total plasma energy is not extreme, amounting to about a factor of 2 between the $a/R_\odot = 1$ and $a/R_\odot = 7$ configurations, but it remains significant.

Figure 5 illustrates the variations with a of the total plasma energy content (i.e., the sum of the thermal, gravitational and kinetic energies plotted on both parts of Figure 4), of the magnetic energy (squares), and of the total energy (thick solid line) for the three cases listed in Table I. The magnetic energy increases as a decreases, i.e., as the field opens up, and reaches its maximal value for a fully open field ($a \to 1$). Most interesting is that the *total* energy, i.e., plasma plus magnetic, is a non-monotonic function of a, showing a minimum for $a/R_\odot \approx 3.5$, 2.5 and 1.5 for cases I, II and III, respectively. The second value is similar to the typical radial extent of the cusp point associated with the equatorial streamer belt that is so characteristic of the quiet, solar minimum corona (see Fig. 1

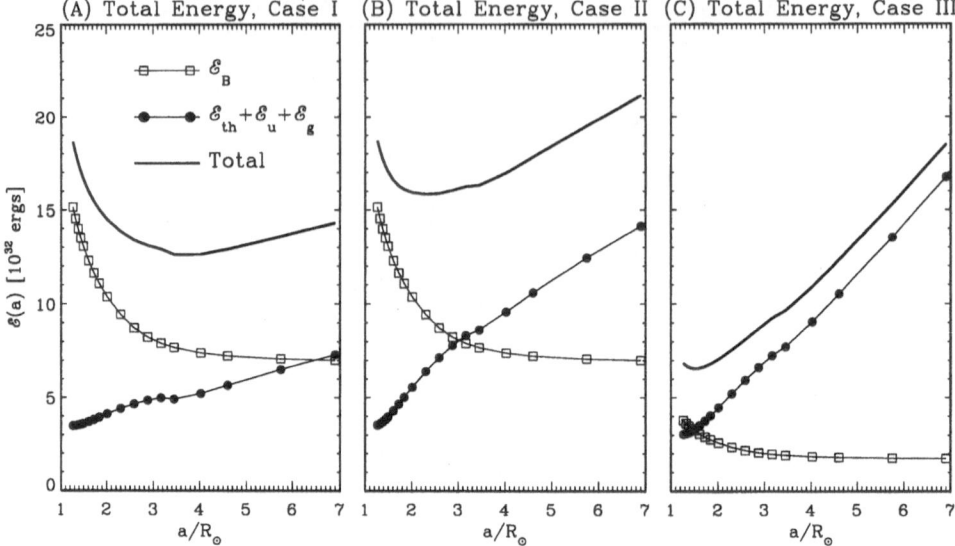

Figure 5. A sample of minimum energy curves for the three sequences of partially open coronal models defined in Table 1. The magnetic and plasma contributions to the total energy (thick solid line) are shown separately. Note the presence of a minimum in each of the total energy curves. Curve coding is identical in all panels.

herein). This result, while interesting, should not be overinterpreted. There are different combination of parameters that yield minima at comparable heliocentric distances, and there are some parameter values that lead to minima for $a/R_{\odot} \to 1$ and/or $a/R_{\odot} \to \infty$ (see Figs. 5 and 6 of CH96 for more examples).

It is of course possible to adjust any (or all) of the model parameters so as to produce an energy minimum at a value of a exactly equal to the observed average location of the equatorial streamer belt cusp during quiet, solar minimum conditions, but such an exercise is clearly meaningless. What is significant is that along the sequence of coronal models constructed here and for reasonable coronal parameters, there exists a minimum in the total energy curve; when the presence of plasma is taken into account, a completely closed magnetic configuration is *not* a minimum total energy state. Furthermore, there can exist configurations (cf. case II, $a \gtrsim 5$; case III, $a \gtrsim 2$) for which the total energy exceeds that associated with the fully open magnetic field ($a \to 1$). Despite the simplistic nature of the partially opened hybrid corona/wind models used to develop the argument, we believe that the presence of a minimum in the total energy curve can be expected to remain a robust feature of more realistic coronal models.

3. Implications

3.1. CORONAL MASS EJECTIONS

The sequence of events leading to the triggering of CMEs continues to attract considerable attention on the part of both observers and numerical modelers. Broadly speaking, two classes of scenario have been investigated in detail. The first relies on the sudden deposition of momentum and/or energy, such as may be caused by a large flare, in a helmet streamer-like initial configuration (see, e.g., Wang *et al.* 1995 for representative results for this class of models). While the simulations show that ejection events can be produced in this way under certain circumstances, growing observational evidence is indicating that this scenario does not apply to the majority of CMEs (see, e.g., Harrison 1991; Kahler 1992; Hiei *et al.* 1993; Hundhausen 1994) The second class of models attempts to trigger CMEs by slowly perturbing a helmet streamer-like structure, until it reaches a state where loss of equilibrium and spontaneous transition to a dynamical state can occur. At the onset, this approach must deal with a rather stringent energetic constraint: Aly's conjecture informs us that for a given boundary flux distribution, a force-free closed or partially closed magnetic configuration always has less magnetic free energy than the fully open magnetic field associated with the same boundary flux; no amount of field aligned currents in the pre-CME configuration can store enough energy to allow a spontaneous transition to the fully open field presumably corresponding to the immediate post-event magnetic configuration.

While this is clearly a stringent constraint, to argue on that basis that a coronal mass ejection cannot be magnetically driven remains a premature conclusion. One should first note that the pre-CME magnetic configuration may well not be force-free, and/or that is may violate some of the topological requirements of Aly's theorem (see Smith & Low 1993; Low, this volume). Such considerations notwithstanding, the plasma contribution to the total energy *must* be taken into account in the energy budget. In the context of the minimum energy corona conjecture, the results presented in the preceding section suggest that a helmet streamer-like configuration, energetically speaking, makes a *worse* pre-CME candidate than a fully closed magnetized corona (contrast this with the views expressed in Linker & Mikić 1995, last paragraph of their §1).

Nevertheless, various suggestive scenarii for the onset of CMEs have been put forth by numerous authors on the basis of quasi-steady and/or fully time-dependent numerical simulations. The recent simulations of Linker & Mikić (1995) are a nice example of the current state of the art. These authors apply a large-scale shear at the footpoints of a helmet streamer, until the streamer is disrupted by an ejection-like event. The ejection opens the bulk of the magnetic fieldlines initially threading the streamer, but they close again following the ejection event proper. Continuous application of the shear leads to a sequence of eruption events characterized by the opening and closing of magnetic fieldlines, with which is associated a parallel increase and subsequent decrease of magnetic energy (see their Fig. 3). These authors note that the opening of the originally closed streamer-like structure occurs distinctly before the magnetic energy reaches the value associated with a fully open magnetic configuration, but do not elaborate on the source of the "missing energy". It is clear from their Fig. 2(A) and (B) that the pre-ejection configuration is characterized by a closed region of significant radial and latitudinal extent, in which the plasma presumably remains in a state close to hydrostatic equilibrium. The thermal and gravitational potential energy of the plasma contained therein may well be enough to provide the additional contribution required for the *total* energy of the pre-ejection configuration to exceed the *total* energy of their post-ejection configuration. Figure 5 herein would certainly suggest that transitions from (nearly fully) closed to (nearly fully) open magnetic configurations are energetically possible, over a reasonably wide range of model parameters. Of course, the dynamical and topological aspects of such a transition lie outside of the purview of the minimum energy conjecture.

3.2. NUMERICAL SIMULATIONS

At this point it may be worth reflecting on the fundamental difference between global energetics and local force balance. Solving the steady MHD (differential) equations amounts to enforcing *local* balance for quantities such as momentum flux, mass flux, etc, at all locations in space. This is quite different from a statement of total energy minimization, which implies something about *global* stability. It is a characteristic of closed systems that stable equilibria correspond to energy minima. The situation is nowhere as obvious for open systems. Following Prigogine (e.g., 1947, §6; 1961, chap. 6), it is believed that open systems in equilibrium should minimize entropy rather than energy (see also deGroot & Mazur 1962, chap. 5). The crux of the matter is then to come up with a suitable definition of entropy for the system (and, particularly critical, of entropy flux through its boundaries), something that turns out to be considerably trickier than one may initially imagine. This is clearly an intriguing avenue for future exploration. Nevertheless, the possibility that the corona effectively behaves like a closed system may hold some important clues concerning the nature and mode of operation of the ever elusive coronal heating mechanism(s).

The fact remains that one can "predict" helmet streamers using arguments based either on local dynamics or on global energetics. In view of this, could not one then argue that the minimum energy corona conjecture is a mere tautology? Should not any coronal configuration in a state of force balance automatically be in a minimum energy state? The answer is emphatically no, and this is most easily seen by asking whether one could actually conceive of a magnetized corona that is everywhere in force balance while *not* being in a minimum energy state. A split-monopole corona, with an equatorial current sheet extending all the way to the photosphere (cf. Fig. 3A), is a obvious example; because the field is everywhere open, the corona simply expands outward in the form of a wind. Force balance —radial and latitudinal— is everywhere satisfied, but the resulting config-uration is in general far from a minimum total energy state (cf. Fig. 5). Numerical simulations have amply illustrated the fate of such a config-uration if used as initial condition in a time-dependent calculation (e.g., Endler 1971; R. Steinolfson, personal communication); magnetic fieldlines gradually reconnect across the equator starting near the coronal base un-til a helmet streamer-like configuration is produced, with a radial extent depending on the adopted boundary conditions, namely coronal base tem-perature, plasma β, and so on. This is precisely the behavior expected from global energy consideration, *but this behavior cannot be predicted from the initial state on the basis of dynamical considerations alone.* This is a dif-ferent situation from initiating the simulation using a fully closed dipolar configuration, and letting the dynamical pressure associated with the so-lar wind open a subset of magnetic fieldlines, leading again to a helmet streamer-like end state (e.g., Steinolfson *et al.* 1982; Wang *et al.* 1993). In the language of our simple model, this latter evolution is also a transition from a higher to a lower total energy state, but proceeds from $a = \infty$ to $a \sim 2$ along the total energy curves of Fig. 5 and start with a configuration that is not in force balance.

4. Applications

Adopting a more pragmatic viewpoint, one may wonder whether our min-imal energy conjecture can be put to practical use, as opposed to simply remaining a thermodynamical curiosity. We believe that the kind of global constraints associated with minimal energy arguments can prove useful in the empirical modelling of coronal structure. This section elaborates on one specific example.

Ground-based and space borne coronagraphs and coronameters have provided and continue to provide systematic observations of the white light corona. Figures 1 and 2 herein provides an example of such observations. The analysis of such data in terms of the physical properties of the emitting plasma is fundamentally an inverse problem. The coronal brightness at a given position in the plane of the sky is a measure of the integrated line-of-sight emission; in practice the polarization brightness (PB) is used, in order to eliminate unwanted contributions from the background sky. In white light, the primary coronal emission process is the scattering of sunlight into the line of sight by free electrons, so that PB is effectively a measure of the electron density (n_e). The appeal of the PB analysis goes well beyond the mere determination of the electron density distribution. Because the coronal plasma is imbedded in and structured by the coronal magnetic field, in theory modelling the PB allows to reconstruct not only the plasma density, but also the large scale magnetic field of the solar corona, the latter being largely inaccessible to direct observation.

It is crucial to realize that the observed PB data represents a *global* and *non-local* measure of the density. Mathematically, the modelling problem is expressed as an integral equation:

$$\text{PB} = \int_{\text{l.o.s.}} C(r) n_e(r, \theta, \phi) \mathrm{d}s, \qquad (6)$$

where the integration is carried out along the line of sight, and $C(r)$ is the scattering function, a quantity of known functional form that represents the so-called kernel of the integral equation. As with most inverse problems, the smoothing properties of the kernel and of the integral operator translate into an irretrievable loss of information, making the problem formally ill-posed (see Craig & Brown 1986 for an accessible and insightful discussion of these issues). In practice, this means that there will exists multiple n_e distributions that produce PB profiles that are equally compatible with the data. In other words, The mapping from PB data space to solution space is not one-to-one, *even for noise-free data*. In the presence of data noise, a related pathology emerges; very closely spaced "points" in data space (these could be, for example, two PB profiles obtained by adding two distinct noise realizations to the same synthetic PB profile) may well map to widely separated "points" in n_e space. In other words, noise and errors are *amplified* in going from data space to solution space. This is obviously a serious difficulty for inverse modelling that aims at reconstructing n_e unambiguously and with useful accuracy.

The non-uniqueness of inverse problems can be accommodated by imposing additional *constraints* on the solution. Smoothness constraints are a classical example; this may involve fitting the data not by simply minimizing the χ^2, but instead minimizing the sum of χ^2 and some functional of the solution that is sensitive to smoothness (e.g., spatial derivatives of various orders). Another approach, effectively equivalent to a regularization technique, consists in using a parametric, physically-based model for n_e and $\mathbf{B}$ to fit the data. The inversion problem then reduces to determining the parameter values that produce the best fit to the PB data. This can be recast in terms of a minimization problem, where the quantity to be minimized is, for example, a χ^2 measure of goodness of fit between the data and the PB predicted by the model for a given set of defining parameters. This can greatly reduces the dimension and size of the search space, but does not, in itself, necessarily guarantee uniqueness.

The strategy outlined in the preceding paragraph is precisely that adopted by Gibson & Bagenal (1995; see also Bagenal & Gibson 1991 and Gibson *et al.* 1995). These authors invert PB data obtained by cross-calibrating and combining SMM coronagraph and MkIII coronameter data (cf. Fig. 1). They then attempt to fit this data using 6-parameters coronal models constructed from a class of magnetostatic solutions due to Bogdan & Low (1986). The resulting reduced inverse problem is non-linear in nature, in view of the dependency of the coronal solutions on some of these parameters. The analysis of Gibson & Bagenal shows that some level of degeneracy remains in the inversion, in that there exists long, narrow χ^2 "valleys" in some hyperplanes of 6-D parameter space (see their Fig. 11). This lack of contrast along some directions in parameter space means that noise in the PB data can lead to large jumps in solution space, exemplifying the fundamentally ill-posed nature of the inverse problem in all its glory.

Evidently, one requires additional physical constraints to lift this degeneracy and isolate a unique solution. Gibson & Bagenal use the net photospheric magnetic flux in each hemisphere to extract bounds on some of the parameters in their models. This reduces somewhat the aforementioned degeneracy, but they find that the resulting (reduced) range of acceptable solutions remains overly wide. Clearly, stronger constraints are required. It may well be that the minimum energy conjecture can provide such a constraint; among all candidate low-χ^2 solutions, one simply selects the solution that minimizes the total energy of the coronal configuration. In fact, such a *global* constraints may well be physically more meaningful that *local* smoothness-like constraints, given the inherently global nature of the inversion problem. This is a strategy we plan to explore in the near future.

References

Aly, J.J.: 1984, *Astrophys. J.* **283**, 349.
Aly, J.J.: 1991, *Astrophys. J. (Letters)* **375**, L61.
Bagenal, F., & Gibson, S.E.: 1991, *J. Geophys. Res.*, **96**, 17663.
Bogdan, T.J., & Low, B.C.: 1986, *Astrophys. J.* **306**, 271.
Charbonneau, P., & Hundhausen, A.J.: 1996, *Sol. Phys.*, in press. (CH96)
Craig, I.J.D., & Brown, J.C.: 1986, *Inverse Problems in Astronomy* (Bristol, UK: Adam Hilger)
de Groot, S.R., & Mazur, P.: 1962, *Non-equilibrium thermodynamics* (New York: 1984 Dover reprint).
Endler, F.: 1971, *Zur Wechselwirkung zwischen Sonnenwind und koronalen Magnetfeldern*, PhD Thesis, Universitätssternwarte Göttingen.

Gibson, S., & Bagenal, F.: 1995, *J. Geophys. Res.*, in press.

Gibson, S., Bagenal, F., & Low, B.C.: 1995, submitted to *J. Geophys. Res.*

Harrison, R.A.: 1991, *Adv. Space Res.* **11**, 1, 25.

Hiei, E., Hundhausen, A.J., & Sime, D.G.: 1993, *Geophys. Res. Letters* 20, 2785.

Hundhausen, A.J.: 1994, in (eds. K.T. Strong, J.L.R. Saba, & B.M. Haisch) *The Many Faces of the Sun: Scientific Highlights of the Solar Maximum Mission*, in press.

Hundhausen, J.R., Hansen, R.T., & Hansen, S.F.: 1981, JGR, 86, 2079

Kahler, S.: 1992, *Ann. Rev. Astr. Ap.* **30**, 113..

Kopp, R.A., & Holzer, T.E.: 1976, *Sol. Phys.* 49, 43.

Linker, J.A., & Mikić, Z.: 1995, *Astrophys. J. (Letters)* **438**, 45.

Low, B.C.: 1986, *Astrophys. J.* **310**, 953.

Prigogine, I.: 1947, *Étude thermodynamique des phénomènes irréversibles* (Liège: Desoer).

Prigogine, I.: 1961, *Introduction to thermodynamics of irreversible processes (second ed.)* (New York: John Wiley & Sons).

Smith, D.F., & Low, B.C.: 1993, *Astrophys. J.* **410**, 412.

Steinolfson, R.S., Suess, S.T., & Wu, S.T.: 1982, *Astrophys. J.* **255**, 730.

Wang, A.-H., Wu, S.T., Suess, S.T., & Poletto, G.: 1993, *Sol. Phys.* 147, 55.

Wang, A.-H., Wu, S.T., Suess, S.T., & Poletto, G.: 1995, *Solar Phys.*, in press

ULYSSES OBSERVATIONS OF THE SOLAR WIND OUT OF THE ECLIPTIC PLANE

W.C. FELDMAN, J. L. PHILLIPS, B. L. BARRACLOUGH AND
C. M. HAMMOND
Los Alamos National Laboratory,
Los Alamos, NM 87545, USA
wfeldman@lanl.gov

Abstract. This chapter presents a summary of *Ulysses* plasma observations of the global structure of the solar wind in the inner heliosphere (1.3 AU < R < 5.4 AU) made between the maximum and minimum phases of solar cycle 22. Salient features are the presence of a low speed, variable flow that marks the extension of the heliomagnetic streamer belt into interplanetary space. A high speed, relatively structure-free wind flows from the large-area coronal holes that cover both solar polar caps. The streamer belt was roughly planar and tilted by about $30°$ relative to the ecliptic plane just after the *Ulysses* Jupiter encounter. It evolved to an approximately $42°$-wide disk parallel to the equatorial plane at the time of perihelion. The $30°$ tilt at the beginning of the high latitude traverse of *Ulysses* caused the generation of a series of strong corotating interaction regions that were bounded by equatorward-propagating forward shocks at their leading edges and poleward-propagating reverse shocks at their trailing edges. A new class of forward-reverse shock pairs driven by the over-expansion of coronal mass ejections was also observed. Both types of shocks were observed up to $-58°$ heliographic latitude. The radial and latitudinal dependences of many basic quantities that characterize the high-latitude solar wind are documented and discussed.

1. Introduction

Most of what is known about the solar wind inside of 5 AU comes from in-situ measurments made within a $\pm 7.25°$ band of heliographic latititude centered on the solar equatorial plane. These data have been supplemented by remote-sensing observations using the interplanetary scintillation (IPS) technique. Two general categories of flow have been identified, those char-

K. C. Tsinganos (ed.), Solar and Astrophysical Magnetohydrodynamic Flows, 265–300.
© *1996 Kluwer Academic Publishers.*

acterized by low speed and high speed wind (Feldman et al., 1977; Schwenn, 1991). Intercomparison with simultaneous optical, EUV, and X-ray observations of the solar disk and limb have shown that these two states emanate from different parts of the corona. The low-speed flows emanate from the magnetically-open portions of the corona near the streamer belt and the high-speed flows from coronal holes (e.g. Withbroe et al., 1991, and references therein). Altogether, a global-scale picture of solar wind structure has emerged that varies with solar activity (e.g., Hundhausen, 1977, and references therein). Whereas streamers generally cap a belt that encircles the Sun at its magnetic equator (which is closely parallel to the ecliptic plane near solar minimum), this belt tilts relative to the ecliptic plane at other phases of the solar cycle. Streamers can also occur at all latitudes near solar maximum. In contrast, coronal holes generally cover both polar caps throughout the solar cycle but their size is smallest at solar maximum and is largest at solar minimum. Coronal holes can additionally occur for short periods of time (relative to the 11-year solar-cycle period) at all latitudes.

MHD models of the coronal expansion have relied almost exclusively on detailed flow parameters provided by in-situ observations at low latitudes in interplanetary space. IPS observations at all latitudes have only been able to provide gross estimates of flow conditions because they necessarily integrate over large volumes of plasma along the line of sight. Resultant gaps in our knowledge have been filled, to some extent, by registering in-situ observations made in the ecliptic plane to the heliomagnetic equator, which can tilt by as much as $30°$ relative to the ecliptic plane (Zhao and Hundhausen, 1981). However, a complete in-situ survey of solar wind conditions at high heliographic latitudes has only recently been made possible by the first complete latitude scan by *Ulysses*, from $-80.2°$ to $+80.2°$ heliographic latitude. It occurred between September, 1994, and August, 1995, which was close to the minimum of solar cycle 22. This circumstance is fortuitous from an MHD modeling point of view (Lima and Tsinganos 1995) because the global structure of the solar wind was consequently in its most simple configuration, that corresponding to a thin, equatorial low-speed streamer belt capped on both sides by high-speed solar wind. In this chapter, we review observations from the ion and electron plasma spectrometers of the *Ulysses* solar wind experiment (Bame et al., 1992) to provide a detailed global survey of the solar wind.

Ulysses was launched in October 1990 into an orbit that used a gravity assist from Jupiter on 8 February 1992 to tilt its orbital plane from the ecliptic to one that was close to meridional. After passage by Jupiter at 5.4 AU, the spacecraft traveled south where it reached its maximum heliographic latitude of $-80.2°$ at 2.29 AU on 12 September, 1994, proceeded to its perihelion where it crossed the ecliptic plane at 1.34 AU on 12 March,

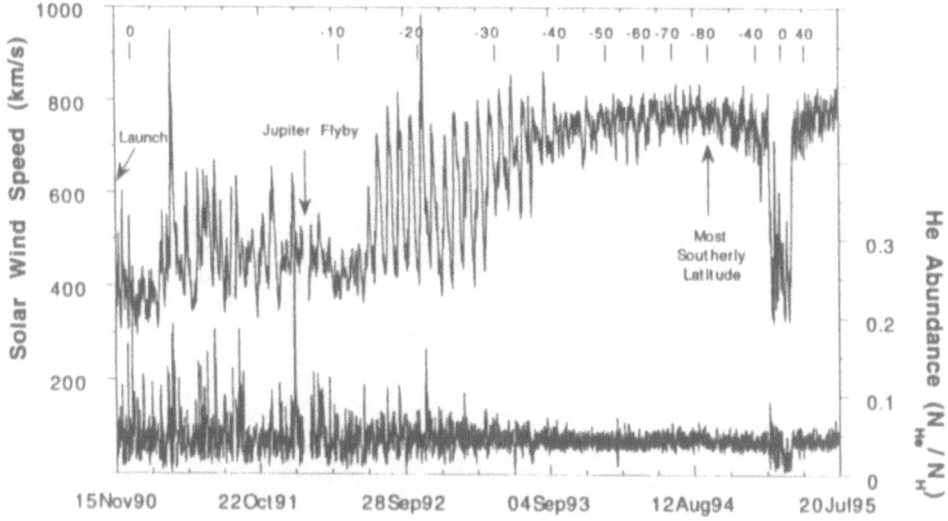

Figure 1. 1. A global overview of the solar wind observed by *Ulysses* provided by two hourly averages of the solar wind bulk speed and He abundance, at top and bottom, respectively (from Barraclough et al., 1995).

1995, and then swung up to + 80.2° at 2.02 AU on 31 July, 1995. Analyses of measurements of the photospheric magnetic field suggest that the maximum and minimum spacecraft heliographic latitudes of this orbit corresponded to about − 88° and + 85° heliomagnetic latitudes, respectively (B.E. Goldstein, private communication, 1995).

2. Observations

2.1. OVERVIEW OF SOLAR WIND FLOW STRUCTURE

The simplest way to survey the various plasma regimes traversed by *Ulysses* is through a plot of solar wind speed as a function of time, as shown in Figure 1 (from Barraclough et al., 1995). The He abundance is also shown at the bottom of Figure 1 for later reference. The ecliptic portion of the trajectory occurred just after solar maximum (the polar fields reversed in early 1990) and the streamer belt overlying many individual active regions

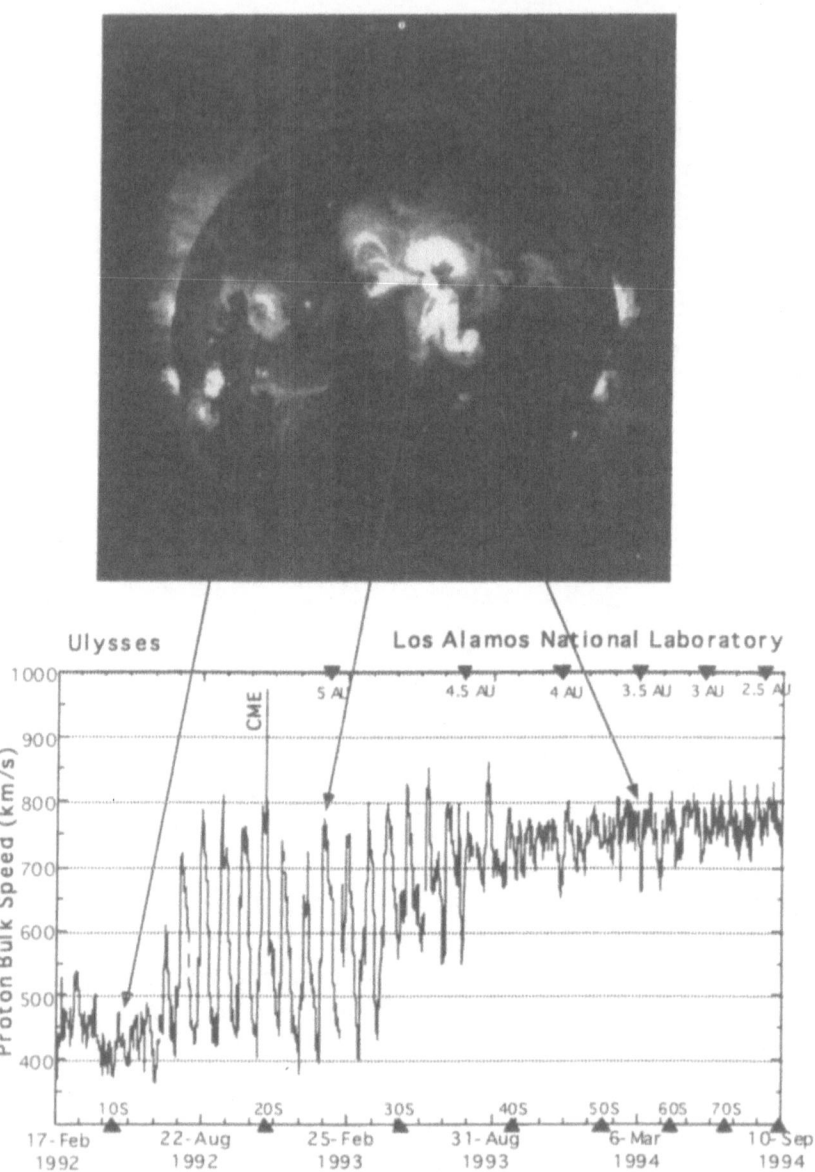

Figure 2. Identification of the coronal origins of the two different flow states of the solar wind observed by *Ulysses*. The X-ray image of the corona was kindly provided by the Yohkoh mission of ISAS, Japan.

was tilted relative to the ecliptic plane by about $30°$ (Smith et al., 1993), as shown by the Yohkoh image of the Sun measured on 8 January, 1993 shown in Figure 2 (the X-ray image was kindly provided by the Yohkoh mission of ISAS, Japan). *Ulysses* was immersed in generally low-speed wind at latitudes equatorward of about 13.5°, but at this point it began to encounter a series of corotating high speed streams from the equatorward edge of the southern polar coronal hole. Observation of high speed flows before Jupiter encounter were due primarily to encounters with coronal mass ejection (CME) events. During the post Jupiter portion of the mission, a magnetic sector reversal was detected during each return to low speed conditions (which marked entry into the outward extension of the heliomagnetic streamer belt) until *Ulysses* passed poleward of – 30° latitude (Smith et al., 1993). No sector boundary crossings were detected poleward of this point. The minimum solar wind speed jumped from about 450 km s^{-1} to about 600 km s^{-1} here also (Bame et al., 1993). Corotating intervals of low- and high-speed flows ranging between about 600 km s^{-1} and 800 km s^{-1}, were then observed by *Ulysses* until it passed poleward of about – 36° latitude (Phillips et al., 1994, 1995a,b; McComas et al., 1995a). From this position until its orbit returned to – 21° latitude, *Ulysses* remained continuously immersed in the high-speed solar wind coming from the large-area coronal hole that covered the southern polar cap of the Sun. After a relatively brief latitude scan of the streamer belt in February and March, 1995 (which at this time was relatively thin and parallel to the ecliptic plane), *Ulysses* penetrated into, and remained within, the high- speed solar wind coming from the northern polar cap of the Sun (Phillips et al., 1995c,d; Gosling et al., 1995a) through the end of the period discussed here.

2.2. COROTATING INTERACTION REGIONS (CIRS)

Forward and reverse shock pairs generally bounded the corotating interaction regions (CIRs) that formed the interface between slow and fast solar wind streams encountered by *Ulysses* (Gosling et al., 1993; 1994a; 1995b; Gosling, 1995; Phillips et al., 1995a,b,e). A summary of CIR-related shock encounters is given in the top panel of Figure 3 (from Phillips et al., 1995a). A close inspection shows that the most polar location of a forward shock occurred at a latitude of about – 34°, while reverse shocks continued to beyond – 58°. This difference is readily explained by a three-dimensional model of solar wind flow and is a natural consequence of the tilt of the heliomagnetic dipole relative to the solar equatorial plane during this period. This model assumes two large-area holes that cover both polar caps, separated by a planar streamer belt tilted relative to the equatorial plane by 30°, as shown in the bottom panel of Figure 3 (Gosling et al., 1993, Pizzo and Gosling, 1994). This tilt ensures that fast wind from both polar-caps overtake slower wind from the streamer belt at all heliographic latitudes equatorward of ± 30°. The ensuing interaction causes regions of enhanced pressure that are

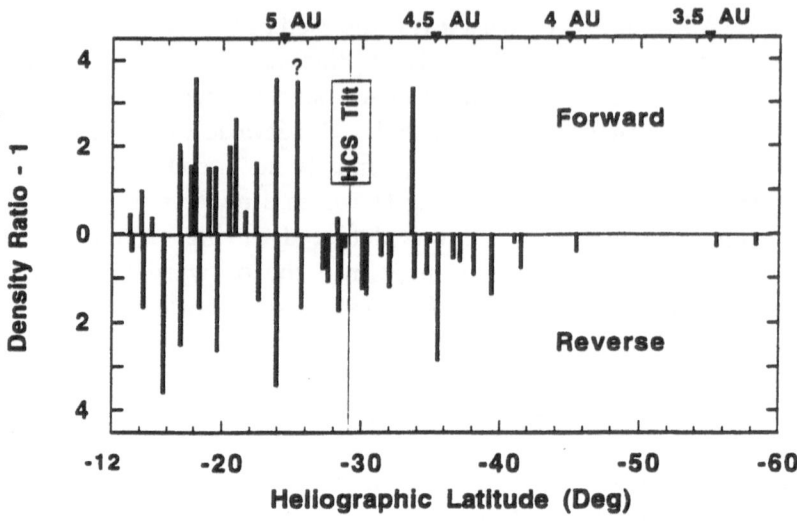

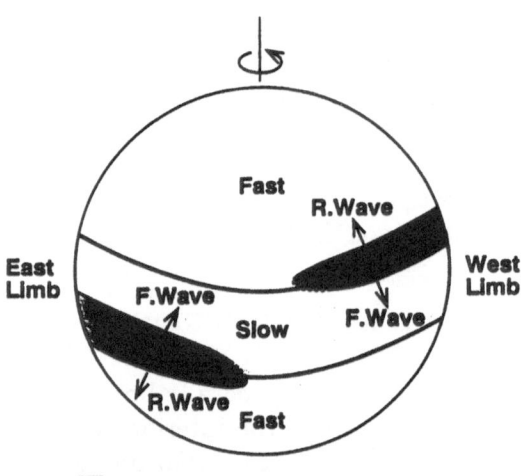

Interaction Region Far From Sun

Figure 3. Summary of the fast-mode shocks observed by *Ulysses* that were driven by CIRs (from Phillips et al., 1995a). The sketch at the bottom illustrates a global model of solar wind structure that is consistent with the pattern of CIRs observed by *Ulysses*. See the text for a detailed explanation.

bounded by compressive waves, which steepen into forward and reverse shocks by 2 AU at their leading and trailing edges, respectively. While the forward waves travel equatorward into the slow solar wind as shown in Figure 3, the reverse waves travel poleward. This model therefore provides a natural expanation of the poleward displacement of reverse shocks relative to the forward shocks observed by *Ulysses*. This poleward displacement has

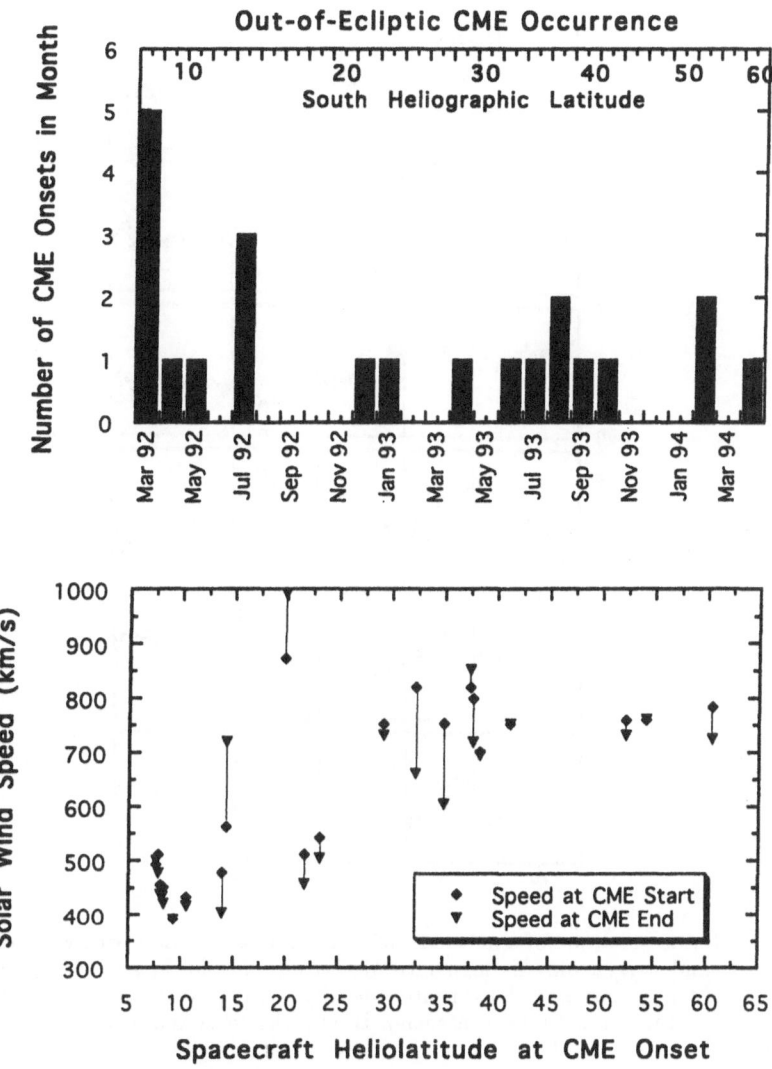

Figure 4. A summary of occurrences of CMEs observed by *Ulysses* is shown at the top and their bulk speeds are shown at the bottom.

necessitated a reevaluation of classical theories of cosmic-ray transport and of the energization of solar wind particles in the heliosphere.

2.3. CORONAL MASS EJECTIONS (CMES)

Ulysses encountered many CMEs from launch until passing poleward of $-58.2°$ latitude. They were identified in the plasma data through observation of counter-streaming halo electrons (e.g., Gosling, 1990, and references

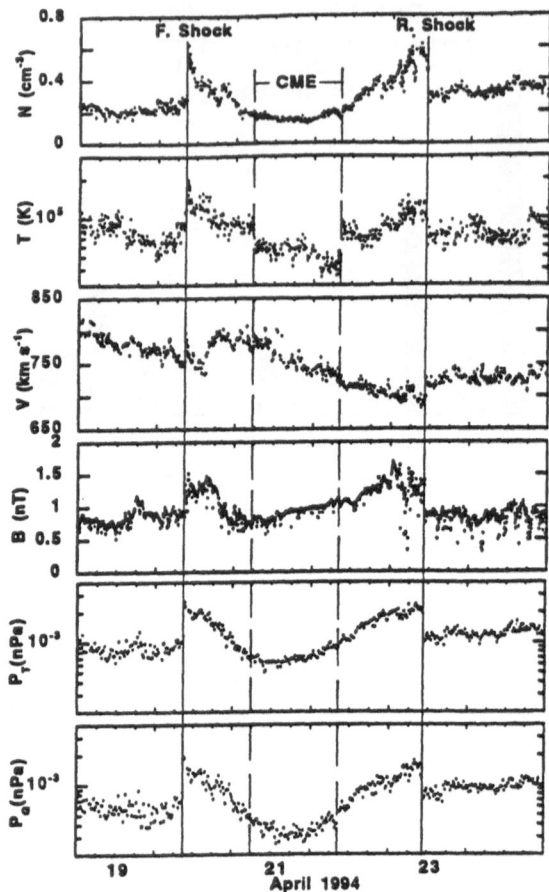

Figure 5. Internal structure of a new class of CME interaction that is driven by a central overpressure, and is seen only at high latitudes (from Gosling et al., 1994b). From top to bottom the traces correspond to, proton density, temperature and speed (N, T, and V, respectively), the magnetic field intensity, B, and the total and gas pressures (P_t, and P_g, respectively).

therein). Verification that an event was indeed a CME came from checking for abnormally low proton temperatures, and for evidence of a magnetic cloud (e.g., Burlaga, 1991, and references therein). A latitude survey of CME encounters between March 1992 and April 1994 yielded 22 events, as summarized in the top panel of Figure 4. Nearly half the total (12) were detected while *Ulysses* was within $25°$ heliolatitude of the ecliptic plane, and the rest were evenly distributed in latitude up to $-58.2°$. Analyses of these events revealed two new characteristics that were not known before the *Ulysses* high-latitude survey. First, the speeds of CMEs generally follow those of the ambient solar wind through which they propagate (Gosling

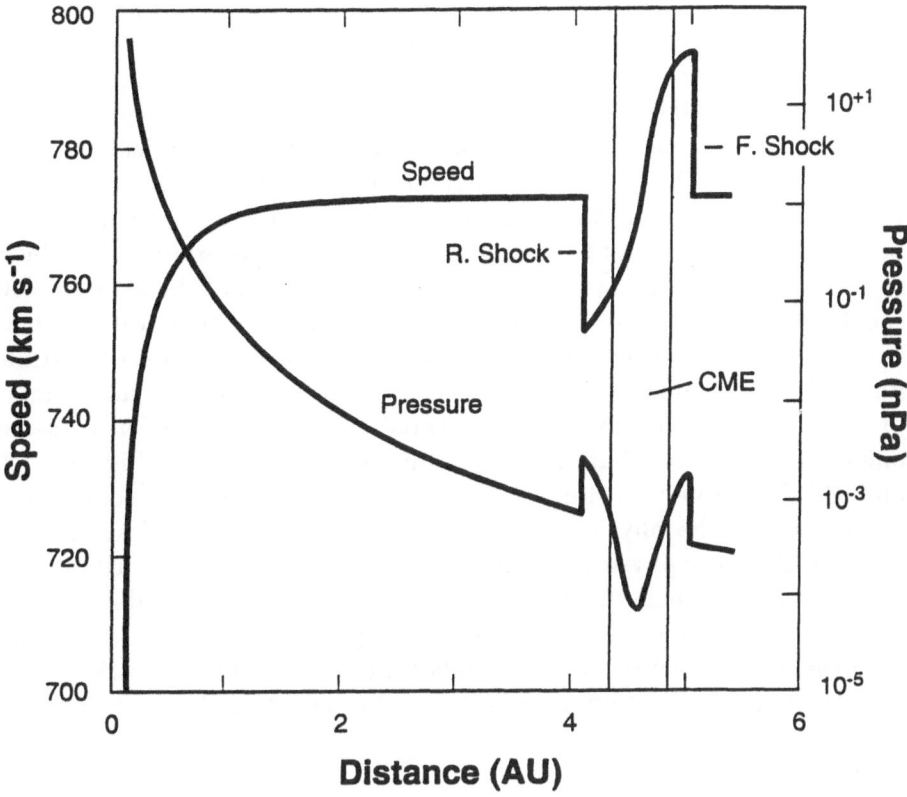

Figure 6. A 1-D hydrodynamic model of a CME driven by a time-dependent bell-shaped density enhancement introduced at 0.14 AU (From Gosling et al., 1994b).

et al., 1994a, 1995c; Gosling, 1994; Hammond et al., 1995a). This fact is seen by comparing the CME speeds shown in the bottom panel of Figure 4 with the solar wind speed profile given in Figure 2. Equatorward of – 27° heliolatitude most of the CMEs had speeds between 400 and 550 km s⁻¹, overlapping the range of speeds observed for the low- speed component of the ambient solar wind at these latitudes. In contrast, all CMEs observed poleward of – 27° had speeds between about 700 and 850 km s⁻¹, overlapping the range of speeds of solar wind observed at these latitudes.

A second characteristic was the discovery of a new class of CME event, unique to high latitudes. These events drive forward and reverse shock pairs through CME expansion in the ambient flow caused by a thermal and/or magnetic over-pressure supplied to the wind at the point of its release close to the Sun (Gosling et al., 1994b, c, 1995 c, d; Gosling, 1994; Pizzo and Gosling, 1994). A clean example of this type of event is shown in Figure 5. The data from top to bottom give the time sequence of proton density, temperature, and speed, followed by the magnetic field intensity, the total pressure (gas plus field), and the gas pressure (protons, alpha particles, and electrons) (Gosling et al., 1994b). The portion of the event containing bistreaming electrons occurred between the central two vertical dashed lines labeled CME. It is symmetrically flanked by a forward and reverse shock identified by the outer two solid vertical lines. These shocks could not have been driven by a drift of CME material relative to the ambient flow because its speed is comparable to that of the wind upstream of the leading forward shock, as can be seen in the third panel from the top. However, the bottom two panels clearly reveal enhanced pressure in the gas contained within the forward-reverse shock pairs. This situation can be driven by an initial over-pressure close to the Sun, as illustrated through application of a simple one-dimensional, adiabatic, gas dynamic code, as shown in Figure 6 (Pizzo and Gosling, 1994; Gosling et al., 1994b). Although this code does not contain all of the physical interactions present in the real solar wind, it contains the basic physics, thereby making it useful for illustrative purposes. Introduction of a bell-shaped density profile at 0.14 AU results in the radial dependence of speed and pressure profiles plotted in Figure 6. Comparison with the data shown in Figure 5 lends credibility to the suggestion that this new type of CME interaction is driven by an internal over-pressure near the Sun.

2.4. FLOW STRUCTURE AT HIGH LATITUDES

An overview of the global structure of the solar wind close to solar minimum is afforded by the profiles of speed and density during the *Ulysses* rapid latitude scan, shown in Figure 7 (Phillips et al., 1995c,d; Gosling et al., 1995a). The traces give 6-hour averages of proton speed at the top and density scaled to 1 AU at the bottom. Inspection shows a generally high-speed flow that covered both solar polar caps separated by a low-speed flow that marked the streamer belt. The speed in the high-speed portion slowly increased from about 750 km s^{-1} at the edge of the low-latitude streamer belt to about 800 km s^{-1} at $\pm$ 8.2° heliolatitude. The boundaries of this belt were at about $\pm$ 21° centered on the ecliptic plane (Gosling et al., 1995a). Within the streamer belt, the solar wind speed was generally low, its density was high, and both were variable. The only exception occurred during a brief encounter with the streamer belt at about $-$ 35° just south of the streamer belt. Superimposed on the velocity profile are relatively small variations ($\sim \pm$ 35 km s^{-1}), which for the most part, reflect the

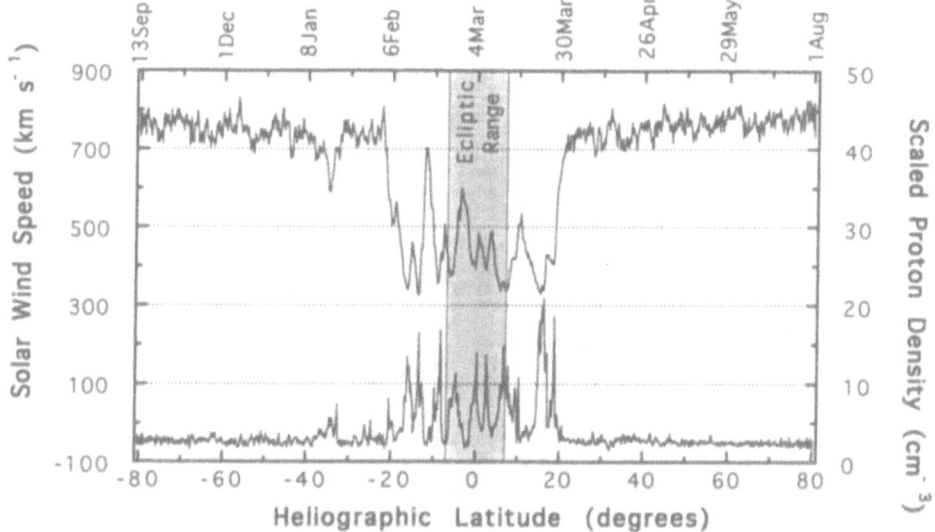

Figure 7. An overview of solar wind conditions during the *Ulysses* rapid latitude scan between – 80.2° and + 80.2° (From Phillips et al., 1995c). The top and bottom traces gives the bulk speed and density of the solar wind, respectively. The shaded stripe at the center indicates the latitude range sampled by *in-situ* observations in the ecliptic plane.

presence of large-amplitude Alfvén waves (Smith et al., 1995). This fact is documented in Figure 8 through plots of the hourly normalized variance of magnetic field components and magnitude as a function of south heliographic latitude (Balogh et al., 1995). The large ratio of directional to field magnitude variances evident in the Figure is one of the primary defining properties of Alfvén waves. Poleward of about – 35°, after *Ulysses* penetrated permanently into the high-speed solar wind, the amplitude of Alfvén waves remained high, but increased slowly to a maximum at the highest heliolatitude. In contrast, the variance of normalized magnetic field magnitude was sensibly constant and low throughout the polar cap.

Properties of the solar wind that control the interaction of the heliosphere with the local interstellar medium are its mass and momentum flux densities. The latitudinal variation of solar-rotation averages (∼ 26 days) of these parameters are shown in Figure 9 (Phillips et al., 1995c). Inspection shows relatively constant, low-variance flux densities at high latitudes separated by higher mass flux and lower momentum flux densities at low latitudes. Both flux densities at low latitudes are characterized by large variations as seen by the 5% and 95% traces, which bracket the median traces. Mean, median, 5%, and 95% values of these flux densities and a

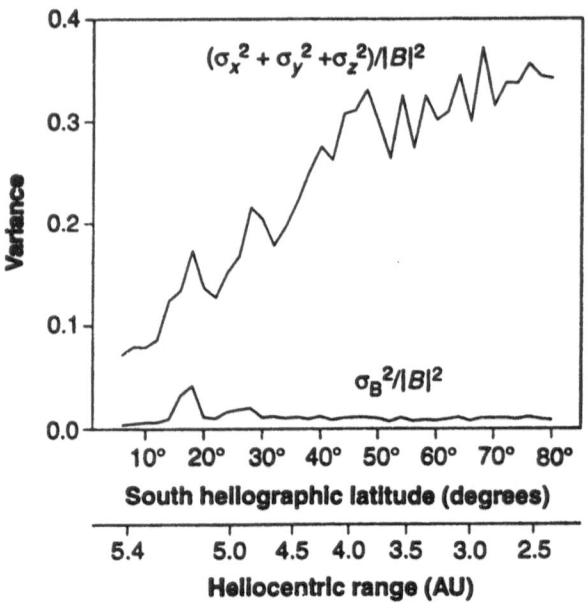

Figure 8. Hourly variances of the normalized components (top) and magnitude (bottom) of the interplanetary magnetic field averaged over successive 27-day intervals (from Balogh et al., 1995).

range of other parameters that define the state of the high latitude wind are collected in Table I. These parameters were averaged over all plasma measurements made poleward of oτ 70° heliolatitude. Comparison with a similar collection made using IMP data at 1 AU (Feldman et al., 1976a, 1977) and Helios data between 0.3 AU and 1 AU (Schwenn, 1991) reveals only relatively minor differences. This close similarity reinforces a conclusion reached previously using IMP data (Bame et al., 1977), that the high-speed solar wind is indeed a relatively structure-free, fundamental flow state of the solar wind.

Although the salient feature of the solar wind at polar latitudes is its relatively low level of parameter fluctuations (with the exception of Alfvén waves), fluctuations are nevertheless present. When viewed on a time scale of days (rather than the hourly period used for the normalized magnetic field variances shown in Figure 8), the intensity of interplanetary magnetic field, $|B|$, is also variable, as shown in Figure 10 (Balogh et al., 1995). The 10-day interval displayed, corresponds to the maximum southerly latitude reached by *Ulysses*. Inspection of the bottom panel shows that variations in $|B|$ exceed 50% over a 10-day time period, corresponding to a greater than 100% variation in the magnetic pressure.

Other variability of note at high helio-latitudes is pressure balanced and

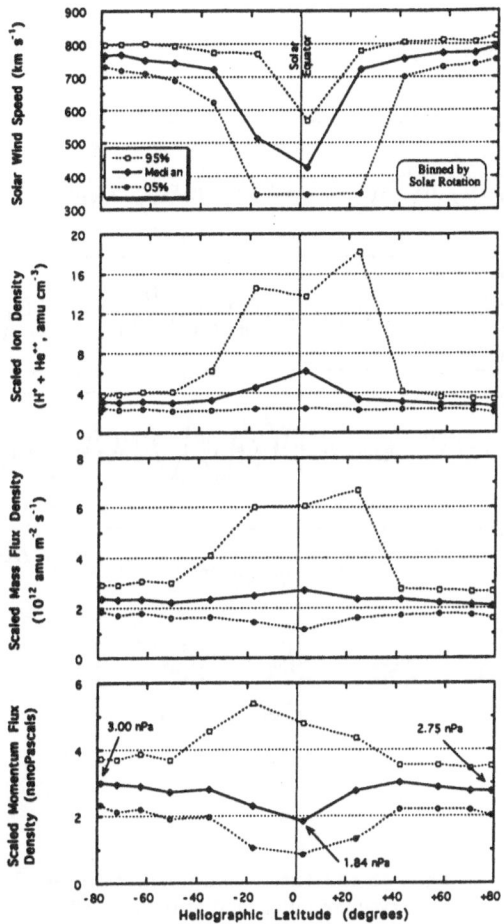

Figure 9. Successive solar-rotation averages of the solar wind speed, density (scaled to 1 AU using an R^{-2} gradient law), scaled mass flux density, and scaled momentum flux during the rapid, pole-to-pole latitude scan (from Phillips et al., 1995c).

compressional structures (McComas et al., 1995b; Neugebauer et al., 1995; Phillips et al., 1995a). An example of each is shown in Figure 11. The defining feature of pressure balanced structures is an anticorrelated, generally small-scale variation of the gas and magnetic pressures that have a sensibly constant sum, as shown at the left in Figure 11. These events are not accompanied by sensible variations in the bulk speed and are most evident in plots of plasma beta (the ratio of gas to magnetic pressures, not shown here, but see McComas et al., 1995b). In contrast, compressional structures are characterized by a correlated increase in gas and magnetic pressures associated with positive speed gradients in the plasma flow, as shown at the right in Figure 11. These structures are observed throughout the high-latitude solar

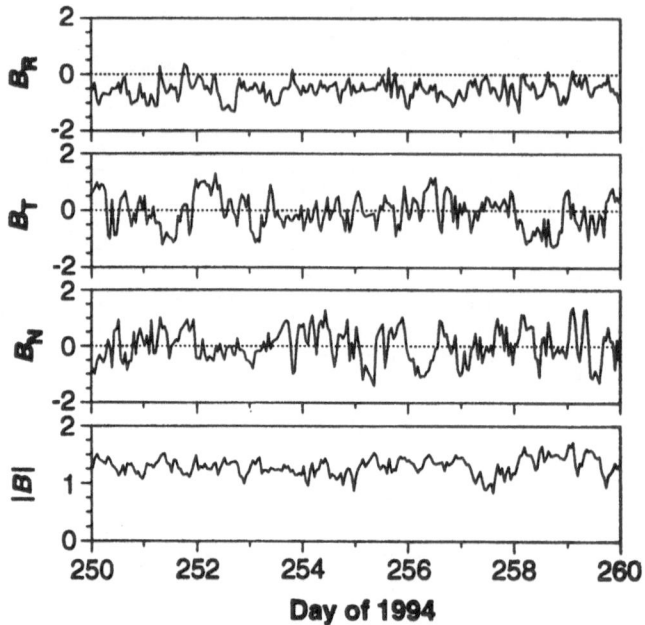

Figure 10. Hourly averages of the three components (in R, T, N coordinates) and magnitude of the interplanetary magnetic field during a 10-day period bracketing the highest southerly heliographic latitude of the *Ulysses* orbit (From Balogh et al., 1995).

wind and are characterized by minimum-to-maximum speed amplitudes of about 50 km s^{-1} (Neugebauer et al., 1995).

The relative constancy in latitude of the flow state of the high speed solar wind allows a determination of the radial variation of proton density, proton temperature, and electron temperature using the *Ulysses* polar-pass data. Different heliocentric distances were sampled at each latitude because the *Ulysses* orbit is elliptical. Binning data in $\pm$ 2° intervals about $\pm$ 50°, $\pm$ 60° and $\pm$ 70° yields a determination of the radial variation of these parameters between about 1.3 AU and 3 AU, shown in Figure 12. The bin corresponding to $-$ 50° from the initial excursion to high southerly latitudes (corresponding to heliocentric distances of about 3.5 AU) was not included in this plot because heating by reverse pressure waves from CIRs was detected at this time caused by the $\sim$ 30° tilt of the heliomagnetic

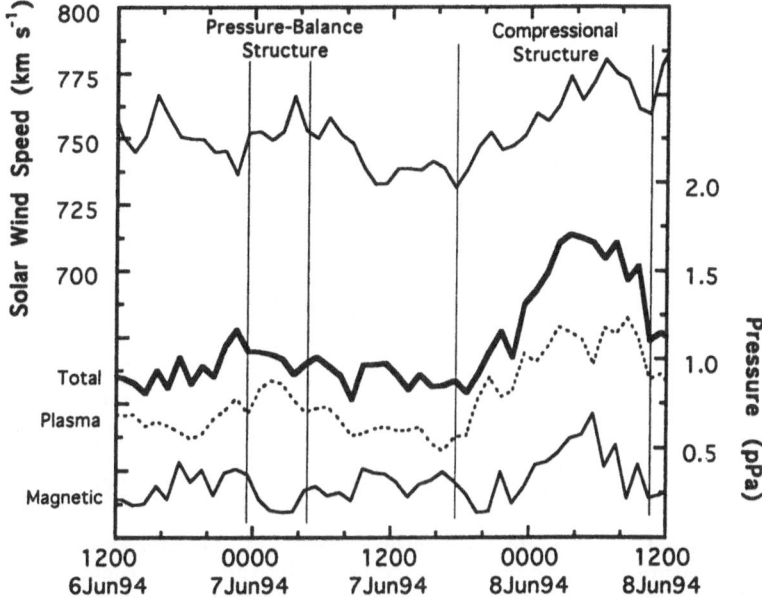

Figure 11. Hourly averages of solar wind speed, and the total, plasma, and magnetic pressures during a two day period in the high latitude wind that illustrates both a pressure- balanced structure and a compressional structure driven by a microstream (from Phillips et al., 1995b).

equator (Phillips et al., 1995e). Variations of all three parameters were fit with power law functions, given by the overlays in the figure. Inspection of the left-hand panel shows that the density is well described by the function $N = 2.4 \times R^{-1.9}$. The closeness of this regression curve to the polar average, $N = 2.4 \times R^2$, listed in Table I, leads us to believe that the technique (and implicit assumptions) used here to determine radial gradients in the high-latitude wind are valid. Proceding to the radial variations of proton and electron temperatures shown in the right- hand panel of Figure 12, we see that the proton temperature data can be adequately described by a power law, $T_p = 2.9 \times R^{-0.57}$ (10^5 K). In contrast, the electron temperature is close to isothermal, $T_e = 0.93 \times R^{-0.098}$. Both yield radial variations that are considerably less than the adiabatic scaling law ($R^{-1.33}$) that is applicable for a spherical expansion such as that indicated by the measured radial variation of proton density. We note, though, that the 1 AU value of the proton temperature (2.9×10^5 K) is somewhat higher than, but close to, that measured using IMP and Helios (both were 2.3 x 10^5 K, Feldman et al., 1976a; Schwenn, 1991). The 1 AU value of the electron temperature (0.91 x 10^5 K) is likewise very close to that measured using IMP and Helios at 1 AU (0.99 x 10^5 K and 1.0 x 10^5 K, respectively). These close comparisons

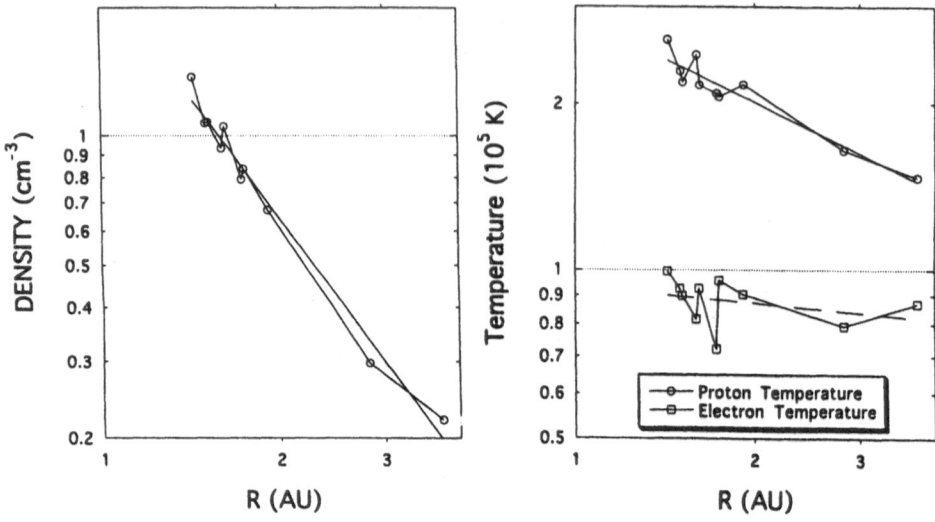

Figure 12. Radial variation of the proton density (at the left) and the proton and electron temperature (at the right) within the high-latitude solar wind. The power law fits to the density, proton temperature, and electron temperature are given by, $N = 2.4 \times R^{-1.9}$, $T_p = 2.9 \times R^{-0.53}$, and $T_e = 0.91 \times R^{-0.098}$, respectively. Details are given in the text.

strengthen our conclusions that the radial variation of proton temperature in the polar wind is less than adiabatic and that of the electron temperature is nearly isothermal between 1 AU and 3.7 AU. A similar isothermal electron temperature gradient was observed within a large, corotating coronal-hole-associated high-speed stream observed by MVM (during its inward trajectory to Mercury), and IMP (at 1 AU, Feldman et al., 1979)

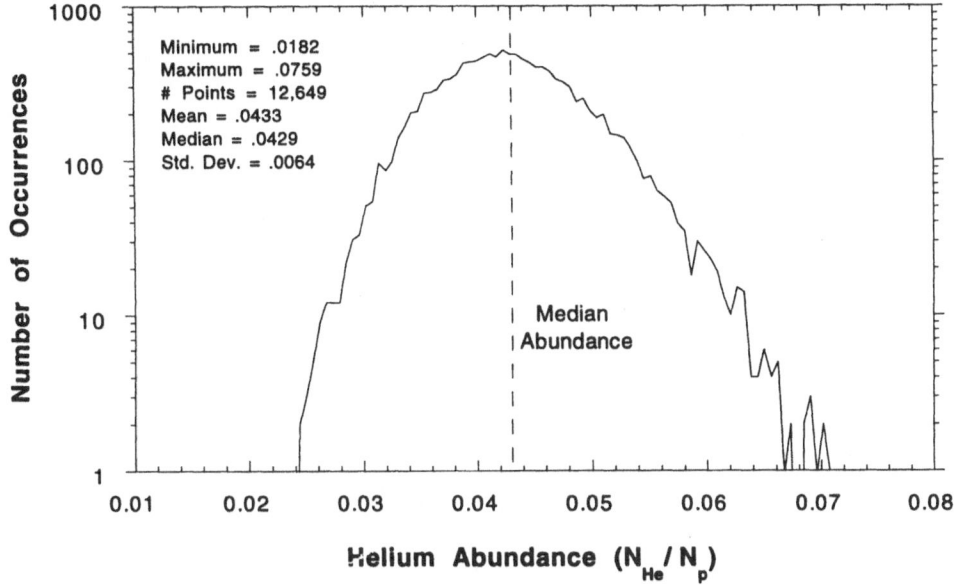

Figure 13. Histogram of hourly averages of Helium abundances in the high-latitude solar wind. Measurements were accepted only if the solar wind speed exceeded 700 km s^{-1}. Seven-day periods bracketing encounters with six CMEs were not included in the histograms.

in the ecliptic plane.

2.5. HELIUM ABUNDANCE

The character of the solar wind Helium abundance ([He] = NH_e/N_p) differs markedly between the low-speed and high-speed solar wind. The bottom trace in Figure 1 shows that [He] varies widely at low speeds and equatorial latitudes, while it is relatively constant at high speeds and polar latitudes (Barraclough et al., 1995). This impression is quantified for the high-latitude solar wind in Figure 13, which presents a histogram of hourly-averaged He abundances for all flows having $V_{sw} > 700$ km s^{-1} observed poleward of $\pm$ or 4° heliographic latitude. Seven-day intervals that bracketed each of six CMEs observed in this latitude range were also excluded from the histogram. The distribution is seen to be quite narrow, having [He] = 0.043$\pm$ 0.0064. This finding compares favorably with similar determinations in the high-speed solar wind near the equatorial plane using both IMP and Helios data, both giving [He] = 0.048 $\pm$ 0.005 (Feldman et

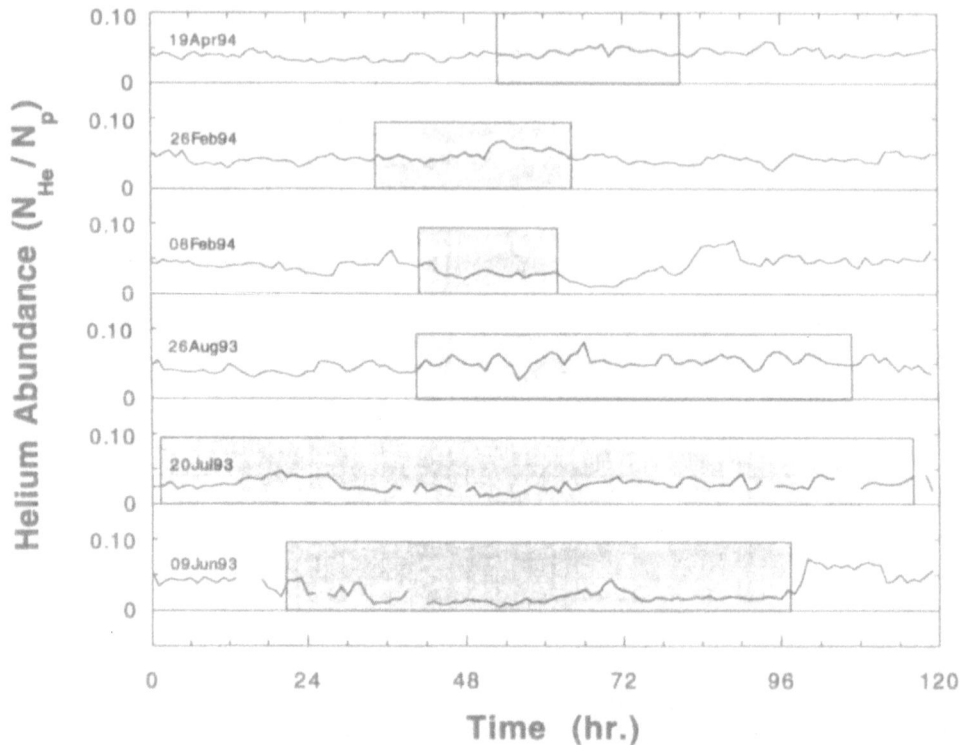

Figure 14. Hourly averages of the Helium abundance during five-day intervals containing each of the six CME events encountered by *Ulysses* within the high-latitude solar wind (from Barraclough et al., 1995).

al., 1977) and 0.038 (Schwenn, 1991), respectively. Also important to note is that the maximum hourly-averaged abundance was 0.076. We note that this value of [He] is considerably smaller than the primordial solar Helium abundance preferred by the most current models of the internal structure of the Sun, [He] = 9.6% (Bahcall and Pinsonnearult, 1992). It is also less than the cosmic He abundance inferred from a variety of measurments of objects within our galaxy (e.g., Iben, 1969, and references therein).

The foregoing discrepancy is not changed by observations within the six high-latitude CMEs observed by *Ulysses* but excluded from the high-latitude histogram of [He] in Figure 13. This fact is seen by the time-sequence plots of [He] during each of the CMEs shown in Figure 14. The shaded portions of each trace delineate the times when counter-streaming halo electrons were observed. A scan by eye (verified by histograms of [He] that are not shown) shows no anomalously high He abundances, as is often apparent in comparable CMEs observed in the ecliptic plane (e.g., Borrini et al., 1982, and references therein). Indeed, the CME observed by *Ulysses* on

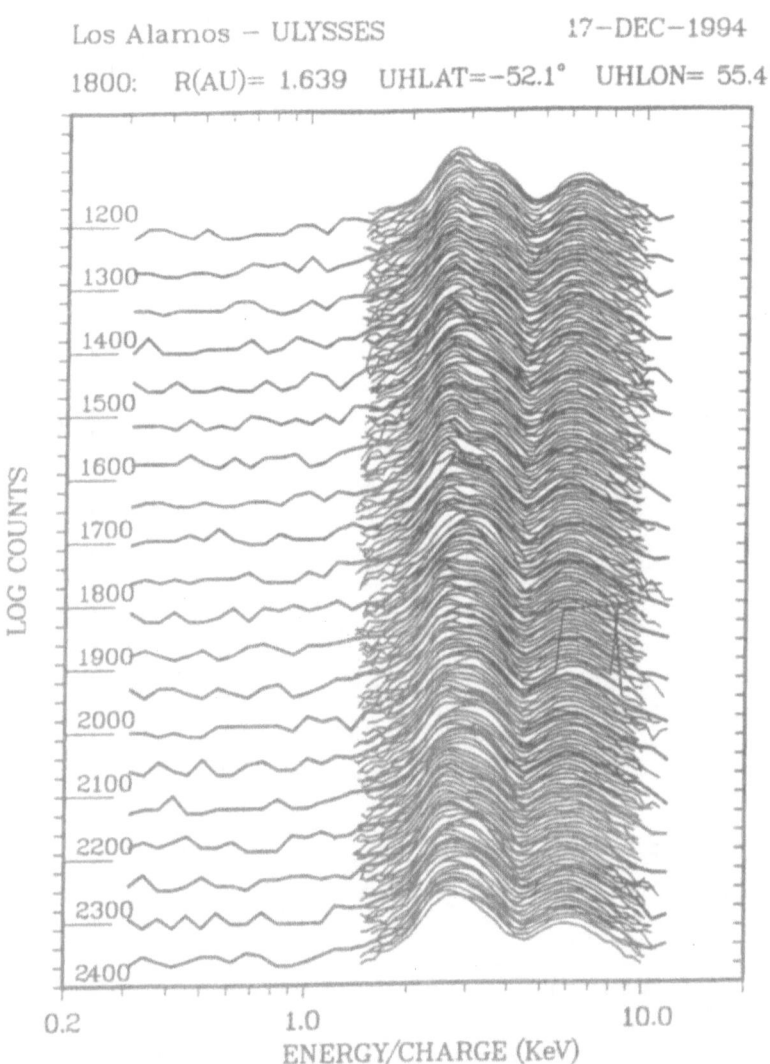

Los Alamos – ULYSSES   17–DEC–1994

1800:   R(AU)= 1.639   UHLAT=−52.1°   UHLON= 55.4°

Figure 15.   Stacked plots of one-dimensional cuts through *Ulysses* ion velocity distributions during a 12-hour period on 17 December, 1994. *Ulysses* was in the high-latitude wind during this period at a heliocentric distance of 1.64 AU and a heliographic latitude of − 52.1°.

26 February, 1994 was also observed by IMP 8 in the ecliptic plane (Gosling et al., 1995d). While the *Ulysses* data show no evidence for enhanced Helium abundances during this event, IMP 8 observed abundances ranging up to 21%. Because the He abundance of the solar wind is frozen in to the flow at low heliocentric distances, this difference in [He] within the February, 1994 CME implies that conditions at low altitudes within the high-latitude

284

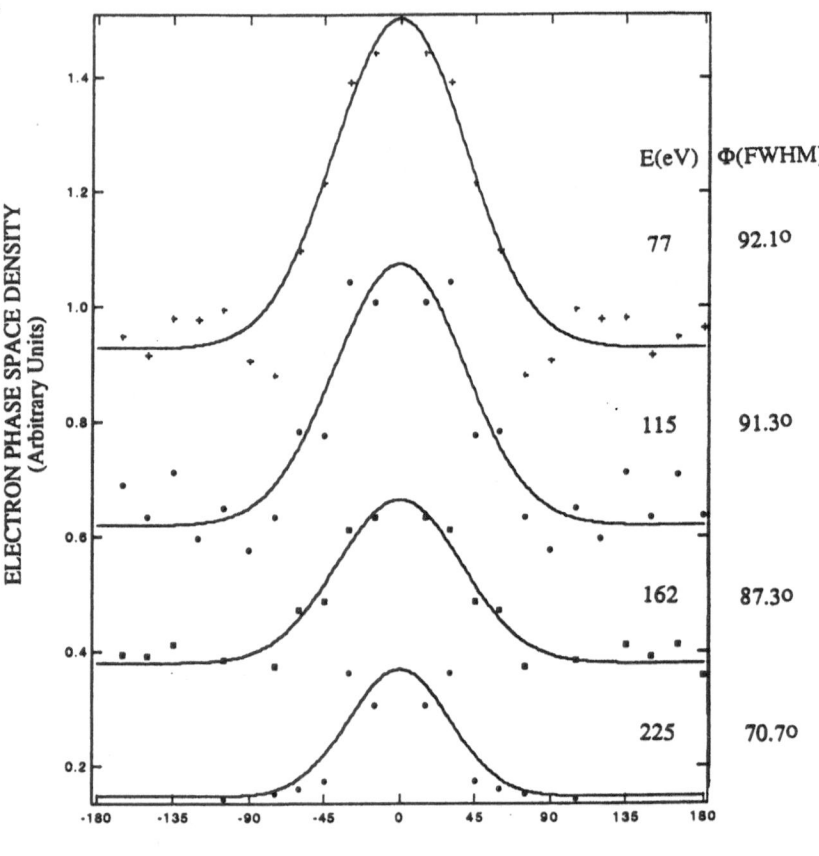

Figure 16. Pitch-angle distributions of suprathermal electrons measured near the most southerly latitude of the *Ulysses* orbit. Each point gives the average of individual phase-space samples contained within successive 15° angular bins relative to the interplanetary magnetic field. The nested curves give fits to the respective pitch-angle scans (from top to bottom corresponding to electron energies of 77, 115, 162, and 225 eV, respectively) using a model of a Gaussian function superimposed on a constant background. Respective full width at half maximum (FWHM) of each of the Gaussians are given at the right. See the text for more details.

polar coronal hole differ markedly from those at equatorial latitudes within coronal mass ejections.

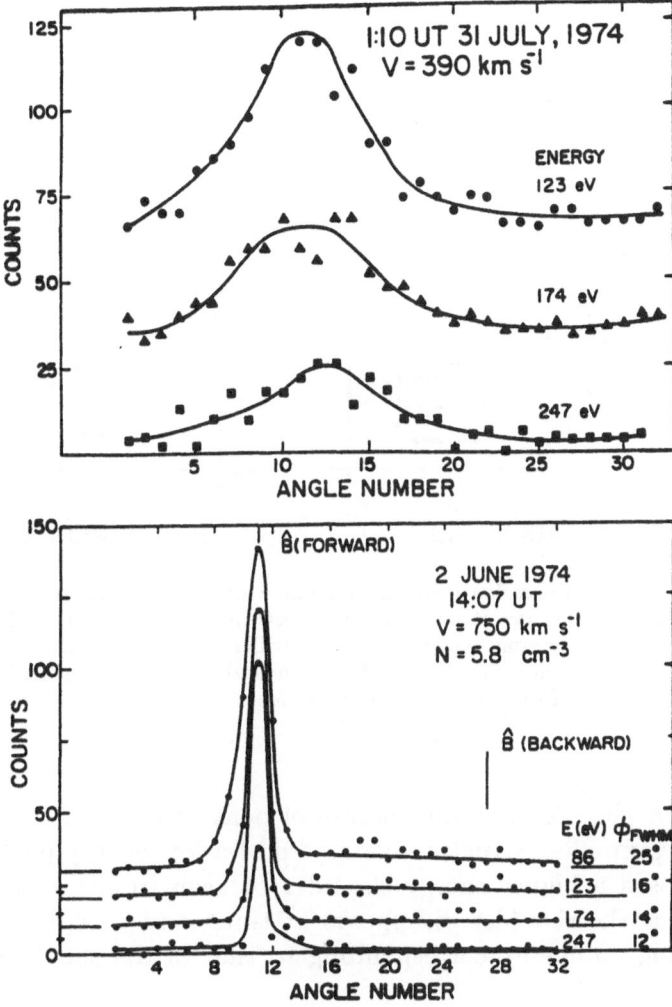

Figure 17. Pitch-angle distributions of suprathermal electrons measured using the LANL IMP plasma experiment in the low- (top) and high- (bottom) speed solar wind. Individual samples are spaced by 11.25°.

2.6. INTERNAL STATE OF THE HIGH-LATITUDE WIND; PROTONS

Scans of stacked spectra showing one-dimensional cuts through ion velocity distributions in the high-latitude solar wind reveal non-Maxwellian velocity distributions. A typical example of 12 hours of data measured on 17 December, 1994 is reproduced in Figure 15. Two peaks are clearly visible. The one at the left is due to protons and the one at the right to alpha

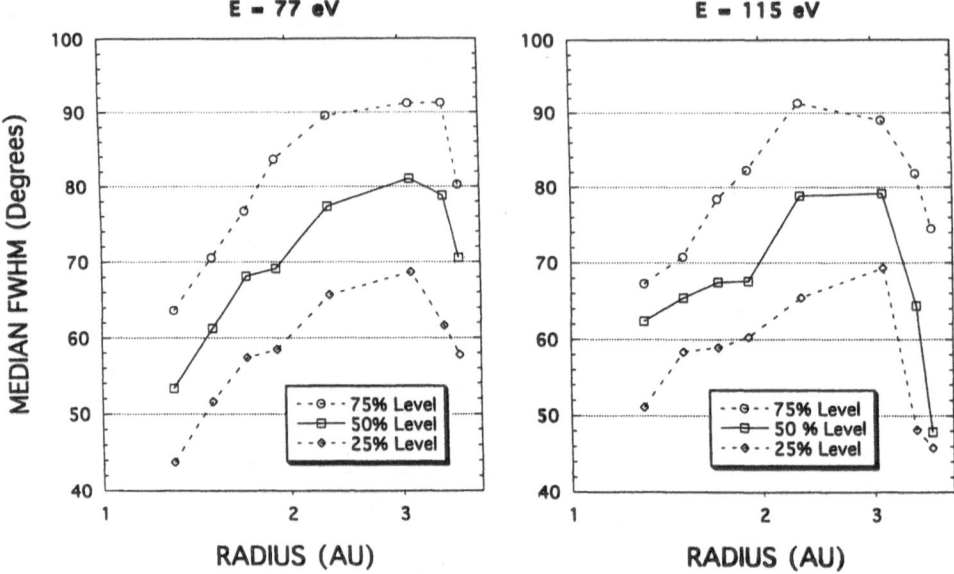

Figure 18. Radial dependence of median full widths at half maxima of Gaussian fits to pitch-angle distributions of suprathermal electrons at 77 eV (left) and 115 eV (right), constructed from histograms accumulated over widely spaced week intervals when *Ulysses* was in the high- latitude solar wind. The upper and lower traces in each panel give the 75% and 25% levels of the weekly histograms, respectively.

particles. A close look at the column of peaks on the left reveals a generally non-Maxwellian structure composed of a low energy peak on which is superimposed a higher energy shoulder. A scan from top to bottom also reveals that the shoulder disappears at times and is replaced by a single peak having an energy corresponding to that of the shoulder. No similar

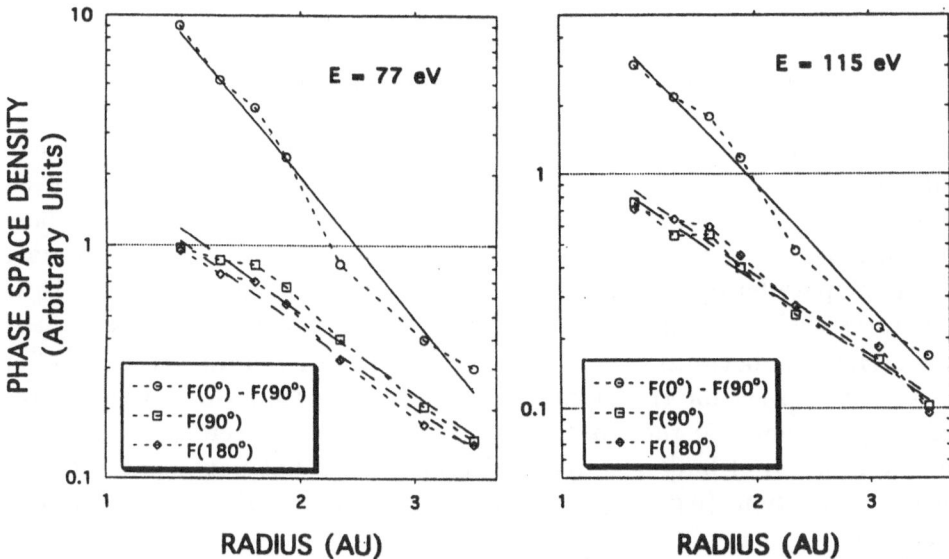

Figure 19. Radial dependence of median values of the phase-space densities of suprathermal electrons in the strahl at $0°$ (given by $F(0°)$ - $F(90°)$), and at $90°$ and $180°$ relative to $\vec{B}$. Individual values are constructed from histograms accumulated over widely spaced week intervals when *Ulysses* was in the high-latitude solar wind. The electron energy is 77 eV for the curves at the left and is 115 eV at the right. The straight lines are best power law fits to each data set using the method of linear least squares. Parameters of these fits are given in the text.

features are evident in the run of alpha-particle peaks at the right. Indeed, these peaks form a neatly nested set that do not vary in energy with time.

Similar behavior has been observed in high speed streams in the ecliptic plane (e.g., Feldman et al., 1976a, 1993; Marsch et al., 1982a). These types of proton spectra have been interpreted (Feldman et al., 1976a, 1993) in terms of two unresolved, interpenetrating proton streams, travelling relative to one another along the interplanetary magnetic field. The variation in shapes of the one-dimensional cuts through the proton distributions is caused by fluctuations in the direction of B associated with the field of large-amplitude Alfvén waves that is a prominant feature of the high-speed solar wind. These directional changes alter the view afforded by a one-dimensional cut through the velocity distribution so that they alternate from one parallel to $\vec{B}$ (revealing the shoulder) to one perpendicular to $\vec{B}$ (characterized by no shoulder but an apparent increase in V_{sw} caused by the correlation between $\delta\vec{B}$ and $\delta\vec{V}_{sw}$ inherent in Alfvén waves). No similar behaviour is seen for the alpha particles because they surf on the waves (at a speed equal to the wind bulk speed plus the Alfvén speed) relative to the solar wind bulk velocity, and therefore do not participate in bulk velocity variations (e.g., Marsch et al., 1982b, Marsch, 1991, and references therein).

2.7. INTERNAL STATE OF THE HIGH-LATITUDE WIND; ELECTRONS

Pitch-angle (relative to B) distributions of suprathermal electrons ranging in energy between about 70 eV and 320 eV were constructed using the 3-D spectra contained in *Ulysses* plasma-electron data (Phillips et al., 1995b; Hammond et al., 1995). Ten one-week intervals, spread over the high-latitude portion of the *Ulysses* orbit between January, 1994 and April, 1995, were processed. Several constraints were imposed before a spectrum was accepted for analysis. First was a requirement that B did not change more than $10°$ in angle during the electron spectrum accumulation time, 120 s. A second criterion required at least one pitch angle sample within 15o of B after the raw data were filtered for spacecraft potential and light-leak effects. And lastly, all pitch-angle samples were required to contain more than 5 counts.

A typical example of pitch-angle distributions of electron phase-space density measured in the high latitude solar wind is presented in Figure 16. From top to bottom, the four superimposed spectra correspond to electron energies of 77, 115, 162, and 225 eV. Data points consist of averages of all individual pitch-angle samples of phase-space density contained within successive 15o angular bins at a given electron energy. Directions along B ,but generally directed away from the Sun, correspond to $0°$ and those directed toward the Sun to $180°$. The distributions as displayed were mirrored about $0°$ to make a better presentation. The curves that overlay the data show the best fitting model, comprising a Gaussian function superimposed on a constant background. This background was estimated using the average phase-space density of the three 15° angle bins centered on 90°. Also calculated but not displayed in Figure 16 were average phase-space densities of the 165° and 180° angle bins.

Inspection shows the model fits the data quite well at all energies. Angular widths are seen to vary from 92° FWHM at 77 eV to 71° at 225 eV. Although the widths in this spectrum decrease monotonically from low to high energies, such behavior is not consistently observed in the *Ulysses* data set. Comparison with similar spectra measured by IMP in the high-speed solar wind at 1 AU reveals a marked difference. Inspection of IMP pitch-angle distributions for electron energies between 86 eV and 247 eV at the bottom of Figure 17 shows full widths that decrease monotonically with energy from 25° at 86 eV to 12° at 247 eV (Feldman et al., 1978). In contrast to the general case observed by *Ulysses*, the monotonically decreasing widths of the IMP spectra is the general rule in the high-speed solar wind at 1 AU. Most important, though, is that the IMP widths are far less than the *Ulysses* widths. Indeed, the *Ulysses* widths agree better with those measured by IMP in the low-speed solar wind, as seen by comparing the *Ulysses* spectra with the IMP spectra in the top panel of Figure 17.

Pitch-angle widths of *Ulysses* spectra seem to vary in a regular way with heliocentric distance as shown in Figure 18. A heliographic latitude explanation for this variation is ruled out by a similar plot of width as a function of latitude (not shown here). The three curves presented in Figure 18 correspond (from top to bottom) to full widths at half maxima (FWHM) at the 75%, 50%, and 25% levels in histograms constructed within each of the 0.2 AU-wide radial bins of data. Some radial bins are missing in the figure because of gaps in our samples of week-long intervals. Inspection shows that the widths increase with increasing heliocentric distance from 1.3 to about 2.5 AU, where they attain a relative maximum. They then decrease at larger heliocentric distances. This variation is in qualitative agreement with the much lower widths measured by IMP at 1 AU.

The radial variation of separate phase-space densities associated with the halo and strahl components of suprathermal electrons can be determined using the same data set of pitch-angle distributions just presented. We quantify the strahl by the peak phase-space density of the Gaussian functions that fit the measured electron distributions (given by $F(0°)$ - $F(90°)$), and the halo by the average phase-space densities at 90° and 180° to the anti-sunward direction along B. Variations of these parameters with radial distance for E = 77 eV and 115 eV are shown in Figure 19. Inspection shows the $F(90°)$ and $F(180°)$ curves to be nearly identical and describable by power laws. For both energies and angles, the radial variation of phase-space density is well characterized by an $R^{-\alpha}$ law, with α close to 1.9. We note that this dependence is close to that determined for the total proton density (see Figure 12). In contrast, the peak phase space density of the strahl (given by F(0°) - F(90°)) is seen to fall off more steeply with increasing heliocentric distance. The strahl data for both E = 77 eV and 115 eV can likewise be fit with a power law characterized by an $R^{-\alpha}$ dependence, but here, α = 3.4 for E = 77 eV and 3.0 for E = 115 eV.

3. Summary and Discussion

Salient features of the structure of the solar wind out of the ecliptic plane observed by *Ulysses* were presented. This first high-latitude survey of solar wind conditions occurred between February, 1992 and August, 1995, between the maximum and minimum activity phases of solar cycle 22. During this time the corona changed from one containing coronal holes that covered both polar caps, separated by an active streamer belt that was tilted by about 30° with respect to the ecliptic plane, to one consisting of two larger-area polar holes separated by a much less active belt in a plane that was nearly parallel to the solar equator. The gross latitudinal structure of the solar wind conformed to this coronal structure, with generally low-speed, variable flows mapping to extensions of the streamer belt into interplanetary space, and a high-speed, generally structure-free flow mapping into the polar coronal holes. The initial tilt of the streamer belt combined with solar rotation ensured the occurrence of strong corotating interaction regions (CIRs) equatorward of about 30° heliographic latitude. These CIRs were driven by the overtaking of slow solar wind from the streamer belt, by fast wind from the heliomagnetic equatorward edges of the polar-cap holes. They were bounded by equatorward-propagating forward shocks at their leading edges and poleward-propagating reverse shocks at their trailing edges. In consequence, *Ulysses* encountered only reverse shocks from CIRs at latitudes poleward of – 35°. These shocks extended to latitudes as high as – 58°.

Ulysses also encountered many shocks driven by coronal mass ejections (CMEs) up to latitudes as high as – 60°. The speeds of the CMEs generally matched that of the ambient flow, which was low equatorward of – 30° latitude and high poleward of – 30° latitude. A new class of interaction between CMEs and the ambient solar wind was discovered at high latitudes. It is characterized by a forward-reverse shock pair driven by a central overpressure in the CME.

The state of the high-latitude solar wind is similar to that observed in the ecliptic plane during the last two solar cycles. An assortment of bulk flow parameters measured at the highest southerly latitude of the *Ulysses* orbit is collected in Table I. Although the bulk speed within the high-speed

TABLE 1. Bulk Flow Parameters of the High Latitude Solar Wind

Parameter	Mean	Median	5% Level	95% Level
$N_p R^2$ (cm^{-3})	2.42	2.38	1.78	3.19
V_{sw} $(km\ s^{-1})$	776	776	738	813
T_p $(10^5\ K)$	1.89	1.86	1.44	2.41
T_e $(10^5\ K)$	0.782	0.744	0.565	1.063
T_a/T_p	4.71	4.69	3.97	5.57
N_a/N_p	0.0450	0.0444	0.0366	0.0579
$\sum_i N_i V_i m_i R^2$ $(10^{12}\ AMU\ m^{-2} s^{-1})$	2.21	2.19	1.64	2.85
$\sum_i (N_i V_i^2 m_i R^2$ $(10^{18} AMU\ m^{-1} s^{-1})$	1.72	1.70	1.26	2.22
$\sum_i 0.5 N_i V_i^3 m_i R^2$ $(mW\ m^{-2})$	1.11	1.10	0.80	1.45
Kin. + Grav. Energy Flux Density $(mW\ m^{-2})$	1.81	1.80	1.33	2.35
Helio Dinstance (R, AU) $(R_{min}, R_{max} = 1.73, 2.83\ AU)$	2.20	2.15	1.79	2.74

polar cap gradually increases with poleward-increasing heliographic latitude, all other parameters of the flow show only relatively minor latitude
dependences. In consequence, the only notable global variation of flow parameters observed by *Ulysses* just reflects the difference between the two
known flow states of the solar wind, a high-speed flow that comes from the
polar holes and a low-speed flow that comes from the equatorial streamer
belt.

Of interest to global MHD models of the solar wind are implications of

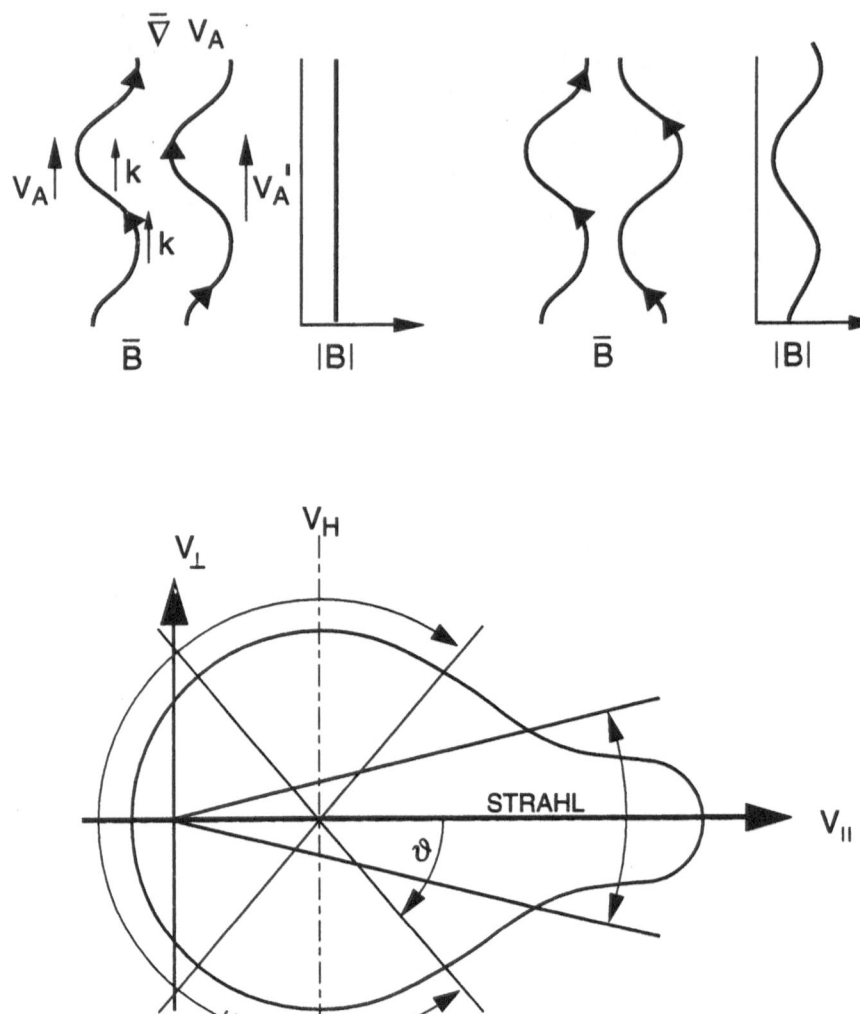

Figure 20. A schematic diagram illustrating a possible explanation the broad electron pitch- angle distributions observed by *Ulysses* in the high-latitude solar wind such as those shown in Figure 16. See the text for details.

the new *Ulysses* results to boundary conditions at large and small heliocentric distances. Starting first with the large distances, we note that the scaled (to 1 AU) momentum flux density of the solar wind is minimum within the streamer belt and is maximum throughout the polar caps (see Figure 9). This configuration should, in turn, impose an hour-glass shape to the interface between the heliosphere and the local interstellar medium, one that is spherical over the heliomagnetic poles and pinched at the equator

(Phillips et al., 1995b,c).

Another important consequence of *Ulysses* observations is that they allow the most accurate estimate to date of the energy and mass fluxes at the coronal base needed to support coronal losses due to the outward expansion of gas from coronal holes. Combining the facts that the holes covering both polar caps are observed to occupy only about 13% of the low corona during the *Ulysses* rapid latitude scan, with the measured size of the polar caps at large distances, covering about 63% of the heliosphere, allows an estimate of about 4.8 for the expansion factor of the flow from low to asymptotic distances (Gosling et al., 1995a). Combining this factor with the scaled value of the total energy flux density of the polar wind listed in Table I (1.8 mW m^{-2} (= 1.8 ergs cm^{-2} s^{-1}) at 1 AU) yields an energy flux of 4.0×10^5 ergs cm^{-2} s^{-1} needed to drive the coronal expansion at its base. The combination of this energy flux with the observed mass flux of 2.2×10^{12} AMU m^{-2} s^{-1} at 1 AU listed in Table I (which translates to 4.8×10^{13} AMU$cm-2s-1$ at the base of the corona), forces the conclusion that classical heat conduction alone cannot drive the high-speed solar wind (Barnes et al., 1995). Extended energy deposition by some other mechanism is essential.

A final consequence can be drawn from the overview of He abundances provided by Figures 1 and 13. Throughout its entire southern and northern polar passes, *Ulysses* observations reveal a remarkably constant He abundance relative to H, averaging 4.3% with a standard deviation of $\pm$ 0.64%. We note that this abundance is less than half the cosmic He abundance, estimated to be between 8.3% and 10.7% by number, (e.g., Iben, 1969, and references therein) and the primordial Helium abundance of the Sun, inferred to be 9.6% from the most current models of the solar interior that are consistent with all measurable solar parameters (Bahcall and Pinsonneault, 1992). It is also considerably less than the value inferred to exist throughout the solar convective zone based on helioseismic measurements and models that include Helium diffusion (see Guzik and Cox, 1993, and references therein), which is 8.1%.

All of the foregoing estimates of [He] are larger than the highest one-hour averaged value of [He] measured by *Ulysses* in the high-latitude solar wind, which is 7.6%. This difference has been explained (see, e.g., von Steiger and Geiss, 1989; Hansteen et al., 1993) as due to fractionation driven by thermal and gravitational diffusion in the region between the chromosphere and the low solar corona. Yet numerous transient phenomena such as spicules, coronal bright point flares (producing, at times, X-ray jets and H$_a$ surges), polar plumes, macrospicules, and a variety of explosive events (observed as plasma jets in the EUV), are known to coexist and course into and through the low corona from below (see, e.g., Beckers, 1972; Habbal, 1992, and references therein). Specifically, spicular eruptions having time scales of 5 to 10 minutes supply more than 20 times the mass to the corona that is carried away from the Sun by the solar wind (Beckers, 1972). Resultant upward surges of chromospheric material thoroughly mix gas from the photosphere to altitudes as high as 9000 km in the corona. This mixing occurs at a rate of about 130 times a day on every magnetic field line that connects the photosphere to the corona (Beckers, 1972). It is supplemented by explosive events at transition zone altitudes, which transport gas from altitudes as low as 2000 km to altitudes in the corona within the solar wind source region. A well documented subset of such transients seen at transition-zone temperatures ($\sim 10^5$ K) have spatial scales of 1500km and birth rates of about $10^{-20} cm^{-2} s^{-1}$ (Dere, 1992). Translation of these measurements alone to large heliocentric distances leads to the expectation that *Ulysses* should encounter ejecta from explosive events at a rate of more than 10 per day. But the net result of these transients is to temporally mix low altitude photospheric material with high altitude coronal material. Local, time-dependent enchancements of the He abundance in the overlying solar wind must then necesarily occur at a rate of more than 10 per day on every magnetic field line that connects the corona to asymptotic distances in the solar wind if a quasi-steady state Helium fractionation does indeed exist in the solar atmosphere, as theory predicts (Hansteen et al., 1993). No such enhancements were observed during the more than 500 days that *Ulysses* spent immersed in the high-speed wind magnetically connected to both solar poles before the end of August, 1995. This null result is reinforced by the absence of enhanced [He] within the six CME events encountered by *Ulysses* at high heliographic latitudes (Barraclough et al., 1995).

The foregoing results can only be understood if the entire solar atmosphere above the photosphere is thotoughly mixed by convective transients on a time scale shorter than diffusion can gravitationally settle Helium to lower altitudes. This conclusion is reinforced by the recent finding that a first- ionization-potential fractionation effect for all ions heavier than Helium is not observed in the high-latitude solar wind (von Steiger et al., 1995). The Helium abundance of the high-speed solar wind measured both by *Ulysses* and IMP must then provide a faithful measure of its abundance throughout the entire solar atmosphere above the photosphere. If the results of current helioseismological measurements stand (which apply to that portion of the solar convection zone below the first Helium ionization zone, located $\sim$ 14,000 km below the photosphere), then processes operating between the photosphere and the first Helium ionization zone must be responsible for the fractionation of Helium from its $\sim$ 8% abundance throughout the solar convection zone (see, e.g., Antia and Basu, 1994; Kosovichev, 1995, and references therein) to its $\sim$ 4% abundance in the source region of the solar wind.

A major new result that was not anticipated prior to *Ulysses*, is the gradual dissipation of the electron strahl with increasing heliocentric distance. Observations in the high-speed solar wind at, and inside of 1 AU, revealed suprathermal electron distributions that are narrowly focussed along B (Rosenbauer et al., 1977; Feldman et al., 1978, Pilipp et al., 1987). These distributions are quantitatively consistent with models of a nearly exospheric expansion (that include Coulomb collisions) beyond about 10 to 20 solar radii (e.g., Lemons and Feldman, 1983, and references therein). Although this theory predicts a continued focussing of the strahl with increasing heliocentric distance, the angular widths of strahl electrons and their peak phase space densities parallel to B, increase and decrease, respectively. These observational results imply a dissipation of the strahl and the heat flux it carries, by some unknown mechanism. A prime candidate is the whistler heat flux instability, which has been shown to predict a radial gradient in electron heat flux (Gary et al., 1994) that is very close to that measured by *Ulysses* in the ecliptic plane between 1 and 5 AU (Scime et al., 1994). Inference of a similar, faster than Coulomb scaling of the heat flux with radial distance in the high-latitude wind (Scime et al., 1995), suggests that this instability may also actively dissipate the heat flux there as well.

The foregoing agreement between theory and observation begs the question as to why the whistler heat flux instability works. It is driven by the convection of the halo electrons relative to the plasma frame, only under conditions of near isotropy. However in the absence of the instability, halo electrons should focus to a narrow beam parallel to $\vec{B}$ through conservation of their magnetic moments and total energy, thereby moving the plasma further from instability. In addition, even if the instability is triggered, the resultant whistler waves should propagate parallel to $\vec{B}$ (i.e. $\vec{k} \| \vec{B}$) and resonate with electrons having large perpendicular velocities and small parallel velocities. Not only haven't such waves been observed in the solar wind, but the electrons that resonate with them do not carry the heat flux.

A possible resolution of this paradox is sketched in Figure 20. It makes use of the fact that a very prominent feature of the high-speed solar wind is its associated field of large-amplitude Alfvén waves. Any small gradient in conditions perpendicular to $\vec{B}$, such as produced by dissipation of the microstreams mentioned earlier (Neugebauer et al., 1995), will cause a lateral gradient in the Alfvén speed, V_A, as shown at the left in the top panel of Figure 20. This condition will cause a lateral shift in the phase of the waves with increasing heliocentric distance as pictured at the right in the top panel, which will induce variations in $|\vec{B}|$. Perhaps this effect is the cause of the large amplitude variations in $|\vec{B}|$ that are so prominent in the bottom panel of Figure 10. If indeed this explanation is correct, the high-speed solar wind should be permeated by time-variable magnetic mirrors that have sufficient amplitude to trap more than half of the halo electron population, forcing them to convect at about the Alfvén speed faster than the bulk solar wind speed. Indeed, inspection of the bottom panel of Figure 10 shows $B_{min}/B_{max} = 0.65$, which is sufficient to trap electrons having pitch angles between $54°$ and $126°$, bounding a solid angle that contains 59% of all suprathermal electrons if they are distributed istropically. This fraction is probably realistic because of offseting effects caused by the generally increasing interplanetary magnetic field (IMF) with decreasing distance. Some of the electrons having pitch angles greater than $126°$ will be reflected by moving mirrors at lower heliocentric distances that are strengthened by the large-scale structure of the IMF, and some with pitch angles less than $54°$ will not be reflected because the mirrors will be weakened at larger distances. A general isotropy of most suprathermal electrons (and, for that matter, of a large fraction of all suprathermal and energetic paricles as well) will be enforced by the magnetic mirrors because the trapped electrons will travel along B at the speed of the mirror, which is considerably slower than the physical speed of the electrons. This condition should be sufficient to drive whistler waves unstable, as predicted by Gary et al., (1994).

Once the whistler waves exist, though, the large amplitude Alfvén waves carried by the ambient medium will act to broaden the resonance to include all electrons. As the whistler waves drift through the wind the large directional fluctuations of $\vec{B}$ associated with Alfvén waves (which produce angular deflections greater than $90°$) will decrease the parallel component of k in places, allowing the waves to resonate with electrons having increased parallel velocities. The resultant scattering should result in a broadening of the strahl and a reduction of its phase-space density, as observed. It should also result in a reduction of the electron heat flux, which is consistent with the nearly isothermal behavior of electrons (which is dominated by the lower- energy, core electrons) shown in Figure 12. It is also consistent with the higher than Coulomb gradient of electron heat flux observed by *Ulysses* in the high-latitude solar wind (Scime et al., 1995). The net effect of the field of large-amplitude Alfvén waves that premeate both polar caps of the Sun, then, is to place a lid on the transport of suprathermal and energetic particles through the high-latitude heliosphere.

Acknowledgments. This review is based on an intense effort, primarily at Los Alamos and JPL, to reduce the *Ulysses* solar wind plasma data and to organize it into files that are accessible by a large number of research workers for ready analysis. We especially owe much thanks to B. Goldstein for his exceptional effort in this regard. We have also benefited from many insightful conservations with J. Gosling, D. McComas and S.P. Gary, to place the new data reported here into its proper physical context. Work at Los Alamos was performed under the auspices of the U.S. Department of Energy with financial support from NASA.

References

Antia, H.M., and Basu, S. (1994) Measuring the Helium abundance in the solar envelope: the role of equation of state, Astrophys. J., **426**, 801-811.

Bahcall, J.N. and Pinsonneault, M.H. (1992) Helium diffusion in the Sun, Astrophys. J. **395**, L119-L122.

Beckers, J.M. (1972) Solar spicules, Ann. Rev. Astron. Astrophys., 9, 73-100.

Balogh, A., Smith, E.J., Tsurutani, B.T., Southwood, D.J., Forsyth, R.J. and Horbury, T.S. (1995) The heliospheric magnetic field over the south polar region of the Sun, Science 268, 1007- 1010.

Bame, S.J., Asbridge, J.R., Feldman, W.C. and Gosling, J.T. (1977) Evidence for a structure-free state at high solar wind speeds, J. Geophys. Res., 82, 1487-1492.

Bame, S.J., McComas, D.J., Barraclough, B.L., Phillips, J.L., Sofaly, K.J., Chavez, J.C., Goldstein, B.E. and Sakurai, R.K. (1992) The *Ulysses* Solar Wind Plasma Experiment, Astron. and Astrophys. Suppl. Ser., 92, 237.

Bame, S.J., Goldstein, B.E., Gosling, J.T., Harvey, J.W., McComas, D.J., Neugebauer, M. and Phillips, J.L. (1993) *Ulysses* Observations of a Recurrent High-Speed Solar Wind Stream and the Heliomagnetic Streamer Belt, Geophys. Res .Letters, 20, 2323, 1993.

Barnes, A., Gazis, P.R., Phillips, J.L. (1995) Constraints on solar wind acceleration mechanisms from *Ulysses* plasma observations: the first polar pass, Geophys. Res. Lett., in press.

Barraclough, B.L., Gosling, J.T., Phillips, J.L., McComas, D.J., Feldman, W.C. and-Goldstein, B.E. (1995) He Abundance Variations in the Solar Wind: Observations from *Ulysses*, Proceedings of the Eighth International Solar Wind Conference, sub.

Borrini, G., Gosling, J.T., Bame, S.J., Feldman, W.C. (1982) Helium abundance enhancements in the solar wind, J. Geophys. Res. 87, 7370-7378.

Burlaga, L.F.E. (1991) Magnetic Clouds, in *Physics of the Inner Heliosphere; 2 Particles, Waves and Turbulence*, R. Schwenn, E. Marsch, eds., Springer-Verlag, Berlin, pp.1-22.

Dere, K.P. (1994) Explosive events, magnetic reconnection, and coronal heating, Adv. Space Res., **14**, 13-22.

Feldman, W.C., Asbridge, J.R., Bame, S.J. and Gosling, J.T., High-speed solar wind flow parameters at 1 AU, J. Geophys. Res., 81, 5054-5060, 1976a.

Feldman, W.C., Abraham-Shrauner, B., Asbridge, J.R. and Bame, S.J. (1976b) The internal plasma state of the high speed solar wind at 1 AU. Physics of Solar Planetary Environments, Vol. 1, Williams, D.J., ed. (Amer. Geophys. Union Publ.) pp 413-427.

Feldman, W.C., Asbridge, J.R., Bame, S.J. and Gosling, J.T. (1977) Plasma and magnetic fields from the sun. The Solar Output and its Variations, White, O.R., ed. (Boulder:Col. Assoc. Univ. Press), pp. 351-382.

Feldman, W.C., Asbridge, J.R., Bame, S.J., Gosling, J.T. and Lemons, D.S. (1978) Characteristic electron variations across simple high-speed solar wind streams, J. Geophys. Res., 83, 5285-5295.

Feldman, W.C., Asbridge, J.R., Bame, S.J., Gosling, J.T. and Lemons, D.S. (1979) The core electron temperature profile between 0.5 and 1 AU in the steady-state high speed solar wind. J. Geophys. Res., 84, 4463-4467.

Feldman, W.C., Gosling, J.T., McComas, D.J. and Phillips, J.L. (1993) Evidence for ion jets in the high-speed solar wind. J. Geophys. Res. 98, 5593-5605.

Gary, S.P., Scime, E.E., Phillips, J.L. and Feldman, W.C. (1994) The Whistler Heat Flux Instability: Threshold Conditions in the Solar Wind, J. Geophys. Res., 99, 23391.

298

Geiss, J. (1982) Processes affecting abundances in the solar wind, Space Sci. Rev., 33, 201-217.

Gloeckler, G., Galvin, A.B., Ipavich, F.M., Hamilton, D.C., Bochsler, P., Geiss, J., Fisk, L.A. and Wilken, B. (1995) Elemental and charge state composition of the fast solar wind observed with SMS instruments on WIND, in Proceedings of the International Solar Wind 8 Conference, D. Winterhalter, D. McComas, N. Murphy, J. Phillips, eds., Sub.

Gosling, J.T. (1990) Coronal mass ejections and magnetic flux ropes in interplanetary space, in Physics of Magnetic Flux Ropes, Russell, C.T., Priest, E.R. and Lee, L.C. eds., Geophys. Monogr. Ser., Vol. 58, AGU, Washington D.C., 343-364.

Gosling, J.T. (1994) Coronal Mass Ejections in the Solar Wind at High Solar Latitudes: An Overview, Proceedings of the Third SOHO Workshop, Kluwer, p. 275.

Gosling, J.T. (1995) Solar Wind Corotating Interaction Regions: The Third Dimension, U.S. National Report to IUGG, Reviews of Geophysics, Supplement, 33, 597.

Gosling, J.T., Bame, S.J., McComas, D.J., Phillips, J.L., Pizzo, V.J., Goldstein, B.E. and Neugebauer, M. (1993) Latitudinal Variation of Solar Wind Corotating Stream Interaction Regions, Geophysical Research Letters, 20, 2789.

Gosling, J.T., Bame, S.J., McComas, D.J., Phillips, J.L., Goldstein, B.E. and Neugebauer, M.. (1994a) The Speeds of Coronal Mass Ejections in the Solar Wind at Mid Heliographic Latitudes: Ulysses, Geophysical Research Letters, 21, 1109.

Gosling, J.T., McComas, D.J., Phillips, J.L., Weiss, L.A., Pizzo, V.J., Goldstein, B.E. andForsyth, R.J. (1994b) A New Class of Forward-Reverse Shock Pairs in the Solar Wind, Geophysical Research Letters, 21, 2271.

Gosling, J.T., Bame, S.J., McComas, D.J., Phillips, J.L., Scime, E.E., Pizzo, V.J., Goldstein, B.E. and Balogh, A. (1994c) A Forward-Reverse Shock Pair in the Solar Wind Driven by Over- Expansion of a Coronal Mass Ejection: Ulysses Observations, Geophysical Research Letters, 21, 237.

Gosling, J.T., Bame, S.J., Feldman, W.C., McComas, D.J., Phillips, J.L., Goldstein, B.E., Neugebauer, M., Burkepile, J.T., Hundhausen, A.J. and Acton, L. (1995a) The Band of Solar Wind Variability at Low Heliographic Latitudes Near Solar Activity Minimum: Plasma Results from the Ulysses Rapid Latitude Scan, Geophysical Research Letters, in press.

Gosling, J.T., Bame, S.J., McComas, D.J., Phillips, J.L., Pizzo, V.J., Goldstein, B.E. and Neugebauer, M. (1995b) Solar Wind Corotating Interaction Regions Out of the Ecliptic Plane: Ulysses, Space Science Reviews, 72, 99-104.

Gosling, J.T., Bame, S.J., McComas, D.J., Phillips, J.L., Balogh, A. and Strong, K.T. (1995c) Coronal Mass Ejections at High Heliographic Latitudes: Ulysses, Space Science Reviews, 72, 133-136.

Gosling, J.T., McComas, D.J., Phillips, J.L., Pizzo, V.J., Goldstein, B.E., Forsyth, R.J. and Lepping, R.P. (1995d) A CME-Driven Solar Wind Disturbance Observed at Both High and Low Heliographic Latitudes, Geophysical Research Letters, 22, 1753

Gosling, J.T., Feldman, W.C., McComas, D.J., Phillips, J.L., Pizzo, V.J. and Forsyth, R.J. (1995e) Ulysses Observations of Opposed Tilts of Solar Wind Corotating Interaction Regions in the Northern and Southern Solar Hemispheres, Geophysical Research Letters, (in press).

Guzik, J.A., and Cox, A.N. (1993) Using solar p-modes to determine the convection zone depth and constrain diffusion-prodused composition gradients, Astrophys. J., 411, 394-401.

Habbal, S.R. (1992) Coronal energy distribution and X-ray activity in the small scale magnetic field of the quiet sun, Annal. Geophysic., 10, 34-46.

Hammond, C.M., Crawford, G.K., Phillips, J.L., Kojima, H., Matsumoto, H., Farnk, L.A., Balogh, A., Kokubun, S. and Yamamoto, T. (1995a) Latitudinal structure of a coronal mass ejection inferred from Ulysses and Geotail observations, Geophys. Res. Lett., 22, 1169- 1172.

Hammond, C.M., Feldman, W.C., McComas, D.J. and Phillips, J.L. (1995b) The electron strahl during the *Ulysses* southern polar pass, Abstract, EOS, sub.

Hansteen, V.H., Holzer, T.E., and Leer, E. (1993) Diffusion effects on the Helium abundance of the solar transition rgion and corona, Astrophys. J. **402**, 334-343.

Hundhausen, A.J. (1977) An interplanetary view of coronal holes, in Coronal Holes and High Speed Wind Streams, Zirker, J., ed., Boulder, Co., Col. Assoc. Univ. Press, pp 225-329.

Hyashi, C. (1966), Evolution of protostars, Ann. Rev. Astron. Astrophys., 4, 171-192.

Iben, I, Jr., (1969), The Cl37 solar neutrino experiment and the solar Helium abundance, Annals of Phys., 54, 164-203.

Kosovichev, A.G. (1995) Helioseismic measurements of elememtn abundances in the solar interior, Adv. Space Res., **15**, 95-99.

Lemons, D.S., Feldman, W.C. 1983. Collisional modification to the exospheric theory of solar wind halo electron pitch angle distributions, J. Geophys. Res., 88, 6881-6687.

Lima, J. and Tsinganos, K., (1995), The heliolatitudinal gradient of the solar wind during solar minimum conditions modelled by exact hydrodynamic solutions, Gephysical Research Letters, (in press).

Lemons, D.S., Feldman, W.C. 1983. Collisional modification to the exospheric theory of solar wind halo electron pitch angle distributions. J. Geophys. Res., 88, 6881-6687.

Marsch, E. 1991. Kinetic physics of the solar wind plasma. Physics of the Inner Heliosphere 2 Particles, Waves and Turbulence, Schwenn, R., Marsch, E., eds. (Berlin:Springer Verlag), pp. 45-133.

Marsch, E., Mühlhäuser, K.-H., Schwenn, R., Rosenbauer, H. Pilipp, W., Neubauer, F.M. 1982a. Solar wind protons: Three-dimensional velocity distributions and derived plasma parameters measured between 0.3 and 1 AU. J. Geophys. Res., 87, 52-72.

Marsch, E., Mühlhäuser, K.-H., Rosenbauer, H., Schwenn, R., Neubauer, F.M. 1982b. Solar Wind Helium ions: Observations of the Helios solar probes between 0.3 and 1 AU. J. Geophys. Res., 87, 35-51.

McComas, D.J., Phillips, J.L., Bame, S.J., Gosling, J.T., Goldstein, B.E. and Neugebauer, M., *Ulysses* Solar Wind Observations to 56 South, Space Science Reviews, 72, 93, 1995a.

McComas, D.J., Barraclough, B.L., Gosling, J.T., Hammond, C.M., Phillips, J.L., Neugebauer, M., Balogh, A., and Forsyth, R.J., Structures in the Polar Solar Wind: Plasma and Field Observations from *Ulysses*, Journal of Geophysical Research, in press, 1995b.

Neugebauer, M., Goldstein, B.E., McComas, D.J., Suess, S.T. and Balogh, A., *Ulysses* Observations of Microstreams in the Solar Wind from Coronal Holes, Journal of Geophysical Research, submitted, 1995.

Phillips, J.L., Balogh, A., Bame, S.J., Goldstein, B.E., Gosling, J.T., Hoeksema, J.T., McComas, D.J., Neugebauer, M., Sheeley, Jr., N.R. and Wang, Y.-M., *Ulysses* at 50 South: Constant Immersion in the High-Speed Solar Wind, Geophysical Research Letters, 21, 1105, 1994.

Phillips, J.L., Bame, S.J., Feldman, W.C., Gosling, J.T., Hammond, C.M., McComas, D.J., Goldstein, B.E. and Neugebauer, M., *Ulysses* Solar Wind Plasma Observations During the Declining Phase of Solar Cycle 22, Advances in Space Research, 16, (9)85, 1995a.

Phillips, J.L., Bame, S.J., Feldman, W.C., Goldstein, B.E., Gosling, J.T., Hammond, C.M., McComas, D.J., Neugebauer, M., Scime, E.E. and Suess S.T., *Ulysses* Solar Wind Plasma Observations at High Southerly Latitudes, Science, 268, 1030, 1995b.

Phillips, J.L., Bame, S.J., Barnes, A., Barraclough, B.L., Feldman, W.C., Goldstein, B.E., Gosling, J.T., Hoogeveen, G.W., McComas, D.J., Neugebauer, M. and Suess, S.T., *Ulysses* Solar Wind Plasma Observations from Pole to Pole, Geophysical Research Letters, submitted, 1995c.

Phillips, J.L., Bame, S.J., Feldman, W.C., Gosling, J.T., McComas, D.J., Goldstein, B.E., Neugebauer, M. and C.M. Hammond, *Ulysses* Solar Wind Plasma Observations from

Peak Southerly Latitude Through Perihelion and Beyond, Proceedings of the Eighth International Solar Wind Conference, submitted, 1995d.

Phillips, J.L., Goldstein, B.E., Gosling, J.T., Hammond, C.M., Hoeksema, J.T., and McComas, D.J., Sources of Shocks and Compressions in the High-Latitude Solar Wind: *Ulysses*, Geophysical Research Letters, in press, 1995e.

Pilipp, W. G., Miggenrieder, H., Montgomery, M.D., Moξhlhoγuser, K.-H., Rosenbauer, H., Schwenn, R. 1987. Characteristics of electron velocity distribution functions in the solar wind derived from the Helios plasma experiment. J. Geophys. Res., 92, 1075-1092.

Pizzo, V.J., Gosling, J.T., 3-D Simulation of High-Latitude Interaction Regions: Comparison with *Ulysses* Results, Geophysical Research Letters, 18, 2063, 1994.

Rosenbauer, H., Schwenn, R., Marsch, E., Meyer, B., Miggenrieder, H., Montgomery, M.D., Moξhlhoγuser, K.H., Pilipp, W., Voges, W., Zink, S.M., A survey on initial results of the Helios plasma experiment, J. Geophys., 42, 561, 1977.

Schwenn, R. 1991. Large-scale structure of the interplanetary medium. Physics of the Inner Heliosphere 1 Large-Scale Phenomena. Schwenn, R., Marsch, E., eds. (Berlin:Springer- Verlag), pp. 99-181.

Scime, E.E., Bame, S.J., Feldman, W.C., Gary, S.P., Phillips, J.L. and Balogh, A., Regulation of the Solar Wind Electron Heat Flux from 1 to 5 AU: *Ulysses* Observations, Journal of Geophysical Research, 99, 23401, 1994.

Scime. E.E., Bame, S.J., Phillips, J.L., and Balogh, A., Latitudinal Variations in the Solar Wind Electron Heat Flux, Space Science Reviews, 72, 105, 1995.

Smith, E.J., Neugebauer, M., Balogh, A., Bame, S.J., Erdxs, G., Forsyth, R.J., Goldstein, B.E., Phillips, J.L. and Tsurutani, B.T., Disappearance of the Heliospheric Sector Structure at *Ulysses*, Geophysical Research Letters, 20, 2327, 1993.

Smith, E.J., Neugebauer, M., Balogh, A., Bame, S.J., Lepping, R.P. and Tsurutani, B.T. (1995) *Ulysses* observations of latitude gradients in the heliospheric magnetic field: radial component and variances, Science, 72, 165-170.

Ulrich, R.K. and Cox, A.N. (1991) The computation of standard solar models, in Solar Interior and Atmosphere. Cox, A.N., Livingston, W.C., Matthews, M.S., eds. (Tucson:Univ. Ariz. Press), pp. 162-191.

Withbroe, G.L., Feldman, W.C., and Ahluwalia, H.S. (1991) The solar wind and its origins. Solar Interior and Atmosphere. Cox, A.N., Livingston, W.C., Matthews, M.S., eds. (Tucson:Univ. Ariz. Press) pp. 1087-1106.

von Steiger, R., and Geiss, J. (1989) Supply of fractionated gases to the corona, Astron. Astrophys., **225**, 222-238.

von Steiger, R., Wimmer Schweingruber, R.F., Geiss, J., and Gloeckler, G. (1995) Abundance variations in the solar wind, Adv. Space Res., **15**, 3-12.

Zhao, X.P. and Hundhausen, A.J. (1981) Organization of solar wind plasma properties in a tilted, heliomagnetic coordinate system, J. Geophys. Res., 86, 5423-5430.

STELLAR WINDS

K. B. MACGREGOR
High Altitude Observatory, NCAR[†]
P.O. Box 3000, Boulder, CO 80307, USA
mac@hao.ucar.edu

Abstract.
We consider some of the ways in which magnetohydrodynamical (MHD) processes can contribute to the acceleration of wind-type outflows from stars. We first summarize the measured properties of the 'average' solar wind, and review the evidence for the existence of analogous flows from solar-type stars in general. The influence of magnetic fields on wind dynamics is then studied by considering how a simple, stationary, thermally driven wind model is modified by the inclusion of several different MHD effects. Specifically, we examine how the radial acceleration of such a wind is influenced by the incorporation of (*i*) magnetically controlled, non-spherical expansion, (*ii*) the Lorentz force associated with a large-scale, stellar magnetic field, and (*iii*) the force arising from outwardly propagating, short-wavelength Alfvén waves into the basic model. We subsequently consider how these processes might affect the dynamical structure of a radiatively driven wind from a luminous, hot, OB star.

1. Introduction

More than 30 years have passed since Parker's prediction of the existence of the solar wind was confirmed by *in situ* measurements of the plasma and magnetic properties of the interplanetary medium, obtained from instruments on board the first generation of satellites in Earth orbit. During this time interval, much has been learned about both the basic morphology of the Sun's expanding corona and the physical mechanisms responsible

---

[†]The National Center for Atmospheric Research is sponsored by the National Science Foundation.

K. C. Tsinganos (ed.), Solar and Astrophysical Magnetohydrodynamic Flows, 301–335.
© 1996 *Kluwer Academic Publishers.*

for its continued existence in a dynamically outflowing state. We know (or think we know) something about the nature of the source regions of the wind in the solar atmosphere, about the processes that contribute to the wind's acceleration and thermodynamic condition, about the longitudinal structuring of the wind into high- and low-speed streams, and about the properties of the distant flow, both in and (from very recent observations) out of the ecliptic plane.

The period of time between the earliest, direct observations of the solar wind and the present has also witnessed rapid growth in the amount of knowledge obtained from the study of large-scale, dynamical phenomena in the atmospheres of stars other than the Sun. Advances in instrumentation together with the development of new observing facilities and techniques have made it possible to detect and study a variety of signatures of mass loss in the radiative fluxes emitted by stars located throughout the Hertzsprung-Russell (HR) diagram. Alternatively, in the absence of measurements of such direct diagnostics of the atmospheric dynamical state, compelling circumstantial evidence for the presence of stellar wind-type outflows is often derived from observations designed to study the consequences of ongoing mass loss. Over the years, these diverse approaches to the task of inferring the existence and properties of winds from stars have yielded an abundance of data, the interpretation of which suggests that mass loss is relatively common, occurring among stars of virtually all types, irrespective of effective temperature or evolutionary status.

In the present paper, we conduct a systematic examination of a few of the possible ways in which a magnetic field can exert a dynamical influence on the flow of a steady, stellar wind. Adopting the perspective of ideal magnetohydrodynamics (MHD), we investigate potential magnetic modifications to the interplay of radial force components that determines the net outward acceleration experienced by a wind emanating from a spherical star. Specifically, we use simple, stationary flow models to comparatively study how the physical properties of an unmagnetized wind are affected by: (i) magnetic definition of the flow geometry; (ii) the Lorentz force associated with the field formed from the outward extension of the stellar magnetic field by the flow from a rotating star; and (iii), the time-averaged force exerted on the flow by outwardly propagating Alfvén waves. Initially, we consider the manner in which each of these effects operates within the winds from the Sun and solar-type stars (*i.e.*, low-mass, main sequence stars with spectral types similar to that of the Sun [G2V]). The temperature of the gas in the outer atmospheres of these stars is known to attain values comparable to or in excess of typical solar coronal temperatures, an indication that apart from any magnetic effects, the thermal pressure gradient force supplies a significant portion of the overall acceleration of any outflowing material. We subsequently extend the purview of the study and perform a similar analysis of the dynamical structure of winds from stars of distinctly non-solar spectral type, namely, the hot, luminous OB stars. Many of the observed properties of the winds from these stars are well-accounted for by a model in which the flow results from the action of the radiative force. That is, the outward acceleration of atmospheric material derives from the momentum imparted to it through the scattering and absorption of photons from the stellar surface layers in the ultraviolet spectral lines of abundant ions in the flow. Magnetic modifications to the dynamical structure of a wind produced by this radiative acceleration mechanism can differ markedly from the changes that occur when MHD processes are included in the description of a flow that is basically thermally driven.

2. The Winds From the Sun and Solar-type Stars

2.1. THE SOLAR WIND: SOME PRELIMINARY PHYSICAL CONSIDERATIONS

In the three decades during which it has been possible to observe the solar wind directly from space, numerous experiments have acquired an enormous amount of information concerning the values of plasma parameters for the flow in the plane of the ecliptic. On the basis of even the earliest such measurements, it was apparent that the solar wind was hardly a steady, structureless outflow emanating from a spherically symmetric corona. Instead, observations revealed the wind structure to be complex and highly variable, changing both spatially and temporally on a variety of scales (see, e.g., Hundhausen 1972). Small amplitude fluctuations in velocity and magnetic field components having time scales in the range between a few minutes and a few hours are now recognized as propagating hydromagnetic waves and discontinuities, presumably generated near the coronal base of the wind. Likewise, the presence of persistent, large-scale structures in the flow (e.g., high speed solar wind streams) is presently understood to be a manifestation of the spatially non-uniform nature of the source regions of the wind in the corona.

In view of these complexities, it is perhaps surprising that the description of stationary, spherical coronal expansion developed by Parker (1958, 1963) is capable of accounting for many of the gross characteristics of the solar wind. Indeed, Parker's basic theory even explains why the corona should expand at all, rather than existing in a state of hydrostatic equilibrium. That this is so must reflect the fact that such a model captures the physical essence of the phenomena it attempts to elucidate. In particular, in some suitably averaged sense, the flow of the solar wind *can* be treated as the steady, radial, hydrodynamical expansion of high-temperature, coronal plasma, driven (for the most part) by the force that results from the thermal pressure gradient in the outwardly streaming gas (see, e.g., Leer, Holzer, & Flå 1982 for a detailed discussion of solar wind acceleration). Because of its mathematical and conceptual simplicity, such a formulation of the Sun's outer atmospheric dynamics makes it possible to understand the consequences of including additional processes in basic physical terms. It is in this spirit that we adopt it as the starting point for an investigation of MHD effects in the winds from the Sun and solar-type stars.

Portions of the sizeable observational data set alluded to at the outset of this section have been analyzed by many authors in an effort to establish 'average' values for solar wind parameters. Taken together, the results from several such analyses (Schwenn 1983; Neugebauer 1983; Withbroe 1988, and references therein) indicate that the low-speed wind at 1 AU is characterized by an average flow speed of about 350 km s^{-1} and an average proton number density of about 12 cm^{-3}. The flux densities of protons and kinetic energy corresponding to these values are approximately 4×10^8 cm^{-2} s^{-1} and 0.4 erg cm^{-2} s^{-1}, respectively. For high-speed wind at 1 AU, the average flow speed and proton number density are approximately 680 km s^{-1} and 3 cm^{-3}, respectively, which values yield a proton flux density of about 2×10^8 cm^{-2} s^{-1} and a kinetic energy flux density of about 0.8 erg cm^{-2} s^{-1}. If either of these sets of average values is assumed to be representative of the wind emanating from the corona at all heliographic latitudes, then the Sun is estimated to be losing mass at a rate $\sim 10^{12}$ g s^{-1} (i.e., $\sim 10^{-14}$ M$_\odot$ y^{-1}) through the action of an outflow whose kinetic energy flux at the orbit of the Earth is $\sim 10^{27}$ erg s^{-1} (i.e., $\sim 10^{-7}$ L$_\odot$).

2.2. WINDS FROM SOLAR-TYPE STARS: OBSERVATIONAL EVIDENCE

In 1960, Parker used a combination of theoretical reasoning and inference from the then-available observations to conclude that '...we expect at least all main sequence stars later than class F to possess *stellar winds*, in analogy with the solar wind...' (see also Parker 1963). At the present time, the number of late-type, dwarf stars with directly detected, wind-type outflows remains the same as it has been since shortly after the paper containing that expectation appeared: namely, one (the Sun). The lack of direct observational evidence pertaining to the existence of winds from main sequence stars of solar and later spectral type can be attributed, in part, to difficulties inherent to the task of detecting and measuring spectroscopic diagnostics of a tenuous, high-temperature flow like the solar wind. By far the largest contribution to the column density of emitting and/or absorbing material in the wind comes from the inner, subsonic portion of the flow, located just beyond the nominal coronal base of the wind at (say) $r = r_0$.

Because of its dynamical state, the density distribution within this region is nearly hydrostatic, and the mass column density is approximately $\rho_0 h$ where ρ_0 is the mass density and h is the scale height at the base of the (assumed) isothermal atmosphere. Using values typical of the solar corona (i.e., gas temperature $T \sim 10^6$ K, proton number density $N \sim 10^8$ cm^{-3} with $r_0 = 1.25$ R$_\odot$) to estimate these quantities for a fully ionized hydrogen plasma, it follows that $\rho_0 h \sim 10^{-6}$ g cm^{-2}. The smallness of this result, together with the low intrinsic abundances of those species capable of forming spectral lines under coronal conditions (e.g., N, O, Ne, Mg, Si) suggests that Doppler-shifted features indicating the presence of an outflow are likely to be quite weak and hard to measure reliably.

In the case of the Sun, spectroscopic identification of the solar wind is further complicated by the structural inhomogeneity of the transition region and corona. Both atmospheric regions contain discrete, loop-like magnetic structures whose collective emission at ultraviolet and X-ray wavelengths dominates that of the magnetically open regions that appear to be the source of most of the outflowing wind material. Thus, spatially unresolved observations are more likely to detect mass motions associated with activity in regions containing such complex, closed magnetic fields. Because of this, if the Sun is observed as a star would be (i.e., without spatial resolution), evidence of persistent, outward bulk motion indicative of the solar wind flow is not present in ultraviolet spectra (see the reviews by Dupree [1986] and Drake [1987] for additional discussion). Hence, the use of radiative signatures to infer the dynamical state of the outer corona-inner solar wind requires either high spectral and spatial resolution (see, e.g., Rottman, Orrall, & Klimchuk 1982) or coronagraphic techniques (Withbroe et al. 1982; Kohl & Withbroe 1982). Neither approach is feasible as a method for the direct observation of winds from distant, unresolved, solar-type stars.

Despite such a pessimistic assessment of the likelihood of obtaining direct observations of stellar coronal expansion, there is a significant amount of indirect evidence which can be used to construct a persuasive (if circumstantial) case for the prevalence of mass loss from Sun-like stars. For example, X-ray emission from main sequence stars has been extensively surveyed and studied, using satellite-borne observing facilities such as *EINSTEIN* (see, e.g., Vaiana et al. 1981), *EXOSAT* (e.g., Pallavicini 1989), and, more recently, *ROSAT* (e.g., Schmitt 1993). Numerous measurements have established that as a group, G dwarf stars exhibit a wide range of emission levels, with soft X-ray luminosities L_x between about 10^{26} and 10^{30} erg s^{-1}. Among late-type dwarf stars with sub-photospheric convection zones and (presumably) dynamo-generated magnetic fields, such emission is interpreted as originating in a high-temperature, magnetically structured corona, as in the case of the Sun. If this analogy is appropriate, then G stars with X-ray luminosities near the lower end of the range given above must have coronae in which the conditions of temperature and density are more like those of solar coronal holes than the active, magnetically confined solar corona (Vaiana & Rosner 1978).

The conjecture made at the close of the preceding paragraph can be combined with elements of the theory governing the flow of a steady, spherical, thermally driven wind to obtain an estimate of the mass loss rate $\dot{M}$ from a G dwarf star with a low level of coronal X-ray emission. Assuming isothermality, the radiative flux from the optically thin material located near the coronal base of such a wind is approximately $N_e N_H h \, P(T)$, where N_e and N_H are, respectively, the electron and total hydrogen number densities, h is (as above) the pressure scale height, and $P(T)$ is the radiative loss function (units: erg cm^3 s^{-1}). As noted by Withbroe & Noyes (1977), the energy radiated by material in coronal holes is a minor component of the overall energy balance of such regions, contributing a flux of magnitude 10^4 erg cm^{-2} s^{-1} (see also Maxson & Vaiana 1977). If the radiative energy losses from the entire solar corona took place at this the rate, the corresponding soft X-ray luminosity would be about 6×10^{26} erg s^{-1}. Adopting this value as typical of low activity G stars in general and equating it to the flux estimate given above, the total number density N_* at the base of the outflow can be expressed as

$$N_* \approx 10^2 \left[\frac{GM_*\mu}{x_e x_H kT R_*^2 P(T)} \right]^{1/2}, \tag{1}$$

where M_* and R_* are the stellar mass and radius, μ is the mean mass per particle in the coronal plasma, and $x_{e,H} \equiv (N_{e,H}/N_*)$ are the fractional abundances of electrons and hydrogen. For a fully ionized gas composed of hydrogen and helium (with $N_{He}/N_H = 0.1$), $\mu = 0.61 \, m_H$, $x_e x_H = 0.23$, so that $N_* \approx 2.26 \times 10^8$ cm^{-3}, assuming $M_* = M_\odot$, $R_* = R_\odot$, and $T = 1.5 \times 10^6$ K. For the asumed coronal temperature, the optically thin, collisional equilibrium loss function $P(T)$ has been evaluated using the fit provided by Rosner, Tucker, & Vaiana (1978), according to which $P(T) \approx 1.15 \times 10^{-22}$ erg cm^3 s^{-1}.

The stationary expansion of an isothermal stellar corona should begin with a subsonic outflow speed at the star and be characterized by a gas pressure which tends toward zero at large distances. As demonstrated by Parker (e.g., 1963; see also Hundhausen 1972), the thermally driven wind solution that satisfies these constraints becomes supersonic at the radius

$$r_s = \frac{GM_*\mu}{2kT}, \tag{2}$$

and, at $r = R_*$, has initial flow velocity

$$u_* \approx a \left(\frac{r_s}{R_*}\right)^2 \exp\left[\left(\frac{3}{2}\right) - 2\left(\frac{r_s}{R_*}\right)\right], \tag{3}$$

where $a \equiv (kT/\mu)^{1/2}$ is the sound speed. Using the parameter values given previously, it is readily found that $(r_s/R_*) \approx 4.7$ and $(u_*/a) \approx 8.2 \times 10^{-3}$, from which it follows that

$$\dot{M} = 4\pi R_*^2 \mu N_* u_* \approx 1.64 \times 10^{12} \text{ g s}^{-1} = 2.60 \times 10^{-14} \text{ M}_\odot \text{ y}^{-1}. \tag{4}$$

We conclude that the observed levels of coronal X-ray emission from solar-type stars are consistent with the occurrence of thermally driven mass loss at rates similar to the rate at which the Sun loses mass due to the solar wind.

A second piece of evidence pointing toward the pervasiveness of mass loss as a dynamical condition of solar-type stellar atmospheres is derived from systematic surveys of rotation among low-mass stars. The logical connection between stellar winds and stellar rotation stems not only from the fact that (as will be seen in subsequent sections) rotation-related effects can significantly influence radial flow dynamics. Instead, note that the wind emanating from a rotating star necessarily has both radial and non-radial flow velocity components. Material motions of the latter kind imply that in addition to transporting mass, momentum, and energy, the wind from a rotating star also carries angular momentum. Since the source of this angular momentum flux is the stellar rotation, the wind effectively exerts a torque on the star, thereby reducing its angular momentum content and (other things being equal) slowing its rate of rotation.

The efficacy of wind-related angular momentum loss as a rotational deceleration mechanism can be assessed by estimating the time required for significant stellar spin down to occur. Consider a spherical star of radius R_* and moment of inertia I_* that rotates rigidly with angular velocity Ω_* while losing mass at the rate $\dot{M}$. In the absence of any forces acting in the azimuthal direction, wind material outflowing in the rotational equatorial plane carries a constant specific angular momentum $\Omega_* R_*^2$. Assume that the wind mass flux density is uniform over the stellar surface, and that the specific angular momentum of the flow emanating from colatitude θ can be approximated by $\Omega_*(R_*\sin\theta)^2$. It then follows that the time rate of change of the stellar angular momentum is $\dot{J}_* \approx -(2/3)\Omega_* R_*^2 \dot{M}$, from which result the spin down timescale is estimated to be $\tau_J = (I_*\Omega_*)/|\dot{J}_*| \approx (3I_*/2\dot{M}R_*^2)$. Using solar parameter values to evaluate this quantity, we find $\tau_J \approx 6.2 \times 10^{12}$ years (!), assuming $I_\odot = 6.3 \times 10^{53}$ g cm^2.

In view of the fact that the estimated spin down timescale is considerably longer than (among other things) the Sun's main sequence lifetime, one might be tempted to conclude that wind-induced torques are incapable of affecting the rotation of solar-type dwarf stars. However, an important component of the wind-rotation interaction has been omitted from the preceding analysis, that being the extension of the stellar magnetic field into circumstellar space by the ionized, electrically conducting outflow. It was Parker (1958) who noted that coronal expansion, when combined with the general solar rotation, transports the Sun's magnetic field into space in such a way that the magnetic lines of force describe a spiral shape when projected into the solar equatorial plane. The magnetic field is not just passively advected by the outflow, but can influence the wind dynamically through the force it exerts on the conducting material in which it is embedded. In the case of the Sun, the effect of this Lorentz force is greatest near the base of the wind where the magnetic energy density is larger than the kinetic energy density of the flow. In a simple spiral field model for the interplanetary magnetic field in the solar equatorial plane (see, e.g., Weber & Davis 1967), the field in this inner portion of the wind is primarily radially directed, so that the Lorentz force acts in the azimuthal direction (i.e., in the sense of the Sun's rotation). As a result of this interaction, the flow is maintained in an approximate state of corotation with the Sun, over a region extending from the coronal base of the wind to the Alfvén radius (that is, the location at which the radial expansion speed becomes equal to the Alfvén speed based on the radial component of the magnetic field). *In situ* measurements of the azimuthal components of the solar wind velocity and magnetic fields in the ecliptic plane yield an average Alfvén radius of about 12 $R_\odot$ (Pizzo et al. 1983).

In addition to the increase in the specific angular momentum of the wind material that results from this effect, the overall solar angular momentum loss rate is further enhanced (and, in fact, dominated by) the direct torque exerted on the Sun by the interplanetary magnetic field itself. If the rotation axis is taken to coincide with the polar axis of a spherical coordinate system, this torque is approximately given by

$$\dot{J}_* \approx \frac{1}{2} r_0^2 \int_0^\pi d\theta \, \sin^2\theta \, B_{r0} B_{\phi0} \approx -\frac{8\pi}{3} \frac{\rho_0 \Omega_* r_0^6 \, B_{r0}^2}{\dot{M}}, \tag{5}$$

where B_{r0} and $B_{\phi0}$ are, respectively, the radial and azimuthal components of the magnetic field, ρ_0 is the mass density, and the subscript '0' denotes evaluation at a reference level $r = r_0$ in the stellar corona. Using values characteristic of the outer solar corona ($N_0 \approx 10^7$ cm^{-3}, $B_0 \approx 1$ G, $\dot{M} \approx 10^{12}$ g s^{-1} with $r_0 \approx R_\odot$), the magnitude of this magnetic torque is estimated to be about $|\dot{J}_*| \approx 3 \times 10^{31}$ dyne cm for $\Omega_* = 3 \times 10^{-6}$ s^{-1}. This result implies that the time for spin down by magnetic braking $(I_*\Omega_*)/|\dot{J}_*|$ has a value ($\approx$ few $\times 10^9$ y) that is quite comparable to the main sequence lifetime ($\approx 10^{10}$

y) of the Sun, and suggests that the present-day Sun (age $\approx 4.5 \times 10^9$ y) must already have suffered considerable angular momentum loss.

The suggestion that a process similar to the one described above might be operative among main sequence stars of solar and later spectral type appears to have first been made by Schatzman (1962). Shortly thereafter, Wilson (1966) sought observational verification of some of these ideas through a comparative study of the rotational and chromospheric properties of field dwarf stars with ages $\gtrsim 10^9$ years. He noted that in proceeding down the main sequence toward later spectral types, an abrupt decrease in rotational velocity (from values ~ 50 km s^{-1} to < 10 km s^{-1}) occurs at a location (about F5V) that is nearly coincident with the point at which Ca II emission is first detected. Wilson reasoned that these behaviors could be simultaneously explained if F5 is the earliest main sequence spectral type for which stars have relatively deep, sub-photospheric, hydrogen convection zones. According to his picture, the mechanical energy flux generated by convection supplies the heating required to produce both the chromospheric Ca II emission and a thermally driven wind. The observed slow rotation of stars cooler than F5V relative to their upper main sequence counterparts is then interpreted as being the result of the torque exerted on individual stars by their respective magnetized winds.

Additional evidence in support of this hypothesis was acquired by Kraft (1967; see also Kraft 1970) who obtained higher dispersion observations of solar-type field stars, and found that, on average, the rotational velocities of stars with Ca II emission were higher than those of stars without. Kraft argued (correctly) that since the stars that exhibit such chromospheric emission are probably younger than those that do not, the rotation rates of solar-type stars must decrease with increasing age on the main sequence. He further substantiated this conclusion by investigating the rotational velocities of solar-type stars in two well-studied open clusters whose known nuclear ages are less than that of the Sun. Kraft showed that the average rate of rotation for Sun-like stars in the Pleiades (age $\approx 7 \times 10^7$ y) is higher than that of similar stars in the Hyades (age $\approx 6 \times 10^8$ y), and that these cluster stars rotate more rapidly than either the field emission or field non-emission stars with $M_* \approx M_\odot$. These results were subsequently quantified by Skumanich (1972), who demonstrated that among G dwarf stars, the observed declines in both the strength of Ca II emission and the velocity of rotation are consistent with an approximate dependence on stellar age t of the form $t^{-1/2}$, over the period between the age of the Pleiades and the age of the Sun. Additional work by Durney (1972) delineated the physical conditions under which such a dependence of rotation on age could be produced using a simple model for the magnetic torque exerted on a star by a wind emanating from its corona.

In recent years, observing capabilites have advanced considerably, making it possible to accurately measure the rotational velocities of large numbers of relatively faint stars by spectroscopic and, in some cases, photometric means. As a result, there now exists a substantial body of observational data pertaining to the angular momentum evolution of low-mass, main sequence stars (see the reviews by Stauffer 1991, 1994, and Soderblom 1991 for detailed discussions). The youngest open cluster which has been extensively studied is α Persei which, at age 5×10^7 years, is only about 10^7 years older than a 1 $M_\odot$ star is at the time of its arrival on the zero-age main sequence. The distribution of rotational velocities among G stars in this cluster is extremely broad, containing a large number ($\sim 2/3$ of the total) of stars with $v \sin i < 30$ km s^{-1} (i is the angle of inclination of the rotation axis to the line of sight), and a smaller but still significant number ($\sim 1/3$ of the total) of stars with $v \sin i > 30$ km s^{-1}, including several with $v \sin i > 100$ km s^{-1}. By the time the age of the Pleiades (about 7×10^7 y) has been attained, almost 90 % of the G stars have rotational velocities < 30 km s^{-1}, with about 50 % of these having $v \sin i < 10$ km s^{-1}. These measurements suggest that considerable spin-down must have occurred within a few $\times 10^7$ years. Finally, at the age of the Hyades cluster (6×10^8 y), the distribution of observed rotational velocities for G stars appears much like a delta function, with the bulk of such stars having $6 \lesssim v \sin i \lesssim 10$ km s^{-1}. Taken together with the measured surface rotation rate of the present-day Sun ($v \approx 2$ km s^{-1}, using the equatorial angular velocity $\Omega_\odot = 2.9 \times 10^{-6}$ s^{-1} obtained by Snodgrass 1983), all of these observational results imply that wind-related braking operates continuously throughout the main sequence lifetimes of low-mass stars.

To summarize the conclusions reached from consideration of the two independent lines of observational evidence presented in this sub-section: The detection of coronal X-ray emission at levels $L_x > 10^{26}$ erg s^{-1} from stars like the Sun suggests that at least some portion of the atmospheres of these objects is characterized by conditions of temperature and density that are conducive to the formation and maintenance of a thermally driven stellar wind with physical properties similar to those of the solar wind. Furthermore, the well-established fact that the rotational velocities of stars of solar spectral type decrease during the course of evolution on the main sequence is reasonably interpreted as being a consequence of the angular momentum loss that results from the flow of such a wind in the presence of a stellar magnetic field. The generality of these conclusions leads us to believe that mass loss as physical phenomenon must be relatively widespread among dwarf stars of solar and later spectral types. We now proceed in our analysis of magnetohydrodynamical modifications to the basic thermally driven wind model outlined above.

3. Magnetic Modifications to Thermally Driven Flow

3.1. MAGNETIC DEFINITION OF FLOW GEOMETRY

As noted in a previous section, the solar corona, as revealed (for example) in X-ray images of the Sun, is quite inhomogeneous spatially. Portions of the corona are composed of loop-like magnetic structures whose uppermost regions contain high-temperature gas that emits at ultraviolet and X-ray wavelengths, and whose footpoints are rooted in the cooler, denser photosphere (see Low 1990, and references therein). Alternatively, other portions of the corona are magnetically open, containing weaker, unipolar, rapidly diverging fields that extend outward into interplanetary space. It is these latter coronal hole regions that have been identified as the sources of recurrent high-speed wind streams during solar minimum, and that are the likely source of much of the low-speed wind as well. Observations have established (e.g., Munro & Jackson 1977) that over a limited height range in the corona, the cross-sectional area of such a region can increase with distance from the solar surface at a rate that is significantly faster than that associated with spherical divergence of the hole. Because the energy density of the coronal magnetic field at these heights is generally larger than the thermal pressure of the not yet supersonically expanding coronal plasma, the geometry of the flow that emanates from a hole region of this type is magnetically controlled near its base.

The dynamical modifications to spherically symmetric expansion arising from a region of faster than r^2 areal divergence were first considered by Parker (1963), and were later systematically investigated in the context of flow from a coronal hole by Kopp and Holzer (1976). We follow the treatment given by these latter authors and consider (for simplicity) a steady, isothermal wind from a spherical, non-rotating star of mass M_* and radius R_*. In the following analysis, we confine our attention to the flow along the central field line of an open magnetic region. The field line in question is assumed to be oriented in the radial (e_r) direction, and the magnetically defined flow tube is taken to be axisymmetric about it. Physically, the cross-sectional area A of the region depends upon the field strength, and varies with height in accord with conservation of magnetic flux; here, A is taken to be a specified function of the radial coordinate r (in spherical polar coordinates).

Throughout the domain $r_0(\geq R_*) \leq r < \infty$, the flow velocity is of the form $\mathbf{u} = u(r)\, e_r$, and is obtained as a function of r by solving the gas dynamic equations expressing conservation of mass

$$\frac{\mathrm{d}}{\mathrm{d}r}(A\rho u) = 0, \qquad (6)$$

and momentum

$$u\frac{du}{dr} = -\frac{1}{\rho}\frac{dp}{dr} - \frac{GM_*}{r^2},\tag{7}$$

where ρ is the mass density and p is the gas pressure. The model is completed by adopting a prescription for the area function A; for the purpose of illustration, we choose the expression given by Kopp and Holzer (1976), according to which $A(r) = (r/r_0)^2 f(r)$ with

$$f(r) = \frac{f_{max}\ \exp[(r - r_1)/\sigma] + f_1}{\exp[(r - r_1)/\sigma] + 1},\tag{8}$$

and

$$f_1 = 1 - (f_{max} - 1)\ \exp[(r_0 - r_1)/\sigma].\tag{9}$$

The constants f_{max}, r_1, and σ appearing in equations (8) and (9) describe, respectively, the net nonspherical divergence, the location at which $df(r)/dr$ is a maximum, and the distance about r_1 over which most of the variation in $f(r)$ takes place.

To proceed, assume that the pressure, mass density, and temperature of the outflowing material are related by the ideal gas equation of state, $p = \rho a^2$, where $a = (kT/\mu)^{1/2}$ is the isothermal sound speed. The differential equation of motion for the flow is then obtained by combining equations (6) and (7) and eliminating ρ, with the result

$$\frac{r}{u}\frac{du}{dr} = \frac{[2 + (d\ \ln f/d\ \ln r)]a^2 - (GM_*/r)}{u^2 - a^2}.\tag{10}$$

As discussed briefly in section 2.2, for specified values of M_*, r_0, T, μ, f_{max}, r_1, and σ, an acceptable wind solution to equation (10) has subsonic expansion speed at r_0 and asymptotically vanishing gas pressure (that is, $p \to 0$ as $r \to \infty$). Parker (1958, 1963) demonstrated that the single solution having these properties is the one that passes smoothly through the sonic critical point at $r = r_s$, where $u = a$ and the numerator and denominator of equation (10) simultaneously have the value zero. In the numerical computation of the wind solutions described below, this criterion has been used as a proxy for the imposition of physical boundary conditions at $r = r_0$ and ∞.

In Figure 1, we depict velocity profiles corresponding to wind solutions derived for the following physical parameter values: $M_* = M_\odot$, $r_0 = 1.25$ $R_\odot$, $T = 1.5 \times 10^6$ K, and $\mu = 0.5$ m_p. All of the curves shown in the Figure have $r_1 = 1.50\ r_0$ and $\sigma = 0.1\ r_0$, but differ in the amount of nonspherical divergence that characterizes the flow tube geometry; the profiles labelled $(a) - (h)$ were computed assuming $f_{max} = 1.00, 2.00, 3.00, 3.55, 3.56, 4.00,$ $5.00, 6.00$, respectively. As is apparent in the Figure, the velocity u_0 of

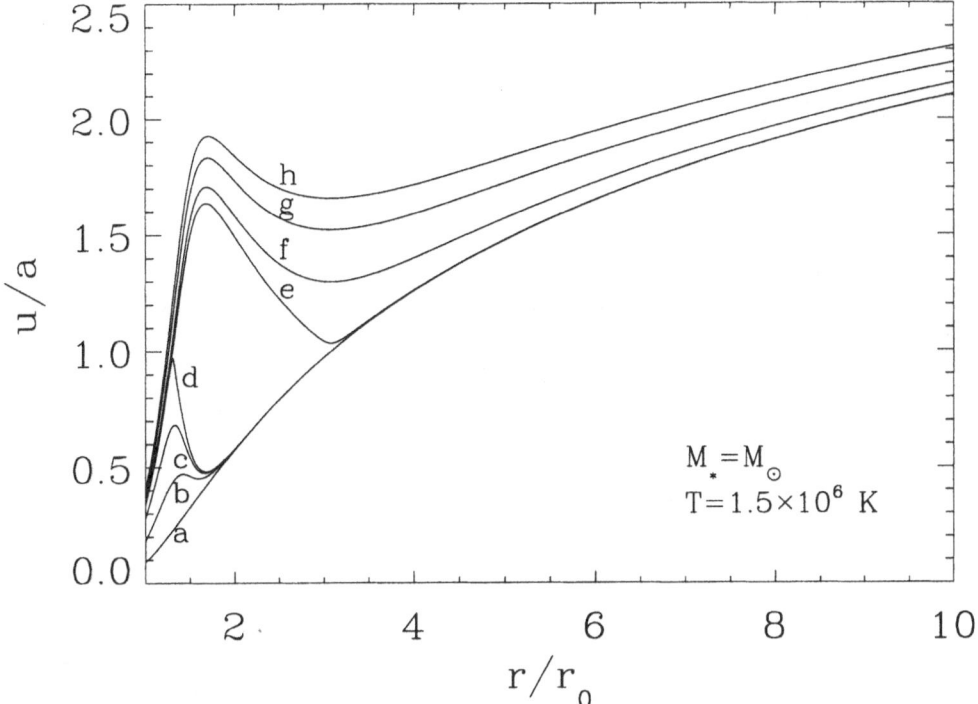

Figure 1. Wind velocity as a function of distance along the central streamline of a magnetically defined, nonspherical flow tube, as described in the text.

the flow at the reference level is an increasing function of the nonspherical expansion parameter f_{max}. The origin of this behavior can be deduced from equation (10), which, in the subsonic portion of the flow $r_0 \leq r \leq r_s$, can be integrated analytically to obtain the approximate result

$$\frac{u_0}{a} \approx \left(\frac{r_s}{r_0}\right)^2 f_s \exp\left[-\frac{1}{2} - \frac{GM_*}{a^2 r_0}\left(1 - \frac{r_0}{r_s}\right)\right], \qquad (11)$$

where $f_s \equiv f(r_s)$. Note that in the case of spherical expansion ($f_{max} = 1$), this expression for the initial flow speed becomes identical to that given in equation (3). For each of solutions (b), (c), and (d) (i.e., $f_{max} = 2.00$, 3.00, and 3.55), the sonic point has essentially the same location as in the spherical solution (a), (r_s/r_0) ≈ 3.1. Inspection of equation (11) indicates that in this case, the expansion speed u_0 should increase in rough proportion to f_s ($\approx f_{max}$), as observed in Figure 1. Evidently then, for $f_{max} \leq 3.55$, the flow interior to r_s experiences an enhanced (relative to the case $f_{max} = 1$), non-monotonic, outward acceleration that produces an increase in u_0 and a local maximum in u, but does not significantly modify the flow properties for $r > r_s$.

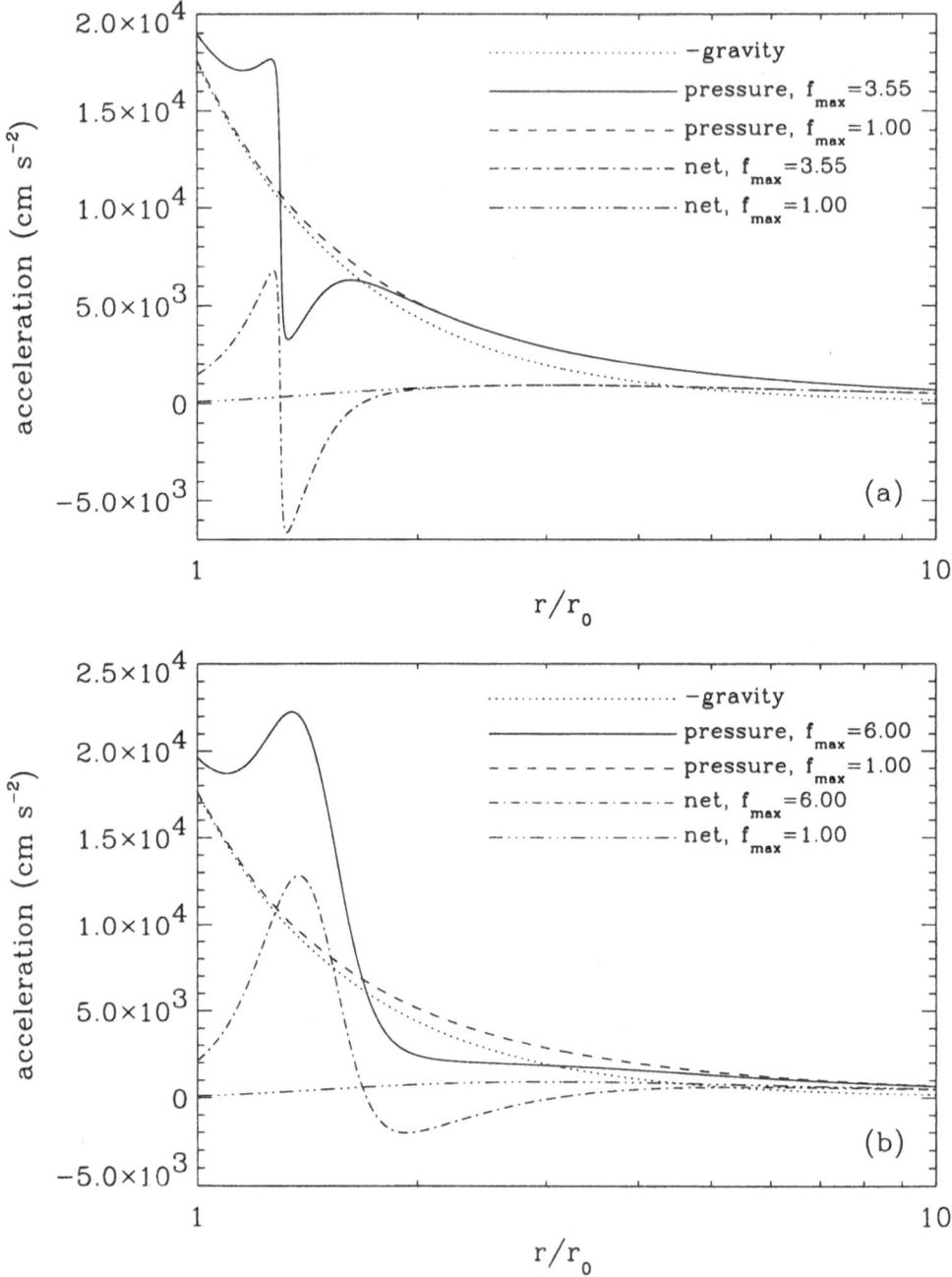

Figure 2. Profiles of the accelerations produced by the gravitational and thermal pressure gradient forces for wind solutions with $f_{max} = 3.55$ (*a*) and $f_{max} = 6.00$ (*b*). The accelerations for the case of spherical expansion ($f_{max} = 1$) are shown in each panel for comparison.

When the amount of nonspherical expansion is such that the value $f_{max} = 3.55$ is exceeded, an abrupt change in flow dynamics occurs. For solutions (e)-(h) (i.e., $f_{max} = 3.56, 4.00, 5.00, 6.00$), the sonic point location lies within the range $1.24 \lesssim (r_s/r_o) \lesssim 1.30$, with r_s now a decreasing function of increasing f_{max}. As equation (11) again indicates, in this case the growth in u_0 for larger values of f_{max} is moderated by the concomitant reduction in r_s noted above. For $f_{max} > 3.55$ then, the additional acceleration arising from the increasingly nonspherical geometry of the flow is of sufficient magnitude to not only induce a transition to supersonic velocities much closer to the star, but to enhance the asymptotic speed of the wind as well.

Our deductions concerning the flow dynamics of the wind solutions presented above are confirmed by examination of Figures 2a and 2b. There are shown the accelerations due to the gravitational and thermal pressure gradient forces as functions of r in the solutions corresponding to $f_{max} = 3.55$ and 6.00. For comparison, each panel also contains the analogous acceleration profiles for the case of a spherically expanding wind. As an aid to the interpretation of these results, note that because the flow tube cross-section increases outward at a rate that is faster than r^2, the pressure gradient force,

$$-\frac{1}{\rho}\frac{dp}{dr} = \frac{a^2}{r}\left(2 + \frac{r}{f}\frac{df}{dr} + \frac{r}{u}\frac{du}{dr}\right), \tag{12}$$

is augmented through the inclusion of an additional term proportional to df/dr (> 0). For the flow tube model used to compute the results presented in the Figures, df/dr is non-monotonic, attaining a maximum value $\approx (f_{max} - 1)/4\sigma$ at the location $r = r_1$. As exemplified by the model of Figure 2a, for mildly nonspherical expansion, the outward acceleration provided by the pressure gradient is increased in the subsonic region $(r/r_0 \lesssim 3.10)$, but remains unchanged from the $f_{max} = 1$ model in the supersonic flow. However, as indicated in Figure 2b, for larger values of the flow tube parameter f_{max}, the enhancement of the pressure gradient force can extend somewhat beyond the sonic critical point $(r_s/r_0 \approx 1.24)$, thereby increasing the velocity of the distant, supersonic wind. These results are consistent with the findings of Leer & Holzer (1980), who noted that the addition of momentum or energy to the subsonic portion of a wind-type outflow would increase the mass loss rate, while the same modifications applied in the supersonic region would result in an increased asymptotic flow speed.

3.2. MAGNETIC/CENTRIFUGAL ACCELERATION

In section 2.2, it was pointed out that the interplanetary magnetic field, representing the extension of the Sun's magnetic field into space through the agency of advection by the ionized solar wind, affects the dynamics of coronal expansion as a result of the Lorentz force it exerts on the outflowing material in which it is entrained. It was furthermore noted that the modifications to flow properties arising from this interaction were largely confined to the region around the coronal base of the wind where the (for the most part) azimuthally directed Lorentz force acts to produce approximate corotation of the flow with the Sun. Also alluded to in that section were observational surveys of rotation among main sequence stars of solar spectral type, which indicate that the youngest, most rapidly rotating such stars (i.e., with ages $\sim 5 - 7 \times 10^7 \mathrm{y}$) spin down rather more quickly than the present-day Sun, with angular momentum loss time scales $\sim 10^7$ years as opposed to $\sim 10^9$ years for the Sun. If this rotational deceleration is the result of magnetic braking by a stellar wind of coronal origin, then the wind and magnetic properties of young solar-type stars must differ considerably from those of the Sun. Among other things, the higher rates of rotation and (possibly) higher magnetic field strengths that prevail among Sun-like stars at the beginning of their lifetimes on the hydrogen-burning main sequence suggest that, in contrast to the solar case, a significant portion of the acceleration of the winds from these objects might be provided by magnetic and rotational (rather than thermal) forces.

In order to ascertain how the radial dynamical structure of the wind from a solar-type star might change over the course of its nearly 10^{10} year residence on the main sequence, we adopt the MHD wind model of Weber & Davis (1967; see also Belcher & MacGregor 1976). The model describes a stationary, axisymmetric, polytropic outflow in the equatorial plane of a star that rotates uniformly with angular velocity Ω. In a spherical coordinate system with polar axis aligned along the stellar rotation axis, the wind velocity and magnetic fields are assumed to have only radial and azimuthal components, $\mathbf{u} = [u_r(r), u_\phi(r)]$, $\mathbf{B} = [B_r(r), B_\phi(r)]$. Upon substitution of these forms into the equations governing conservation of mass and magnetic flux, it follows that the quantities $(r^2 \rho u_r)$ and $(r^2 B_r)$ both have constant values everywhere in the wind. Similarly, with the assumptions given above, the radial and azimuthal components of the momentum equation for the flow are analytically integrable, yielding expressions for the conserved energy per unit mass

$$E = \frac{1}{2}\left(u_r^2 + u_\phi^2\right) + \frac{a^2}{\alpha - 1} - \frac{GM_*}{r} - \frac{\Omega r A_r A_\phi}{u_r}, \qquad (13)$$

and angular momentum per unit mass

$$L = r\left(u_\phi - \frac{A_r A_\phi}{u_r}\right). \qquad (14)$$

In equations (13) and (14), $A_{r,\phi} = B_{r,\phi}/\sqrt{4\pi\rho}$ are the Alfvén speeds based on the radial and azimuthal components of the magnetic field, and $a = (\alpha p/\rho)^{1/2}$ is the sound speed for a gas in which $p \propto \rho^\alpha$ with α the (constant) polytrope index.

The model is completed by further assuming that the gas is perfectly electrically conducting, so that Ohm's law takes the form $\mathbf{E} = -(\mathbf{u} \times \mathbf{B})/c$. Application of Faraday's law of induction then leads to an expression for the azimuthal magnetic field

$$B_\phi = \frac{B_r}{u_r}\left(u_\phi - \Omega r\right), \tag{15}$$

which can be combined with equation (14) to obtain the azimuthal velocity

$$u_\phi = \Omega r \left[\frac{u_r^2(L/\Omega r^2) - A_r^2}{u_r^2 - A_r^2}\right]. \tag{16}$$

Regularity of equation (16) at the Alfvén radius r_A (i.e., the point at which $u_r = A_r$) requires the constant L to have the particular value Ωr_A^2. Hence, the specific angular momentum carried by the wind is equivalent to that of a flow whose azimuthal motion is strict corotation throughout the region between r_0 and r_A.

The wind equation of motion is derived by solving the linear system composed of equations (13)-(16) in differential form for du_r/dr, with the result

$$\frac{r}{u_r}\frac{du_r}{dr} = \frac{(u_r^2 - A_r^2)\,(2a^2 + u_\phi^2 - \frac{1}{2}u_e^2) + 2u_r u_\phi A_r A_\phi}{(u_r^2 - A_r^2)\,(u_r^2 - a^2) - u_r^2 A_\phi^2}, \tag{17}$$

where $u_e = (2GM_*/r)^{1/2}$ is the local gravitational escape speed. For specified values of the quantities α, a_0, $(\Omega r_0/a_0)$, (u_{e0}/a_0), and (A_{r0}/a_0) equation (17) contains two critical points; that is, there are two distinct locations in the (r, u_r)-plane at which the numerator and denominator on the right-hand side of the expression for du_r/dr simultaneously vanish. As can be verified from detailed examination of equation (17), these singularities correspond to the points in the flow at which the radial expansion speed u_r becomes equal to, in turn, the phase speeds of the slow and fast magnetosonic waves (see the chapter by Tsinganos et al. for a more complete discussion of the nature of these critical points). Analogously to the gas dynamic flow described by equation (10), the single MHD wind solution to equation (17) that obeys physical boundary conditions at r_0 and ∞ is the one that passes smoothly through both of these critical points. A number of different techniques have been developed for the purpose of obtaining numerical wind solutions to the model described herein (Weber & Davis 1967; Belcher & MacGregor 1976; Charbonneau 1995).

318

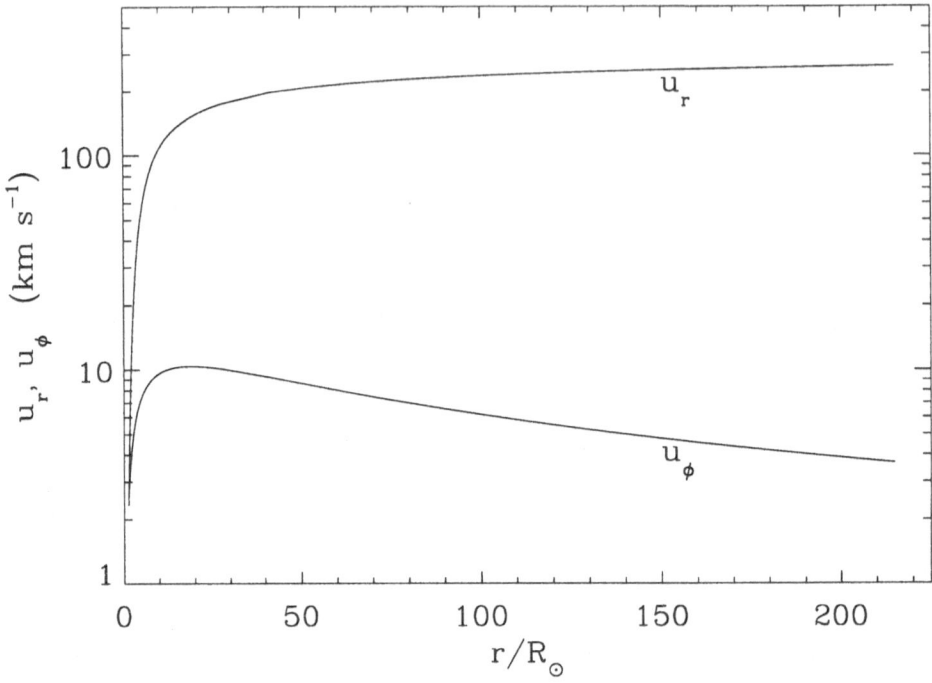

Figure 3. Radial and azimuthal velocity profiles for the present-day, low-speed solar wind solution described in the text.

Some of the dynamical differences between the present-day solar wind and the wind from a solar-mass star on the zero-age main sequence are illustrated in Figures 3-6. In Figure 3, we show the radial and azimuthal components of the flow velocity as functions of $(r/R_\odot)$ for a model having properties that are similar to those of the quiet solar wind model given by Withbroe (1988). At a coronal reference radius $(r_0/R_\odot) = 1.25$ located at the base of the wind, the gas temperature, proton number density, and radial magnetic field strength have the values $T = 1.5 \times 10^6$ K, $N_{p0} = 5 \times 10^7$ cm^{-3}, and $B_{r0} = 2$ G, respectively, with $\Omega = \Omega_\odot = 3 \times 10^{-6}$ s^{-1} and $\alpha = 1.13$. At 1 AU [$(r/R_\odot) \approx 215$], the wind velocity, density, temperature, and proton flux have the values $u_r = 264$ km s^{-1}, $N_e = 15$ cm^{-3}, $T = 2.1 \times 10^5$ K, and $N_p u_r = 3.9 \times 10^8$ cm^{-2} s^{-1}, in good agreement with the results $u_r = 285$ km s^{-1}, $N_e = 14$ cm^{-3}, $T = 2.3 \times 10^5$ K, and $N_p u_r = 3.8 \times 10^8$ cm^{-2} s^{-1} from the Withbroe (1988) model at the same location.

Figure 3 provides graphical substantiation of our previous statement (see section 2.2), to the effect that the interplanetary magnetic field primarily influences the azimuthal dynamics of the solar wind by attempting to sustain an approximate corotational state for $r \lesssim r_A$ (which for this

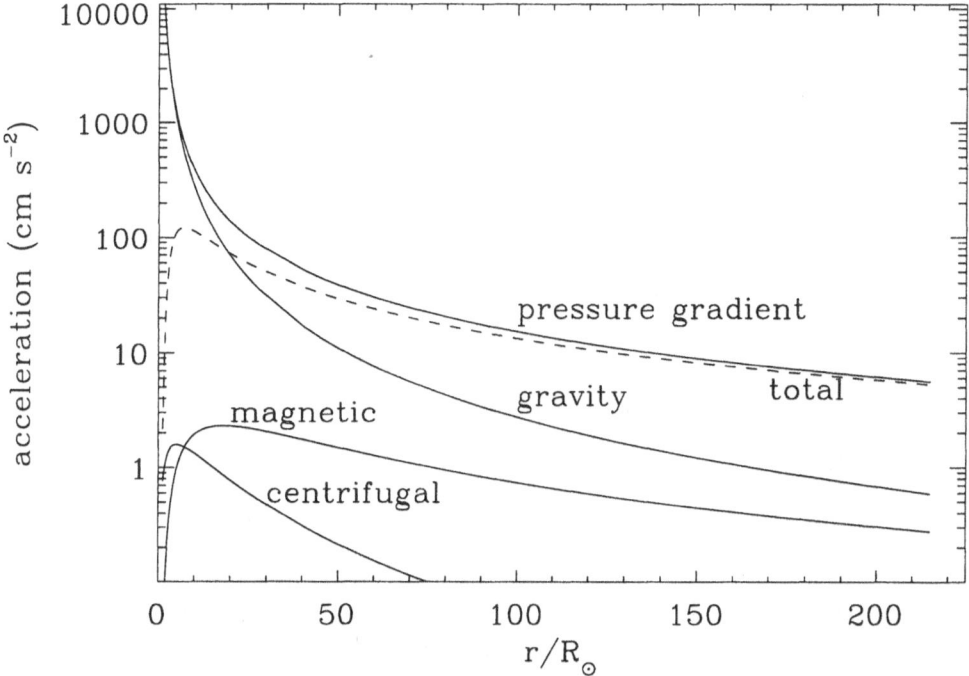

Figure 4. Profiles of the radial accelerations produced by the magnetic, centrifugal, gravitational (depicted with the opposite sign), and thermal pressure gradient forces in the wind solution of Figure 3.

solution has the unrealistically high value $r_A \approx 36\ R_\odot$). Further evidence of the relative unimportance of magnetic and rotational forces for wind dynamics in the radial direction is given in Figure 4, where profiles of the radial accelerations produced by the pressure gradient, gravitational, centrifugal, and Lorentz forces are shown for the solution of Figure 3. As is apparent from the Figure, most of the outward acceleration of the flow in this model is provided by the gradient in thermal pressure, with perhaps a few percent of the initial expansion driven by the combined efforts of the Lorentz and centrifugal forces (see also Barnes 1974). Under these conditions, the asymptotic speed u_∞ of the flow is approximately expressible in terms of reference level quantities according to

$$u_\infty \approx \left(\frac{2a_0^2}{\alpha - 1} - u_{e0}^2 \right)^{1/2}, \qquad (18)$$

a form which makes obvious the thermal origin of the kinetic energy of the wind at large distances. It can likewise be shown (see, e.g., Durney 1972; Kawaler 1988) that for the thermally dominated wind from a 'slow

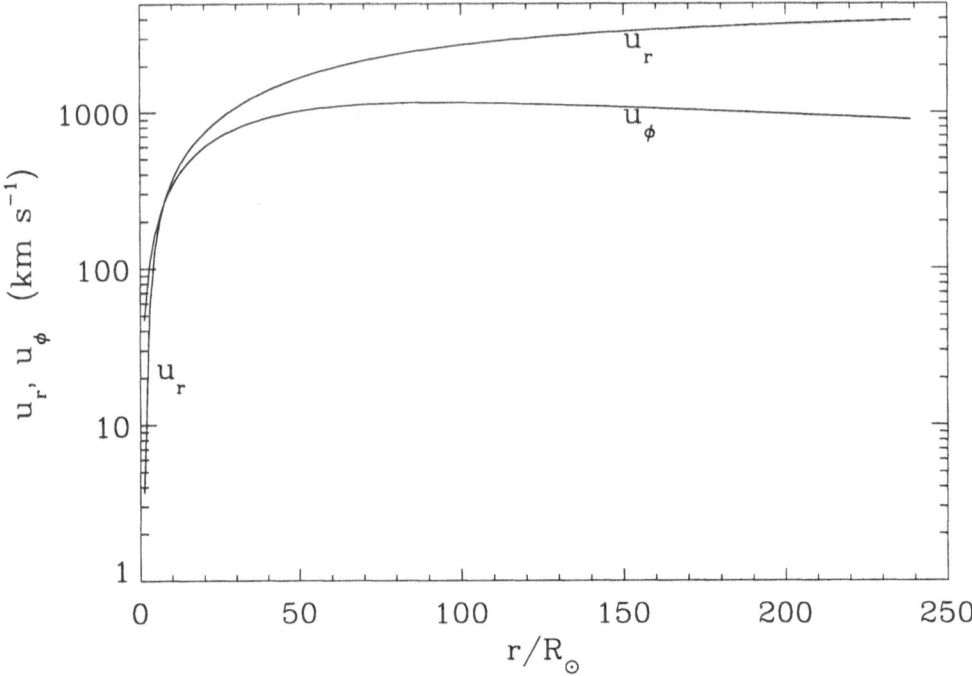

Figure 5. Radial and azimuthal components of the flow velocity as functions of radius in a model for the wind from a solar-type star of age 5×10^7 years.

magnetic rotator' such as the Sun, the Alfvén radius is approximately

$$r_A \approx r_0^2 B_{r0}/(\dot{M} u_\infty)^{1/2}, \tag{19}$$

from which it follows that the e-folding time for angular momentum loss is

$$\tau_J = (3I_*/2\dot{M} r_A^2) \approx (3I_* u_\infty/2r_0^4 B_{r0}^2), \tag{20}$$

where I_* is the stellar moment of inertia.

Wind solutions obtained for rapid rates of rotation and strong magnetic fields exhibit a radial dynamical structure that differs considerably from that seen in the model considered above. In Figure 5, we show the radial and azimuthal velocity profiles corresponding to a model that is intended to represent the wind from a star with $M_* = M_\odot$ at age 5×10^7 years, the putative age of solar-type stars belonging to the α Persei cluster (see section 2.2). The model was constructed using the radius $R_* = 0.885$ $R_\odot$ appropriate to a solar-mass star of this age, as given in unpublished evolutionary tracks computed by D. VandenBerg. In addition, the surface angular velocity was taken to be 20 $\Omega_\odot$, a value that yields a rotational velocity that is roughly equal to the average of measured $v \sin i$ values for

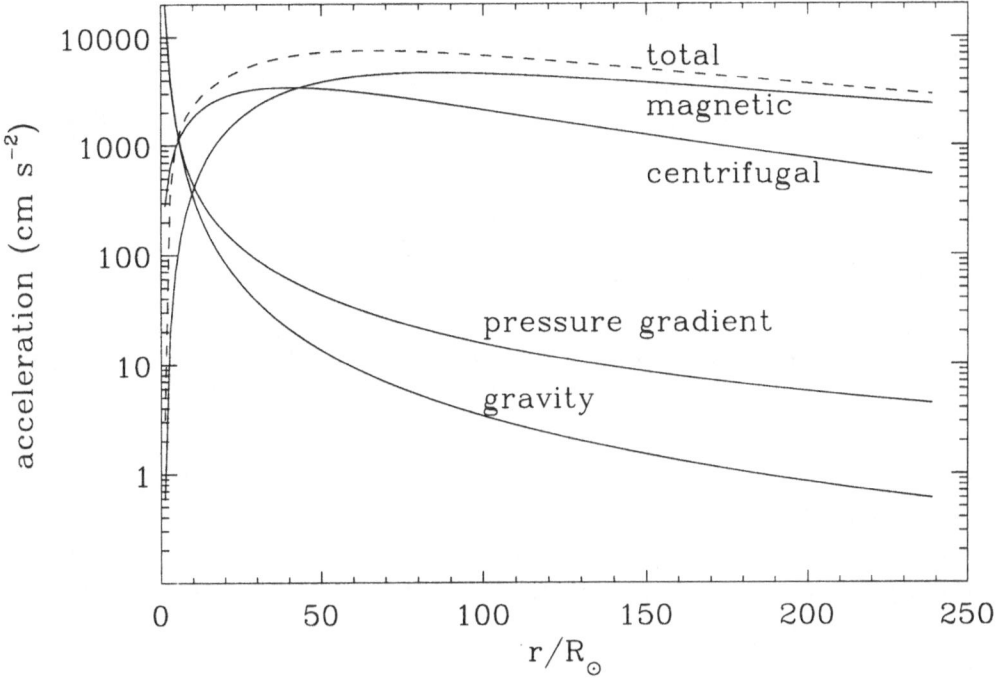

Figure 6. Radial acceleration profiles for the wind model depicted in Figure 5

G dwarfs in α Persei (see, e.g., Keppens et al. 1995, and references therein). The coronal magnetic field strength for the model was evaluated by scaling the adopted solar value ($B_{r\odot} = 2$ G) using the phenomenological dynamo relation $B_{r0} = B_{r\odot}(\Omega/\Omega_\odot)(R_\odot/R_*)^2$. All other reference level parameters have the same values as those given for the model of Figures 3 and 4.

As is evident in Figure 5, the zero-age main sequence solar wind attains a significantly higher radial expansion speed at 1 AU, $u_r = 3917$ km s^{-1}, with $N_e = 1.24$ cm^{-3}, $T = 1.54 \times 10^5$ K, and $N_p u_r = 4.84 \times 10^8$ cm^{-2} s^{-1}. In fact, if the flow is followed to large distances, its asymptotic speed is found to be $u_\infty \approx 5300$ km s^{-1} in contrast to the value $u_\infty \approx 350$ km s^{-1} obtained for the model of Figure 3. In Figure 6, it can be seen that the additional acceleration required to produce this enhancement is provided by the centrifugal and Lorentz forces, both of which are dominant contributors to the radial force balance of the distant flow. Because of this, the mass mass flux in the wind of the model presently under consideration differs little from that of the preceding model, as can be seen by comparing the proton fluxes $N_p u_r$ at 1 AU (see also Leer & Holzer 1980).

Unlike the largely thermally driven present-day solar wind, the flow emanating from the 'fast magnetic rotator' discussed above derives the bulk of its ultimate streaming kinetic energy from the stellar magnetic and rotational energies. In the limit of rapid rotation and strong magnetic fields (cf., Belcher & MacGregor 1976; Hartmann & MacGregor 1982), it can be demonstrated that the initial radial velocity of an isothermal wind is approximately

$$u_{r0} \approx a \left(\frac{r_s}{r_0}\right)^2 \exp\left(-\frac{1}{2} - \Phi\right), \tag{21}$$

where r_s is now the location of the slow magnetosonic critical point and

$$\Phi = \frac{GM_*}{r_0 a_0^2}\left(1 - \frac{r_0}{r_s}\right) - \frac{1}{2}\left(\frac{\Omega r_0}{a_0}\right)^2\left[\left(\frac{r_s}{r_0}\right)^2 - 1\right]. \tag{22}$$

In the absence of rotational and magnetic effects, $\Omega = 0$, r_s becomes equal to the sonic radius $(GM_*/2a_0^2)$, and equation (21) reduces to the expression (cf., equation [3]) appropriate to a thermally driven wind.

In a similar fashion, it is possible to derive an approximate result for the asymptotic speed of a magnetically/centrifugally accelerated wind,

$$u_\infty \approx (\Omega^2 r_0^4 B_{r0}^2/\dot{M})^{1/3}, \tag{23}$$

in terms of which the Alfvén radius is given by

$$r_A \approx (3/2)^{1/2}(u_\infty/\Omega). \tag{24}$$

Equations (23) and (24) are valid for polytropic flows in general; the fact that neither result depends explicitly on the reference level sound speed a_0 is an indication of the subordinate role played by the thermal pressure gradient in establishing the terminal velocity of the wind. Finally, the spin down time scale due to the torque applied to the star by the magnetized outflow can be shown to be

$$\tau_J \approx (I_*\Omega^2)/(\dot{M}u_\infty^2). \tag{25}$$

It is readily verified by using the stellar and solution properties given above that spin down times $\sim 10^7 - 10^8$ years can be attained in the limit that equation (25) holds, as observations of young solar-type stars seem to indicate (see section 2.2; also MacGregor & Brenner 1991)

3.3. ALFVÉN WAVE-DRIVEN WINDS

Yet a third way in which MHD processes can influence the dynamical structure of wind-type outflows from stars like the Sun is through the force associated with propagating hydromagnetic waves. At the outset of section

2.1, it was noted that much of the spatial and temporal variability seen in *in situ* measurements of solar wind properties has been identified as being comprised of travelling MHD disturbances (see, e.g., Belcher & Davis 1971; Roberts & Goldstein 1991; Marsch 1991). In this regard, it is frequently the case that the observed vector fluctuations in plasma velocity are well correlated with those in the magnetic field, implying that Alfvén waves are a major constituent of the microscale structure of the solar wind at 1 AU. These perturbations are probably introduced into the flow at its base in the corona, having been generated by mass motions at lower altitudes within the magnetized solar atmosphere.

That outwardly propagating Alfvén waves can affect the dynamics of solar coronal expansion was first shown by Parker (1965). By solving for the amplitudes of short wavelength disturbances contained in a stationary, spherically symmetric solar wind, he found that the time-averaged wave energy flux density decreased with increasing distance from the Sun. Parker attributed this behavior to the fact that the waves performed work on the flow, contributing to its acceleration by exerting a force (in a time-averaged sense) on the material through which they travelled. Subsequent model calculations confirmed that this was indeed the case (e.g., Belcher 1971), and suggested that the wave force might be partially responsible for the acceleration of high-speed streams in the solar wind.

To quantitatively explore the nature of the dynamical interaction between propagating Alfvén waves and an expanding stellar atmosphere, we consider the steady, radial flow of a fully ionized, perfectly conducting gas from a spherical, non-rotating star of mass M_* and radius R_*. The stellar magnetic field is assumed to have only a radial component, along which linear Alfvénic disturbances with azimuthally directed velocity and magnetic amplitudes ($\delta \mathbf{u}$ and $\delta \mathbf{B}$, respectively) travel. Throughout the domain $r_0 \leq r < \infty$, then, the total velocity and magnetic fields are of the form $u(r)\mathbf{e}_r + \delta u(r,t)\mathbf{e}_\phi$ and $B(r)\mathbf{e}_r + \delta B(r,t)\mathbf{e}_\phi$, where wave-like perturbations with propagation vector $\mathbf{k}(r)$ have time averages $\langle \delta u \rangle = \langle \delta B \rangle = 0$ and orientations $\mathbf{k} \cdot \delta \mathbf{u} = \mathbf{k} \cdot \delta \mathbf{B} = 0$. With these assumptions, it is readily shown that the fluxes $r^2 \rho u$ and $r^2 B$ are both conserved, and that the time-averaged radial component of the perturbed momentum equation can be written

$$u\frac{\mathrm{d}u}{\mathrm{d}r} = -\frac{1}{\rho}\frac{\mathrm{d}p}{\mathrm{d}r} - \frac{GM_*}{r^2} + \frac{1}{\rho}\langle f_w \rangle, \qquad (26)$$

where $\langle f_w \rangle$ is the force per unit volume exerted on the background flow by the waves.

To complete the description of a wave-driven wind, it is necessary to provide for the computation of the wave force density. Formally (see, e.g.,

MacGregor & Charbonneau 1995, and references therein),

$$\langle f_w \rangle = \frac{\rho \langle \delta u^2 \rangle}{r} - \frac{\langle \delta B^2 \rangle}{4\pi r} - \frac{\mathrm{d}}{\mathrm{d}r} \left(\frac{\langle \delta B^2 \rangle}{8\pi} \right), \qquad (27)$$

from which it can be seen that the total force is the sum of: (i) the centrifugal force arising from the oscillatory, azimuthal motion of the gas in the wave and (ii), the Lorentz force $(\delta \mathbf{j} \times \delta \mathbf{B})/c$ produced by the interaction between the perturbation $\delta \mathbf{B}$ and the current density fluctuation $\delta \mathbf{j} = (c/4\pi)(\nabla \times \delta \mathbf{B})$ associated with it. Evaluation of $\langle f_w \rangle$ requires knowledge of the amplitudes δu and δB as functions of r in the inhomogeneous, outflowing stellar atmosphere. In the present brief exposition, we consider only disturbances with wavelengths λ such that $\lambda/\ell \ll 1$, where ℓ is the scale length for variations in ρ, u, and B. Such waves can be treated within the WKB approximation, and solutions for the velocity and magnetic amplitudes can be derived either from the use of perturbation series expansion methods (e.g., Belcher 1971; Barnes 1992, and references therein), or from the application of the principle of wave action density conservation (e.g., Jacques 1977, and references therein). Wind models containing very low frequency Alfvén waves with $\lambda/\ell \gtrsim 1$ have been constructed by MacGregor & Charbonneau (1994, 1995).

According to the WKB analysis, waves travelling in the outward direction have phase speed $(u+u_A)$, and satisfy the relation $(\delta u/u_A) = -(\delta B/B)$ where $u_A = B/\sqrt{4\pi\rho}$ is the Alfvén speed. Using these results together with the explicit WKB solutions for the wave amplitudes, expression (27) for the force density becomes

$$\langle f_w \rangle = -\frac{1}{2} \frac{\mathrm{d}}{\mathrm{d}r} \langle \varepsilon_w \rangle = \frac{\langle \varepsilon_w \rangle}{4} \left(\frac{1 + 3M_A}{1 + M_A} \right) \left(\frac{2}{r} + \frac{1}{u} \frac{\mathrm{d}u}{\mathrm{d}r} \right), \qquad (28)$$

where $\langle \varepsilon_w \rangle = \langle \delta B^2 \rangle /4\pi$ is the wave energy density and $M_A = u/u_A$ is the Alfvénic Mach number. Substitution of this result for $\langle f_w \rangle$ in equation (26) yields, after some rearrangement,

$$\left[u^2 - a^2 - \frac{\langle \varepsilon_w \rangle}{4\rho} \left(\frac{1 + 3M_A}{1 + M_A} \right) \right] \frac{r}{u} \frac{\mathrm{d}u}{\mathrm{d}r} = \left[2a^2 - \frac{GM_*}{r} + \frac{\langle \varepsilon_w \rangle}{2\rho} \left(\frac{1 + 3M_A}{1 + M_A} \right) \right], \qquad (29)$$

for the wind equation of motion, in which it has been assumed that $p = \rho a^2 (= \rho kT/\mu)$. As was the case for the wind models considered in preceding subsections, equation (29) contains a critical point, occuring where the flow velocity equals the propagation speed of a radially travelling sound wave in the presence of short wavelength Alfvén waves (Holzer, Leer, & Flå 1983).

Some properties of isothermal winds containing short-wavelength Alfvén waves are illustrated in Figures 7 and 8. The solutions depicted were obtained for $M_* = M_\odot$ and $r_0 = 1.25\,R_\odot$, with $\mu = m_p/2$, $N_0 = 2 \times 10^7\,\mathrm{cm}^{-3}$,

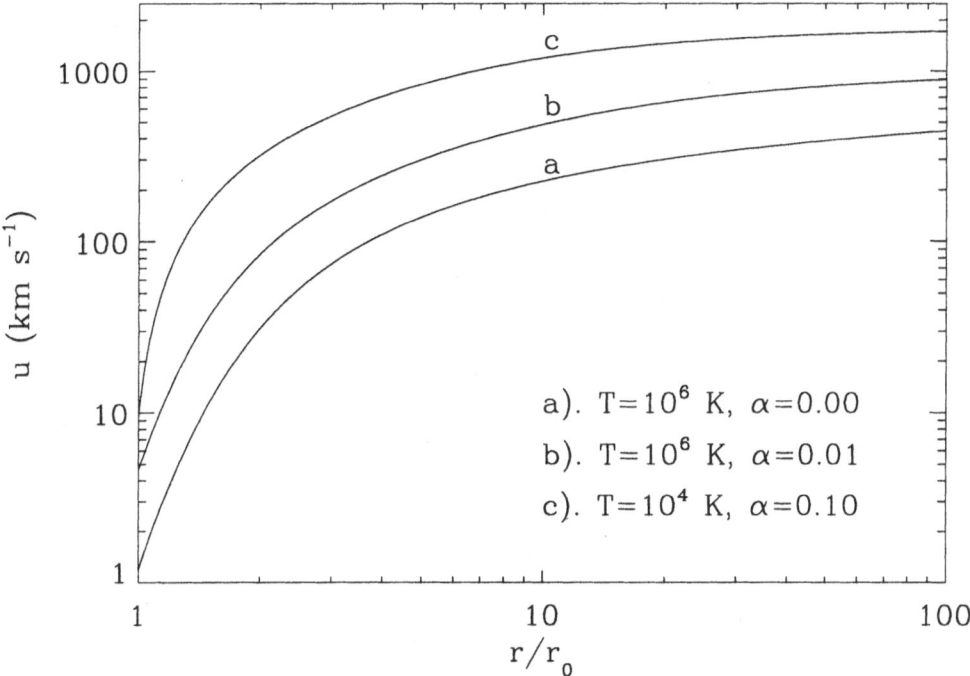

Figure 7. Radial flow velocity as a function of radius for each of the three Alfvén wave-driven wind models discussed in the text.

and $B_0 = 1$ G. In each model, the amount of wave energy present in the wind at its base was controlled by varying the magnitude of the magnetic amplitude through the value of the quantity $\alpha = (\delta B_0/B_0)^2$. In Figure 7, we show the radial expansion speed as a function of distance from the star for wave-driven wind models having $T = 10^6$ K, $\alpha = 0.01$ and $T = 10^4$ K, $\alpha = 0.10$. Also shown is the velocity profile of the thermally driven wind that is present when $\alpha = 0.0$ (i.e., $\delta u = \delta B = 0$) and $T = 10^6$ K. Comparison of curves (a) and (b) indicates that for the same isothermal coronal temeperature, the expansion speed of the wind containing waves is everywhere greater than that of the wind whose acceleration derives from the thermal pressure gradient force alone. For the latter solution, the initial and asymptotic flow speeds are $u_0 = 1.19$ km s^{-1} and $u_\infty = 444.3$ km s^{-1}, respectively, while for the former, which includes effects arising from an Alfvén wave energy flux of magnitude 2.77×10^4 erg cm^{-2} s^{-1} at $r = r_0$, $u_0 = 4.63$ km s^{-1} and $u_\infty = 888.5$ km $^{-1}$.

Note that while the velocity profile corresponding to the $\alpha = 0.01$ wind solution clearly reflects the additional acceleration supplied by waves, the thermal pressure is nonetheless an important component of the wind dynamics in the sub-Alfvénic portion of the flow. Alfvén waves can, however,

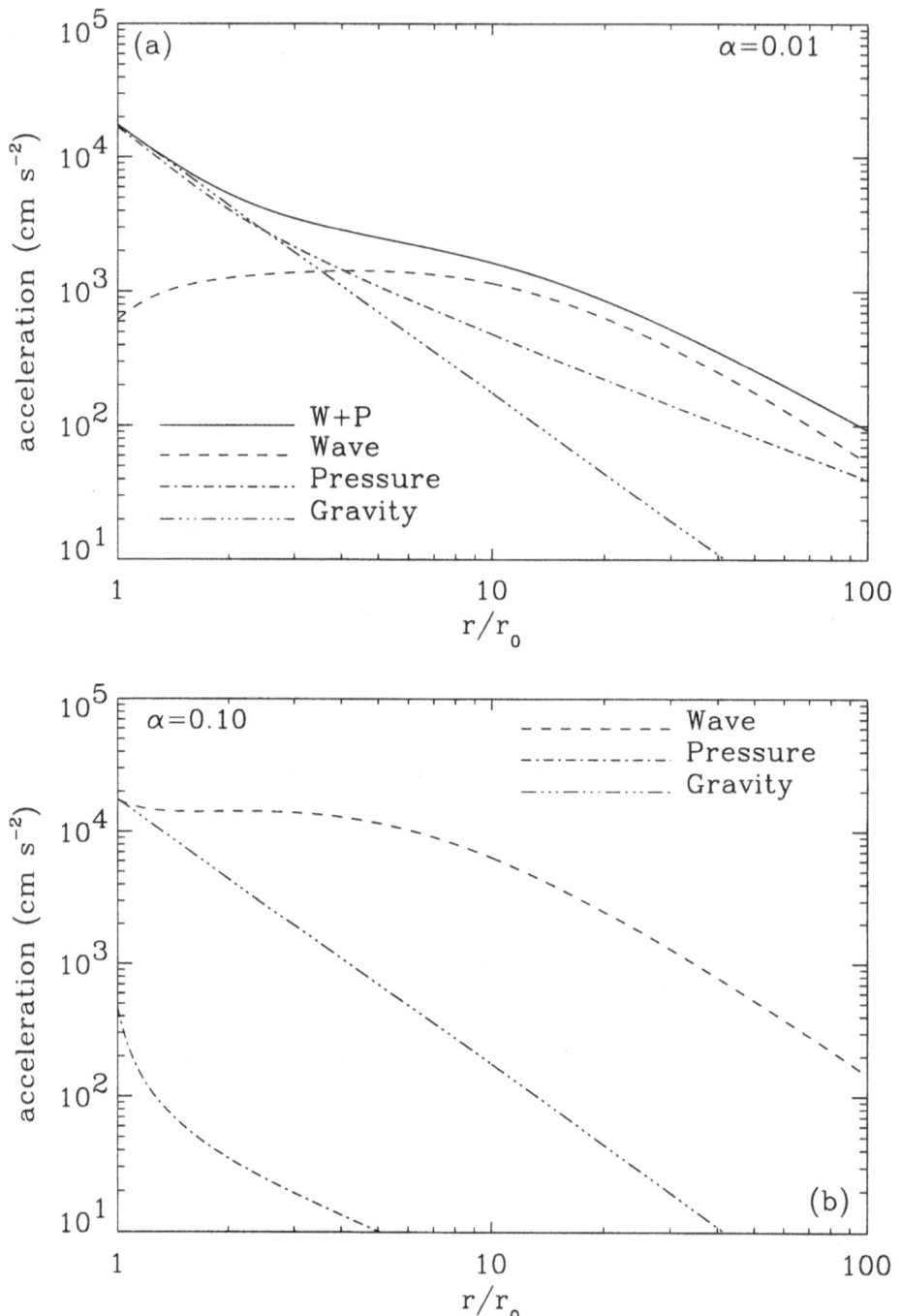

Figure 8. Radial acceleration profiles for wave-driven wind solutions with $T = 10^6$ K, $\alpha = 0.01$ (a) and $T = 10^4$ K, $\alpha = 0.10$ (b). The gravitational acceleration is shown with the opposite sign.

exert considerable dynamical influence even under conditions for which little (if any) of the atmospheric expansion stems from thermal sources. Such a 'wave-dominated' flow is exemplified by the profile labelled (c) in Figure 7. In this case, the assumed coronal temperature ($T = 10^4$ K) is sufficiently low that negligible thermally driven mass loss would take place in the absence of waves. For this solution, $u_0 = 9.56$ km s^{-1}, $u_\infty = 1714$ km s^{-1}, and the value $\alpha = 0.10$ implies a wave energy flux of magnitude 2.80×10^5 erg cm^{-2} s^{-1} at the base of the wind. The dynamical distinctions between models (b) and (c) can be appreciated from inspection of Figure 8, in which the accelerations produced by the thermal pressure gradient, gravitational, and wave forces are shown as functions of r for both models. In model (c), the wave force dominates all others virtually everywhere in the flow, while in model (b), the thermal contribution to the expansion of the stellar corona is evident, particularly for $r \approx r_0$.

For flows that are wave-dominated in the sense described above, it is possible to obtain an approximate, analytic solution to the wind equation of motion. In particular, for a wind whose outward acceleration arises principally from the interaction between Alfvén waves and the flow, it can be shown (see, e.g., Leer, Holzer, & Flå 1982; MacGregor 1983) that the critical point location, the inital expansion speed, and the asymptotic speed of the wind are given by

$$r_c/r_0 \approx \frac{7}{4(1 + \beta/2)}, \tag{30}$$

$$u_0/u_{e0} \approx \frac{1}{8}\beta^2 (r_c/r_0)^{7/2}, \tag{31}$$

and

$$u_\infty/u_{e0} \approx (\beta/M_{A0} - 1)^{1/2}, \tag{32}$$

respectively. In these expressions, $u_{e0} = (2GM_*/r_0)^{1/2}$ is the gravitational escape speed at $r = r_0$, and the validity of equation (32) requires that both the quantity $\beta \equiv \langle \varepsilon_w \rangle / (\rho_0 u_{e0}^2 / 2)$ and u_0/u_{e0} be $\ll 1$.

4. Discussion: Extension to Winds From Non-Solar-type Stars

4.1. WINDS FROM EARLY-TYPE STARS

We have, to this point, confined our attention to the dynamical modifications that result from the inclusion of several well-studied MHD processes in the basic mathematical and physical description of a thermally driven wind. On the basis of observations yielding a significant amount of circumstantial evidence both for the commonality of mass loss from solar-type stars and for its similarity to the solar wind, it seems reasonable to suppose that such a model can plausibly account for the gross characteristics of winds from late-type, low-mass stars. However, as noted in the introductory section of this paper, mass loss as a physical phenomenon is widespread among stars located nearly everywhere in the HR diagram. In view of this fact, we conclude the present analysis by briefly considering how the previously studied MHD processes might operate in the wind from a massive, early-type star; that is, in a flow that is radiatively rather than thermally driven.

The luminous, hot stars having spectral types O through early B and belonging to all luminosity classes display unambiguous radiative signatures of continuous mass loss in the form of high-velocity winds (see, e.g, Cassinelli & Lamers 1987). Most frequently, these indicators take the form of the so-called P-Cygni profiles (consisting of a strongly blue-shifted absorption component together with a red-shifted emission feature) in the ultraviolet resonance lines of species such as C, N, O, and Si. Detailed analyses of measured line shapes and strengths yield typical asymptotic flow speeds $u_\infty \approx$ few $\times 10^3$ km s^{-1} and rates of mass loss in the range $10^{-8} \lesssim \dot{M} \lesssim 10^{-5}$ M$_\odot$ y^{-1}. Winds of this type are thought to be driven by the force produced by the absorption and scattering of radiation from the stellar photosphere in ultraviolet spectral lines formed in the flow (see, e.g., Kudritzki 1988, and references therein).

4.2. THE FORCE OF RADIATION IN SPECTRAL LINES

The force per unit volume produced by a single line having mass absorption coefficient $\kappa_L(\mathbf{r})$ (units: cm^2 g^{-1}) at position $\mathbf{r}$ in the expanding atmosphere of an early-type star is (Rybicki & Hummer 1978)

$$\mathbf{f}_L(\mathbf{r}) = \frac{\rho \kappa_L(\mathbf{r})}{c} \int d\Omega \ \mathbf{n} \int\limits_0^\infty d\nu \ \phi(\mathbf{r},\mathbf{n},\nu) I(\mathbf{r},\mathbf{n},\nu), \qquad (33)$$

where $I(\mathbf{r},\mathbf{n},\nu)$ is the specific intensity (units: erg cm^{-2} s^{-1} Hz^{-1} sr^{-1}) of the radiation at frequency ν traveling in the direction $\mathbf{n}$, $\phi(\mathbf{r},\mathbf{n},\nu)$ is the line profile function, and $d\Omega$ is the element of solid angle. Formally, evaluation of $\mathbf{f}_L$ requires that I be known, a task which in turn requires a complete solution to the non-local transfer problem for line photons in a moving, geometrically extended, circumstellar envelope. In practice, however, rapid radiative acceleration of the atmosphere to supersonic velocities justifies the use of a method commonly referred to as the Sobolev approximation (cf., Rybicki & Hummer 1978), by means of which the line transfer problem is simplified considerably.

Consider a spectral line whose central rest-frame frequency is ν_L. Because of the Doppler shift introduced by atmospheric expansion, material at position $\mathbf{r}$ with velocity $\mathbf{u}(\mathbf{r})$ will interact with radiation that originated in the stellar photosphere with frequency

$$\nu = \nu_L \left[1 + \frac{\mathbf{n} \cdot \mathbf{u}(\mathbf{r})}{c} \right], \tag{34}$$

having travelled to the point in question along a ray with direction given by the unit vector $\mathbf{n}$. If the acceleration of the flow is sufficiently large that $\Delta r \equiv u_{th}/(\mathrm{d}u/\mathrm{d}r) \ll l$ where l is a typical length scale in the medium and u_{th} is the ion thermal speed, then the radial extent of the region in which photons of frequency ν can be absorbed or scattered is small. Under these conditions, the line profile function (here assumed to be Gaussian for simplicity) at the frequency ν tends toward a δ-function of position, and the transfer of radiation at frequency ν becomes purely local. For radially streaming radiation in a spherically symmetric flow, the force due to the single line acts in the radial direction and is approximately given by (e.g., Castor 1974; Rybicki & Hummer 1978)

$$f_L \approx \frac{\pi F_*(\nu) \Delta \nu_D}{c} \left(\frac{R_*}{r} \right)^2 \rho \kappa_L \left[\frac{1 - \exp(-\tau_L)}{\tau_L} \right], \tag{35}$$

where $F_*(\nu)$ is the radiative flux from the stellar photosphere, $\Delta \nu_D$ is the Doppler width of the line, and $\tau_L \equiv \rho \kappa_L u_{th} (\mathrm{d}u/\mathrm{d}r)^{-1}$ is the optical depth parameter for the line.

The winds from massive, hot stars are accelerated by the radiation force produced by a large number of resonance and subordinate lines, formed by ions of a variety of chemical elements. As shown by Castor, Abbott, & Klein (1975), the collective, local force exerted on the flow by many such lines, some of which are optically thick ($\tau_L \gg 1$) and some of which are optically thin ($\tau_L \ll 1$), can be represented by the formula

$$f_{L,total} \approx \frac{\rho \sigma_e L_*}{4 \pi r^2 c} \times k t^{-\alpha}, \tag{36}$$

where σ_e is the electron mass scattering coefficient, L_* is the stellar luminosity, $t \equiv (\sigma_e/\kappa_L) \tau_L$, and k and α are prescribed constants. A total line force of the form given in equation (36) can be formally derived using the Sobolev theory if it is assumed that the force producing lines (i) do not overlap in frequency, (ii) have constant opacities κ_L, and (iii) are distributed in opacity according to a power law.

4.3. MAGNETIC MODIFICATIONS TO RADIATIVELY DRIVEN FLOW

In order to discuss some of the dynamical changes arising from the presence of a magnetic field in a radiatively accelerated wind, it is first necessary to summarize a few of the properties of an unmagnetized flow of this type. Although many such wind models containing varying amounts of detail have been constructed, for this comparison we choose the basic description of a line driven wind presented by Castor, Abbott, & Klein (1975). In this model, the flow is acted upon by gravitational, thermal pressure gradient, and radiation forces, with the later given quantitatively by equation (36). When this expression is used for the line force, the radial component of the momentum equation becomes a transcendental equation which must be solved to obtain the velocity gradient du/dr at a given position in the wind. The flow speed at an incrementally larger radius is then derived by integration of the result, and the procedure is is repeated to extend the solution outward from the star. For specified values of the stellar and line force parameters, a critical point exists in the flow, occurring at the radius and velocity for which two distinct solution branches for du/dr coincide. As in the case of the wind models studied in section 3, the physical properties implied by the boundary conditions are manifest only in the single solution that passes smoothly through the critical point. Imposition of this requirement makes it possible to evaluate the mass loss rate $\dot{M}$ and asymptotic flow speed u_∞ of the wind. For isothermal flow from a star with mass M_* and radius R_*, these quantities are expressible in the form (cf., Castor, Abbott, & Klein 1975)

$$\dot{M} = \left(\frac{4\pi G M_*}{\sigma_e u_{th}}\right) \alpha \, (k\Gamma)^{1/\alpha} \left(\frac{1-\alpha}{1-\Gamma}\right)^{(1-\alpha)/\alpha}, \tag{37}$$

and

$$u_\infty = \left[\frac{\alpha}{1-\alpha} \frac{2 G M_*(1-\Gamma)}{R_*}\right]^{1/2}, \tag{38}$$

where $\Gamma = (\sigma_e L_*/4\pi G M_* c)$ is the ratio of the (optically thin) electron scattering force to the gravitational force.

Consider the application of the wind model described above to a star having $M_* = 60 \, M_\odot$, $R_* = 13.8 \, R_\odot$, and $T_{eff} = 50,000$ K, from which values it follows that $L_* = 1.075 \times 10^6 \, L_\odot$ and $\Gamma = 0.428$. Adopting the force multiplier constants $k = 1/30$ and $\alpha = 0.7$ (see, e.g., Castor, Abbott, & Klein 1975), equations (37) and (38) yield the estimates $u_\infty = 1487$ km s^{-1} and $\dot{M} = 4.73 \times 10^{20}$ g s^{-1} ($\approx 7.5 \times 10^{-6} \, M_\odot$ y^{-1}) for the terminal speed and mass loss rate of a spherical, isothermal ($T = T_{eff}$) wind. If the flow is instead along the central streamline of a magnetically defined, non-spherical flow tube (cf., section 3.1), it is found that even small deviations from spherically symmetric expansion can lead to significant changes in the magnitudes of both the radiative acceleration and the asymptotic speed of the wind (MacGregor 1988). As an example, consider a flow within a tube whose cross-sectional area increases with r according to equations (8) and (9). Detailed, numerical wind solutions (cf., MacGregor 1988) obtained for the parameter values given previously with $r_0 = R_*$, $r_1 = 2.00\ r_0$, and $\sigma = 0.1\ r_0$ have asymptotic flow speeds $u_\infty = 2316,\ 2934,\ 3541,$ and 4151 km s^{-1} for $f_{max} = 1.25,\ 1.50,\ 1.75,$ and 2.00, respectively. In general, for fixed values of r_1 and σ, the dependence of u_∞ on f_{max} can be shown to be (approximately) $u_\infty \propto f_{max}^{\alpha/[2(1-\alpha)]}$. Examination of the balance of forces that prevails throughout each of these models reveals that the increased flow speeds at large distances from the star are a direct result of an enhancement of the radiation force acting on the wind. In particular, the radiative acceleration for $r \approx r_1$ in the solution having $f_{max} = 2$ .exceeds that at the same location in a spherical ($f_{max} = 1$) solution by more than an order of magnitude. Because of the nature of its dependence on ρ and du/dr (see equation [36] and the discussion thereof), the line force is significantly increased in the region of most rapid flow tube divergence. This augmentation is instigated by the (in this case) perturbative influence of the non-monotonic thermal pressure gradient, as in Figure 2. Since, for the solutions under consideration, the additional acceleration occurs beyond the critical point location (cf., MacGregor 1988), the mass flux density at the base of the wind is essentially unchanged from the case of spherical expansion.

When magnetic and rotational forces are incorporated into the basic model of line-driven atmospheric expansion, further modifications to the dynamical structure of the flow are possible. Consider an isothermal wind emanating from a star having $M_* = 50\ M_\odot$, $R_* = 19.7\ R_\odot$, $T_{eff} = 65,000$ K, $L_* = 6.76 \times 10^5\ L_\odot$, and $\Gamma = 0.323$. Assuming again that the constants characterizing the line force law (36) have the values $k = 1/30$ and $\alpha = 0.7$, it follows from equations (37) and (38) that an unmagnetized wind from a non-rotating star has mass loss rate $\dot{M} = 5.2 \times 10^{-6}\ M_\odot\ y^{-1}$ and asymptotic speed $u_\infty = 1240$ km s^{-1}. Now consider the wind from a star that is both uniformly rotating and magnetized, but which is otherwise identical to the object described above. As in section 3.2, consider only that portion of the flow contained in the equatorial plane, wherein the velocity and magnetic fields are assumed to have radial and azimuthal components, each of which is a function solely of r. Numerical solutions to this radiative analogue of the MHD wind model of Weber & Davis (1967) have been derived by Friend & MacGregor (1984; see also, MacGregor & Friend 1987), for a broad range of values of both the surface radial magnetic field strength B_{r0} and equatorial rotation speed v. For constant v, the asymptotic speed u_∞ is found to increase as the specified value of B_{r0} is made larger. For example, models having $v = 350$ km s^{-1} and $B_{r0} = 200,\ 400,\ 800,$ and 1600 G have terminal speeds $u_\infty = 1256,\ 1564,\ 2057,$ and 2877, respectively.

This behavior is not just a consequence of the increasing importance of the radial component of the Lorentz force for higher surface field strengths (cf., section 3.2), but also of the effect this additional magnetic acceleration has on the magnitude of the line force. As was seen for radiation-driven flow in a magnetically defined flow tube, an increase in the radial velocity gradient induced by a non-radiative (in this case magnetic) force can lead to a sizeable enhancement of the force produced by spectral lines, thereby raising u_∞. In addition, it is also found that the mass loss rate $\dot{M}$ is an increasing function of B_{r0} for fixed v, with the magnitude of the increase being larger for larger values of v. To illustrate this, note that the wind solutions presented above have mass loss rates (in units of 10^{-6} $M_\odot$ y^{-1}) $\dot{M} = 6.04, 6.37, 6.62$, and 6.81 while solutions obtained for the same values of B_{r0} but with $v = 400$ km s^{-1} have $\dot{M} = 6.57, 7.13, 7.56$, and 7.86. These changes reflect the increasing magnitude of the centrifugal force at the base of the flow, since, as B_{r0} is made larger, the azimuthal motion of the gas there tends more toward corotation with the star. Unlike the wind solutions studied in section 3.2, in the model presently under consideration, sufficiently strong, outward acceleration of the flow by the line force can cause the azimuthal component of the magnetic force near the base of the wind to reverse direction and act counter to the sense of the stellar rotation. Under these conditions, the azimuthal velocity u_ϕ as a function of r decreases at a rate that is faster than r^{-1} for $r \approx r_0$. A careful examination of the wind dynamics in the corotating frame of reference reveals that this behavior is a consequence of the balance of forces perpendicular to magnetic field lines required by the steady state model.

The azimuthal dynamical structure of a line-driven wind from a rotating star can also be influenced directly by the radiation force. In the wind model described in the preceding paragraph, the vector flow velocity at location r in the equatorial plane of the star has both radial and azimuthal components, $\mathbf{u} = u_r(r)\mathbf{e}_r + u_\phi(r)\mathbf{e}_\phi$. Now consider two rays originating at the leading and trailing limbs of the star, respectively, that intercept the point r. Describe these rays by the unit vectors $\mathbf{n}_\pm = \cos\theta_*\mathbf{e}_r \pm \sin\theta_*\mathbf{e}_\phi$ with $\cos\theta_* = [1 - (R_*/r)^2]^{1/2}$. It then follows from equation (34) that the photons that travel along the directions $\mathbf{n}_\pm$ and are resonant with a spectral line of rest frequency ν_L at position r have frequencies

$$\nu_\pm = \nu_L \left[1 + \frac{u_r}{c}\cos\theta_* \pm \frac{(u_\phi - \Omega_* r)}{c}\sin\theta_*\right]. \tag{39}$$

Hence, the net momentum $h(\nu_+\mathbf{n}_+ + \nu_-\mathbf{n}_-)/c$ gained by the gas upon absorbing a photon from each of the directions $\mathbf{n}_\pm$ has a non-zero azimuthal component, a result of the fact that since $\mathbf{u}$ is not solely in the radial direction, it projects differently onto the unit vectors $\mathbf{n}_\pm$. It can be shown that because of this asymmetry, the material at r experiences an azimuthal radiation force per unit volume given by,

$$f_{L,\phi} = \frac{\rho \kappa_L}{c} \cdot q(r/R_*) \cdot \frac{\nu_L(u_\phi - \Omega_* r)}{c} \cdot \frac{dI_*(\nu)}{d\nu}, \tag{40}$$

where q is a geometrical factor, I_* is the intensity of the photospheric radiation field, and the line has been assumed to be optically thin. Note that the force given by equation (40) is a differential effect, depending on $dI_*/d\nu$ rather than on I_*. If the locally absorbed radiation is unattenuated photospheric continuum radiation, then $dI_*/d\nu$ is of order I_*/ν, and the azimuthal force is small ($\sim u_\phi/c$) relative to the radial force. If, however, the radiation absorbed at r originates in a photospheric absorption feature, then $dI_*/d\nu \sim I_*/\Delta \nu_D$ and the azimuthal force can have a magnitude that is comparable ($\sim u_\phi/u_{th}$) to that of the radial force.

The preceding analysis can be applied to the determination of the transverse radiative force experienced by an Alfvén wave that propagates along the magnetic field in a static, uniform, plane-parallel atmosphere. Consider a small amplitude Alfvénic disturbance that travels in the direction of mean radiation flow with wave vector $\mathbf{k} = k\mathbf{e}_z$, and assume that the velocity and magnetic fields have the forms $\mathbf{u} = \delta u(z,t)\mathbf{e}_x$ and $\mathbf{B} = B_0\mathbf{e}_z + \delta B(z,t)\mathbf{e}_x$ where B_0 is constant and δu and δB are the wave amplitudes. Under these conditions, the frequency of a photon that travels along the ray $\mathbf{n}$ and is resonant with a line of rest frequency ν_L is

$$\nu = \nu_L \left(1 + \frac{\mathbf{n} \cdot \delta \mathbf{u}}{c} \right). \tag{41}$$

As before, it can be shown that there is now a fluctuating component of the line force δf_L, acting in the $\mathbf{e}_x$-direction and given by

$$\delta f_L = \frac{2\pi \rho \kappa_L}{3c} \cdot \frac{\nu_L \delta u}{c} \cdot \frac{dI_*(\nu)}{d\nu}. \tag{42}$$

When this transverse radiation force is included in the perturbation analysis, the linearized momentum and induction equations, after substituting solutions of the form $(\delta u, \delta B) \propto \exp[i(kz - \omega t)]$, yield the dispersion relation

$$k^2 u_A^2 = \omega^2 - i\omega \omega_0, \tag{43}$$

where (as in section 3.3) u_A is the Alfvén speed and $\omega_0 = \delta f_L/(\rho \delta u)$ is a characteristic frequency. For complex k, inspection of equation (43) reveals that the wave can either grow ($\text{Im}(k) < 0$) or decay ($\text{Im}(k) > 0$) depending upon whether ω_0 ($\propto dI_*/d\nu$) is > 0 or < 0, respectively. From equation (43), it is apparent that wave growth occurs when $dI_*/d\nu > 0$ and δf_L and

334

δu are in phase, while for $\mathrm{d}I_*/\mathrm{d}\nu < 0$, the phase difference between δf_L and δu is π radians and the wave is damped. The nature of any changes to the dynamical structure of a line-driven wind that might result from the presence of such radiatively modified Alfvén waves in the flow has yet to be explored.

I am grateful to P. Charbonneau and T. E. Holzer for numerous helpful discussions and comments on the manuscript. I thank K. Tsinganos for his kind hospitality during the course of the workshop, and for carefully reading and commenting on this paper. I am also deeply indebted to E. N. Parker for sharing his insights on many of these subjects with me and to the late H. S. Bridge for providing me with both the opportunity and the motivation to pursue my scientific interests in this area.

References

Barnes, A. 1974, ApJ, 188, 645
Barnes, A. 1992, J. Geophys. Res., 97, 12,105
Belcher, J.W. 1971, ApJ, 168, 509
Belcher, J.W., & Davis, L. 1971, J. Geophys. Res., 76, 3534
Belcher, J.W., & MacGregor, K.B. 1976, ApJ, 210, 498
Cassinelli, J.P., & Lamers, H.J.G.L.M. 1987, in Exploring the Universe with the IUE
    Satellite, ed. Y. Kondo (Dordrecht: Reidel), 139
Castor, J.I. 1974, MNRAS, 169, 279
Castor, J.I., Abbott, D.C., & Klein, R.I. 1975, ApJ, 195, 157
Charbonneau, P. 1995, ApJS, 101, 309
Drake, S.A. 1987, in Proceedings of the Sixth International Solar Wind conference, eds.
    V.J. Pizzo, T.E. Holzer, & D.G. Sime (NCAR Technical Note 306), 129
Dupree, A.K. 1986, ARA&A, 24, 377
Durney, B. 1972 in Solar Wind, eds. C.P. Sonett, P.J. Coleman, & J.M. Wilcox (Wash-
    ington: NASA Scientific and Technical Information Office), 282
Friend, D.B., & MacGregor, K.B. 1984, ApJ, 282, 591
Hartmann, L., & MacGregor, K.B. 1982, ApJ, 259, 180
Holzer, T.E., Leer, E., & Flå, T. 1983, ApJ, 275, 808
Hunhausen, A.J. 1972, Coronal Expansion and Solar Wind (New York: Springer-Verlag)
Jacques, S.A., 1977, ApJ, 215, 942
Kawaler, S.D. 1988, ApJ, 333, 236
Keppens, R., MacGregor, K.B., & Charbonneau, P. 1995, A&A, 294, 469
Kohl, J.L., & Withbroe, G.L. 1982, ApJ, 256, 263
Kopp, R.A., & Holzer, T.E. 1976, Solar Phys., 49, 43
Kraft, R.P. 1967, ApJ, 150, 551
Kraft, R.P. 1970, in Spectroscopic Astrophysics, ed. G.H. Herbig (Berkeley: University
    of California Press), 385
Kudritzki, R.P. 1988, in Radiation in Moving, Gaseous Media, eds. Y. Chmielewski & T.
    Lanz (Geneva: Geneva Observatory), 1
Leer, E., & Holzer, T.E. 1980, J. Geophys. Res., 85, 4681
Leer, E., Holzer, T.E., & Flå, T. 1982, Space Sci. Rev., 33, 161
Low, B.C. 1990, ARA&A, 28, 491
MacGregor, K.B. 1983, in Solar Wind Five, ed. M. Neugebauer (NASA Conference Pub-
    lication 2280), 241
MacGregor, K.B. 1988, ApJ, 327, 794
MacGregor, K.B., & Friend, D.B. 1987, ApJ, 312, 659

MacGregor, K.B., & Brenner, M. 1991, ApJ, 376, 204

MacGregor, K.B., & Charbonneau. P. 1994, ApJ, 430, 387

MacGregor, K.B., & Charbonneau, P. 1995, in Cosmic Winds and the Heliosphere, eds. J.R. Jokipii, C.P. Sonett, & M.S. Giampapa (Tucson: University of Arizona Press), in press

Marsch, E. 1991, in Physics of the Inner Heliosphere, Vol. II, eds. R. Schwenn & E. Marsch (Heidelberg: Springer-Verlag), 159

Maxson, C.W., & Vaiana, G.S. 1977, ApJ, 215, 919

Munro, R.H., & Jackson, B.V. 1977, ApJ, 213, 874

Neugebauer, M. 1983, in Solar Wind Five, ed. M. Neugebauer (NASA Conference Publication 2280), 135

Pallavicini, R. 1989, A&A Review, 1, 177

Parker, E.N. 1958, ApJ, 128, 664

Parker, E.N. 1960, ApJ, 132, 821

Parker, E.N. 1963, Interplanetary Dynamical Processes (New York: Interscience)

Parker, E.N. 1965, Space Sci. Rev., 4, 666

Pizzo, V.J., Schwenn, R., Marsch, E., Rosenbauer, H., & Mühlhäuser, K.-H. 1983, ApJ, 271, 335

Roberts, D.A., & Goldstein, M.L. 1991, Rev. Geophys. Suppl., 29, 932

Rosner, R., Tucker, W.H., & Vaiana, G.S. 1978, ApJ, 220, 643

Rottman, G.J., Orrall, F.Q., & Klimchuk, J.A. 1982, ApJ, 260, 326

Rybicki, G.B., & Hummer, D.G., 1978, ApJ, 219, 654

Schatzman, E. 1962, Ann. d'ap., 25, 18

Schmitt, J.H.M.M. 1993, in Physics of Solar and Stellar Coronae, eds. J.L. Linsky & S. Serio (Dordrecht: Kluwer), 327

Schwenn, R. 1983, in Solar Wind Five, ed. M. Neugebauer (NASA Conference Publication 2280), 489

Skumanich, A. 1972, ApJ, 171, 565

Snodgrass, H.B. 1983, ApJ, 270, 288

Soderblom, D.R. 1991, in Angular Momentum Evolution of Young Stars, eds. S. Catalano & J.R. Stauffer (Dordrecht: Kluwer), 151

Stauffer, J.R. 1991, in Angular Momentum Evolution of Young Stars, eds. S. Catalano & J.R. Stauffer (Dordrecht:Kluwer), 117

Stauffer, J.R. 1994, in Cool Stars, Stellar Systems, and the Sun, ed. J.P.Caillault (ASP Conference Series 64), 163

Vaiana, G.S., & Rosner, R. 1978, ARA&A, 16, 393

Vaiana, G.S. et al. 1981, ApJ, 245, 163

Wilson, O.C., 1966, ApJ, 144, 695

Weber, E.J., & Davis, L. 1967, ApJ, 148, 217

Withbroe, G.L. 1988, ApJ, 325, 442

Withbroe, G.L., & Noyes, R.W. 1977, ARA&A, 15, 363

Withbroe, G.L., Kohl, J.L., Weiser, H., & Munro, R.H. 1982, Space Sci. Rev., 33, 17

SOME IMPLICATIONS OF SOLAR AND STELLAR ACTIVITY

EUGENE PARKER
The University of Chicago,
Enrico Fermi Institute and Depts. of Physics and Astronomy
933 East 56th Street, Chicago, Illinois 60637 U.S.A.
parker@odysseus.uchicago.edu

Abstract. The astonishing variability and suprathermal activity of the Sun and similar stars is presented as a problem in physics, about which there is currently more conjecture than hard fact. Attention is directed to the coronal hole, the active X-ray corona, and to the overall variation in solar brightness. The heat supply to the coronal hole and the quiet corona is the physical cause of the solar wind, providing the loss of mass and angular momentum from a solitary late main sequence star like the Sun. The heat supply to the active corona is the cause of the X-ray emission from such stars. The variation of solar brightness has profound effect on terrestrial climate. The best guess for the causes of these phenomena is the microflaring in the network fields, the spontaneous discontinuities in the bipolar fields, and the enhanced convective heat transport associated with emerging flux bundles, respectively. The observational difficulty is the inability to see below the surface of the Sun and the inability to resolve the 10^2 km internal structure and motions of the faculae, plages, microflares, and magnetic fibrils at the visible surface. A solar microscope with resolution of 0.1" is essential for probing these basic characteristics. Observational study of these solar phenomena is critical for understanding the associated century long warming and cooling of the terrestrial climate, now complicated by the accumulating anthropogenic greenhouse gases, and for understanding the X-ray astronomy of solitary stars.

1. Suprathemal Activity and Variability

The astonishing aspect of the ordinary star like the Sun is the active suprathermal outer atmosphere, or corona, at $1\text{-}10 \times 10^6$K and the variability of the whole system from the photosphere to the corona and the stellar wind. The flare in the corona is the extreme example. The corona

337

K. C. Tsinganos (ed.), Solar and Astrophysical Magnetohydrodynamic Flows, 337–354.
© *1996 Kluwer Academic Publishers.*

consists of fast particles (100-300 eV) with enough energy to stand outward from the star without, however, enough energy (1.5 keV) to escape immediately to infinity. The general variability attests to the convective magnetic activity deeper in the stellar envelope. It was pointed out fifty years ago (Biermann, 1946; Schwarzchild, 1948; Schatzman, 1949; Alfven, 1950; Piddington, 1956; Osterbrock, 1961; Billings, 1966) that the suprathermal energy that creates the corona comes from the mechanical work of the sub-surface convection. However, the precise nature of the connection between convection and coronal heating has proved to be surprisingly elusive.

Observational studies show that the corona exists in two distinct states, the active X-ray corona and the coronal hole, each with its own particular form of heat supply. The two states depend upon the local mean magnetic field strength, which usually lies at 10^2 gauss or more, creating the active regions, or 10 gauss or less, making the quiet corona and the coronal hole. The fields of 10^2 gauss or more form the bipolar magnetic active regions at the surface of the Sun. The bipolar fields confine the coronal gas within them, and the gas density builds up to about 10^{10} atoms/cm^3 at 2-6×10^6K to achieve a quasi-steady balance between the heat input of approximately 10^7 ergs/cm^{2}sec and the thermal X-ray emission (Withbroe and Noyes, 1977). This is the basis for the X-ray astronomy of solitary late main sequence stars. On the other hand, the regions of weak field, far from bipolar regions and largely at high latitudes, are not able to confine the coronal gas, so that the gas expands away to infinity, producing the supersonic solar wind and extending the weak magnetic field to "infinity" to form the heliosphere. The temperature in the coronal hole is of the order of 1.5 - 2 ×10^6K and the density is limited by the free expansion to about 10^8 atoms/cm^3, so that the X-ray emission is negligible.

Coronal holes have an energy input of the order of 0.5-1 ×10^6 ergs/cm^3, most of which goes into maintaining the temperature as the coronal gas expands gently away from the Sun (Withbroe and Noyes, 1977). Withbrow (1988) points out that the major portion of the heat input occurs in the first 1R$_\odot$ above the surface in order to maintain the temperature in the continuing expansion. On the other hand, the high velocity (500-800 km/sec) solar wind from coronal holes indicates a substantial energy, and perhaps momentum, input beyond the sonic point in the wind (r$\gtrsim$3R$_\odot$) (see also Wang, 1994a,b).

One assumes that the corona of the Sun is the paradigm for most stars, carrying on their X-ray emission and stellar winds unseen by the terrestrial telescope. So how does it work? What is the physical process that supplies the energy to the expanding coronal hole and to the active X-ray corona of the typical solitary late-main sequence star? The energy supply is the key in both cases, with the mass loss and X-ray emission a direct consequence of the heat input and the resulting upward expansion of the heated gas from the top of the chromosphere. The heat input, then, is the basis for the solar wind and consequent heliosphere of the Sun, and, by inference, of the other stars. It provides the mass loss and angular momentum loss from solitary late-main sequence stars. The heat input is the basis for the X-ray emission and hence the basis for the X-ray astronomy of late-main sequence stars. Curiously the X-ray atronomy community shows little active interest in the cause of the X-ray corona, evidently content with vague reference to waves from the convective zone. On the other hand, the physicist can hardly avoid the scientific challenge.

The implications of the extensive forms of solar activity are broad, showing the restless nature of the Sun and, by inference, most other late main-sequence stars, and producing a continuing variation of conditions in the magnetosphere and atmosphere of Earth. The X-ray emission of the Sun varies by factors of 10-10^2 in association with the general level of magnetic activity. The variability declines with increasing wavelength into the UV where it is only several percent. The X-rays and the UV drive the temperature and photochemistry of the ionosphere, thermosphere, and upper stratosphere of Earth, with the temperature in the thermosphere at the altitude of low orbiting space craft falling to 600 K during years of minimum solar activity and rising to 1200 K at activity maximum. The associated variation in total brightness of the Sun, mainly in the visible and infra-red, was observed to be about two parts in 10^3 between the maximum in 1979 and the minimum in 1986 (Fröhlich, et al, 1991; Hoyt, et al, 1992; Lean, et al, 1994; Fröhlich, 1994; see also the early reports of Chapman, 1987; Hudson, 1988). Ground based observations of the brightness of other solitary main-sequence stars shows the same generally variability, by one to five parts in 10^3 in step with the cyclic level of magnetic activity (Radick, Lockwood, and Baliunas, 1990; Baliunas, 1991; Baliunas, et al, 1991, 1995; Lockwood, et al, 1992).

There is, then, the complication that the 11 year cyclic variation of the activity and brightness of the Sun (and other solar-type stars) sometimes vanishes, leaving the star in a profoundly inactive state for 10^2 years, e.g. the period 1645-1715 AD now known as the Maunder Minimum (Maunder, 1894, 1922). To add to the confusion, there are centuries of greatly enhanced activity, e.g. the Medieval Maximum when the Sun was hyperactive during the 12th and 13th centuries (Eddy, 1976, 1977a,b, 1983a,b). The important point first made by Eddy is that the Sun has been essentially without activity in ten of the last seventy centuries and has been hyperactive in eight of the last seventy centuries. Eddy's second point is that for at least the last 2000 years, for which information on terrestrial climate is available (Beer, et al, 1990; Damon and Sonell, 1991; Friis-Christensen and Lasen,

1991; Hameed and Wyant, 1982; Hameed and Gong, 1994) the mean annual temperature in the northern temperate zone of Earth tracks the general level of solar activity.

The mean annual temperature in the northern hemisphere varies 1-2 degrees above and below the average, with the cold climate of the 15th and 17th centuries (the Spörer and Maunder Minima, respectively) causing agricultural calamity in northern Europe and China, and the warm climate of the Medieval Maximum causing agricultural calamity in the semi-arid regions, e.g. southwest United States. Unfortunately the terrestrial climate is not sufficiently well understood to allow one to deduce the change $\Delta L_\odot$ in solar brightness $L_\odot$ that caused the changes in the mean annual temperature. More recent studies (Friis-Christensen and Lasser, 1991) show the climate tracking solar activity over periods of a decade as well as a century, so the connection is well established even if the physics of $\Delta L_\odot$ and of the terrestrial atmosphere is not fully understood. Detailed studies of the Sun and other solar-type stars suggests that during the Maunder Minimum the Sun was fainter on the average by at least two, and probably four, parts in 10^3 than the present average (Baliunas and Jastrow, 1990; Foukal and Lean, 1990; Baliunas, 1991; Schuurman, 1991; Schlesinger and Ramankutty, 1992; Soon, Baliunas, and Zhang, 1994; Zhang, et al, 1994; Hout, Schatten and Nesmes-Ribes, 1994; Lean, et al, 1994; Baliunas and Soon, 1995).

It would be interesting to know the state of the solar wind during the abnormal centuries. One may guess that there were extended coronal holes and associated high speed solar wind during the Maunder Minimum. There were only a few sunspots and these appeared only in the southern hemisphere within about 20° of the equator (Ribes and Nesmes-Ribes, 1993). There may have been few, if any, high speed solar wind streams during the Medieval Maximum. But that is only a guess (see Geiss and Bochsler, 1991 for long term trends in the solar wind).

The obvious question is why the brightness and general level of activity of the Sun should vary at all, apart from the nuclear evolution of its core over times of $10^9 - 10^{10}$ years. The answer seems to be that the internal hydrodynamics of the Sun is not in a steady state. The hydrodynamics takes on the characteristics of a magnetohydrodynamic dynamo, which gets itself tangled up with the magnetic fields that it creates, and the net result is a nonsteady system. Which is not to say that we understand the process. In view of the terrestrial impact of the solar variability, pursuit of the problem should be motivated by more than purely scientific curiosity (Parker, 1994a).

2. Energy Supply

At the present time the best guess for the energy input to a coronal hole is a combination of microflaring in the network fields (i.e. in the magnetic hedgerows in the boundaries between supergranules) to heat the near corona (Martin, 1988; Porter and Moore, 1988) and Alfven waves generated by the turbulent granules in the mean field ($\sim$ 10 gauss) to heat and push the corona beyond the sonic point (Parker, 1991). There is no direct observational support for sufficient energy in the microflaring nor compelling theoretical reason to expect enough Alfven waves from the granules. Indeed, the required Alfven wave amplitude is extreme, comparable to the wavelength. But what else can accomplish the necessary heat input to the coronal hole? Postulating exotic and unexpected wave forms (e.g. a powerful Alfven wave flux at 1Hz) for the task merely transforms the mystery of coronal heating into the mystery of the origin of exotic waves. It is a step to be taken only if forced by the facts. Thus a supersonic solar wind is an inevitable consequence of the 10^6 K temperatures in the open corona, but it is not clear why the temperature is so high. Some mechanism, or mechanisms, probably driven by the subsurface convection, accelerates ions to about 200 km/sec and/or electrons to 10^4km/sec (i.e. to 10^6K or more) over an extended radius, out to $5R_\odot$ at the very least. But that is not to say that we understand why a star like the Sun is "compelled" to do this.

This brings to mind the somewhat similar situation for the theory of sunspots. We know a lot about the structure of sunspots, but we do not know enough to state why a star is "compelled" by the laws of nature to form sunspots.

Consider, then, the variation of the brightness of the Sun in association with magnetic activity. Foukal and Lean (1986) show that perhaps half or a little more of the varying brightness of the Sun arises from the competition between the bright faculae and plages on the one hand and the dark sunspots on the other. The enormously greater area of the faculae outweighs the sunspots so that the Sun is brighter on the average during the years of high activity. On the other hand, the observations show that there are other effects as well (Foukal, Harvey, and Hill, 1991; Kuhn and Libbrecht, 1991; Sprüit, 1994). So the nature of the enhanced electromagnetic emissions from the visible surface of the Sun can be described at least superficially, but there is no theory for the component parts, e.g. sunspots, faculae, plages, the calcium network, etc. The basic question is why more energy is transported upward across the convective zone when the Sun is active? For it can be shown that the heat content in the excess of the temperature gradient above the adiabatic gradient is not sufficient to maintain a century of enhanced or suppressed solar brightness.

In fact the answer to that question may be straightforward. For the surface activity of the Sun is a consequence of the continuing emergence of Ω-loops, creating and maintaining the long lived bipolar magnetic regions. It appears that the Ω-loops are formed in the azimuthal magnetic field of the Sun at the bottom of the convective zone. The point is that the energy radiated from the visible surface is supplied to the surface from the bottom of the convective zone by convective mixing. The convective transport is relatively inefficient because of the vertical density stratification which limits the vertical mixing length to something of the order of the pressure scale height. The pressure scale height at any depth z below the surface of the Sun is approximately $0.3\, z$, (cf. Sprüit, 1974) so the convective heat transport to the surface is a slow process of many successive relays. In contrast, the vertical wakes created by the upward passage of successive Ω-loops are driven all the way from the bottom to somewhere near ($\sim 10^9$cm) the upper surface of the convective zone, providing a much more effective local heat transport than the ambient convection. Then if the convective updrafts associated with the emerging Ω-loops occupy a fraction $\epsilon \sim 2 \times 10^{-3}$ of the solar surface during the years of maximum activity, then doubling the heat transport within the updrafts provides a brightness increase of two parts in 10^3 (Parker, 1994b,c; 1995a-d).

There may, of course, be other contributing mechanisms. For instance, the X-rays and UV also increase with the activity level. However, they are only a small part of the total and do not fully account for more than a modest fraction of the total observed brightness variations of one part in 10^3 or more.

As a final comment, we note that the temperature in the northern hemisphere showed a small cyclic variation with a period of 23 years through the Maunder Minimum (Hameed and Wyant, 1992), suggesting that the solar dynamo and the associated effects continued to operate in spite of the near absence of effects at the surface (cf. Ribes and Nesme-Ribes, 1993)

3. The X-ray Corona

Consider, then, the problem posed by the X-ray corona. The popular explanation has been the propagation of Alfven waves upward along the mean magnetic field ($B \sim 10^2$ gauss) into the corona where the waves dissipate and supply the estimated 10^7 ergs/cm^{2}sec that maintains the X-ray emitting gas at $N = 10^{10}$ atoms/cm^3 and $T = 2 - 3 \times 10^6$K. Other wave forms, e.g. compressional waves and internal gravity waves tend to be refracted away from the vertical as well as developing large amplitudes in the upward declining density, so that they dissipate at chromosphere levels, and, indeed, may be the major heat source of the chromosphere (but see the recent suggestion by Lou, 1995 that gravity waves may be trapped in the lower corona). Alfven waves, on the other hand, propagate upward along the nearly vertical magnetic field and their velocity amplitudes Δv diminishes with increasing height relative to the Alfven speed C ($\Delta v \sim \rho^{-\frac{1}{4}}$, $C \sim \rho^{-\frac{1}{2}}$ so that $\Delta v / C \sim \rho^{\frac{1}{4}}$). So Alfven waves in the corona are of small amplitude, i.e. they are linear waves, with $\Delta v \sim 20$ km/sec and $C \sim 2 \times 10^3$ km/sec (for $B = 10^2$ gauss and $N = 10^{10}$ atoms/cm^3). The relative magnetic amplitude is $\Delta B / B \sim \Delta v / C \sim 10^{-2}$, so there is no significant nonlinear steepening. The important point is that one expects that the principal waves have periods comparable to the 300 sec coherence time and turnover time of the granules. The characteristic wavelength λ is then 6×10^5 km $\sim 1 R_\odot$. Such waves do not fit into the typical bipolar magnetic field with an overall length of $2 - 20 \times 10^9$ cm. Even a period as short as 10^2 sec yields a wavelength $\lambda = 2 \times 10^{10}$ cm, comparable to the characteristic length L of the largest bipolar regions and ten times L for the smaller bipolar regions.

Now the high Reynolds numbers ($10^6 - 10^7$) of the individual granule would produce Alfven waves of shorter wavelength. However a Kolmogoroff spectrum predicts that the kinetic energy density in eddies with period τ is proportional to τ. So we would expect the energy input to the corona to vary directly with $\tau = L/C$ where C is the coronal Alfven speed. That is to say, the X-ray brightness would be expected to vary more or less in proportion to the length L of the bipolar region. Yet it is an observed fact (Rosner, Tucker, and Valana, 1978) that the brightness varies more like $L^{\frac{1}{6}}$, i.e. hardly at all.

Further investigation turns up more difficulties. The Alfven speed C is approximately 0.6 km/sec in a mean field of 10^2 gauss at the visible surface, from which it follows that the rms fluid velocity Δv within the waves would have to be 0.3 km/sec in order that the energy flux $\rho(\Delta v)^2 C$ be equal to the estimated 10^7 ergs/cm^2 sec. The granules, with characteristic scale $\ell \sim 5 \times 10^7$ cm, correlation times of 300 sec, and rms fluid velocities of 1 km/sec may perhaps produce waves of enough intensity, but there is no excess (Collins, 1992) to overcome reflection at the chromosphere-corona boundary, and the power falls off dramatically at shorter periods, as already noted. Resonance effects have been explored to reduce the reflection and increase the amplitude and dissipation in the corona (cf. Davila, 1987; Hollweg, 1987 and references therein), but the observed independence of the heat input and the length L does not look like a resonance phenomenon which would favor a particular L. The conclusion is simply that the observed X-ray corona does not look like a result of wave heating.

Consider, then, the fact that observations show the X-ray brightness ($\lesssim 10^7$ ergs/cm^2 sec) to vary approximately with the plasma pressure and the magnetic pressure and to decline with increasing time since emergence of the flux bundle through the visible surface of the Sun (Rosner, Tucker, and Valana, 1978). The decline with age is not expected for wave heating, of course. The two facts together suggest that the principal heating arises from the dissipation of magnetic free energy in the emerging quasi-static field, which gradually declines over characteristic times of $1\text{-}2 \times 10^5$ sec. That is to say, the heat input is a direct magnetic effect. So the question is how the magnetic energy is converted to heat. The high temperature ($T \sim 2\text{-}3 \times 10^6$ K) provides only a small resistive diffusion coefficient $\eta \sim 10^3$ cm^2/sec so that the resistive dissipation time ℓ^2/η over a scale ℓ is large. For $\ell \sim 5 \times 10^7$ cm the dissipation time is approximately 10^5 years. The question is whether there is any circumstance under which magnetic free energy with scale comparable to the characteristic granule scale $\ell \sim 5 \times 10^7$ cm can be dissipated in a characteristic time of the order of $10^4 - 10^5$ sec so as to provide the observed direct association between magnetic field and X-ray brightness. The fact of the observed X-ray corona indicates that there is an effective dissipation mechanism.

4. Dissipation of Magnetic Energy

The bipolar magnetic fields in which the X-ray corona resides are subject to continual internal deformation by the subsurface convection in which the fields are anchored. The characteristic length of the deformation is suggested by the coherence scale ℓ of the granules ($\ell \sim 5 \times 10^7$cm). The characteristic speed of the convection in the granules is 0.5×10^5 cm/sec or more (cf Sprüit, 1974) and the characteristic density is 10^{-6} gm/cm^3 over the same characteristic depth below the surface. The equipartition magnetic field $B_{eq} = (4\pi\rho v^2)^{\frac{1}{2}}$ is 170 gauss, suggesting that the magnetic fibrils of the mean 10^2 gauss field are freely moved around by the granules. The conclusion is simply that, with the expected shuffling and intermixing of the photospheric footpoints of the bipolar field, one expects that the field lines are wound and interwoven throughout the bipolar field, providing a substantial magnetic free energy. That is to say, the field lines wander stochastically about the mean field direction throughout the bipolar field above the visible surface. Presumably the field line topology is already stochastic as the successive Ω-loops emerge through the surface to form the bipolar region, which an observer with adequate telescopic resolution sees as a random motion of the footpoints during emergence. Further interweaving upon emergence is expected as a consequence of the convective transport of individual magnetic fibrils by the granules and supergranules.

Then note the basic theorem of magnetostatics which asserts that the Maxwell stresses in a stochastic field push the field toward internal surfaces of tangential discontinuity as the field relaxes to static equilibrium. That is to say, in the absence of resistivity, the ultimate asymptotic static equilibrium of the stochastic bipolar field involves current sheets of vanishing thickness and infinite current density (Parker, 1972, 1981a,b 1994d). This is the principal effect responsible for dissipating magnetic free energy in the bipolar fields to produce the active X-ray corona of the Sun and similar stars. (Parker, 1983a,b). The essential point is that the small but nonvanishing resistivity prevents the field from reaching the final ideal equilibrium state. The Maxwell stresses continually push the field toward the tangential discontinuity and the transverse field components are dissipated as rapidly as they are pressed into the current layer. The lower the resistivity the thinner and more intense the current layer.

To put it in different words the finite regions of continuous field on either side of the current layer are not parallel (because of the general stochastic deformation of the field) so that the two regions undergo continual rapid reconnection where they press together at the current layer. If the two misaligned regions of field press together over a width λ, a simple analysis of the fluid dynamics and field dissipation (Sweet, 1958; Parker, 1957) indicates that the characteristic thickness δ of the current sheet, which the Maxwell stress is striving to reduce to zero, is limited by the resistivity η to something of the order of $\lambda/N_L^{\frac{1}{2}} = (\lambda\eta/\Delta C)^{\frac{1}{2}}$, where $N_L = \lambda\Delta C/\eta$ is the Lundquist number for the Alfven speed $\Delta C = \Delta B/(4\pi\rho)^{\frac{1}{2}}$ in the transverse field component $\pm\Delta B$ on either side of the discontinuity. As an example suppose that $\lambda = \ell = 5 \times 10^7$ cm, $\Delta B = \frac{1}{4}B = 25$ gauss, $N = 10^{10}$ atoms/cm^3 and $\eta = 10^3$ cm/sec (for $T = 3 \times 10^6$ K). Then $\Delta C = 5 \times 10^7$ cm/sec and $N_L = 2.5 \times 10^{12}$, with the result that $\delta = 30$ cm, comparable to the ion cyclotron radius of the thermal ion in the mean field of 10^2 gauss.

The speed u at which the fields on either side are carried into the current sheet follows from energy considerations (Parker, 1957). The fluid is ejected along the current sheet of thickness δ at the velocity ΔC, so that conservation of fluid requires

$$u\lambda = \Delta C\delta$$

in order to magnitude. Thus, for instance, if $\delta = 30$ cm, the result is $u = 30$ cm/sec, cutting across λ in a characteristic time $\lambda/u = 1.6 \times 10^6$ sec.

In fact the thickness δ probably never gets as small as 30 cm because when δ declines to about 10^3 cm, the electron conduction velocity in the current sheet reaches the characteristic ion thermal velocity (200 km/sec) and plasma turbulence is generated by the interaction between the ions and the streaming electrons. This provides anomalous resistivity some 10^2 or more times larger than η and greatly enhances the dissipation, preventing δ from getting much smaller than 10^3 cm. At this point quantitative theoretical calculations become complicated, if not intractable, requiring the simultaneous computation of the anomalous resistivity, fluid motion, and magnetic field. It is interesting to note that the numerical simulations presently available show that a localized region of anomalous resistivity fosters development of the Petschek mode, in which the two regions of field come into contact only over a width λ' small compared to the characteristic scale λ. The result is an enhanced dissipation speed u characterized by $\Delta C / \ell n N_L \sim 0.01 \Delta C$. Even without anomalous resistivity $0.01\ \Delta C$ has the large value 5×10^5 cm/sec. Experience with laboratory Tokamak devices and with numerical simulations indicates that the dissipation becomes intermittent, with explosive bursts of merging interspersed with slower reconnection $u \sim \Delta C / N_L^{\frac{1}{2}}$. There is no way to provide a direct theoretical estimate of the mean dissipation or reconnection speed u. The observed characteristics of the corona indicate a dissipation time in the range $10^4 - 10^5$ sec, again suggesting the mean rate $u \sim 10^3$ cm/sec already noted. It is instructive, therefore, to examine the dissipation from another point of view.

5. Energy Input and Reconnection Rates

Consider the problem from the view that under steady state conditions the convective motions of the footpoints of the field must do work on the field at the average rate of 10^7 ergs/cm^2 sec at which the field dissipates to heat the X-ray corona. Suppose, then, that the fibril footpoints of a bipolar magnetic field are carried about at random, intermixing with individual velocities of the order of v. In the bipolar field above the photosphere the flux bundles associated with each magnetic fibril at the photosphere are intertwined along the length L of the bipolar field. After a time t the footpoint of an individual elemental flux bundle has traveled a distance vt along a random path among the neighboring flux bundles. Therefore after a time t the individual flux bundle makes an angle of the order of θ to the mean field where $tan\theta \sim vt/L$, i.e. the transverse field ΔB is of the order of Bvt/L. The Maxwell stress opposing the forward wandering v of the footpoints of the bundle is $B\Delta B/4\pi$ so the rate W at which v does work on the field is

$$W = vB\Delta B/4\pi$$
$$= (B^2/4\pi)v^2 t/L$$
$$= (B^2/4\pi)vtan\theta.$$

With $B = 10^2$ gauss and $v \sim 0.5$ km/sec, the necessary rate $W = 10^7$ ergs/cm^2 sec is achieved with $tan\theta = \frac{1}{4}$, for which $\Delta B \sim \frac{1}{4}B \sim 25$ gauss ($\theta \cong 14°$). For $L = 10^{10}$ cm, the time to build up to $vt/L = \frac{1}{4}$ is 5×10^4 sec, and correspondingly less for smaller L.

Alternatively the field may be in a stochastic topological state prior to emergence, in which case the motion of the footpoints during emergence is determined by the speed of rise of the fixed topology through the surface of the Sun. There is the additional effect that the expansion of the field upon rising above the visible surface draws further winding up from below as the field pushes toward a stable static equilibrium concentrating the winding in the region of weakest field, i.e. in the neighborhood of the apex of the bipolar field (Parker, 1974, 1975, 1979). We expect both of these effects to some unknown degree in addition to the continuing intermixing of the footpoints by the granules. It is clear upon reflection that the observation of the motion of the footpoints alone can establish the interweaving of the field lines of the bipolar field but cannot readily distinguish the contributions of these three different input phenomena.

Now the essential point is that the reconnection speed u is initially (if $\Delta B \sim 0$) too small to provide significant dissipation of the emerging interwoven magnetic fields. The cutting speed u across the developing tangential discontinuities increases as the Maxwell stresses decrease the thickness δ of the current sheets. As already noted, δ does not decline much below 10^3 cm (for which the cutting speed $u \sim \eta/\delta \sim 1$ cm/sec and the characteristic cutting time λ/u is $\sim 5 \times 10^7$ sec for $\lambda \sim 5 \times 10^7$ cm). For as δ declines below 10^3 cm the electron conduction velocity exceeds the ion thermal velocity and the ensuing plasma turbulence provides an enormous localized increase in the effective resistivity, with a tendency for the reconnection to evolve toward the rapid Petschek form. The result is a value of u perhaps as large as 10^6 cm/sec for brief periods. One expects that, in place of a quasi-steady state with $1 \lesssim u << 10^6$ cm/sec, there are bursts of high speed reconnection (nanoflares). The bursts are quenched as ΔB falls below some critical threshold for maintaining the anomalous resistivity. Following the burst there is a subsequent slow increase of ΔB to a level that again initiates anomalous resistivity and the rapid reconnection again, etc.

The energy input $P \sim 10^7$ ergs/cm^{2}sec inferred from the observed brightness combined with the foregoing estimate of 25 gauss for ΔB requires dissipation of the magnetic free energy density $L(\Delta B)^2/8\pi$ ergs/cm^2 in a characteristic time τ, given by the relation

$$P = \frac{L\,(\Delta B)^2}{\tau} \frac{}{8\pi}$$

so that $\tau \sim 10^4$ sec for $L = 5 \times 10^9$ cm, equivalent to a cutting speed $u = \lambda/\tau \sim 5 \times 10^3$ cm/sec. As already noted this lies between the extreme values of 1 cm/sec in the absence of anomalous resistivity and 10^6 cm/sec or more during a burst of reconnection fostered by anomalous resistivity.

Hence the heat input per unit volume may be comparable, with most of the total heat going into the corona.

An obvious question is the variation of the cutting speed u with height, particularly between the chromosphere (T$\sim$ 7×10^3 K, N $\sim$ 4×10^{12} atoms/cm^3) and the corona. Then the ratio α of the electron conduction velocity w to the ion thermal velocity $(kT/M)^{\frac{1}{2}}$ appears to be the critical parameter. It follows from Ampere's law that

$$w \sim c\Delta B/4\pi e N \delta$$

so that

$$\alpha = M^{\frac{1}{2}} c\Delta B/4\pi e \delta N^{\frac{1}{2}} (NkT)^{\frac{1}{2}}.$$

It is apparent that there is no strong variation of α upward through the corona. The transverse field component ΔB in a simple twisted flux bundles varies in proportion to $B^{\frac{1}{2}}$ (Parker, 1975), so that $\Delta B, N$, and NkT decrease upward through the chromosphere and only a detailed model can suggest the overall variation of α with height.

Between the corona and chromosphere the principal effect is the approximate continuity of NkT and ΔB and the smaller value of N and δ, with $\alpha \sim 1/N^{\frac{1}{2}} \delta$, in the corona. The first point is that one has no reason to expect anomalous resistivity to appear in the high electron density at chromosphere levels. In that case we estimate δ to be of the order of $\lambda/N_L^{\frac{1}{2}}$ in the chromosphere where the characteristic width λ of the flux bundle is about the same as in the corona and $N_L = \lambda \Delta C/\eta$ again. At 10^4 K the resistivity is $\eta \sim 10^7$ cm^2/sec. Chromospheric densities of N = 5×10^{12} atoms/cm^3 provide $\Delta C = 2.5 \times 10^6$ cm, with $\Delta B = 25$ gauss again. Putting $\lambda = 5 \times 10^7$ cm yields $N_L = 1.2 \times 10^6$ so that $\delta \cong 5 \times 10^4$ cm and $u = \Delta C/N_L^{\frac{1}{2}} = 2.5 \times 10^3$ cm/sec. The characteristic dissipation time λ/u is then 2×10^4 sec. This is to be compared with the estimated mean dissipation speed $u \sim 5 \times 10^3$ cm/sec and dissipation time of 10^4 sec estimated for the corona. It appears, then, that the reconnection in the chromosphere goes nearly as fast as reconnection in the corona in spite of the absence of anomalous resistivity. Hence the heat input per unit volume may be comparable with the input to the corona, with most of the heat going into the greater height of the corona.

Recalling that the magnetic field at photospheric levels consists mostly of widely separated magnetic fibrils of $1 - 2 \times 10^3$ gauss, it would appear that reconnection at photospheric levels is limited to occasional collisions between flux bundles. In fact, the individual flux bundles expand and come into contact with their neighboring bundles only at some distance up in the chromosphere (cf. Athay, 1981). So reconnection takes place only above some elevation in the chromosphere. There is not sufficient information available to determine that height, which probably has no unique fixed value in any case. The essential point is that chromospheric reconnection takes place over a height of the order of 10^8 cm, which is small compared to the height or length of the order of $10^9 - 10^{10}$ cm in the corona. This simple estimate indicates that the dissipation of magnetic free energy occurs principally in the corona.

6. Observational Tests

It is apparent from the foregoing discussion of solar variability, heating coronal holes, and heating the X-ray corona that there is a lot of theoretical conjecture backed up by relatively little direct observational evidence. The conjectures are not implausible, and few if any alternatives are presently available. But experience repeatedly shows nature to be more imaginative than ourselves, so it is essential that observational tests be carried out before it can be said that we understand the cause of the X-ray emission, brightness variations, etc. The serious physicist does not confuse conjecture with experimentally established fact.

The basic difficulties in providing an observational foundation for solar variability and suprathermal activity are the fact that there are no direct observations yet of conditions below the visible surface of the Sun and the fact that the action at the visible surface occurs on scales of the order of 10^2 km, which is well below the limit of resolution of ground based telescopes. It appears that most of the magnetic fibrils in the photosphere have diameters of the order of 10^2 km, and the internal structure of faculae plages, and sunspots has important characteristic lengths of the same order. So if we are to understand the dynamical nature of any of these structures, telescopic resolution to 0.1" is essential.

To treat the X-ray corona first, note that the individual current sheets and their nanoflares are too transparent to observe directly. So the available crucial test is the assumed intermixing of the footpoints of the bipolar magnetic fields at the visible surface of the Sun. For if the small fibrils move among each other with a speed v at some fraction of 1 km/sec, then the stochastic nature of the bipolar field above is established. This would guarantee the dynamical trend toward surfaces of discontinuity. The remaining observational question would be whether ΔB is large enough (i.e. reconnection rate u small enough) that $vB\Delta B/4\pi$ reaches 10^7 ergs/cm^2 sec. The crude theoretical estimates of §IV clearly need observational confirmation or correction.

Ideally one would wish to measure the magnitude of ΔB and the inclination $\Delta B/B$ directly with a vector magnetograph (cf. Bernasgoni, et al, 1995). However, it is easy to show that the intense photospheric fibril fields B_f are inclined only by the small fraction $B/B_f \lesssim 0.1$ of the inclination $\theta \sim 10° - 15°$ of the mean field in the corona above. So it appears doubtful that the detection of the various inclinations of individual fibrils is tractable.

Recent work by Alan Title has provided a ground based movie of the filigree (the unresolved bright points associated with each small magnetic fibril) showing rapid (0.5-2 km/sec) intermixing, as well as frequent splitting and regrouping. This observational milestone was shown unofficially at the Memphis meeting of the Solar Physics Division of the American Astronomical Society, 5-8 June, 1995. It clearly establishes the essential intermixing, thereby establishing the principal coronal heating as a consequence of the spontaneous current sheets and nanoflares of one form or other. These preliminary observations of filigree motion open up the possibility for extensive future observational studies and more quantitative theoretical treatment.

Ultimately the basic observation is simply the detection and telescopic resolution of the individual fibrils at photospheric levels to establish whether they do, or do not, intermix to provide the assumed stochastic field line topology in the bipolar fields on the Sun. Any such observations require clear resolution of the individual fibrils in order to follow the motions through occasional close approaches, where the two fibrils may or may not circle around each other before they part.

The essential observations require telescopic resolution of 75 km or better (≤ 0.1"). That means a diffraction limited mirror 1.2 m or more in diameter operating from a balloon or spacecraft above the terrestrial atmosphere. The telescopic observations in the visible and infra-red should be supplemented by at least occasional high resolution UV and X-ray observations, most easily accomplished by co-aligned optical and X-ray telescopes functioning as a single unit.

In fact such a solar "microscope" would provide basic information on several important aspects of solar activity. In particular the variations in the brightness of the Sun are in large part associated with the competition between sunspots and plages (Foukal and Lean, 1986). Yet the nature of the plage is not presently understood nor are the reasons for the formation of sunspots known. It has been suggested (Parker, 1994b,c, 1995a-b) that the general increase in brightness is supplied by the enhanced convection associated with the emerging Ω-loops of azimuthal magnetic field. These emerging fields provide the bipolar magnetic fields in which the X-ray corona forms, so there is a direct association between X-ray emission and a general increase in brightness. The point is that the structure and behavior of plages, sunspots, and emerging magnetic fields lies at scales of 100 km or less. So the dynamics can be studied, and hopefully understood, only with a suitable solar microscope. There is a whole new world of physics with direct implications for X-ray astronomy, solar variability, and the terrestrial climate waiting to be discovered by the solar microscope. That world is waiting for its Anton van Leeuwenhoek to produce the microscope to discover it.

Acknowledgements. This work was supported in part by the National Aeronautics and Space Administration through NASA grant NAGW 2122.

References

Alfven, H. (1950) *Cosmical Electrodynamics*, Oxford, Clarendon Press, p. 151.

Athay, R.G. (1981) Chromospheric-corona transition region models with magnetic field flow, *Astrophys. J.* **249**, 340.

Baliunas, S.L. (1991) *The past, present, and future of solar magnetism: Stellar magnetic activity,* in *The Sun in Time*, Tucson, University of Arizona Press, ed. C.P. Sonett. M.S. Giampapa, and M.S. Matthews, p. 809.

Baliunas, S.L. and Jastrow, R. (1990) Evidence for long-term brightness changes of solar-type stars, *Nature* **348**, 520.

Baliunas, S. and Soon, W. (1995) Are variations in the length of the activity cycle related to changes in brightness of solar type stars? *Astrophys. J.* (in press).

Baliunas, S.L. et al, (1985) Time-series measurements of chromospheric Ca II. H and K emission in cool stars and the search for differential rotation, *Astrophys. J.* **294**, 310.

Baliunas, S.L. et al, (1995) Chromospheric variations in main-sequence stars. II, *Astrophys. J.* **438**, 269.

Beer, J., Blinov, A. Bonani, G. Finkel, R.C., Hofmann, H.J., Lehmann, B., Oeshger, H., Sigg, A., Schwander, J., Staffelbach, T., Stauffer, B., Suter, M., and Wölfli, W. The 11-year cycle of solar activity. *Nature* **347**, 164.

Bernasconi, P.N., Keller, C.U., Povel, H.P. and Stenflo, J.O. (1995) Direct measurements of flux tube inclinations in solar plages. *Astron. Astrophys.* (in press).

Biermann, L. (1946) The meaning of chromospheric turbulence and the UV excess of the Sun, *Naturwischenschaften* **33**, 118.

Billings, D.E. (1966) *A Guide to the Solar Corona*, New York, Academic Press.

Chapman, G.A. (1987) Variations of solar irradiance due to magnetic activity, *Ann. Rev. Astron. Astrophys.* **25**, 633.

Collins, W. (1992) The theory of magnetohydrodynamic wave generation by localized dissipation of far-field waves, *Astrophys. J.*, **384**, 319.

Damon, P.E. and Sonett, C.P. (1991) *Solar and terrestrial components of the atmospheric ^{14}C variation spectrum,* in *The Sun in Time*, Tucson, University of Arizona Press, ed. C.P. Sonett, M.S. Giampapa, and M.S. Matthews, p. 360.

Davila, J.M. (1987) Heating of the solar corona by the resonant absorption of Alfven waves, *Astrophys.J.* **317**, 514.

Eddy, J.A. (1976) The Maunder Minimum, *Science*, **192**, 1189.

Eddy, J.A. (1977a) Climate and the changing Sun, *Clim. Change* I, 173.

Eddy, J.A. (1977b) *Historical evidence for the existence of the solar cycle,* in *The Solar Output and Its Variation*, Boulder, Colorado Associated University Press, ed. O.R. White, p. 51.

Eddy, J.A. (1983a) The Maunder Minimum, A reappraisal, *Solar Phys.*, **89**, 195.

Eddy, J.A. (1983b) *An historical review of solar variability, weather, and climate,* in *Weather and Climate Responses to Solar Variations*, Boulder, Colorado Associated University Press, ed. B.M. McCormack, p. 1.

Foukal, P., Harvey, K. and Hill, F. (1991) Do changes in the photospheric magnetic network cause the 11-year variation of total solar irradiance? *Astrophys. J. Lett.* **383**,

L89.

Foukal, P. and Lean, J. (1986) The influence of faculae on total solar irradiance and luminosity, *Astrophys. J.* **302**, 826.

Foukal, P. and Lean, J. (1990) An empirical model of total solar irradiance variation between 1874 and 1988, *Science* **247**, 556.

Fris-Christensen, E. and Lassen, K. (1991) Length of the solar cycle: An indicator of solar activity closely associated with climate, *Science* **254**, 698.

Fröhlich, C. (1994) *Irradiance observations of the Sun, in The Sun as a Variable Star*, Cambridge, Cambridge University Press, ed. J.M. Pap, C. Fröhlich, H.S. Hudson, and S.K. Solanki, p. 28.

Fröhlich, C. Foukal, P.V., Hickey, J.R., Hudson, H.S. and Willson, R.C. (1991) *Solar irradiance variability with modern instruments, in The Sun in Time*, Tucson, University of Arizona Press, ed. C.P. Sonett, M.S. Giampapa, and M.S. Matthews, p. 11.

Geiss, J. and Bochsler, P. (1991) *Long time variations in solar wind properties: Possible causes versus observations, in The Sun in Time*, Tucson, University of Arizona Press, ed. C.P. Sonett, M.S. Giampapa, and M.S. Matthews, p. 98.

Hameed, S. and Gong, G. (1994) Variation of spring climate in lower-middle Yangtse River Valley and its relation with solar-cycle length *Geophys. Res. Lett.* **21**, 2693.

Hameed, S. and Wyant, P. (1982) Twenty three year cycles in surface temperatures during the Maunder Minimum, *Geophys. Res. Lett.* **9**, 83.

Hollweg, J.V. (1987) Resonance absorption of magnetohydrodynamic surface waves: Viscous effects, *Astrophys. J.* **320**, 875.

Hoyt, D.V., Kyle, H.L., Hickey, J.R. and Maschhoff, R.H. (1992) Nimbus-7 solar total irradiance: A new algorithm for its derivation, *J. Geophys. Res.* **97**, 51.

Hoyt, D.V., Schatten, K.H. and Nesmes-Ribes, E. (1994) *A new reconstruction of solar activity, 1610-1993, in The Solar Engine and its Influence on Terrestrial Atmosphere and Climate*, Berlin, Springer-Verlag, ed. E.Nesmes-Ribes, p. 57.

Hudson, H.S. (1988) Observed variability of the solar luminosity, *Ann. Rev. Astron. Astrophys.* **26**, 473.

Kuhn, J.R. and Libbrecht, K.G. (1991) Nonfacular solar luminosity variations, *Astrophys. J. Lett.* **381**, L35.

Lean, J., Skumarich, A., White, O.R. and Rind, D. (1994) *Estimating solar forcing of climate change during the Maunder Minimum, in The Sun as a Variable Star*, Cambridge, Cambridge University Press, ed. J.M. Pap, C. Fröhlich, H.S. Hudson, and S.K. Solanki, p. 236.

Lockwood, C.W., Skiff, B.A., Baliunas, S.L. and Radick, R.R. (1992) Long term solar brightness changes estimated from a survey of Sun-like stars *Nature*, **360**, 653.

Lou, Y.Q. (1995) Gravity waves in the lower solar corona, *Astrophys. J.* **442**, 401.

Martin, S. (1988) The indentification and interaction of network, intranetwork, and ephemeral-region magnetic fields *Solar Phys.* **117**, 243.

Maunder, E.W. (1984) A prolonged sunspot minimum, *Knowledge* **17**, 173.

Maunder, E.W. (1922) The prolonged sunspot minimum, 1645-1715, *J. Brit. Astron. Assoc.* **32**, 140.

Osterbrock, D.E. (1961) Heating of the solar chromosphere, plages and corona by magnetohydrodynamic waves, *Astrophys. J.* **134**, 347.

Parker, E.N. (1957) Sweet's mechanism for merging magnetic field in conducting fluids, *Geophys. Res.* **62**, 509.

Parker, E.N. (1972) Topological dissipation and the small-scale fields in turbulent gases, *Astrophys. J.* **174**, 499.

Parker, E.N. (1974) The dynamical properties to twisted ropes of magnetic field and the vigor of active regions on the Sun, *Astrophys. J.* **191**, 245.

Parker, E.N. (1975) X-ray bright spots on the Sun and the non-equilibrium of a twisted flux tube in a stratified atmosphere, *Astrophys. J.* **201**, 494.

Parker, E.N. (1979) *Cosmical Magnetic Fields*, Oxford, Clarendon Press, pp. 175-200.

Parker, E.N. (1981a) The dissipation of inhomogeneous magnetic fields and the problem of coronas. I. Dislocation and flattening of flux tubes, *Astrophys. J.* **244**, 631.

Parker, E.N. (1981b) The dissipation of inhomogeneous magnetic fields and the problem of coronas. II. The dynamics of dislocated flux tubes, *Astrophys. J.* **244**, 644.

Parker, E.N. (1983a) Magnetic neutral sheets in evolving fields. I. General theory, *Astrophys. J.* **264**, 635.

Parker, E.N. 91983b) Magnetic neutral sheets in evolving fields. II. Formation of the solar corona, *Astrophys. J.* **264**, 692.

Parker, E.N. (1991) Heating solar coronal holes, *Astrophys. J.* **372**, 719.

Parker, E.N. (1994a) Summary comments, in *The Solar Engine and its Influence on Terrestrial Atmosphere and Climate*, Berlin, Springer-Verlag, ed. E. Nesme-Ribes, p. 527.

Parker, E.N. (1994b) *Theoretical interpretation of magnetic activity, in The Sun as a Variable Star, Solar and Stellar Irradiance Variations*, Proc. IAU Symp. No. 143, Boulder, Colorado, 1993, Cambridge, Cambridge University Press, ed. J.M. Pap, C. Fröhlich., H.S. Hudson, and S.K. Solanki, p. 264.

Parker, E.N. (1994c) Theoretical properties of Ω-loops in the convective zone of the Sun. I. Emerging bipolar magnetic regions, *Astrophys. J.* **433**, 867.

Parker, E.N. (1994d) *Spontaneous Current Sheets in Magnetic Fields*, New York, Oxford University Press.

Parker, E.N. (1995a) Theoretical properties of Ω-loops in the convective zone of the Sun. II. The origin of enhanced solar irradiance, *Astrophys. J.* **440**, 415.

Parker, E.N. (1995b) Theoretical properties of Ω-loops in the convective zone of the Sun. III. Extended updrafts,; *Astrophys. J.* **442**, 405.

Parker, E.N. (1995c) Theoretical properties of Ω-loops in the convective zone of the Sun. IV. Stability of updrafts, *Astrophys. J.*, 1 August.

Parker, E.N. (1995d) Theoretical properties of Ω-loops in the convective zone of the Sun. V. Coriolis force and the centrifugal potential barriers, *Astrophys. J.*, (submitted for publication).

Piddington, J.H. (1956) Solar atmospheric heating by hydromagnetic waves, *Mon. Not. Royal Astron. Soc.* **116**, 314.

Porter, J.G. and Moore, R.L. (1988) in Proc. 9th Sacramento Peak Summer Symposium, 1987, August 17-21, *Sunspot, NM: NSO/ Sacramento Peak*, ed. R.C. Altrock, p. 30.

Radick, R.R., Lockwood, C.W. and Balunas, S.L., (1990) Stellar activity and brightness variations: A glimpse of the Sun's history, *Science* **247**, 39.

Ribes, J.C. and Nesmes-Ribes, E. (1993) The solar sun-spot cycle in the Maunder Minimum, AD 1645 to AD 1715, *Astron. Astrophys.* **276**, 549.

Rosner, R., Tucker, W.H. and Valana, G.S. (1978) Dynamics of the quiescent solar corona, *Astrophys. J.* **220**, 643.

Schatzman, E. (1949) The heating of the solar corona and chromosphere, *Ann. d. Astrophys.* **12**, 203.

Schlesinger, M.E. and Ramankutty, N. (1992) Implications for global warming of intercycle solar irradiance variations, *Nature* **360**, 330.

Schuurman, C.J.E. (1991) Changes of coupled troposphere and lower stratosphere after solar activity events, *J. Geomag. Geo. electri. suppl.* **43**, 767.

Schwarzschild, M. (1948) Noise arising from solar granulation, *Astrophys. J.* **107**, 1.

Soon, W.H., Baliunas, S.L. and Zhang, Q. (1994) *A technique for estimating long-term variations of solar total irradiance: Preliminary estimates based on observations of the Sun and solar-type stars*, in *The Solar Engine and its Influences on Terrestrial Atmosphere and Climate*, Berlin, Springer-Verlag, ed. E. Nesme-Ribes, p. 133.

Sprüit, H.C. (1974) A model of the solar convection zone, *Solar Phys* **34**, 277.

Sprüit, H.C. (1994) *Theories of radius and luminosity variations, in The Solar Engine and its Influence on Terrestrial Atmosphere and Climate*, Berlin, Springer-Verlag, ed. E. Nesmes-Ribes, p. 107.

354

Sweet, P.A. (1958) The production of high energy particles in solar flares, *Nuovo Cimento Suppl.* **8**, Ser. X, 188.

Wang, Y.M. (1994a) Polar plumes and the solar wind, *Astrophys J. Lett.* **435**, L153.

Wang, Y.M. (1994b) Two types of slow solar wind, *Astrophys. J. Lett.* **437**, L67.

Willson, R.C. and Hudson, H.S. (1991) The Sun's luminosity over a complete cycle, *Nature*, **351**, 42.

Withbroe, G.L. (1988) The temperature structure, mass, and energy flow in the corona and inner solar wind, *Astrophys. J.*, **325**, 442.

Withbroe, G.L. and Noyes, R.W. (1977) Mass and energy flow in the solar chromosphere and corona, *Ann. Rev. Astron. Astrophys.*, **15**, 363.

Zhang, Q., Soon, W.H., Baliunas, S.L., Lockwood, G.W., Skiff, B.A. and Radick, R.R. (1994) A method for determining possible brightness variations of the Sun in past centuries from observations of solar-type stars, *Astrophys, J. Lett.*, **427**, L111.

PART II:
ASTROPHYSICAL MAGNETOHYDRODYNAMIC FLOWS

HUBBLE SPACE TELESCOPE OBSERVATIONS
OF GALACTIC AND EXTRAGALACTIC JETS

JOHN A. BIRETTA
Space Telescope Science Institute
Baltimore, Maryland, USA
biretta@stsci.edu

1. Introduction

Jets, or narrow, high-velocity plasma streams, are ubiquitous in the universe. Our own galaxy, just in the region near Earth, contains over a dozen jets which are formed by very young or old stars accreting nearby material. These "galactic jets" typically consist of neutral or ionized atomic material, and are visible due to emission lines of common elements such as hydrogen, oxygen, and sulfur. They extend over distances up to of order 10^4 AU (1 AU $= 1.5 \times 10^{13}$ cm) before dissipating or being stopped by the tenuous interstellar medium. Beyond the reaches of our own galaxy are numerous "extragalactic jets" which are formed in the centers of distant galaxies. Accretion on to super-massive black holes (mass of order 10^9 solar masses) are thought to power these jets, though until recently there has been little direct evidence for this. These jets are thought to consist of ionized plasma containing electrons and positive particles (either protons or positrons) in a magnetic field. They shine via synchrotron emission from the most energetic electrons, which have Lorentz factors $\gamma > 10^3$. The extragalactic jets are very energetic, and have flow speeds approaching the velocity of light (c). They may extend over distances exceeding 10^6 parsecs (1 pc $= 3 \times 10^{18}$ cm) from the host galaxy.

The Hubble Space Telescope (*HST*) is a 2.4 meter aperture telescope in near-earth orbit. Since it is unaffected by atmospheric turbulence, it is able to provide diffraction-limited imaging with a resolution of ~ 0.06 arcseconds in visible light, which is more than an order of magnitude better than is attainable from ground-based telescopes. Furthermore, it is unaffected by atmospheric absorption, and can observe at wavelengths as short as $\lambda \sim 1200$ Å in the ultraviolet, with a diffraction limited resolution ~ 0.03

357

K. C. Tsinganos (ed.), Solar and Astrophysical Magnetohydrodynamic Flows, 357–382.
© *1996 Kluwer Academic Publishers.*

arcseconds. After its initial launch, the primary mirror of *HST* was found to suffer spherical aberration. Corrective optics were installed in December 1993, and these restored the telescope to its full "as designed" capabilities.

There are two cameras on board *HST* which provide the images discussed here. The Faint Object Camera covers a wavelength range from 1200 to 5000Å. Due to its very fine pixel scale (0.014 arcseconds per pixel) and UV capability, it provides the highest angular resolution images of all the instruments on HST. It is an imaging photon-counting device, with a multi-stage image-intensifier whose output is read by a television camera tube. The second camera is the Wide Field / Planetary Camera, which was replaced by the Wide Field Planetary Camera 2 during the service mission. These both provide imaging from 3000 to 9000Å, with either 0.046 or 0.1 arcsecond pixels. Due to the larger pixel size and large area of its Charge-Coupled Device (CCD) detectors, it offers a much wider field of view than the FOC. With its unprecedented resolution, and its imaging capability at UV wavelengths, *HST* is providing insights to the formation, structure, and kinematics of both galactic and extragalactic jets.

The balance of this paper is divided into the following sections: (2) Observations of *galactic jets*, including jets from young stellar objects and older, symbiotic stars; (3) Observations of *extragalactic jets*, principally those in M87 and 3C273, as well as several optical jets discovered by *HST*; (4) Problems posed by the observations; and (5) Future prospects for *HST* observations.

2. Galactic Jets

The high resolution of *HST* allows us to probe the region near the jet's origin and determine the properties of the jet-forming object. The properties of the jet itself, its width or degree of collimation, the detailed morphology and motions of its knots, and its emission line species, all give us additional clues about the process of jet formation. Finally, the interaction of the outer reaches of the jet with the surrounding medium gives us further constraints on the jet's velocity and kinetic energy.

2.1. JETS FROM YOUNG STELLAR OBJECTS

Young Stellar Objects (YSOs) are a common site of jet formation. These newly formed stars are still imbedded in their parent molecular cloud. Accretion of this molecular material onto the star is a probable energy source for jet formation. Usually a pair of oppositely directed jets is seen spanning the YSO (i.e. these jets are two-sided). Bright emission line regions, known as Herbig-Haro objects, form where the outer reaches of the jet impinge

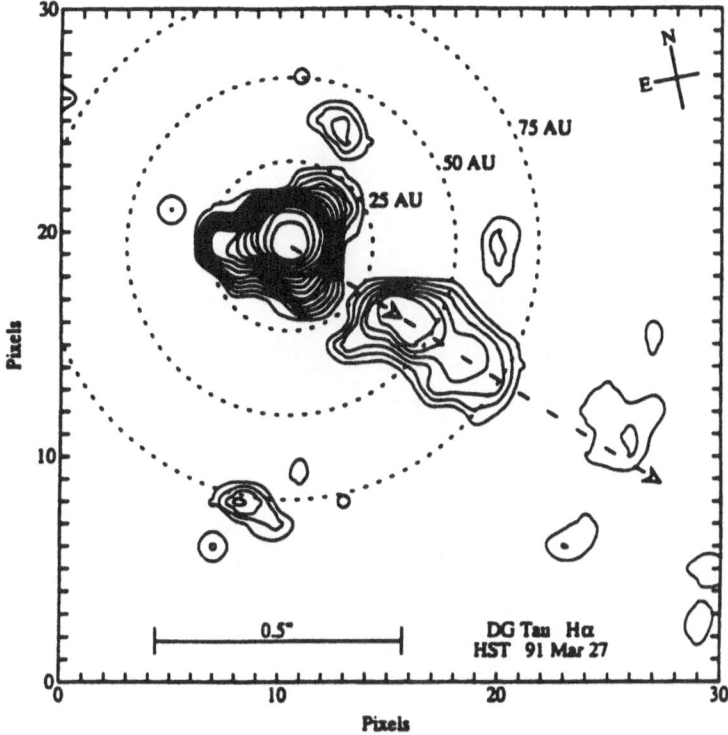

Figure 1. Contour plot of WF/PC-1 image of DG Tau and jet in Hα light. Dotted circles indicate radii from central star in AU. From Kepner et al. 1993.

on surrounding interstellar material. These Herbig-Haro objects typically contain bow shocks which are visible in collisionally excited emission lines.

Different lines are excited under different conditions, and hence the lines provide a diagnostic of the physical conditions within the jets and in their surroundings. In particular, [SII] 6716 and 6731Å lines are excited at weak shocks, with collision speeds in the range 15 - 30 km sec^{-1}. In contrast, [OIII] 5007Å emission is excited in strong shocks with collision speeds exceeding 100 km sec^{-1}. Emission in Hα 6563Å is seen at intermediate shock velocities. A separate review by Prof. Ray in this volume discusses the physics of YSO jets in detail; here we will concentrate on observational material. (See also Ray and Mundt 1993; Reipurth and Heathcote 1993).

We will review six young stellar jets where *HST* has made important contributions, proceeding from the smallest structures to the largest.

The first object is DG Tau, a bright and highly active T-Tauri star. At a relatively nearby distance of 150 parsecs, it offers an opportunity to study the innermost regions of the jet at high linear resolution. Figure 1

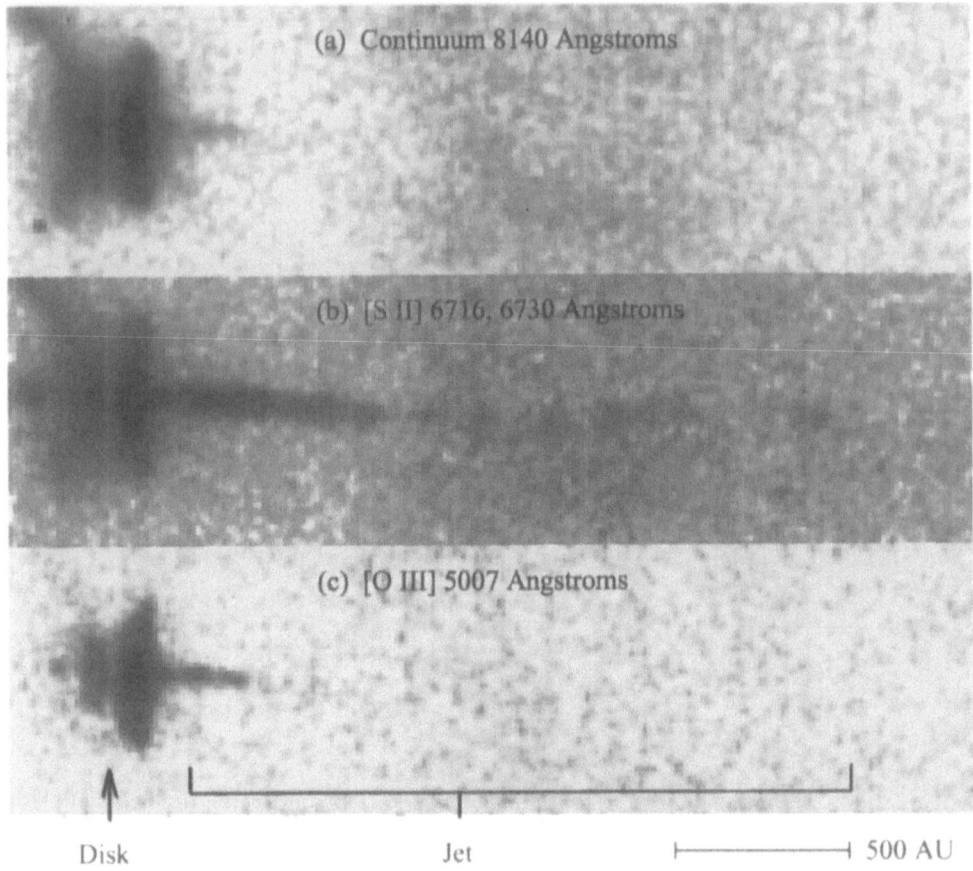

Figure 2. WFPC2 image of HH-30 disk and jet in: (A) ∼8140Å continuum light, (B) Light of low excitation [SII] lines, and (C) Light of high excitation [OIII] lines. The disk faces are visible, while the center of the disk is obscured by dust. The jet is seen primarily in the low excitation [SII] emission. From Burrows et al. 1996.

shows a contour plot of an Hα image obtained with WF/PC. This image clearly shows a jet-like structure which is well collimated at 40 AU from the central star (Kepner et al. 1993). The jet is unresolved across its width, indicating an opening angle of < 10°. Hence, most of the jet collimation must occur very near to the central star.

Herbig-Haro 30 (HH-30) is at a similar distance and is shown in Figure 2. This image shows, for the first time, an accretion disk surrounding the newly forming star (Burrows et al. 1996). The disk appears nearly edge-on, and its central plane is darkened due to dust absorption. The surfaces of the disk are illuminated by the central star and are highly visible. Accretion processes have long been thought to be at the root of jet formation; this image clearly supports these theories.

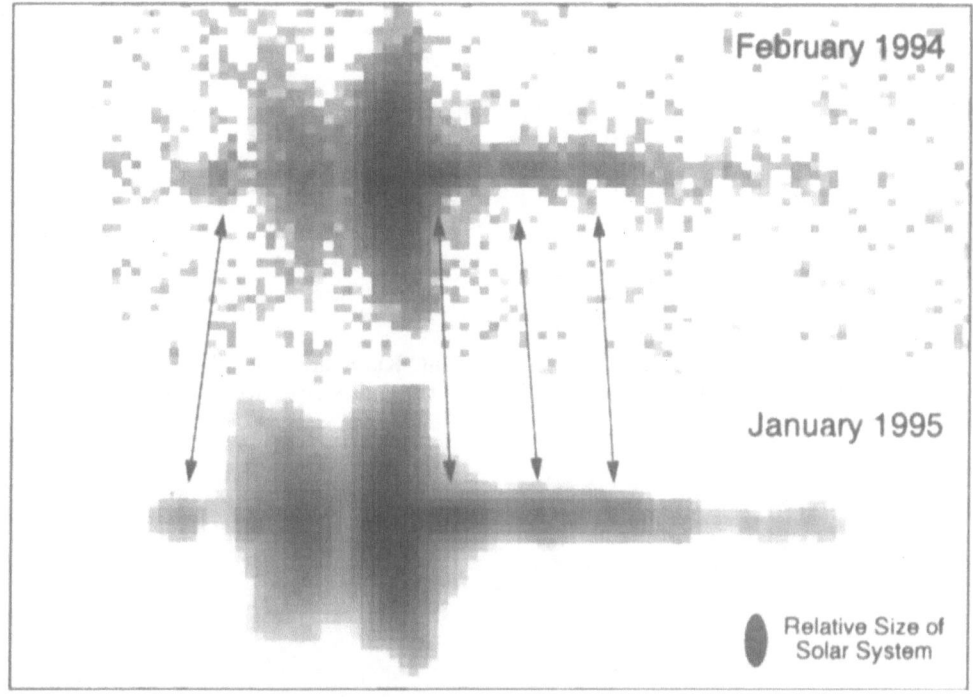

Figure 3. Comparison of [SII] images of the HH-30 jets taken in February 1994 and January 1995. Knots in the jet appear to move outward at ~ 200 km sec^{-1}. From Burrows et al. 1996.

On either side of the disk we see a narrow, knotty jet which can be traced to about 1500 AU. The jet shines mostly in the light of [SII] 6716 and 6731Å emission, indicating only very weak shocks within the jet, with collision velocities of a few dozen km sec^{-1}. Very little high excitation [OIII] 5007Å emission is seen from the jet. The jet also has a slight curvature, perhaps due to precession of the engine's rotation axis.

A second set of observations, taken a year later, show that condensations in the jet and opposite "counter-jet" are moving outward at 200 km sec^{-1} (Figure 3). The lack of high excitation [OIII] 5007Å emission shows that the flow within the jet must be relatively smooth and coherent. The [SII] emission must arise from relatively weak shocks in this rapid flow.

Herbig-Haro 34 (HH-34) is about three times more distant, at ~ 450 parsecs. Again, this jet shines mostly in the light of low excitation [SII] emission. Here, for the first time, we are able to see the detailed structure of condensations within the jet (Figure 4; Hester et al. 1996). The bright condensations appear to be miniature bow shocks, such as would be formed by dense clouds moving through slower, more tenuous jet material ejected at an earlier time. This indicates the flow in the jet is not steady, but instead

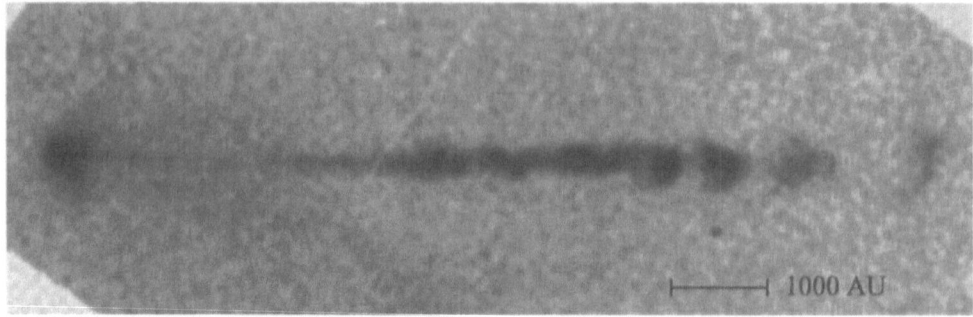

Figure 4. WFPC2 image of HH-34 jet in [SII] light. Knots in the jet contain bow shocks, indicating they consist of dense material moving through slower material of less density. From Hester et al. 1996.

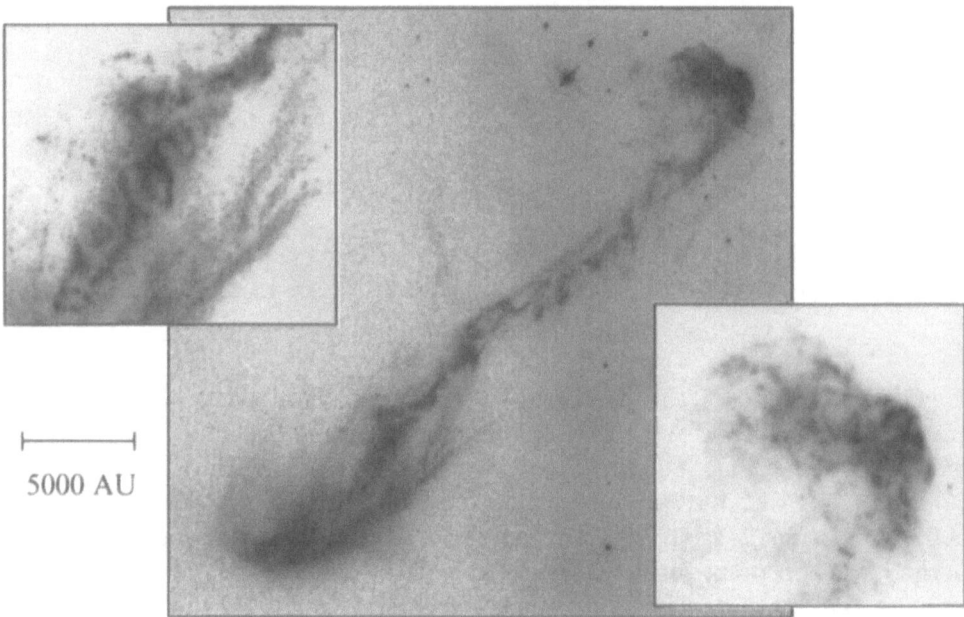

Figure 5. Images of HH-47 jet taken with WFPC2 in [SII] light. Jet appears to be twisted in a helical pattern. The central star is located near the bottom left, but is obscured by dust. Insets show details of jet kinks and terminus. From Morse et al. 1995.

undergoes brief episodes of higher density and/or velocity. (See also Morse 1992.)

Herbig Haro 47 (HH-47) is at a similar distance and shows a very complicated jet (Figure 5; Morse et al. 1995). Here the jet is not straight, but appears to wiggle from side-to-side in a helical pattern. This might be caused by wobbling of the engine's rotation axis, perhaps due to the influence of a companion star. Again, the jet shines primarily in low excitation [SII] lines. High excitation [OIII] emission is seen at the end of the jet, where it burrows into surrounding interstellar gas, indicating a jet velocity exceeding of 100 km sec^{-1}.

Finally, HH-1 and HH-2 illustrate many details of the jet/interstellar medium interaction. Here, a narrow jet is seen emanating from the engine, which is obscured by a dark dust cloud near the middle of the source. (Figures 6 and 7; Hester et al. 1996). Strong bow shocks are seen in [OIII] emission at the ends of both jets, where the high-speed jet material encounters slow-moving or stationary material. (See also Schwartz et al. 1993; Schwartz 1994.)

2.2. JETS FROM OLDER, SYMBIOTIC STARS

While jets are common in young stars, they are relatively rare in older stars. This is presumably because the molecular material that fuels accretion has been blown away during early stages of the star's evolution. An exception is symbiotic stars, where two stars orbit each other and exchange material. Late in their lifetimes, stellar envelopes expand to many times their "main sequence" diameter, and hence aged stars provide material which can be accreted onto a companion. Stars can also produce winds late in their lives, and these winds are an alternate source of accretion material.

The best example of a jet from an older symbiotic star is R Aquarii (Hollis and Michalitsianos 1993; Paresce and Hack 1994; Hack 1996). This system contains a Mira-type star, and is located at a distance of only $\sim$ 200 parsecs. Images taken with the *HST* Faint Object Camera show a two-sided jet with an opening angle of < 15°. The highest resolution images taken in the ultraviolet (Figure 8) show the jet is well collimated within $\sim$ 20 AU of the central star. This is the closest distance to the central source that a jet has ever been detected.

To summarize this section, *HST* observations provide information about the engine, structure, and terminus of jets from YSOs and symbiotic stars. *HST* has provided the first detection of an accretion disk (in HH-30), which confirms the basic picture of an accretion-driven jet. Jets from both YSOs and symbiotic stars are already well collimated at distances of 20 to 40 AU from the central star, indicating the engine is responsible for nearly all of the collimation; external pressures must contribute little to the collimation. The YSO jets show low velocity, collisionally excited lines of [SII] and yet exhibit motions $\sim$ 200 km sec^{-1}; this indicates the shock velocities are small compared to flow speeds, and implies relatively coherent flow with only weak internal shocks. Curvature and "wiggles" in jets (e.g. HH-30 and HH-47) indicate precession of the ejection axis, perhaps due to a companion

364

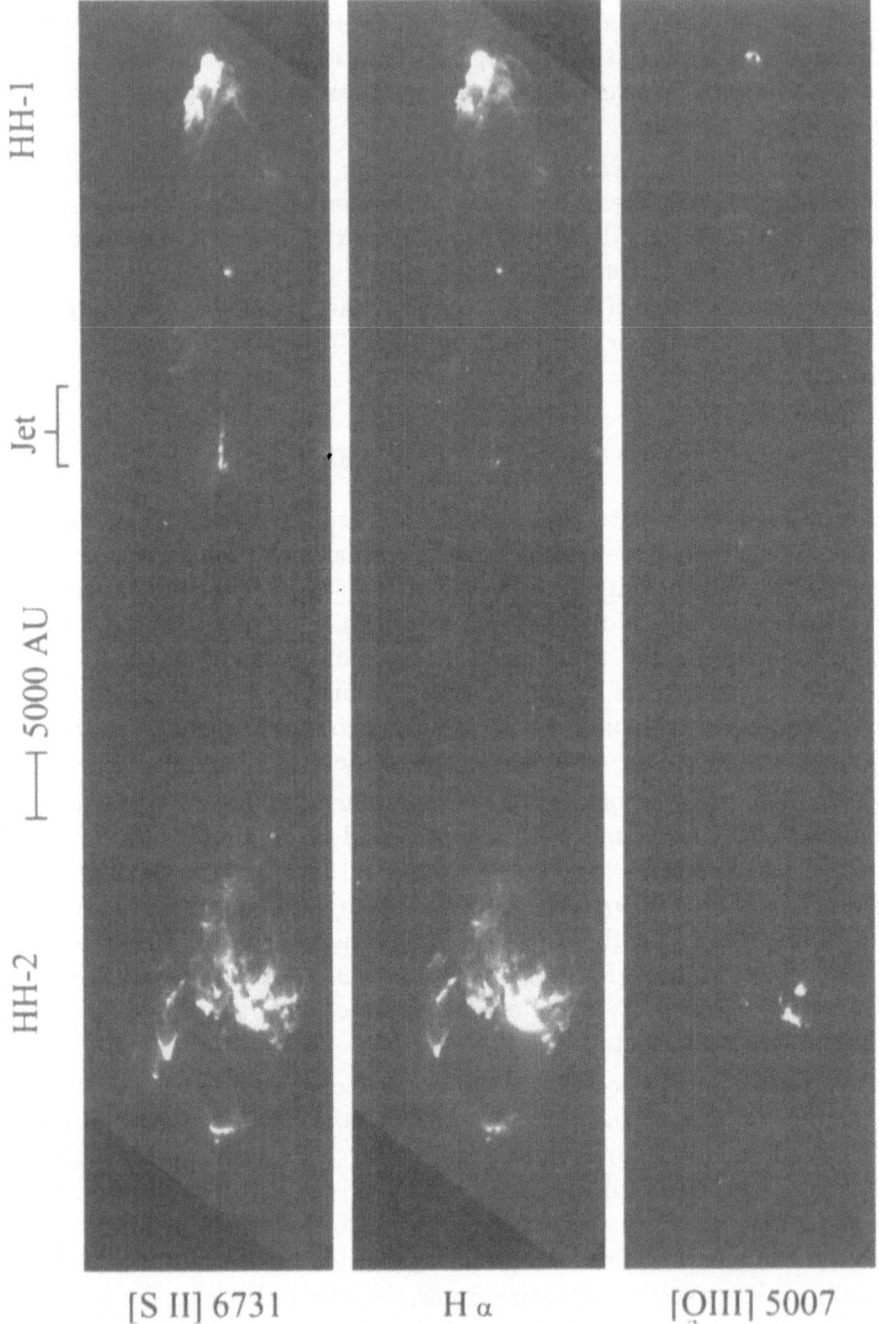

[S II] 6731   H α   [OIII] 5007

Figure 6. Images of HH-1 and HH-2 taken with WFPC2 in [SII], Hα, and [OIII] light. The jet near the center shines predominantly in low-excitation [SII] lines. Bow shocks at jet terminae are bright in [OIII] emission, indicating high-velocity shocks. The YSO is located below the jet, but is obscured by dust. Original data from Hester et al. 1996.

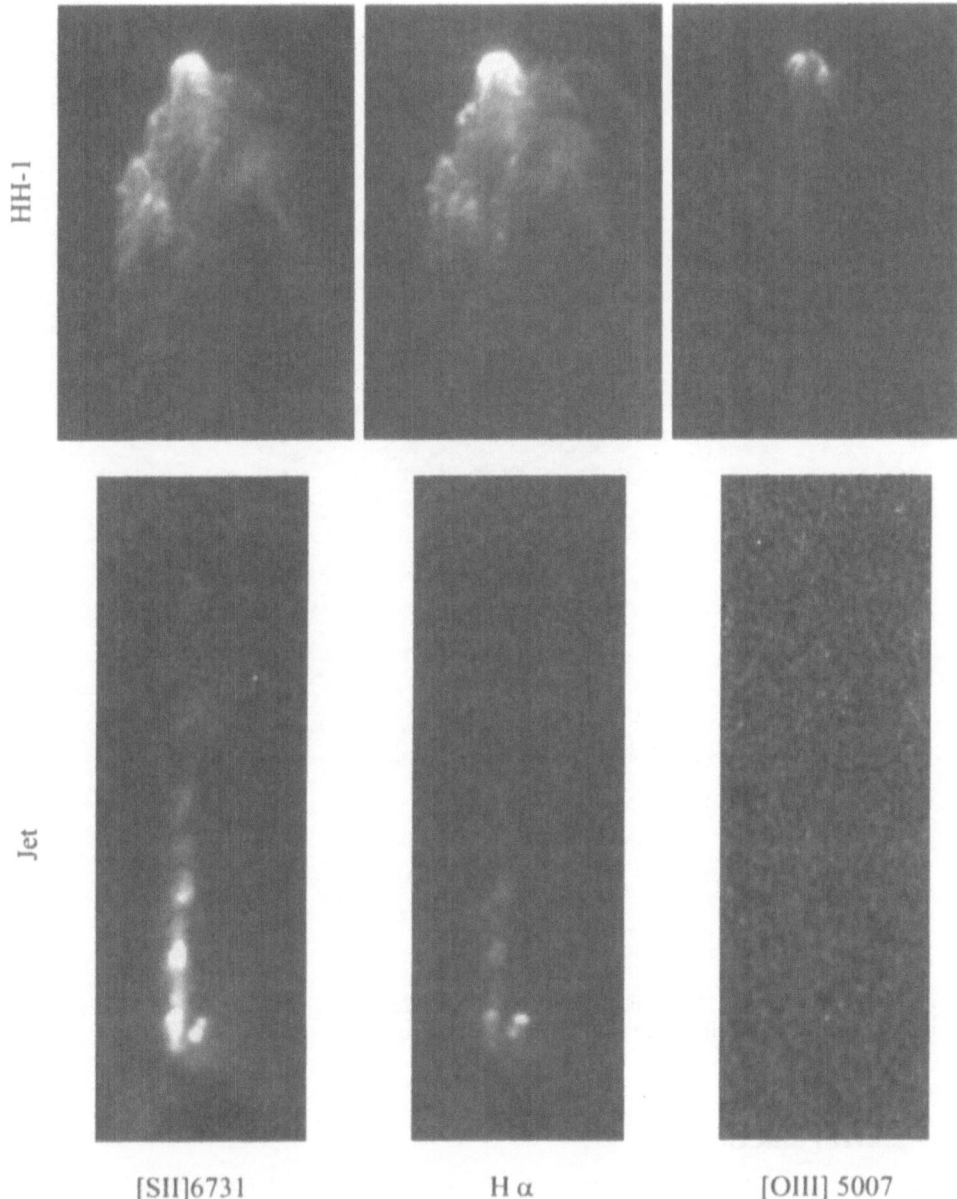

HH-1

Jet

[SII]6731   H α   [OIII] 5007

Figure 7.   Detail of HH-1 and jet (see previous Figure). Intensity scale is logarithmic. Original data from Hester et al. 1996.

of the central star. Finally, observations of the jet terminae give evidence for classical bow shock morphology, with high excitation lines [OIII] consistent with jet flow speeds exceeding ~ 100 km sec^{-1} in YSOs.

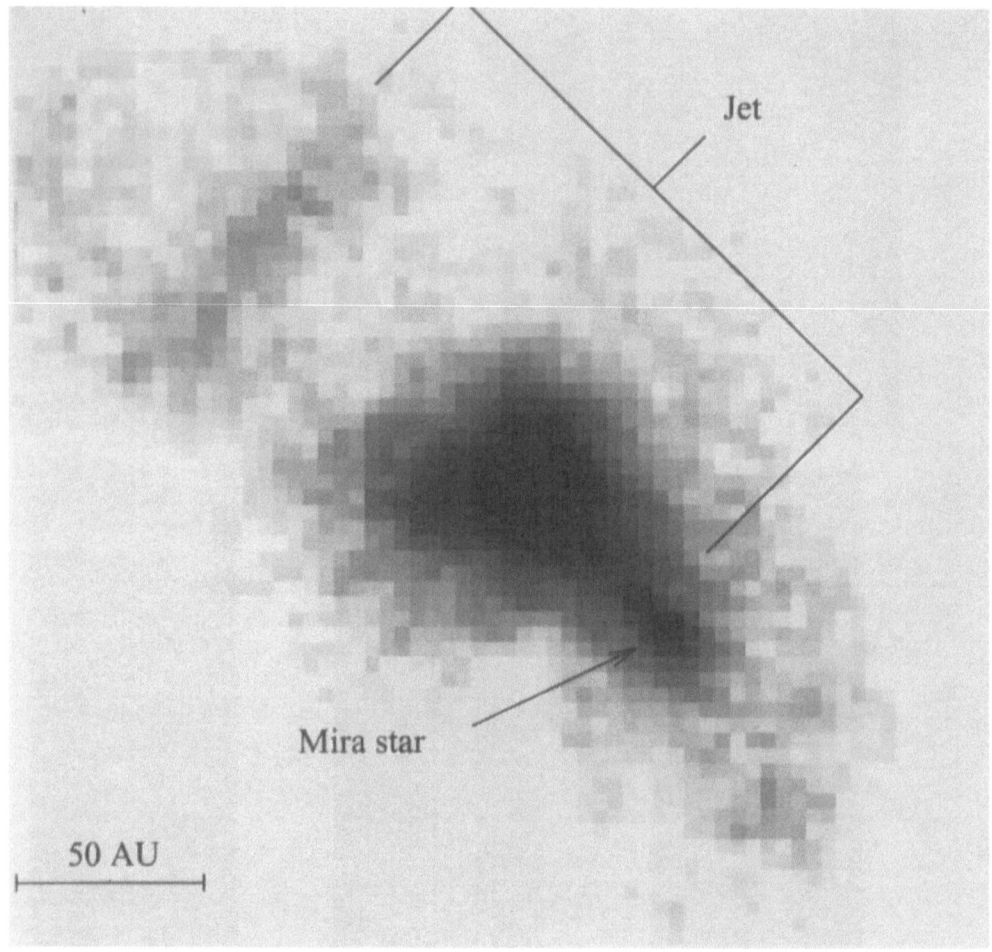

Figure 8. Image of R Aquarii obtained with the *HST / FOC* in the C III] emission line. The location of the Mira-type star is indicated. From Paresce and Hack 1994.

3. Extragalactic Jets

Massive galaxies produce jets which are in many ways similar to the stellar jets, but are greatly scaled-up in terms of energy. Extragalactic jets have speeds which approach the velocity of light, and can extend for up to 10^6 parsecs from the host galaxy. Jets produced in the centers of "active" galaxies vary in length from 2 to > 50 kpc. They shine by synchrotron radiation, presenting a featureless continuum caused by high energy electrons (Lorentz factor $\gamma > 1000$) spiraling in weak magnetic fields ($B \sim 10^{-4}$ Gauss).

HST is well-suited to the study of synchrotron jets, as it permits observation of very high energy electrons ($\gamma \sim 10^6$) which emit in the optical and ultraviolet; by comparison, radio emission arises from electrons with $\gamma \sim 10^4$. (Electrons with energy $E = \gamma mc^2$, in field B, emit at frequency $\nu \propto BE^2$). Optical observations also allow us to clearly delineate sites of electron acceleration in the jet. Electrons lose half their energy in time τ, where

$$\tau \propto B^{-3/2}\nu^{-1/2}$$

Therefore, optical and UV emitting electrons cannot diffuse far from their production sites, as compared to electrons emitting at radio frequencies (Sparks, Biretta and Macchetto 1994).

Optical observations can also reveal the structure of the jet's magnetic field (e.g., degree of order, direction), since the B field of radiation is parallel to emission region B. Faraday rotation by intervening plasma can corrupt the polarization measurements at radio frequencies, but this is not a problem at optical frequencies.

Finally, *HST* observations can provide evidence regarding the mass and nature of the jet's engine. Radial velocities of spectral lines from gas very near the engine provide constraints on the engine's mass.

We now consider *HST* observations of two well-studied jets – that in the nearby galaxy M87 (redshift z=0.0043), and that in 3C273 (z=0.158), the nearest quasar. Next we review three optical jets discovered by *HST*: 3C78 (z=0.029); 3C264 (z=0.021); and 3C346 (z=0.161). And finally we briefly mention results on two other optical jets, 3C66B (z=0.0215) and PKS 0521-36 (z=0.055).

3.1. THE JET IN M87

The giant elliptical galaxy M87 (3C 274, NGC4486) contains one of the best studied synchrotron jets, and is one of the nearest examples of this phenomenon (distance 16 Mpc; Tonry 1991). The host galaxy is perhaps the most massive in the Virgo cluster of galaxies. The 20 arcsecond long (2500 pc) jet is a prominent source of radio, optical, and X-ray emission, and has been well studied from the ground and with X-ray satellite observatories (Biretta, Stern, and Harris, 1991; Biretta 1993).

The M87 jet can be divided into several regions on a morphological basis: the first 1000 pc or "inner jet," which is relatively faint and straight; the "transition" region where it is brightest and shows slight bending; and the "outer jet" beyond 1500 pc where it is strongly bent, and fades rapidly. Figure 9 shows an image made with the Faint Object Camera at near 3400Å. The inner jet contains prominent knots D, E, F and I which

368

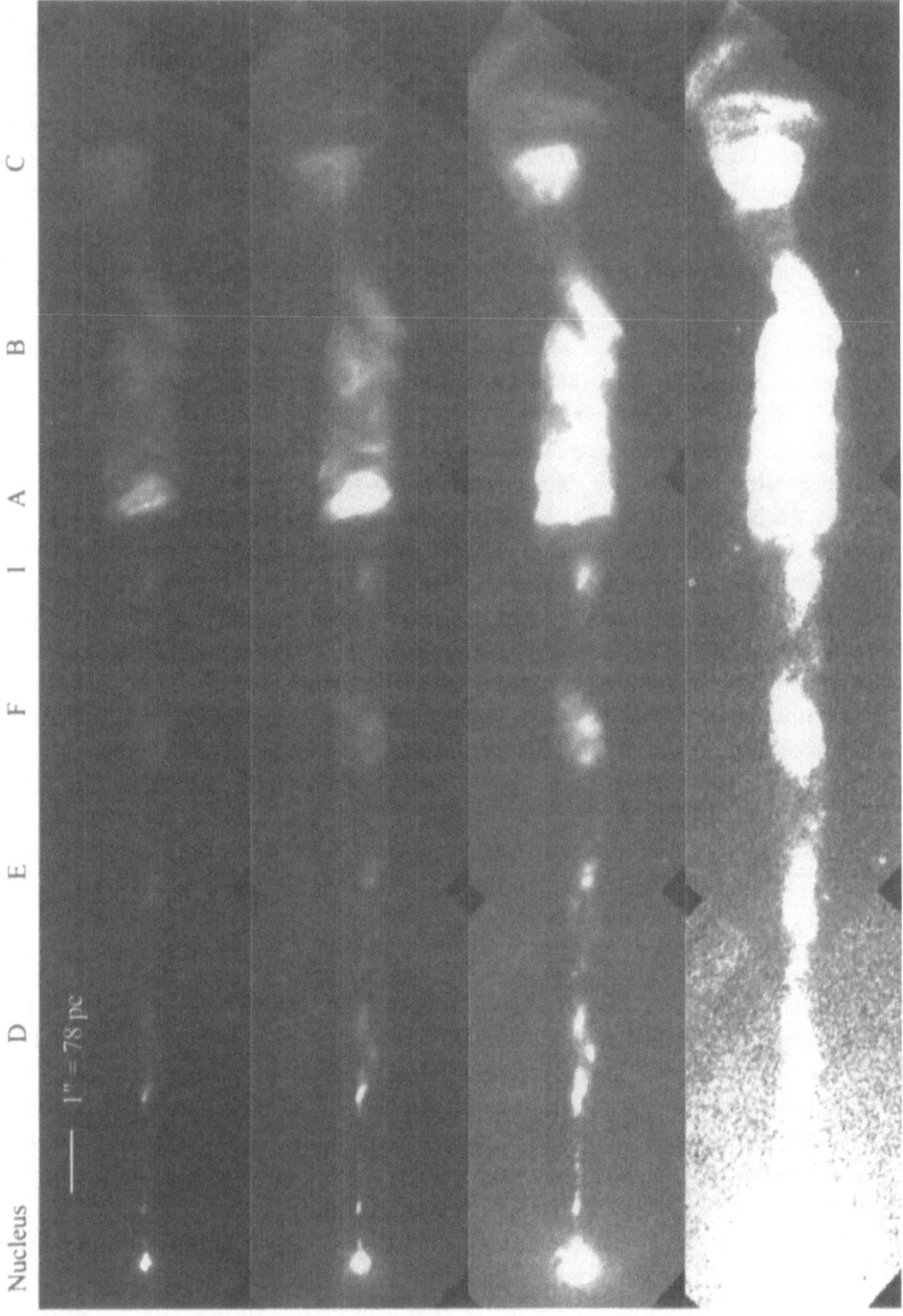

Figure 9. Mosaicked *HST/FOC* image of the M87 jet at three different display contrasts. Resolution is ~ 0.03 arcseconds. Knots are labeled along the bottom. From Biretta, et al. 1996.

have quasi-periodic spacing. These knots have relatively complex, dissimilar, morphologies: knot D has several bright condensations; knot E is a faint complex structure; knot I is dominated by a single bright feature. There are also several compact knots where the jet first emerges from the galaxy's nucleus.

It is interesting to note that "diagonal" emission features are common along the inner jet. The brightest regions of both knots D and I appear elongated along lines at angles 10° to 20° from the jet axis. These structures might be portions of helical structures (magnetic field lines) within the jet. Knots E and F also show faint hints of diagonal structures.

The morphology of the jet changes dramatically at knot A where the transition region starts. While the inner jet is quite straight, in this region containing knots A, B, and C, the entire jet appears to bend and oscillate from side to side. At knot A it is deflected towards the south; in knot B it is deflected back towards the north.

The expansion angle of the jet also changes at knot A. Between the nucleus and knot A the (radio) jet appears as a uniformly expanding cone with an opening angle of $\sim 7°$. In this transition region, however, the jet appears to become nearly cylindrical, perhaps even with some decrease in the diameter between knots A and B. Between knots B and C, the jet seems to narrow considerably and then expand again.

The transition region contains numerous narrow, linear emission features which are oriented approximately normal to the jet axis. Knot A contains several very bright transverse features which run nearly perpendicular to the jet axis; these are likely to be associated with shock waves in the jet where the magnetic field, and hence synchrotron emission, is greatly enhanced. There are also interesting structural details, which are first seen in these *HST* images: the side of knot A nearest knot I shows a faint loop of emission precisely on the jet axis; there's also a faint bar exactly perpendicular to the jet axis. Knot C contains a pattern of linear features which is symmetric about the jet axis; these are similar to X-pattern shocks seen in numerical jet models (c.f. Norman, Smarr, and Winkler 1985; Falle and Wilson 1986). The prominent linear features indicate a very different morphology from the inner region of the jet, probably due to the jet slowing dramatically in the region.

Polarization observations of the jet (Capetti et al. 1996; Figure 10) indicate the magnetic field lines run predominantly parallel to the jet axis over the inner region, but is more complex in the transition region. In knots A and C, we see that the field direction always runs along the bright, linear, transverse features, further supporting the idea that they are shock waves where the normal component of the B-field is enhanced. Fractional polarizations are quite high, in the range 20% to 40%, indicating highly

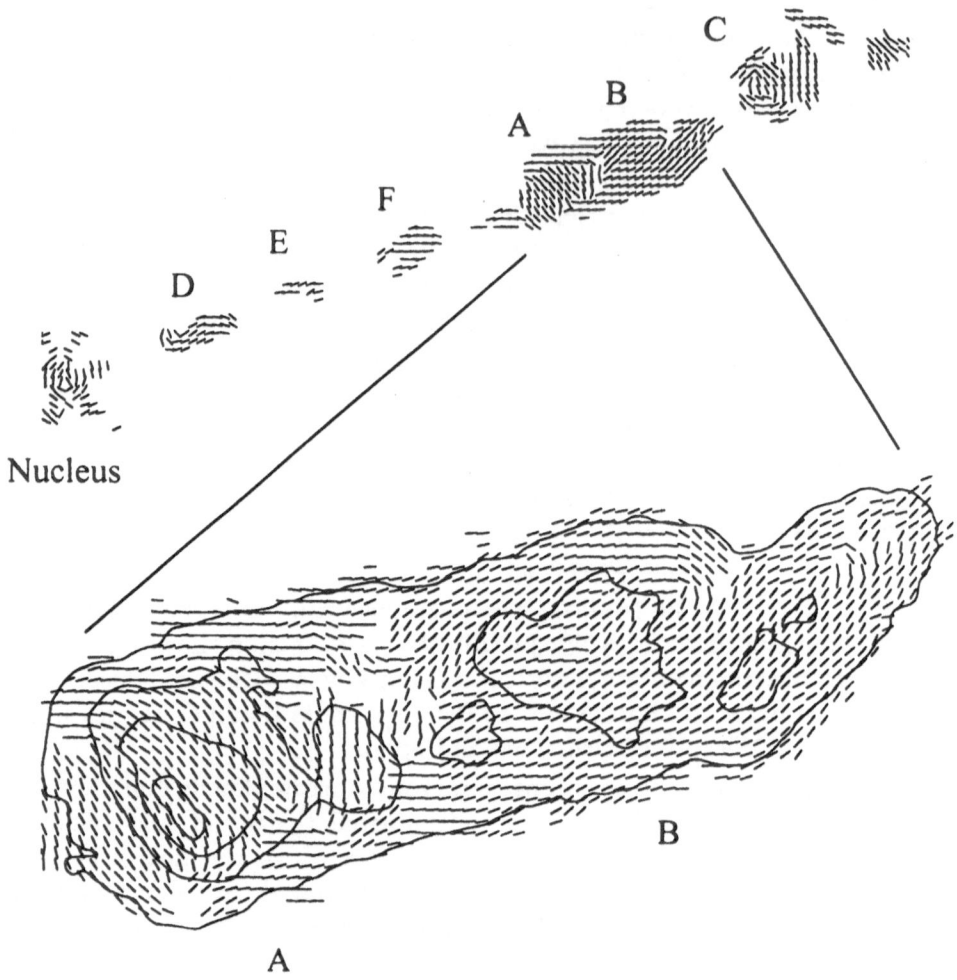

Figure 10. Magnetic field directions derived from FOC polarization data. Top panel show entire jet, while bottom panel shows detail of knot A-B region. From Capetti, et al. 1996.

waves where the normal component of the B-field is enhanced. Fractional polarizations are quite high, in the range 20% to 40%, indicating highly ordered fields. The highest polarization fractions, up to 60%, are seen near the jet edges where shear may tend to align the field lines.

While the jet appears generally similar in the radio and optical bands (Figure 11; Biretta, Stern, and Harris 1991; Boksenberg, et al. 1992), there are striking differences in the detailed morphology. The optical emission is

For example, in knots E and I, the radio image shows significant emission all the way across the jet, but the optical emission is strongly concentrated in one or two small features near the jet axis. Put another way, the radio band shows a diffuse emission component which seems to fill the jet. This component is also visible in the optical, but it is much weaker, implying a relative deficiency of high energy electrons. Clearly, we cannot think of the jet as a homogeneously filled cylinder; instead there must be various layers with respect to the jet axis, each with differing properties.

The velocity field of the jet has been derived from radio interferometer observations at multiple epochs (Biretta, Zhou, and Owen 1995). Knots D and F both show average speeds approaching c, the speed of light, with velocity vectors pointing parallel to the jet axis (within uncertainties). Separate measurements on different condensations within knot D give a more complex picture, with speeds ranging from $-0.2c \pm 0.1c$ to $2.5c \pm 0.3c$. These largest speeds, which exceed c, arise from light travel time effects, and require the jet to have motions with Lorentz factor $\gamma > 3$, and be oriented $\sim 40°$ to the line of sight. Early results of monitoring with the FOC appear to confirm these large speeds; Figure 12 shows the knot D region in mid 1994 and 1995. Several features are seen to move outwards at apparent speeds in the range $2c$ to $4c$.

Motion is also seen very close to the nucleus. The FOC images in Figure 13 show the very bright nucleus and a second bright feature, M, which is located ~ 0.13 arcseconds from the nucleus. This feature appears to move outward at $2.5c \pm 0.5c$, again implying a Lorentz factor $\gamma > 3$. This is especially interesting, since radio observations of this region show very little motion (Biretta and Junor 1995). Apparently the radio emission may come from slower regions nearer the jet's surface. We note that these *HST* observations of motion represent the first detection of superluminal motion in the optical waveband.

Figure 12 also shows a pair of compact features (FOC 2 and 3) which vanish between the *HST/FOC* epochs, indicating rapid evolution is possible; in fact much more rapid than the canonical optical synchrotron lifetimes in this source (~ 100 years).

While the jet in M87 is quite prominent, there is very little evidence of a "counter-jet" on the opposite side of the nucleus. This is rather different from stellar jets, where two-sided jets are most common. Evidence has been discovered for an optical "hot spot" of synchrotron opposite the jet, and about 25 arcseconds from the nucleus (Stiavelli et al. 1992; Sparks et al. 1992; Neumann et al. 1995). This feature is remarkable in that the lifetimes of the optically emitting electrons is expected to be of order 10^3 years in the feature. This implies that an unseen "counter-jet" must be supplying energy to this feature.

372

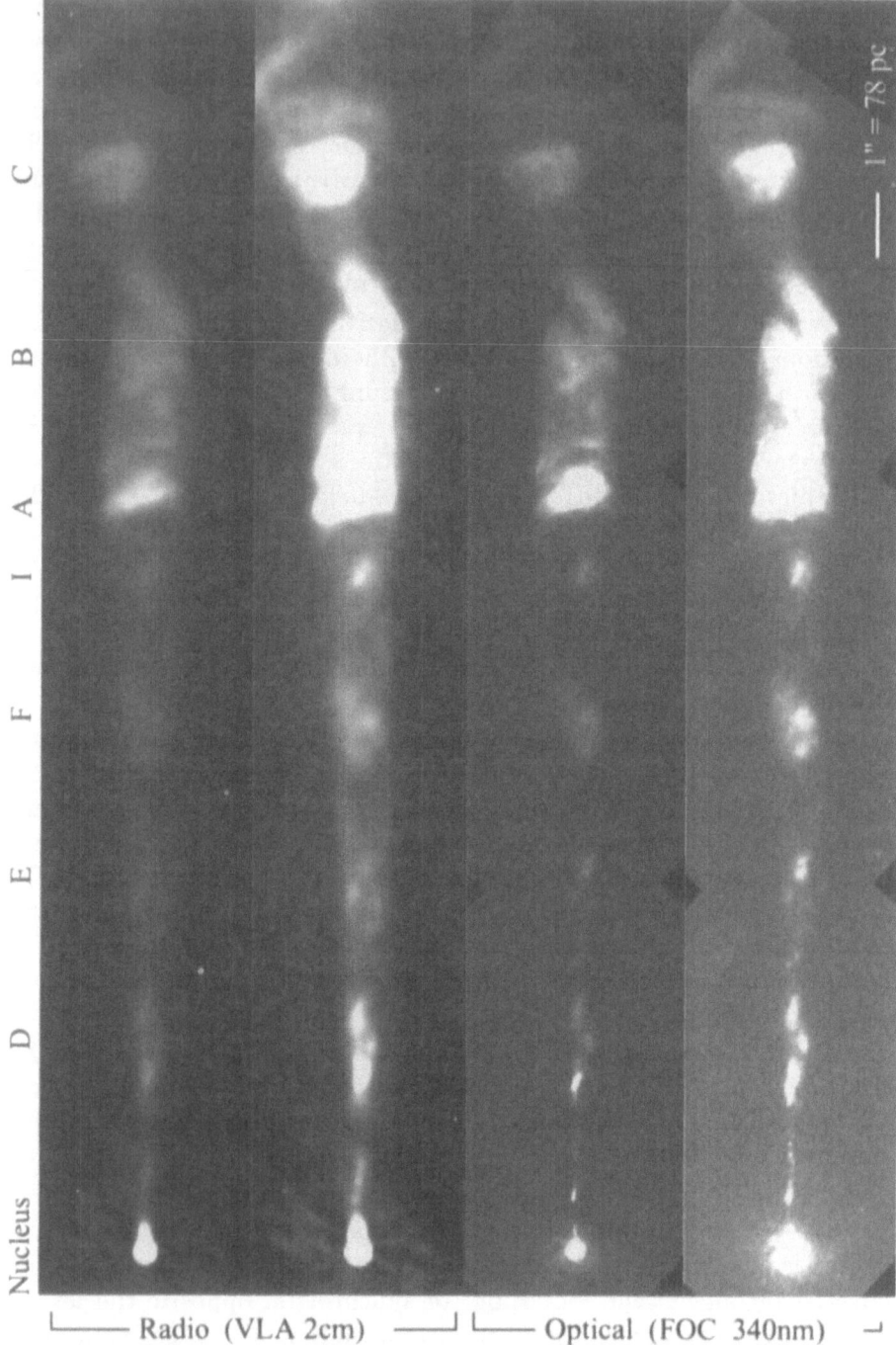

Figure 11. Radio and optical images of the M87 jet. Top two panels show $\lambda = 2$ cm VLA observations in May 1994 at 0.15 arcsecond resolution at different contrast settings. Bottom two panels show $\lambda = 3400$ Å HST Faint Object Camera observations from August 1994 at 0.03 arcsecond resolution. From Biretta, et al. 1996.

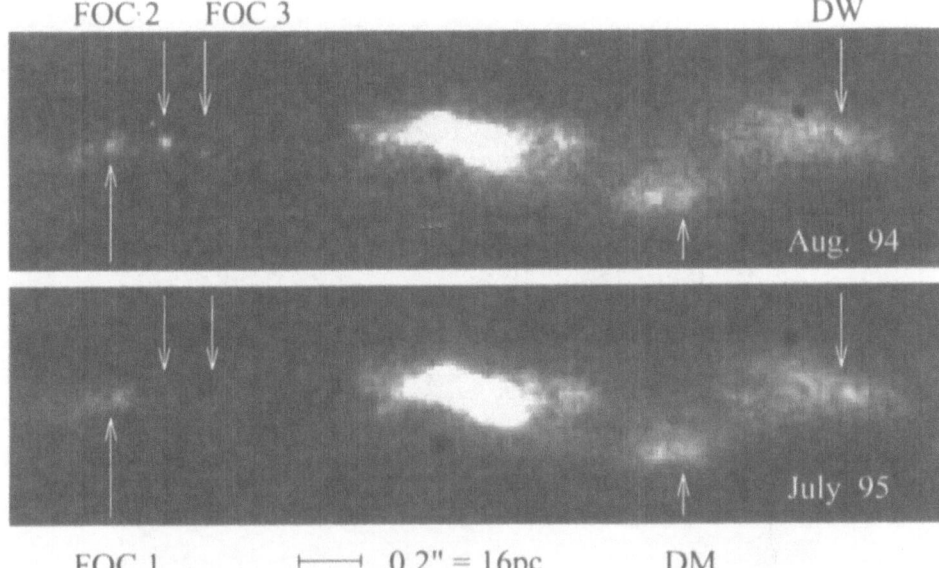

Figure 12. Comparison of HST/FOC $\lambda = 3400$ A observations of knot D region in August 1994 and July 1995 at 0.03 arcsecond resolution. Features FOC1, DM, and DW appear to move at speeds in range 2.5c to 4c. FOC2 and FOC3 fade from view. Arrows are at fixed locations relative to the nucleus; dark spots are Reseau marks inside the camera. From Biretta, et al. 1996.

A final important contribution of *HST* in M87 has been the discovery of a gas disk orbiting the nucleus (Figure 14). This disk is presumably the very outer regions of an accretion disk which produces the jet. *HST* has allowed spectroscopic observations of the inner regions of the disk, ~ 18 pc from the nucleus, which show a velocity gradient of 1000 km sec^{-1} across the nucleus (Ford et al. 1994; Harms et al. 1994). When interpreted as Keplerian orbital motion, these velocities imply a central mass of 2.4×10^9 solar masses. This requires a very large mass-to-light ratio for the nucleus, and strongly supports the idea the supermassive black holes are responsible for forming extragalactic jets.

3.2. THE JET IN 3C273

At a distance of 506 Mpc (z=0.158), 3C273 is the nearest quasar. Radio observations find superluminal motion in the nucleus of the quasar, implying a relativistic flow with $\gamma > 5$ (Pearson et al. 1981). Ground-based radio and optical observations show a narrow jet extending approximately 25 arcseconds towards the south-west from the nucleus (Conway et al. 1993). There

374

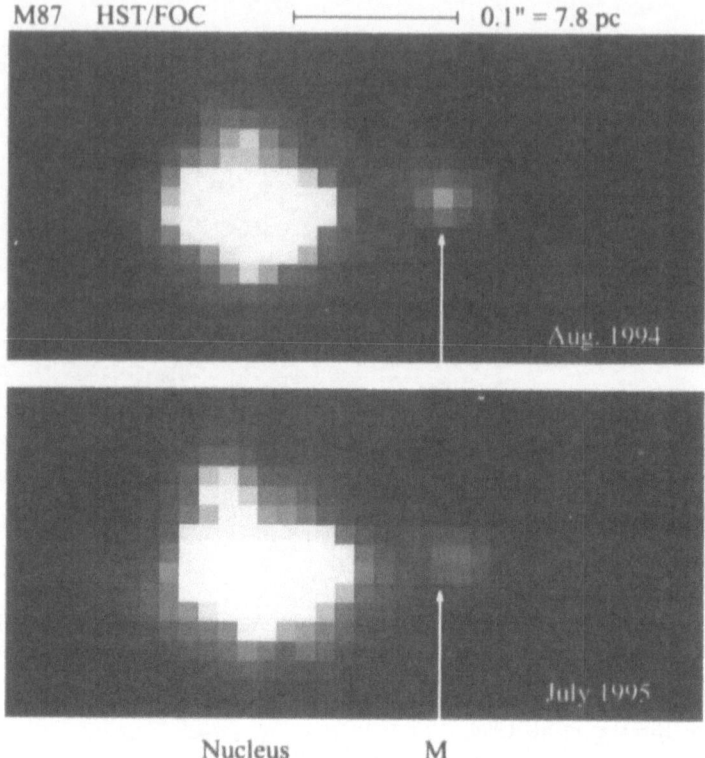

Figure 13. Comparison of HST/FOC $\lambda = 3400$ Å observations of the nucleus in August 1994 and July 1995 at 0.03 arcsecond resolution. Feature "M" appears to move outward by about 0.6 pixel (0.008 arcsecond) corresponding to $2.5c \pm 0.5c$. From Biretta, et al. 1996.

is no evidence whatsoever for a second jet (or "counter-jet") north-east of the nucleus; any counter-jet must be extremely faint.

Observations taken with the *HST* Faint Object Camera show the jet is quite narrow, with an opening angle of $\sim 1°$ (Figure 15; Thompson, Mackay, and Wright 1993). In addition, the jet is not smooth and continuous, but rather appears to consist of many discrete condensations. Superposing the radio contours onto the FOC image indicates that the radio and optical morphologies are roughly similar. However, close examination reveals systematic differences. Much as in M87, optical emission is more concentrated into discrete knots, and is also concentrated along the jet's centerline. Apparently the highest-energy electrons reside near the jet's axis.

The optical emission appears highly polarized, with polarization fractions of 20 to 50% in the bright knots (Figure 16). As in M87, the magnetic field is oriented parallel to the inner jet and becomes more complex at the jet's end, with the brightest knots showing fields normal to the jet axis.

Active Galaxy M87

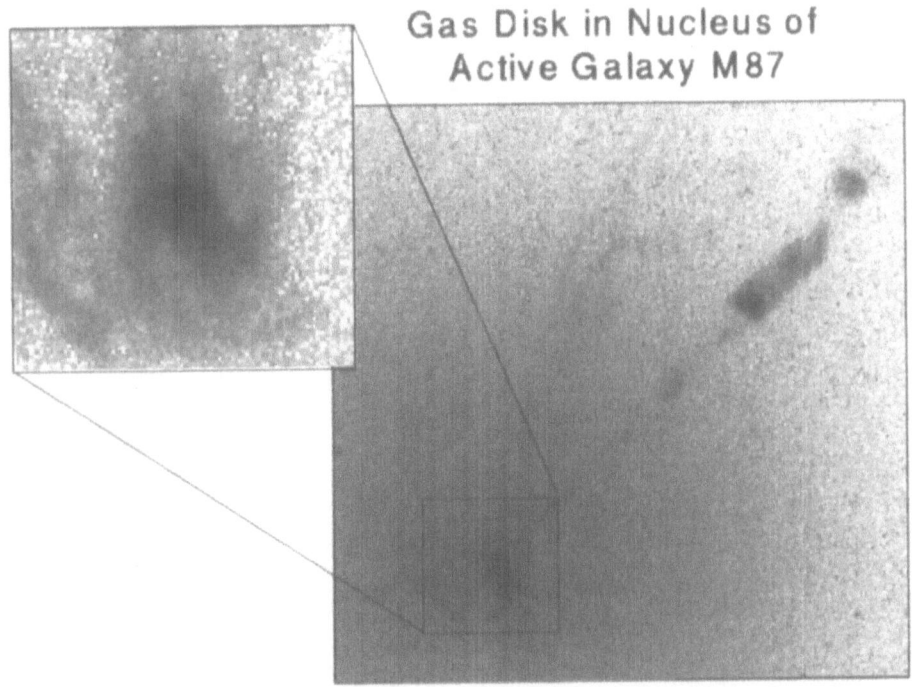

Figure 14. WFPC2 image of M87 showing gas disk surrounding galaxy's nucleus. The disk is oriented normal to the jet, as expected for an accretion driven jet. From Ford et al. 1994.

3.3. OPTICAL SYNCHROTRON JETS DISCOVERED BY HST

Due to its high resolution *HST* has been able to discover a number of optical jets that were previously obscured by stellar light in the host galaxy (Figure 17). This is significant, since only a handful of optical jets were previously known.

Images of the radio galaxy 3C78 (z=0.029) taken with the WFPC2 revealed a short (500 pc long) optical jet with a single prominent knot (Sparks et al. 1995). Similarly, WFPC2 observations of radio galaxy 3C264 (z=0.021) (Gavazzi et al. 1981) found a short jet with four knots having quasi-periodic spacing (Crane et al. 1993, Baum et al. 1996).

Finally, observations of 3C346 (z=0.161) reveal an unusual curved jet with a sharp bend. (Biretta 1996; Dey and vanBreugel 1994; vanBreugel et al. 1992). This is the only optical jet to exhibit significant curvature, and as we will see later, the sharp curve raises interesting questions about jet confinement and collimation. It is also one of the very longest optical jets.

It is likely that very many more optical jets will be discovered now that

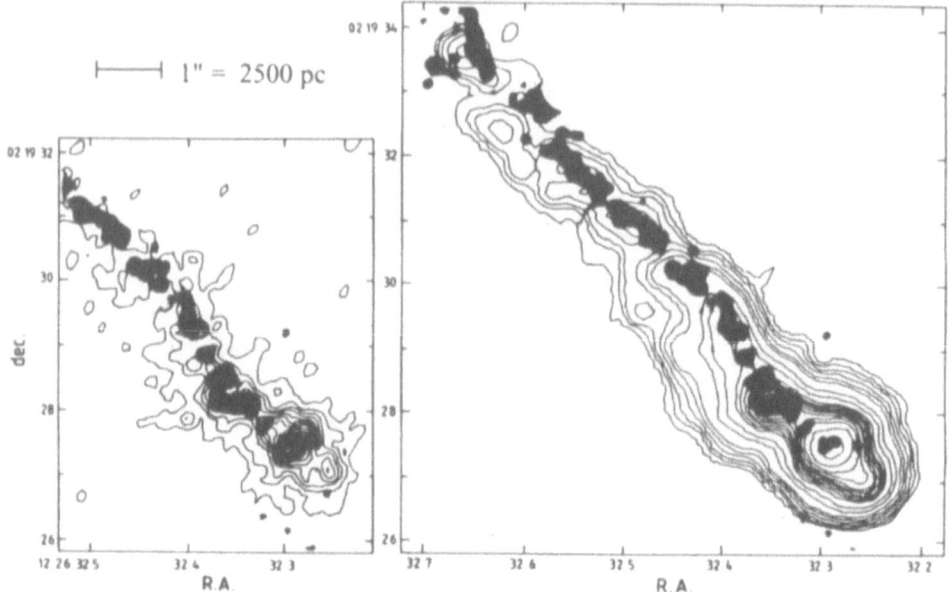

Figure 15. FOC image of 3C273 jet at 4300Å (grey scale) superposed on $\lambda = 2$ cm radio image contours (left) and $\lambda = 6$ cm radio image contours (right). Resolution of *HST* image is 0.15 arcseconds, while radio images are 0.22 and 0.35 arcseconds, respectively. The quasar nucleus is approx. 15 arcseconds off top left of image. From Thompson, Mackay, and Wright 1993.

the corrected optics have been put in place.

Finally we briefly note *HST* results on two other optical synchrotron jets – 3C66B and PKS 0521-36. The jet in 3C66B (z=0.0215; Macchetto et al. 1991) is notable for a double filamentary structure. These seem likely to be magnetic structures. PKS 0521-36 (z=0.055; Macchetto et al. 1991) shows a relatively smooth optical jet, possibly indicating continuous particle accelerate along its length.

3.4. SUMMARY OF HST OBSERVATIONS OF EXTRAGALACTIC JETS

To summarize, *HST* observations of extragalactic jets have given important insights to their production, structure, and kinematics. The discovery of a gas accretion disk with high velocities in M87 tends to confirm the presence of a super massive black hole, and its role in jet production.

It is possible to generalize about the structure of the jets observed by *HST*. They appear to be well collimated, with prominent knots (sometimes with periodic spacing). Filamentary features are common. The radio and optical morphologies are generally similar ($\gamma = 10^4$ for radio vs. $\gamma = 10^6$ for optical emitting elecrons). However, both M87 and 3C273 show a tendency

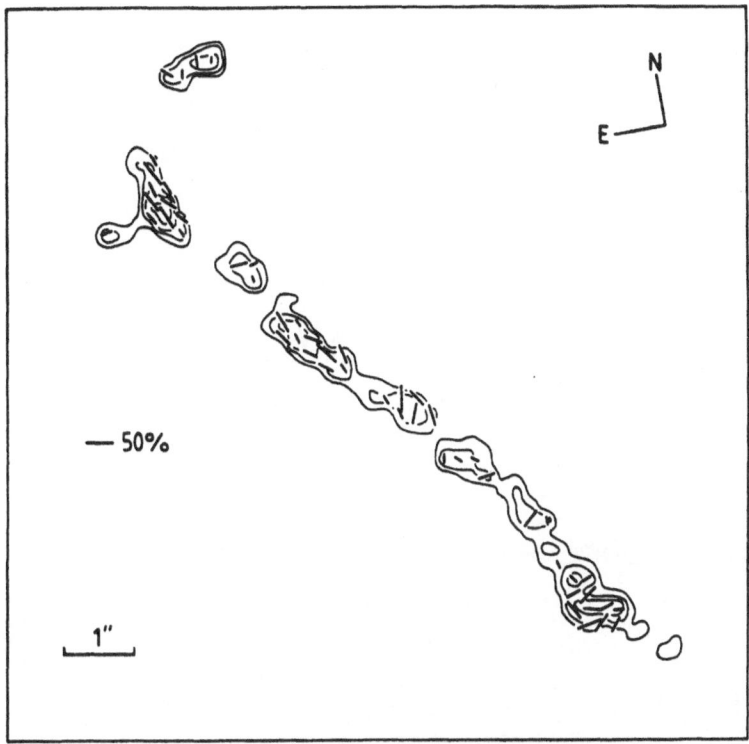

Figure 16. Polarization map of the 3C273 jet as derived from *HST/FOC* images. The orientation of the magnetic field vectors is shown. Vector length is proportional to fractional polarization, with the longest vectors indicating approximately 50% polarization.

for the optical emission to be more concetrated on the jet axis and in the knots.

Highly ordered magnetic fields are often seen, with polarization fractions of 20% to 60%. In both M87 and 3C273 the field lines are oriented along the inner jet, but become more complex near the jet terminae. Also, B tends to run along any filaments or shock fronts. The filaments may be magnetic flux tubes, and shock fronts may compress B perpendicular to the flow.

4. Problems Posed by Observations of Extragalactic Jets

The interesting results obtained by HST observations raise many questions:

1. First, how are the jets confined and collimated? This question is especially interesting for 3C346, the unusual curved jet.
2. There is a particle lifetime problem for optical/UV emitting electrons. While the lifetime of optical / UV emitting electrons is only about

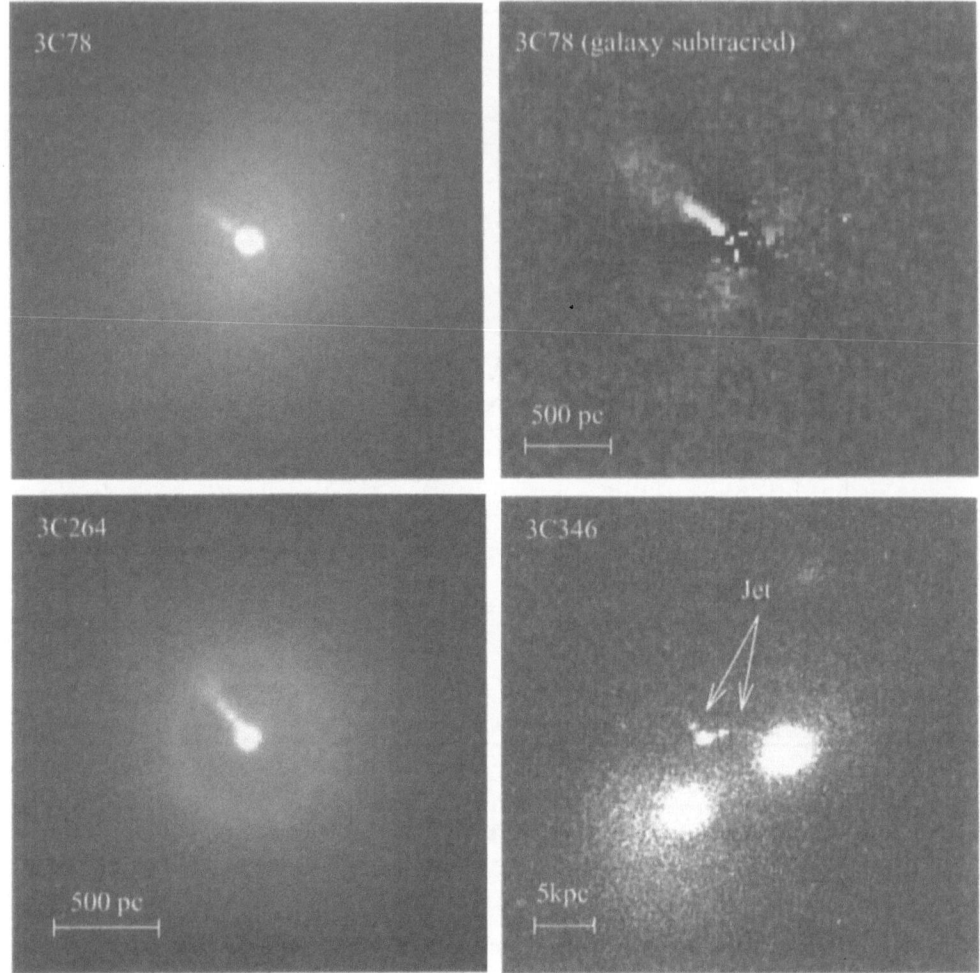

Figure 17. WFPC2 images of extragalactic jets discovered by HST in 3C78, 3C264, and 3C346. The top right panel shows 3C78 with a model of the stellar light subtracted.

100 years, these emissions are seen up to 10,000 light years from the engine.

3. Finally, why are all the optical jets one-sided, when models produce two-sided jets?

While the answers to these questions will occupy astronomers for years to come, a number of possibilities are discussed below.

Jet Confinement / Collimation. Thermal X-ray observations show that jets are surrounded by gas pressure (e.g. in M87; White and Sarazin 1988). The synchrotron emission region's minimum pressure is 10 to 20 times external pressure. So, what keeps jets confined to a narrow, well-collimated cone? And what keeps the jet of 3C346 together over a curved and sharply bent path? One potential solution is that emitting regions are not in static pressure balance, but rather contain shocks. However, many regions don't look like shocks. It is also possible that the jets are held together by B fields wrapped around the current carrying the jet Benford (1978).

Particle Lifetime Problem for Optical / UV Emitting Electrons. For the typical knot, magnetic fields are $B \sim 10^{-4}$ Gauss. Hence the lifetime of optical / UV emitting electrons is only ~ 100 years, and yet optical / UV emission is seen 1,000 to 10,000 light years from the engine. There are two leading explanations of this phenomenon. The first involves particle acceleration at shocks by a first-order Fermi process. Under this theory, moving magnetic walls at the shock front accelerate the high-energy particles. But again, many features don't look like shocks. Furthermore, it is difficult to get similar particle energy spectra from shocks, mild turbulence, etc., as needed to explain their similar radio and optical appearance. The other theory posits a low-loss pipe at the center of the jet. This pipe would transport high energy particles from the engine to the outer reaches of the jet. The question then arises as to what might provide this pipe: a low B field channel (Owen, Hardee, and Cornwell 1989)? A high Lorentz factor flow ($\gamma = 10$ to 100?; Meisenheimer 1991) at the center, where energy losses are slowed by time-dilation? More study of jet structure and kinematics are needed to resolve this.

Why Are Jets One-Sided? As noted above, all optical jets are one-sided, but models for the accretion disk / black hole engine produce two-sided jets. One potential solution: Relativistic flow in the jet causes strong Doppler boosting of the jet's near side, and concomitant dimming of other side. But then, why is there so little evidence of the counter jet? One might expect sub-relativistic flow where the counter-jet interacts with, or impinges on, the external medium.

5. Future Prospects

We are just beginning to see results *HST* can yield when operating at its full design potential. Monitoring programs on well-known jets should give direct evidence on jet kinematics over the next few years; it may even be possible to derive detailed velocity fields for the jets. *HST* is also looking for new optical synchrotron jets, and the statistical results should have important influences on relativistic beaming models for extragalactic jets. Better information about the engines will be obtained by imaging the initial collimation region of YSOs with the Near Infrared Camera (to be installed in 1997), and by obtaining the masses of more galactic nuclei from ongoing imaging / spectroscopy programs on galactic nuclei.

References

Baum, S. (1996) in preparation

Benford, G. (1978) Current-carrying Beams in Astrophysics: Models for Double Radio Sources and Jets, *Mon. Not. R. Astron. Soc.*, **Vol. no. 183**, pp. 29–48

Biretta, J.A., Stern, C.P. and Harris, D. (1991) The Radio to X-Ray Spectrum of the M87 Jet and Nucleus, *Astronomical Journal*, **Vol. no. 101, no. 5**, pp. 1632–1646

Biretta, J. A. (1993) The M87 Jet. In *Astrophysical Jets*, eds. D. Burgarella, M. Livio, C. P. O'Dea, Cambridge Univ. Press, Cambridge: pp. 263–304

Biretta, J. A., Zhou, F. and Owen, F. N. (1995) Detection of Proper Motions in the M87 Jet, *Astrophysical Journal*, **Vol. no. 447**, pp. 582–596

Biretta, J. A., and Junor, W. (1995) The Parsec-scale Jet in M87, *Proc. Natl. Acad. Sci.*, in press

Biretta, J. A., Sparks W. B., Macchetto, F. and Capetti A. (1996), HST Observations of Fine Structure and Motions in the M87 Jet *Astrophysical Journal*, submitted

Biretta, J. (1996); in preparation

Boksenberg, A. Macchetto, F., Albrecht, R., Barbieri, C., Blades, J. C., Crane, P., Deharveng, J. M., Disney, M. J., Jakobsen, P., Kamperman, T. M., King, I. R., Mackay, C. D., Paresce, F., Weigelt, G., Baxter, D., Greenfield, P., Jedrzejewski, R., Nota, A., Sparks, W. B. (1992) Faint Object Camera Observations of M87: The Jet and Nucleus, *Astronomy and Astrophysics*, **Vol. no. 261**, pp. 393–404

Burrows, C. et al. (1996); in preparation

Capetti, A., Macchetto, F., Sparks, W.B. and Biretta, J. (1996) HST Polarization Observations of the Jet of M87, *Astronomy and Astrophysics*, in press

Capetti, A., Macchetto, F., Sparks, W.B. and Miley, G.K. (1994) HST Observations of 3C 449: Discovery of an Extended Nuclear Disk and of Possible Optical Jets, *Astronomy and Astrophysics*, **Vol. no. 289**, pp. 61–66

Conway, R.G., Garrington, S.T., Perley, R. and Biretta, J. (1993) *Astronomy and Astrophysics*, **Vol. no. 267**, pp. 347–362

Crane, P., Peletier, R., Baxter, D., Sparks, W.B., Albrecht, R., Barbieri, C., Blades, J.C., Boksenberg, A., Deharveng, J.M., Disney, M.J., Jakobsen, P., Kamperman, T.M., King, I.R., Macchetto, F., Mackay, C.D., Paresce, F., Weigelt, G., Greenfield, P., Jedrzejewski and Nota, A. (1993) Discovery of an Optical Synchrotron Jet in 3C 264, *Astrophysical Journal*, **Vol. no. 402**, pp. L37–L40

Dey, A. and van Breugel, W.J.M. (1994) Blue Optical Continuum Associated with a Radio Knot in 3C346, *Astronomical Journal*, **Vol. 107, no. 6**, pp. 1977–1983

Falle, S. A. E. G. and Wilson, M. J. (1986) M87: An Example of a Reconfining Jet, *Canadian Journal of Physics*, **Vol. no. 64**, pp. 470

Ford, H.C., Harms, R.J., Tsvetanov, Z.I., Hartig, G.F., Dressel, L.L., Kriss, G.A., Bohlin, R., Davidsen, A.F., Margon, B., and Kochhar, A.K. (1994) Narrowband HST Images of M87: Evidence for a Disk of Ionized Gas Around a Massive Black Hole, *Astrophysical Journal*, **Vol. no. 435**, pp. L27–L30

Gavazzi, G., Perola, G.C. and Jaffe, W. (1981) Observations of the Head-Tail Radio Galaxy NGC 3862 (3C 264), *Astronomy and Astrophysics*, **Vol. no. 103**, pp. 35–43

Hack, W. (1996); in preparation

Harms, R.J., Ford, H.C., Tsvetanov, Z.I., Hartig, G.F., Dressel, L.L., Kriss, G.A., Bohlin, R., Davidsen, A.F., Margon, B., and Kochhar, A.K. (1994) HST FOS Spectroscopy of M87: Evidence for a Disk of Ionized Gas Around a Massive Black Hole, *Astrophysical Journal*, **Vol. no. 435**, pp. L35–L38

Hester, J. et al. (1996); in preparation

Hollis, J.M. and Michalitsianos, A.G. (1993) Evidence for Precession of the R Aquarii Jet, *Astrophysical Journal*, **Vol. no. 411**, pp.235–238

Kepner, J., Hartigan, P., Yang, C. and Strom, S. Hubble Space Telescope Images of the Subarcsecond Jet in DG Tauri (1993) *Astrophysical Journal*, **Vol. no. 415**, pp. L119–L121

Macchetto, F., Albrecht, R., Barbieri, C., Blades, J.C., Boksenberg, A., Crane, P., Deharveng, J.M., Disney, M.J., Jakobsen, P., Kamperman, T.M., King, I.R., Mackay, C.D., Paresce, F., Weigelt, G., Baxter, D., Greenfield, P., Jedrzejewski, R., Nota, A., Sparks, W.B. and Miley, G.K. (1991a) HST Observations of 3C 66B: A Double-Stranded Optical Jet, *Astrophysical Journal*, **Vol. no. 373**, pp. L55–L58

Macchetto, F., Albrecht, R., Barbieri, C. Blades, J.C., Boksenberg, A., Crane, P., Deharveng, J.M., Disney, M.J., Jakobsen, P., Kamperman, T.M., King, I.R., Mackay, C.D., Paresce, F., Weigelt, G., Baxter, D., Greenfield, P., Jedrzejewski, R., Nota, A and Sparks, W.B. (1991b) First Results from the Faint Object Camera: Observations of PKS 0521-36, *Astrophysical Journal*, **Vol. no. 369**, pp. L55–L57

Meisenheimer, K. (1991) The Spectra of Extended Radio Jets and Hot Spots, in *Physics of Active Galactic Nuclei*, eds. W. Duschl and S. Wagner, Springer-Verlag, Berlin, p. 525

Morse, J.A., Hartigan, P., Cecil, G., Raymond, J.C. and Heathcote, S. (1992) The Bow Shock and Mach Disk of HH 34, *Astrophysical Journal*, **Vol. no. 339**, pp.231–245

Morse, J.A. et al. (1996); in preparation

Neumann, M., Meisenheimer, K., Rosner, H.-J. and Stickel, M. (1995) Detection of Near Infrared Emission from the Eastern Lobe of M87, *Astronomy and Astrophysics*, **Vol. no. 296**, pp. 662–664

Norman, M. L., Smarr, L. and Winkler, K.-H. A. (1985) Fluid Dynamical Mechanisms for Knots in Astrophysical Jets, in *Numerical Astrophysics*, eds. J. M. Centrella, J. M. LeBlanc, and R. L. Bowers, Jones and Bartlet, Boston. 88-125

Owen, F. N., Hardee, P. E. and Cornwell, T. J. (1989) High-Resolution, High Dynamic Range VLA Images of the M87 jet at 2 Centimeters, *Astrophysical Journal*, **Vol. no. 340**, pp. 698–707

Paresce, F. and Hack, W. (1994) New HST Observations of the Core of R Aquarii, *Astronomy and Astrophysics*, **Vol. no. 287**, pp. 154–162

Pearson, T.J., Unwin, S.C., Cohen, M.H., Linfield, R.P., Readhead, A.C.S., Seielstad, G.A., Simon, R.S., and Walker, R.C. (1981) Superluminal Expansion of Quasar 3c273, *Nature*, **Vol. no. 290**, pp. 365-368

Ray, T.P. and Mundt, R. (1993) Interpreting Observations of Jets from Young Stars, in *Astrophysical Jets*, eds. D. Burgarella, M. Livio, C. P. O'Dea, Cambridge Univ. Press, Cambridge: pp. 146–175

Reipurth, B. and Heathcote, S. (1993) Observational Aspects of Herbig-Haro Jets, in *Astrophysical Jets*, eds. D. Burgarella, M. Livio, C. P. O'Dea, Cambridge Univ. Press, Cambridge: pp. 35–71

Schwartz, R.D., Cohen, M., Jones, B.F., Bohm, K.-H., Raymond, J.C., Hartmann, L.W., Mundt, R., Dopita, M.A. and Schultz, A.S.B. (1993) Hubble Space Telescope Imaging of Herbig-Haro Object No.2, *Astronomical Journal*, **Vol. no. 106, no.2**, pp. 740–746

Schwartz, R.D (1994) The Herbig-Haro Object 1-2 System Revisited, in *Stellar and Circumstellar Astrophysics, ASP Conference Series*, G. Wallerstein and A. Noriega-Crespo, eds., **Vol. 57**

Sparks, W. B., Fraix-Burnet, D., Macchetto, F., and Owen, F. N. (1992) A Counterjet in the Elliptical Galaxy M87, *Nature*, **Vol. no. 355**, pp. 804–806

Sparks, W.B., Biretta, J.A. and Macchetto, F. (1994) Hubble Space Telescope Observations of Synchrotron Jets, *The Astrophysical Journal Supplement Series*, **Vol. no. 90**, pp. 909–916

Sparks, W. B., Golombeck, D., Baum, S., Biretta, J. A., deKoff, S., Macchetto, F., McCarthy, P., and Miley, G.K. (1995) Discovery of an Optical Synchrotron Jet in 3C 78,*Astrophysical Journal*, **Vol. no. 450**, pp. L55-58

Sparks, W. B., Biretta, J. A. and Macchetto, F. (1996) The Jet of M87 at Tenth Arcsecond Resolution: Optical, UV, and Radio Observations, *Astrophysical Journal*, submitted

Stiavelli, M., Biretta, J., Möller, P. and Zeilinger, W. W. (1992) Optical Counterpart of the East Radio Lobe of M87, *Nature*, **Vol. no. 355**, pp. 802–804

Thomson, R.C., Mackay, C.D., and Wright, A.E. (1993) Internal Structure and Polarization of the Optical Jet of the Quasar 3C273, *Nature*, **Vol. no. 365**, pp. 133–135

Tonry, J. L. 1991.Surface Brightness Fluctuations: A Bridge from M31 to the Hubble Constant, *Astrophysical Journal*, **Vol. no. 373**, pp. L1–L4

Uomoto, A., Caganoff, S., Ford, H.C., Rosenblatt, E.I., Antonucci, R.R.J., Evans, I.N. and Cohen, R.D. (1993) The Optical Jet in Markarian 463, *Astronomical Journal*, **Vol. no. 105, no. 4**, pp. 1308–1312

vanBreugel, W.J.M., Fanti, C., Fanti, R., Stanghellini, C., Schilizzi, R.T. and Spencer, R.E. (1992) Compact Steep-Spectrum 3CR Sources: VLA Observations at 1.5, 15 and 22.5 GHz, *Astronomy and Astrophysics*, **Vol. no. 256**, pp. 56–78

White, R. E. and Sarazin, C. L. (1988) Star Formation in the Cooling Flows of M87/Virgo and NGC 1275/Perseus, *Astrophysical Journal*, **Vol. no. 335**, pp. 688–702

EXACT MHD SOLUTIONS FOR SELF-SIMILAR OUTFLOWS

EDOARDO TRUSSONI
Osservatorio Astronomico di Torino
Strada dell'Osservatorio 20
I-10025 Pino Torinese, ITALY, trussoni@ph.unito.it

CHRISTOPHE SAUTY
Observatoire de Paris, DAEC, F-92195 Meudon, FRANCE
sauty@gin.obspm.fr

AND

KANARIS TSINGANOS
Department of Physics, University of Crete
Foundation for Research and Technology Hellas (FORTH)
GR-710 03 Heraklion, GREECE, tsingan@physics.uch.gr

Abstract. Any effort to self-consistently model MHD flows encounters prohibitive mathematical difficulties and several simplifications need to be adopted. One common such technique is the so-called *self-similar* approach where the physical variables are factorized and a suitable scaling law is chosen along a prescribed direction. For example, the well known axisymmetric, and *radially* self-similar solution of Blandford and Payne (1982) has been widely used to model winds from accretion disks, although it is pathogenic at the flow axis. Here we review a complementary wide class of axisymmetric MHD solutions for rotating magnetized flows which are *latitudinally* self-similar and are appropriate to model flows from central gravitating bodies, while they do not have the limitations of the radially self-similar ones. In one subclass the streamline pattern is selfconsistently calculated while in the other it is *a priori* prescribed to correspond to outflows of various degrees of collimation. The main features of these two subclasses of latitudinally self-similar solutions are presented. A quantitative energetic criterion is also derived saying roughly that when the magnetorotational energy density exceeds the thermal energy density, the flow cross sectional area undergoes oscillations before it finally collimates in a jet of several Alfvén radii width and finite Alfvén number in the range of about 1–100. If on the other hand, the thermal energy density exceeds the magnetorota-

K. C. Tsinganos (ed.), Solar and Astrophysical Magnetohydrodynamic Flows, 383–410.
© 1996 *Kluwer Academic Publishers.*

384

tional one, the asymptotic channel cross sectional area and Alfvén number are much larger and the outflow resembles more to the classical picture of a thermally driven wind. A simple relation is given relating the width of the beam with the terminal Mach number in realistic jet-type flows found for a wide range of the values of the parameters. The evolutionary history of the morphology of outflows from the protosun to the present date Ulysses-measured solar wind, is also sketched within the model. The advantages and limits of the self-similar approach are also discussed.

1. Introduction

Since the discovery of the solar wind, outflows have been observed in a widespread class of astrophysical objects: stars of all spectral types, regions of star formation, collapsed galactic objects and AGN. The existence of jets and bipolar flows in galactic and extragalactic environments indicates that quite often the mass ejection can be highly asymmetric and collimated. Since the pioneering work of Parker (1958) on the solar wind, several theoretical models have been proposed for the acceleration of astrophysical outflows. It has been found that in solar-like and cool stars the wind from a hot corona can be effectively thermally driven, while in luminous objects (e.g. early type stars) it can be radiatively driven. On the other hand, since almost all celestial bodies are magnetized and rotating, the combination of rotation and magnetic field gives rise to a flux of the Poynting vector which in turn can drive a wind to supersonic terminal speeds. Furthermore, the presence of a toroidal magnetic component can easily collimate the outflow and thus jets are likely to be associated with this driving mechanism (see also the chapters by Macgregor and Heyvaerts & Norman in this volume).

The study of magnetized winds is usually performed in the framework of a steady MHD approach, that implies the treatment of a system of 3-D partial, nonlinear differential equations. In the case of axially symmetric geometry, suitable for collimated jets, the set of the MHD equations is reduced to the coupled Bernoulli and transfield (or Grad-Shafranov) equations, which fully describe the dynamics of the wind along and across its streamlines, respectively. However, due to prohibitive mathematical difficulties some basic assumptions are generally adopted in order to solve these two equations, with three different ones having been so far followed.

In the first, the problem is reduced to an ordinary differential equation by simplifying, or even totally ignoring the most difficult transfield equation, in which case the results are valid only in restricted regions. For example, very close to the equatorial plane we obtain an ordinary first order differential equation with three singularities wherein the wind speed is equal to the fast/slow magnetoacoustic speeds and the poloidal Alfvén equation is obtained to model hollow flows in thin shells by performing a preliminary averaging of the physical variables in the direction perpendicular to this axis (Lovelace *et al.* 1991).

In a second class of studies the analysis of the complete MHD equations is performed only in asymptotic regions without studying full solutions from the base of the flow to infinity. In this view, a rather important result has been obtained by Heyvaerts and Norman (1989), who showed that in the superAlfvénic region at least the inner part of the wind must be collimated into cylinders because of the effect of the toroidal component of the magnetic field.

Finally, a third approach is based on the *self-similar* technique where the physical variables are separated and a specific scaling is prescribed along one of the coordinates. This technique allows to reduce the partial differential MHD equations to a system of ordinary differential equations yielding fully 2-D solutions. The crucial point here is the assumption of the scaling law, that depends on the specific scenario that is to be studied. For example, in the framework of flow acceleration from a magnetized accretion disk, Blandford and Payne (1982) presented an analysis for radially self-similar cold winds. Their results for a polytropic flow were extended assuming more general scaling laws by Contopoulos and Lovelace (1994) and Rosso and Pelletier (1994).

However, this radially self-similar description fails at the system rotational axis; instead, in this region the flow can be nicely studied by assuming latitudinal self-similarity, i.e. the colatitude is the scaling coordinate for the physical variables and not the spherical radius. A second important difference between these latitudinally self-similar and the previous radially self-similar solutions is that in the former case the scaling laws for the gas pressure and the geometrical structure of the streamlines are not independent. This means that there is <u>no</u> polytropic relation between the gas pressure and density for any prescribed constant value of the polytropic index γ. For example, it is well known that a polytropic equation of state with such given constant value of γ cannot yield correct values of the wind variables both close and far from our Sun (Parker 1963). If the distribution of the pressure is assumed such that its gradient has unrelated components parallel and perpendicular to the streamlines, then the streamline structure must be provided *a priori*. This subclass of latitudinally self-similar solutions has been analyzed in previous studies for collimated or noncollimated winds with or without rotation (Low and Tsinganos 1986, Tsinganos and Low 1989, Hu and Low 1989, Tsinganos and Trussoni 1991, Tsinganos *et al.* 1992, Trussoni and Tsinganos 1993). If, conversely, the two components of the pressure gradient are correlated, then the structure of the streamlines is free to be deduced from the equations. This second subclass of latitudinally self-similar solutions has been recently studied by Tsinganos and Sauty (1992), Sauty and Tsinganos (1994), Sauty et al. (1995).

In the framework of the latitudinal self-similar approximation, in this study we outline the main features of these solutions for the two cases of prescribed and free streamlines, according to the assumptions on the pressure. In the following section the general properties of the axisymmetric MHD equations are summarized, while in section 3 the technique for generating latitudinally self-similar solutions is outlined with the assumed scaling for the variables and the final equations are deduced and discussed in section 4. The general solutions obtained assuming free streamlines are studied in section 5, while solutions with prescribed collimated streamlines are presented in section 6. The main implications of our results are discussed and summarized in section 7.

2. MHD equations for axisymmetric configurations

The steady motion of an inviscid magnetized plasma with large conductivity in the spherically symmetric potential well of a central body with mass m is governed by the following well known ideal MHD equations

$$\nabla \cdot \mathbf{B} = \nabla \cdot (\rho \mathbf{V}) = \nabla \times (\mathbf{V} \times \mathbf{B}) = 0 \qquad (2.1a)$$

$$\rho(\mathbf{V} \cdot \nabla)\mathbf{V} - (\mathbf{B} \cdot \nabla)\mathbf{B}/4\pi = -\nabla(P + B^2/8\pi) - \rho\frac{\mathcal{G}m}{r^2}\mathbf{e}_r \qquad (2.1b)$$

$$\rho\mathbf{V} \cdot \left(\nabla h - \frac{\nabla P}{\rho}\right) = \rho\sigma \qquad (2.1c)$$

where $\mathbf{B}$ and $\mathbf{V}$ are the magnetic field and velocity, P, ρ and T are the gas pressure, density and temperature, h the enthalpy, $\mathcal{G}$ the gravitational constant, and r the radial distance from the central body (a spherical frame of reference r, θ, ϕ is assumed) while $\rho\sigma$ is the volumetric rate of heating/cooling. Note that since in astrophysical plasmas $\rho\sigma$ is poorly known, the energy equation is usually replaced by a polytropic relation between the gas pressure and density, $P = C\rho^\gamma$, with $1 \leq \gamma \leq \Gamma$, where Γ is the ratio of the specific heats, while γ is the constant polytropic index. In such a case the volumetric rate of heating/cooling is simply

$$\rho\sigma = \rho\mathbf{V} \cdot \vec{\nabla}\rho\frac{\gamma - \Gamma}{\Gamma - 1}C\rho^{\Gamma-1} . \qquad (2.1d)$$

Some basic properties of Eqs. (2.1) under axisymmetry conditions have been extensively discussed in the literature (e.g. Tsinganos 1982, Heyvaerts and Norman 1989, Heyvaerts and Norman in this volume) and are quickly summarized in the following only for establishing the notation. Thus, by defining the poloidal and toroidal components of the magnetic field and velocity

$$[\mathbf{B}, \mathbf{V}] = [B_r, V_r]\mathbf{e}_r + [B_\theta, V_\theta]\mathbf{e}_\theta + [B_\phi, V_\phi]\mathbf{e}_\phi$$
$$= [\mathbf{B}_p, \mathbf{V}_p] + [B_\phi, V_\phi]\mathbf{e}_\phi \tag{2.2}$$

it is possible to treat separately the toroidal and poloidal components of Eqs. (2.1) in terms of $\mathbf{B}_p$, $\mathbf{V}_p$ and B_ϕ, V_ϕ.

2.1. POLOIDAL COMPONENTS

The structure of the system (2.1) allows to define the poloidal field components through the magnetic flux function $A(r, \theta)$ and the mass flux $\Psi(r, \theta)$:

$$\mathbf{B}_p = \nabla \times \left[\frac{A(r, \theta)}{r \sin\theta}\mathbf{e}_\phi\right], \qquad \rho\mathbf{V}_p = \nabla \times \left[\frac{\Psi(r, \theta)}{r \sin\theta}\mathbf{e}_\phi\right]. \tag{2.3}$$

From the induction equation (2.1a) it follows immediately that $\Psi = \Psi(A)$ and $\mathbf{B}_p \parallel \mathbf{V}_p$

$$\mathbf{V}_p = \frac{\Psi_A}{4\pi\rho}\mathbf{B}_p \tag{2.4}$$

where $\Psi_A = d\Psi/dA$.

It is clear that the poloidal streamlines (or fieldlines, the two terms are equivalent) are defined from the intersection of the surfaces $A(r, \theta) = \text{const}$ with every meridional plane crossing the symmetry axis. From a physical point of view the magnetic flux function is proportional to the number of streamlines included between the two surfaces labelled A_1 and A_2.

From the previous expressions we can define a new variable as the ratio of the poloidal flow velocity and the poloidal Alfvén velocity

$$M^2 = 4\pi\rho\frac{V_p^2}{B_p^2} = \frac{\Psi_A^2}{4\pi\rho}. \tag{2.5}$$

In the following M will be simply called the *Alfvén number* although in reality it is the poloidal Alfvén Mach number.

2.2. TOROIDAL COMPONENTS

The ϕ− component of the induction equation can be integrated, giving

$$\frac{V_\phi}{r \sin\theta} - \frac{\Psi_A}{4\pi\rho}\frac{B_\phi}{r \sin\theta} = \Omega(A). \tag{2.6a}$$

This relation is also known as the *isorotation law* since a streamline labelled with A has at its footpoint an angular speed equal to $\Omega(A)$ if there the Alfvén number is vanishingly small.

Also the toroidal component of the momentum equation can be integrated, providing a conservation law for the total angular momentum

$$r \sin\theta \left(V_\phi - \frac{B_\phi}{\Psi_A} \right) = L(A) \qquad (2.6b)$$

where the quantity $L(A)$ is the total angular momentum of the fluid and of the magnetic field on the particular streamline A.

From the above relations we can get the following expressions for the toroidal components of the velocity and magnetic field:

$$B_\phi = -\frac{L(A)\Psi_A}{r \sin\theta} \frac{1 - r^2 \sin^2\theta\, \Omega(A)/L(A)}{1 - M^2}$$

$$V_\phi = \frac{L(A)}{r \sin\theta} \frac{r^2 \sin^2\theta\, \Omega(A)/L(A) - M^2}{1 - M^2} \qquad (2.7)$$

in terms of the Alfvén number, defined in Eq. (2.5).

2.3. ALFVÉN SINGULARITY

The azimuthal components depend on the three free integrals $L(A)$, $\Psi(A)$ and $\Omega(A)$, which are not independent. A regularity condition in Eqs. (2.7) requires that at the Alfvén point the following relation must hold between $L(A)$ and $\Omega(A)$

$$r^2 \sin^2\theta|_{M=1} = \frac{L(A)}{\Omega(A)}. \qquad (2.8)$$

The ratio between the two integrals gives the radius of the streamline labelled A at the Alfvén point. Furthermore, from the definition of the Alfvén number M we get the following expression for the density at the Alfvén singularity

$$\rho(r,\theta)|_{M=1} = \frac{\Psi_A^2}{4\pi} \qquad (2.9)$$

which means $\rho(r,\theta)|_{M=1} \equiv \rho_a(A)$ and $M^2 = \rho_a(A)/\rho$.

3. Scaling laws for self-similar solutions

We can now proceed to treat Eqs. (2.1) following the θ−self-similar technique. As a first step we express the magnetic flux function in separable form

$$A(r,\theta) \propto f(r)\, g(\theta). \qquad (3.1)$$

The choice of the self-similar direction and the scaling law along that direction depends on the wind structure to be modelled, and must lead the original equations to a system of ordinary differential equations.

In r–self-similar models for winds from accretion disks it is effectively assumed that

$$f(r) = r^x \qquad (3.2a)$$

where the exponent x is related to the rotation law of the disk (Blanford and Payne 1982, Contopoulos and Lovelace 1994, Rosso and Pelletier 1994). Furthermore, it is assumed that the Alfvén number is only a function of the colatitude, $M(\theta)$. The contour surfaces of M are conical, with their axis coinciding with the rotational axis of the wind. As a consequence, the solutions are singular on this axis, where the r–self-similarity assumption is not valid.

The θ–self-similar approximation, followed in the present study, can be considered as complementary with the previous one and it is suitable to model the wind emerging from the central body in the region around the rotational axis. In Eq. (3.1) we assume the following scaling with the colatitude

$$g(\theta) = \sin^2 \theta \qquad (3.2b)$$

while the Alfvén number is assumed to depend on the radial distance only, $M(R)$: the contour surfaces of constant M are spherical in this case. We can now proceed to define the scaling and the functional form for the other physical variables. The interested reader is referred to Tsinganos *et al.* in the present volume for an extended discussion of some general properties of the MHD equations under such self-similar conditions.

It is convenient next to define in nondimensional form the variables, by introducing a reference level r_o. In the following discussion r_o is chosen to coincide with the Alfvén distance r_* and the dimensionless radial distance used is $R = r/r_*$. Also, $V_*^2 = B_*^2/(4\pi\rho_*)$ and $\alpha = 2A/(r_*^2 B_*)$ are used, where B_*, V_* and ρ_* are the poloidal magnetic field, velocity and density at the Alfvén distance r_* on the rotational axis, respectively.

Density distribution. The assumption of a spherical Alfvén surface automatically implies that the density can be expressed in a separable form; from Eq. (2.5) we have $\rho(R, \theta) = \Psi(\alpha)/4\pi M^2(R)$, and from Eq. (2.9) $\rho(R, \theta) = \rho_a(\alpha)\rho'(R)$. If we expand $\rho_a(\alpha)$ in terms of α up to the first order, we get the following expression for the density

$$\rho(R, \alpha) = \frac{\rho_*}{M^2(R)}(1 + \delta\alpha) \qquad (3.3)$$

where the parameter δ rules the nonspherically symmetric distribution of the density; for $\delta > 0$ ($\delta < 0$) the density increases (decreases) as we move away from the rotational axis.

Poloidal and toroidal velocity and magnetic field. The previous expression for the density implies that Ψ_A has the functional form (see Eq. 2.5)

$$\Psi_A = \frac{4\pi\rho_* V_*}{B_*}\sqrt{1+\delta\alpha}.\tag{3.4}$$

From conservation of the mass and magnetic fluxes and taking into account Eqs. (2.3), (2.4) and (3.4) we get the following expressions for the two components of the poloidal velocity and magnetic field

$$V_r = V_* M^2(R)\frac{f(R)}{R^2}\frac{\cos\theta}{(1+\delta\alpha)^{1/2}},\qquad V_\theta = -\frac{V_*}{2}\frac{M^2(R)}{R^2}\frac{\mathrm{d}f(R)}{\mathrm{d}R}\frac{\sin\theta}{(1+\delta\alpha)^{1/2}}$$

$$B_r = B_*\frac{f(R)}{R^2}\cos\theta,\qquad B_\theta = -\frac{B_*}{2}\frac{1}{R^2}\frac{\mathrm{d}f(R)}{\mathrm{d}R}\sin\theta.\tag{3.5}$$

The expressions for V_ϕ and B_ϕ require some specification of of the integrals $L(\alpha)$ and $\Omega(\alpha)$. Inside a flux tube with label α, the axial electric current enclosed at a cylindrical distance $r\sin\theta$ is $\propto r\sin\theta B_\phi \propto L(\alpha)\Psi(\alpha)$. Therefore the quantity $L(\alpha)\Psi(\alpha)$ must be suitably chosen to avoid some unphysical current distribution, as it is the case with the available radially self-similar models. It occured to us to assume that the electric current vanishes at the rotational axis being linearly dependent on α through some constant λ

$$L(\alpha)\Psi_A(\alpha) = \lambda r_* B_* \alpha.\tag{3.6}$$

Taking into account Eqs. (3.4) and (2.8) we obtain then

$$L(\alpha) = \lambda r_* V_*\frac{\alpha}{\sqrt{1+\delta\alpha}},\qquad \Omega(\alpha) = \frac{\lambda V_*}{r_*}\frac{1}{\sqrt{1+\delta\alpha}}\tag{3.7}$$

while the toroidal velocity and magnetic field (Eqs. 2.7) become

$$V_\phi = V_*\lambda\frac{f(R)}{R^2}\frac{R^2/f(R) - M^2(R)}{1 - M^2(R)}\frac{R\sin\theta}{(1+\delta\alpha)^{1/2}}\tag{3.8a}$$

$$B_\phi = -B_*\lambda\frac{f(R)}{R^2}\frac{1 - R^2/f(R)}{1 - M^2(R)}R\sin\theta.\tag{3.8b}$$

We note that the parameter λ is proportional to the ratio between the azimuthal velocity at the equator and the poloidal velocity on the rotational axis at the Alfvén distance $R = 1$.

The gas pressure. The last variable to be modelled, the gas pressure P, is also expressed through a formal expansion in α truncated at the first order

$$P(R,\theta) = P_o(R)[1 + K(R)\alpha]\tag{3.9a}$$

that can be written as function of the colatitude θ in non dimensional form

$$Q(R,\theta) = Q_0(R) + Q_1(R)\sin^2\theta\,, \qquad Q_1(R) = Q_0(R)\,K(R)\,f(R) \quad (3.9b)$$

where $Q_{0,1} = 2P_{0,1}/(\rho_* V_*^2)$.

According to Eqs. (3.9) the gradient of the pressure has two components. One is directed in the radial direction partly balancing the gravitational pull of the central body, while the other is directed perpendicular to the streamlines, contributing to the dynamical balance across them.

4. Equations of the wind

Following the scaling laws discussed in the previous section, we obtain from Eqs. (2.1)

$$\frac{f f'}{2R^2}\frac{\mathrm{d}M^2}{\mathrm{d}R} = Q_1 + \frac{f}{2R^2}\left(f'' - \frac{2f}{R^2}\right) - \frac{M^2}{2R^2}\left[f f'' - \frac{(f')^2}{2}\right]$$

$$-\frac{\lambda^2}{R^2(1-M^2)^2}\left[\frac{1}{M^2}(fM^2 - R^2)^2 - 2(f - R^2)^2\right] \quad (4.1a)$$

$$\frac{\mathrm{d}Q_1}{\mathrm{d}R} = \frac{2f^2}{R^4}\frac{\mathrm{d}M^2}{\mathrm{d}R} - \frac{\delta\nu^2}{M^2 R^2}f - \frac{f'}{2R^2}\left(f'' - \frac{2f}{R^2}\right) + \frac{M^2}{R^3}\left[\frac{(f')^2}{2} - \frac{4f^2}{R^2} + \frac{f f'}{R}\right]$$

$$+\frac{2\lambda^2}{R^2(1-M^2)}\left[\frac{1}{M^2 R}\frac{(M^2 f - R^2)^2}{1-M^2} - (f - R^2)\frac{\mathrm{d}}{\mathrm{d}R}\left(\frac{f - R^2}{1 - M^2}\right)\right] \quad (4.1b)$$

$$\frac{\mathrm{d}Q_o}{\mathrm{d}R} = -\frac{2f}{R^2}\left[\frac{f}{R^2}\frac{\mathrm{d}M^2}{\mathrm{d}R} + \frac{M^2}{R^2}\left(f' - \frac{2f}{R}\right)\right] - \frac{\nu^2}{M^2 R^2}\,. \quad (4.1c)$$

With the primes in f we indicate its derivatives with respect to R, and also have defined $\nu^2 = V_{esc}^2/v_*^2 = 2\mathcal{G}m/r_* v_*^2$.

We remark that Eq. (4.1a) is deduced from the $\theta-$ component of the momentum conservation law, while the radial component provides the two Eqs. (4.1b) and (4.1c). With the four unknowns f, M, Q_o, Q_1, the system appears dynamically overconstrained. This is related to our choice to split the pressure gradient into two terms, so that the dynamics of the flow along the radial direction is provided by two independent relations. This in turn means that the gas pressure, i.e. the thermal energy, is important for our model. In fact, along the polar axis the flux of the Poynting vector and the centrifugal force vanish, so that the only driving force is the thermal energy of the flow.

Another important point to remark is that our equations have been deduced without any polytropic relation between the gas pressure and the density. The relations for P and ρ have been derived independently, as the result of the self-similar approach. We are dealing with *non polytropic* winds where the gas is far from adiabatic conditions and heating processes must be present. As we do not know the heating function σ, we can anyway deduce *a posteriori* from Eq. (2.1c) the functional form of $\sigma(R, \theta)$ consistent with our model (Tsinganos *et al.* 1992).

Altogether, we have the three Eqs. (4.1) for the four variables M, f, Q_o and Q_1 and as no extra relation exists to close this system, we are allowed to freely impose one last assumption to our model. There are two possibilities.

1) The two components of the pressure are related. In Eq. (3.9b) we may assume for simplicity that $K(R) = \kappa = \text{const}$, such that the unknowns are M, f and Q_o; the streamline structure $f(R)$ is deduced in this case.

2) The two components of the pressure are left independent. The shape of the streamlines, i.e. the functional form of $f(R)$, is prescribed and the unknown variables are M, Q_o and Q_1 instead.

5. Solutions with prescribed pressure distribution

Since in Eqs. (4.1) terms involving $f''(R)$ are present, these three equations can be expressed as a system of four equations of the first order. For convenience, let us define the two new variables (Sauty and Tsinganos 1994)

$$G(R) = \frac{R}{\sqrt{f}}, \qquad F(R) = \frac{R}{f}\frac{\mathrm{d}f}{\mathrm{d}R}. \tag{5.1}$$

The quantity $G(R)$ represents the transversal cross sectional area of a flux tube in units of its cross sectional area at the Alfvén point, while $F(R)$ governs the opening of the streamlines. For $F(R) = 0$ we have radially expanding poloidal streamlines, while for $F(R) = 2$ the streamlines are collimated parallel to the polar axis $[f(R) \propto R^2$ and $G(R) = \text{const}]$. For $F(R) \leq 2$ we have an expanding flux tube that does not attain full collimation but to infinity, while for $F(R) > 2$ the radius of the flux tube shrinks and the streamlines are bent towards the polar axis.

From Eqs. (4.1), (5.1) and (3.9b) with $K(R) = \kappa$, the following system of four equations for the variables Q_o, G, F, and M is deduced

$$\frac{\mathrm{d}Q_o}{\mathrm{d}R} = -\frac{2}{G^4}\left[\frac{\mathrm{d}M^2}{\mathrm{d}R} + \frac{M^2}{R}(F-2)\right] - \frac{\nu^2}{R^2 M^2} \tag{5.2a}$$

$$\frac{\mathrm{d}G^2}{\mathrm{d}R} = \frac{G^2}{R}(2-F) \tag{5.2b}$$

$$\frac{\mathrm{d}F}{\mathrm{d}R} = \frac{N_F(R, G, F, M^2, Q_o; \kappa, \delta, \nu, \lambda)}{D(R, G, F, M^2; \kappa, \lambda)} \tag{5.2c}$$

$$\frac{dM^2}{dR} = \frac{N_M(R,G,F,M^2,Q_o;\kappa,\delta,\nu,\lambda)}{D(R,G,F,M^2;\kappa,\lambda)} \quad (5.2d)$$

where

$$D = \frac{F^2}{4} + (M^2 - 1)\left(1 + \kappa\frac{R^2}{G^2}\right) + R^2\lambda^2\frac{(1-G^2)^2}{(1-M^2)^2} \quad (5.3)$$

and N_F and N_M are two rather lengthy expressions (see Sauty and Tsinganos 1994). In the following we discuss some general properties of these equations.

5.1. ENERGETICS OF THE OUTFLOW

The Bernoulli integral can be expressed in its generalized form as

$$\frac{1}{2}V_p^2 + \frac{1}{2}V_\phi^2 + h - \frac{\mathcal{G}m}{r} - \frac{\Omega(A)}{\Psi_A}r\sin\theta B_\phi - \int_{s_o}^{s}\frac{\rho\sigma}{\rho V_p}ds = H(A) \quad (5.4)$$

where $H(A)$ is a streamline constant. In our case this expression does not provide any practical information since σ is not known before a solution of the system has been found. However, expressing the various terms in Eq. (5.4) with our self-similar variables we see that, if $K(R) = \kappa$, the following *new* integral emerges which is a global constant for all streamlines (Sauty and Tsinganos 1994)

$$\epsilon = \frac{M^4}{R^2G^2}\left(\frac{F^2}{4} - 1 - \kappa\frac{R^2}{G^2}\right) - (\delta - \kappa)\frac{\nu^2}{R}$$

$$+ \frac{\lambda^2}{1-M^2}\left[\frac{(M^2-G^2)^2}{G^2(1-M^2)} + 2(1-G^2)\right]. \quad (5.5a)$$

As we shall see in the following, this new integral ϵ measures the excess of the volumetric energy along a given streamline as compared to that along the pole, normalized to the corresponding volumetric magnetic rotator energy $\rho(r,A)L(A)\Omega(A)$

$$\frac{\epsilon}{2\lambda^2} = \frac{\rho(r,A)H(A) - \rho(r,\text{pole})H(\text{pole})}{\rho(r,A)L(A)\Omega(A)} \quad (5.5b)$$

and determines the asymptotic character of the solution.

5.2. ASYMPTOTIC STRUCTURE OF THE OUTFLOW

An analytical study of Eqs. (5.2) can be carried out in the asymptotic region $(R \gg 1)$ for a cylindrically collimated flow $(F = 2)$. In this case the dynamical equilibrium in the cylindrical radius direction $\varpi\ (= R\sin\theta)$ is given by

$$\rho\frac{V_\phi^2}{\varpi} = \frac{d}{d\varpi}\left(\frac{B_\phi^2}{8\pi}\right) + \frac{B_\phi^2}{4\pi\varpi} + \frac{dP}{d\varpi} \qquad (5.6a)$$

assuming that asymptotically the pressure is a function only of ϖ, $Q(\varpi) = Q_{0,\infty}(1 + \kappa\varpi^2)$. Then, Eq. (5.6a) gives

$$(G_\infty^2 - M_\infty^2)^2 = 2M_\infty^2(1 - G_\infty^2)^2 + \frac{\kappa}{\lambda^2}G_\infty^4 M_\infty^2(1 - M_\infty^2)^2 Q_{0,\infty}. \qquad (5.6b)$$

This relation can be combined with (Eq. 5.5a) which becomes for $R \gg 1$

$$\epsilon = \frac{\lambda^2}{(M_\infty^2 - 1)^2}\left[\frac{(M_\infty^2 - G_\infty^2)^2}{G_\infty^2} + 2(G_\infty^2 - 1)(M_\infty^2 - 1)\right] - \kappa\frac{M_\infty^4}{G_\infty^4} \qquad (5.7)$$

such that M_∞ can be obtained as a function of the parameters λ, κ, ϵ and $Q_{0,\infty}$.

In particular, for a pressure vanishing asymptotically, $Q_{0,\infty} = 0$, M_∞ is given as the root of a fourth degree polynomial

$$\frac{\kappa}{\lambda^2}M_\infty^4 + \sqrt{2}\frac{\kappa}{\lambda^2}M_\infty^3 + \frac{\kappa + \epsilon}{2\lambda^2}M_\infty^2 + \sqrt{2}\left(\frac{\epsilon}{\lambda^2} - 1\right)M_\infty + \frac{\epsilon}{\lambda^2} - 2 = 0. \qquad (5.8)$$

The asymptotic cross sectional area is given by Eq. (5.6b), by taking the positive root which is the only physically accepted (Sauty and Tsinganos 1994),

$$G_\infty^2 = M_\infty\frac{M_\infty + \sqrt{2}}{1 + \sqrt{2}M_\infty}. \qquad (5.9a)$$

We find then a simple relation between the asymptotic cross sectional area and Alfvén number for the usually expected large values of M_∞, namely that the asymptotic cross sectional area is approximately equal to the square root of the asymptotic Alfvén number

$$G_\infty^2 = \frac{M_\infty}{\sqrt{2}}. \qquad (5.9b)$$

Consider for a moment further the case where the gas pressure is assumed to be spherically symmetric, $\kappa = 0$. As we see from Figs. 1, cylindrically collimated solutions with M_∞ and G_∞ finite exist only for positive values of ϵ. And, when we approach negative values of ϵ (for $\kappa = 0$), both M_∞ and the asymptotic cross section (see Eqs. 5.9) become infinitely large, a situation corresponding to a radial opening of the streamlines. This simple criterion was used by Sauty and Tsinganos (1994) to distinguish between strongly collimated jets where rotation, i.e. λ, was large enough (see Eq. 5.5a) to ensure positive values of ϵ, from ordinary noncollimated winds emerging from slower magnetic rotators such as the Sun.

Consider next positive values of κ corresponding to an underpressured jet in relation to its exterior. As we can see from Figs. 1, this external overpressure helps collimation through the positive pressure gradient in Eq. (5.6a), similarly as the magnetic pinching does. Note that such an increasing of the pressure with latitude is expected in many astrophysical jets and winds (Tsinganos and Sauty 1992). Let us have a closer look then to one of the tracks of constant κ/λ^2 in Figs. 1, say, $\kappa/\lambda^2 = 10^{-3}$. For positive values of ϵ/λ^2 we have $M_\infty \approx 10$ corresponding also to a value of the same order for G_∞^2, i.e., a rather tightly collimated jet. Then, as we move upwards along this track in the direction of decreasing ϵ/λ^2 we see that we have a transition from the tightly collimated jet of low G_∞ and M_∞ for ϵ roughly positive, towards a loosely collimated wind with large asymptotic Alfvén number and cross sectional area. In this sense the criterion between winds and jets still holds but even ordinary winds should be collimated into cylinders at very large asymptotic distances. This may be actually the case in the solar wind probed by Ulysses (see the chapter by Feldman *et al.* in this volume) and we may conjecture that the relatively constant magnetic field across the south pole may result from some bending of the magnetic fieldlines of the wind that starts to weakly collimate at relatively large distances. Such a track along a constant κ/λ^2 may thus correspond to the evolutionary history of the Sun as it started from a protoplanetary T Tauri stage with a presumably collimated jet-type outflow, to its present stage with a loosely collimated wind. The value of κ/λ^2 remained roughly constant througout this period because both λ and κ decreased keeping their ratio roughly constant, with the value of λ dropping because of angular momentum loss and subsequent slowdown and the value of κ decreasing because of mixing and more homogenisation with the surrounding medium.

5.3. OSCILLATIONS IN THE WIDTH OF COLLIMATED OUTFLOWS

Before the asymptotic collimated region is reached, the outflow develops a series of pinchings in its cross section. By a formal expansion of $G(R)$ and $M(R)$, $G^2 = G_\infty^2(1 + \xi)$ and $M^2 = M_\infty^2(1 + \mu)$ where $(\xi, \mu) \ll 1$, we find from the equations of motion an equation similar to the classical harmonic oscillator problem: $\ddot{\xi} + \omega^2(G_\infty, M_\infty)\xi = 0$. The frequency ω is related to the oscillations wavelength Λ_{osc} by $4\pi^2 r_*^2/\Lambda_{osc}^2 = \omega^2$ with Λ_{osc} given in terms of G_∞ and M_∞ (Sauty and Tsinganos 1994).

Evidently, these oscillations are a natural consequence of the interplay along the cylindrical radius distance ϖ of the inwards magnetic and pressure

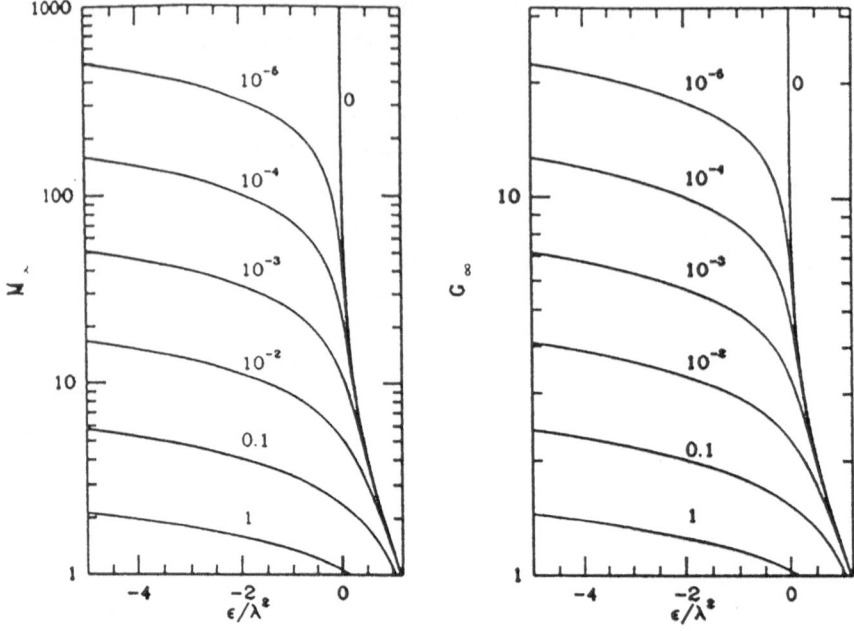

Figure 1. Plot of M_∞ (left panel) and G_∞ (right panel) vs ϵ/λ^2, for some values of κ/λ^2.

forces (for $\kappa > 0$) with the outward centrifugal force. Then, the centrifugal force acts as a restoring force when the cross section of the jet is reduced beyond some limit and the magnetic/pressure gradient force acts as the restoring force when the cross section of the jet is increased beyond the same limit.

5.4. ALFVÉN AND X-TYPE SINGULARITY

For the numerical integration of Eqs. (5.2) we must take into account the presence of singularities. The smooth crossing of these critical points constrain the values of the parameters and select unique solutions. First, the Alfvén singularity at $R = 1$ (where $M = G = 1$) appears explicitly. A second singularity is also present, as it can be seen from the analytical expressions of the derivatives of F and M^2 in Eqs. (5.2c) and (5.2d). There, a singularity appears when $N_F = N_M = D = 0$.

Alfvén singularity. It is well known that if the transfield equation is neglected, the Alfvén transition is <u>not</u> a true singularity (see the Chapter by Tsinganos *et al.* in this volume); all solutions cross this point smoothly without any additional constraint on their slope. In the present case however where the transfield equation is indeed taken into account, the Alfvén

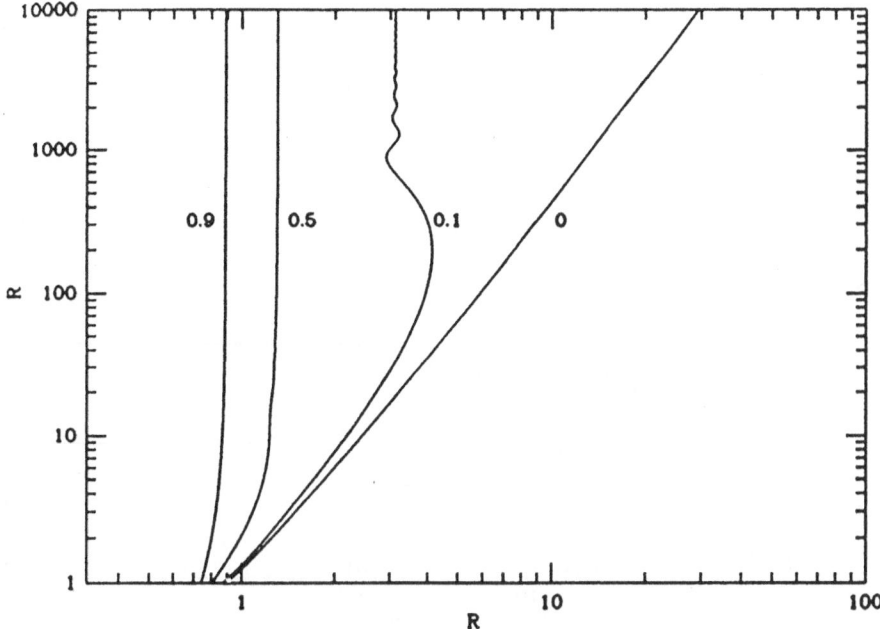

Figure 2. Plot of a poloidal streamline for $\kappa = 0$, $\lambda = 3$, $\delta = 0.01$ and some values of ϵ/λ^2. The streamline with $\theta = 45°$ at $R = 1$ is followed.

transition is a critical point. By applying de l'Hospital's rule, we obtain from Eq. (5.2d) at $R = 1$ that the slope $\mathrm{d}M^2/\mathrm{d}R|_{R=1} = p$ is given by a third degree polynomial in τ

$$\tau^3 + 2\tau^2 - \left[1 - \frac{1}{\lambda^2}\left(\kappa Q_{o,*} + \frac{F_*^2}{4} - 1\right)\right]\tau - \frac{F_*}{2\lambda^2}(2 - F_*) = 0 \qquad (5.10)$$

where $\tau = (2 - F_*)/p$ and $Q_{o,*} \equiv Q_o(R = 1)$, $F_* \equiv F(R = 1)$ (see Sauty and Tsinganos 1994).

X-type singularity. Considering the case with $\kappa \geq 0$ only, a critical point is expected to be present in the subAlfvénic region, where the expression in Eq. (5.3) changes sign. It has been shown (Tsinganos and Sauty 1992, Sauty and Tsinganos 1994) that this is a classical X-type point, whose position depends on the strength of the azimuthal components. By lowering λ the position of this singularity is moved closer to the Alfvén point. The interpretation of this singularity in physical terms is further explained elsewhere in this volume (Tsinganos *et al.*). It is actually shown there that this transition corresponds in fact to the limiting characteristic (separatrix) of slow MHD waves.

398

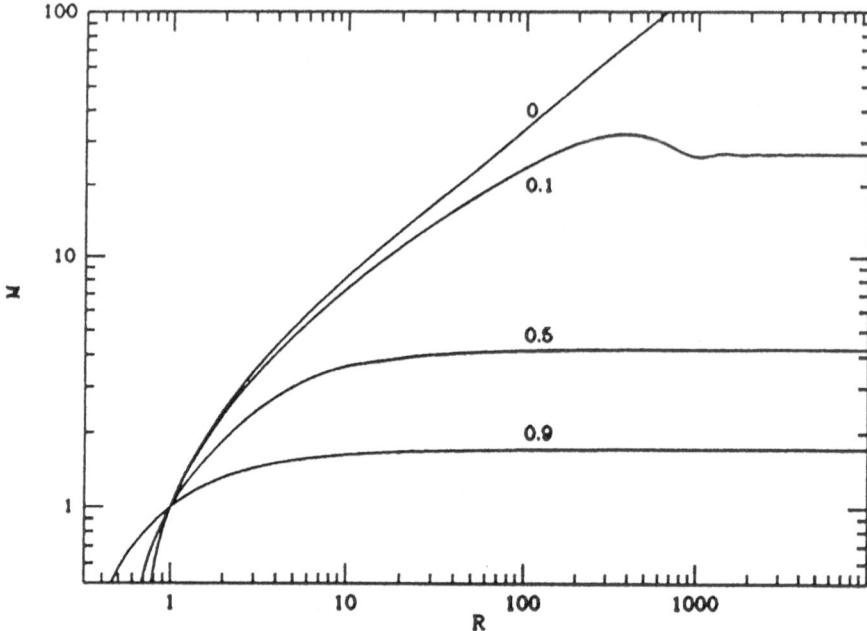

Figure 3. Plot of M vs R along the polar axis ($\theta = 0$) for $\kappa = 0$, $\lambda = 3$, $\delta = 0.01$ and some values of ϵ/λ^2.

5.5. NUMERICAL RESULTS

The solution of the system of Eqs. (5.2) depends on the five parameters λ, κ, ϵ, ν and δ, which however are not all independent due to Eqs. (5.5). To investigate the general physical properties of the solutions, we shall fix λ, κ, ϵ and δ, deducing thence ν from Eq. (5.5a). Results will be presented here for $\delta = 0.01$ and $\lambda = 3$, for $\kappa \geq 0$ and for some values of ϵ.

The numerical integration is performed downwind and upwind of the Alfvén point. First we start from the position $R = 1 \pm dR$ with $M = 1 \pm p\,dR$, where p is given by Eq. (5.10) for an arbitrary set of the parameters F_*, κ, $Q_{o,*}$ and λ. This automatically fulfills the boundary conditions for M and G at $R = 1$ and also the Alfvén regularity condition. This set of parameters however does not guarantee crossing of the X-type slow critical point in the subAlfvénic region, unless the second regularity condition there is fulfilled too. This second regularity condition will then tune the value of, say, F_* such that the solution crosses this critical point too. Finally, the value of $Q_{o,*}$ at $R = 1$ is also tuned in order to obtain a positive pressure all along the flow. This last condition may be regarded as the requirement that the solution crosses another critical point at infinity.

In Fig. 2 are plotted the poloidal streamlines for $\kappa = 0$ and some values of ϵ. In agreement with the previous asymptotic analysis, collimated solutions can be obtained only for $0 < \epsilon/\lambda^2 < 2(2 - \sqrt{2})$; for larger values of ϵ we do not have superAlfvénic solutions ($M < 1$), while for $\epsilon \leq 0$ the streamlines become asymptotically radial ($F_\infty \to 0$). The collimation distance from the Alfvén point increases by reducing ϵ, and, for $\epsilon \to 0$, the width of the jet reaches a maximum, before attaining its asymptotic collimation ($F_\infty \to 2$). Downstream of this maximum the cylindrical flux tube undergoes a series of periodic pinching of its cross section, with decreasing amplitude. The onset of these oscillations was predicted in the asymptotic analisis of section 5.3.

In Fig. 3 we plotted M vs R for different values of ϵ/λ^2. In solutions with asymptotically radial streamlines, the Alfvén number diverges, while for $\epsilon > 0$ a constant asymptotic value of M is attained, which decreases with the increase of ϵ, consistently with the previous asymptotic results. Furthermore, for $\epsilon \to 0$, in the region of the largest extend of the flux tube the Alfvén number has a maximum before attaining its asymptotic value through small amplitude oscillations.

If $\kappa > 0$ the pressure gradient is directed inwards, contributing to the wind collimation, and solutions with cylindrically collimated streamlines can be found also for $\epsilon < 0$, as discussed in the previous asymptotic analysis (see also Figs. 1). Accordingly, for $\epsilon > 0$ the effect of κ is marginal, but it plays a major role for $\epsilon < 0$ for obtaining asymptotically collimated streamlines (see Figs. 4). For $\kappa \ll \delta$, in the inner region the streamlines are basically similar to those obtained for $\kappa = 0$, while farther away they reach their maximum width and then they drastically shrink to quite narrow, oscillating channels. For $\kappa \to \delta$ the oscillations almost disappear.

The plots of M vs R are shown in Figs. 5. Again for $\epsilon > 0$ the effect of κ is not crucial, while for $\epsilon < 0$ the Alfvén number has a maximum at the pinching distance before reaching its asymptotic value. This maximum disappears for $\kappa \to \delta$. In these solutions, for $R \gg 1$ the pressure is much lower than its value at $R = 1$, therefore the asymptotic values of M are consistent with those obtained from Eq. (5.8). By comparing Figs. 2 - 5 we note that the parameter κ has a similar effect as ϵ when the nonradial component of the gradient of the pressure is included.

The above conclusions are not basically modified for other values of our set of parameters. Thus, collimated solutions can be easily constructed which may be appropriate for different astrophysical scenarios. With suitable choices for κ and ϵ, more or less collimated outflows can be modelled. They correspond to jets of various asymptotic Alfvén numbers and cross sectional areas, or to asymptotically open winds.

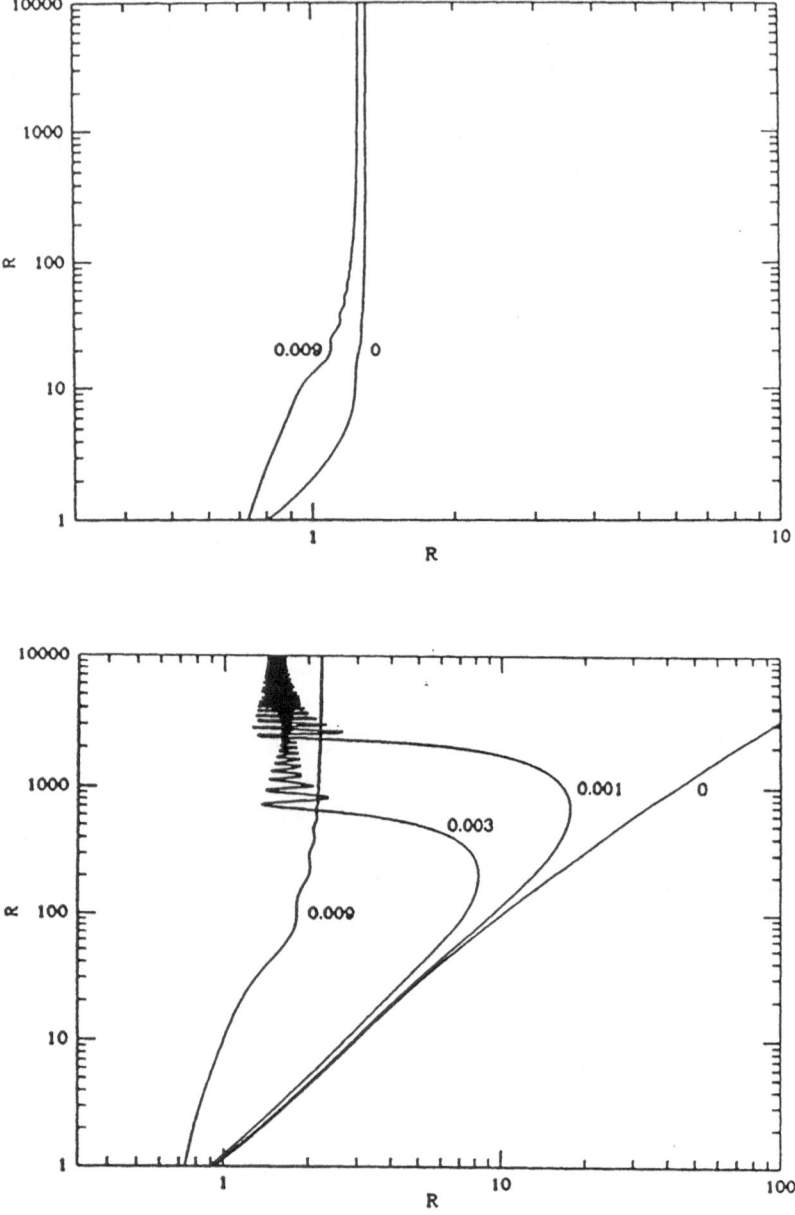

Figure 4. Plot of a poloidal streamline for $\lambda = 3$, $\delta = 0.01$ and some values of κ/λ^2, for $\epsilon/\lambda^2 = 0.5$ (upper panel) and $\epsilon/\lambda^2 = -0.2$ (lower panel). The representative streamline with $\theta = 45°$ at $R = 1$ is plotted.

6. Solutions with prescribed streamlines

If the structure function $f(R)$ of the poloidal streamlines is assumed *a priori* and the two components Q_1 and Q_o of the pressure are independent,

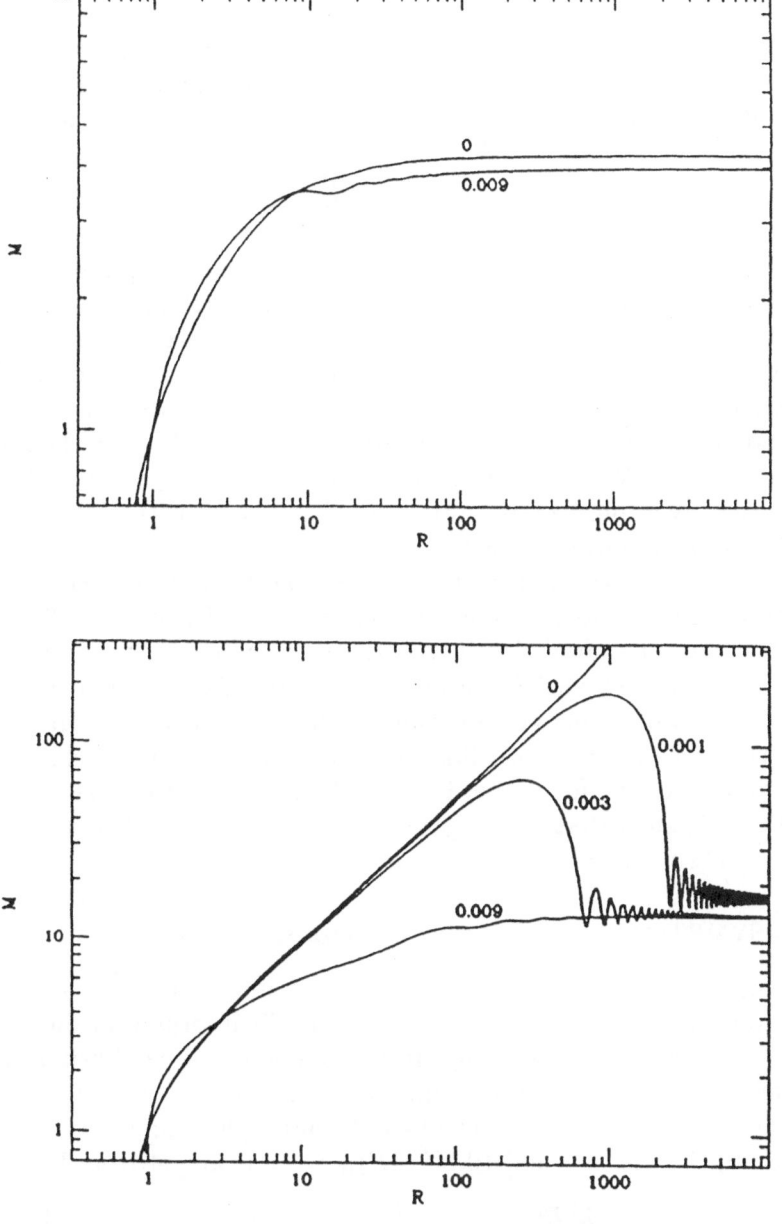

Figure 5.   Plot of M vs R along the polar axis ($\theta = 0$) for $\lambda = 3$, $\delta = 0.01$ and some values of κ/λ^2, for $\epsilon/\lambda^2 = 0.5$ (upper panel) and $\epsilon/\lambda^2 = -0.2$ (lower panel).

the original Eqs. (4.1) can be solved for M, Q_1 and Q_o. In particular, for $f' = 0$, Eqs. (4.1) reduce to a first order differential equation for $M(R)$ and an analytical expression for $Q_1(R)$ (Tsinganos and Trussoni 1991).

6.1. PRESCRIPTION OF THE STREAMLINES

The shape of the streamlines is determined by the particular form of the function $f(R)$. The following simple expression allows to describe several different configurations for axisymmetric outflows (Trussoni and Tsinganos 1993)

$$f(R) = \frac{1 + (R/R_c)^n + (R_e/R)^m}{1 + 1/R_c^{\,n} + R_e^{\,m}} \tag{6.1}$$

in terms of the positive exponents n and m and the parameters R_c and R_e. The parameters R_c and R_e determine the collimation distance such that for $R_c \longrightarrow \infty$ and $R_e = 0$, the fieldlines are everywhere radial, while for $R_c \longrightarrow \infty$ and $R_e \neq 0$ they expand towards the equatorial plane and become radial at large distances. In the general case with R_c and R_e finite, the fieldlines converge initially towards the equatorial plane, then polewards asymptotically. We will restrict our analysis to the case of streamlines monotonically converging to the rotational axis, $R_e = 0$, with R_c finite. We shall call R_c *the collimation radius*.

We note that the parameter n plays the same role with the function F introduced in the previous section. As with F, for $n = 2$ the flow is cylindrically collimated, for $n < 2$ the streamlines collimate for $R \to \infty$, and for $n > 2$ they shrink towards the polar axis. On the other hand, R_c rules the distance where collimation is effective; the larger is R_c the farther away the streamlines are collimated, and for cylindrical fieldlines it fixes the diameter of the flux tube. For $n = 2$ and $R \gg 1$ we have $G = \sqrt{R_c^2 + 1}$ (see Eq. 5.1) such that the larger is R_c the wider is the beam. For large R_c is $G_\infty \approx R_c$.

6.2. ASYMPTOTIC STRUCTURE OF THE OUTFLOW

Since in the present case Q_o and Q_1 are uncorrelated, no energy integral equivalent to ϵ exists and we can obtain information on the asymptotic structure of the wind only from the dynamical equilibrium across the streamlines. In the case of cylindrically collimated flows, $n = 2$, we obtain, from the condition of transversal equilibrium, the equivalent expression of Eq. (5.9b). For $R \gg 1$ and $(dM^2/dR)_{R\to\infty} = 0$ we get from Eq. (4.1a)

$$Q_1 = \frac{\lambda^2 R^2}{M^2(1-M^2)^2(1+R_c^2)^2} \left[\left(R_c^2 + 1 - M^2\right)^2 - 2R_c^4 M^2 \right]. \tag{6.2a}$$

By defining the quantity $q = Q_1/R^2\lambda^2$, this expression leads to a third degree polynomial for M_∞^2

$$M_\infty^6 - \left[2 + \frac{1}{q(R_c^2 + 1)^2} \right] M_\infty^4 +$$

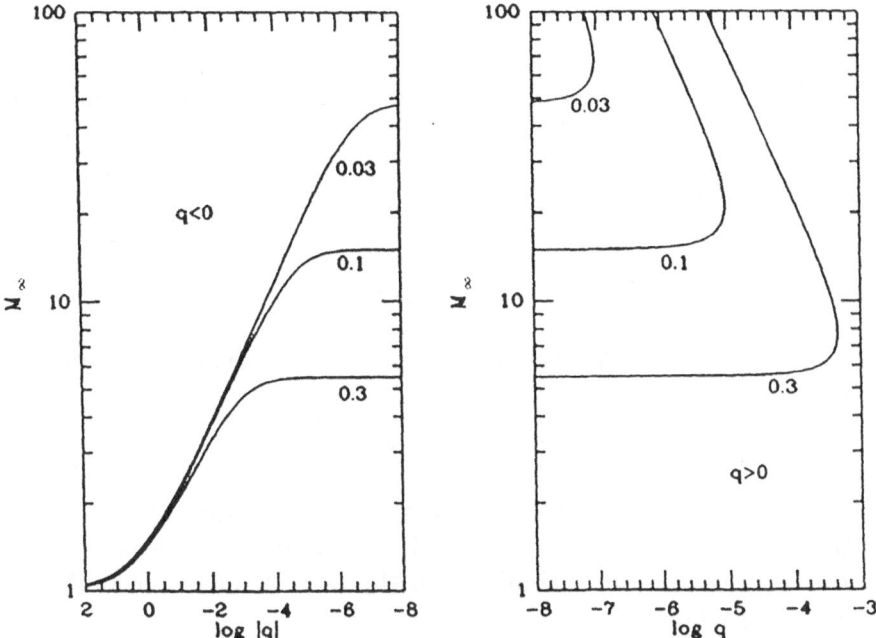

Figure 6. Plot of M_∞ vs q (left panel $q < 0$, right panel $q > 0$) for some values of $1/R_c^2$.

$$+\left[1 + \frac{2}{q(R_c^2 + 1)}\left(1 + \frac{1}{R_c^2 + 1}\right)\right] M_\infty^2 - \frac{1}{q} = 0 \qquad (6.2b)$$

whose real and positive roots provide the asymptotic Alfvén number as function of the parameters q and R_c. In Fig. 6 the value of M_∞ vs q is plotted for some values of R_c. For negative q the Alfvén number increases, while for $q > 0$ two branches exist, that join for a maximum value of q. Which of these branches is attained by the complete solution, depends on the regularity conditions at the Alfvén point, as we will discuss later.

For $q = 0$, Eq. (6.2a) reduces to the simpler form

$$M_\infty^2 \pm \sqrt{2}R_c^2 M_\infty - R_c^2 - 1 = 0 \qquad (6.2c)$$

while for $R_c \gg 1$, it gives simply

$$M_\infty \approx \sqrt{2}\, R_c^2 = \sqrt{2}\, G_\infty^2\,. \qquad (6.2d)$$

This expression is identical to Eq. (5.9b) obtained also for $Q_{o,\infty} = 0$ and thus in both cases the Alfvén number increases with the square of the collimation radius R_c of the jet.

6.3. NUMERICAL SOLUTIONS

The same procedure discussed in section 5.4 is followed now to integrate Eqs. (4.1), taking into account that equations have different singularities. *Alfvén singularity.* The Alfvén critical point is the same in both Eqs. (4.1) and (5.2) and therefore the slope of M at $R = 1$ is always given by Eq. (5.10) with $\kappa Q_{o,*} \to Q_{1,*}$ and $F_* \to f'$, and

$$p = \frac{1}{\tau} \frac{2R_c^2 + (2 - n)}{R_c^2 + 1} . \qquad (6.3a)$$

When $n = f' = 0$, Eq. (5.13) reduces to a second degree polynomial with positive root

$$\tau = \left[2 + \frac{1}{\lambda^2}(1 - Q_1^*) \right]^{1/2} - 1 . \qquad (6.3b)$$

Since Q_1^* in this case is given by an analytical expression, a family of infinite solutions exists (provided that $Q_1^* < 1 + 2\lambda^2$), all of which cross smoothly the Alfvén singular point, similarly to the case of the classical Weber and Davis (1967) model.

Other singularities. We see from the structure of Eqs. (4.1) that for $n > 0$ apparently no other critical point is present since the derivatives of the variables are not expressed through as the ratio of two quantities that can go simultaneously to zero (unless $f' = 0$, a case that we do not consider here). However, from the topologies it appears that for $R \to \infty$ all solutions become similar to those nearby an X-type critical point and therefore a singularity must be present there asymptotically. This singularity is not related to rotation, since it has been also found in the case of non-rotating flows (see Trussoni and Tsinganos 1993).

The topology is different for radial streamlines ($n = 0$) since the equation for M is given through the ratio of two expressions (Trussoni and Tsinganos 1991). In this case an X-type critical point is found downwind of the position of the Alfvén point, corresponding to the limiting characteristic of the fast magnetosonic waves (see Tsinganos *et al.*, this volume). *Numerical integration and boundary conditions.* For $n > 0$ the integration is performed upwind and downwind from the Alfvén point, as for Eqs. (5.2). However the different structure of the equations requires a slightly modified technique. The integration in the upstream region is straightforward, since no singularity is present there. We cannot integrate downstream from the Alfvén point due to the presence of the singularity for $R \to \infty$. The solutions have a *flaring* topology with the variables diverging to ∞ or $-\infty$ unless the parameters are tuned very carefully. Therefore, the integration in the superAlfvénic region is carried out with a finite difference code, suitable to treat stiff systems (see Trussoni and Tsinganos 1993). For $n = 0$, the equation for M is integrated numerically upwind and downwind of the X-type critical point, such that the Alfvén singularity can be crossed with the correct slope (Tsinganos and Trussoni 1991).

Concerning the boundary conditions, we start with $M = 1$ at the Alfvén point, and its slope is given by Eq. (6.3a). A unique value of Q_1 must be selected at $R = 1$, such that asymptotically the solution does not diverge, because of the presence of the asymptotic singularity. Again Q_o must be chosen at $R = 1$ such that the pressure is positive all along the flow. In the particular case with $n = 2$ (cylindrical streamlines) a range of values of Q_o is allowed and in fact collimated solutions exist with a non vanishing pressure asymptotically.

In the following we present some typical solutions, restricting ourselves to the cases of radially expanding and cylindrically collimated winds, by fixing $\lambda = 3$ and $\nu^2 \delta = 16$.

Solutions with radial streamlines. The behaviour of the outflow depends mainly on λ, that rules the magnetorotational contribution to the acceleration. For $\lambda \to 0$ the flow is basically thermally driven. The critical point almost coincides with the Alfvén point while V_p increases very rapidly in the subAlfvénic and transAlfvénic region, reaching asymptotically a constant value, appropriate for a thermally driven Parker-type wind, albeit with a nonspherically symmetric pressure distribution,

$$V_{p,\infty} = V_o\sqrt{1 + \delta\nu^2} \tag{6.4}$$

(Tsinganos *et al.* 1992). By rising λ the X-type critical point is shifted downwind, the flow velocity is lowered in the intermediate region but increases slowly for $R \gg 1$, without reaching a formally constant asymptotic value (see Fig. 7).

This behavior is understood by the following asymptotic expression of the poloidal velocity V_p along the polar axis, that can be obtained analytically from Eq. (5.2d) for $R \gg 1$ (see Tsinganos *et al.* 1992),

$$V_{p,\infty}(r) = v_o \left[\left(\frac{F_B \Omega_o^2}{4\pi F_m} \right)^{1/3} \right] [6\ln(r/r_o)]^{1/3} \tag{6.5}$$

where F_B and F_m are the magnetic and mass fluxes (by "o" we indicate the base of the flow). The quantity in the square brakets is the well known *Michel velocity* corresponding to Michel's (1969) minimum energy solution where 1/3 of the energy escapes in kinetic form while the other 2/3 in Poynting flux form. In one-dimensional winds the flow attains asymptotically the Parker or the Michel velocity depending on whether we have effectively a slow magnetic rotator SMR (basically a thermally driven wind) or a

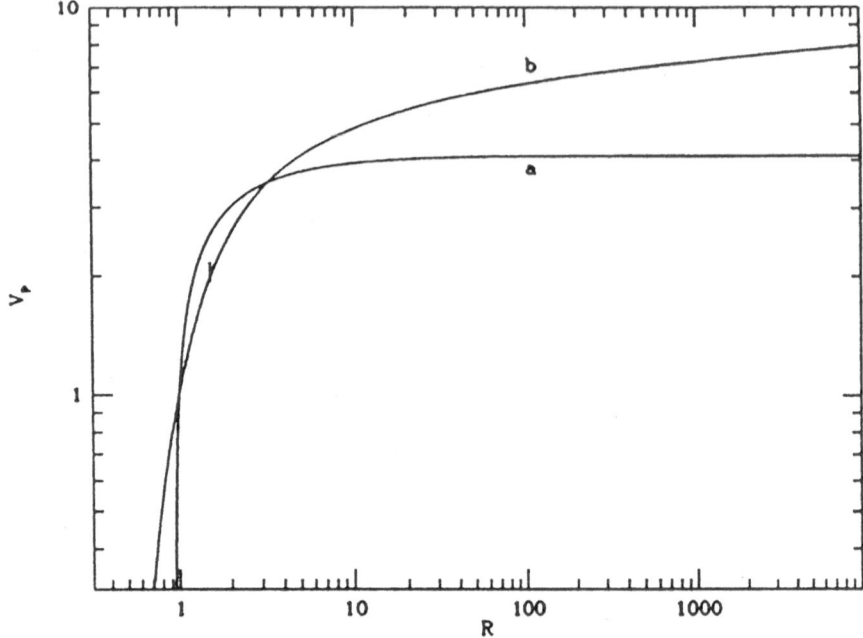

Figure 7. Plot of V_p (solid line) vs R along the polar axis ($\theta = 0$) for a radially expanding wind ($n = 0$) with $\delta\nu^2 = 16$, and $\lambda = 0$ (a) and $\lambda = 3$ (b, the small bar marks the position of the X-type critical point).

fast magnetic rotator FMR (basically a magnetocentrifugally driven wind). A similar trend is also found for outflows from accretion disks, following a 1.5-D treatment of the MHD equations wherein parabolically expanding poloidal streamlines are assumed (Kudoh and Shibata 1995). The difference in our case is the logarithmic increase of the velocity in the case of FMR which is consistent with the heating processes that keep asymptotically the wind close to isothermal conditions.

We further remark that an increase of λ implies that the pinching magnetic force increases too, and, to preserve the dynamical balance in the flow, the transversal gradient of the gas pressure is strong enough to lead to negative values of Q_1. Therefore these solutions are valid only in a small conical region around the polar axis, where $Q_o \geq |Q_1|\sin^2\theta$ (Tsinganos and Trussoni 1991).

Solutions with collimated streamlines. In Fig. 8 are shown plots of M vs R for $n = 2$, $R_c = \sqrt{10}$, and different values of the quantity $q = Q_{1,\infty}/(R^2\lambda^2)$. The assumed value of R_c means that the width of the flux tube is asymptotically ≈ 3.3 times larger than that at the Alfvén point.

The main feature of the solutions is that for negative values of $q = -|q|$, the Alfvén number rapidly increases across the transAlfvénic region, reach-

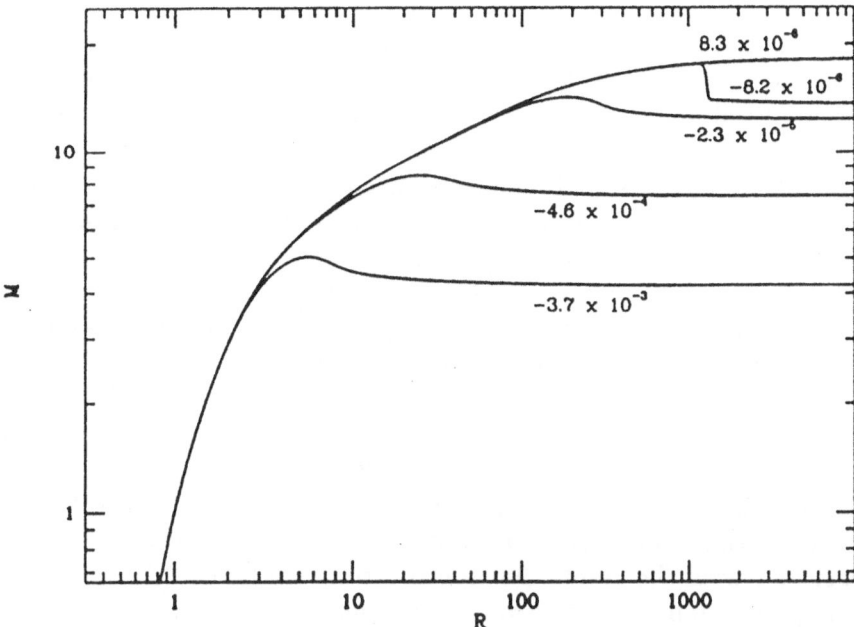

Figure 8. Plot of M vs R along the polar axis ($\theta = 0$) for a cylindrically collimated wind ($n = 2$) with $\lambda = 3$, $\delta\nu^2 = 16$, $a = 0.1$ and some values of q.

ing a maximum ("bump") before attaining its asymptotic constant value. By increasing the value of $|q|$, the position of the bump is shifted upwind, with an increasing amplitude. Consistently with this bump in velocity, the density and pressure have a minimum before they reach further downstream constant values. In order to avoid negative values of the quantities at these minima, the asymptotic pressure and density need to be nonvanishing, and increase by lowering $|q|$.

Two different classes of solutions exist whether $q < 0$ or $q > 0$ and therefore we shall discuss them separately. In the case of $q < 0$ we find that the value of M decreases with q; numerical solutions that fulfill the boundary conditions are found for each value of q without any constraint. However, from Eq. (3.5) it follows that, for $M^2 \leq R_c^2 + 1$, it is $V_r < V_*$, i.e. the flow is superAlfvénic but with a lower velocity with respect to its value at $R = 1$. These solutions can be considered as unphysical, accordingly a lower limit exists for q below which $M_\infty < \sqrt{R_c^2 + 1}$. The structure of these solutions is not strongly affected by the values of the parameters; in particular the amplitude of the bump increases by reducing λ.

For $q = 0$ the nature of the solutions appears to be rather different, since only for a single value of $q = q'$ an acceptable solution exists from $R = 1$ to infinity. The main solution feature in this case is that the asymptotic value of M is reached without any bump, while the density and the pressure decrease monotonically, and vanish asymptotically. This means that there must be an upper limit in the values of q for the class of solutions with $q < 0$, as we discussed above although it is rather difficult to isolate numerically this upper limit.

We see from Figs. 6 and 8 that the asymptotic value of M in this case lies on the lower branch, and we have tested that the upper branch can be attained by changing the parameters. By reducing λ the value of q' increases on the lower branch, reaches its maximum value (where the two branches merge) and then slides back along the upper branch, with the asymptotic Alfvén number sharply increasing by lowering q'.

Since no singularity is present, the integration in the upwind region is straightforward. The solution here is quite independent of λ and it is ruled mainly by the parameters δ and ν; the higher is $\delta\nu^2$ the steeper is the acceleration in the subAlfvénic region. Accordingly, there is a value of $\delta\nu^2$ below which the solution has a minimum, and M increases for $R \to 0$. This is consistent with the results for non rotating winds. In this zone the flow is mainly thermally driven, and the acceleration process from the poles becomes more efficient by reducing the weight of the wind, i.e. rising δ (Tsinganos and Trussoni 1991).

Referring to the pressure, as in the case with radial streamlines Q_1 can have negative values and diverge asymptotically for the two classes of solutions ($Q_1 \propto R^2$, see Eq. 6.2a). However along a streamline the total pressure is $Q_o + Q_1\alpha/f \approx Q_o + q\lambda^2 R_c^2\alpha$. Therefore, now we can easily tune the value of Q_o at the Alfvén point such that the pressure is positive everywhere inside the collimated wind.

7. Concluding remarks

Preliminary applications of this study to various types of astrophysical outflows and jets have been already presented elsewhere (Sauty and Tsinganos 1994, Sauty et al 1995). We only note briefly here a possible relation of this self-similar modelling to the remarkable jet associated with the elliptical galaxy M 87 (see Figs. 9, 11 of previous chapter by Dr. J. Biretta). The inner jet of M 87 (up to knot I) apparently expands radially while its outer part (from knot A to C) switches rather suddenly to a cylindrical pattern. It is tempting then to associate the distance to knot A (~ 1 kpc) to the collimation distance R_c (Eq. 6.1) used in our self-similar models, wherein the outflow expands almost radially up to R_c, switching to the cylindrical shape after, exhibiting also some oscillations in the cross-sectional area of the channel, Figs. (2–5). In the following we summarize some of the main features of this approach, as well as its advantages and limitations.

Even though the present analytical study is not yet complete, a conclusion already emerging is that collimated and superAlfvénic MHD solutions can be obtained with reasonable enough assumptions on the parameters. In the case of self-consistently determined streamline shape, it was found that a *nonrotating* magnetized wind can only expand asymptotically radially (Tsinganos and Sauty 1992). However, cylindrical collimation has been found for a wide range of parameters for *rotating* magnetized outflows, in particular if the non-spherically symmetric component of the pressure gradient is included ($\kappa > 0$). And, in this case of external additional confining pressure, we were able to distinguish *two* kinds of outflows, one being strongly collimated, the other loosely collimated far from the source. Those two classes may indeed correspond to jet-like structures and ordinary solar-like winds.

We have seen that, if the fieldlines are prescribed to be radial, the wind never reaches an asymptotic velocity, and, to avoid negative pressures, the solutions must be arbitrarily limited to a narrow region close to the axis. On the other hand, if the streamlines are bent polewards in a nonrotating wind, a cylindrical collimation cannot be attained because the heating of the flow diverges (Trussoni and Tsinganos 1993). These problems do not appear in collimated and rotating flows, suggesting that magnetorotational acceleration indeed plays a major role in collimated outflows.

It should be remarked that our results are consistent with those obtained by other studies following a self-similar approach (Blandford and Payne 1982, Pelletier and Pudritz 1992, Contopoulos and Lovelace 1994), or, through an asymptotic study of the full set of the MHD equations (Heyvaerts and Norman 1989). It is worth noting in particular the presence of oscillations in the collimated region, found for solutions with free streamlines. This peculiar behaviour which may be very interesting from the observational point of view, was also found in other studies (Pelletier and Pudritz 1992, Contopoulos and Lovelace 1994). Obviously these oscillations do not appear in the solutions with prescribed fieldlines where cylindrical collimation was assumed. However the bump found in the intermediate region (Fig. 8) can be related to such an oscillation in the outflow.

Several points of these classes of solutions can be further developed. A more complete analysis requires a wider parametric study, in particular solutions must be studied for $\kappa < 0$ and $0 < n < 2$ (i.e. partially collimated fieldlines). A more complex scaling law of the stream function with the co-latitude, than the one given in Eq. (3.2b), could be also pursued to analyze specific structures. Such an undertaking has already been started by Lima and Priest (1995) for radially expanding winds.

We conclude by pointing out that this approach of studying axisymmetric MHD outflows has the advantage of providing a class of exact 2-D solutions. Such solutions in closed form can be used, for example, for a stability analysis of the jet. However, as radially self-similar models break down along the rotational axis, it may happen that these meridionally self-similar solutions are approriate to describe properly the flow dynamics not far from the polar axis, if at the equator the velocity is nonzero. Finally, we should keep in mind that the energy input in the plasma is deduced *a posteriori*; this means that 'ad hoc' heating processes must be present in the plasma. An important future direction is then to examine which kind of physical mechanism can be consistent with those solutions. We note however that a similar problem one confronts also in polytropic winds, where the energy processes that maintain γ fixed are also assumed *a priori*.

Aknowledgements. We are indebted to Profs. Jean Heyvaerts, Attilio Ferrari, Robert Rosner and Drs. Boon Chye Low, Sergei Bogovalov, John Contopoulos, Joao Lima, and George Surlantzis for stimulating discussions. Two of the authors (E.T. and C.S.) acknowledge the financial support of the ASI. Part of this research has been developed in the framework of bilateral scientific agreements of Italy and France with Greece.

References

Belcher J.W. and MacGregor K.B. 1976, *Ap. J.*, **210**, 498.
Blandford R.D. and Payne D.G., 1982, *M.N.R.A.S.*, **199**, 883.
Contopoulos J. and Lovelace R.V.E., 1994, *Ap. J.*, **429**, 139.
Heyvaerts J. and Norman C., 1989, *Ap. J.*, **347**.
Hu Y.Q. and Low B.C., 1989, *Ap. J.*, **342**, 1049.
Kudoh T. and Shibata K., 1995 *Ap. J. (Lett.)*, **452**, L41.
Lima J.J. and Priest E., 1995, *Astr. Ap.*, in press.
Lovelace R.V.E., Berk H.L. and Contopoulos J., 1991, *Ap. J.*, **379**, 696.
Low B.C. and Tsinganos K., 1986, *Ap. J.*, **302**, 163.
Michel, F.C., 1969, *Ap. J.*, **158**, 727.
Parker E., 1958, *Ap. J.*, **128**, 664.
Parker E., 1963, *Interplanetary Dynamical Processes*, Interscience, New York.
Pelletier G. and Pudritz R.E., 1992, *Ap. J.*, **394**, 117.
Rosso F. and Pelletier G., 1994, *Astr. Ap.*, **287**, 325.
Sauty C., 1994, *Ann. Phys. Fr.*, **19**, 469.
Sauty C. and Tsinganos K., 1994, *Astr. Ap.*, **287**, 893.
Sauty C., Tsinganos K. and Trussoni E., 1995, in *Disks and Outflows around Young Stars*, in press.
Trussoni E. and Tsinganos K., 1993, *Astr. Ap.*, **269**, 589.
Tsinganos K., 1982, *Ap. J.*, **253**, 775.
Tsinganos K. and Low B.C., 1989, *Ap. J.*, **342**, 1028.
Tsinganos K. and Sauty C., 1992, *Astr. Ap.*, **257**, 790.
Tsinganos K. and Trussoni E., 1991, *Astr. Ap.*, **249**, 156.
Tsinganos K., Trussoni E. and Sauty C., 1992, in *The Sun: A Laboratory for Astrophysics*, T.E. Schmeltz and J.C. Brown Eds., Kluwer Academic Publishers, Dordercht, p. 349.
Weber E.J. and Davies L., 1967, *Ap. J.*, **148**, 217.

NUMERICAL SIMULATION OF PLASMA FLOW IN THE MAGNETOSPHERE OF AN AXISYMMETRIC ROTATOR

SERGEI V. BOGOVALOV

Moscow Engineering Physics Institute
Kashirskoje shosse 31, Moscow, 115409, RUSSIA
bogoval@mx.iki.rssi.ru

Abstract. The boundary conditions and the nature of the critical surfaces of the stationary flow of magnetized plasma ejected by rotating objects are discussed from the causality point of view. To obtain the solution of the problem of the stationary plasma ejection by a magnetized rotating star, the three dimensional time-dependent axisymmetric problem is solved numerically. The solution for a cold nonrelativistic plasma ejected by a star having a monopole-like magnetic field is obtained. An analysis of the plasma flow near the critical surfaces is performed. It is shown that in the nonrelativistic case almost 50% of the Poynting flux can be transformed in to the kinetic energy of the plasma.

1. Introduction

Axisymmetric flows in the magnetosphere of a rotating object are widely considered as the simplest example of plasma flow in the environment of several astrophysical objects, such as stars (Parker, 1958; Weber & Davis, 1967; Mestel, 1968; Sakurai, 1985; Tsinganos & Trussoni, 1991), active galactic nuclei (AGN) (Camenzind, 1986; Blandford & Payne, 1982; Heyvaerts & Norman, 1989; Takahashi et al 1990), radio pulsars (Goldreich & Julian, 1969; Kennel & Coroniti, 1984; Emmering & Chevalier, 1987; Arons, 1990; Michel, 1991; Beskin et al 1983; Bogovalov, 1991) and young stellar objects (Lada, 1985; Pelletier & Pudritz, 1992). However, even for the simplest of those cases, there are unavailable standard methods for the solution of the stationary problem. One of the ways to understand various intriguing aspects of the stationary problem is to follow the evolution in time of some initial configurations. This method has been used in vari-

411

K. C. Tsinganos (ed.), Solar and Astrophysical Magnetohydrodynamic Flows, 411–425.

ous recent works (Shibata & Uchida, 1990; Washimi & Shibata, 1993a; Washimi & Shibata, 1993b; Ustugova et al, 1995). Thus, the main goals of our present work have been, the development of a regular method of solution of the stationary problem from the nonstationary problem, a detailed investigation of the plasma flow near the Alfven surface (AS) and the fast magnetosonic surface (FMS) in the stationary flow, an investigation of the collimation of the plasma along the axis of rotation, and the centrifugal acceleration of the plasma in the flow under consideration.

2. Equations of stationary MHD flow

The equations describing stationary ejection of plasma by an axisymmetric magnetized rotator have been widely studied starting with the work (Weber & Davis, 1967). Here we follow the nonrelativistic formalism presented in a previous paper (Bogovalov, 1994). An equivalent version of these equations was considered in (Sakurai, 1985), while the relativistic version of the same equations was investigated in (Ardavan, 1979; Kennel et al, 1983; Bogovalov, 1991). Finally, the equations in the general relativistic formalism were considered in (Mobarry & Lovlace, 1986; Camenzind, 1986; Takahashi et al 1990; Beskin & Par'ev, 1993).

In the axisymmetric flow the magnetic field $\mathbf{H}$ can be presented as a sum of the poloidal magnetic field $\mathbf{H_p}$ and azimuthal magnetic field $\mathbf{H}_\varphi$. The poloidal magnetic field can be expressed as

$$\mathbf{H_p} = \frac{\nabla\psi \times \mathbf{e}_\varphi}{\rho}, \tag{1}$$

where ρ is the distance to the axis of rotation. $\mathbf{e}_\varphi$ is the unit vector corresponding to the rotation around the axis z. The function ψ is proportional to the total flux of the poloidal magnetic field through a surface at radius ρ. In the frozen in approximation the relationship between the electric field $\mathbf{E}$ and the poloidal magnetic field is $\mathbf{E} = -\frac{\rho\Omega}{c}\mathbf{q}(\psi)\mathbf{H_p} \times \mathbf{e}_\varphi$ (Weber & Davis, 1967), where Ω is the angular velocity of the central object. This relationship follows directly from frozen in condition and two equations $(\nabla \cdot \mathbf{H}) = 0$ and $\nabla \times \mathbf{E} = 0$. The function $q(\psi)$ is constant along the poloidal field force line and describes the differential rotation of the lines of force.

The first equation defining the dynamics of the plasma along the lines of the poloidal field is the conserved energy flux

$$\frac{u^2}{2} + \mu + \phi - f(\psi)\rho\Omega q(\psi)H_\varphi = W(\psi). \tag{2}$$

The second equation is the conserved angular momentum flux

$$\rho\Omega u_\varphi - f(\psi)\rho\Omega H_\varphi = M(\psi). \tag{3}$$

Projection of the frozen-in condition on the electric field gives

$$\rho \Omega q H_p + u_p H_\varphi = u_\varphi H_p. \tag{4}$$

Here, $u^2 = u_p^2 + u_\varphi^2$, u_p is the velocity of plasma along a field line, u_φ is the azimuthal velocity of plasma, μ is the enthalpy of plasma per one particle, ϕ is the gravitational potential of the star, $f = \frac{H_p}{4\pi m n u_p}$. The functions $W(\psi)$ and $M(\psi)$ are proportional to the energy and angular momentum flux per particle. These functions are constant along field lines. Therefore they depend only on ψ. The last equation is the adiabatic condition

$$s = s(\psi), \tag{5}$$

where s is the entropy per particle. Non adiabatic flows were considered in the range of works by Tsinganos (Trussoni et al, 1996; Sauty & Tsinganos, 1994).

The transfield equation for ψ in the nonrelativistic limit is as follows (Bogovalov, 1994)

$$(u_p^2 - u_a^2)[(u_p H_\rho^2 - H_p^2 U_m)\psi_{\rho\rho} + 2H_\rho H_z u_p \psi_{\rho z} + (u_p H_z^2 - H_p^2 U_m)\psi_{zz} +$$

$$+ H_z H_p^2 \times \{u_m + \frac{u_\varphi^2}{u_p}\} + \frac{H_p^4 k}{u_p}\rho^2(u_p^2(\ln f)'_\psi + \frac{v'_\psi u_\varphi}{\rho\Omega q}) - \frac{H_p^2 c}{\Omega q u_p}(\mathbf{E} \cdot \nabla\phi) +$$

$$+ \frac{\rho^2 k H_p^4 u_\varphi^2}{u_p}(\ln q)'_\psi] = -u_a^2 H_p \rho\Omega q H_\varphi\{2H_z - \frac{1}{(q\Omega)^2}(\frac{R H_p U_m}{f} -$$

$$- \frac{V'_\psi H_p}{u_p}\mathbf{U} \cdot \mathbf{H}) + (\ln q)'_\psi \frac{\rho H_p}{q\Omega}(cE + \frac{u_\varphi}{u_p}\mathbf{U} \cdot \mathbf{H})\}, \tag{6}$$

where $V = W - Mq$, $k = (\frac{C_s}{u_p})^2$, $U_m = k\frac{(u_p^2 - u_a^2)}{u_p} + f\frac{H^2}{H_p}$, $R = W'_\psi + \rho\Omega H_\varphi(fq)'_\psi$, $\mathbf{U} \cdot \mathbf{H} = u_p H_p + u_\varphi H_p$. Symbol $()'$ denotes the derivative with respect to ψ. Equation (6) is a mixed elliptic -hyperbolic equation. In the regions $0 < u_p < u_c$ and $u_{sl} < u_p < u_f$, it is an elliptic equation. In the regions $u_c < u_p < u_{sl}$ and $u_p > u_f$ equation (6) is hyperbolic. u_c is the cusp velocity (Polovin & Demutskii, 1990), u_{sl} and u_f are the slow and fast magnetosonic velocities, C_s is the adiabatic sound velocity.

3. Boundary conditions and critical surfaces for mixed-type equations

Equation (6) is a second order equation in partial derivatives. To solve this equation it is necessary to specify some boundary conditions. This problem is not trivial even for the simplest mixed-type equations (Guderley, 1957). Singularities of equations (2-6) and the existence of a priori unspecified functions such as the velocity of plasma and two components of the magnetic field on the surface of the star make the boundary value problem for equation (6) especially complicated (Bogovalov, 1994).

Let's start our analysis of the problem trying to answer the question: how many conditions must be specified on any boundary of the stationary flow in order to obtain a physically reasonable solution? The answer on this question follows from the causality principle. According to this principle the boundary conditions of the stationary problem are well specified if the temporal evolution of a perturbation of the flow has a unique solution without the formation of discontinuities on the boundary. This principle has been formulated in (Gelfand, 1959) and was used for investigation of the stability of shock fronts in relation to their splitting under the name of "evolutionarity" (Polovin & Demutskii, 1990; Achiezer et al, 1958). According to this principle the time dependent problem must have a unique solution in terms of small amplitude perturbations, if the original perturbation of the stationary flow was small. In particular, it must be correct in relation to the perturbation in the form of arbitrary plane wave incident on the boundary at arbitrary angle. In other words the boundary conditions of the stationary problem are well specified if the problem of reflection of a small amplitude plane wave incident on the boundary has a unique solution. This problem is reduced to the solution of a system of linear equations in relation to the amplitudes of waves outgoing from the boundary. The number of the equations equals to the number of independent boundary conditions. The system has a solution if the number of independent boundary conditions is equal to the number of outgoing waves generated at the reflection of some arbitrary plane wave.

It is important that the number of outgoing waves generated at the reflection of an MHD wave incident on the boundary at arbitrary angle, always equals to the number of waves outgoing from the boundary perpendicular to its surface. This remarkable statement was proved in (Kontorovich, 1958). It follows from this statement that the number of boundary conditions in the problem satisfying the causality principle must be equal to the number of waves outgoing perpendicularly to the surface of the boundary.

In the frame system connected with plasma plane waves are distinguished by polarization, wave vector and frequency. At one wave vector the number of different waves equals the number of physical parameters characterizing plasma. In magnetized plasma they are density, pressure, three components of the velocity and two components of the magnetic field. The magnetic field has only two freedom levels because all the components are connected by relationship $(\nabla \cdot \mathbf{H}) = 0$. Total number of waves is equal 7. Waves with the same polarization and wave number can differ by the sign of the frequency. These are the waves propagating in opposite directions. In moving plasma the velocity of a wave is equal the sum of the own velocity of the wave in the plasma and the velocity of the plasma. The direction of propagation of the wave in the moving plasma finally depends on the relationship between the own group velocity of the wave and the velocity of plasma (Kontorovich, 1958).

Below we consider the boundary conditions on a stellar surface for several cases typical for MHD. In the first one, we assume that the velocity of the plasma is less than the slow magnetosonic velocity (u_{sl}) (hot plasma) on the stellar surface. This case is typical of stellar winds. In this case the number of waves outgoing from the stellar surface is equal to 4. They are the entropy wave, the slow, the fast and the Alfven waves (Achiezer et al, 1958). Correspondingly, the density and temperature of the plasma, together with the normal component of the magnetic field and the angular velocity of the star ought to be prescribed on the star surface in this case (Sakurai, 1985).

The second case corresponds to the cold plasma when the starting velocity exceeds the slow magnetosonic velocity. This is typical for the problem of plasma ejection by radio pulsars. In this case the number of outgoing from the star surface waves is equal 5. Along with slow mode wave with the wave vector directed from the boundary the slow mode wave having the wave vector directed to the boundary becomes outgoing. Total velocities of these waves are directed from the boundary. Correspondingly an additional boundary condition has to be added in compare to the first case. From the physical point of view the velocity of plasma on the stellar surface has to be taken as the additional boundary condition. Analysis shows that the problem of perturbation of the stationary solution in this case has really reasonable solution in terms of small perturbations (Bogovalov, 1992).

Now we consider the important case of pure hydrodynamic wind. The magnetic field equals to zero. At first sight it looks natural that three parameters have to be specified on the stellar surface. They are temperature, density of plasma and the angular velocity of the star. But more careful analysis shows that they are not sufficient to peak out unique solution.

It was mentioned above that the number of different waves propagating in the plasma is equal to the number of physical parameters defining the plasma. In nonmagnetized plasma this number is equal 5. Along with the well known entropy and two sound waves propagating in opposite directions there are two vorticity waves. They have own velocity of propagation in the plasma equal to zero and are polarized perpendicular to each other ((Landau & Lifshitz, 1959), the problem to Section 82). The perturbation in the vorticity wave propagates with the velocity of the plasma.

If the velocity of the plasma on the surface of the star is less than the sound velocity, four waves out go from the surface. They are the entropy, sound and two vorticity waves. Three boundary conditions are not sufficient to specify correctly the problem. Direct solution of this problem really confirms that at these boundary conditions there are infinitely large amount of solutions satisfying to them ((Bogovalov, 1996)). We have to add some additional condition characterizing the vorticity of the plasma on the surface of the star to specify correctly boundary conditions in pure hydrodynamical case. For more detailes see (Bogovalov, 1996).

Boundary conditions specified a priory on the stellar surface together with equations (2-6) allow in principle to determine a stationary solution. Note however that the problem cannot be treated as a Cauchy problem. The reason is that the number of boundary conditions that can be given is less than the number of the parameters defining the state of the magnetized plasma. These parameters are the density and pressure, the three components of the flow speed and the two components of the magnetic field, i.e., the total number of problem parameters equals to 7 (Achiezer et al, 1958). Among these 7, if we subtract the number of boundary conditions we are left with the free parameters. The number of these parameters is equal 3 if the plasma velocity on the surface of the star does not exceed the slow mode velocity and is equal 2 if it does that. These free parameters should be specified in the process of the solution of the problem.

The problem is similar to a boundary value problem. The difference is only that we do not know *a priori* the boundary conditions on the outer boundary of the flow. Moreover we even do not know where this outer boundary is placed.

Experience shows that the outer boundary conditions are the conditions of regularization of the stationary solution on some surfaces named "critical surfaces". The basic property of these surfaces is that some boundary conditions can be applied on them. This property allows to determine the critical surfaces in the stationary flow.

Let us introduce artificially some surface surrounding the star in a known stationary MHD flow. We can ask then the question. How many boundary conditions must be specified on this surface to reproduce the flow in the region from this surface to infinity? The answer can be derived from the above discussion. This number equals the number of waves outgoing normally from the surface to the direction of infinity. Let's now move this artificial boundary downstream the flow. At the immediate vicinity of a critical surface the following situation appears. The number of boundary conditions specified on our surface placed infinitesimally close to the critical surface but upstream to it will differ from the number of conditions which have to be specified on the surface if it is placed infinitesimally close to the critical surface but downstream this critical surface. This follows directly from the definition of the critical surface. Simultaneously it means that the number of waves outgoing normally from the surface to the direction of infinity differs in dependence on whether the surface is placed up or down the flow in relation to the critical surface.

In astrophysical winds the velocity of plasma increases downstream in comparison with the MHD wave velocities. Therefore in the vicinity of the critical surface the following relationship holds place: $v_{wave} < v_n$ downstream from the critical surface and $v_{wave} > v_n$ upstream from the critical surface. On the critical surface we have

$$v_{wave} = v_n, \tag{7}$$

where v_{wave} is the velocity of one of types of MHD waves propagating perpendicular to the surface, v_n is the normal to the surface component of the plasma speed. It follows from equality (7) that the slow and fast critical surfaces are placed in the regions where equation (6) is hyperbolic (this statement is trivial for fast mode critical surface but is not so trivial for slow mode critical surface). Critical surfaces coincide with one of the characteristics. Only the Alfven critical surface coincides with the so called Alfven surface (AS), where the poloidal velocity of the plasma equals the local Alfven velocity.

Slow and fast critical surfaces are the characteristics of equation (6) which are closed lines. This property picks out them from other characteristics. These characteristics were named as slow magnetosound separatrix surface (SMSS) and fast magnetosound separatrix surface (FMSS) (Bogovalov, 1994). In more detail the structure of the characteristics and their separatrices is discussed elsewhere in this volume (Tsinganos et al, 1996).

So, we distinguish three critical surfaces in the stationary MHD flow. They are the slow mode critical surface coinciding with the SMSS, the Alfven mode critical surface coinciding with the AS and the fast mode critical surface coinciding with the FMSS. The slow and fast critical surfaces do not coincide with the classical slow and fast magnetosound surfaces (Weber & Davis, 1967) found in the analysis of equations (2-5) describing the dynamics of plasma in prescribed poloidal magnetic fields. It is clear why. Classical surfaces are also critical but in relation to signals propagating exactly along field lines which are not perturbed. They are critical in relation to the artificially limited amount of MHD signals. The singularity of the stationary solutions on the FMSS (Blandford & Payne, 1982; Lovelace et al, 1991; Li et al 1992; Contopoulos, 1994; Tsinganos & Trussoni, 1991) and on the SMSS (Tsinganos, 1992; Tsinganos et al, 1993; Sauty & Tsinganos, 1994) was found directly in self-similar solutions.

It follows from the analysis made above that the number of free functions on the boundary of the star always equals the number of critical surfaces in the flow. Regularization of the solution on the critical surfaces can in principle be performed by variation of these free functions. This equality is the necessary condition of the existence of stationary continuous solution of the problem. But this is not the sufficient condition. It does not warrant us that the stationary solution supersonic at the infinity exist. In particular it is seen very well on the example of self-similar solutions. It is impressing fact that no self-similar solutions passing smoothly through all critical surfaces have been obtained up to now. In this situation the numerical simulation of the stationary flows which has no self-similarity like limitations is important to investigate the flow near the critical surfaces and to clarify the problem.

4. Time dependent model

Following to (Sakurai, 1985) we assume that the poloidal magnetic field of the nonrotating star has a monopole like structure. We consider the cold nonrelativistic wind ejected from the surface with the velocity u^0 exceeding the slow magnetosound velocity. We assume that the poloidal magnetic field, the plasma velocity and the plasma density of the nonrotating star are spherically symmetrical. The AS coincides with the FMS and is placed on the constant distance R_a from the star center. This distance is defined by the relationship $u_0 = H_p(R_a)/\sqrt{4\pi mn(R_a)}$. Let us assume that the magnetic field and the plasma density on the AS of the nonrotating star are H_a, n_a. Magnetic field can be presented as a sum of poloidal and azimuthal ones. The poloidal magnetic field is expressed through function ψ' according to formula (1). Now ψ' depends on not only coordinates but time also. We introduce dimensionless variables expressed through the values of these variables on the AS of the nonrotating magnetosphere.

$$\psi = \frac{\psi'}{H_a R_a^2}, U_x = \frac{u_x}{u_0}, U_z = \frac{u_z}{u_0}, U_\varphi = \frac{u_\varphi}{u_0}$$

$$n = \frac{n'}{n_a}, x = \frac{\rho}{R_a}, z = \frac{z'}{R_a}, H_p = \frac{H_p'}{H_a}, H_\varphi = \frac{H_\varphi'}{H_a}, t = \frac{t' u_0}{R_a}$$

The symbol $'$ marks usual dimension variables. The system of equations defining the dynamics of plasma in cylindrical system of coordinates in new variables is as follows,

$$\frac{\partial \psi}{\partial t} = -U_x \frac{\partial \psi}{\partial x} - U_z \frac{\partial \psi}{\partial z}, \tag{8}$$

$$\frac{\partial H_\varphi}{\partial t} = \frac{\partial (U_\varphi H_z - U_z H_\varphi)}{\partial z} - \frac{\partial (U_x H_\varphi - U_\varphi H_x)}{\partial x}, \tag{9}$$

$$\frac{\partial U_\varphi}{\partial t} = -U_x \frac{\partial x U_\varphi}{x \partial x} - U_z \frac{\partial U_\varphi}{\partial z} + \frac{1}{n}(H_x \frac{\partial x H_\varphi}{x \partial x} + H_z \frac{\partial H_\varphi}{\partial z}), \tag{10}$$

$$\frac{\partial U_x}{\partial t} = -U_x \frac{\partial U_x}{\partial x} - U_z \frac{\partial U_x}{\partial z} - \frac{1}{2x^2 n} \frac{\partial (x H_\varphi)^2}{\partial x} + \frac{U_\varphi^2}{x} + \frac{H_z}{n} (\frac{\partial H_x}{\partial z} - \frac{\partial H_z}{\partial x}), \quad (11)$$

$$\frac{\partial U_z}{\partial t} = -U_x \frac{\partial U_z}{\partial x} - U_z \frac{\partial U_z}{\partial z} - \frac{1}{2x^2 n} \frac{\partial (x H_\varphi)^2}{\partial z} - \frac{H_x}{n} (\frac{\partial H_x}{\partial z} - \frac{\partial H_z}{\partial x}), \quad (12)$$

$$\frac{\partial n}{\partial t} = -\frac{\partial n x U_x}{x \partial x} - \frac{\partial n U_z}{\partial z}, \quad (13)$$

Equations (8,9) are the frozen in condition. These equations are obtained from the equation $\frac{\partial \mathbf{H}}{\partial t} = \mathbf{u} \times \mathbf{H}$. Equation (8) ensures conservation exactly of the poloidal magnetic field flux.

Two step Lax-Wendroff differencing scheme was used for the simulation (Press et al, 1988). It was performed in the quarter of the total box of simulation. It was assumed that the solution is symmetrical in relation to the equator and to the axis of rotation. The radius of the star was taken 0.5. The outer boundary of the box of simulation was placed far enough beyond the FMS.

On the stellar surface we accept boundary conditions corresponding to the formulation of the boundary - value problem for stationary flow of cold plasma.

1. The function ψ is specified a priori on the stellar surface and does not depend on time. This boundary condition specifies the normal component of the poloidal magnetic field on the stellar surface. The tangential component is left free.

2. The tangential component of the electric field is continuous on the stellar surface. This condition specifies a priori the law of rotation of the field lines $q(\psi)$.

3. The velocity of the plasma on the stellar surface in the frame of reference rotating with the star is kept constant. Mathematically it means that the relationship $U_x^2 + U_z^2 + (U_\varphi - \alpha x)^2 = 1$ takes place on the surface. Here $\alpha = \frac{\Omega R_a}{u_0}$ is the dimensionless angular velocity of the star. Under this condition the velocity of the plasma on the stellar surface is specified a priori.

4. The density of the plasma is specified a priori on the stellar surface and is kept constant in time.

The simulation was performed in several dimensionless boxes with sizes 4×4 (the lattice 100×100) and 2×8 (the lattice 50×200). Simulation was terminated when the stationary solution was achieved.

5. Structure of the stationary solutions

Figure 1 displays the structure of one quarter of the magnetosphere. The star is placed in the left lower corner of the figure. Scales in X and Z directions are different (that is why the star has nonspherical form). Circles show the AS. Crosses show the surface where the right hand part of equation (6) goes to zero. In the stationary solution these surfaces must merge together. Coincidence of these surfaces in the numerical solution proves that the stationary solution is really achieved and this solution is described by equation (6).

The strong deflection of the flow to the axis of rotation due to the compression by the azimuthal magnetic field is the most prominent result. Earlier similar results in numerical calculations was pointed out in the works (Washimi & Shibata, 1993a; Washimi & Shibata, 1993b). Figure 1 clearly shows that the rotation of the magnetized object ejecting plasma leads to the collimation of the flow along the axis of rotation. Analysis shows that this collimated flow forms a jet at large distances from the star (Heyvaerts & Norman, 1989; Bogovalov, 1995).

Another interesting feature of the solution is shown in figure 2. The plot of the function xU_φ is presented there. This figure shows half of the total flow. It is obtained by pasting together two symmetrical quarters of the flow along the equator. The center of the star is placed in the point with coordinates x,z (0,0). The solution is **irregular** in the point P. As it can be seen in figure 1, this point is placed on the equator where the FMS is perpendicular to the field lines.

This result provides evidence that the stationary solution of the problem has a singular point. There are no doubts that an additional numerical simulation on the lattice with much more higher spacial resolution is necessary to make final conclusion about physical reality of the irregularity found in the numerical solution. But there are two arguments that this irregularity is likely not a numerical effect. Firstly, the effect is very stable. The simulation was performed for several angular velocities: α=1, 2.5, 3.03, 4. For $\alpha = 1$ no irregularity was found. In other cases every time the irregularity appears in the characteristic point P. Secondly, the effect is certainly connected with physically particular point P. This point is the point of divergence of one of the families of fast mode characteristics. For the numerical scheme this is an ordinary point.

The nature of the irregularity of the solution in the point P can finally be clarified in future numerical experiments. If this irregularity is real it is likely connected with the structure of fast mode characteristics near this point. Below we will refer on the point where the velocity of plasma is perpendicular to the FMS as to the characteristic point. Qualitatively the structure of the characteristics near the characteristic point depends on whether the flow is converging or diverging in this point. For example in the Laval nozzle the flow converges in the characteristic point. The structure of characteristics is not particular in this case (Landau & Lifshitz, 1959). If the flow diverges in the characteristic point as in the point P, it appears

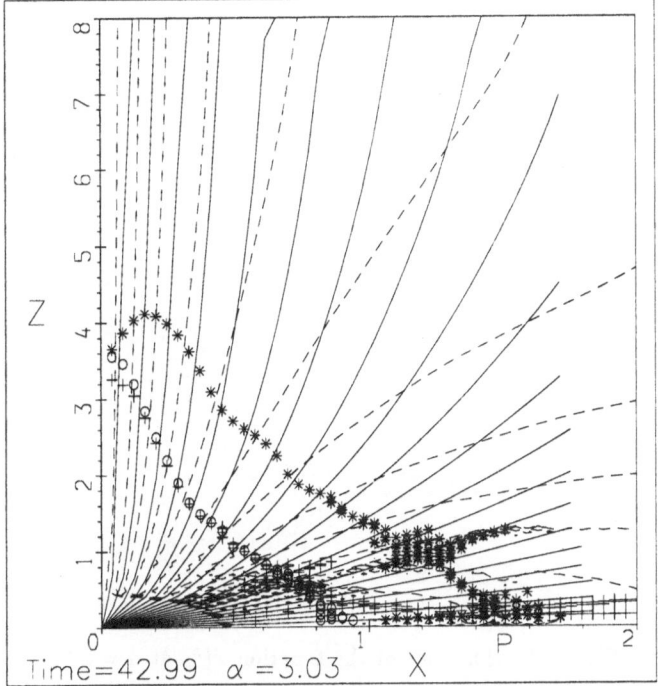

Figure 1. The structure of the stationary magnetosphere of the axisymmetric rotator at $\alpha = 3.03$. The scales in Z and X directions are different. Solid lines are the field lines of the poloidal magnetic field. Dashed lines are the lines of the poloidal electric currents. Circles o mark points on the AS. Stars ⋆ mark points on the FMS. Crosses † mark points where the right hand side of equation (6) goes to zero.

particular point of the set of characteristics (Bogovalov, 1996). A family of characteristics outgoes from this point. This can lead to the irregularity of the solution.

It is necessary to keep in mind that the irregularity can also be the result of impossibility of the supersonic flow of plasma to fill all space (Bogovalov, 1995) and especially near the equator (Camenzind, 1989).

6. Centrifugal acceleration of plasma

Michel (Michel, 1969) was apparently the first who applied ideal MHD to the investigation of the centrifugal acceleration of plasma by rotation powered pulsars in the model of an axisymmetrical rotator. He considered the dynamics of cold plasma in the monopole like poloidal magnetic field. Deformation of the poloidal field by the moving plasma was not taken into account. The FMS was placed on the infinity in his solution. As a result the effectivity of transformation of the Poynting flux carrying the main part of

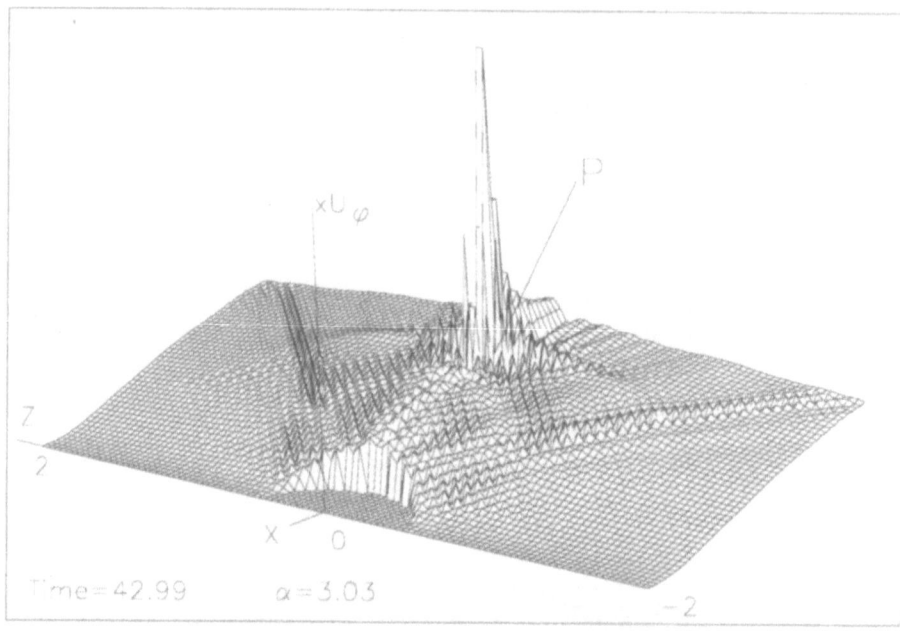

Figure 2. The plot of the function xU_φ at $\alpha = 3.03$

the rotational losses of the star in to the kinetic energy of plasma appeared very low. This conclusion does not depend on the angular velocity of the star and the energy of plasma.

The model used in this work is exactly similar to the Michel's model. But here the full problem was solved. Plasma moves in the poloidal magnetic field deformed by the plasma. The result drastically differs from that obtained in(Michel, 1969). First of all it is evident from figure 1 that the FMS is placed on the finite distance from the star. The transformation of the Poynting flux in to the kinetic energy of plasma appears very effective. The effectivity of transformation can be characterized by the coefficient $\mu = \frac{((xH_\varphi)_s - (xH_\varphi)_{x=2})}{(xH_\varphi)_s}$. The index s marks the value on the surface of the star, the index x=2 marks the value on the boundary of the box of simulation. This coefficient shows what part of the Poynting flux is transformed in to the kinetic energy of plasma. The dependence of μ on the angular velocity of the star is presented in figure 3. This coefficient was calculated for equatorial field line. It increases with increase of α and achieves 46% at $\alpha = 4$.

This result can be of crucial importance for the physics of radio pulsars if to extrapolate it on the flow of relativistic plasma ejected by fast rotating star. Existing models of electrostatic acceleration of plasma in inner (Ruderman & Sutherland, 1975; Arons, 1983) or outer gaps (Cheng et al, 1986)

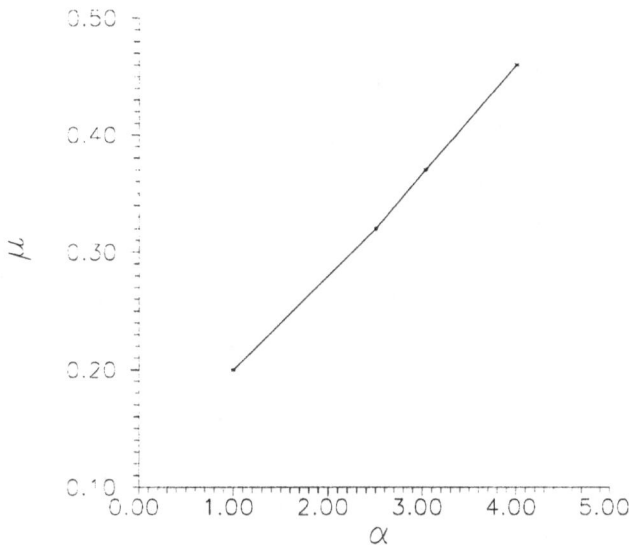

Figure 3. The dependence on α of the coefficient of the transformation of the Poynting flux in to the kinetic energy of plasma.

ensure the effectivity of transformation of the rotational energy of the neutron star in to the kinetic energy of plasma less than 1%. At the same time the analysis of the Crab Nebular observations (Kennel & Coroniti, 1984; Emmering & Chevalier, 1987) and pulsars in binary systems show that at least half of this energy is transformed in to the kinetic energy of plasma. It follows from observations that some basic very effective mechanism of plasma acceleration in pulsars exists. According to the results of the numerical simulations this mechanism is more likely the centrifugal acceleration.

It is important however to point out that it is impossible to transform 100% of Poynting flux in to the kinetic energy of plasma as it was concluded in (Chiueh et al, 1991). Our analysis of asymptotical behavior of stationary MHD solutions shows that the net poloidal electric current flowing in to the infinity is never equal to zero (Bogovalov, 1995). According to the energy conservation law (2) it means that some part (may be small) of the energy flux is always concentrated in the electromagnetic field.

Acknowledgments. The author is grateful to the participants of the ASI confernce: Prof. K.Tsinganos, Prof. M.Camenzind, Dr. J.Contopoulos, Dr. J.Ferreira, Dr. E.Trussoni and Dr. C. Sauty for stimulating discussions related to the problems in this report.

References

Parker E. (1958) Astrophysical Journal, **128**, 664

424

Weber E.J., Davis L. (1967) Astrophysical Journal, **148**, 217
Mestel L. (1968) MNRAS, **138**, 359
Sakurai T. (1985) Astronomy and Astrophysics, **152**, 121
Tsinganos K., Trussoni E., (1991) Astronomy and Astrophysics, **249**, 156
Camenzind M. (1986) Astronomy and Astrophysics, **162**, 32
Blandford R.D., Payne D.G. (1982) MNRAS, **199**, 883
Heyvaerts J., Norman C. (1989) Astrophysical Journal, **347**, 1055
Takahashi M., Nitta S., Totematsu Ya., Tomimatsu A., (1990) Astrophysical Journal, **363**, 206
Goldreich P., Julian W.H. (1969) Astrophysical Journal, **157**, 869
Kennel C.F., Coroniti F.V. (1984) Astrophysical Journal, **283**, 710
Emmering R.T., Chevalier R.A. (1987) Astrophysical Journal, **321**, 334
Arons J. (1990) in Hankins T.H., Rankin J.M., Gil J.A., eds,Proc. IAU Colloquium 128, The magnetospheric structure and emission mechanisms of radio pulsars. Pedagogical Univ. Press, Zielena Gora, p.56
Michel F.C. (1991) Theory of Neutron Star Magnetosphere. Chicago: The university of Chicago Press
Beskin V.S., Gurevich A.V., Istomin Ya.N. (1983) Sov. Phys. JETP, **58**, 235
Bogovalov S.V. (1991) AZh, **68**, 1227
Lada C.J. (1985) ARA&A, **23**, 267
Pelletier G., Pudritz R.E. (1992) Astrophysical Journal, **394**, 117
Shibata K., Uchida Y. (1990) PASJ, **42**, 39
Washimi H., Shibata S. (1993a) MNRAS, **262**, 936
Washimi H., Sakurai T. (1993b) Solar Phys., **143**, 173
Ustugova G.V., Koldoba A.V., Romanova M.M., Chechetkin V.M., Lovelace R.E. (1995) Astrophysical Journal(Letters), **439**, L39
Bogovalov S.V. (1994) MNRAS, **270**, 721
Ardavan H. (1979) MNRAS, **189**, 397
Kennel C.F., Fujimura F.S., Okamoto I. (1983) Geophys., Astrophys.Fluid Dynamics, **26**,147
Mobarry C.M., Lovlace R.V.E. (1986) Astrophysical Journal, **309**, 455
Beskin V.S., Par'ev V.I. (1993) Uspechi Phys. Nauk. **163**, 96.
Sauty C., Tsinganos K. (1994) Astronomy and Astrophysics, **287**, 893
Polovin R.V., Demutskii V.P. (1990) Fundamentals of Magnetohydrodynamics, Consultants Bureau, New York
Guderley K.G. (1957) Theorie schallnaher stromungen. Springer- Verlag,Berlin
Gelfand I.M. (1959) Uspechi Matematicheskich Nauk, **14**, 87
Achiezer A.I., Lubarskii G.Ya., Polovin R.B. (1958) JETP, **35**, 731
Kontorovich V.M. (1958) JETP, **35**, 1216
Landau L.D., Lifshitz E.M. (1959) Fluid Mechanics, Pergamon, London
Bogovalov S.V. (1996) Astronomy and Astrophysics, in preparation
Bogovalov S.V., (1992) Pis'ma Astron. Zh, **18**, 832
Lovelace R.V.E., Berk H.L., Contopoulos J. (1991) Astrophysical Journal, **379**, 696
Li Z., Chiuen T., Begelman M.C. (1992) Astrophysical Journal, **394**, 459
Contopoulos J. (1994) Astrophysical Journal, **432**, 508
Tsinganos K., Sauty C., Surlantzis, G., Trussoni, E., Contopoulos, J. (1996), this volume
Tsinganos K., Sauty C. (1992) Astronomy and Astrophysics, **257**, 790
Tsinganos K., Trussoni E., Sauty C. (1993) Outflows Focusing in Rotationing Stellar Magnetospheres. In:Linsky J., Serio S. (eds.) Advances in Stellar and Solar Coronal Physics. Kluwer, Dordrecht
Trussoni E., Sauty, C., Tsinganos K., this volume
Press W. H., Flannery B.P., Teukolsky S.A., Vetterling W.T. (1988) Numerical Recipes, Cambridge University Press, Cambridge
Bogovalov S.V. (1995) Pis'ma Astron. Zh, **21**, 633
Bogovalov S.V. (1996) MNRAS, in press

Camenzind M. (1989) in Belvedere G., ed., Accretion Disks and Magnetic Fields in Astrophysics. Kluwer, Dordrecht. p.139.
Michel F.C. (1969) Astrophysical Journal, **158**, 727.
Ruderman M., Sutherland P.C., (1975) Astrophysical Journal, **196**, 51
Arons J. (1983) Astrophysical Journal, **266**, 215
Cheng K.S., Ho C., Ruderman M.A. (1986) Astrophysical Journal, **300**, 500
Chiueh T., Li Z., Begelman M.C. (1991) Astrophysical Journal, **377**, 462

CRITICAL POINTS AND SEPARATRIX CHARACTERISTICS IN SOLAR AND ASTROPHYSICAL MHD FLOWS

KANARIS TSINGANOS

Department of Physics, University of Crete
Foundation for Research and Technology Hellas (FORTH)
GR-71409 Heraklion, GREECE, tsingan@physics.uch.gr

CHRISTOPHE SAUTY

Observatoire de Paris, DAEC, F-92195 Meudon, FRANCE,
sauty@gin.obspm.fr

GEORGE SURLANTZIS

Department of Physics, University of Crete
GR-71409 Heraklion, GREECE, sourl@physics.uch.gr

EDOARDO TRUSSONI

Osservatorio Astronomico di Torino
I-10025 Pino Torinese, ITALY, trussoni@ph.unito.it

AND

JOHN CONTOPOULOS

NASA Goddard Space Flight Center, Greenbelt
MD 20771, USA conto@jets.uchicago.edu

Abstract. One of the main difficulties in solving the MHD equations is due to the fact that a physically accepted solution is determined by the requirement that it passes through some critical points which are not known *a priori*, but are only determined simultaneously with the complete solution. In this chapter the relation of such critical points to separatrix characteristics is established, through examples of several classes of analytical, axisymmetric and self-similar solutions for flows in a uniform or spherical gravitational field. Besides the well known polytropic cases with constant polytropic index, more general flows with a variable polytropic index are also examined. These examples are the only available classes of exact 2-D dynamical equilibria which give us a chance to understand some novel and subtle properties of the nonlinear set of the MHD equations encountered in

K. C. Tsinganos (ed.), Solar and Astrophysical Magnetohydrodynamic Flows, 427–458.
© *1996 Kluwer Academic Publishers.*

the modelling of various astrophysical flows, such as in winds, jets, coronal loops, etc.

In the reviewed examples of exact solutions, the various families of characteristics of the MHD partial differential equations are plotted; it is then found that one member of each family of characteristics is asymptotically tangent to the corresponding separatrix. Furthermore, we also find that these separatrices coincide with the location of novel X-type critical points which control the topology of the solutions and wherein nevertheless the flow speed does not correspond to any of the known speeds for MHD wave propagation in a plasma (cusp/tube speed, or fast/slow and sound/Alfven speeds). Instead, it is shown that at the critical points of all axisymmetric and self-similar solutions examined, the component of the flow velocity perpendicular to the directions of axisymmetry and self-similarity, equals the characteristic slow/fast MHD wave speed in that direction. As a byproduct of this understanding of the nature of the critical points, the sound speed can be calculated locally even in flows where no polytropic relation exists.

1. Introduction

The nonlinear set of the steady MHD equations reduces to the much simpler pair of the Bernoulli and Grad-Shafranov (or transfield) equations in case there exists a translational, axial, or helical symmetry in the system under consideration and a polytropic equation of state is assumed (see e.g., Tsinganos 1982). However, even after this reduction in the number of equations, the mathematical difficulty for obtaining exact solutions remains at prohibitive high levels and in order to tackle the problem some further restrictive approximations need to be introduced. At the beginning, only one-dimensional problems were considered, such that the two resulting coupled equations are reduced to one ordinary differential equation. This is for example in the case of the well known 1-D (spherically symmetric) transonic wind (Parker 1958), or in the case of the 1-D magnetic stellar wind on the equatorial plane of a star (Weber & Davis 1967). A single equation is also obtained in the "quasi two-dimensional" approach followed to study collimated outflows (Ferrari et al. 1985, Koupelis 1990, Lovelace et al. 1991).

In the fully 2-D case the general properties of the Bernoulli and transfield equations have been analyzed for axisymmetric geometries only and in asymptotic regimes (Heyvaerts & Norman 1989, Appl & Camenzind 1993). To make further progress the assumption of self-similar symmetry is often employed, wherein various classes of exact solutions for the MHD system exist for different self similarity configurations. Following this approach, it has been possible to analyze the solutions of the MHD equations in different astrophysical environments, e.g. for flows along magnetic arcades in the solar corona (Tsinganos, Surlantzis and Priest 1993), the magnetic structure of cylindrical jets (e.g. Chan & Henriksen 1980, Bacciotti & Chiuderi 1993), the configuration of non-rotating magnetospheres (Low & Tsinganos 1986, Tsinganos & Low 1989, Low & Hu 1989, Tsinganos & Sauty 1992, Trussoni & Tsinganos 1993), the acceleration of winds from accretion disks (Bardeen & Berger 1978, Blandford & Payne 1982, Henriksen 1987, Henriksen & Valls-Gabaud 1993, Li et al 1993, Rosso & Pelletier 1994, Contopoulos & Lovelace 1994), or outflows from spherical rotating bodies (Tsinganos & Vlastou 1988, Tsinganos & Trussoni 1991, Sauty & Tsinganos 1994). The interested reader is also encouraged to see the complementary chapters by Macgregor, Trussoni et al and Heyvaerts & Norman in this volume.

One of the main problems encountered in working out the MHD system is the presence of critical points, where the equations become singular. These singularities are present both in the 1-D and 2-D geometries: a physically acceptable solution must fulfill the boundary conditions passing smoothly across these points. In one dimensional flows, the critical point appears wherever the flow speed becomes equal to one of the characteristic speeds of propagation of a disturbance in the medium (the sound speed in the HD case, and the fast and slow magnetoacoustic speeds in the MHD case). In flows of higher than one dimension however, the problem is much more complex and, in general, the critical points are not found wherever the plasma bulk flow velocity equals the aforementioned characteristic speeds.

This feature clearly appears in MHD self-similar solutions, where the assumed geometry implies constraints on the propagation of the disturbances in the plasma. This strict relation between the symmetry properties of the flow and the properties of the critical points has been given a rather limited attention in the related literature (Bardeen & Berger 1978, Blandford & Payne 1982, Li et al. 1993, Bogovalov 1994). However, some confusion oftentimes seems to arise, as to which critical point in various astrophysical problems does select a physically important MHD solution. The purpose of this review is to clarify some aspects of this question also through the examination of representative examples in various self-similar symmetries and coordinates, related to particular astrophysical scenarios.

The outline of our chapter is thus as follows. In the next Section 2 and in order to establish notation we shall quickly go through the general structure of 2-D steady MHD equations. In particular, we shall focus our attention on general properties of these equations and on how they are modified under the self-similarity assumption. In Section 3 we discuss some polytropic exact solutions in case of translational or radial self-similarity. The simpler first case refers to a stratified flow in a uniform gravitational field while the later case is typical for outflows from accretion disks. In Section 4 we discuss nonpolytropic meridionally self-similar solutions, typical for outflows from spherically symmetric gravitational bodies. The main results are discussed and summarized in Sec. 5.

2. Critical Points in Symmetric MHD Flows

In this section we summarize quickly the general properties of the reduced MHD equations for symmetric systems with emphasis on the true critical points that select the physically interesting solutions.

2.1. BERNOULLI AND TRANSFIELD EQUATIONS FOR SYMMETRIC POLYTROPIC FLOWS

The well known set of steady ideal MHD equations for an inviscid flow in a gravitational field that expresses conservation of mass, momentum, energy and magnetic flux, as well as Ampere's law is,

$$\vec{\nabla} \cdot \rho \vec{V} = 0 , \quad \vec{\nabla} \cdot \vec{B} = 0 , \tag{2.1a}$$

$$\rho(\vec{V} \cdot \vec{\nabla})\vec{V} = -\vec{\nabla}P + \frac{(\vec{\nabla} \times \vec{B}) \times \vec{B}}{4\pi} - \rho\vec{\nabla}\Phi , \tag{2.1b}$$

$$\vec{\nabla} \times \left(\vec{V} \times \vec{B} \right) = 0 , \tag{2.1c}$$

$$q = \rho\vec{V} \cdot \left[\frac{\Gamma}{\Gamma - 1}\vec{\nabla}\left(\frac{P}{\rho}\right) - \frac{\vec{\nabla}P}{\rho} \right] , \tag{2.1d}$$

where $\vec{V}$, P and ρ are the flow velocity, pressure and density, $\vec{B}$ the magnetic field, $\vec{\Phi}$ the gravitational field, Γ the adiabatic index and q the heating source. When $q = 0$ the energy equation reduces to the adiabatic relation $P \propto \rho^{\Gamma}$ ($\Gamma = 5/3$). Since oftentimes q is a complicate or unknown function, instead of eq. (2.1d) a similar relation is usually assumed, $P \propto \rho^{\gamma}$, where γ is the familiar polytropic index ($1 \leq \gamma \leq \Gamma$). Note that equations (2.1) are assumed to govern a wide variety of phenomena, such as, MHD structures with flows in a low stellar corona (prominences, Evershed flows in sunspots), or, material ejected from a star or an accretion disc (winds or jets).

The set of Eqs. (2.1) may be reduced to two coupled partial differential equations when one coordinate is ignorable (Tsinganos 1982, see also Heyvaerts & Norman and Trussoni et al, in this Volume). In fact, assuming a generalized curvilinear *orthogonal* coordinate system (x_1, x_2, x_3) wherein the coordinate x_3 is ignorable, the fields $\vec{B}$ and $\vec{V}$ can be expressed through the magnetic flux function $A(x_1, x_2)$ as,

$$\vec{B}(x_1, x_2) = \vec{\nabla} A(x_1, x_2) \times \vec{\nabla} x_3 + \vec{\nabla} x_3 \frac{h_3^2 \Omega \Psi_A - L\Psi_A}{1 - \Psi_A^2/4\pi\rho}\,, \qquad (2.2a)$$

$$\vec{V}(x_1, x_2) = \frac{\Psi_A}{4\pi\rho} \vec{\nabla} A(x_1, x_2) \times \vec{\nabla} x_3 + \vec{\nabla} x_3 \frac{h_3^2 \Omega - L\Psi_A^2/4\pi\rho}{1 - \Psi_A^2/4\pi\rho}\,, \qquad (2.2b)$$

where $h_i(x_1, x_2)$ $i = 1, 2, 3$ are the corresponding line elements of our curvilinear coordinates, while an integration of the partial differential equations (2.1a), (2.1c) and the $\hat{x}_3$-component of Eq. (2.1b) leads to the result that the angular frequency of rotation $\Omega(A)$, the total angular momentum $L(A)$ and the mass flux per unit magnetic flux $\Psi_A(A) \equiv d\Psi/dA$, are three arbitrary functions of the magnetic flux function A.

If we further assume that the flow is polytropic, $P = K\rho^\gamma$ (with constant γ), the set of Eqs. (2.1) can be reduced to two equations, representing force balance along and across the field lines (for details see Tsinganos 1982, Agim & Tataronis 1985). The first one is the well known Bernoulli equation expressing force balance along each streamline ($A = const.$),

$$\frac{1}{2}|\vec{V}|^2 + \frac{\gamma}{\gamma - 1}\frac{P}{\rho} + \Phi - \frac{\Omega h_3}{\Psi_A} B_3 =$$

$$\frac{1}{2}|\vec{V}_p|^2 + \frac{1}{2}(V_\phi - \Omega h_3)^2 + \frac{\gamma}{\gamma - 1}\frac{P}{\rho} + \Phi - \frac{1}{2}\Omega^2 h_3^2 + \Omega L = E(A)\,, \qquad (2.3)$$

Force balance across streamlines on the other hand, yields a second order partial differential equation of mixed elliptic/hyperbolic type, the so-called transfield, or Grad-Shafranov equation,

$$\left[1 - \frac{\Psi_A^2}{4\pi\rho}\right]\left[\vec{\nabla} \cdot \left(\frac{\vec{\nabla} A}{h_3^2}\right)\right] - \Psi_A\left[\frac{\vec{\nabla} A}{h_3^2}\right] \cdot \left[\vec{\nabla}\left(\frac{\Psi_A}{4\pi\rho}\right)\right] +$$

$$+\frac{1}{2(1 - \Psi_A^2/4\pi\rho)}\left[\frac{1}{h_3^2}\frac{d(L^2\Psi_A^2)}{dA} + 4\pi\rho h_3^2\frac{d\Omega^2}{dA} - 8\pi\rho\frac{d(L\Omega)}{dA}\right] +$$

$$+\frac{1}{2(1 - \Psi_A^2/4\pi\rho)^2}\left[\frac{L^2\Psi_A^2}{h_3^2} + 4\pi\rho h_3^2\Omega^2 - 8\pi\rho L\Omega\right]\frac{1}{4\pi\rho}\frac{d\Psi_A^2}{dA} +$$

$$+4\pi\rho\frac{dE}{dA} - 4\pi\frac{\rho^\gamma}{\gamma - 1}\frac{dK}{dA} = 0\,. \qquad (2.4)$$

Along each streamline the Alfvén point is located wherein the poloidal Alfvén number $M_a = V_p/V_{ap}$ equals unity, the density takes the value $\rho_a(A) \equiv \Psi_A^2(A)/4\pi$, while the otherwise free functions Ω and L are related as $h_3\Omega(A) = L(A)$.

2.2. CHARACTERISTICS OF SYMMETRIC MHD FLOWS

The notion of characteristics is rather familiar in the literature of partial differential equations (PDE) indicating some special curves/surfaces in their solution space. Even in the nonpolytropic case, the system of the MHD equations can be written down conveniently in a matrix form as a quasilinear set of mixed-type PDE's (Contopoulos 1996). Thus, denoting by $[S]$, the 7×7 column vector of the unknown quantities $\rho, \mathbf{V}, \mathbf{B}$ and by $[M]$ and $[N]$, 7×7 matrices of the MHD set, we can write

$$[M][S_{,x_i}] = [N],\qquad(2.5)$$

where the index x_i indicates partial differentiation with respect to the nonignorable coordinates x_i. For a solution in some volume bounded by a curve/surface S_o, we need to prescribe boundary conditions on S_o and solve Eq. (2.5) provided that the determinant of this system is nonzero. In other words, the requirement is that this boundary is different from any surface along which the determinant of the system [2.5] is zero. Such a surface/curve along which the determinant is zero defines a so-called characteristic surface/curve of the system, $\phi(x_1, x_2) = 0$ (Jeffrey & Taniuti 1964). In terms of the classical Alfven and sound speeds, $V_A = B/(4\pi\rho)^{1/2}$ and C_s, respectively, the equation of the characteristics can be written in the following form,

$$\{\mathbf{V} \cdot \nabla\phi\}\{(\mathbf{V} \cdot \nabla\phi)^2 - (\mathbf{V}_A \cdot \nabla\phi)^2\}$$

$$\{(\mathbf{V} \cdot \nabla\phi)^4 - |\nabla\phi|^2(C_S^2 + V_A^2)(\mathbf{V} \cdot \nabla\phi)^2 + |\nabla\phi|^2 C_S^2 (\mathbf{V}_A \cdot \nabla\phi)^2\} = 0.\quad(2.6)$$

Seven characteristics emerge from this equation. The first pair of brackets corresponds to an entropy wave propagating along the streamlines of the flow, the second pair of brackets corresponds to the two Alfvén characteristics while the third to the four slow/fast characteristics. Evidently, the component of the flow speed perpendicular to the Alfven characteristics equals the Alfvén speed while the flow component perpendicular to the slow/fast characteristics is equal to the slow/fast magnetoacoustic speeds.

2.3. SEPARATRIX CHARACTERISTICS OF BERNOULLI AND TRANSFIELD EQUATIONS IN POLYTROPIC FLOWS

In this section we explore the nature of the differential equations describing polytropic flows. Consider first the case where the function $A(x_1, x_2)$ giving the shape of the poloidal streamlines is regarded as a *known* function. This means that x_2 can be regarded as some function of A and x_1, $x_2 = x_2(A, x_1)$ while similarly $h_3 = h_3(A, x_1)$. The question we want to explore then is whether there exist any characteristic speeds emerging from

433

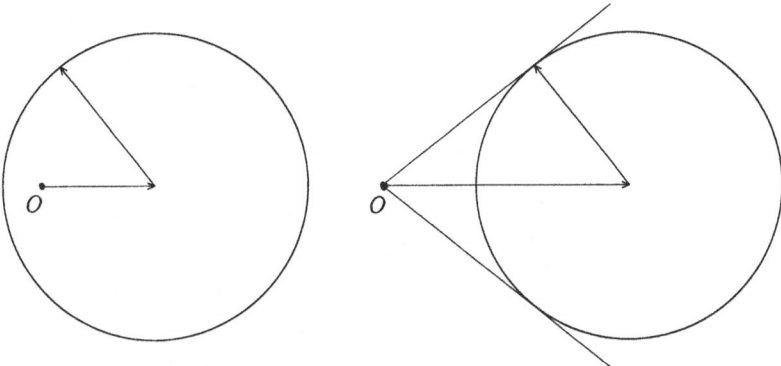

Figure 1. A uniform flow of gas with sound speed C_s and velocity $\vec{V}$, subject to a small perturbation at some fixed in space point O. The velocity $\vec{V} + C_s\hat{n}$ with which the perturbation has propagated from O, relative to the fixed system of coordinates, has its end on the circle, having thus different magnitudes for different directions of the unit vector $\hat{n}$ and also for the two cases of subsonic (a), or, supersonic hydrodynamic flow, (b). Note that in the supersonic flow (b), a disturbance starting from any point is propagating only downstream within the Mach cone of aperture $\mu = \arcsin(C_s/V)$ formed by the two characteristics C_-, C_+ (tangents from O to the wave front).

the Bernoulli equation considered <u>alone</u>, without taking into account the transfield equation [since $A(x_1, x_2)$ is considered as <u>given</u> in this case].

With a polytropic equation of state, the enthalpy is a function of ρ, while the velocity can be also regarded as a function of the two independent coordinates x_1 and ρ on each given A,

$$|\vec{V}(x_1, \rho, A)|^2 = \frac{\Psi_A^2(A)}{16\pi^2\rho^2}|\vec{\nabla}A|^2 + \left[\frac{h_3\Omega(A) - L\Psi_A^2(A)/4\pi\rho}{1 - \Psi_A(A)^2/4\pi\rho}\right]^2. \tag{2.7}$$

In the Bernoulli integral (2.3) then, along each poloidal streamline A = const. the density ρ is a function of x_1 alone, $E = E(x_1, \rho)$. Thus, the density is a level contour of the Bernoulli function E in the (x_1, ρ) plane. In this plane there are critical points wherein the Bernoulli constant has an extremum,

$$\vec{\nabla}E(\rho, x_1) = 0. \tag{2.8}$$

It follows from Eq. (2.8) $[\partial E(\rho, x_1)/\partial\rho = 0]$ that the poloidal speed $V_p = \sqrt{V_1^2 + V_2^2}$ at these characteristic points is

$$V_p^2 = C_s^2 - \frac{(L - h_3^2\Omega)^2 M^4}{h_3^2(1 - M_a^2)^3}, \tag{2.9}$$

where $C_s = \sqrt{\gamma P/\rho}$ is the sound speed, while in combination with the relation $\partial E(\rho, x_1)/\partial x_1 = 0$, their location can be determined. It is evident from Eq. (2.9) that if $M_a < 1$ we have $V_p < C_s$ while if $M_a > 1$ it is $V_p > C_s$. As a matter of fact it is straightforward to write down Eq. (2.9) in the more familiar form,

$$V_p^4 - V_p^2(C_s^2 + V_a^2) + C_s^2 V_{a,p}^2 = 0, \qquad (2.10)$$

in terms of the Alfvén speed associated with the poloidal velocity component, $V_{a,p} = \sqrt{V_{a,1}^2 + V_{a,2}^2}$ and the total Alfvén velocity, V_a. In other words, the poloidal speed equals the slow/fast MHD wave speeds at the points along each streamline $A = $ const. wherein the Bernoulli function $E(A)$ considered as function of x_1 and ρ has an extremum. These are precisely the only two critical points encountered for example in the Weber & Davis (1967) analysis of equatorial magnetic winds, wherein the requirement that $E(x_{1s}, \rho_s) = E$ and $E(x_{1f}, \rho_f) = E$ at these slow/fast critical points $(x_{1s}, \rho_s), (x_{1f}, \rho_f)$ provides the two needed conditions for a unique determination of the physical solution. Note that the Alfvén point *is not* a critical point in such an approach and therefore it plays no role in selecting a particular solution; all transAlfvenic solutions pass automatically through the point (x_{1a}, ρ_a) in the (x_1, ρ) plane and there is no criticality condition there, as there is in the case of a self-consistent determination of the streamline shape, that we shall pursue in the following. Note also that the same critical points have been found by Sakurai (1985) and in some sense by Pneumann and Kopp (1971), although there only the sonic velocity appears in the absence of rotation, because of their numerical technique. In fact the numerical scheme is an iterative process where the Bernoulli and the transfield equations are solved successively until the system converges. As a result critical points of the Bernoulli equation are always considered for a <u>given</u> shape of the streamlines and the previous discussion is relevent to these cases.

On the other hand, let us consider the transfield equation (2.4) in addition to the Bernoulli equation (2.3). Note that both these two equations should be taken into account simultaneously for a self-consistent determination of the shape of the poloidal streamlines. In the Bernoulli equation (2.3) we may substitute V_ϕ in terms of the MHD integrals as in (2.2b) and then take its divergence. From this expression we substitute the value of $\vec{\nabla}\rho/\rho$ in the transfield equation (2.4) to obtain,

$$\left\{ \frac{1 - M_a^2}{h_3^2} \right\} \left\{ \nabla^2 A - \frac{\vec{\nabla}A \cdot \vec{\nabla}(\vec{\nabla}A)^2}{2(\vec{\nabla}A)^2} \frac{V_p^4}{V_p^4 - V_p^2(C_s^2 + V_a^2) + C_s^2 V_{ap}^2} \right\} = F_o, \qquad (2.11)$$

where F_o is a function of $A, \vec{\nabla}A$ and ρ.

This combination of the transfield and Bernoulli equations, Eq. (2.11), contains crucial information. *First*, it shows that the Alfvénic transition

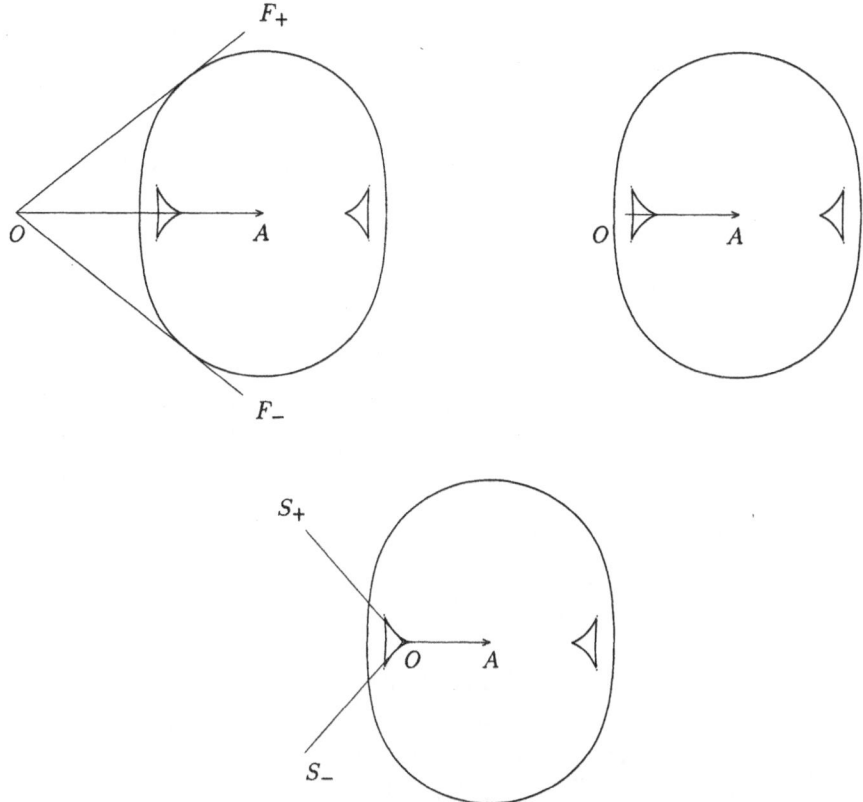

Figure 2. Construction of the slow and fast magnetoacoustic characteristics in a MHD flow on the poloidal plane aligned with a magnetic field, $\vec{V}_p \parallel \vec{B}_p$. The magnetized gas has fast (slow) speed $\vec{V}_s(\theta_B)$ $(\vec{V}_f(\theta_B))$ where θ_B is the angle a perturbation makes with the horizontal magnetic field $\vec{B}_p$. The flow is subject to a small perturbation at the fixed in space point O. The velocity $\vec{V}_p + \vec{V}_f(\theta_B)$ with which this perturbation has propagated from O with the fast(slow) speed, relative to the fixed system of coordinates, has its end on the ellipse (two spherical triangles). Thus, its magnitude depends on the different directions of propagation (θ_B) relative to $\vec{B}_p$ and also on the three possibilities (a), (b), or, (c) wherein the velocity vector $\vec{V}_p = \vec{OA}$ has its origin O outside the wavefront of the fast wave, (a), between the wavefront of the fast and slow waves, (b), and finally inside the wavefront of the slow wave, (c). Note that in the superfast flow (a), a disturbance starting from any point O is propagating only downstream within the fast Mach cone of aperture $\mu_f = \arcsin(V_f/V_p)$ formed by the two fast characteristics F_-, F_+ (tangents from O to the fast wave front) facing downstream. On the other hand, in case (c), a disturbance starting from O is propagating now only upstream within the slow Mach cone of aperture $\mu_s = \arcsin(V_s/V_p)$ formed by the two slow characteristics S_-, S_+ (tangents from O to the spherical triangle representing the slow wave front). Finally, in case (b) no fast or slow characteristics exist since we cannot draw tangents neither to the fast nor to the slow wavefronts, as in the previous cases (a), (c). For propagation along the magnetic field, the classical cusp speed is $V_c = \vec{AC}$, the slow speed $V_s = \vec{AS}$ and the fast speed $V_f = \vec{AF}$. Note that more pairs of characteristics exist in nonaligned MHD flows (Polovin & Demutskii 1990).

$M_a = 1$ is indeed a critical point, with two conditions emerging consecutively there : $F_o(M_a = 1) = 0$, together with the corresponding regularity condition which appears after applying L'Hospital's rule into Eq. (2.11). The *second* interesting feature of Eq. (2.11) has to do with its nature. Let us choose a local orthogonal coordinate system (ξ, η) with ξ along the poloidal magnetic field such that $\vec{\nabla} A = |\vec{\nabla} A| \hat{\eta}$. Linearizing Eq. (2.11) such as $A' = A + \delta A$, with $\delta A \ll A$ we obtain,

$$\left\{\frac{1 - M_a^2}{h_3^2}\right\} \left\{\frac{\partial^2}{\partial^2 \xi^2} - \frac{(V_p^2 - V_c^2)(C_s^2 + V_a^2)}{(V_p^2 - V_s^2)(V_p^2 - V_f^2)}\frac{\partial^2}{\partial^2 \eta^2}\right\} \delta A = \delta F_o, \quad (2.12)$$

where V_s, V_f are the slow/fast MHD speeds which satisfy the familiar quartic (2.10) and V_c is the cusp speed (equivalent in cartesian geometry to the tube speed), $V_c^2 = C_s^2 V_{a,p}^2/(V_a^2 + C_s^2) = V_s^2 V_f^2/(V_s^2 + V_f^2)$ with $V_c < min(V_s, V_f)$. Therefore the transfield equation is elliptic in the region $V_p < V_c$, hyperbolic for $V_c < V_p < V_s$, elliptic in the zone $V_s < V_p < V_f$ and hyperbolic for $V_p > V_f$. Note that the cusp speed V_c is not related to any singularity of this equation, even though the equation changes nature for $V_p = V_c$.

The characteristic curves of Eq. (2.12) have slopes,

$$\frac{d\eta}{d\xi} = \pm\sqrt{\frac{(V_p^2 - V_c^2)(C_s^2 + V_a^2)}{(V_p^2 - V_s^2)(V_p^2 - V_f^2)}}. \quad (2.13)$$

Therefore, at the *neighborhood* of the velocities $V_p = (V_s, V_f)$ and inside the hyperbolic region we may write

$$V_p^2(\xi, \eta) = V_{s,f}^2 + \frac{\partial V_p^2}{\partial \xi}\bigg|_{V_p = V_{s,f}} \xi + \frac{\partial V_p^2}{\partial \eta}\bigg|_{V_p = V_{s,f}} \eta + \cdots, \quad (2.14)$$

such that the equation of the characteristics is obtained by integrating

$$\frac{d\eta}{d\xi} = \pm\sqrt{\frac{1}{q_\xi \xi + q_\eta \eta}}, \quad (2.15)$$

i.e.,

$$\eta = \pm\frac{2}{q_\xi}\sqrt{q_\xi \xi + q_\eta \eta} - \frac{q_\eta}{q_\xi}\log\left(1 \pm \frac{q_\xi}{q_\eta}\sqrt{q_\xi \xi + q_\eta \eta}\right)^2, \quad (2.16a)$$

where $q_\xi \equiv \partial V_p^2/\partial \xi$, and $q_\eta \equiv \partial V_p^2/\partial \eta$. It follows that of the two families of characteristics the one with the lower sign goes asymptotically to the curve

$$q_\xi \xi + q_\eta \eta = q_\eta^2/q_\xi^2. \quad (2.16b)$$

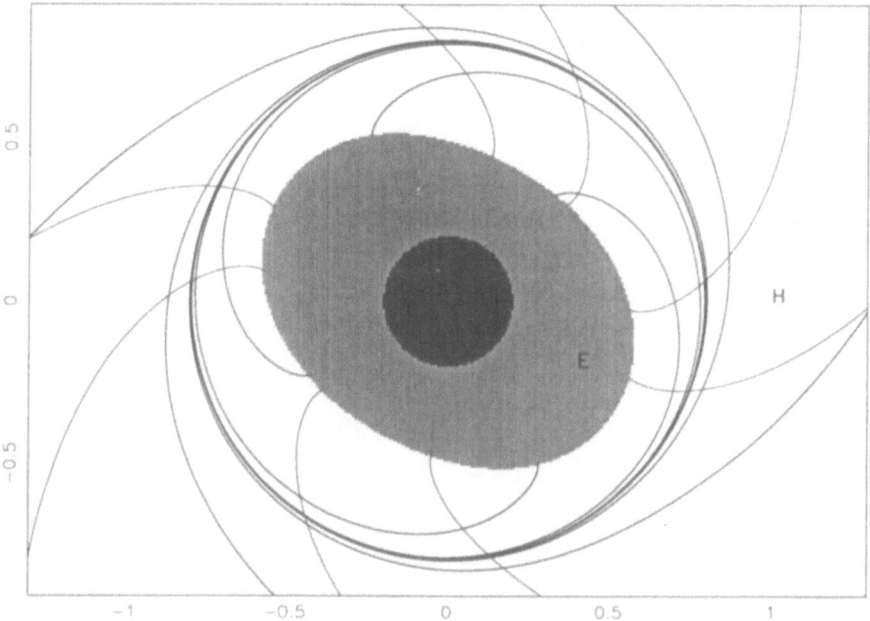

Figure 3.   Characteristics in an example HD wind type flow from a spherical source (dark area). Two families of characteristics are shown (thin and thick solid curves), with the thick family spiraling towards a spherical separatrix characteristic. The gray area E is the elliptic domain with the flow speed reaching the sound speed at its outer boundary. The exterior H area is the hyperbolic domain.

This last equation is precisely the definition of a 'limiting characteristic' (e.g. Guderley 1962), or a separatrix characteristic (Bogovalov 1994) for our problem. We note at this point that, if the condition which determines the one family of characteristics which becomes tangent to the separatrix characteristic is continuously differentiable, the separatrix characteristic is itself one characteristic of that family. If this is not the case, the above curve appears as an envelope of characteristics of that family, in which case it is also called an 'edge of regression' (e.g. Courant & Hilbert, 1962, Volume II).

We claim that the true critical surfaces coincide with those of the limiting characteristics. This conjecture will be shown to predict the same critical points like those found in exact solutions. The structure of characteristics in spherical coordinates and in the neighborhood of the fast and slow critical points in the hyperbolic regimes $V_p > V_f$ and $V_c < V_p < V_s$ is presented in Sakurai (1990), and a schematic plot of the characteristics is shown in his Fig. 4. However, no discussion of the implications of the exis-

438

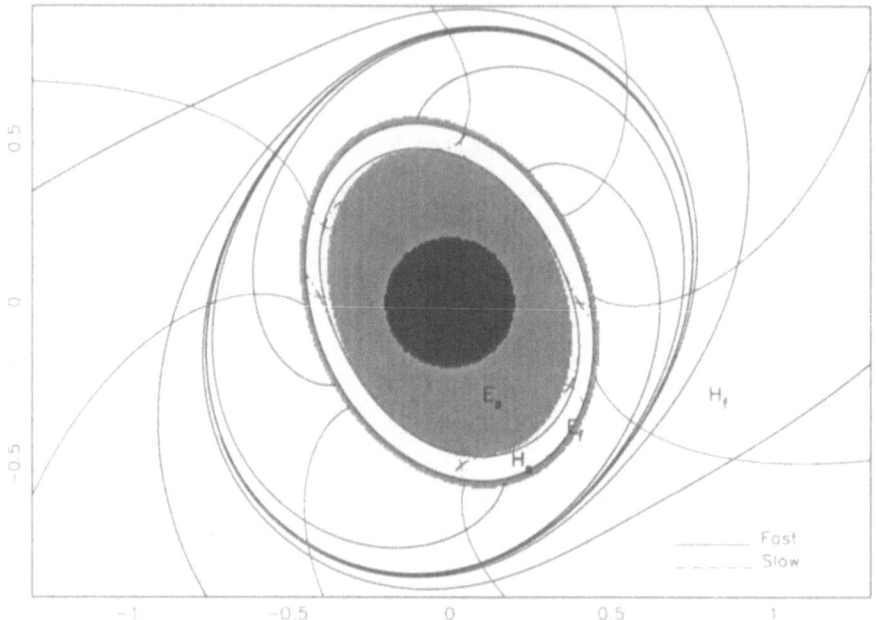

Figure 4.   Characteristics in an example MHD wind type flow from a spherical source (dark area) and with aligned poloidal velocity and magnetic fields. Two families of fast (thin and thick solid curves) and slow (thin and thick dot-dashed curves) characteristics are shown, with each thick family spiraling toward the corresponding elliptical separatrix. The gray areas E_s and E_f are the slow and fast elliptic domains with the flow speed reaching the cusp speed at the outer boundary of E_s, the slow speed at the inner edge of the gray ring E_f and the fast speed at the outer edge of the gray ring E_s. The white areas H_s and H_s are the corresponding hyperbolic domains.

tence of such separatrix characteristics and their connection to the correct boundary conditions of the problem is given there. Such a connection was attempted by Bogovalov (1994) and the term fast and slow magnetoacoustic separatrix surfaces was given; it is also claimed there that boundary conditions should be applied at the fast separatrix characteristic occuring downstream of the fast critical point, for a correct solution of the problem.

¿From the above discussion it turns out that the true critical surfaces are inside the hyperbolic regions, while evidently the positions of the true critical points do not coincide with the positions where the flow velocity equals one of the slow/fast characteristic speeds in the plasma, where the equation changes from elliptic to hyperbolic (or vice-versa). Another important point to be remarked is that the separatrix characteristics in general do not coincide with the surfaces where the equation changes regime, *unless* the flow is perpendicular to these surfaces. In that case only, the surfaces

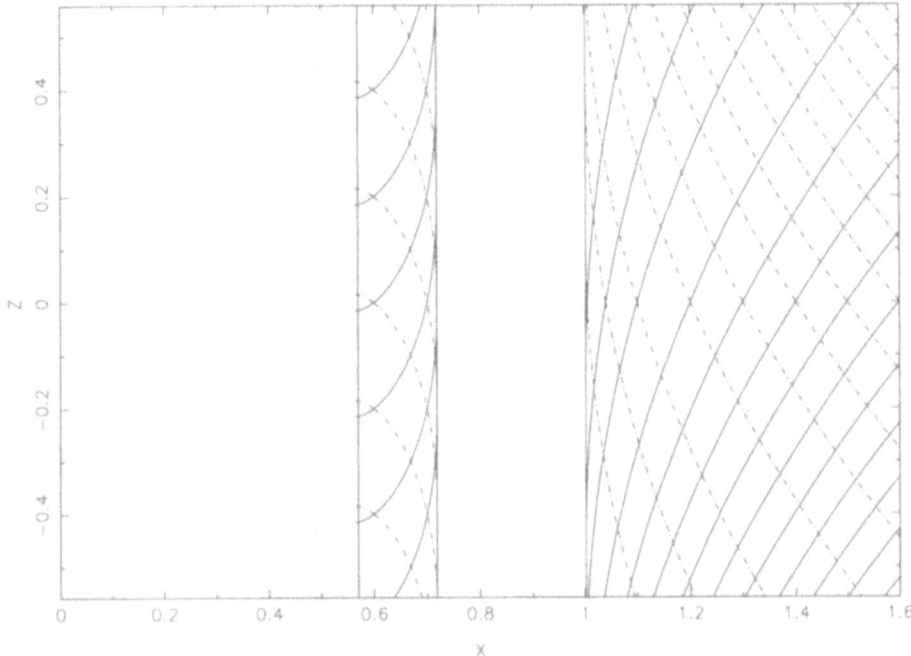

Figure 5. *Plot of the two families of characteristics in the Weber & Davis (1967) model of a magnetized wind from a "cylindrical" star, $\vec{V} = V_{\varpi}\hat{\varpi} + V_{\varphi}\hat{\varphi}$. The characteristics exist in the poloidal plane only in the domains where the governing MHD equation is hyperbolic. This is the case when the radial speed exceeds the cusp speed but it is smaller than the slow MHD speed and in the super-fast domain. Note that in this degenerate case the separatrix characteristics coincide with the slow and fast surfaces because the poloidal (radial) speed is perpendicular to these surfaces there. The three boundaries represent the radial distances where the radial speed equals the cusp, slow and fast MHD speeds.*

where the equation changes regime are themselves characteristics (by the definition that a characteristic curve makes an angle equal to the corresponding Mach angle with the flow, which in that special case is equal to $90°$). This last point explains why, e.g., Anderson (1992) didn't encounter any of the above complications: he considered only flows which cross the elliptic/hyperbolic boundary perpendicularly.

In Fig. 3 we plot in the meridional plane the two families of characteristics of an example HD flow, in Fig. 4 the characteristics of an example MHD flow, while in Fig. 5 the two families of characteristics of the classical Weber & Davis (1967) wind model. Their construction is graphically shown in Figs. 1 and 2. In the hydrodynamic case of Fig. 3 there is one separatrix towards which one of the family spirals, while in the MHD case there are two such characteristics, the slow and the fast one. In the Weber & Davis (1967) case of Fig. 5 they exist only in the domains where the radial speed exceeds the cusp speed but it is smaller than the slow MHD speed and in the super-fast domain, where the MHD equation is hyperbolic. In this degenerate case the separatrix characteristics coincide with the slow and fast surfaces because the poloidal speed is perpendicular to these surfaces there.

The above points will become more transparent in the following paragraph, where we discuss the nature of critical points and separatrix characteristics in the case of self-similar solutions.

2.4. CRITICAL POINTS IN SELF-SIMILAR FLOWS

Before we proceed, let us rewrite Eq. (2.11) in the form,

$$\left[1 - \frac{\Psi_A^2}{4\pi\rho}\right]\left[\frac{a}{h_1^2}\frac{\partial^2 A}{\partial x_1^2} + \frac{2b}{h_1 h_2}\frac{\partial^2 A}{\partial x_1 \partial x_2} + \frac{c}{h_2^2}\frac{\partial^2 A}{\partial x_2^2} + d\right] + e = 0, \quad (2.17a)$$

with

$$a = \frac{V_{a,p}^2}{V_{a,1}^2}[V_1^4 - V_1^2(C_s^2 + V_a^2) + C_s^2 V_{a,1}^2], \quad (2.17b)$$

$$b = V_1 V_2 V_p^2, \quad (2.17c)$$

$$c = \frac{V_{a,p}^2}{V_{a,2}^2}[V_2^4 - V_2^2(C_s^2 + V_a^2) + C_s^2 V_{a,2}^2], \quad (2.17d)$$

where $(V_1, \ V_{a,1})$ and $(V_2, \ V_{a,2})$ are the flow and Alfvén speeds in the directions $\hat{x}_1$ and $\hat{x}_2$, respectively. The coefficients d and e are complicated expressions containing only the free functions of A but not its derivatives; if there is no component of the fields parallel to the ignorable direction $\hat{e}_3$ we have $e = 0$. On the other hand, when in Eq. (2.17a) $e \neq 0$, the Alfvénic transition is found for $\Psi_A^2/4\pi\rho \equiv V_p^2/V_{a,p}^2 = 1$, in a region where the transfield equation remains elliptic. It has been shown (Heyvaerts & Norman 1989, Sakurai 1990, Sauty & Tsinganos 1994) that not satisfying this Alfvén regularity condition leads to the formation of current sheets, with consequent kinks in the fieldlines, related to a discontinuity in the second derivatives of A. Note in passing that in a recent paper, Li & Melrose (1994) discuss which is the correct Alfven point in relativistic flows, which should not be confused with 'fake' singularities that might appear in the equations.

A mixed partial differential equation like Eq. (2.17) is hyperbolic whenever the sign of the determinant of the coefficients of its second order terms is positive: $D = b^2 - ac > 0$. In the present case we have

$$D = (V_p^2 - V_f^2)(V_p^2 - V_s^2)(V_p^2 - V_c^2)(V_a^2 + C_s^2). \quad (2.18)$$

In each one of the hyperbolic regimes ($D > 0$), characteristics exist and can be defined as follows. Through each point in the domain (x_1, x_2) define two families of curves with slopes $h_2 dx_2 / h_1 dx_1$ given by

$$\left(\frac{h_2 dx_2}{h_1 dx_1} \right)_\pm = \frac{b \pm (b^2 - ac)^{1/2}}{a}. \qquad (2.19)$$

Clearly, wherever $a = 0$, one root for the slope of the characteristics is $h_2 dx_2 / h_1 dx_1 = c/2b$ whereas the second one is simply $dx_1 = 0$, i.e. one direction of characteristics becomes the local direction $\hat{x}_2$ of the coordinate x_2. According to Eq. (2.17b), the condition $a = 0$ implies that $V_1^4 - V_1^2(C_s^2 + V_a^2) + C_s^2 V_{a,1}^2 = 0$, which is equivalent to say that the component of the flow speed on the poloidal plane (x_1, x_2) *perpendicular* to the direction $\hat{x}_2$ of the one family of characteristics (i.e. V_1, since the directions x_1 and x_2 are by definition orthogonal), is equal to the component of the slow/fast magnetosonic speeds *perpendicular* to that direction. Note that this last equality describes mathematically the graphical construction of the characteristics in figures 1 and 2.

Let us now further restrict our discussion to consider only self-similar flows. The self-similar approach for the description of 2-D MHD flows is based on the following separate assumptions : (i) separation of the variables in some key physical quantities, (ii) prescribed scaling of all functions with one of the two variables, (iii) the scaling is such as to remove all dependence from the scaling variable. In the present case let us assume $\hat{x}_2$ as the self-similarity direction, so that the solutions become x_2 self-similar. Accordingly the physical variables can be defined in separable forms, for example the magnetic flux function A can be assumed as,

$$A = f(x_1) f_A(x_2), \qquad (2.20)$$

with $f_A(x_2)$ a known function of x_2. The same definition can hold for other quantities of the same physical dimensions, for example, all velocities scale with the same function $f_v(x_2)$. Therefore, in these models critical surfaces can only be surfaces of self-similarity, i.e., are given by functions of x_1 alone.

The assumption in Eq. (2.20) allows to simplify the transfield equation (Eq. 2.17a) in an ordinary differential equation for $f(x_1)$:

$$f_A(x_2) \frac{a}{h_1^2} \frac{d^2 f(x_1)}{dx_1^2} + \frac{df_A(x_1)}{dx_2} \frac{2b}{h_1 h_2} \frac{df(x_1)}{dx_1} + f(x_1) \frac{c}{h_1^2} \frac{d^2 f_A(x_2)}{dx_2^2} + d' = 0.$$
$$(2.21)$$

This ordinary differential equation is clearly directly integrable, *unless* the coefficient of the second order derivative term becomes zero at some point, i.e. if $a = 0$. Consistently with the previous discussion, the condition $a = 0$

for the singularity corresponds to *the symmetry direction $\hat{x}_2$ becoming locally a characteristic direction* (at the particular value of x_1). This statement is equivalent to the equality of the flow speed component V_1 (in the direction *perpendicular* to the self-similarity direction $\hat{x}_2$), with the component of the slow/fast MHD mode wave along the same direction ($V_{s,1}, V_{f,1}$). We need to note that although the singularity condition is easily defined in a self-similar flow, we know of no simple algebraic relation between the coefficients of the second order partial differential equation (2.17) that would determine the singular points of a general (non self-similar) flow.

The above "strange looking" relation between the typical plasma speeds can be easily understood taking into account that the direction along $\hat{x}_3$ is ignorable, while along $\hat{x}_2$ the system is self–similar. Therefore the only allowed direction for the propagation of a perturbation is along $\hat{x}_1$, and consequently, it is physically reasonable to expect that a singularity will arise whenever V_1 (and not V or V_p) becomes equal to either $V_{s,1}$, or $V_{f,1}$ (and not simply V_s or V_f). In conclusion, the self-similarity modifies the mathematical problem in the sense that it converts the PDE's to ODE's, and also modifies the positions of the actual critical points, which (in general) are different from the points where the flow velocity becomes equal to the characteristic slow/fast MHD speeds.

At this point, note a similar discussion in Williams & Dyson (1994), for the special case of axisymmetric accretion disks where angular momentum is conserved, and therefore one component of the velocity (namely V_ϕ) becomes a prespecified function of r. In that case, considering only the Bernoulli equation, the critical point appears at the position where $V_r^2 = C_s^2$ (and *not* where $V_r^2 + V_\phi^2 = C_s^2$). In the context of limiting or separatrix characteristics in flows which exhibit some kind of symmetry (in the present trivial case axisymmetry), this relation is equivalent to the statement that the component of the flow velocity perpendicular to the directions of symmetry, i.e. V_r, becomes equal to the speed of the characteristic wave of the problem, the speed of sound. The circle where $V_r = C_s$ is indeed a limiting characteristic (*c.f.*, Fig. 3).

We will discuss in detail in the following two sections the above important points for three typical configurations. In the first case we shall consider a constant vertical gravitational field which leads naturally to assume a vertical stratification of the isothermal atmosphere (*z-self-similarity*). In the second example of winds originating from Keplerian discs, the spherical gravitational field and the lack of any intrinsic scale of the Keplerian law leads to the choice of a model where everything scales with the spherical radius (*r-self-similarity*). In the third example a formal expansion around the axis is assumed where the stellar source is spherical, which gives rise to a natural scaling with the colatitude θ (*θ-self-similarity*). It is important to point out that in the first two examples the polytropic relation between the gas pressure and density still holds, while it is released in the third case.

3. Examples of Critical Points and Separatrix Characteristics in Symmetric and Self-similar Polytropic MHD Flows

In this section we shall discuss briefly two exact solutions for polytropic flows wherein the critical points found correspond to setting the coefficient a=0 in the combined transfield and Bernoulli equation (2.17a).

3.1. VERTICALLY SELF-SIMILAR FLOWS IN LOOPS EMBEDDED IN A PLANAR ATMOSPHERE

Tsinganos, Surlantzis & Priest (1993) recently constructed self-consistent z-self-similar solutions for an isothermal atmosphere with $x_1 = x, x_2 = z, x_3 = -y$, following the general scheme described in Sec. 2.1. The ignorable coordinate is $x_3 = -y$ (direction of symmetry), while $x_2 = z$ is the self-similar coordinate. Their assumptions actually can be reduced to giving the dependences of M_a, A, P and the free integral $\Psi_A(A)$, as follows,

$$M_a^2(z, x) = M_a^2(x), \quad A(z, x) = A_o f(x) e^{-\frac{\xi z}{2H}}, \tag{3.1a}$$

$$\Psi_A = 4\pi k A, \quad P = V_s^2 \rho = 4\pi \frac{k^2}{M_a^2(x)} A, \tag{3.1b}$$

with H the constant scale height of the atmosphere, $H = V_s^2/g$, and the positive parameter $\xi > 0$ controlling the decline of the density with the height z. Note that the Alfvén surfaces given by (Eq. 3.1a) are *planar* with their normal along $\hat{x}$ while the direction of self-similarity is $\hat{z}$. Furthermore, these assumptions correspond to the density and pressure having exponential z-dependences, similarly to the density and pressure in a vertically stratified atmosphere in a uniform gravitational field in the absence of magnetic fields and flows. The flow speed and Mach number are z-independent because of the choice of the free function $\Psi_A(A)$ which is linear in A. By substituting the z-dependence of the density ρ and A in the Bernoulli and transfield Eqs. (2.3)-(2.4) we obtain for the Mach number M and Alfvén number M_a a set of ordinary differential equations. A topological analysis of this system of equations shows that besides the trivial mathematical singularity at $M_a = M = 0$, there exists in the $(M_a - M)$-plane a *new* X-type critical point at the vanishing of the numerator and denominators of the equations. The corresponding critical condition at this point can be understood by noting that it defines a curve in the (M_a, M)-plane where the coefficient of the second order partial derivative of $A(z, x)$ with respect

444

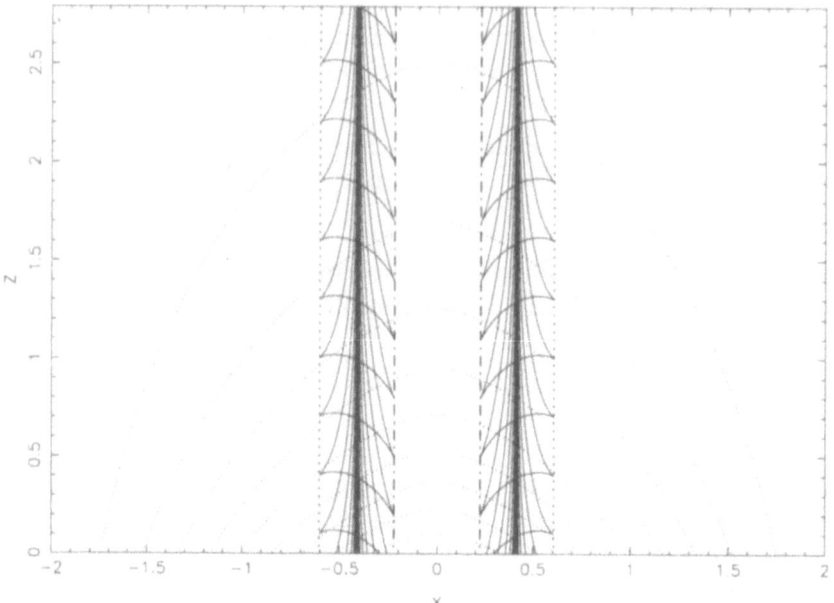

Figure 6. Plot of the characteristics (solid curves) in the vertically self-similar model of Tsinganos, Surlantzis and Priest (1993) for a magnetized and vertically stratified loop with flows, in the uniform gravitational field of the lower solar atmosphere. The characteristics are plotted in the domains where the transfield equation is hyperbolic ($V_{cusp} < V < V_{slow}$) together with the path of the plane arcade-type field lines and streamlines (dotted curves) in all space. The cusp line is indicated with thick dots while the slow magnetoacoustic line with thick dots-dashes. Characteristics are parallel to the flow speed when $V = V_{cusp}$ and perpendicular to it when $V = V_{slow}$. The two thick solid lines correspond to the slow separatrix where one of the two families of the characteristics becomes tangent. There, the x-component of the flow speed equals the slow MHD speed, $V_x = V_{slow}$.

to x in the PDE (2.17a) becomes zero. Furthermore, it is easily shown that the critical condition may take the more familiar form,

$$V_x^4 - V_x^2(V_s^2 + V_a^2) + V_s^2 V_{a,x}^2 = 0. \qquad (3.2)$$

This means that the x-component of the flow speed along the critical curve is equal to the characteristic speed of the fast/slow MHD waves propagating in the x-direction. So the critical points found by Tsinganos, Surlantzis & Priest (1993) correspond indeed to the self-similarity modified critical points that we introduced at the end of the last section. In other words, since the flow is z-self-similar and symmetric in y, only those waves can propagate that preserve those two symmetries, i.e., they propagate in the $\hat{x}$-direction which is perpendicular to the $\hat{y}$ and $\hat{z}$ symmetry directions.

The previous critical conditions arise because the strength of the magnetic field along the path s of a flow tube is affected by the magnitude of the flow and therefore $B = B(s, V)$. In flows through a rigid flow tube, $B = B(s)$ only, the Bernoulli equation for force balance along s is similar to that for flow through a nozzle with its effective throat located at the sonic transition $M = 1$. However, in flows interacting with the magnetic field, i.e., when lateral force balance (transfield equation) is also taken into account in addition to longitudinal force balance (Bernoulli equation), the nozzle of the effective throat is located along the critical curve.

In Fig. 6 as the flow accelerates from left along the magnetic field lines climbing upwards in the uniform vertical gravitational field, the governing equation of the MHD flow remains elliptic until it reaches the position of the cusp or tube speed wherein $V^2 = V_c^2 = V_a^2 C_s^2 / (C_s^2 + V_a^2)$. Then the equation becomes hyperbolic and characteristics which are parallel to the flow speed emanate at the horizontal distance of the first dashed vertical line. At the X-type critical point one of the families of the characteristics becomes tangent to the separatrix characteristic (thick vertical line) where the x-component of the flow speed equals the slow MHD speed, $V_x = V_s$, Eq. (3.2). Further downstream where the total speed is equal to the slow magnetoacoustic speed (the sound speed in this case), $V = C_s$, the characteristics become perpendicular to the flow velocity and stop there (dot-dashed vertical line). The MHD equation for the shape of the streamlines becomes again elliptic further downstream and the situation is symmetric in the downflowing leg of the coronal loop. This is our first example for plotting the characteristics in self-similar and polytropic MHD flows which shows all new features discussed in the previous section.

3.2. RADIALLY-SELF-SIMILAR POLYTROPIC DISC WINDS AND JETS

A second broad class of self-similar solutions are the radially self-similar models for winds from accretion discs. Bardeen & Berger (1978) initiated such models in the context of hydrodynamic polytropic galactic winds. Yet, their generalization to a cold but magnetized plasma by Blandford & Payne (1982) remains more known because of its success in showing that magnetically driven flows can be related to astrophysical jets. The hypotheses used by Blandford & Payne (1982), Königl (1989), Safier (1993a,b) and Rosso & Pelletier (1994) were recently generalized by Contopoulos & Lovelace (1994) who proposed a more general extension of the original Bardeen & Berger's (1978) model using a parametrized distribution of the magnetic flux on the disc with the following assumptions,

$$M_a = M_a(\theta) , \qquad A \propto f(\theta) r^x , \qquad (3.3a)$$

$$\Psi_A \propto A^{1-3/(2x)} , \qquad L\Psi_A \propto A^{1-1/x} , \qquad \frac{L}{\Omega} = \varpi_a^2 \propto A^{2/x} , \qquad (3.3b)$$

$$P \propto A^{2-2\gamma+(3\gamma-4)/x}\rho^{\gamma} \propto A^{2-4/x}\frac{1}{M_a(\theta)^{2\gamma}}, \qquad (3.3c)$$

where, $L\Psi_A$ is the total angular momentum per unit of magnetic flux related to the electrical current as shown hereafter. L/Ω must be equal at the Alfvén surface to the square of the cylindrical radius of the flux tube, hence we introduce this radius ϖ_a while x is a constant index. The special case $x = 3/4$ corresponds to the original Blandford & Payne (1982) model.

If $\gamma = \Gamma = 5/3$ (the polytropic index corresponding to the adiabatic one) then the special case $x = 3/4$ corresponds to the case of constant entropy throughout the disc, $S(A) \propto A^{2-2\gamma+(3\gamma-4)/x}=$const. This justifies the original choice made by Bardeen & Berger (1978). Blandford & Payne's (1982) case is indeed justified by recovering the Bardeen & Berger's hydrodynamical model although they assumed a cold plasma. Rosso & Pelletier (1994) did make such a restriction requiring that $\gamma = 1 - 1/(2x - 3)$, considering thus a constant entropy distribution throughout the disc although finally they neglected the enthalpic term. The analysis of Contopoulos & Lovelace (1994) is more general since a warm plasma is considered, while in order to concentrate on the effect of the magnetic field, and for reasons of computational simplification, they started the solutions beyond the slow magnetoacoustic point.

The spherical coordinates (r, θ, ϕ) are scarcely used in the usual presentation of those models. A more convenient way is to use the cylindrical/spherical coordinates (ϖ, θ). The cylindrical distance ϖ is a key quantity in terms of which we can calculate the spherical distance $r = \varpi/\sin\theta$ and the axial distance $z = \varpi \cot\theta$. Denoting by $y = (z/\varpi) = \cot\theta$, Eq. (3.3a) can be written as $A \propto f[\theta(y)]\sin^{-x}\theta(y)\varpi^x$. Rosso & Pelletier (1994) use this system of coordinates (ϖ, y), introducing a function $f_{\mathrm{RP}}(y) = f(\theta(y))\sin^{-x}\theta(y)$ $[A = f_{\mathrm{RP}}(y)\varpi^x]$.

Another natural way to describe the same system is by using the streamline coordinates, (A, θ, ϕ). Then reversing Eq. (3.3a), we can express the spherical radius as a function of A and θ, $r \propto A^{1/x}f^{-1/x}(\theta)$. This immediately shows that the cylindrical coordinates $\varpi = r\sin\theta$ and $z = r\cos\theta$ of a fieldline split in terms of a function of A and θ,

$$\varpi(A, \theta) = \varpi_o(A)R(\theta), \qquad z(A, \theta) = \varpi_o(A)Z(\theta), \qquad (3.4a)$$

where

$$\varpi_o(A) \propto A^{1/x}, \quad R(\theta) = \sin\theta f^{-1/x}(\theta), \quad Z(\theta) = \cos\theta f^{-1/x}(\theta), \quad (3.4b)$$

$Z(\theta)$ and $R(\theta)$ are arbitrary well behaved functions of θ subject only to the constraint that $R(\theta = \pi/2) = 1$ and $Z(\theta = \pi/2) = 0$. Evidently, in the context of outflows from an accretion disk, $\varpi_o(A)$ is the distance where the fieldline A intersects the disk at $z = 0$ or $\theta = \pi/2$, since this distance depends solely on the streamline label A.

Instead of using (r, θ) as variables, Blandford & Payne (1982), Königl (1989), Safier (1993) and Contopoulos & Lovelace (1994) prefered to use (ϖ_o, Z) which is equivalent as shown by Eq (3.4b). The surfaces $Z = const.$ are identical to the surfaces $\theta = const.$ Hence, the unknown function $f(\theta(Z)) = (Z^2 + R^2)^{-x/2}$ is replaced by the unknown function $R(Z)$. The original scaling with the distance on the disc can now be recovered. In particular, it follows immediately that the magnetic field on the disk level $z = 0$ scales as $B \propto B_o \varpi_o^{x-2}$ while the flow speed at the disk level $Z = 0$ $(\theta = \pi/2)$ scales as the Keplerian rotational speed $V \propto \varpi_o^{-1/2}$. Note that the Alfvén and the sound speeds V_a and C_s at the disk level $Z = 0$ $(\theta = \pi/2)$ also scale as the Keplerian rotational speed.

It is important to note here that in the present r-self-similar case, the scaling of all quantities in the problem depends on the scaling of only two: the magnetic field (or equivalently the density), and the velocity. In the scaling of the magnetic field, we introduced the free exponent x. The scaling of the velocity is however uniquely determined by the form of the gravity term in the Bernoulli equation. This is the reason why, in the context of r-self-similar scaling with Newtonian gravity from a central compact object, velocities can scale only as $r^{-1/2}$ along the accretion disk. If however we are allowed to assume a more general form for the gravity term (i.e. $\propto r^{-a}$), velocities will then have to scale as $r^{-a/2}$ (e.g. Bardeen & Berger 1978, Chakrabarti & Bhaskaran 1992).

With the above assumptions (3.3) it is easy to calculate the form of the MHD integrals. Thus, the angular velocity of the roots of the streamlines, Ω, and the total angular momentum per unit mass, L, scale independently of the parameter x, as expected from a Keplerian disc,

$$\Omega(A) \propto A^{-\frac{3}{2x}} \propto \varpi_o^{-\frac{3}{2}} , \qquad L(A) \propto A^{\frac{1}{2x}} \propto \varpi_o^{\frac{1}{2}} . \qquad (3.5)$$

Note that in the present r-self-similar model, critical surfaces are conical with their normal perpendicular to the direction of self-similarity, similarly to the previous case of z-self-similarity.

On the other hand consider the total axial current I_z enclosed by a flux tube at a cylindrical radius ϖ_o and the rate of the total angular momentum flux per unit of magnetic flux $dJ/dA = L\Psi_A$ carried by the flow tube of the outflow,

$$I_z = -\frac{c}{2} L\Psi_A \frac{1 - \varpi_o^2/\varpi_a^2}{1 - M^2(\theta)} \propto \varpi_o^{x-1} \propto \frac{dJ}{dA} . \qquad (3.6)$$

This explain the roles played by the parameter x and the quantity $L\Psi_A$ used in Eqs. (3.3b). In fact, it follows that when $x < 1$ infinite current and angular momentum rates are flowing along the axis $\varpi = 0$. Thus the axial current and angular momentum rate are finite everywhere only for $x > 1$. The solutions are evidently not valid close to the rotation axis due to the Keplerian law. In addition to that, the case $x < 1$ is characterized by a strong *infinite* current along the axis which is compensated by the return current flowing back to the disc through the wind. Conversely the case $x > 1$ corresponds to a distributed current flowing everywhere through the disc. In that case, the electric circuit is assumed to close outside the jet.

In the above r-self-similar solutions the self-similarity direction is $\hat{r}$. At the same time the solutions are axisymmetric and the axisymmetry direction is $\hat{\phi}$. Therefore, a wave that preserves those two symmetries should propagate along $\hat{\theta}$ *on the poloidal plane* (Blandford & Payne 1982). The incompressible Alfvén mode propagates along the magnetic field $\vec{B}$ with velocity V_a and in an inclined direction $\hat{\theta}$ along the poloidal plane, with a phase velocity $V_{a,\theta} = \vec{B}_p \cdot \hat{\theta} / \sqrt{4\pi\rho}$. Therefore, at the Alfvén point we should have

$$\frac{\vec{B}_p \cdot \hat{\theta}}{\sqrt{4\pi\rho}} = \vec{V}_p \cdot \hat{\theta} \implies M_a = 1, \tag{3.7}$$

since, in ideal MHD, $\vec{V}_p \propto \vec{B}_p$. The compressible slow/fast MHD modes propagate in a *poloidal* direction $\hat{\theta}$ with a phase velocity $V_{(s,\theta;f,\theta)}$ satisfying the quartic

$$V^4_{(s,\theta;f,\theta)} - V^2_{(s,\theta;f,\theta)}(V_a^2 + C_s^2) + C_s^2 V^2_{a,\theta} = 0, \tag{3.8}$$

and therefore at the slow/fast MHD critical points we should have

$$V_{(s,\theta;f,\theta)} = \vec{V}_p \cdot \hat{\theta}, \tag{3.9}$$

as expected from the discussion in Sect. 2.3. We expect therefore that when conditions (3.7) or (3.9) are satisfied, the transfield and Bernoulli equations should have singularities.

In Fig. 7 we plot the characteristics in such a model of a disk-wind, where the magnetic field has for simplicity only a toroidal component. All critical surfaces are conical, i.e. functions of θ only. Thus, characteristics start to exist at meridional angles greater than the angle of the fast magnetoacoustic surface at $\theta = \theta_{\rm fm}$ (at $\theta_{\rm fm}$ the Mach number for the fast magnetoacoustic mode n=1, dashed line). The separatrix characteristic (dot-dashed line) appears at $\theta = \theta_{\rm m.fms}$ where t=1 and one of the two families of characteristics becomes tangent there while the other crosses it.

The special case of a cold plasma with $C_s = 0$ and $V_f = V_a$ was first considered by Blandford & Payne (1982). The more general case of polytropic self-similar outflows has been considered by Contopoulos & Lovelace

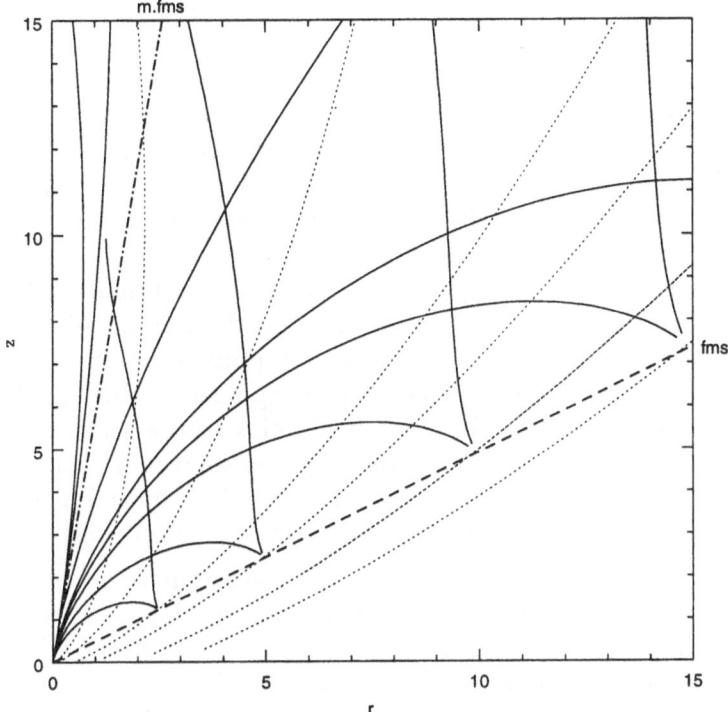

Figure 7. Plot of the characteristics in the model of Contopoulos (1995) for a radially self-similar disk-wind, with the magnetic field having only a toroidal component.
With dotted lines the poloidal flux/flow lines are indicated, with solid the two families of
the characteristics, with dashed the fast magnetoacoustic surface (fms) and finally with
dash-dotted the modified by self-similarity fast magnetoacoustic surface, or fast separatrix
(m.fms). Note that one of the two families of characteristics becomes tangent to the fast
separatrix characteristic while the other simply crosses it.

(1994). The transfield equation has the Alfvén singularity at $M_a = 1$ (m=1
in Blandford & Payne's 1982 notation) and the modified by self-similarity
slow/fast MHD critical point (the self-similarity modified fast MHD critical
point corresponds to $t = 1$ in Blandford & Payne's notation). In fact, the
expression which determines these so called 'modified' slow and fast singularities, is also present in equation (4.35) of Contopoulos (1992). Solutions
presented in the above literature pass through the Alfvén singularity (m=1)
at a finite distance, while the modified by self-similarity fast MHD critical
point (t=1) is never approached. In other words solutions that have been
plotted are either some kind of breeze solutions or terminated solutions
in Parker's terminology. Fig. 7 shows however that recollimating solutions
that do cross the $t = 1$ FMSS exist by tuning adequately the parameters.

At this point, we need to clarify a point of misunderstanding in the literature. Blandford & Payne (1982), Li, Chiueh & Begelman (1992), Contopoulos & Lovelace (1994), Contopoulos (1994) and others give emphasis to the ratio of the poloidal flow speed to the speed of the fast mode MHD wave in the poloidal direction (which they denote by n), and consider the critical point at t=1 (the so called 'modified' fast magnetoacoustic point) as being an *unphysical artifact* of the assumption of self-similarity. It is interesting however, that in all of the above works, the point where $V_p = V_f$ (i.e. $n = 1$) presented no numerical complications in the numerical integration of the transfield equation (it only presented minor problems in the Bernoulli equation, but this is well understood from our discussion in the beginning of Sect. 2.2). In other words, the point where $n = 1$ imposes no restrictions in determining uniquely the one physical solution which crosses it smoothly, it is not therefore a critical point in the sense that the sonic point in the solar wind 'chooses' the physical solar wind solution. On the other hand, the point where $V_\theta = V_{f,\theta}$ (i.e. $t = 1$) is indeed a singular point of the equations, as shown in the discussion of Section (2.3).

Similarly, were anyone to obtain a self-consistent r-self-similar solution around the slow magnetoacoustic point, the singular point will appear at the point where $V_\theta = V_{s,\theta}$, and not at the point where $V_p = V_s$ This was also actually obtained by Ferreira & Pelletier (1993a,b, 1995) in their ejection-accretion model for Keplerian discs. They used similar assumptions as Contopoulos & Lovelace (1994) except that the gravitational field is cylindrical instead of being radial because they restrict themselves to the vicinity of the disc. They also include non ideal MHD but only in the disc itself. Once the ejection starts and the plasma becomes lighter, ideal MHD is operant. They find a critical point just above the disc that is indeed a modified slow magnetoacoustic point.

4. Examples of Critical Points and Separatrix Characteristics in Axisymmetric and Meridionally-Self-similar Nonpolytropic MHD Flows

In meridionally self-similar flows the meridional dependence of the flow quantities is prescribed while their radial dependence is derived self-consistently from the MHD equations. The simplest such meridionally self-similar flow is the Parker (1958) solar wind treatment, where there isn't any meridional dependence of the wind quantities. The next broad class of meridionally self-similar flows of interest in the present paper has been recently studied in the context of nonspherically symmetric wind and jet-type outflows from a spherical gravitational field (see the Chapter by Trussoni et al in this Volume). The flows are expanding from a spherical source and from the disc close to its axis of symmetry in a region where a Keplerian law does not necessarily hold. Their meridional-self-similarity corresponds to a *formal* expansion in the magnetic flux around the axis.

As before, such models rest on six assumptions for the Alfvén Mach number, magnetic flux, the free functions and the pressure. Here we choose to normalize everything at the Alfvén point in the polar axis ($\theta = 0, r = r_\star$) where $M_a = 1$; $\rho_\star$, $V_\star$ and $B_\star$ are respectively the density, the velocity and the magnetic field at this point. Note that by construction $B_\star^2 = 4\pi\rho_\star V_\star^2$ and $A_\star = r_\star^2 B_\star/2$. Thus,

$$M_a^2 = M_a^2(r), \quad A = A_\star f(r)\sin^2\theta, \quad \Psi_A = \frac{4\pi\rho_\star V_\star}{B_\star}(1 + \delta\frac{A}{A_\star})^{1/2},$$

$$(4.1a)$$

$$L\Psi_A = r_\star B_\star \lambda \frac{A}{A\star}, \quad \frac{L}{\Omega} = \varpi_a^2 = r_\star^2 \frac{A}{A\star}, \quad (4.1b)$$

$$P = P_o(r) + P_1(r)\frac{A}{A_\star}, \quad \rho = \frac{\rho_\star}{M_a^2}(1 + \delta\alpha). \quad (4.1c)$$

The axial electrical current $L\Psi_A$ is zero on the axis and increases linearly with the magnetic flux. All quantities are well defined along the polar axis contrarily to the models of the previous section. Surfaces of constant Alfvén Mach numbers are *spherical* like the central stellar source instead of being conical.

The above assumptions combined with the original equations of Section 2 lead to the following magnetic and velocity fields,

$$B_r = B_\star \frac{f(R)}{R^2}\cos\theta, \quad V_r = V_\star \frac{M_a^2 f(R)}{R^2}\frac{\cos\theta}{\sqrt{1 + \delta\alpha}}, \quad (4.2a)$$

$$B_\theta = -B_\star \frac{1}{2R}\frac{df(R)}{dR}\sin\theta, \quad V_\theta = -V_\star \frac{M_a^2}{2R}\frac{df(R)}{dR}\frac{\sin\theta}{\sqrt{1 + \delta\alpha}}, \quad (4.2b)$$

$$B_\phi = -B_\star \frac{\lambda}{R}\frac{f(R) - R^2}{1 - M_a^2}\sin\theta, \quad V_\phi = -V_\star \frac{\lambda}{R}\frac{M_a^2 f(R) - R^2}{1 - M_a^2}\frac{\sin\theta}{\sqrt{1 + \delta\alpha}}, \quad (4.2c)$$

with (R, θ) the familiar spherical coordinates, the radial distance R measured in units of the Alfvén radius $r_\star$ and $\alpha = A/A_\star = f(R)\sin^2\theta$ the dimensionless magnetic flux function. Two classes of solutions can be distinguished according to the latitudinal dependence of the pressure $P(R,\theta)$. In one, $f(R)$ is prescribed to be 1 such that the field/streamlines on the meridional plane are conical and the pressure is written as (Tsinganos & Trussoni 1991):

$$P(R,\theta) = \frac{1}{2}\rho_\star V_\star^2 \left\{Q_o(R) + Q_1(R)\sin^2\theta\right\}. \quad (4.3)$$

In this case the pressure is splitted into two independent terms: the first is the spherically symmetric part of the pressure Q_o while the second Q_1 accounts for deviations from spherical symmetry. In the other (Sauty & Tsinganos 1994), $f(R)$ is unknown and the pressure is written as,

$$P(R,\theta) = \frac{1}{2}\rho_* V_*^2 Q_o(R)\left\{1 + \kappa f(R)\sin^2\theta\right\}. \tag{4.4}$$

Both classes are θ-self-similar because in all physical quantities the θ-dependence is prescribed while a solution of the equations of motion determines the remaining unknown radially dependent functions, $[M_a(R),$ $Q_o(R), Q_1(R)]$ in the first case of prescribed meridional streamline geometry and $[M_a(R), Q_o(R), f(R)]$ in the second case of deduced meridional streamline geometry. In the following, we shall examine the characteristics of the deduced meridional streamline geometry. The solutions of prescribed streamline geometry may be less interesting for our present purposes, because the separatrix characteristic coincides with the fast surface in that case wherein $V_p = V_r$.

In the free streamline situation, momentum balance results in three independent equations, for $Q_o(R)$, $F(R)$ and $M_a^2(R)$ where $F(R)$,

$$F(R) = \frac{R}{f}\frac{df}{dR}, \tag{4.5a}$$

is the logarithmic derivative of the streamlines flux and gives the flux tube geometry. The derivatives of $Q_o(R)$, $F(R)$ and $M_a^2(R)$ are then given in terms of the functions and variables R, G, F, M_a^2 and Q_o where

$$G(R) = \frac{R}{\sqrt{f}}, \tag{4.5b}$$

measures the cylindrical radius of the jet. The calculations show then that the expressions for $F(R)$ and $M_a^2(R)$ possess the same critical points whose nature is revealed by putting their respective denominators equal to zero [see Eqs. (4.1) in Sauty & Tsinganos 1994],

$$(M_a^2 - 1)\left(1 + \kappa\frac{R^2}{G^2}\right) + \frac{F^2}{4} + R^2\lambda^2\frac{(1-G^2)^2}{(1-M_a^2)^2} = 0. \tag{4.6}$$

In order to evaluate the sound speed we have to calculate the infinitesimal variation of the pressure and the density at a *fixed* point of a given streamline, (R,α). The variation in the density is due to the infinitesimal variation in the Alfvén number and we may write,

$$d\rho|_{(R,\alpha)} = -\frac{\rho_*}{M_a^4}(1 + \delta\alpha)dM_a^2. \tag{4.7a}$$

Considering the variation of the pressure, due to relation (4.4) we have,

$$dP|_{(R,\alpha)} = -\frac{1}{2}\rho_* V_*^2(1 + \kappa\alpha)dQ_o|_{(R,\alpha)}. \tag{4.7b}$$

Thus the sound speed at a given point may be expressed in terms of the local pressure variation with the Alfvén number,

$$C_s^2 = \frac{\partial P}{\partial \rho}\Big|_{(R,\alpha)} = -\frac{V_*^2}{2}\frac{\partial Q_o(M_a^2, R)}{\partial M_a^2}\Big|_{(R,\alpha)} M_a^4\frac{1 + \kappa\alpha}{1 + \delta\alpha}. \tag{4.7c}$$

However, since it holds that

$$\frac{dM_a^2}{dR} = \frac{-\dfrac{\nu^2}{M_a^2 R^2} - \dfrac{2}{G^4}\dfrac{M_a^2}{R}(F - 2) - \dfrac{\partial Q_o(M_a^2, R)}{\partial R}}{\dfrac{\partial Q_o(M_a^2, R)}{\partial M_a^2} + \dfrac{2}{G^4}}, \tag{4.8}$$

at the critical point (R_x, M_x), the numerator and denominator of the above expression should be identically zero. We may then calculate the sound speed by substituting the quantity $\partial Q_o(R, M_a^2)/\partial M_a^2$ at the critical point (R_x, M_x) with its equivalent $-2/G^4$ to obtain for the sound speed in the neighbour of the critical point,

$$C_s^2 = V_*^2\frac{M_a^4}{G^4}\frac{1 + \kappa\alpha}{1 + \delta\alpha}. \tag{4.9}$$

Let's now go back to the single X-type critical point $(M = M_x, R = R_x)$ as calculated by putting to zero the denominator of the expressions giving the derivatives of the Alfvén number and the expansion factor, Eq. (4.6). This expression can be written down in terms of the various Alfvén and characteristic speeds, as to

$$V_r^4 - V_r^2 V_a^2 - V_r^2(V_r^2 - V_{a,r}^2)\frac{(1 + \kappa\alpha)}{\cos^2\theta} = 0, \tag{4.10}$$

and is valid at the neighborhood of the critical point where we have calculated the value of the sound speed. Combining the above expression with the one of the definition of the sound speed, Eq. (4.7c), we recover immediatel the usual expression holding for the MHD modes in the radial direction

$$V_r^4 - V_r^2(V_a^2 + C_s^2) + C_s^2 V_{a,r}^2 = 0. \tag{4.11}$$

The above equation is the equation for the speed of an MHD wave propagating in the r-direction. The self-similarity direction is $\hat{\theta}$, because all quantities are prescribed functions of θ. Also, the symmetry direction is $\hat{\phi}$.

Therefore, a fast/slow MHD wave that keeps those two symmetries should propagate along $\hat{r}$ (which is perpendicular to both, $\hat{r}$ and $\hat{\phi}$). At the X-type critical point the r-component of the flow speed, equals the slow/fast MHD mode wave speed in that direction, $V_r = V_{\text{fms},r}$. This confirms again the expected modification of the critical speed due to the self-similarity. Moreover, we have been able to calculate the sound speed of this non polytropic flow at least localy where the critical point is present.

In Fig. 8 we plot the characteristics of the present self-similar model of a jet/wind. Although the effective sound speed is well defined only locally around the critical surface, we may extend its definition (Eq. 4.7c) everywhere to see the role of characteristics. Characteristics start to exist at angles and distances above the surface where the poloidal speed equals the cusp speed (thick dots). The separatrix characteristic is at the spherical distance R≈ 0.7845 where Eq. (4.6), or, (4.11) hold and one family of the characteristics is tangent there while the other crosses it. An interesting point is that the flow is almost radial at this distance R_x and almost perpendicular to the separatrix characteristic. As a result, the separatrix and the slow magnetoacoustic surface seem to merge in Fig. 8a and all the characteristics are tangent to it as in the Weber & Davis case (Fig. 5). However, by making a magnification of the crucial region in the immediate neighborhood of the separatrix, Fig. (8b), we see that the flow is not strictly radial and perpendicular to the direction of self-similarity; this separatrix does not coincide with the slow magnetoacoustic surface (dot-dashes in Fig. 8b) and one of the family of characteristics (the one with thin solid lines) is tangent to the critical surface while the second family (with thick solid lines) crossed it.

A similar calculation has been performed in the case of prescribed geometry of Tsinganos & Trussoni (1991). It is shown again that in this θ-self-similar model, Eq. (4.11) is equivalent to the condition at the critical point found in the original paper where the sound speed is $V_s(R_x)$ at the critical point ($M_a = M_x$, $R = R_x$),

$$V_s^2(R_x, \theta) = V_*^2 \frac{M_x^4}{R_x^4} \frac{1 + \kappa(M_x, R_x)\sin^2\theta}{1 + \delta\sin^2\theta}, \qquad (4.12a)$$

where

$$\kappa(M_x, R_x) = \frac{\lambda^2 R_x^2(1 - R_x^2)^2}{(1 - M_x^2)^3} - 1, \qquad (4.12b)$$

Note that for this case, the characteristic lines would be very much like in the one dimensional case as the flow is everywhere radial thus meaning perpendicular to the self-similarity direction. In such a case the separatrix characteristic defining the critical surface is also the fast magnetoacoustic surface.

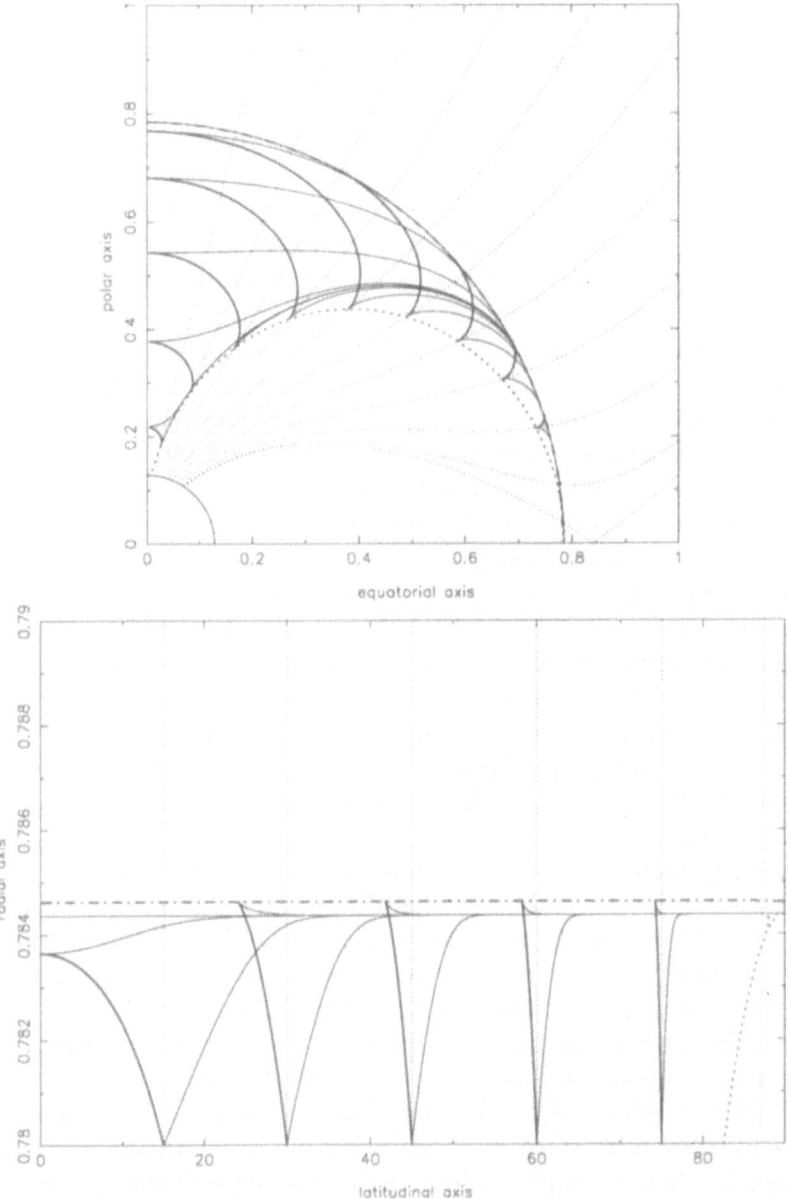

Figure 8. Plot of the characteristics in the model of Sauty and Tsinganos (1994) for a meridionally self-similar wind/jet. The dotted lines are the poloidal flux/flow lines while the solid ones the two families of the characteristics. In (a) a global view of the hyperbolic region is shown while in (b) a zoom of the characteristics around the slow separatrix showing that one of the two families (the one with thick lines) cross the separatrix while the other (the one with the thin lines) becomes tangent to it. The cusp line is drawn with thick dots and the slow magnetoacoustic one with dot-dashes.

5. Summary

The aim of the present paper is to explain the observed relocation and modification of the critical surfaces which appear in the domain of the solutions of the PDE's which govern stationary solar and astrophysical MHD flows in case the symmetry of self-similarity is imposed on the system, in addition to the symmetry of an ignorable coordinate. We recall that although axi-symmetry or planar-symmetry allows to reduce the system of the partial MHD differential equations to the coupled Bernoulli and transfield equations, the problem still remains complicated to be solved analytically as well as numerically. Self-similarity on the other hand, allows a simplified treatment of the basic equations governing ionized flows in a gravitational field, by converting the system of PDE's into a system of ODE's. Such a reduction is possible if it is assumed that all quantities can be scaled in a given direction which defines thus the direction of self-similarity. Furthermore, we pay special attention to the role of the so-called limiting or separatrix characteristics (Guderly 1962) which appear in the hyperbolic regimes of the combined Bernoulli and transfield equation. We find that they coincide with the location of the novel X-type critical points which control the topology of the solutions. At these critical surfaces one of the family of characteristics is asymptotically tangent while the component of the flow velocity perpendicular to the directions of axisymmetry and self-similarity is equal to the characteristic slow/fast MHD wave speeds in that direction.

Three representative examples have been analyzed in detail: vertically, or z-self-similar isothermal flows in uniform gravity applied to prominences and Evershed flows in the low solar corona; radially, or r-self-similar polytropic and axisymmetric flows from discs in spherical gravity which are the basis of standard disc wind theories; and finally, meridionally, or θ-self-similar, axisymmetric but non polytropic winds from a spherical central object. In sections 3 and 4 it has been shown that despite the wide variety of self-similar models, all assumptions can be put in a similar and comparable form. Basically, all these models assume a separation of variables in the magnetic flux, a simple scaling of the magnetic flux in the self-similar direction and a poloidal Alfvén Mach number that depends only on the variable in the direction perpendicular to the symmetry and the self-similarity direction.

For isothermal or polytropic flows with a direction of symmetry, the equation of motion can be integrated along each fieldline giving rise to the so-called Bernoulli's integral. Then the remaining part of momentum balance or transfield equation takes a specific form of a second order equation for the magnetic flux. This equation is of mixed type and changes from hyperbolic to elliptic at the classical MHD speeds. Yet these speeds do not necessarily correspond to critical ones. Conversely, if self-similarity is assumed, we show that critical points appear to avoid the formation of shocks or current sheets which are discontinuities in the second derivatives of the magnetic flux. Furthermore, we show that independently of the specific model, at the critical points, it is the component of the bulk velocity in the perpendicular direction which is equal to one of the speeds of the perturbation waves propagating in this specific direction. In an ionized plasma three types of waves may be considered, the slow magnetosonic, the Alfvén and the fast magnetosonic. Thus the relation between the flow velocity and the velocity of the mode is not as simple as in the spherically symmetric case or when equatorial flows are considered.

We further show that the same result occurs in non-polytropic flows. The problem there is that the sound speed in this more general case is ill defined. In other words, the speed of sound cannot be formally calculated *a priori*, because of the lack of an explicit relation between density and pressure. Nevertheless, it is remarkable that the structure of the dynamical equations allows to calculate explicitly the value of the sound speed at each critical point. It is then shown in the different contexts of fixed and free streamline geometry that again critical points occur when the component of the flow speed perpendicular to both the symmetry and the self-similar directions is equal to one of the three speeds of the normal modes.

The question of the number of the appropriate boundary conditions required in order to obtain a well behaved unique solution to the problem is directly related to the above remarks on critical points and characteristics. Bogovalov in the appropriate chapter in this volume, argues that the number of the conditions we need to prespecify at the stellar boundary of the outflow in order to get a solution equals the number of the waves that can be emitted (or reflected) by this boundary. In Sec 2.2, Eq. (2.6), we have shown that in an MHD flow there are 4 such waves (entropy, Alfvén, slow, fast) which can be sent downstream from a boundary placed in the subslow part of the MHD outflow. As we also discussed in Sec 2.2, the number of the unknowns of the problem is 7 and therefore there are 3 free parameters that need to be determined by the criticality conditions at the respective true critical points. As we have shown here, these are the critical points associated with the slow and fast magnetoacoustic separatrix characteristics (see Fig. 4) together with the Alfvén critical point at the corresponding Alfvén separatrix characteristic. In the case of a HD flow on the other hand, there are 4 such waves propagating downstream from a stellar surface placed in the subsonic part of the flow (entropy, sound and the two vorticity waves). Thus, out of the 5 unknown variables of the problem the one free parameter left should be determined by the critical

458

condition at the sonic separatrix characteristic (see Fig. 3). As we stated earlier, the main difficulty of producing exact MHD solutions lies in the fact that these free boundary conditions should be specified in the process of the solution of the problem, beeing unknown *a priori*.

Acknowledgements. We are indebted to Profs. Eugene Parker, Jean Heyvaerts, Eric Priest and Dr Sergei Bogovalov for sharing with us their invaluable insights on these aspects of the fascinating subject of MHD flows.

References

Agim Y.Z., Tataronis J.A., 1985, J. Plasma Phys. 34(3), 337
Anderson M. R., 1989, MNRAS, 239, 19
Bardeen J.M., Berger B.K., 1978, ApJ, 221, 105
Blandford R.D., Payne D.G., 1982, MNRAS, 199, 883
Bogovalov S.V., 1994, MNRAS, 270, 721
Chakrabarti S.K. & Bhaskaran P., 1992, MNRAS, 255, 255
Contopoulos J., 1992, PhD thesis, Cornell Univ., p. 54
Contopoulos J., 1994, ApJ, 432, 508
Contopoulos J., 1996, ApJ (in press).
Contopoulos J. and Lovelace R.V.E., 1994, ApJ, 429, 139
Courant R., Hilbert D., 1962, Methods of Mathematical Physics, Volume II, John Wiley.
Ferreira J., Pelletier G., 1993b, A&A, 276, 637
Ferreira J., Pelletier G., 1995, A&A, 293, 665
Guderley K. G., 1962, The Theory of Transonic Flow, Addison–Wesley: Reading
Heinemann M., Olbert S., 1978, J. of Geophys. Res., 83-A6, 2457
Heyvaerts J., Norman C.A., 1989, ApJ, 347, 1055
Holzer T.H., 1977, J. Geophys. Res., 82, 23
Li Z., Chiueh T., Begelman M., 1992, ApJ, 394, 459
Li J., Melrose D.B., 1994, MNRAS, 270, 687
Lima J.J., Priest E., 1993, A&A, 268, 641
Lovelace R.V.E., Berk H.L., Contopoulos J., 1991, ApJ, 379, 696
Low B.C., Tsinganos K., 1986, ApJ, 302, 163
Michel F.C., 1969, ApJ, 158, 727
Parker E., 1958, ApJ, 128, 664
Pneuman G.W., Kopp R.A., 1971, Sol. Phys. 18, 258
Polovin R.V., Demutskii V.P., 1990, Fundamentals of Magnetohydrodynamics, Consultants Bureau, New York
Roberts, B., 1976, ApJ, 204, 268
Rosso F., Pelletier G., 1994, A&A, 287, 325
Sakurai T., 1985, A&A, 152, 121
Sakurai T., 1987, PASJ, 39, 821
Sakurai T., 1990, Computer Physics Reports, 12, 247
Sauty, C. and Tsinganos K., 1994, A&A, 287, 893
Trussoni E., Tsinganos K., 1993, A&A, 269, 589
Tsinganos K., 1982, ApJ, 252, 775
Tsinganos K., Low B.C., 1989, ApJ, 342, 1028
Tsinganos K., Sauty C., 1992a, A&A, 255, 405
Tsinganos K., Sauty C., 1992b, A&A, 257, 790
Tsinganos K., Trussoni E., 1990, A&A, 231, 270
Tsinganos K., Trussoni E., 1991, A&A, 249, 156
Tsinganos K., Surlantzis G., Priest, E. 1993, A&A, 275, 613
Weber E.J., Davis L., 1967. ApJ, 148, 217
Williams R.J.R., Dyson J.E., 1994, MNRAS, 270, L52

COLLIMATION OF MHD OUTFLOWS

JEAN HEYVAERTS
Observatoire de Strasbourg, Université Louis Pasteur
11, Rue de l'Université, 67000 Strasbourg, FRANCE
heyvaert@cdsxb6.u-strasbg.fr

AND

COLIN A. NORMAN
Space Telescope Science Institute
and Johns Hopkins University
3700 San Martin Drive, Baltimore, MD 21218, USA

Abstract. This paper discusses the magnetic focusing of MHD winds, in particular the possibility that they asymptotically converge to a cylindrical geometry. Basics of magnetized rotating MHD winds are quickly reviewed. The condition for the solution to remain regular at the Alfvèn surface is discussed and explicitly expressed in terms of surface functions which determine the flow solution. Theorems are established, which constrain the asymptotic shape of rotating polytropic MHD winds. In particular it is proved that winds which carry a non-vanishing Poynting flux and poloidal current to infinity must contain a cylindrically collimated core, whereas other winds focus parabolically.

1. Introduction

Collimated winds are found in nature, blown by such different objects as active galactic nuclei, compact galactic objects and young stellar objects. Stellar winds sometimes show some degree of anisotropy, and the anisotropy of the solar wind itself is presently under study. Anisotropy in wind outflow may have different origins, the most obvious one being a non uniformity of the wind source itself. A non homogeneous intervening medium may also cause anisotropic outflow, for example if a much denser equatorial medium diverts the outflow polewards. MHD forces can also focus winds. Such forces

K. C. Tsinganos (ed.), Solar and Astrophysical Magnetohydrodynamic Flows, 459–474.
© *1996 Kluwer Academic Publishers.*

may act at large distances from the wind source and can produce significant focusing. Whether or not they do so depends on properties of the flow to be reviewed in this communication, which discusses the model of a stationary, axisymmetric, differentially rotating MHD wind. We first present the basic properties of such flows in sections 2, 3 and 4 and then discuss some general asymptotic properties of MHD winds in section 5.

2. Basics of axisymmetric stationary MHD flows

2.1. THE FOCUSING HOOP STRESS

The force which is responsible for MHD wind focusing is the so-called hoop stress. Consider a certain cylindrical region of space carrying a net current $\vec{j}$, parallel to its axis. This current creates an azimuthal field $\vec{B}_\theta$. A moment's thought reveals that the force density $\vec{j} \times \vec{B}_\theta$, the hoop stress, is always oriented such as to constrict the current carrying region towards the axis. So, MHD focusing is exerted on winds which carry axial poloidal currents. However, the hoop stress fights other forces, such as the pressure gradient and the gradient of poloidal magnetic pressure, most often outwardly oriented, or the centrifugal force, which is always radially oriented outwards. The question is then whether the hoop stress can dominate over these other forces to asymptotically focus the wind.

2.2. THE AXISYMMETRIC STATIONARY WIND MODEL

Let us model the wind as stationary and axisymmetric, and simplify the thermodynamics to a polytropic relation between pressure and mass density

$$P = Q\rho^\gamma. \tag{1}$$

The factor Q is constant following the fluid, but can assume different values for different fluid elements.

Cylindrical coordinates r and z are used. Any vector $\vec{V}$ can be split into a poloidal component, $\vec{V}_P$, in the (r-z) plane, and a toroidal component, $\vec{V}_T$, in the θ-direction. Because of axisymmetry, the toroidal part of the magnetic field has no divergence and the poloidal part satisfies $\operatorname{div}\vec{B}_P = 0$. So, it derives from a vector potential, which can be chosen in the azimuthal direction and can be written as

$$\vec{B}_P = \vec{\nabla} \times \left(a(r,z)\frac{\vec{e}_\theta}{r} \right) = \frac{\vec{\nabla}a \times \vec{e}_\theta}{r}. \tag{2}$$

This implies that the function $a(r, z)$ is constant along poloidal field lines, because its gradient is perpendicular to $\vec{B}_P$: these field lines are lines of constant a. If the convention is taken that a vanishes on the polar axis, the flux through a circle of axis z passing at point (r,z) is $2\pi a(r,z)$.

Field lines of the total field $\vec{B}$ are drawn on the axisymmetric surfaces of constant $a(r,z)$, generated by the rotation about the polar axis of a poloidal field line. Indeed, the tangent plane to such a surface contains both $\vec{B}_P$ and the toroidal component $\vec{B}_\theta$. These surfaces are called magnetic surfaces.

2.3. SURFACE FUNCTIONS

The equations which describe an axisymmetric MHD wind have a number of first integrals. The associated quantities are constant on a magnetic surface. As will be seen, this is equivalent to saying that they are constant following the motion of the fluid. These functions then depend on r and z through the value assumed by the function a on this particular magnetic surface. They are called "surface functions".

Let us give an example. Because of stationarity, the electric field is irrotational, which implies that it is the gradient of an electric scalar potential ϕ. Because of axisymmetry any derivative with respect to θ vanishes. So, the electric field has no toroidal component. Let $\vec{v}$ be the plasma velocity. The toroidal component of the flux-freezing equation, $\vec{E} + \vec{v} \times \vec{B} = 0$ reduces to $\vec{v}_P \times \vec{B}_P = 0$. This shows that the poloidal velocity and magnetic field must be aligned and implies that a function α exists such that

$$\rho \vec{v}_P = \alpha \vec{B}_P, \tag{3}$$

where ρ is the mass density. It follows from the alignement of the poloidal velocity to the poloidal field that the wind flows on magnetic surfaces, because the sum of the poloidal and toroidal components of the velocity is in their tangent plane. Now, take the divergence of equation (3), noting that for a stationary flow $\mathrm{div}(\rho \vec{v}_P)$ vanishes. It results that

$$\vec{B}_P . \vec{\nabla} \alpha = 0 \tag{4}$$

So, α does not vary following the poloidal field, which means that a line of constant α is also a line of constant a, or, stated anotherway, that α is a function of a alone:

$$\alpha = \alpha(a) \tag{5}$$

It is a surface function: the mass flux to magnetic flux ratio. We now enumerate, without detailed proof, the other surface functions which appear.

The second one results from the poloidal part of the flux-freezing equation $\vec{\nabla}\phi = \vec{v} \times \vec{B}$, from which it can be deduced that the electric potential ϕ is itself a surface function, $\phi(a)$. Its derivative with respect to a, $\Omega(a)$, has dimension of the inverse of a time. The poloidal part of the induction equation leads to the relation:

$$\rho v_\theta = \rho r \Omega(a) + \alpha(a) B_\theta \tag{6}$$

Equations (3) and (6) show that the fluid velocity is the sum of a field-aligned component of amplitude $(\alpha B/\rho)$ and of a rotation about the polar axis at angular velocity $\Omega(a)$. The flow can be viewed as if the fluid were moving along "pipes" (the magnetic field lines) helically coiled on the magnetic surfaces, each of which rotates like a solid body at its own rotation frequency $\Omega(a)$. Equation (6) is for this reason sometimes refered to as Ferraro's isorotation law. Note again that it implies that the fluid flows on magnetic surfaces. A consequence of this is that the factor Q in equation (1) is constant on a magnetic surface, because fluid elements flow on them:

$$Q = Q(a) \tag{7}$$

The toroidal part of the equation of motion

$$\rho \left(\vec{v}.\vec{\nabla} \right) \vec{v} = -\vec{\nabla} P + \rho \vec{g} + \vec{j} \times \vec{B} \tag{8}$$

leads to the following first integral:

$$r v_\theta - \frac{r B_\theta}{\mu_0 \alpha} = L(a) \tag{9}$$

In the absence of a magnetic field, $L(a)$ would reduce to the (conserved) angular momentum of a unit mass of material flowing on magnetic surface a. In the presence of MHD forces, the material specific angular momentum is not conserved anymore. This is because MHD stresses exert a torque on the flowing material. It is possible, though, to define a (conserved) material and electromagnetic momentum flux. The electromagnetic part is associated with the moment with respect to the polar axis of a component of the Maxwell stress tensor. This conserved total angular momentum flux is just $L(a) \, \rho \, \vec{v}_P$. $L(a)$ can then be regarded as the specific angular momentum of matter and field together.

The poloidal part of the equation of motion is still a vector equation, which can be projected along $\vec{B}_P$ or perpendicular to it. The perpendicular projection will be considered later. The projection on $\vec{B}_P$ yields the following, so-called Bernoulli first-integral:

$$\frac{v^2}{2} + \frac{\gamma}{\gamma - 1} Q \rho^{\gamma-1} + G(r, z) - \frac{r \Omega B_\theta}{\mu_0 \alpha} = E(a) \tag{10}$$

In this equation v is the total fluid velocity and G the gravitational potential. The gravitational field is $\vec{g} = -\vec{\nabla} G$. In the absence of MHD forces, this first integral would express the well known Bernoulli's theorem, i.e. that the sum of the kinetic, enthalpy and gravitational energy fluxes is constant. The magnetic field introduces another energy flux, the Poynting flux, which adds to the other three to form a conserved total energy flux, comprising a material and a magnetic part. This energy flux, divided by mass flux, defines a conserved total specific energy of escaping matter, E. The fourth term on the left-hand side of equation (10) is the Poynting flux per unit of escaping mass flux, which we may call the Poynting specific energy. Note that $2\pi r B_\theta/\mu_0$ is, by Ampere's theorem, the electric current I at altitude z through a circle of radius r. So, the Poynting specific energy is $I\Omega/2\pi\alpha$.

2.4. ALFVÈN RADIUS AND DENSITY

By the isorotation law, equation (6), and the specific angular momentum conservation law, equation (9), the toroidal components of the velocity and of the magnetic field can be expressed in terms of the density and radius:

$$v_\theta = r\Omega + \frac{\mu_0\alpha^2}{r}\frac{L - r^2\Omega}{\mu_0\alpha^2 - \rho},\tag{11}$$

$$B_\theta = \frac{\mu_0\alpha\rho}{r}\frac{L - r^2\Omega}{\mu_0\alpha^2 - \rho}.\tag{12}$$

The denominator of these expressions vanishes when $\rho = \mu_0\alpha^2$. For regularity of v_θ and B_θ, r^2 must become equal to L/Ω when this happens. It can easily be checked that the poloidal velocity, given by equation (3), becomes in this case equal to the local Alfvèn velocity calculated with the poloidal component of the magnetic field, v_{PA}. Said anotherway, it becomes equal to the phase speed of the torsional Alfvèn mode propagating in the poloidal direction. For this reason $\mu_0\alpha^2$ is named the Alfvèn density

$$\rho_A \equiv \mu_0\alpha^2.\tag{13}$$

and L/Ω is named the square of the Alfvèn radius,

$$r_A^2 \equiv \frac{L}{\Omega}\tag{14}$$

More generally, the Alfvènic Mach number, $M_A = v_P/v_{PA}$, turns out to be given by

$$M_A^2 = \frac{\rho_A}{\rho}\tag{15}$$

The specific angular momentum L of escaping matter is then Ωr_A^2, where r_A appears as a lever arm. Generally r_A is larger than the size of the wind source. This means that the torques exerted by MHD forces are more effective than the angular momentum carried away by the escaping matter in spinning down the wind source.

By equations (3), (11) and (12), the Bernoulli equation (10) can be written in terms of the density ρ and flux function a as

$$\frac{1}{2}\frac{a^2(\nabla a)^2}{\rho^2 r^2} = E - G - \frac{\gamma Q \rho^{\gamma-1}}{\gamma - 1} + \rho \Omega^2 \frac{r_A^2 - r^2}{\rho_A - \rho} - \frac{\Omega^2}{2r^2}\left(r_A^2 + \rho\,\frac{r_A^2 - r^2}{\rho_A - \rho}\right)^2.$$
$$(16)$$

3. Field-aligned flows

3.1. THE BERNOULLI INTEGRAL

The Bernoulli equation (16) is not self-contained, since the left-hand side can only be known when the function $a(r, z)$ is. The projection of the equation of motion perpendicular to the poloidal field gives an equation, to be examined later, which determines the shape of magnetic surfaces. Assume provisionally this shape to be known. Models where the shape of magnetic surfaces is assumed have been the standard description of winds in the mid 1960's, as exemplified by the Weber and Davis (1967) model for the equatorial solar wind, where magnetic surfaces were taken to be cones.

For known magnetic surfaces the Bernoulli equation is simply an algebraic relation relating the density ρ to the radius r at fixed a. The solution of this equation, however, is generally singular when the flow velocity passes the characteristic magnetosonic mode velocities, unless specific conditions are imposed on the first integrals. It is however possible to do so because not all of them are defined by the boundary conditions.

Actually, $Q(a)$ is known because the state of the plasma at the base of the wind is known, and $\Omega(a)$ is essentially the rotation of the matter at the base of the wind, because when ρ becomes much larger than ρ_A, (v_θ/r) becomes approximatly equal to Ω. However the other first integrals are not similarly determined by boundary conditions.

We seek a solution that can somehow be matched, at large distances along the field lines, to virtually empty space, i.e. with a plasma of arbitrarily small density and pressure threaded by a known potential magnetic field, generally taken as uniform.

The early work of Parker (1958) has shown that such a matching can only be achieved through shocks, with the wind density vanishing at infinity for flaring flow surfaces and the velocity becoming supersonic. In the more general situation where the geometry of magnetic surfaces is not a-priori known, the necessity for the wind to reach a velocity at infinity exceeding that of all the three MHD modes should be discussed in each particular case.

From equation (15), however, it is seen that the flow must become superalfvènic if the density is to become vanishingly small. The flow should probably also become faster than the fast mode speed, because the magnetic field in the wind which, when magnetic surfaces widen off axis, is dominated at large distances by its toroidal component, has to match through a system of discontinuities to the uniform field at infinity. This should involve a fast MHD-shock.

Therefore solutions are sought which reach a velocity larger than the Alfvèn speed and the fast mode speed at large distances on each magnetic surface.

Since the flow starts with a very small velocity at the source, where the plasma is very dense, it must accelerate to pass first the slow mode speed, v_s, then the Alfvèn speed, and finally the fast mode speed, v_f. The condition that the flow be transalfvènic introduces at this stage no constraint. Actually, multiplying equation (16) by $(\rho_A - \rho)^2$, it can be seen that any solution which reaches $\rho = \rho_A$ does it at $r = r_A$.

3.2. CRITICAL POINTS AND CRITICALITY RELATIONS

Most solutions of the Bernoulli equation (16), however, are singular when the poloidal velocity reaches the slow or fast mode speed. It takes some algebra to prove this. The derivative $(d\rho/dr)$ is obtained by differentiating equation (16) at constant a with respect to ρ and r. If we write it as

$$B(\rho, r) = E, \tag{17}$$

we obtain

$$\frac{d\rho}{dr} = -\frac{\partial B/\partial r}{\partial B/\partial \rho}. \tag{18}$$

It can be shown that the condition for the denominator to vanish is just the dispersion relation of MHD modes, with the poloidal fluid velocity replacing the mode phase speed. The direction of propagation to consider is that of the poloidal velocity with respect to the total field $\vec{B}_P + \vec{B}_T$. More details can be found, for example, in Heyvaerts (1996). This is the reason why most solutions of the Bernoulli equation become singular when the poloidal velocity passes one of the MHD mode speeds. The only regular solutions are those for which the numerator of the right-hand side of equation (18) vanishes at the same time as its denominator. Equivalently, it can be said that the Bernoulli function $B(r, \rho)$ has a vanishing differential at these particular places of the (r, ρ) plane. This statement does not depend on the variables used in expressing the specific energy. It will also hold true if the velocity v_P, or the Alfvèn Mach number, M_A, is used instead of ρ in equation (17).

Those points of the (r, ρ) plane where the differential of the Bernoulli function B vanishes are called "critical points". The condition that the numerator and the denominator of the right hand side of equation (17) vanish simultaneously gives for each critical point a pair of relations between its position and density, r_{cr} and ρ_{cr}. The first integrals Q, L, Ω and α stand as parameters in these relations.

Since there are two critical velocities to pass, the slow and the fast one (subscript $cr = s$ and $cr = f$ resp.), we obtain altogether four such equations. They can, at least in principle, be solved to give the position and density at the fast and slow critical point as a function of the first integrals, but for E, which eliminates in differentiating equation (16).

$$r_s = R_s(\Omega, Q, \alpha, L),$$ (19.a)

$$\rho_s = D_s(\Omega, Q, \alpha, L),$$ (19.b)

$$r_f = R_f(\Omega, Q, \alpha, L),$$ (19.c)

$$\rho_f = D_f(\Omega, Q, \alpha, L).$$ (19.d)

We then demand, for regularity, that the solution passes both through the fast and slow critical point. This requires that the Bernoulli function $B(r, \rho)$ has the same value, $E(a)$, at these two points:

$$B(r_s, \rho_s) = B(r_f, \rho_f) = E(a).$$ (20)

This gives two relations, the so-called criticality relations, between the five surface functions, E, α, L, Ω and Q. Two of them being fixed by the boundary conditions, one remains, at this stage, arbitrary. In those models which disregard the transfield equation, this arbitrariness usually appears as a further imposed boundary condition, for example that the density be known at some point in the wind.

4. The transfield equation

4.1. FORCE BALANCE PERPENDICULAR TO FIELD LINES

At this point, all the equations of the model have been written, except for the projection of the poloidal part of the equation of motion perpendicular to the magnetic field. This equation determines the shape of magnetic surfaces. Some tedious algebra reduces it to the following form, one among many different possible ones:

$$\frac{\alpha}{\rho r}\left(\frac{\partial}{\partial z}\frac{\alpha}{\rho r}\frac{\partial a}{\partial z} + \frac{\partial}{\partial r}\frac{\alpha}{\rho r}\frac{\partial a}{\partial r}\right) - \frac{1}{\mu_0 \rho r}\left(\frac{\partial}{\partial z}\frac{1}{r}\frac{\partial a}{\partial z} + \frac{\partial}{\partial r}\frac{1}{r}\frac{\partial a}{\partial r}\right) = E' - \frac{Q'\rho^{\gamma-1}}{\gamma - 1}$$

$$+\frac{\alpha'}{\alpha}\ \frac{\mu_0\alpha^2\rho}{r^2}\ \frac{(L-r^2\Omega)^2}{(\mu_0\alpha^2-\rho)^2} - \frac{\rho}{r^2}\ \frac{(L'-r^2\Omega')(L-r^2\Omega)}{(\mu_0\alpha^2-\rho)} - \frac{LL'}{r^2}. \quad (21)$$

In this so-called transfield equation, primes denote derivatives with respect to a, i.e. $E' = dE/da$. The expressions (11) and (12) for the toroidal components of the velocity and magnetic field have been used. This is a nonlinear partial differential equation for the function $a(r, z)$. It is coupled to the Bernoulli equation because the density is still present in equation (21). The non-linearities appear on the right-hand side of the equation, as well as on the left-hand side, where terms quadratic in the first order derivatives of a appear.

The equation is however linear in the second order derivatives, so it is of the quasi-linear type. It can be shown (Heinemann and Olbert 1978) that it is of a mixed elliptic/hyperbolic type. It is elliptic from the base of the wind to the cusp surface, which is the locus of points where the poloidal fluid velocity reaches the so-called cusp velocity, v_c, defined by

$$v_c^2 = \frac{v_f^2\ v_s^2}{v_f^2 + v_s^2}. \quad (22)$$

It becomes hyperbolic between the cusp surface and the slow surface, elliptic again between the slow and the fast mode surface, to finally become hyperbolic from the fast surface on. These properties of the transfield equation specify which boundary conditions determine the solution. In the case of no motion, $\alpha = 0$, it is possible, using (11) and (12) again, to prove that equation (21) reduces to the well known Grad-Shafranoff equation of axisymmetric magneto-hydrostatics.

4.2. ALFVÈN REGULARITY CONDITION

The transfield equation (21) is singular at the Alfvèn surface where all its second order terms vanish (Heinemann and Olbert 1978). A somewhat more direct proof is as follows. In deriving equation (21), it is found that it can also be written as

$$\frac{1}{\mu_0\rho}\left(\vec{\nabla}\times\vec{B}_P\right)\times\vec{B}_P - \left(\vec{\nabla}\times\vec{v}_P\right)\times\vec{v}_P = \vec{\nabla}a\left(E' - \frac{Q'\rho^{\gamma-1}}{\gamma-1} - \Omega L'\right)$$

$$+\vec{\nabla}a\left(\frac{\alpha'}{\alpha}\ \frac{B_\theta^2}{\mu_0\rho} - \frac{rB_\theta\Omega'}{\mu_0\alpha} - \frac{\alpha B_\theta L'}{\rho r}\right). \quad (23)$$

The Bernoulli equation gives information on the modulus of $\vec{v}_P$. Then it must be possible to turn the transfield equation into an equation for its direction. Let us write

$$\vec{v}_P = v_P\vec{t}. \quad (24)$$

with

$$\vec{t} = cos\psi \ \vec{e}_r + sin\psi \ \vec{e}_z, \tag{25}$$

and define the normal to magnetic surfaces as

$$\vec{n} = -sin\psi \ \vec{e}_r + cos\psi \ \vec{e}_z. \tag{26}$$

Inserting this in equation (23) we obtain an equation for the curvature of poloidal field lines which can be written as :

$$\left(1 - \frac{\rho}{\rho_A}\right) \left(v_P^2 \frac{d\psi}{ds} - \frac{\alpha^2}{\rho^2}(\vec{n}.\vec{\nabla}) \left(\frac{B_P^2}{2}\right)\right) = \frac{B_P^2}{2}(\vec{n}.\vec{\nabla}) \left(\frac{\alpha^2}{\rho^2}\right) + |\nabla a|E'$$

$$-|\nabla a| \left(\frac{Q'\rho^{\gamma-1}}{\gamma-1} + \Omega L' + \frac{rB_\theta\Omega'}{\mu_0\alpha} + \frac{\alpha B_\theta L'}{\rho r} - \frac{\alpha'}{\alpha} \frac{B_\theta^2}{\mu_0\rho}\right). \tag{27}$$

In this equation $d\psi/ds$ is the derivative of the direction angle with respect to the curvilinear absissa along the poloidal field line. This follows from the Fresnet curvature formulae, noting that $\vec{t}.\vec{\nabla} \equiv d/ds$. It is seen that the left hand side, which contains the second order derivative terms, vanishes at the Alfvèn surface. It can be shown by explicit calculations that the second order terms present in the first term on the right-hand side of equation (27) also vanish at the Alfvèn point.

Equation (27) can then be turned into an expression for the curvature $d\psi/ds$, involving a denominator which vanishes at the Alfvèn surface. Magnetic surfaces would have infinite curvature there (a kink), unless the numerator of this expression also vanishes.

At the Alfvèn surface, the transfield equation then reduces to a first order differential equation, the so-called Alfvèn regularity condition, which can be written as:

$$\frac{B_P^2}{2|\nabla a|}(\vec{n}.\vec{\nabla}(\frac{\alpha^2}{\rho^2})) + E' - \frac{Q'\rho^{\gamma-1}}{\gamma-1} - \Omega L' - \frac{rB_\theta\Omega'}{\mu_0\alpha} - \frac{\alpha B_\theta L'}{\rho r} + \frac{\alpha'B_\theta^2}{\alpha\mu_0\rho} = 0 \tag{28}$$

We worked out (Heyvaerts and Norman 1989) a more explicit form of this regularity condition. It involves the direction angle ψ at the Alfvèn point on magnetic surface a, $\psi_A(a)$, the poloidal velocity there, $v_{PA}(a)$, and the slope at the Alfvèn point of the relevant solution of the Bernoulli equation, which can be represented by the parameter q:

$$q = \left(\frac{dlog\rho}{dlogr}\right)_{a,A}. \tag{29}$$

This parameter can be calculated in terms of the values of the first integrals which determine the solution of the Bernoulli equation for fixed a. Finally, the Alfvèn regularity condition can be written as:

$$\frac{\alpha'}{\alpha} - q \ \frac{r'_A}{r_A} + q \ sin\psi_A \ \frac{1}{r|\nabla a|_A} + \frac{E'}{v_{PA}^2} - \frac{Q'\rho_A^{\gamma-1}}{(\gamma-1)v_{PA}^2}$$

$$+\frac{\Omega^2 r_A^2}{v_{PA}^2}\left(\frac{4}{q^2}\frac{\alpha'}{\alpha} + \frac{4+2q}{q}\frac{r'_A}{r_A} - \frac{\Omega'}{\Omega}\right) = 0. \tag{30}$$

5. General results on the asymptotic behaviour of polytropic rotating MHD winds

5.1. ROTATING POLYTROPIC MHD WINDS
DO NOT ASYMPTOTICALLY BEND TOWARDS THE EQUATOR

We now discuss asymptotic properties of MHD winds. An interesting bound on the asymptotic behaviour of the density on a given magnetic surface is obtained by considering the Bernoulli equation (16). Assume that when following a poloidal field line to infinity the radius r approaches infinity. No generality is lost by assuming this, since otherwise cylindrical focusing would be certain. Then, gather all terms in equation (16) which might be dominant in this limit, knowing that the density must be smaller than the Alfvèn density, i.e. remain finite. The behaviour of ρr^2, however, cannot be anticipated. Equation (16) then reduces to the following approximate form:

$$\frac{1}{2}\frac{\alpha^2(\nabla a)^2}{\rho^2 r^2} \approx E - \frac{\gamma Q\rho^{\gamma-1}}{\gamma-1} - \frac{\rho\Omega^2 r^2}{\rho_A - \rho} - \frac{1}{2}\frac{\rho^2\Omega^2 r^2}{(\rho_A - \rho)^2}. \tag{31}$$

The negative terms on the right hand side cannot exceed E, since the sum has to remain positive. This implies that ρr^2 must remain bounded following the poloidal field line a to infinity, a limit which we represent by the symbol $\lim_{\infty a}$, or simply lim. So,

$$\lim_{\infty a}(\rho r^2) < \infty. \tag{32}$$

This allows us to further reduce expression (31) to

$$\frac{1}{2}\frac{\alpha^2(\nabla a)^2}{\rho^2 r^2} \approx E - \frac{\rho r^2\Omega^2}{\rho_A} \tag{33}$$

and to improve our bound on ρr^2:

$$\lim_{\infty a}\left(\frac{\rho r^2\Omega^2}{\rho_A}\right) < E. \tag{34}$$

This relation only expresses the fact that the Poynting flux at infinity cannot exceed the total energy flux. Since the specific poloidal kinetic energy also cannot exceed E, we can write:

$$\lim_{\infty a} v_P \leq \sqrt{2E}. \tag{35}$$

By equations (2), (3) and (34) this can be converted into a bound on $r|\nabla a|$:

$$r|\nabla a| < \frac{\sqrt{2}\mu_0 \alpha E^{3/2}}{\Omega^2} = \lambda(a). \tag{36}$$

This can be further used to deduce a lower bound on the value of z at radius r on poloidal field line a, $z_a(r)$. Indeed,

$$|\nabla a| \geq |\frac{\partial a}{\partial z}|. \tag{37}$$

Since the magnetic surfaces are nested, z grows at fixed r when the magnetic surface becomes more "polar", whatever the sense of variation of the flux function a from equator to pole. So, the inequality (37) can be converted, at fixed r, into

$$|dz| \geq \frac{r}{\lambda(a)}|da|, \tag{38}$$

which can be integrated to obtain a lower bound on z. To be specific, assume that a vanishes at the pole and increases equatorwards to a value A_{eq} (generalizations are easy). In this case equation (37) integrates into

$$z_a(r) \geq r \int_a^{A_{eq}} \frac{da'}{\lambda(a')}. \tag{39}$$

This shows that the magnetic surface a is situated above a certain cone. We then conclude that for a polytropic wind no magnetic surface can be such that $\lim_{\infty a}(z/r) = 0$. Or, stated another way, the magnetic surfaces of a polytropic wind cannot asymptotically bend towards the equator.

5.2. WINDS WHICH CARRY NO POYNTING FLUX AT INFINITY ALSO CARRY NO POLOIDAL CURRENT AND FOCUS PARABOLICALLY

It has been shown above that a wind carries no Poynting flux to infinity (inside a certain magnetic surface a) if $\lim(\rho r^2)$ vanishes. The Poynting specific energy at z on some magnetic surface a is $I\Omega/2\pi\alpha$, where the electric current I enclosed between the axis and this magnetic surface at this z is $(2\pi r B_\theta/\mu_0)$. For large (r/r_A), and small (ρ/ρ_A) this current approaches (see equation (12)) $2\pi\rho r^2\Omega/\alpha$. The asymptotic Poynting flux then vanishes when $\lim_{\infty a}(\rho r^2) = 0$.

We now show that if $\lim(\rho r^2)$ is zero, so is $\lim(z/r)$. Note that this may only happen if $\lim(r)$ is infinite, for otherwise both ρ and r would approach finite limits, and the assumption that ρr^2 approaches zero could not be satisfied. From equation (33), we see that

$$r|\nabla a| \leq \frac{\sqrt{2E}}{\alpha} \; \rho r^2. \tag{40}$$

If $lim_{\infty a}(\rho r^2)$ vanishes, as assumed, a function $\lambda(a, r)$ exists, which approaches zero as $r \to \infty$, such that

$$r|\nabla a| < \lambda(a, r). \tag{41}$$

Repeating the reasoning made in the preceding section, we find that:

$$z > r \int_a^{A_{eq}} \frac{da'}{\lambda(a', r)} = r \, \Lambda(a, r), \tag{42}$$

with

$$\Lambda(a, r) \to \infty \qquad \text{as} \qquad r \to \infty \tag{43}$$

So,

$$\lim_{\infty a} \left(\frac{z}{r} \right) = \infty. \tag{44}$$

This means that such magnetic surfaces have, at infinity, the shape of paraboloids.

We have thus proved the statement made in the title of this subsection, that magnetic surfaces on which no Poynting flux is carried to infinity are asymptotically parabolic.

Note however that the inverse proposition, i.e. that parabolic magnetic surfaces should necessarily enclose an asymptotically vanishing poloidal current, is not true. Magnetic surfaces which are parabolic at infinity enclose finite current if $\log(z/r)$ approaches a constant independant of a. If, on the other hand, the surfaces can be asymptotically represented as $z = K(a)r^{p(a)}$ with a power index $p(a)$ which does depend on a, then $r|\nabla a|$ approaches zero, as can be checked by direct differentiation, and this implies that ρr^2 approaches zero too.

5.3. MAGNETIC SURFACES WHICH DIVERGE OFF AXIS ENCLOSE NO POLOIDAL CURRENT BETWEEN THEM AT INFINITY

Up to now, our conclusions have been based on the properties of the Bernoulli equation only. We now turn to the transfield equation, in the form which appears in equation (22). The first term on the left-hand side is negligible when $r \to \infty$, because its ratio to the second one goes as M_A^{-2}, which, from equation (15), approaches zero when the density goes to zero. In the same limit, the gas pressure term on the right hand side can be neglected as compared to finite terms. One easily calculates that $(\vec{\nabla} \times \vec{v}_P) \times \vec{v}_P = v_P^2 d\psi/ds - (\vec{n}.\vec{\nabla})(v_P^2/2)$, and that the asymptotic limits, for large r and small ρ, of (rB_θ) and v_P^2 are $(-\rho r^2\Omega/\alpha)$ and $v_\infty^2 = E - (\rho r^2\Omega^2/\mu_0\alpha^2)$ respectively. Introducing the notation

$$R_\infty^2(a) = \lim_{\infty a} \left(\frac{\rho r^2}{\mu_0\alpha^2} \right) \tag{45}$$

the asymptotic form of equation (22) can be written as:

$$r\frac{d\psi}{ds} = \frac{\mu_0 \; \alpha \; \Omega^2 \; R_\infty^4}{v_\infty} \left(\frac{\Omega'}{\Omega} + 2\frac{R_\infty'}{R_\infty} + \frac{\alpha'}{\alpha} \right). \tag{46}$$

Obviously, the curvature $d\psi/ds$ should decrease to zero at infinity, but it is not obvious that its product with r should do. This is nevertheless true and can be shown as follows. The right hand side of equation (46) is a certain function of a, call it $k(a)$. The asymptotic shape of a poloidal field line of a certain a-value is then given by the curvature equation

$$r\frac{d\psi}{ds} = k, \tag{47}$$

where k is a constant on this field line. Equation (47) can be solved, associated to the two auxiliary equations

$$dr = cos\psi \; ds \qquad dz = sin\psi \; ds \tag{48}$$

to give

$$r = r_0 \; exp\left(\frac{sin\psi}{k} \right) \tag{49.a}$$

$$z = z_0 + \frac{r_0}{k} \int^\psi exp\left(\frac{sin\psi'}{k} \right) sin \; \psi'd\psi' \tag{49.b}$$

So, r is a periodic function of the parameter ψ. It is bounded from above, and cannot reach infinity, as assumed, unless the constant k vanishes. In this case an higher order expansion of the transfield equation is required to calculate the curvature of the poloidal field lines and find their actual shapes. To lowest order in $(1/z)$, the transfield equation then reduces, asymptotically, to the condition that the constant k vanishes, which can be written as

$$\frac{\alpha \; \Omega^2 \; R_\infty^4}{v_\infty} \left(\frac{\Omega'}{\Omega} + 2\frac{R_\infty'}{R_\infty} + \frac{\alpha'}{\alpha} \right) = 0 \tag{50}$$

This asymptotic form of the transfield equation can be satisfied in two different ways. On the one hand, the factor R_∞ may vanish, which, from its definition given by equation (45), means that $\lim(\rho r^2) = 0$. In this case the magnetic surfaces become asymptotically parabolic and enclose an asymptotically vanishing poloidal current. On the other hand, the factor in parenthesis on the left-hand side of equation (50) may vanish, which implies that

$$\lim_{\infty a} \frac{d}{da}\left(\frac{\rho r^2 \Omega}{\alpha}\right) = 0. \tag{51}$$

This means that the poloidal current enclosed in the magnetic surface a becomes a constant independant of a when z approaches infinity. An immediate consequence of this is that the current enclosed between two such surfaces asymptotically vanishes.

We have thus proved that in either case there is an asymptotically null current between magnetic surfaces which diverge off the rotation axis, i.e. for which $\lim_{\infty a}(r) = \infty$.

5.4. WINDS WHICH CARRY NON VANISHING POYNTING FLUX AT INFINITY CONTAIN A CYLINDRICALLY COLLIMATED CORE

If a wind conveys a non-zero Poynting flux to infinity on at least one of its magnetic surfaces, the poloidal current enclosed in it does not approach zero at large distances. If it happens that r does not diverge on this surface, nothing more needs to be proved. If not, then, there is no current between this surface and other similar surfaces for which r also diverges, which are nested in it closer to the polar axis. However, all these surfaces must enclose the same, non-zero, poloidal current. So, this current must flow in a "core" of the wind made of magnetic surfaces on which r remains bounded as $z \to \infty$. This completes the proof of the statement made in the title of this subsection.

References

Blandford, R. D., and Payne, D.G., (1982), Hydromagnetic flows from accretion discs and the production of radio jets, *Month. Not. Roy. Astr. Soc.*, **199**,pp. 883–903

Heinemann, M. and Olbert, S., (1978), Axisymmetric ideal MHD stellar wind flows, *J. Geophys.Res.*, **83**, pp. 2457-2460

Heyvaerts, J. and Norman, C.A., (1989) The collimation of magnetized winds, *Astrophys.J.*, **347** , pp. 1055-1081

Heyvaerts, J., (1996) *Plasma Astrophysics*, Rotating MHD winds, in Proceedings of the EADN Summer School, San Miniato (Italy), G.Einaudi, ed., "Lecture Notes in Physics" series, Springer Verlag, in press.

Parker, E.N., (1958) Dynamics of the interplanetary gas and magnetic field, *Astrophys.J.*, **128**, pp. 664–676

474

Weber, E.J. and Davis, L. Jr, (1967) The angular momentum of the solar wind *Astrophys.J.*, **148** , pp. 217-227

THE PARKER INSTABILITY IN THE INTERSTELLAR MEDIUM

TELEMACHOS CH. MOUSCHOVIAS
University of Illinois at Urbana-Champaign
Departments of Physics and Astronomy
1002 West Green street
Urbana, IL 61801, U. S. A.   E-mail: tchm@astro.uiuc.edu

Abstract. We review the elementary formalism and basic physics of the instability in a galactic disk, first in two dimensions (defined by the vertical component of the galactic gravitational field and by the essentially horizontal interstellar magnetic field). Then we discuss how the nature of the instability changes by the presence of three-dimensional perturbations and by a spatially varying gravitational field. "Final" equilibrium states are obtained in two dimensions, reachable from Parker's stratified initial state through continuous deformations of the field lines. In the process, the inherently open system of magnetohydrostatic equations is closed in a self-consistent fashion. The equilibrium states represent giant clouds (or cloud complexes) in valleys of the field lines. The variation of the calculated column density along spiral arms is in agreement with observations. An energy integral for an isothermal plasma is also discussed. Finally, we review calculations which have recently followed numerically the nonlinear evolution of the instability. Work which questions the significance of the instability for the interstellar medium or attributes to the instability unattainable consequences is also reviewed critically.

1. Introduction

Obvious as it may appear to us today that the interstellar magnetic field is confined to the Galaxy by the weight of the gas in the Galactic gravitational field, the subject was one of considerable controversy in the 1960s. Although Biermann and Davis (1960) had arrived at the correct conclusion, it was Parker (1966) who first excluded other suggested possibilities (such as force-free twisted flux tubes, or interstellar field lines passing through and confined by the Galactic nucleus) and explored the dynamical consequences of the gas-confined field. The magnitude of the field was yet another controversial issue (see review by Spitzer 1968). The then existing observational evidence (from optical polarization of starlight, synchrotron radiation, 21-cm Zeeman measurements, and Faraday rotation) was interpreted by some workers as implying "strong" fields ($B > 10 \ \mu G$; e.g., Woltjer 1965), while it was used by others to argue for "weak" fields ($< 5\mu G$; e.g., Davies *et al.* 1960; Parker 1966). Chandrasekhar and Fermi (1953) used the mean angular deviation of the observed polarization vectors from the direction of the local spiral arm and the assumption that this is due to Alfvén waves along the arm (thought to be a cylindrical flux tube), and predicted a $B = 7 \ \mu G$. By assuming that a spiral arm of stars and gas is a self-gravitating, primarily magnetically supported flux tube, they also found that $B = 6 \ \mu G$. The arguments in the literature are fascinating. However, this is not a review of the history of the origins of our current knowledge of the interstellar magnetic field. It will suffice to state here that sometimes correct arguments had led to incorrect conclusions because of one or more

K. C. Tsinganos (ed.), Solar and Astrophysical Magnetohydrodynamic Flows, 475–504.
© *1996 Kluwer Academic Publishers.*

incorrect assumptions, while incorrect arguments had led to correct conclusions by accident. The focus of this review is what has come to be known as the Parker instability, in a sense a magnetic Rayleigh-Taylor instability but with sufficiently distinct features to justify its current name (see § 2.2). We examine the instability and its consequences in the interstellar medium at a level accessible to graduate students beginning research in astrophysics, and certainly to researchers in solar and astrophysical magnetohydrodynamics. Readers not interested in the formalism can skip virtually every equation and still get a clear physical picture of the linear instability as well as its nonlinear evolution and observable consequences. Readers interested in a review of the observational evidence for the structure and magnitude of the interstellar magnetic field are referred to Mouschovias (1981, § 2) and Heiles (1987).

The observed magnitude of the interstellar magnetic field (3 - 5 μG) implies that the magnetic energy density is comparable in magnitude to the other forms of energy present in the interstellar medium (e.g., thermal, gravitational, cosmic-ray, and turbulent). The physical conditions in most regions of the interstellar medium (outside dense molecular clouds) are such that ohmic dissipation of magnetic flux is negligible; $\tau_{ohmic} = 4\pi\sigma L^2/c^2 \sim 10^{19} - 10^{22}$ yr, for the Spitzer (1962) conductivity $\sigma_c \sim 10^7 \, T^{3/2}$, $T \simeq 10^2 - 10^4$ K, and a typical lengthscale $L = 1$ pc. Redistribution of mass in the various flux tubes of the system by ambipolar diffusion is also negligible, at least for phenomena occurring over timescales $\lesssim 10^8$ yr and lengthscales greater than a few parsecs [$\tau_{AD} \sim \tau_{A,i}^2/\tau_{in} \sim 10^8$ yr, where $\tau_{A,i}$ is the Alfvén crossing time in the ions and τ_{in} the ion-neutral collision time (see Mouschovias 1991, § 2.3); we have used $B \simeq 4 \, \mu$G, $L \simeq 3$ pc, $n_i \simeq 3 \times 10^{-2}$ cm^{-3}, $n_n \simeq 1$ cm^{-3}, and an ion-neutral collisional rate $\langle \sigma w \rangle_{in} \sim 10^{-9}$ cm^3 s^{-1}]. One may then use the single-fluid magnetohydrodynamic (MHD) and magnetohydrostatic (MHS) equations to study dynamical processes and equilibrium structures, respectively, in the interstellar medium. (For a quantitative review of the issues related to the reduction of the three-fluid to the single-fluid MHD equations, see Mouschovias 1991, § 2.2.) The basic physics and elementary formalism of the instability are presented in § 2; the description of cosmic rays as a fluid is given in § 2.1; the instability itself is discussed first in two and then in three dimensions in § 2.2. Final equilibrium states reachable by the system in which the instability has developed are the subject of § 3; the question of why they should exist at all is addressed in § 3.1; the equilibrium problem is formulated in a self-consistent manner in § 3.2; a representative final equilibrium state is presented in § 3.3, with emphasis on observable features. Energy integrals and work that employed an energy principle to study the stability of the gas-field system are reviewed in § 4. Finally, recent work that followed numerically the nonlinear evolution of the instability is summarized in § 5. The main conclusions are stated in § 6.

2. Basic Equations and Physics of the Instability

Intuitively, the conditions that allow us to treat the three-component interstellar gas (electrons, ions, and neutrals of density n_e, n_i, and n_n, respectively) as a single compressible fluid, can be stated as follows. For long-wavelength (or low-frequency) hydromagnetic disturbances, for which an electron collides many times with ions and neutrals before it is forced to reverse its direction of motion by the oscillating electric field of the disturbance, charge neutrality ($n_i ze = n_e e$) is an excellent approximation and, therefore, collective plasma effects, being relatively high-frequency phenomena (the plasma frequency is $\omega_p \sim 10^3$ s^{-1}), can be ignored. Although collision frequencies ($\omega_{ei} \sim 10^{-7}$ s^{-1} and $\omega_{in} \sim 10^{-9}$ s^{-1}) are relatively large compared to the frequency of the hydromagnetic wave ($\omega \sim 10^{-14}$ s^{-1}), they are nevertheless much smaller than gyrofrequencies ($\omega_c \sim 10$ s^{-1} for electrons). Hence, diffusion across the magnetic field, which would result from collisions between opposite charges or between a charged and a neutral particle before a gyration is completed, may be neglected. In summary, we have the inequalities $\omega_p \gg \omega_c \gg \omega_{coll} \gg \omega$, where ω_{coll} denotes collision frequencies. For lengthscales much greater than collision mean free paths, viscosity and thermal conduction may also be neglected.

We thus consider a conducting fluid of density ρ, thermal pressure P, and temperature T, threaded by a frozen-in magnetic field B, and affected by a gravitational field g derivable from a potential ψ. Both the gas and stars, of density ρ_*, may contribute to ψ. The velocity of a fluid element is denoted by v. An electric current density j maintains the magnetic field, which is derivable from a vector potential A. The entropy per gram of matter is denoted by S, and $\mathcal{L}$ represents the net rate of energy loss (losses minus gains) per gram of matter. The MHD equations, then, are

mass conservation
$$\frac{d\rho}{dt} + \rho \, \nabla \cdot \, v \, = \, 0 \,, \tag{1a}$$

force equation
$$\rho \frac{dv}{dt} \, = \, - \nabla P - \rho \nabla \psi + \frac{j}{c} \times B \,, \tag{1b}$$

energy equation
$$\rho T \frac{dS}{dt} \, = \, - \rho \mathcal{L}(\rho, T) \,, \tag{1c}$$

flux freezing
$$\frac{\partial B}{\partial t} \, = \, \nabla \times (v \times B) \,, \tag{1d}$$

Ampere's law
$$\nabla \times B \, = \, \frac{4\pi}{c} \, j \,, \tag{1e}$$

Poisson equation
$$\nabla^2 \psi \, = \, 4\pi G \, (\rho_* + \rho) \,, \tag{1f}$$

ideal-gas law
$$P \, = \, \frac{\rho}{m} \, k_B T \,, \tag{1g}$$

definitions
$$g \, = \, - \nabla \psi \,, \tag{2a}$$
$$B \, = \, \nabla \times A \,, \tag{2b}$$
$$S \, = \, \frac{1}{\gamma - 1} \frac{k_B}{m} \, \ln\left[\frac{P}{\rho^\gamma}\right] + const \,. \tag{2c}$$

where m is the mean mass per particle ($m = \mu m_H$), and γ is the ratio of specific heats.

A hydromagnetic disturbance in the interstellar medium which occurs over a relatively short timescale may be considered as an adiabatic process, in which case the energy equation becomes simply

adiabatic process
$$\frac{d}{dt}\left(\frac{P}{\rho^\gamma}\right) \, = \, 0 \,. \tag{3a}$$

At the other extreme, a slow process may allow thermal equilibrium to be maintained, in which case an isothermal description is appropriate and the energy equation reduces to $T = const$ or, by using equation (1g),

isothermal process
$$\frac{d}{dt}\left(\frac{P}{\rho}\right) \, = \, 0 \,. \tag{3b}$$

Note that, if one views equation (3a) as the operational definition of the exponent $\gamma \equiv$

$(\partial \ln P / \partial \ln \rho)_\theta$, where θ is a thermodynamic variable (not necessarily the entropy) kept fixed, then equation (3a) can be used to describe processes in addition to adiabatic ones. For example, an isothermal process will be characterized by $\gamma = 1$, while an adiabatic process in an ideal monoatomic gas will have $\gamma = 5/3$.

2.1. Cosmic Rays as a Fluid

The dynamical effects of cosmic rays in the Galaxy have been the subject of intensive investigation for many years. A series of papers has examined the conditions under which cosmic rays, considered as a very hot, collisionless plasma, may be described in the MHD approximation. Three excellent reviews point out what phenomena are excluded when such a description is adopted, but they argue that the MHD description is the proper one for cosmic rays in the Galactic environment (Parker 1968b, 1969; Lerche 1969). Here we give a physically motivated, brief but adequate for our purposes, MHD description of the cosmic-ray gas, so that the conspiracy between magnetic fields and cosmic rays to destabilize the interstellar medium and lead to the formation of cloud complexes can be investigated.

The interstellar magnetic field, being frozen in the matter, couples the cosmic rays to the interstellar matter because the relativistic-particle gyroradii are much smaller than typical lengthscales in the interstellar medium. For example, the gyroradius of a 10^{11} eV proton is only about 10^{14} cm. In the absence of a magnetic field, the cosmic-ray pressure is maintained isotropic to within $\sim 1\%$ by various rapidly growing ($\tau \sim 10^3$ yr) relativistic microinstabilities (e.g., see Lerche 1969). In introducing the magnetic field, we restrict our attention to very long-wavelength (much greater than gyroradii) hydromagnetic waves. In this regime and for slow bulk motions ($v_{rel}^2 \ll c^2$), Parker argues that an isotropic cosmic-ray pressure is a fair approximation in most astrophysical situations even in the presence of a magnetic field. In the extreme relativistic case, the relation between the cosmic-ray pressure P_{rel} and density ρ_{rel} is simply

$$P_{rel} = \frac{1}{3} \rho_{rel} c^2 \equiv \rho_{rel} C_{rel}^2 . \tag{4}$$

Because the cosmic rays and the thermal gas are tied to the magnetic field, their bulk velocities v_{rel} and v (v_{rel}^2, $v^2 \ll c^2$) will have equal components in a direction perpendicular to the field; i.e.,

$$v \times B = v_{rel} \times B . \tag{5a}$$

Thus, an equation identical to (1d), with v replaced by v_{rel}, holds for the cosmic-ray gas as well. Note that, as in the case of the thermal gas, P_{rel} will, in general, respond to changes in the volume of a flux tube; it is *not* a constant of the motion. Hence, the right-hand side of the equation

$$\frac{dP_{rel}}{dt} = ? , \tag{5b}$$

will not, in general, vanish. Under the assumptions that (i) the motion of cosmic rays along field lines is completely uncoupled from that of the thermal gas (cf. Kulsrud and Pearce 1969), (ii) the cosmic-ray gas is not subject to the galactic gravitational field (because the ratio of gravitational and cosmic-ray-pressure forces $\sim v_{esc}^2 / C_{rel}^2 \ll 1$; the quantity v_{esc} is the escape speed in the galactic gravitational field), and (iii) inertial effects of the relativistic gas are negligible (in a typical H I region $\rho_{rel} / \rho = 3 P_{rel} / \rho c^2 \sim 10^{-10}$), one may write the force equation for cosmic rays in a direction parallel to the field simply as

$$\nabla_\| P_{rel} \equiv \frac{B}{B} \cdot \nabla P_{rel} = 0 . \tag{5c}$$

Physically, the nearly instantaneous communication of cosmic rays along field lines establishes pressure equilibrium in the cosmic-ray gas over a distance L in a time $L/C_{rel} \simeq L/c$. Even if L is as large as 1 kpc, this time is 10^{11} s, which is much smaller than the timescale of the hydromagnetic phenomena of interest to us here ($\simeq 10^{14}$ s). Hence, the inertial effects of the cosmic-ray gas can indeed be ignored. *Equation (5c) states that the cosmic-ray pressure is constant on a field line, but it does not determine its value, which is different for different field lines.* (Mouschovias 1975a shows how to calculate P_{rel} at the position of any field line in a quasi-equilibrium state.) Equation (5c) will be exact at equilibrium insofar as the hot ($T_{rel} \sim 10^{13}$ K) and tenuous ($n_{rel} \sim 10^{-10}$ cm^{-3}) cosmic-ray gas is not affected by the galactic gravitational field ($g \sim 10^{-9}$ cm s^{-2}).

It remains to specify how cosmic rays will modify the force equation (1b) in a direction normal to B. Under the approximations discussed above, the main contribution of the cosmic rays to the thermal-gas force equation (1b) is to add a term $- \nabla_\perp P_{rel}$ to the right-hand side. The electric current density j may, in principle, have a contribution from the cosmic rays, but as long as one uses the *total* current density on the right-hand side of Ampere's law (1e) and denotes it by j, the magnetic force term in equation (1b) does not change. All together, then, the combined force equation becomes

$$\rho \frac{d\mathbf{v}}{dt} = - \nabla P - \nabla P_{rel} - \rho \nabla \psi + j \times B/c , \tag{6}$$

where j is the total current density, which appears on the right-hand side of equation (1e). Note that, because of equation (5c), the term $- \nabla_\perp P_{rel}$ has been replaced by $- \nabla P_{rel}$ in equation (6).

In summary, the presence of cosmic rays adds the term $- \nabla P_{rel}$ on the right-hand side of the equation of motion (1b) of the thermal gas. Hence, another equation for dP_{rel}/dt must, in general, be specified in order to close the system of the single-fluid MHD equations. In the case of the interstellar gas, which is supported by thermal-pressure, magnetic, and cosmic-ray-pressure forces against the vertical component of the galactic gravitational field, Parker (1966) considered perturbations which leave the volume of each magnetic flux tube unchanged. For such disturbances, the cosmic-ray pressure is a constant of the motion, so that the equation needed to close the system is simply

$$\frac{dP_{rel}}{dt} \equiv \frac{\partial P_{rel}}{\partial t} + (\mathbf{v} \cdot \nabla) P_{rel} = 0 , \tag{7}$$

where d/dt is the time derivative comoving with the thermal gas. Ames (1973) showed that equation (7) is indeed valid (to first order) for the perturbations considered by Parker.

2.2. Magnetic Rayleigh-Taylor (or Parker) Instability

A "light" (low-density) fluid of density ρ_2 can support a "heavy" (denser) fluid of density ρ_1 against a vertical (downward) gravitational field g (assumed to be constant for simplicity) if the interface is perfectly horizontal. However, deformations of the interface grow, as fingers of heavy fluid protrude downward into the light fluid, thus reducing the energy of the system. The growth rate is given by

$$n = [kg (\rho_1 - \rho_2)/(\rho_1 + \rho_2)]^{1/2} , \tag{8}$$

where k is the wavenumber of the horizontal perturbation. Clearly, shorter wavelengths along the interface grow faster than longer wavelengths. This is a classical Rayleigh-Taylor instability. It can also develop if the downward gravitational field is replaced by an upward acceleration of the heavy fluid by the lighter one. This instability changes if a frozen-in, horizontal magnetic field plays the role of the light fluid in supporting the gas against the gravitational field. The light fluid (the field) and the heavy fluid (the gas) now coexist in the same region of space, and analogies with the nonmagnetic case break down. The nature of the magnetic Rayleigh-Taylor instability has been worked out by Parker (1966) in the context of the interstellar medium. Is it a viable mechanism for the formation of interstellar clouds? In other words, are the unstable wavelengths of the proper size (at least a few hundred parsecs), and are the corresponding growth times short enough ($\lesssim 3 \times 10^7$ yr)? In addition, what final densities are achieved by the nonlinear development of the instability?

2.2.1. The Zeroth-Order State

Observations show that the interstellar magnetic field is predominantly parallel to the Galactic plane (Mathewson and Ford 1970), although arches of magnetic field lines rising high above the plane are also revealed. Parker (1966) assumed a zeroth-order equilibrium state which, for simplicity, has field lines exactly parallel to the galactic plane. The galactic gravitational field (due to stars) was taken, again for simplicity, to be constant and to reverse its direction across the galactic plane; i.e., $g = - \hat{z}g(z)$, where $g(z) = - g(-z) = const > 0$. The assumption was also made that the pressure ratios

$$\alpha \equiv B_0^2/8\pi P_0 \quad \text{and} \quad \beta \equiv P_{rel,0}/P_0 \qquad (9a, b)$$

are constant in the zeroth-order, equilibrium state. With the magnetic field in the zeroth-order state taken parallel to the galactic plane and written as $B_0 = \hat{y}B_{0,y}(z) \equiv \hat{y}B_0(z)$, the MHS force equation (eq. [1b] with the inertial term set equal to zero) becomes

$$(1 + \alpha + \beta) \frac{dP_0}{dz} = - \frac{g}{C^2} P_0 . \qquad (10)$$

Because of the symmetry of the problem about the plane $z = 0$, we need only be concerned with the upper half-plane $z \geq 0$. The solution is

$$P_0(z) \equiv \beta^{-1}P_{rel,0}(z) \equiv \frac{B_0^2(z)}{8\pi\alpha} \equiv \frac{[- A_0(z)]^2}{32\pi\alpha H^2} = \rho_0(0) \, C^2 \, \exp\left(- \frac{z}{H}\right), \qquad (11a)$$

where

$$H \equiv (1 + \alpha + \beta) \, C^2/g \qquad (11b)$$

is the total scale height of the gas in the zeroth-order state, and A_0 is the only nontrivial (x-)component of the magnetic vector potential (see eq. [2b], with B having only a y-component), namely, $A_0(z) = \hat{x}A_0(z)$. The thermal pressure contribution to the scale height, for $T \simeq 6000$ K and $g \simeq 3 \times 10^{-9}$ cm s^{-2}, is $C^2/g \simeq 42$ pc. This is smaller than the observed scale height by at least a factor of 3, possibly 4. With $\alpha \simeq \beta \simeq 1$, a value consistent with observations, one finds that $H \simeq 126$ pc --not an unreasonable value.

2.2.2. Two-Dimensional Stability Analysis

Parker showed, through a linear stability analysis, that the above one-dimensional equilibrium state is unstable with respect to deformations of the field lines in the (y, z)-plane. The vertical gravitational field acquires a component along a deformed (nonhorizontal) field line, thus causing gas to slide along the field line from a raised into a lowered portion. The unloading of gas from the raised portion leaves magnetic and cosmic-ray pressure gradients unbalanced in that region, thereby causing further inflation of the already raised portion of a field line. The component of gravity along the now more vertical field line is greater, with the result that gas can be unloaded more effectively into the "valley" of the field line. The process will stop only when field lines have inflated enough for their tension to balance the expansive magnetic and cosmic-ray pressure gradients (Mouschovias 1974, 1975a). In this picture, the matter which accumulates in the valleys of the field lines represents interstellar clouds. Or, does it?

We first obtain Parker's instability criterion and maximum growth rate for two-dimensional (y, z) perturbations, discuss the physics of the instability more quantitatively, and then address the question of whether the "final" states of the system in which the instability has developed are expected to resemble interstellar clouds (see, also, § 3).

The MHD equations (1a) - (1g), with the energy equation (1c) replaced by (3a) and the force equation (1b) modified as in equation (6) so as to accommodate the cosmic-ray gas, are written in an expanded form in terms of the magnetic vector potential $A \equiv \hat{x}A(y, z)$ and then linearized to find

$$\frac{\partial \rho_1}{\partial t} + (v_1 \cdot \nabla)\rho_0 + \rho_0 \nabla \cdot v_1 = 0 , \tag{12a}$$

$$\rho_0 \frac{\partial v_1}{\partial t} + \nabla(P_1 + P_{rel,1}) - \rho_1 g + \frac{1}{4\pi}[(\nabla^2 A_0)\nabla A_1 + (\nabla^2 A_1)\nabla A_0] = 0 , \tag{12b}$$

$$\rho_0 \frac{\partial P_1}{\partial t} + \rho_0(v_1 \cdot \nabla)P_0 - \gamma P_0 \frac{\partial \rho_1}{\partial t} - \gamma P_0(v_1 \cdot \nabla)\rho_0 = 0 , \tag{12c}$$

$$\frac{\partial A_1}{\partial t} - v_1 \times B_0 = 0 , \tag{12d}$$

$$\frac{\partial P_{rel,1}}{\partial t} + (v_1 \cdot \nabla)P_{rel,0} = 0 . \tag{12e}$$

Ampere's law (1e) has been used to eliminate j from the force equation, and then the definition of A (eq. [2b]) to eliminate B. Although equation (7) was taken as an adequate description of the cosmic-ray gas in order to close the system of the MHD equations, it is emphasized that this step is legitimate only because the equations have been linearized --as explained at the end of § 2.1, equation (7) is valid only to first order for the perturbations considered by Parker. We choose to work with A in this geometry, instead of

$$B = \nabla \times A = - \hat{x} \times \nabla A , \tag{13}$$

because $v \times B = - (v \cdot \nabla)A$, so that the flux-freezing equation (1d) becomes

$$\frac{\partial A}{\partial t} - v \times B = \frac{\partial A}{\partial t} + (v \cdot \nabla)A \equiv \frac{dA}{dt} = 0 , \tag{14a}$$

which states that A is a constant of the motion. Since equation (13) implies directly that

$$B \cdot \nabla A = 0 , \tag{14b}$$

A is also constant on a field line (but has different values on different field lines). One may therefore calculate A at some convenient time, for example, in the initial equilibrium state, and then identify different values of A with different field lines with the assurance that those values will always label the corresponding field lines no matter how the field lines are deformed during the evolution of the system --as long as the magnetic field remains frozen in the matter.

We look for a solution of equations (12a) - (12e) having the time dependence $\exp(nt)$, where n is the growth rate and is, in general, complex. A real, positive n implies instability, a real negative n exponential decay of the perturbation, and an imaginary n corresponds to waves. The time derivatives in equations (12a) - (12e) are thus replaced by n. Since the first-order part of the $\mathbf{v} \cdot \nabla$ is $v_z(\partial/\partial z)$, and $\mathbf{v}_1 \times \mathbf{B}_0 = -\hat{x}v_z B_0$, it is seen that the coefficients of all the perturbed quantities in these equations do not depend on y. We may therefore look for solutions of the form (i.e., Fourier analyze in y)

$$a(y, z, t) = f(z) \exp(nt + ik_y y) , \qquad (15)$$

which, in addition, has the effect of replacing $\partial/\partial y$ with ik_y. Because the coefficients of the perturbed quantities in the resulting equations still depend on z, it is not yet possible to Fourier analyze in z. To get constant coefficients, one changes variables to

$$\xi \equiv \frac{B_0}{\rho_0} \rho_1 , \qquad \eta \equiv B_0 v_y , \qquad (16a, b)$$

which then permits Fourier analysis in z as well. (To the best of our knowledge, the transformation [16a, b] was first pointed out by Field 1970±1.) The resulting set of algebraic equations is then reduced as follows. Equation (12d) is solved for the perturbed velocity v_z (note that the subscript 1 has been dropped from the velocities since v_0 vanishes identically), equation (12c) is solved for P_1, equation (12e) for $P_{\text{rel},1}$, and the resulting expressions are substituted in the rest of the equations to obtain a homogeneous system of three algebraic equations for the three unknowns ξ, η, and A_1. By setting the determinant of the coefficients equal to zero, one obtains a dispersion relation for motions in the (y, z)-plane:

$$n^4 + C^2[(\gamma + 2\alpha)(k^2 + k_0^2/4)]n^2 + k_y^2 C^4 \{2\alpha\gamma k^2 + k_0^2[\gamma(1 + \beta + 3\alpha/2) - (1 + \alpha + \beta)^2]\} = 0, \qquad (17)$$

where we have set

$$k_0 \equiv H^{-1} , \qquad (18)$$

and we have let the magnitude of the wavenumber be denoted by k; i.e.,

$$k^2 = k_y^2 + k_z^2 . \qquad (19)$$

Clearly, equation (17) has the form $n^4 + c_1 n^2 + c_2 = 0$, which is a quadratic for n^2, with the constant c_1 being always positive, and the constant c_2 being capable of having either sign.

One can show (Parker did not) that all roots of equation (17) are real --a tedious algebraic proof. Here we are content with finding a criterion that one root $n^2 > 0$, so that one root $n > 0$, corresponding to instability. Let the two roots of the quadratic be denoted by $n_{(1)}^2$ and $n_{(2)}^2$. From the properties of quadratic equations, the sum of the two roots is

$$n_{(1)}^2 + n_{(2)}^2 = -c_1 < 0 . \qquad (20a)$$

Therefore, at least one root (say, $n_{(1)}^2$) is < 0, and the corresponding modes are waves. Hence, *at most* one root (the second one, $n_{(2)}^2$) corresponds to an unstable mode. To determine whether this is so, we look at the algebraic sign of the product of the two roots, which is equal to the constant term of the quadratic, namely,

$$n_{(1)}^2 n_{(2)}^2 = c_2 . \qquad (20b)$$

It follows, since we have already found that $n_{(1)}^2 < 0$, that the second root $n_{(2)}^2$ has the same algebraic sign as $-c_2$. It is therefore sufficient for instability to demand that $c_2 < 0$, so that $n_{(2)}^2$ will be positive. All together, *the criterion for having one unstable mode is*

$$\left[\frac{k}{k_0/2}\right]^2 < \frac{2(1 + \alpha + \beta - \gamma)(1 + \alpha + \beta) - \alpha\gamma}{\alpha\gamma}, \qquad \alpha \neq 0, \qquad (21)$$

which is the instability criterion obtained by Parker (1966).

In terms of wavelengths, the criterion (21) is equivalent to the following two conditions, which must be satisfied simultaneously,

$$\lambda_y > \Lambda_y \equiv 4\pi H \left[\frac{\alpha\gamma}{2(1 + \alpha + \beta - \gamma)(1 + \alpha + \beta) - \alpha\gamma}\right]^{1/2}, \qquad (22a)$$

$$\lambda_z > \Lambda_z(\lambda_y) \equiv \lambda_y \left[1 - \left(\frac{\Lambda_y}{\lambda_y}\right)^2\right]^{-1/2} > \Lambda_y. \qquad (22b)$$

For the interstellar gas, $\gamma \equiv d\ln P/d\ln\rho \simeq 1$. If $\lambda_y < \Lambda_y$, the radius of curvature of a typical, deformed field line is small, hence the tension is large and straightens the field line out by propagating the perturbation away as a (modified Alfvén) wave. If $\lambda_z < \Lambda_z(\lambda_y)$, the system is stable even though λ_y may exceed its critical value Λ_y. The physical reason lies in the fact that the volume available for the field lines to expand in, and thereby decrease the magnetic energy of the system, is limited. The increase in the field strength in the valleys and the pileup of field lines near the first undeformed field line, which forms a natural "lid" to the system below, represent an increase in magnetic energy which suppresses the instability (see Mouschovias 1974, 1975b for details). For a fixed λ_y, the growth rate of the perturbation increases monotonically as λ_z ($> \Lambda_z$) increases. For a fixed $\lambda_z > \Lambda_z$, the growth rate first increases and then decreases as λ_y increases. Equation (22a) shows that the horizontal critical wavelength for $\alpha \simeq \beta \simeq \gamma \simeq 1$ is $\Lambda_y = 1.2\pi H \simeq 477$ pc. The maximum growth rate occurs at $\lambda_y = 1.8\Lambda_y = 2.2\pi H \simeq 868$ pc and $\lambda_z = \infty$, and its inverse (the e-folding or growth time) is given by

$$\tau_{min} = 1.1 \frac{H}{C} = 2.2 \times 10^7 \left[\frac{T}{6000 \text{ K}}\right]^{1/2} \left[\frac{3 \times 10^{-9} \text{ cm s}^{-2}}{g}\right] \quad \text{yr.} \qquad (23a, b)$$

This growth time is short enough to be relevant for cloud formation behind a spiral density shock wave. In fact, it may be smaller than the value given in equations (23a, b) because, in a strict sense, the quantity H is the scale height in the *initial* state; not its value today. It has been shown by exact determination of final equilibrium states for the Parker instability that, in the valleys of field lines, $H_{final} \simeq 1.7 H_{initial}$ (Mouschovias 1974, Fig. 2b). This implies that $\tau_{min} \simeq 1.3 \times 10^7$ yr and $\lambda_y(\tau_{min}) \simeq 511$ pc. A further decrease in τ_{min} can take place because of the fact that the instability is externally *driven* by a spiral density shock wave (Mouschovias 1975b, p. 73). There is yet another reason for which τ_{min} can decrease further. Giz and Shu (1993) took into consideration the variation of g with z, as Kellman (1972) had done earlier, and found that the value of g which enters equation (23b) is greater than the one given above by a factor of 3. The amount of matter involved in a cylinder (along a spiral arm) of length 511 pc and diameter 250 pc (the approximate thickness of a galactic shock, as well as the thickness of the galactic disk in which most of the gas is found) is $8.6 \times 10^5 \ M_\odot$. Thus the Parker instability is most suitable for the formation of large-scale condensations (or cloud complexes), rather than individual interstellar clouds (Mouschovias 1974; Mouschovias, Shu, and Woodward 1974). The implosion of individual clouds by shock waves within these complexes can give rise to OB associations and giant H II regions, all aligned along spiral arms "like beads on a string" and separated by regular intervals of 500 - 1000 pc, in agreement with observations both in our galaxy and in external galaxies (Westerhout 1963; Kerr 1963; Hodge 1969; Morgan 1970).

2.2.3. *Instability in Three Dimensions*

In the direction $g \times B$ (the $\hat{x}$-direction of §§ 2.2.1 and 2.2.2), wavelengths λ_x in the approximate range $H/25$ - $20H$ (for the reasonable values $\alpha = \beta = \gamma = 1$) can grow with almost identical growth rates; very short wavelengths in this direction are stabilized by ambipolar diffusion (Parker 1967a, b). The presence of perturbations in the x-direction (essentially the radial direction in a galactic disk) introduces a new effect. Along the x-axis, at a distance $\lambda_x/2$ away from the sinking valley of a field line, there now exists a rising mountain top of another field line. If λ_x is actually near the lower limit ($\simeq H/25$) of the above unstable range, the rising flux-tube tops may move laterally rapidly enough (in $\lesssim 10^7$ yr) along the x-axis into the (relatively) evacuated regions of the sinking valleys of other flux tubes at a distance $\lambda_x/2$ away. This, Parker suggested, leads to a disordered, turbulent dynamical state, which may cease only when a cloud becomes so dense as to be dominated by self-gravity. The x-direction being essentially the radial direction in a galactic disk, perturbations characterized by large λ_x are stabilized by differential galactic rotation (e.g., see Shu 1974). Hence, an intermediate range of λ_x's, from about 10 pc to about 100 pc can grow with a significant growth rate. Although the physics of the instability in the linear regime is well understood, important questions remain on the nonlinear evolution and the ultimate state of the interstellar gas and field system in which the instability develops. Even the basic issue of whether the instability can lead to the formation of typical interstellar clouds, as opposed to giant clouds or cloud complexes, has not yet been definitively settled.

If the growth rate of the instability itself or some mechanism (other than a galactic shock) selects wavelengths $\lambda_x \simeq 10$ pc in the x-direction, then the mass involved would be only slightly greater than $10^4\ M_\odot$. This begins to approach masses of *individual* clouds. Whether in fact individual clouds can form by the Parker instability will be decided only after realistic nonlinear three-dimensional calculations are carried out. Although there is a wealth of ideas on how phase transitions (e.g., see Field, Goldsmith, and Habing 1969) and conversion of atomic to molecular hydrogen in the valleys of field lines can convert the nongravitating clumps of gas into dense molecular cloud complexes (Field 1969; Mouschovias 1975b, 1978; Blitz and Shu 1980), no quantitative calculation has been produced yet. Initial perturbations which allow field lines to cross the Galactic plane are more likely to initiate the necessary phase transitions. Such perturbations can lead to a gas density (and pressure) in the plane significantly greater (a necessary condition for phase transitions) than its value in the initial (unstable) equilibrium state (see § 5.2.2). Perturbations with reflection symmetry about the Galactic plane (see § 5.2.1) can lead to such phase transitions only under special circumstances (Mouschovias 1975b, pp. 55 - 57).

As mentioned near the end of § 2.2.2, Kellman (1972) considered the effect of the variation of the gravitational field with height z above the galactic plane; i.e., the gaseous disk is taken to be embedded in a disk of stars, which is responsible for $g(z)$. He showed that perturbations in the x-direction on the verge of instability are described by a time-independent, Schrödinger-like equation. Hence, in addition to the continuum of unstable modes, there also exists a discrete set (i.e., only certain values of λ_x can exist, just like the quantum mechanical problem of a particle in a potential well). A similar but more complete analysis has been carried out recently by Giz and Shu (1993), who also find the existence of discrete modes. Despite the theoretical significance of these modes, they are actually irrelevant to the Parker instability in the interstellar medium, because they become important only if $\alpha \ll 1$ --a condition not satisfied in the general interstellar medium, in which $\alpha \gtrsim 1$.

3. Final Equilibrium States

As soon as the Parker instability was discovered, the issue of the ultimate state of the gas-field system arose quite naturally. Lerche (1967a) considered a cold ($P = 0$) interstellar gas and determined an equilibrium state in a constant galactic gravitational field, as originally specified by Parker (see § 2.2.1). Since thermal-pressure forces, which are the only ones opposing gravity along field lines, were ignored, this state consisted of vertical, infinitesimally thin sheets of matter separated by a horizontal distance $y = \lambda_y$ in the direction of the unperturbed magnetic field. It can be reached from a stratified initial equilibrium state through continuous deformations of the field lines. However, the sheets are unstable with respect to horizontal displacements, with the result being a tendency for them to coalesce and form new sheets separated by a distance $y = 2\lambda_y$ (Lerche 1967b).

Parker (1968a) found a restricted class of equilibrium states with curved magnetic field lines and a nonvanishing thermal pressure. Unfortunately, these states cannot be reached from the initial, zeroth-order state through continuous deformations of the field lines (see discussion in § 3.2 below).

The difficulty of determining final equilibrium states for the Parker instability is not so much a mathematical (or numerical) one as it is one of formulating the problem self-consistently in the first place. The MHS equations are obtained from the single-fluid MHD equations (1a) - (1g) by dropping the terms containing time derivatives and velocities:

$$- \nabla P - \rho \nabla \psi + j \times B/c = 0 , \tag{24a}$$

$$\nabla \times B = (4\pi/c) \, j , \tag{24b}$$

$$P = \rho C^2 , \tag{24c}$$

where $C = const$ for an isothermal gas. If the self-gravity of the gas is negligible, the gas responds to the (known) gravitational field of the stars, and $g \equiv - \nabla \psi$ in equation (24a) is not an unknown. If the self-gravity of the gas is important, then the Poisson equation,

$$\nabla^2 \psi = 4\pi G \rho \tag{24d}$$

must also be considered. *The above system of four equations is an inherently open set* --it contains five unknowns, namely, P, ρ, ψ, j, and B. *Ad hoc* assumptions are usually made to close the system and obtain equilibrium solutions. In what follows, we return to first principles and close the system in a *self-consistent* manner. Then we use the equations to determine (final) equilibrium states for the Parker instability (see § 3.3).

It may appear as a puzzle that, although the MHD equations form a closed system, the MHS equations, which are obtained from the former, are inherently open. One may also think that adding the condition $\nabla \cdot B = 0$ would close the system. It does not. What is required is a relation between one of the electromagnetic variables (j or B) and at least one of the two variables P, ρ, and, possibly, the gravitational potential, ψ. In order to find a clue as to how the system may be closed self-consistently, we look for the physical principle(s) that must have gotten lost in going from the MHD to the MHS equations. Equations (1a) and (1d), representing mass and flux conservation, respectively, are the culprits. Since all their terms contain either time derivatives or velocities, they all drop out of the MHS equations. In other words, the problem of closing the system of MHS equations is reduced to one of finding a way to build into them the lost physical principles of conservation of mass and magnetic flux. *It is necessary to discover a new equation.* This was done for the two-dimensional cartesian geometry relevant to the determination of final equilibrium states for the Parker instability (Mouschovias 1974), and for the three-dimensional axisymmetric geometry relevant to the study of equilibria of self-gravitating, magnetic clouds (Mouschovias 1976). We outline below the method and the results relevant to the Parker instability in the interstellar medium (as well as in stellar atmospheres, for which the self-gravity of the gas is negligible).

We first point out that Cowling's integral of the continuity and flux-freezing equations, $B/\rho = [(B_0/\rho_0) \cdot \nabla]r'$, which relates the ratio B/ρ in a fluid element at some time t to its value B_0/ρ_0 in the same fluid element at a previous time t_0 through the displacement vector r' of the fluid element, does not close the system of MHS equations because r' is not known --in fact, to know r' requires a knowledge of the solution.

3.1. Why Final States Exist

Since it has been argued that, once the Parker instability develops in the interstellar medium, the field lines will inflate forever, it is necessary first to give at least a plausibility argument why this is not so. We consider two neighboring field lines labeled by A and $A+\delta A$ (recall eqs. [14a, b], which show that the only nontrivial component of the magnetic vector potential in this two-dimensional geometry is both a constant of the motion and constant on a field line), which have inflated to a distance above the galactic plane ($z = 0$) comparable to the (unstable) horizontal wavelength of the perturbation, λ_y. Let h be the (small) separation between the above two field lines in the "wings" or "mountains" of the field lines, where inflation is taking place. Then the ratio of the magnitudes of the confining magnetic-tension force and the expansive magnetic-pressure force is

$$\frac{B^2/4\pi R_c}{B^2/8\pi h} \propto \frac{1/h^2 R_c}{1/h^3} \propto \frac{1/\lambda_y}{1/h} \propto h , \tag{25a}$$

where we have used $B \simeq \delta A/h \propto 1/h$, $|\nabla_\perp| \sim 1/h$, and for the radius of curvature $R_c \sim \lambda_y \simeq const$. Since the separation h between neighboring field lines increases upon expansion, the magnetic-tension force will eventually overwhelm the magnetic-pressure force and the expansion will not continue indefinitely. After all, this is the reason for which the field lines of a bar magnet do not all recede to infinity: the tension and pressure forces exactly balance.

The expansive cosmic-ray-pressure force in the region where inflation occurs is

$$|\nabla_\perp P_{rel}| \propto h^{-1} n_{rel}^{4/3} \propto h^{-1} V^{-4/3} \propto h^{-7/3} , \tag{26}$$

where the relativistic relation $P_{rel} \propto n_{rel}^{4/3}$ between pressure and density has been used, and the quantity V is the volume of the flux tube (A, $A+\delta A$), which was conservatively taken to increase only as h upon expansion. It therefore follows that the ratio of the magnitudes of the magnetic-tension force and the cosmic-ray-pressure force varies as

$$\frac{B^2/4\pi R_c}{|\nabla_\perp P_{rel}|} \propto \frac{h^{-2}}{h^{-7/3}} \propto h^{1/3} , \tag{25b}$$

which increases upon expansion. Hence, magnetic tension will also overwhelm the expansive tendencies of cosmic rays, and equilibrium becomes possible as long as the number of cosmic rays in each flux tube of the system remains quasi-steady (see Mouschovias 1975a for a more complete picture and for an explanation of a fat Galactic radio disk of half-thickness $\sim$ 1 kpc in a quasi-steady state).

3.2. Self-Consistent Formulation of the Problem

First we reduce the system (24a) - (24c) to one equation. We define the scalar function of position $q(y, z)$ by[1]

$$q = P \exp(\psi/C^2) . \tag{27}$$

By taking the gradient of both sides of equation (27) and using equation (24c), we find that the thermal-pressure force and the gravitational force in equation (24a) combine into a single term if written in terms of q, namely,

$$-\nabla P - \rho \nabla \psi = -\exp(-\psi/C^2)\,\nabla q\,. \qquad (28a)$$

The magnetic force is written in terms of the magnetic vector potential, by using Ampere's law to eliminate j in favor of B and equation (13) to eliminate B in favor of $A \equiv A_x(y, z)$, as

$$\frac{j}{c} \times B = -\frac{1}{4\pi}\nabla^2 A \times (-\hat{x} \times \nabla A) = -\frac{1}{4\pi}(\nabla^2 A)\,\nabla A\,, \qquad (28b)$$

where in the last step we have used the fact that $\nabla^2 A$ is in the $\hat{x}$-direction, and $\hat{x} \cdot \nabla A = 0$. Hence, the MHS equations reduce to

$$-\exp(-\psi/C^2)\,\nabla q - \frac{1}{4\pi}(\nabla^2 A)\,\nabla A = 0\,. \qquad (29)$$

To reduce equation (29) further, we look at the properties of the function q. Since A is constant on a field line (see eq. [14b]), we may take the dot product of B and equation (29) to find that $B \cdot \nabla q = 0$; i.e.,

$$q \equiv P\exp(\psi/C^2) = \text{constant on a field line} = q(A)\,. \qquad (30a)$$

We may therefore write in equation (29) $\nabla q = \nabla A\,[dq(A)/dA]$ and, since ∇A does not vanish identically, finally find that

$$\nabla^2 A(y, z) = -4\pi\frac{dq(A)}{dA}\exp(-\psi/C^2)\,. \qquad (31)$$

Note that, since from Ampere's law written in terms of A in this geometry we have that $\nabla^2 A = -(4\pi/c)j$, in deriving equation (31) we have also determined j:

$$\frac{j}{c}\exp(\psi/C^2) = \frac{dq(A)}{dA} = \text{constant on a field line}\,. \qquad (30b)$$

The physical meaning of what we have done in order to reduce the MHS equations (24a) - (24c) to one equation (31) is as follows. (i) Equation (30a) states that, at hydrostatic equilibrium, the gravitational forces balance the thermal-pressure forces *along* field lines (since in eq. [28a] $\nabla_\parallel q = 0$). (ii) Equation (30b) gives the current density j for which forces balance *perpendicular* to field lines if q is such that forces balance along field lines [i.e., if $q = q(A)$]. However, j and B will not, in general, be consistent with each other unless they satisfy equation (31).

One should note that the gravitational potential in equation (31) has been left unrestricted. It can be due to stars (and therefore externally specified), as we shall assume in this section, or due to the self-gravity of the gas itself, in which case Poisson's equation (24d) must also be considered (see review by Mouschovias 1991, § 5.2.1).

---

[1]For simplicity, we ignore cosmic rays here although their presence can be accounted for in a straightforward fashion: a second function, q_{rel}, is defined which is exactly equal to P_{rel} (see Mouschovias 1975a) because the cosmic-ray gas is not affected by the galactic gravitational field; i.e., $\psi/C_{rel}^2 \simeq (v_{esc}/c)^2 \ll 1$ (see § 2.1).

In order to determine final equilibrium states for the Parker instability, we must solve equation (31) for a pair of *unstable* wavelengths (λ_y, λ_z) with $q(A)$ determined in such a manner that a final state can be reached (by the nonlinear time evolution of the system) from the corresponding initial, one-dimensional state through continuous deformations of the field lines.

What is $q(A)$? In the case of a cold gas $(T = 0)$, Parker (1968a, b) assumed that q is proportional either to A or to A^2, so that equation (31) becomes linear in A and can easily be solved. However, it follows from the definition of q (eq. [30a]) and from the solution of the zeroth-order equilibrium state (eq. [11a] with $P_{rel} = 0$) that

$$q_0(A) = \rho_0(0) \, C^2 \left[- \frac{2HB_0(0)}{A} \right]^{2\alpha} , \qquad (32)$$

which clearly is an *inverse power of A* (since $\alpha > 0$).

Although $q = q(A)$ at equilibrium, $dq/dt \neq 0$; i.e., q is *not* a constant of the motion. Therefore, one cannot calculate q once and for all from the initial equilibrium state and assume that a final state is characterized by the same $q(A)$.

3.2.1. Calculation of the Function q(A)

With $Y \equiv \lambda_y/2$, the mass (δm) in a flux tube between field lines A and $A+\delta A$ is, by definition,

$$\delta m(A) = \int_{-Y}^{+Y} dy \int_{z(y, A)}^{z(y, A+\delta A)} dz(y, A) \, \rho[y, z(y, A)] . \qquad (33)$$

It is natural to consider y and A as the independent variables. Since the integration over z in equation (33) is performed keeping y fixed, we may write $dz = (\partial z/\partial A)dA$ and effect the change of variables from z to A. We eliminate ρ in favor of A by using equations (30a) and (24c), perform the trivial integration over A, and then solve the resulting equation for $q(A)$ to find that

$$q(A) = \frac{C^2}{2} \frac{dm(A)}{dA} \left\{ \int_0^Y dy \, \frac{\partial z(y, A)}{\partial A} \exp\left[- \frac{\psi(y, A)}{C^2} \right] \right\}^{-1} , \qquad (34)$$

where the quantity $z(y, A)$ refers to the z-coordinate of the field line A at y.

The function $dm(A)/dA$ on the right-hand side of equation (34) is the mass-to-flux ratio in a flux tube about field line A in one period of the system in the y-direction and having a unit length in the "third" $(x-)$ direction. By conservation of both mass and flux, dm/dA is a *constant of the motion*. If it is given/known, then $q(A)$ can be calculated for any equilibrium configuration characterized by the *same dm/dA*.

Note that $q(A)$ depends on the shape of the field lines, which, for the final equilibrium state, are not known. Hence, in general, one must solve equations (31) and (34) simultaneously for any given dm/dA. The initial state of the interstellar gas and magnetic field system is not known in reality, for it depends on the mechanism which creates the magnetic flux. Here we take it to be the plane-parallel system proposed by Parker (1966) (see § 2.2.1). We emphasize, however, that the only information needed in order to determine an equilibrium state is the mass-to-flux ratio in each flux tube. As discussed by Mouschovias (1974), if it becomes possible for the distribution of mass among the various flux tubes to be obtained from observations, we can determine an equilibrium state without reference to any particular initial state.

3.3. Determination of Equilibrium States

The system of (self-consistently) closed MHS equations (31) and (34) was solved by a successive underrelaxation method (see Mouschovias 1974, Appendix C) with the differential mass-to-flux ratio calculated from the zeroth-order state of § 2.2.1. For an isothermal gas, a pair of initially unstable wavelengths $\lambda_y \equiv 2Y = 24C^2/g$ and $\lambda_z \equiv 2Z = 50C^2/g$, and $\alpha_0 = 1$ (see eq. [9a]), the final equilibrium state is shown in Figure 1. The unit of length is the thermal-pressure scale height, C^2/g, where $g = -\hat{z}g(z)$, with $g(z) = -g(-z)$ = a positive constant. *Solid curves* represent field lines chosen such that the increment of magnetic flux (per unit length in the x-direction), ΔA, is constant between consecutive field lines; hence, the spacing between consecutive field lines is inversely proportional to the mean strength of the magnetic field in that region. The *dashed curves* represent isodensity contours at which the density decreases to e^{-1}, e^{-2}, and e^{-3} its value on the plane $z = 0$. The number on each curve is the z-coordinate of that curve in the initial state, in which all curves were horizontal.

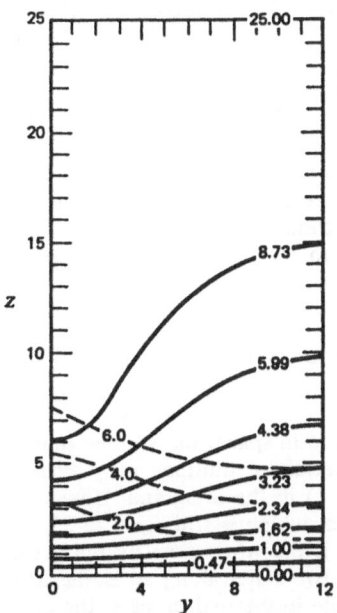

Figure 1. *An equilibrium state of the interstellar gas and magnetic field system in a vertical, uniform galactic gravitational field g.* Distance is measured in units of C^2/g, where C is the isothermal speed of sound in the gas. Only half a wavelength in the y- and half a wavelength in the z-direction are shown. Half the critical wavelength in the y-direction is equal to 7.26. Field lines (*solid curves*) are chosen so that the magnetic flux between any two consecutive ones is constant. The isodensity contours (*dashed curves*) at which the density decreases to e^{-1}, e^{-2}, and e^{-3} its value on the y-axis are shown. The number labeling each curve is the z-coordinate of that curve in the initial state, in which $\alpha = 1$.

The particular horizontal wavelength λ_y chosen for this state is that which corresponds to the maximum growth rate at the chosen (unstable) λ_z. The critical horizontal wavelength is $\Lambda_y = 14.52C^2/g$ (see eqs. [22a] and [11b], with $\alpha = \gamma = 1$ and $\beta = 0$). The increase of the gas density in the "valleys" of the field lines is due primarily to efficient drainage along field lines, rather than to compression perpendicular to the galactic plane. A parameter study found that, for any given λ_z ($> \Lambda_z$), this drainage is more efficient the larger λ_y ($> \Lambda_y$) is. This makes the magnetic field a nearly vacuum field at high z in the region where inflation of field lines has taken place. The gas density in the galactic plane $\rho(y, z = 0)$ remains constant because g has no horizontal component and, therefore, no thermal-pressure gradients can be sustained (at equilibrium) along field lines in the plane. [Perturbations that permit deformation of the field line originally at $z = 0$ have been studied by Mouschovias, Basu, and Paleologou (see § 5.2.2), who found typical final enhancements of the density in the plane by a factor of 2.] Observations of spiral arms of a galaxy seen face-on were predicted to show a column density of gas through the center of a condensation (along the z-axis of Fig. 1) typically twice as large as the column density observed through the "mountains" of the field lines. The galaxy M81 shows precisely this kind of variation of column density along its spiral arms (Rots 1974).

For a detailed discussion of the physics of the Parker instability and the energetics of final equilibrium states, see Mouschovias (1974), where an energy integral for an isothermal gas is also obtained. The effect of cosmic rays on final equilibrium states is discussed in Mouschovias (1975a), while the explanation of the alignment of cloud complexes, OB associations, and giant H II regions "like beads on a string" along spiral arms is given by Mouschovias, Shu, and Woodward (1974). The nonlinear development (time evolution) of the Parker instability is discussed in § 5 below.

4. Energy Integral for Isothermal Plasma

It is well known that the MHD equation, with $P \propto \rho^\gamma$, possess the energy integral

$$\int dV \left[\frac{1}{2}\rho v^2 + \frac{B^2}{8\pi} + \rho\psi + \frac{P}{\gamma - 1} \right] = const, \tag{35}$$

where the volume integral is over all space (Bernstein *et al.* 1958). Clearly, the last three terms in the integral represent the effective potential energy of the system (magnetic + gravitational + thermal). For an isothermal plasma, one cannot simply take the limit $\gamma \to 1$, because of the last term in the integrand. Returning to the MHD equations for such a plasma, one can show that the proper energy integral is

$$\int dV \left[\frac{1}{2}\rho v^2 + \frac{B^2}{8\pi} + \rho\psi + P\ln P \right] = const \tag{36}$$

(Mouschovias 1974, Appendix B). If the system is periodic in space, the volume integral can be performed over only one period, provided that certain conditions are satisfied (see Mouschovias 1974). The term $P\ln P$ has replaced $P/(\gamma - 1)$ in the integrand. However, $P\ln P$ is *not* the thermal energy density of an isothermal plasma, which is still equal to $3P/2$ for an ideal monoatomic gas. The quantity $\Delta W_P \equiv \Delta \int dV \, P\ln P$ is the work done by the gas against thermal-pressure forces in making a transition between two states along an isothermal path. (If $\Delta W_P > 0$, heat is released by the gas.) For a reversible isothermal process, the change in the entropy is given by $\Delta S = - \Delta W_P / T$. Hence, W_P provides a measure of the entropy and is equal to the Helmholtz free energy of the gas, to within an additive constant.

The existence of the energy integrals (35) and (36) allow one to reexamine the Parker instability through an energy principle, both for adiabatic and for isothermal perturbations (Mouschovias 1974). Since the first variation of the energy integral yields the algebraic sum of all the forces (i.e., the force eq. [1b]), which vanishes at equilibrium, the algebraic sign of the second variation determines the stability of the system. If one can find a perturbation which renders this change of potential energy negative, then the system is unstable with respect to this perturbation.

Zweibel and Kulsrud (1975) used the energy principle to study the stability of Parker's initial, one-dimensional equilibrium state, and they recovered Parker's instability criterion. They also investigated the stability of a different initial state, which the authors regard as more realistic. The thermal pressure P in this state is due to random motions of clouds in the model interstellar medium, and the magnetic field is assumed to be tangled. The authors assume that cloud-cloud collisions lead to collapse and star formation, which, in turn, pumps energy into the system, accelerates clouds through the rocket effect, HII-region expansion and supernova remnants, and thereby stiffens the equation of state (to $\gamma = 5/3 - 2$), thus tending to stabilize the system (see eq. [21] or [22]). Since the assumption was made that clouds already exist in the zeroth-order state, the issue addressed is not whether the Parker instability can develop and lead to the formation of interstellar clouds, but whether it can occur in an interstellar medium in which clouds have already formed by some other, unspecified mechanism, and possibly lead to reconnection of expanding field lines and escape of magnetic "bubbles" and cosmic rays from the Galactic disk. Although they concede that the instability may still lead to the formation of cloud complexes and OB associations as suggested by Mouschovias (1974) and Mouschovias, Shu, and Woodward (1974), they conclude that it is irrelevant to the escape of cosmic rays from the disk suggested by Parker (1968b). Its e-folding timescale would be significantly greater than 10^7 yr.

Taking the initial state at face value [actually Parker (1975) argues that that is the *result* of the onset of the instability], the validity of the conclusion hinges on the value of the γ *for the perturbation* (not the zeroth-order state) being in the range 5/3 - 2, deduced from a series of uncertain assumptions. If slow ($\tau \simeq 10^7 - 10^8$ yr) perturbations in the present-day interstellar medium can only be characterized by $\gamma > 5/3$, it would indeed be more difficult for the instability to develop. The critical horizontal wavelength would nearly double in magnitude (see eq. [22a]), the wavelength corresponding to maximum growth rate would correspondingly increase, and the minimum e-folding time would increase to several times 10^7 yr (for $\alpha = \beta = 1$). Parker (1968b) argues, appealing to past work on the thermal properties of the interstellar gas, that such long-period disturbances are more nearly isothermal ($\gamma = 1$). Moreover, in order for the small-amplitude perturbation itself to trigger the sequence of events (cloud collapse, star formation, and cloud reacceleration) envisioned by the authors, clouds not only should pre-exist but a sufficiently large fraction of them must be right on the verge of collapse. For, if the perturbation must grow into the nonlinear regime before it induces cloud collapse, then, by assumption, the instability has set in on a relevantly short timescale, and the argument becomes moot. Present-day observations do not show many clouds to be on the verge of collapse or those that are forming stars to be collapsing from the outside-in. Most star formation is taking place in cloud cores, surrounded by relatively quiescent, massive, envelopes. This is as originally suggested by Mouschovias (1977) for ambipolar-diffusion-induced (or *self-initiated*) star formation in cloud interiors (where the degree of ionization is very low) of magnetically supported clouds. In this picture, supported by recent, detailed, multifluid MHD simulations, typically only about 10% of the mass of a cloud collapses to form stars, while the rest remains magnetically supported, in the form of a massive, extended envelope, in which the magnetic field is nearly frozen in the matter (see review by Mouschovias 1995, or article in this volume). Concerning the escape of cosmic rays, however, as a direct result of the onset of Parker's instability, there may be a problem even without the complications introduced by Zweibel and Kulsrud regarding the value of γ (see § 5.2.3).

Asséo et al. (1978) used the energy principle for isothermal perturbations and obtained an instability criterion for two-dimensional equilibrium states with curved magnetic field lines, reachable from initial stratified states through continuous deformations of the field lines. They found that such states are unstable with respect to certain three-dimensional perturbations. However, in estimating the growth rate, they took an invalid limit, namely, that the radius of curvature of field lines is $R_c \ll H$, where H is the vertical scale height of the gas in the zeroth-order state (see eqs. [11a, b] with $\beta = 0$). On the basis of this, they found that the growth rate is $n \propto (H/R_c)^{1/2}$, which can become very large for $R_c \ll H$. They concluded that the curvature of field lines destabilizes the system. However, the radius of curvature R_c is comparable to the horizontal wavelength λ_y of the perturbation. Since λ_y must exceed the critical value Λ_y for instability to occur in the first place (see eq. [22a]), and since $\Lambda_y \simeq \pi H$ for $\alpha = \beta = \gamma = 1$, one must always have $R_c \sim \lambda_y \gtrsim \pi H$. It is not legitimate to take $R_c \ll H$. Such perturbations are stable. The magnetic-tension force propagates them away as (modified) Alfvén waves. If the above invalid limit is not taken and the growth time of the instability is estimated properly, it exceeds 10^8 yr. This makes its relevance to the dynamics of the interstellar medium of little practical significance.

5. Nonlinear Evolution

5.1. Summary of Past Work

The nonlinear evolution of the Parker instability in the absence of cosmic rays (i.e., $\beta = 0$ in our notation) was followed numerically in two dimensions [the (y, z)-plane of §§ 2.2, 3.2, and 3.3] by Matsumoto et al. (1988, 1990), with the z-component of the gravitational field at a cylindrical polar radius r taken to be that due to a point mass M at the center of coordinates. The assumed gravitational field is appropriate for nongravitating accretion disks around a central mass, but precludes the results from being relevant to the Parker instability in the interstellar medium (see below). We nevertheless review certain features of the results which may have qualitative relevance.

There are three key dimensionless free parameters in the authors' formulation of the problem: in our notation, the wavelength of the perturbation along field lines λ_y, the ratio of magnetic and gas pressures in the initial state α, and the new parameter

$$\mathscr{E} \equiv \frac{GM}{r} \frac{1}{(1 + \alpha)C^2}, \tag{37}$$

which is essentially equal to $(v_K/C)^2$, where v_K is the Keplerian speed in the disk at the radius r. (Although not investigated by the authors, λ_z is a significant free parameter, as we have seen in §§ 2.2.2 and 3.3.) Models are studied with λ_y in the range $(0.67 - 2.0) \times \lambda_{y,max}$, where $\lambda_{y,max}$ is the wavelength corresponding to the maximum growth rate of the linear analysis, $\alpha = 0.1 - 1.0$, and $\mathscr{E} = 6$. The above values of λ_y and the value $\alpha = 1$ are certainly reasonable for the interstellar medium. However, the value $\mathscr{E} = 6$ is much too small. In the solar neighborhood, the Keplerian speed is approximately 250 km s^{-1} while the isothermal sound speed (at $T = 6,000$ K and a mean mass per particle $\mu = 1.27$, to account for a 10% He abundance by number) is $C = (k_B T/\mu m_H)^{1/2} = 6.2$ km s^{-1}. Hence, $\mathscr{E} = 1.6 \times 10^3$. This exceeds the value assumed by the authors by a factor of about 270. Put in a different way, the chosen unit of time is r/C. With $r \simeq 10$ kpc and C as above, this unit exceeds 10^9 yr. Since the typical model requires at least 10 such time units to evolve significantly, 10^{10} yr is much too long a timescale to be of relevance to the interstellar medium.

In any case, for the shorter wavelengths studied, the development of the instability leads to nonlinear oscillations, while for the longer wavelengths it leads to the formation of shallow $\cup$-shaped shocks above (and $\cap$-shaped below) the equatorial plane $z = 0$. It is also found that the nonlinear evolution favors wavelengths λ_y greater than $\lambda_{y,max}$, which is selected by the linear instability. Energetically, this tendency was identified earlier (Mouschovias 1974, §§ VIb and VIc). However, the distances involved are significantly greater than 1 kpc. Thus, even with $\alpha = 1$ (the largest value studied by Matsumoto *et al.*), the Alfvén speed $v_A = (2\alpha)^{1/2}C$ in the initial state is only $\simeq 9$ km s^{-1}, implying that the evolutionary timescale would exceed 10^8 yr. Only a very quiescent interstellar medium could wait for these long-wavelength disturbances to evolve into the nonlinear regime and affect its structure, as originally suggested by Mouschovias, Shu, and Woodward (1974).

An interesting effect was pointed out by Shibata and Matsumoto (1991) which, if true, would have significant implications. As gas flows down a deformed flux tube of decreasing cross section, a cloud will form through essentially spherical, isotropic contraction, which will result in considerable spinup because of conservation of angular momentum. If ρ_1, r_1, and Ω_1 denote, respectively, the density, radius, and angular velocity of the gas high in the flux tube, and the subscript 2 denotes the corresponding quantities near the valley of the flux tube, then a toroidal component $B_{\phi,2}$ will be generated, which will be related to the initial poloidal component B_0 by

$$\frac{B_{\phi,2}}{B_0} = \frac{v_{\phi,2}}{v_A} = \frac{r_2\Omega_2}{v_A} = \frac{r_1\Omega_1}{v_A}\left[\frac{\rho_2}{\rho_1}\right]^{1/3} \tag{38a}$$

$$= 3\left[\frac{r_1}{100 \text{ pc}}\right]\left[\frac{\Omega_1}{10^{-15} \text{ s}^{-1}}\right]\left[\frac{10 \text{ km s}^{-1}}{v_A}\right]\left[\frac{\rho_2}{\rho_1}\right]^{1/3}. \tag{38b}$$

The quantity v_A is the initial mean Alfvén speed due to the poloidal field. This equation is correct for the (unrealistic but useful for illustration purposes) spherical, isotropic contraction of the parcel of gas considered by the authors. Even so, however, the poloidal magnetic field B_0 cannot be assumed to remain constant under the authors' assumptions. Since it is frozen in the matter, spherical, isotropic contraction implies that its final value B_2 will exceed the initial value by the factor $(\rho_2/\rho_1)^{2/3}$, which, for the ratio $\rho_2/\rho_1 = 10^3$ assumed in equation (38b), is equal to 10^2. Hence, the final value of the toroidal magnetic field $B_{\phi,2}$ will not exceed the *final* value of the poloidal field B_2 by a factor of 3. It will, instead, be only 3% of B_2. Altogether, then, the Parker instability cannot generate a significant twist in the field lines as a result of angular-momentum conservation of gas sliding down a deformed (nonhorizontal) flux tube of decreasing cross section.

There is yet a second reason for which a significant toroidal component of the field could not be sustained, even if it were to be generated in the first place: magnetic braking. Mouschovias and Paleologou (1980) showed that the twists in the field lines propagate away as torsional Alfvén waves, and the field lines straighten out in a relatively short time.

Matsumoto and Shibata (1992) also undertook a three-dimensional numerical simulation of the instability in the case that the field lines initially at $z = 0$ remain undeformed at all times. Certain qualitative features of the evolution are probably relevant, but adoption of the value of the free parameter $\mathscr{E} = 6$ (see eq. [37] and associated discussion) makes quantitative applications to the interstellar medium impossible. The authors themselves are aware of this difficulty, but they apply their calculation to the interstellar medium anyhow, by offering heuristic arguments. The results are very similar to those of their two-dimensional simulations, except for the presence of periodic structure in the third (x-)direction as well. The nonlinear evolution is faster (by about a factor of 2) than it was in the two-dimensional case, as one expects from the linear analysis. This conclusion is almost certain to survive in future three-dimensional calculations with more realistic values of the free parameters.

We now turn to recent calculations specifically designed for application to the interstellar medium.

5.2. New Numerical Simulations

5.2.1. *Nonlinear Evolution of the Odd Linear Modes*

We introduce a perturbation in the zeroth-order equilibrium state of § 2.2.1 characterized by a pair of unstable wavelengths (λ_y, λ_z) and having the property that it leaves the field line originally coinciding with the galactic plane ($z = 0$) undeformed at all times. We choose $\lambda_z \equiv 2Z = 25H = 25 (1 + \alpha)C^2/g$, and $\alpha = 1$ ($\beta = 0$, i.e., no cosmic rays), so that $\lambda_z = 50C^2/g$. Hence, the first undeformed field line above the galactic plane is located at $z = 25C^2/g$. We place the upper boundary of the computational region at this $z = Z$, which is high enough not to influence the evolution of the system at lower altitudes, where most of the mass is found. (Note that the density in the initial state at Z is equal to $e^{-12.5} = 3.7 \times 10^{-6}$ its value in the plane $z = 0$. Moreover, about 95% of the mass of the initial state is found below the height $z = 6C^2/g$.) With this choice of λ_z, we set $\lambda_y \equiv 2Y = 24C^2/g$, which is approximately equal to the wavelength corresponding to the maximum growth rate of the linear instability. Because of the spatially periodic nature of the perturbation, the evolution need only be followed in the rectangular region $0 \leq y \leq 12C^2/g$, $0 \leq z \leq 25C^2/g$. The field lines have reflection symmetry about the plane $z = 0$ and the plane $y = 0$ at all times. (Note, however, that the density ρ and the magnetic field and velocity vectors B and v have reflection symmetry only about the plane $z = 0$. The magnetic vector potential does not have even this symmetry; see Mouschovias 1974, eq. [20].) A simple initial perturbation having, and at later times inducing evolution preserving, reflection symmetry about the plane $z = 0$ is given by

$$v_z = - \epsilon C \sin\left[\frac{\pi z}{Z}\right] \cos\left[\frac{\pi y}{Y}\right], \qquad (39)$$

where ϵ is a positive constant $\ll 1$. It is chosen such that the kinetic energy introduced in the system by the perturbation is less than 1% of any other form of energy (gravitational, magnetic, or thermal).

We choose convenient units as follows: [length] = thermal-pressure scale height C^2/g; [speed] = isothermal sound speed C; [density] = initial density in the galactic plane $\rho_0(y, z=0)$ = *const*; and [magnetic field] = initial field strength in the galactic plane $B_0(y, z=0)$ = *const*. The implied unit of time is C/g, which is the time required for a free-falling fluid element to acquire a speed equal to the sound speed. We note that the typical dimensional values of $T = 6000$ K, $g = 3 \times 10^{-9}$ cm s^{-2}, mean density in the plane $n = 1$ cm^{-3} (including a 10% He abundance), and the choice $\alpha = 1$ imply the following typical values of the basic units: [length] = 42 pc; [speed] = 6.2 km s^{-1}; [density] = 2.12 g cm^{-3}; and [magnetic field] = 4.56 μG. Hence, [time] = 6.6×10^6 yr.

Figures 2a, 2b, and 2c exhibit field lines, isodensity contours, and velocity vectors at time $t = 10$. The same quantities are shown in Figures 2d, 2e, and 2f at $t = 20$, in Figures 2g, 2h, and 2i at $t = 30$, and in Figures 2j, 2k, and 2l at $t = 40$. The field lines are chosen so that the magnetic flux between any two consecutive ones is constant (hence, they are the same as those of the equilibrium state of Fig. 1). The density decreases by a factor of e from one isodensity contour to the next higher one. The maximum velocity in Figures 2c, 2f, 2i, and 2l is $v_{max} = 0.74, 2.2, 3.0,$ and 2.1, respectively.

As is the case in the linear theory, the field lines deform at early times in an essentially sinusoidal fashion. The vertical gravitational field acquires a component along the deformed field lines and induces mass motions from the mountains to the valleys of the field lines. By the time $t = 10$, significant downward velocities ($v_{max} \lesssim 0.7C$) occur in the valleys of field lines relatively high above the plane, in a part of the computational region roughly defined by $9 \lesssim z \lesssim 18$ and $0 \leq y \leq 4$ (see Figure 2c); similarly significant upward velocities are found in the region $11 \lesssim z \lesssim 18$, $7 \lesssim y \leq 12$. Figure 2a shows that field lines are hardly deformed at small z's. Deformations are significant only at high z's, where matter begins to accumulate in valleys of the field lines (see Fig. 2b). Soon thereafter, the evolution becomes fully dynamic and nonlinear.

By $t = 20$, an almost vertical shock front is clearly present high above the valleys of the field lines ($z \gtrsim 10$) --see Fig. 2f. Behind it, a slab of matter of half-thickness $y = 1 - 2$ extends high above the plane. It is noteworthy that, for $z > 0$, each isodensity contour is higher at $y = 0$ than at neighboring y's. This reveals that the increase in density in valleys of the field lines at $z > 0$ is due to accumulation of matter there, because of motions along field lines, rather than due to significant compression perpendicular to the field lines, as also found for the final equilibrium states (see Mouschovias 1974). The velocities are very small below $z = 5$, although the field lines and isodensity contours in this region are visibly deformed relative to those of the initial state (see Figs. 2d and 2e). By the time $t = 30$ (see Figs. 2g - 2i), the shock has gained strength ($v_{max} = 3.0$) and has moved to smaller z's, and the vertical slab of matter has become more pronounced, as significant deformation of field lines has taken place above $z \simeq 2$. At even later times, $t = 40$ (see Figs. 2j - 2l), the system below $z \simeq 10$ has essentially settled into a final equilibrium state (the same one determined from the solution of the MHS equations; see Fig. 1). Significant velocities are found only at high z's and intermediate y's (see Fig. 2l). The shock front is still present but has begun to lose strength ($v_{max} = 2.1$) as it moves upward and to the right along field lines, where the slope of the field lines decreases. Although not shown in Figure 2 for economy of space, we have followed the evolution to a time $t = 50$. The system settles into the equilibrium state of Figure 1, with leftover, subsonic oscillations about this state at high z's, where the inertia of the gas is very low. It is mainly a reduction of the gravitational potential energy of the system that has allowed the transition from the initial to the final equilibrium states, as also found by MHS calculations (see Mouschovias 1974, Table 2). The evolution is discussed much more extensively by Mouschovias, Basu, and Paleologou (1996). A summary of features other than those given above is found in Basu, Mouschovias, and Paleologou (1996).

5.2.2. Nonlinear Evolution of the Even Linear Modes
We consider, once again, Parker's zeroth-order, one-dimensional equilibrium state, but we now introduce a perturbation that allows the field line originally coinciding with the galactic plane to deform. The linear instability criterion and growth rate are identical with those of the previous class of perturbations, which leave the field line originally at $z = 0$ undeformed, provided that the pair of unstable wavelengths (λ_y, λ_z) is the same in the two cases. However, we expect that the nonlinear evolution will be substantially different from the one found above. If a final equilibrium state is reached (and we expect that it will be, since the argument of § 3.1 did not assume that the field line initially at $z = 0$ remains undeformed at all times), the density at $z = 0$ will now both exceed its initial value (because field lines cross the galactic plane and, therefore, gravity can compress the gas along them) and vary with y (because, although g still has no horizontal component, horizontal density gradients can be sustained by magnetic forces even when a final equilibrium state is reached). We introduce the velocity perturbation

$$v_z = - \epsilon C \cos\left[\frac{\pi z}{2Z}\right] \cos\left[\frac{\pi y}{Y}\right], \qquad (40)$$

which has the same y dependence as that of equation (39) but, as a function of z, it has its maximum value at $z = 0$ and vanishes at $z = \pm Z$. (Note that λ_z for this even linear mode is chosen to be twice as large as the λ_z of the odd linear mode studied in § 5.2.1, so that the first undeformed field lines above and below the plane are located at $z = \pm Z$ in both cases. This, however, implies that the growth rate of the linear instability is somewhat greater for

496

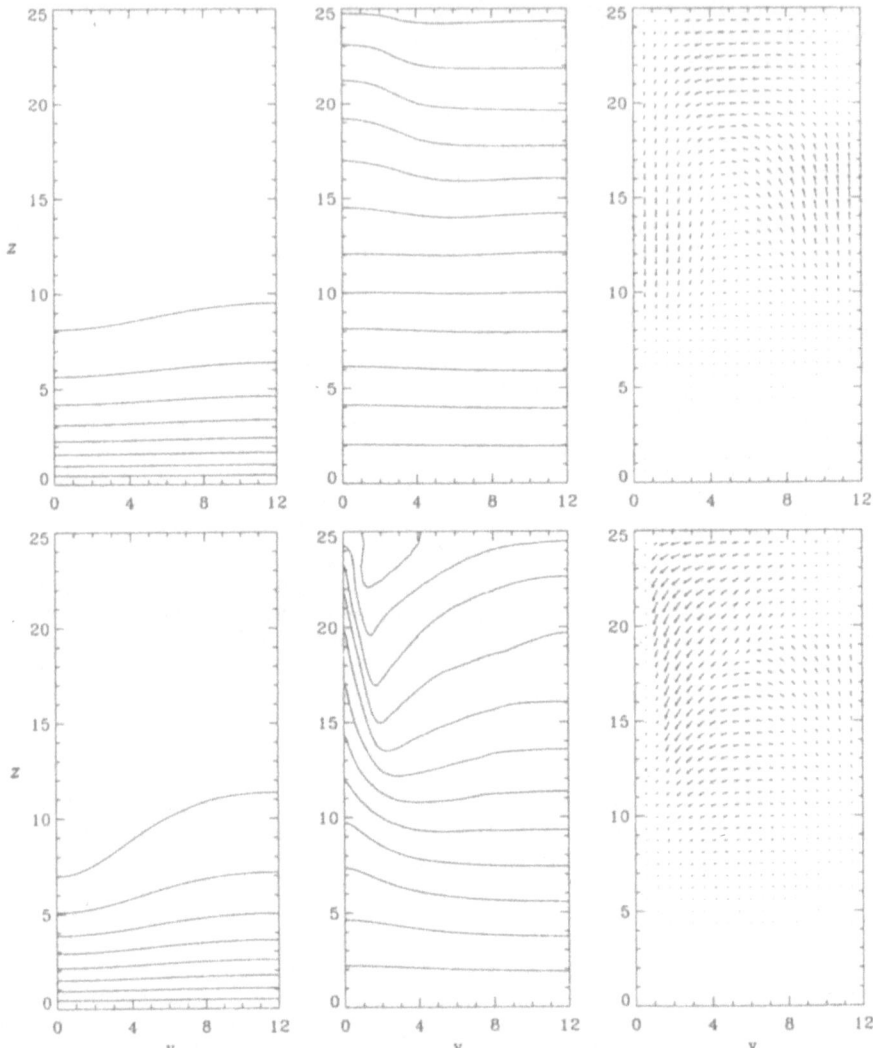

Figure 2. *Nonlinear evolution of the model of* § 5.2.1. The *left, middle* and *right* panels exhibit equiflux magnetic field lines, e-folding isodensity contours, and velocity vectors, respectively. The unit of length is $C^2/g \simeq 42$ pc. The top row of panels (a, b, and c) refers to $t = 10$, and the bottom row (d, e, and f) to $t = 20$. The unit of time is $C/g = 6.6$ Myr. The velocity scale is given in the text.

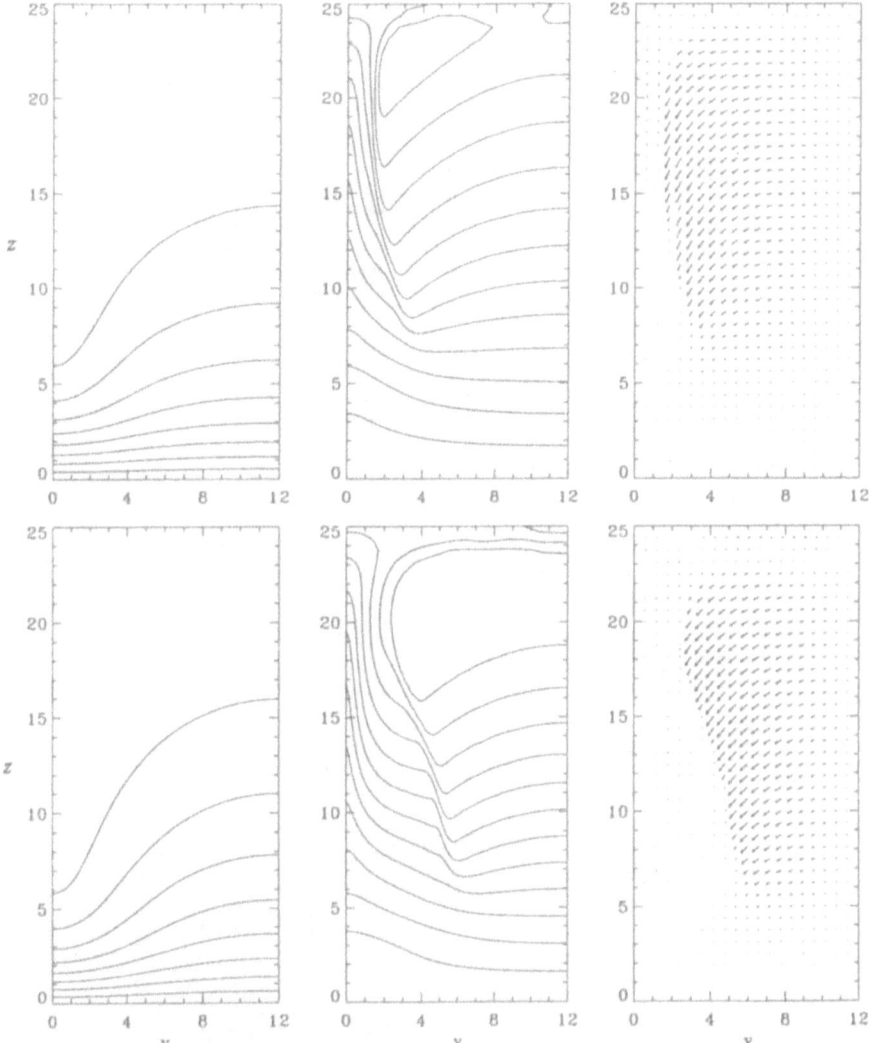

Figure 2, cont'd. The top row of panels (g, h, and i) refers to $t = 30$, and the bottom row (j, k, and l) to $t = 40$.

498

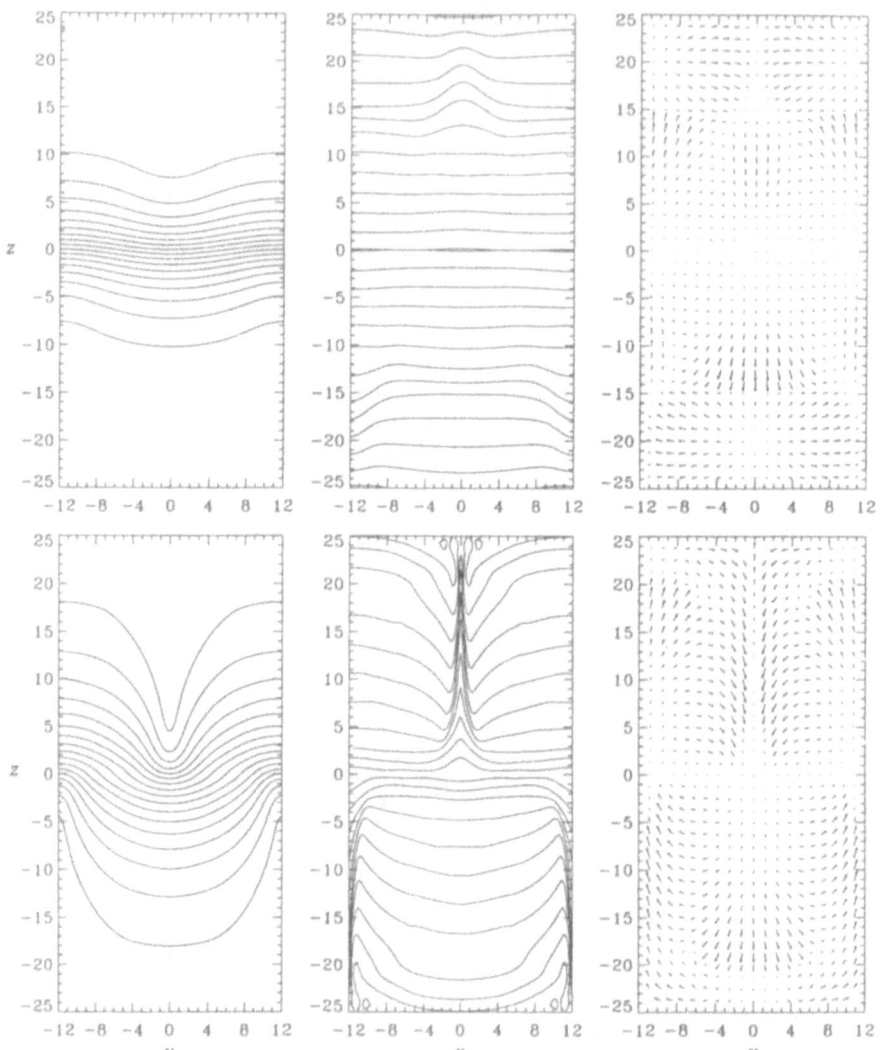

Figure 3. *Nonlinear evolution of the model of* § 5.2.2. The *left*, *middle* and *right* panels exhibit equiflux magnetic field lines, *e*-folding isodensity contours, and velocity vectors, respectively. The units of length and time are as in Figure 2. The top row of panels (*a*, *b*, and *c*) refers to *t* = 10, and the bottom row (*d*, *e*, and *f*) to *t* = 20. The velocity scale is given in the text.

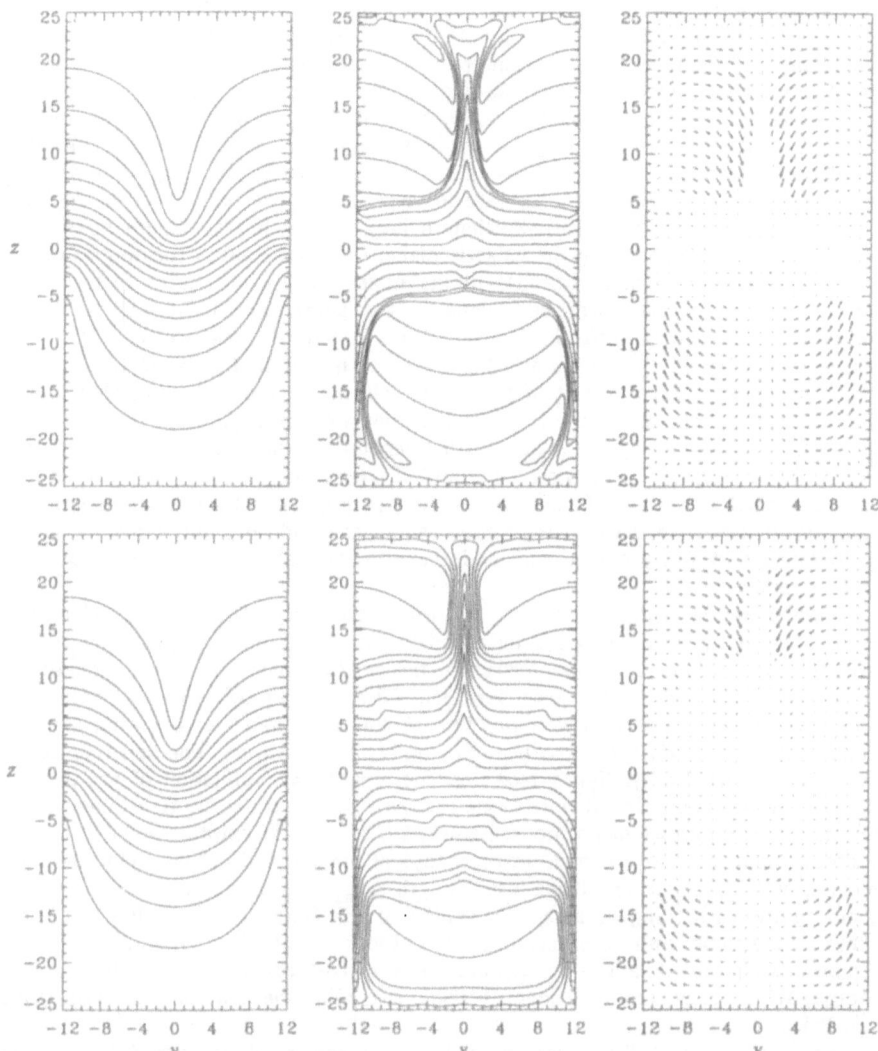

Figure 3, cont'd. The top row of panels (g, h, and i) refers to $t = 30$, and the bottom row (j, k, and l) to $t = 40$.

the present, even mode than it was for the odd mode.) Unlike the case of § 2.2.1, the quantities ρ, B and v in the present case are not invarient with respect to reflection about the plane $z = 0$. They are invarient only with respect to such reflection followed by a translation in the y-direction by an amount $\pm\lambda_y/2$.

The system enters its nonlinear stage of evolution[2] at an earlier time compared to the previous run, it develops larger velocities, and it reaches a final equilibrium state significantly earlier. Figures 3a, 3b, and 3c correspond to Figures 2a, 2b, and 2c in that they refer to time $t = 10$. However, for ease in visualization, we now show the region $-12 \leq y \leq +12$, $-25 \leq z \leq +25$ (i.e., a full wavelength in y- and half wavelength in the z-direction). With field lines now crossing the galactic plane, there are valleys forming at both $y = 0$ and $y = \pm12$. Gas from higher altitudes in the upper half-plane begins to accumulate in the former, and from the lower half-plane in the latter. The field lines in Figure 3a are somewhat steeper than those in Figure 2a, the maximum velocity (= 1.3) in Figure 3c is almost twice that in Figure 2c, and the sheets of matter in Figure 3b (now at $y = 0$ in the upper half-plane and $y = \pm12$ in the lower half-plane) are at a somewhat more advanced (but still early) stage of their formation than they were in Figure 2b. By the time $t = 20$, the field lines shown in Figure 3d are much steeper than those in Figure 2d; the velocities in Figure 3f are greater than those in Figure 2f (now $v_{max} = 2.6$) and (more significantly) extend to smaller z's; and the sheets of matter in Figure 3e are thinner in y, more extended in z, and denser than those in Figure 2e. Matter can and does now move closer to the plane than it could in the previous run, and the gravitational potential energy of the system is further reduced.

Figures 3g, 3h, and 3i show the field lines, isodensity contours, and velocity vectors at $t = 30$. The maximum velocity has increased to 3.6, and there exist prominent and distinctive in shape shock fronts at the boundaries of the vertical sheets of matter extending alternately above and below the plane (see Fig. 3i). *These shocks, if observed, cannot possibly be mistaken for supernova remnants and can therefore be taken as evidence for the nonlinear development of the Parker instability in the interstellar medium.* The deformation of the field lines has increased only slightly (see Fig. 3g), but the accumulation of gas in the valleys of the field lines has continued slowly (see Fig. 3h). By this time, the field lines have essentially reached their equilibrium positions. Figures 3j, 3k, and 3l exhibit the field lines, isodensity contours, and velocity vectors, respectively, at time $t = 40$. The field lines have hardly moved, and the shock front has lost strength ($v_{max} = 2.5$) and moved to higher z's, where some additional settling of matter in valleys of the field lines is taking place. At yet later times, the system undergoes small-scale, subsonic oscillations, mainly in the low-inertia regions. A more detailed description of the features of the evolution is given by Mouschovias, Basu, and Paleologou (1996); see, also, the summary in Basu, Mouschovias, and Paleologou (1996).

The signatures of this mode of evolution of the instability are: (1) gas concentrations half a horizontal wavelength apart ($\lambda_y/2 \equiv Y \simeq 500$ pc); (2) sheets of matter extending high *above* the galactic plane separated by a distance equal to a full horizontal wavelength ($\lambda_y \equiv 2Y \simeq 1$ kpc), accompanied by similar sheets extending *below* the galactic plane and also

---

[2]Readers interested in the numerical techniques used to follow the nonlinear evolution of the instability are referred to the Appendix of the paper by Mouschovias, Basu, and Paleologou (1996). We only note here that the placement of the upper boundary at $z = Z = 25$ affects the evolution of the system only very near that boundary ($z \gtrsim 22$ in the run described in § 5.2.1, and $z \gtrsim 20$ in the run of this section), where there is a negligible amount of matter.

separated by a distance equal to λ_y, but with the former set of sheets shifted in y with respect to the latter set by a distance equal to $\lambda_y/2$; and (3) during the intermediate to late stages of the evolution, prominent, distinctively shaped shocks at the sheet boundaries. Although the density structure in the final state of this run is quite different from that of the run that leaves the field line originally at $z = 0$ undeformed, the *column density* as a function of y (which is an observable quantity along spiral arms of external galaxies seen nearly face-on) has a maximum-to-minimum contrast approximately equal to 2 in both cases. However, as explained above, the maxima are only $\lambda_y/2$ apart in the present case.

5.2.3. *Effect of Cosmic Rays on the Instability and vice versa*

As seen is § 2.2.2, in the linear regime cosmic rays destabilize the gas-field system; they decrease the critical wavelengths in the (y, z)-plane and increase the growth rate. No nonlinear calculation has accounted for the presence of cosmic rays yet, so as to test Parker's (1968b) argument that they will cause a continuous and unlimited inflation of the field lines, limited only by magnetic reconnection and escape of magnetic "bubbles" with entrapped cosmic rays. It is nevertheless the case that Parker's conclusion depends crucially on the assumption that the cosmic-ray pressure in all flux tubes rising above the galactic plane is kept fixed by a copious supply of cosmic rays within the disk. If such a supply of cosmic rays can indeed render P_{rel} a constant of the motion (which implies sources of cosmic rays in all the flux tubes of the system), unlimited inflation of the field lines seems likely. If, however, the *number* (instead of the pressure) of cosmic rays remains fixed (or quasi-steady) in each flux tube as its volume increases upon expansion, then a final equilibrium state is still possible (see Mouschovias 1975a), as we have shown in § 3.1. It is very similar to the ones found in the absence of cosmic rays, but with larger deformation of the field lines. A fat radio disk of half-thickness ~ 1 kpc was predicted on this basis.

The above summary concerns the effect of cosmic rays on the linear instability and its nonlinear evolution. Does the instability itself have the effect of allowing cosmic rays to escape from the Galactic disk? We have seen that, with $\alpha = \beta = \gamma = 1$, the minimum e-folding time of the instability in two dimensions is $\tau_{min} = 2.2 \times 10^7$ yr. Accounting for the third direction and for the variation of the Galactic gravitational field with z can reduce τ_{min} to $\simeq 10^7$ yr. Since Be^{10} observations imply a mean cosmic-ray lifetime $\tau_{CR} > 10^7$ yr (most of which is spent in regions of density ≤ 0.2 cm^{-3}; Garcia-Munoz, Mason, and Simpson 1975, 1977), Parker's suggestion survives a comparison of timescales. There is a difficulty, however, with the assumption that *all* inflating flux tubes maintain a constant cosmic-ray pressure. It is much more likely that a flux tube close to the Galactic plane is suddenly overpressurized with cosmic rays. Then the following sequence of events, which has little to do with a large-scale instability (i.e., the growth of a small-amplitude perturbation), may be initiated. (1) The cosmic-ray gas streams quickly along field lines (mainly in the y-direction) and equalizes its pressure along the flux tube. (2) The flux tube buckles (in the vertical, z-direction) and inflates, with its valley remaining anchored low in the disk, or rises as a whole through overlying field lines, which (instead of inflating themselves) make way by moving laterally (mainly in the x-direction) as the flux tube passes by. The displaced background field lines tend to move back to their original position after the flux tube has risen above them. This, in a sense, is an interchange instability, or a cosmic-ray buoyancy effect. (3) If the flux tube rises as a whole, it will simply escape into the Galactic halo carrying the cosmic rays and the thermal gas with it. (This may feed the "Galactic fountain" of Shapiro and Field 1976.) If it buckles and inflates into a magnetic loop of increasing cross section, the oppositely directed magnetic field lines of its footpoints (unfortunately an unknown distance apart) may suffer reconnection in the plane, as originally suggested by Parker, or may reconnect with background field lines at higher z's. Both kinds of reconnection restore the equilibrium of the gas-field system at small z's,

However, the former allows closed flux tubes with entrapped cosmic rays to escape, while the latter forms a fat radio disk with arched field lines still confined by the weight of the gas in the disk and the tension of the field lines in the region of inflation. The escape of cosmic rays in this scenario is not limited by the growth time of the linear instability. It is driven instead by the excess pressure of cosmic rays in a magnetic flux tube. This variation of Parker's original proposal is nevertheless far from being characterized by the rigor of the discussion in the rest of this paper.

6. Summary

The interstellar gas is supported against the vertical galactic gravitational field ($g = \hat{z}g(z)$, where $g(z) = - g(-z) > 0$) by thermal-pressure, magnetic, and cosmic-ray-pressure gradients (Parker 1966). We have reviewed quantitatively this notion and its consequences in the interstellar medium. With the magnetic field essentially parallel to the galactic plane along a spiral arm [$B_0 = \hat{y}B_0(z)$], long-wavelength perturbations in the plane defined by g and B [the (y, z)-plane] are unstable. Gas drains from raised portions of field lines into field-line valleys, thus allowing the unburdened, inflated portions to rise even further and unload gas even faster into the valleys. For typical parameters of the interstellar medium, the critical wavelength along the field lines is $\lambda_{y,\text{crit}} \equiv \Lambda_y \simeq \pi H$, where H is the vertical scale height of the density in the undisturbed state ($\simeq 126$ pc). The maximum growth rate occurs at $\lambda_{y,\text{max}} \simeq 2\pi H$ and $\lambda_{z,\text{max}} = \infty$, and corresponds to an e-folding time $\tau_{\text{min}} \simeq 2 \times 10^7$ yr. In the third (x-)direction, wavelengths λ_x in the approximate range $H/25 - 20H$ can grow with almost identical growth rates, while long wavelengths are stabilized by differential galactic rotation (this being essentially the radial direction in a galactic disk) and short wavelengths by ambipolar diffusion.

Approximate analytical arguments show that "final" equilibrium states should exist, provided only that the total number of cosmic rays in each flux tube is fixed or quasi-steady; the confining magnetic-tension force decreases *less rapidly* than the expansive magnetic-pressure and cosmic-ray-pressure forces at the "mountains" of inflating field lines. Final equilibrium states reachable from the initial stratified state through continuous deformation of the field lines have been calculated by formulating and solving a self-consistent magnetohydrostatic problem (Mouschovias 1974). One of their important testable predictions is the variation of the column density of the gas along a spiral arm (the y-axis). The maximum-to-minimum column-density contrast is $\simeq 2$. This is consistent with observations of external galaxies seen nearly face-on. The gas accumulations in valleys of the field lines are mainly the result of drainage along field lines, rather than compression perpendicular to the field. They do not (yet) represent typical interstellar clouds, which are much denser objects. They are giant clouds or cloud complexes (with masses $10^4 - 10^6\ M_\odot$), of relatively low mean density, within which star formation is initiated by galactic shocks or other means, leading to the appearance of OB associations, giant HII regions and cloud complexes like "beads on a string" along spiral arms, separated by regular intervals $\simeq 0.5 - 1$ kpc. Although the stability of these two-dimensional equilibrium states with curved field lines has been questioned (with respect to three-dimensional perturbations) by Asséo *et al.* (1978), the properly calculated e-folding time of the presumed instability exceeds 10^8 yr. The equilibrium states are therefore of significance in the interstellar medium.

Thus far, numerical simulations of the nonlinear evolution of the Parker instability have used input parameters relevant to accretion disks about a central point mass but completely inappropriate for application to the interstellar medium. (If scaled to the interstellar medium, the evolutionary timescale would approach or exceed 10^{10} yr.) Simulations by Mouschovias, Basu, and Paleologou (1996), specifically designed for the interstellar medium, find that the bulk of the system settles in $\leq 10^8$ yr into equilibrium states virtually indistinguishable from those determined by magnetohydrostatic calculations. At later times the system undergoes small-scale, subsonic oscillations, mainly in the low-inertia (very high z) regions. Yet the transition from the initial to the final states is quite dynamic, with velocities along field lines (in the case that the horizontal wavelength of the initial perturbation corresponds to that of maximum growth rate of the linear instability) exceeding $3C$, where C is the isothermal sound speed in the initial state. Shock fronts define the boundaries of slabs of matter rising high above valleys of field lines. The details of the evolution depend on whether the initial perturbation leaves the field line originally coinciding with the galactic plane ($z = 0$) undeformed ("odd linear mode") or allows field lines to cross the galactic plane ("even linear mode"). In the former case, slabs of matter rising above and below the plane form in valleys of field lines, separated by a full horizontal wavelength $\lambda_y \simeq 1$ kpc. (The field lines are symmetric with respect to reflection about the galactic plane.) In the latter case, the valleys of field lines above and below the plane alternate. (The field lines are symmetric with respect to reflection about the plane followed by a translation in y by an amount $\lambda_y/2$.) Hence, sheets of matter extending high *above* the plane, separated by a distance λ_y, are accompanied by sheets extending *below* the plane, also separated by λ_y but shifted in y with respect to the former set of sheets by a distance equal to $\lambda_y/2 \simeq 0.5$ kpc. This structure and the distinctive shocks, present during the intermediate stages of the evolution at the boundaries of the sheets, are unmistakable signatures of the nonlinear evolution of the Parker instability.

The nonlinear calculations have not yet accounted for the presence of cosmic rays. Hence, the issue of whether cosmic rays and magnetic flux escape from the galactic disk through magnetic reconnection at the "footpoints" of inflated field lines (Parker 1968b) remains unsettled, and so does the variation of it proposed in this paper.

Acknowledgements. This work was supported in part by the National Science Foundation under grant AST 93-20250. The hospitality of the Max-Planck-Institut für Extraterrestrische Physik in Garching, Germany, and the Alexander von Humboldt Foundation during the writing of part of this paper is gratefully acknowledged.

7. References

Ames, S. 1973, *ApJ*, **182**, 387
Asséo, E., Cesarsky, C. J., Lachièze-Rey, M., and Pellat, R. 1978, *ApJ*, **225**, L21
Basu, S., Mouschovias, T. Ch., and Paleologou, E. V. 1996, *ApLC, in press*
Biermann, L., and Davis, L. 1960, *Z. Astrophys.*, **51**, 19
Blitz, L., and Shu, F. H. 1980, *ApJ.*, **238**, 148
Chandrasekhar, S., and Fermi, E. 1953, *ApJ*, **118**, 113
Davies, R. D., Slater, C. H., Shuter, W. L. H., and Wild, P. A. T. 1960, *Nature*, **187**, 1088
Field, G. B. 1969, in *Interstellar Gas Dynamics*, ed. H. J. Habing (Reidel: Dordrecht), 51
_____ . 1970±1, *unpublished notes*
Field, G. B., Goldsmith, D. W., and Habing, H. J. 1969, *ApJ*, **155**, L149
Garcia-Munoz, M., Mason, G. M., and Simpson, J. A. 1975, *ApJ*, **201**, L141
_____ . 1977, *ApJ*, **217**, 859
Giz, A. T., and Shu, F. H. 1993, *ApJ*, **404**, 185
Heiles, C. 1987, in *Physical Processes in Interstellar Clouds*, ed. G. E. Morfill and M. Scholer (Dordrecht: Reidel), 429
Hodge, P. W. 1969, *ApJ*, **156**, 847
Kellman, S. A. 1972, *ApJ*, **175**, 363

504

Kerr, F. J. 1963, in *The Galaxy and the Magellanic Clouds*, ed. F. J. Kerr (Dordrecht: Reidel), 81

Kulsrud, R., and Pearce, W. 1969, *ApJ*, **156**, 445

Lerche, I. 1967a, *ApJ*, **149**, 395

———. 1967b, *ApJ*, **149**, 553

———. 1969, in *Adv. Plasma Phys.*, ed. A. Simon and W. B. Thompson (New York: Interscience), **2**, 47

Mathewson, D. S., and Ford, V. L. 1970, *Mem. RAS*, **74**, 143

Matsumoto, R., Horiuchi, T., Hanawa, T., and Shibata, K. 1990, *ApJ*, **356**, 259

Matsumoto, R., Horiuchi, T., Shibata, K., and Hanawa, T. 1988, *PASJ*, **40**, 171

Matsumoto, R., and Shibata, K. 1992, *PASJ*, **44**, 167

Morgan, W. W. 1970, in *The Spiral Structure of Our Galaxy*, ed. W. Becker and G. Contopoulos (Dordrecht: Reidel), 9

Mouschovias, T. Ch. 1974, *ApJ*, **192**, 37

———. 1975a, *A&A*, **40**, 191

———. 1975b, *Ph.D. Thesis*, University of California at Berkeley

———. 1976, *ApJ*, **206**, 753

———. 1977, *ApJ*, **211**, 147

———. 1978, in *Protostar and Planets*, ed. T. Gehrels (Tucson: Univ. of Arizona), 209

———. 1981, in *Fundamental Problems in the Theory of Stellar Evolution*, ed. D. Sugimoto, D. Q. Lamb, and D. N. Schramm (Dordrecht: Reidel), 27

———. 1991, in *The Physics of Star Formation and Early Stellar Evolution*, ed. C. J. Lada and N. D. Kylafis (Dordrecht: Kluwer), 61

———. 1995, in *The Physics of the Interstellar Medium and Intergalactic Medium*, ed. A. Ferrara, C. F. McKee, C. Heiles, and P. R. Shapiro (San Francisco: ASP), **80**, 184

Mouschovias, T. Ch., Basu, S., and Paleologou, E. V. 1996, *ApJ, submitted*

Mouschovias, T. Ch., and Paleologou, E. V. 1980, *ApJ*, **237**, 877

Mouschovias, T. Ch., Shu, F. H., and Woodward, R. 1974, *A&A*, **33**, 73

Parker, E. N. 1966, *ApJ*, **145**, 811

———. 1967a, *ApJ*, **149**, 517

———. 1967b, *ApJ*, **149**, 535

———. 1968a, *ApJ*, **154**, 57

———. 1968b, in *Stars and Stellar Systems*, Vol. 7, *Nebulae and Interstellar Matter*, ed. B. Middlehurst and L. H. Aller (Chicago: Univ. of Chicago), 707

———. 1969, *Space Sci. Rev.*, **9**, 651

———. 1975, *ApJ*, **201**, 74

Roberts, W. W. 1969, *ApJ*, **158**, 123

Rots, A. H. 1974, *Ph.D. Thesis*, Groningen University

Shapiro, P. R., and Field, G. B. 1976, *ApJ*, **205**, 762

Shibata, K., and Matsumoto, R. 1991, *Nature*, **353**, 633

Shu, F. H. 1974, *A&A*, **33**, 55

Spitzer, L., Jr. 1962, *Physics of Fully Ionized Gases*, 2nd ed. (New York: Interscience)

———. 1968, in *Stars and Stellar Systems*, Vol. 7, *Nebulae and Interstellar Matter*, ed. B. Middlehurst and L. H. Aller (Chicago: Univ. of Chicago), 1

———. 1978, *Physical Processes in the Interstellar Medium* (New York: Wiley-Interscience)

Westerhout, G. 1963, in *The Galaxy and the Magellanic Clouds*, ed. F. J. Kerr (Dordrecht: Reidel), 78

Woltjer, L. 1965, in *Stars and Stellar Systems*, Vol. 5, *Galactic Structure*, ed. A. Blaauw and M. Schmidt (Chicago: Univ. of Chicago), 531

Zweibel, E., and Kulsrud, R. 1975, *ApJ*, **201**, 63

MULTIFLUID MAGNETOHYDRODYNAMICS AND STAR FORMATION

TELEMACHOS CH. MOUSCHOVIAS
University of Illinois at Urbana-Champaign
Departments of Physics and Astronomy
1002 West Green street
Urbana, IL 61801, U. S. A.

E-mail: tchm@astro.uiuc.edu

Abstract. The theoretical description of the isothermal phases of star formation (densities $\leq$ 10^{10} cm^{-3}) in magnetically supported molecular clouds is reviewed and the main results are summarized. A modern theory of (self-initiated) protostar formation is described, properly accounting for ambipolar diffusion and magnetic braking. The role of interstellar grains both in determining the degree of ionization and in transmitting the magnetic forces to the predominantly neutral matter (H$_2$ and He) in molecular clouds through grain-neutral collisions is explained physically and accounted for in the five-fluid MHD equations. Criteria for direct as well as indirect attachment of charged particles to the magnetic field are obtained and explained physically. Results of analytical calculations and numerical simulations on the formation and contraction of protostellar fragments (or cores), including consequences of nonlinear hydromagnetic waves, are discussed and compared with observations. Excellent quantitative agreement is found. New, testable predictions are made.

1. Introduction - A Modern Theory of Star Formation

Unlike solar and most other astrophysical magnetohydrodynamic (MHD) flows discussed at this meeting, the flows that lead to star formation are driven by *self-gravity*. This introduces an almost unique simplicity in the physical problem of star formation but, at the same time, it complicates its mathematical solution.

Typical, massive ($M \sim 10^2 - 10^5 \ M_\odot$) interstellar molecular clouds, in which most star formation activity is observed, are relatively dense, cold, nearly isothermal objects (number density of neutral particles $n_n \gtrsim 10^3$ cm^{-3}, and temperature $T \simeq 10$ K). High-energy (> 100 MeV) cosmic rays maintain a degree of ionization $x_i \equiv n_i/n_n \lesssim 10^{-7}$ in their deep interiors, and ultraviolet (UV) radiation is responsible for the much greater degree of ionization ($x_i \sim 10^{-4}$) in the outer envelopes. Approximately 1% of the mass of a cloud is in the form of dust particles (or grains), 90% of which carry a single electronic charge $-e$ at any one time, the remaining 10% being electrically neutral. In the density range of interest ($\sim 10^3 - 10^{10}$ cm^{-3}), the dominant ions in dense cores are HCO$^+$, Mg$^+$, Na$^+$, and Fe$^+$, and in cloud envelopes C$^+$, S$^+$, Mg$^+$, Si$^+$, Na$^+$, Fe$^+$, and HCO$^+$. Zeeman measurements yield magnetic field strengths $B \simeq 10 - 150 \ \mu$G, with 30 μG being typical, while polarization observations reveal nicely ordered magnetic fields inside and in the neighborhoods of clouds. Molecular clouds rarely exhibit rotation significantly higher than that of the background medium and, when they do, their measured angular velocities are much smaller than those expected from angular-momentum conservation from an initial angular velocity $\Omega_0 \simeq 10^{-15}$ s^{-1} and density $n \simeq 1$ cm^{-3} (the approximate local Galactic rotation rate and mean interstellar density). The observed spectral linewidths are almost always supersonic but subAlfvénic.

K. C. Tsinganos (ed.), Solar and Astrophysical Magnetohydrodynamic Flows, 505–538.
© 1996 *Kluwer Academic Publishers.*

Despite the fact that the critical mass for collapse of a pressure-bounded isothermal sphere is only 5.8 $M_\odot$ at a density $n_n = 10^3$ cm^{-3} and temperature $T = 10$ K (Bonnor 1956; Ebert 1955, 1957), *molecular clouds are not collapsing; they are magnetically supported* (Mouschovias 1976a, b; Mouschovias and Spitzer 1976; Spitzer 1978; Mouschovias 1978). The need for multifluid MHD as the proper tool for the theoretical description of star formation arises from the fact that *ambipolar diffusion*[1] is an inevitable process in magnetically supported,[2] self-gravitating clouds. Its timescale (with τ_{ff} denoting the free-fall and τ_{ni} the neutral-ion collision time)

$$\tau_{AD} \simeq \frac{\tau_{ff}^2}{\tau_{ni}} \simeq 2 \times 10^6 \left[\frac{n_i/n_{H_2}}{10^{-7}} \right] \quad \text{yr} \tag{1}$$

was calculated and shown to be typically three orders of magnitude smaller in the deep interior of a cloud than in the outermost envelope, thus suggesting that fragmentation and star formation can be *self-initiated* (Mouschovias 1977, 1979b): Ambipolar diffusion allows fragments, each having mass somewhat greater than the critical thermal mass ($\simeq$ Jeans mass) to form and contract *quasistatically* (i.e., with negligible acceleration) through essentially stationary plasma and magnetic field lines. The presence of the magnetic field does not prevent this process. It simply regulates it. That is, it slows down the contraction through plasma-neutral drag and converts the would-be violent free fall into a quasistatic process. The fragmentation (or core-formation) timescale is the mass-redistribution timescale among the central flux tubes of a cloud. No magnetic flux is lost by the cloud as a whole. The essence of ambipolar diffusion in this modern picture of its role in star formation is precisely this kind of redistribution of mass or, equivalently, the increase of the central mass-to-flux ratio $(dm/d\Phi_B)_c$ (Mouschovias 1978, 1979b). Eventually, $(dm/d\Phi_B)_c$ exceeds the critical (central) value for collapse

$$\left[\frac{dm}{d\Phi_B} \right]_{c,crit} = 1.5 \left(\frac{1}{63G} \right)^{1/2} \tag{2}$$

(Mouschovias and Spitzer 1976) and dynamical contraction ensues. For a cloud as a whole and for a core during its quasistatic (or subcritical) phase of contraction, *magnetic braking* is very effective and keeps the cores and the cloud essentially corotating with the background (Mouschovias 1977, 1979a; Mouschovias and Paleologou 1979, 1980; Basu and Mouschovias 1994). Hence, the centrifugal forces have a negligible effect on the evolution. In the case of an *aligned disk rotator* (cloud or fragment) of density ρ_{cl} and half-thickness Z, with field lines "fanning out" away from it, surrounded by an "external" medium (or envelope) of density ρ_{ext}, the magnetic-braking timescale is given by (Mouschovias 1983)

---

[1]Ambipolar diffusion refers to the magnetically driven drift of the plasma relative to the neutrals. It was proposed by Mestel and Spitzer (1956) as a way for a cloud to lose its magnetic flux and thus collapse and fragment to form stars. With the magnetic field frozen in the matter, the critical mass would be $M_{crit} \propto B^3/n^2 = const$ during spherical, isotropic contraction, thought at the time to be the collapse mode; hence, fragmentation would be impossible.

[2]Magnetic support can include a contribution from hydromagnetic (HM) waves. In fact, it was suggested that it is precisely the damping of HM waves by ambipolar diffusion in magnetically supported molecular clouds, rather than the growth of perturbations in collapsing clouds, that initiates fragmentation and star formation (Mouschovias 1987a).

$$\tau_{\parallel\,\text{fan}} = \frac{\rho_{\text{cl}}}{\rho_{\text{ext}}} \frac{Z}{v_{\text{A,ext}}} \left[\frac{R_{\text{cl}}}{R_0}\right]^4 \equiv \frac{\sigma_{\text{n,cl}}}{2\rho_{\text{ext}} v_{\text{A,ext}}} \left[\frac{R_{\text{cl}}}{R_0}\right]^4 , \tag{3a, b}$$

$$\equiv \left[\frac{\pi}{\rho_{\text{ext}}}\right]^{1/2} \frac{M}{\Phi_{\text{B}}} \left[\frac{R_{\text{cl}}}{R_0}\right]^2 , \tag{3c}$$

where $v_{\text{A,ext}}$ is the Alfvén speed in the external medium, $\sigma_{\text{n,cl}} \equiv M/\pi R^2$ the column density of matter along field lines, R the equatorial radius, Φ_{B} the magnetic flux threading the cloud (or fragment), R_0 the original equatorial radius of the flattened cloud, when its density was ρ_0 ($> \rho_{\text{ext}}$), because of its formation by motions along field lines, and its magnetic field was equal to that of the external medium, B_{ext} (see Mouschovias and Morton 1985b, Fig. 2, which shows this geometry, but for several fragments). Note that $\tau_{\parallel\,\text{fan}}$ is equal to the ratio of the moment of inertia of the cloud and the rate of increase of the moment of inertia of that part of the external medium affected by the outward-propagating torsional Alfvén waves. For a *perpendicular rotator* with $\rho_{\text{cl}}/\rho_{\text{ext}} \gg 1$, the magnetic-braking timescale is

$$\tau_{\perp} = \frac{1}{2} \left[\frac{\rho_{\text{cl}}}{\rho_{\text{ext}}}\right]^{1/2} \frac{R}{v_{\text{A}}(R)} \equiv 2 \left[\frac{\pi}{\rho_{\text{cl}}}\right]^{1/2} \frac{M}{\Phi_{\text{B}}} , \tag{4a, b}$$

(Mouschovias and Paleologou 1979; Mouschovias 1985), where $v_{\text{A}}(R)$ is the Alfvén speed just outside the cloud (or core) surface. It follows from equation (3c) that, as a cloud (or core) contracts at constant M/Φ_{B} in an environment whose properties (ρ_{ext}) do not change much as a result of the cloud's (or core's) contraction, $\tau_{\parallel\,\text{fan}}$ decreases as R_{cl}^2 ($\propto Z \propto \rho_{\text{cl}}^{-1/2}$ for isothermal contraction with balance of forces maintained along field lines). A similar contraction for a perpendicular rotator yields that $\tau_{\perp}$ also decreases as $\rho_{\text{cl}}^{-1/2}$ and, moreover, does not depend on ρ_{ext}.

During the collapse phase, force balance is maintained *along* field lines, and both magnetic flux and angular momentum tend to get trapped inside the magnetically and thermally supercritical protostellar cores, with a consequent increase of the magnetic field and angular velocity with gas density as $B_{\text{c}} \propto \rho_{\text{c}}^{1/2}$ and $\Omega_{\text{c}} \propto \rho_{\text{c}}^{1/2}$ (Mouschovias 1976b, 1979a; 1989; Fiedler and Mouschovias 1993; Ciolek and Mouschovias 1994; Basu and Mouschovias 1994). Hence, the leftover angular momentum and magnetic flux can be calculated from a knowledge of the density at which dynamical contraction sets in. For oblate clouds maintaining force balance along field lines, the central density enhancement at which the central mass-to-flux ratio becomes critical is given by

$$\frac{n_{\text{n,c,crit}}}{n_{\text{n,c0}}} \simeq \left[\frac{\sigma_{\text{n,c,crit}}}{\sigma_{\text{n,c0}}}\right]^2 \equiv \left[\frac{B_{\text{c,crit}}}{B_{\text{c0}}}\right]^2 \left[\frac{1}{\mu_{\text{c0}}}\right]^2 \tag{5}$$

(Mouschovias 1991c, eqs. [9], [10], and § 2.4). The quantity σ_{n} denotes the neutral column density along field lines; the ratio $B_{\text{c,crit}}/B_{\text{c0}}$ is the total enhancement of the central field strength during the subcritical phase of contraction (it is typically $\simeq 1.3 - 1.7$); and μ_{c0} is the initial central mass-to-flux ratio in units of its critical value for collapse. If the mass-to-flux ratio in the envelope of a molecular cloud, not very far from a core, is measured, one may compare it to the critical value given by equation (2) to obtain μ_{c0}, and then use equation (5) with $(B_{\text{c,crit}}/B_{\text{c0}})^2 \simeq 2 - 3$ to obtain an approximate but reasonable value for the factor by which the initial central density will need to increase for critical conditions to be achieved in a core.

Recent, axisymmetric numerical simulations have shown the remarkable result that, even during the dynamical phase of contraction, the infall of a core does not evolve into free fall.

(see review by Mouschovias 1995). Even at a central density as high as 10^{10} cm^{-3}, the *maximum* infall acceleration does not exceed 30% that of gravity, and the magnetic force dominates the thermal-pressure force everywhere in the supercritical core, except in a small region of size $\simeq$ 30 AU, in which the two forces are comparable.

The leftover magnetic flux at the end of these, relatively early, isothermal stages of contraction ($n_n \lesssim 10^{10}$ cm^{-3}) will not by any means survive to main-sequence densities. Pneuman and Mitchell (1965) had shown that Ohmic dissipation is effective during the opaque stages of contraction. Yet, calculations accounting for magnetic effects applicable to the opaque phases have so far been confined to estimates and comparisons of timescales (i.e., τ_{ff}, τ_{AD}, and τ_{ohmic}; e.g., see Nakano and Umebayashi 1986a, b; Nishi, Nakano, and Umebayashi 1991). If past experience can be a guide, dynamical calculations will unveil many surprises. As for the angular momenta of main-sequence stars, although we have shown that magnetic braking can resolve the thorny angular momentum problem of star formation during the isothermal phase of contraction and that it can account for the periods of binary stars from 10 hr to 100 yr as well as for the observed rotation of O5 to F5 main-sequence stars, refinements are still needed before the claim can be made that the stellar rotation-mass relation has been understood.

We first summarize in § 2 the five-fluid (neutral molecules, electrons, ions, charged grains and neutral grains) description of the evolution of isothermal, self-gravitating molecular clouds as it relates to star formation. The emphasis is on the physical origin and interpretation of the equations. The conditions for charged-particle attachment to the magnetic field are elucidated, including those for the new effect of partial, indirect attachment of negatively-charged grains due to electrostatic attraction by well-attached ion "quasiparticles". The two-fluid (plasma, i.e., electrons and ions, plus neutral molecules) approximation is described in § 3 for both nonrotating and rotating model clouds; key results are explained physically in § 4. In § 5 we discuss the results and new effects introduced by the presence of grains. For the isothermal phase of evolution, with which we are concerned in this paper, a four-fluid (neutrals, plasma, charged and neutral grains) description is sufficient. A summary of the results of a detailed, quantitative comparison of an evolutionary model with observations is given is § 5.2, while § 5.3 summarizes the effects of UV ionization on the evolution. Some consequences of the presence of hydromagnetic waves are discussed briefly in § 6. The main conclusions are stated in § 7.

2. Five-Fluid MHD Description of Protostar Formation

The three-fluid equations, including their validity conditions, have been reviewed at some length (Mouschovias 1991b), and the five-fluid equations have been developed in Ciolek and Mouschovias (1993). Here we summarize the results.

2.1. Physical Origin of the Five-Fluid Equations

As explained in § 1, in the density range of interest ($\sim 10^3$ - 10^{10} cm^{-3}), a molecular cloud is a nearly isothermal, weakly ionized physical system consisting of neutral molecules (mainly H_2, with a He abundance 20% by number), electrons, ions (each of charge $+e$), singly negatively charged grains, and neutral grains. Electric (E) and magnetic (B) fields affect the charged particles, and gravitational ($g = -\nabla\psi$) fields affect, at least in principle, all species. Elastic collisions transfer momentum primarily between the dominant neutral molecules and each of the other species. However, (inelastic) electron capture by neutral grains and ion capture (and neutralization) by charged grains represent a momentum transfer between the charged- and neutral-grain fluids whose effect on at least the less abundant neutral grains cannot be ignored. For the low-frequency phenomena of interest here, we may assume charge neutrality, i.e.,

$$n_i = n_e + n_{g_-} , \tag{6a}$$

so as to suppress or average over (high-frequency) plasma phenomena, and we may ignore the displacement current in Ampere's law, which is then written as

$$\nabla \times B = (4\pi/c) \, j \, . \tag{6b}$$

The electric current density is defined by

$$j = e(n_i v_i - n_e v_e - n_{g_-} v_{g_-}) , \tag{6c}$$

where n_s and v_s denote the number density and fluid velocity of species s. The time evolution of the magnetic field is given by Faraday's law of induction,

$$\frac{\partial B}{\partial t} = - c \, (\nabla \times E) , \tag{6d}$$

with B satisfying the condition $\nabla \cdot B = 0$ at all times. In quasineutral plasmas the electric field E is not calculated from Maxwell's equation $\nabla \cdot E = 4\pi\rho_q$, where ρ_q is the net charge density. It is obtained instead from the electron or ion force equations in the form of a generalized Ohm's law (e.g., see review by Mouschovias 1991b, § 2.2).

Keeping only the dominant terms, the five momentum (or force) equations are

$$\frac{\partial}{\partial t}(\rho_n v_n) + \nabla \cdot (\rho_n v_n v_n) = - \nabla P_n - \rho_n \nabla \psi + F_{ni} + F_{ne} + F_{ng_-} + F_{ng_0} , \tag{6e}$$

$$0 = - n_e e \left[E + \frac{v_e}{c} \times B \right] + F_{en} , \tag{6f}$$

$$0 = + n_i e \left[E + \frac{v_i}{c} \times B \right] + F_{in} , \tag{6g}$$

$$0 = - n_{g_-} e \left[E + \frac{v_{g_-}}{c} \times B \right] + F_{g_- n} + F_{g_- g_0,\text{inel}} , \tag{6h}$$

$$0 = + F_{g_0 n} + F_{g_0 g_-,\text{inel}} , \tag{6i}$$

where the frictional force on species s due to collisions with species l is denoted by F_{sl}. All these forces arise because of elastic collisions, except $F_{g_- g_0,\text{inel}}$ (and the equal but opposite force $F_{g_0 g_-,\text{inel}}$), which represents the transfer of momentum from the (negatively) charged-grain fluid, because of ion attachment and neutralization, to the neutral-grain fluid (and vice versa for $F_{g_0 g_-,\text{inel}}$). Expressions for these forces are given below.

The thermal pressure and mass density of the neutrals, P_n and ρ_n respectively, are related through the equation of state

$$P_n = \rho_n C_n^2 , \tag{6j}$$

where $C_n = (k_B T/\mu m_H)^{1/2}$ is the isothermal speed of sound in the neutrals, with k_B being the Boltzmann constant and μ the mean mass per neutral particle in units of the atomic-hydrogen mass, m_H. Because the grains contribute only 1% to the mass of a cloud, while the ions and electrons contribute much less than that, the gravitational potential ψ appearing in equation (6e) is calculated from Poisson's equation using only the density of neutrals as a source term,

$$\nabla^2 \psi = 4\pi G \rho_n . \tag{6k}$$

Moreover, ionizations and recombinations have a negligible effect on ρ_n, which permits us to write the continuity equation

$$\frac{\partial \rho_n}{\partial t} + \nabla \cdot (\rho_n v_n) = 0 . \tag{6l}$$

It is also a reasonable approximation to use a mass continuity equation for the grains (charged + neutral) at least in the case of clouds (the ones of interest here) which have not yet given birth to stars:

$$\frac{\partial (\rho_{g-} + \rho_{g0})}{\partial t} + \nabla \cdot (\rho_{g-} v_{g-} + \rho_{g0} v_{g0}) = 0 . \tag{6m}$$

The system of 13 equations (6a) - (6m) contains 15 unknowns. It will be closed once a prescription is given for calculating the number densities of electrons and ions. (The four elastic-collision forces appearing in eqs. [6e] - [6i] will be expressed in terms of known quantities and dependent variables already appearing in the equations, while elimination of the inelastic forces in eqs. [6h] and [6i] will introduce only the velocity of neutral grains as a new unknown, which has already been counted as part of the set of fifteen. The quantities ρ_{g-} and n_{g-} are, of course related by $\rho_{g-} = n_{g-} m_g$, where m_g is the mass of an individual dust particle; a similar relation exists between ρ_{g0} and n_{g0}.)

In our numerical simulations of the formation and evolution of protostellar cores (from 3×10^3 cm^{-3} to 3×10^9 cm^{-3}) we find that the timescales of the microscopic processes (ionization, charge transfer, and recombination, including grain effects) that determine the abundances of charged species are much smaller than the evolutionary timescales of the cores. We may therefore assume that a chemical equilibrium is established and maintained among the charged particles. In such a case, one may adopt the chemistry designed for static clouds (e.g., see Elmegreen 1979). We consider two kinds of ions, molecular (denoted by the subscript m$_+$), such as HCO$^+$, and metallic or atomic (i.e., elements heavier than carbon; subscript a$_+$), such as Mg$^+$ and Na$^+$. The creation/destruction rate equations for electrons, molecular ions, and atomic ions are, respectively,

$$\zeta_{CR} n_n = n_e (\alpha_{dr} n_{m_+} + \alpha_{rr} n_{a_+} + \alpha_{eg0} n_{g0}) , \tag{7a}$$

$$\zeta_{CR} n_n = n_{m_+} (\alpha_{dr} n_e + \beta n_{a0} + \alpha_{m_+g-} n_{g-}) , \tag{7b}$$

$$\beta n_{a0} n_{m_+} = n_{a_+} (\alpha_{rr} n_e + \alpha_{a_+g-} n_{g-}) . \tag{7c}$$

The quantity ζ_{CR} is the cosmic-ray ionization rate ($\simeq 10^{-17}$ - 10^{-16} s^{-1}, Spitzer and Tomasko 1968; Payne, Salpeter, and Terzian 1984); α_{dr} the dissociative recombination rate ($\simeq 10^{-6}$ cm^3 s^{-1}, Langer 1985; Dalgarno 1987); α_{rr} the radiative recombination rate ($\simeq 10^{-11}$ cm^3 s^{-1}, Oppenheimer and Dalgarno 1974); α_{eg0} the electron capture rate by neutral grains (Spitzer 1948; Hollenbach and Salpeter 1970; Watson and Salpeter 1972); β the charge transfer rate from molecular ions to metallic atoms ($\simeq 0.5 - 8 \times 10^{-9}$ cm^3 s^{-1}, Watson 1976); α_{m_+g-} the capture rate of molecular ions by negatively charged grains; and α_{a_+g-} the atomic-ion capture rate by negative grains. The above cited values of the reaction rates refer to a $T \simeq 10$ K plasma. The electron and ion capture rates by neutral and negative grains, respectively, are given by Spitzer (1941, 1948) and modified by Draine and Sutin (1987) to account for grain polarizability (see Ciolek and Mouschovias 1993, eqs. [57a] - [57c]). UV ionization has been neglected in equations (7a) - (7c) only for simplicity in presentation (see

§ 5.3). Also, in equation (7b), cosmic-ray ionization of the dominant neutral molecule (H_2) does not directly lead to the formation of HCO^+. The relevant chain of reactions is: $H_2 + CR \rightarrow H_2^+ + e$, $H_2^+ + H_2 \rightarrow H_3^+ + H$, $H_3^+ + CO \rightarrow HCO^+ + H_2$, but equation (7b) gives a sufficiently accurate molecular-ion abundance for our present purposes because each cosmic-ray ionization of H_2 leads to the formation of a HCO^+ molecule relatively rapidly.[3] (The detailed work of Ciolek and Mouschovias 1995 includes all the relevant intermediate reactions in the chemical model, which also accounts for UV ionization.) The total ion density appearing in equation (6g) is the sum of the molecular and metallic ion densities,

$$n_i = n_{m_+} + n_{a_+} . \tag{8}$$

Although equations (7a) - (7c) refer to electrons and ions, they imply an equilibrium abundance of charged grains at each time during the evolution. By subtracting equations (7b) and (7c) from (7a), we find that

$$\alpha_{eg_0} n_e n_{g_0} = n_{g_-} (\alpha_{a_+g_-} n_{a_+} + \alpha_{m_+g_-} n_{m_+}) . \tag{9}$$

This equation states that the rate at which charged grains are produced by electron capture onto neutral grains is equal to the rate at which they are neutralized by metallic and molecular ion captures. Typically, 90% of the grains are charged and 10% are neutral.

It remains to specify the frictional forces. The force (per unit volume) on species s due to collisions with neutrals is given by

$$F_{sn} = \frac{\rho_s}{\tau_{sn}} (v_n - v_s), \qquad s = i, e, g_-, g_0 , \tag{10a}$$

where the mean (momentum exchange) collision times, accounting for both s-H_2 and s-He collisions, are

$$\tau_{sn} = \frac{\tau_{sH_2}}{a_{sHe}} = \frac{1}{a_{sHe}} \frac{m_{H_2} + m_s}{\rho_{H_2} \langle \sigma w \rangle_{sH_2}} , \qquad s = i, e, g_-, g_0 , \tag{10b}$$

and the quantity

$$a_{sHe} = 1.14 \quad \text{for } s = i, \tag{10c}$$
$$= 1.16 \quad \text{for } s = e, \tag{10d}$$
$$= 1.28 \quad \text{for } s = g_- \text{ or } g_0 \tag{10e}$$

is the factor by which the presence of He reduces the slowing-down time of species s due to s-H_2 collisions alone. The quantity $\langle \sigma w \rangle_{iH_2}$ is the mean collisional rate between ions of mass m_i and hydrogen molecules of mass m_{H_2}; it is equal to 1.69×10^{-9} cm^3 s^{-1} for HCO^+ - H_2 collisions, and to a similar value for Na^+ - H_2 and Mg^+ - H_2 collisions (McDaniel and Mason 1973). In calculating a_{sHe} in equations (10c) - (10e) we have made use of the following facts. The ratio of the ion-He and ion-H_2 collisional rates is $\langle \sigma w \rangle_{iHe}/\langle \sigma w \rangle_{iH2} = 0.3679$. The measured value of the e-H_2 cross section at low energies is 6.4×10^{-16} cm^2,

---

[3] Dissociative recombination of H_3^+, which in principle would compete with the $H_3^+ + CO$ reaction, is negligible despite the recently measured (see Larsson et al. 1993) relatively large rate coefficient. This is so because in the core the electron abundance is small relative to that of CO, and in the envelope atomic ions, such as C^+ and Mg^+, dominate HCO^+ despite the relatively large electron abundance there.

and the electron speed $(8k_BT/\pi m_e)^{1/2} = 2.0 \times 10^6$ cm s^{-1} at $T = 10$ K; hence, $\langle \sigma w \rangle_{eH2} = 1.3 \times 10^{-9}$ cm^3 s^{-1}.[4] The grain-H$_2$ collisional rate is given by

$$\langle \sigma w \rangle_{gh} = \pi a^2 (8k_BT/\pi m_h)^{1/2}, \qquad h = H_2, He, \tag{10f}$$

for the case $|v_n - v_g| < C_n$, a condition fulfilled in the density range of interest here; this rate is the same for both charged and neutral grains of radius $a \geq 10^{-6}$ cm (the kind of grains we consider). The quantity in parentheses on the right-hand side of equation (10f) is the most probable (Maxwellian) speed of a neutral particle s (H$_2$ or He), and πa^2 is the geometric cross section of a grain.

The collisional forces F_{ns} in equation (6e) are obtained from F_{sn}, specified above, by using Newton's third law, i.e., $F_{ns} = -F_{sn}$, which implies that

$$\tau_{ns} = (\rho_n/\rho_s) \tau_{sn}, \qquad s = i, e, g_-, g_0; \tag{11a}$$

$$= a_{He\text{-}s} \tau_{H_2 s} = a_{He\text{-}s} \frac{m_s + m_{H_2}}{\rho_s \langle \sigma w \rangle_{sH_2}}, \qquad s = i, e, g_-, g_0; \tag{11b}$$

$$\simeq a_{He\text{-}s} \frac{1}{n_\alpha \langle \sigma w \rangle_{\alpha H_2}}, \qquad \alpha = i, g_-, g_0. \tag{11c}$$

The quantity $a_{He\text{-}s}$ is the factor by which the presence of He lengthens the slowing-down time relative to the value it would have if only H$_2$-s collisions were considered. We find

$$a_{He\text{-}s} = 1.23 \quad \text{for } s = i, \tag{11d}$$

$$= 1.21 \quad \text{for } s = e, \tag{11e}$$

$$= 1.09 \quad \text{for } s = g. \tag{11f}$$

The approximate form of τ_{ns} in equation (11c) follows from the fact that $m_g \gg m_i \gg m_{H_2}$. (Note that eq. [11c] is not valid for H$_2$-e collisions because $m_e \ll m_{H2}$.) If He-s collisions are ignored but the contribution of He to the inertia of the H$_2$ fluid is accounted for, then $a_{iHe} = a_{eHe} = a_{gHe} = 1$ and, consequently, $a_{He\text{-}i} = a_{He\text{-}e} = a_{He\text{-}g} = 1.4$.

The inelastic forces appearing in equations (6h) and (6i) are written as

$$F_{g_- g_0, inel} = \frac{\rho_g}{\tau_{g_- i, inel}} (v_{g_0} - v_{g_-}), \qquad F_{g_0 g_-, inel} = \frac{\rho_{g_0}}{\tau_{g_0 e, inel}} (v_{g_-} - v_{g_0}), \tag{12a, b}$$

where the ion- and electron-capture timescales by charged and neutral grains, respectively, are given in terms of the corresponding capture rates, introduced in connection with equations (7a) - (7c), by

$$\tau_{g_- i, inel} = 1/n_i \alpha_{ig_-}, \qquad \tau_{g_0 e, inel} = 1/n_e \alpha_{eg_0}. \tag{12c, d}$$

---

[4]If the Langevin approximation were valid not only for ion-neutral but for electron-neutral collisions as well, one would find that $\langle \sigma w \rangle_{eH2} \simeq (m_{H2}/m_e)^{1/2} \langle \sigma w \rangle_{iH2}$ and that, therefore, $\tau_{ni}/\tau_{ne} \simeq (m_e/m_{H2})^{1/2}(n_e/n_i) = 1.65 \times 10^{-2} (n_e/n_i) \ll 1$; i.e., F_{ne} could be neglected compared to F_{ni}. Actually, the Langevin approximation is not valid for e-H$_2$ collisions (see Mott and Massey 1971). With the values of $\langle \sigma w \rangle_{iH2}$ and $\langle \sigma w \rangle_{eH2}$ given above, we find that $\tau_{ni}/\tau_{ne} \simeq 2.0 \times 10^{-4} (n_e/n_i)$, a value $\simeq 100$ times smaller than that found from the Langevin approximation. Hence, electrons are completely insignificant, relative to the ions, in transmitting the magnetic forces to the neutrals.

2.2. Reduction of the Five-Fluid Equations: Flux-Freezing in the Plasma

The sum of the four drag forces appearing on the right-hand side of equation (6e) is eliminated in favor of the magnetic field by adding equations (6f) - (6i) and using the charge-neutrality equation (6a), Newton's third law, and Ampere's law (eq. [6b]) to find that

$$F_{ni} + F_{ne} + F_{ng} + F_{ng_0} = \frac{1}{4\pi}(\nabla \times B) \times B . \tag{13}$$

Then the force equation (6e) for the neutrals becomes

$$\frac{\partial}{\partial t}(\rho_n v_n) + \nabla \cdot (\rho_n v_n v_n) = -C^2 \nabla \rho_n - \rho_n \nabla \psi + \frac{1}{4\pi}(\nabla \times B) \times B , \tag{14}$$

where the equation of state (6j) has also been used to eliminate P_n in favor of ρ_n.

A further simplification occurs by noting that the third term, F_{en}, in the force equation (6f) for the electrons is negligible compared to the magnetic force (the second term) for the physical conditions characterizing the isothermal stage of evolution. The ratio of the magnitudes of the two terms can be written as

$$\frac{c |F_{en}|}{n_e e |v_e \times B|} = \frac{1}{\omega_{c,e} \tau_{en}} \frac{|v_e - v_n|}{|v_e|} , \tag{15a}$$

where $\omega_{c,e} = eB/m_e c$ is the electron cyclotron frequency. We evaluate $\omega_{c,e} \tau_{en}$ to find that

$$\omega_{c,e} \tau_{en} = 1.4 \times 10^8 \left[\frac{B}{30 \ \mu G}\right] \left[\frac{3 \times 10^3 \ cm^{-3}}{n_n}\right] \left[\frac{10 \ K}{T}\right]^{1/2} . \tag{15b}$$

The magnetic field and gas density on the right-hand side have been normalized to values typical of initial conditions at the cloud center, before a protostellar core forms. Our numerical simulations show that, during the quasistatic phase of contraction, B in the core increases by less than a factor of 2, while n_n increases by a factor 10^1 - 10^2 and $|v_e - v_n|/|v_e| \simeq 20$ (because the electrons are essentially held in place by magnetic forces as the neutrals contract through them). So, the ratio of forces in equation (15a) is negligible. By the end of a typical run, the neutral density and magnetic field strength in the core increase by a factor of 10^6 and 10^2, respectively, while $|v_e - v_n|/|v_e| \simeq 1/3$ (i.e., the electrons contract at 2/3 the infall speed of the neutrals). Hence, $|F_{en}|$ is still nearly 10^4 times smaller than that required to detach the electrons from the magnetic field. For the ions, we have that

$$\omega_{c,i} \tau_{in} = 3.2 \times 10^4 \left[\frac{B}{30 \ \mu G}\right] \left[\frac{3 \times 10^3 \ cm^{-3}}{n_n}\right] . \tag{16}$$

A similar argument shows, therefore, that the neutral drag on the ions falls short, by a factor > 10^3 initially, of the magnitude needed to dislodge the ions from the field lines, but that by a central density of 3×10^9 cm^{-3} the ions are on the verge of becoming detached from the field. We may thus obtain the electric field from equations (6f) and (6g),

$$E = -\frac{v_e}{c} \times B = -\frac{v_i}{c} \times B , \tag{17a, b}$$

which is then used to express Faraday's law (6d) as

$$\frac{\partial B}{\partial t} = \nabla \times (v_e \times B) = \nabla \times (v_i \times B) , \tag{18a, b}$$

with equations (17b) and (18b) being valid only for densities $< 10^{10}$ cm^{-3}. In summary, *the magnetic flux is frozen in the plasma during the isothermal phase of star formation.* The electron and ion fluids may thus be treated as a single fluid, the plasma.

2.3. Grain Motion and Attachment to the Magnetic Field

The physics of charged-grain coupling to the magnetic field can be most easily understood if we ignore for the moment the presence of neutral grains. The *direct attachment parameter* of grains to the field, defined by $\Gamma_{dir} = \omega_{c,g} \tau_{gn}$, is evaluated to find that

$$\Gamma_{dir} \equiv \omega_{c,g} \tau_{gn} = 4.4 \times 10^2 \left[\frac{B}{30 \ \mu G} \right] \left[\frac{3 \times 10^3 \ \text{cm}^{-3}}{n_n} \right] \left[\frac{10 \ \text{K}}{T} \right]^{1/2} \left[\frac{10^{-6} \ \text{cm}}{a} \right]^2 , \quad (19)$$

with the typical grain radius used in our numerical simulations being $a = 3.75 \times 10^{-6}$ cm. In a typical model cloud, the subcritical phase of core formation and contraction ends by a central density enhancement $\lesssim 10^2$ and a corresponding increase of the central field strength by a factor < 2. Hence, $\Gamma_{dir} \simeq 1$ at the end of this phase of evolution, and the grains are expected to decouple from the field. This, however, is only part of the story. As the (negatively charged) grains begin to fall in (because of drag by the contracting neutrals) faster than the ions, which are still very well attached to the magnetic field, electrostatic attraction by the ions tends to keep the grains at least partially attached to the field even at densities at which $\Gamma_{dir} \ll 1$. This is an *indirect attachment* mechanism. Because, as we have seen, the electrons are extremely well attached to the field at this stage, they tend to shield the ionic charge, thus reducing the grain-ion electrostatic attraction. (Recall that $n_e = n_i - n_{g^-}$, so that the shielding is never complete.) The electron-shielded ions, to which we refer as "*quasiparticles*", thereby behave as if they carried a fractional electronic charge equal to $[(n_i - n_e)/n_i]e$. Partial coupling occurs if the *indirect attachment parameter*

$$\Gamma_{indir} \equiv [(n_i - n_e)/n_i] \omega_{c,i} \tau_{in} = (n_{g^-}/n_i) \omega_{c,i} \tau_{in} \equiv \omega_{c,q} \tau_{in} , \quad (20)$$

is $\gtrsim 1$. The quantity $\omega_{c,q}$ is the cyclotron frequency of the ion quasiparticles.

The above physical argument is put on a rigorous mathematical footing as follows. For simplicity, we consider a cylindrically symmetric model cloud with the magnetic field lines parallel to the $(z-)$ axis of symmetry. The force equation (6h) for the grains without the last term can be solved for the radial component of the grain velocity (denoted simply by v_{g^-}), by using equation (17b) to eliminate the electric field, equation (10a) and Newton's third law, and the facts that $\omega_{c,g} \tau_{gn} \ll \omega_{c,i} \tau_{in} \ll \omega_{c,e} \tau_{en}$ and $\tau_{ni}/\tau_{ne} \ll 1$:

$$v_{g^-} = \frac{\Theta}{\Theta + 1} v_i + \frac{1}{\Theta + 1} v_n , \quad (21)$$

where the function Θ is given by

$$\Theta = (1 + \tau_{ni}/\tau_{ng^-}) (\omega_{c,g^-} \tau_{g^- n})^2 , \quad (22a)$$

$$= (\omega_{c,g^-} \tau_{g^- n})^2 + (n_{g^-}/n_i) (\omega_{c,i} \tau_{in}) (\omega_{c,g^-} \tau_{g^- n}) . \quad (22b)$$

It is clear from equation (21) that, if $\Theta \gg 1$, then $v_{g^-} = v_i$, i.e., the grains are well attached to the magnetic field. If $\Theta \ll 1$, then $v_{g^-} = v_n$, and the grains fall in with the neutrals, i.e., they are decoupled from the field. *Partial attachment to the field can still take place if*

$$\Theta \equiv (\omega_{c,g^-} \tau_{g^- n})^2 + (n_{g^-}/n_i) (\omega_{c,i} \tau_{in}) (\omega_{c,g^-} \tau_{g^- n}) \gtrsim 1 . \quad (23)$$

For example, if $\Theta = 1$, $v_{g-} = (v_i + v_n)/2$, the arithmetic mean of the ion and neutral infall speeds. Condition (23) can be satisfied even if $\omega_{c,g-}\tau_{g-n} \lesssim 1$, in which case the grains would not be *directly* attached to the field, provided only that $(n_{g-}/n_i)(\omega_{c,i}\tau_{in})(\omega_{c,g-}\tau_{g-n}) \gtrsim 1$. This means that partial attachment can be achieved *indirectly*, through grain-quasiparticle attraction, even if $n_{g-}/n_i \ll 1$ as long as $\Gamma_{indir} \equiv (n_{g-}/n_i)(\omega_{c,i}\tau_{in}) > 1$. We find in our numerical simulations in axisymmetric geometry that the last condition is satisfied (because $\omega_{c,i}\tau_{in} \gg 1$) even at densities high enough that the direct attachment parameter (see eq. [19]) is $\Gamma_{dir} \ll 1$ (see § 5.1, Fig. 7e).

Neutral grains behave as if they are attached to the field if the electron-capture timescale $\tau_{g_0 e,inel}$ is much smaller than $\tau_{g_0 n}$, the momentum-transfer timescale to the neutral grains by collisions with neutral molecules. If the opposite is true, i.e., if $\tau_{g_0 e,inel} \gg \tau_{g_0 n}$, the neutral grains fall in with the neutrals. Equation (6i) can be solved for v_{g_0} in terms of v_g and v_n to obtain an equation analogous to (21), and thus quantify the degree to which neutral grains are attached to the field (see Ciolek and Mouschovias 1993, eq. [44]).

During the opaque phases of star formation ($n_n \gtrsim 10^{10}$ cm^{-3}), not reviewed here, the ions will not be directly attached to the magnetic field. However, we predict that they will be at least partially attached indirectly, because of electrostatic attraction by well-attached electrons (see eq. [15b]). The shielding of the ionic charge by negatively-charged grains will tend to shortcircuit the electron-ion electric field, but this shielding will not be complete unless n_e becomes insignificant ($\ll 10^{-3}n_i$) because of electron attachment onto grains. The indirect ion attachment will be weakened when ohmic dissipation reduces the field strength (and, therefore, $\omega_{c,e}$) significantly so that $(n_e/n_i)\omega_{c,e}\tau_{en} < 1$.

3. Two-Fluid Description of Protostar Formation

3.1. Basic Equations

The modern theory of star formation outlined in § 1 is mainly based on analytical and numerical results obtained over the last twenty years on the basis of a two-fluid (neutrals and electron-ion plasma) MHD description of the process of ambipolar-diffusion-induced fragmentation and star formation. With the macroscopic effects of grains ignored, but their role in the microscopic processes that determine the degree of ionization retained, the five-fluid equations reduce to

$$\frac{\partial \rho_n}{\partial t} + \nabla \cdot (\rho_n v_n) = 0 , \qquad (24a)$$

$$\frac{\partial}{\partial t}(\rho_n v_n) + \nabla \cdot (\rho_n v_n v_n) = -C_n^2 \nabla \rho_n - \rho_n \nabla \psi + \frac{1}{4\pi}(\nabla \times B) \times B , \qquad (24b)$$

$$v_i = v_n + \frac{\tau_{ni}}{\rho_n} \frac{1}{4\pi}(\nabla \times B) \times B , \qquad (24c)$$

$$\tau_{ni} = a_{He-i} \frac{m_i + m_{H_2}}{\rho_i \langle \sigma w \rangle_{iH_2}} , \qquad (24d)$$

$$\frac{\partial B}{\partial t} = \nabla \times (v_i \times B) , \qquad (24e)$$

$$\nabla^2 \psi = 4\pi G \rho_n . \qquad (24f)$$

As shown in § 2.1, $a_{\text{He-i}}$ is equal to 1.23 if He-ion collisions are accounted for, and to 1.4 if they are ignored but the contribution of the He inertia to that of H_2 is included. These are six equations for seven unknowns (ρ_n, v_n, ρ_i, v_i, τ_{ni}, B, and ψ). Equations (7a) - (7c) can be used to determine the ion density and thus close the system. In the past, calculations of ionization equilibrium in *static* clouds (Elmegreen 1979; Nakano 1979) have been used to parametrize the ion density in terms of the neutral density as

$$\rho_i = m_i K \left[\frac{\rho_n / m_n}{10^5 \text{ cm}^{-3}} \right]^k , \tag{25}$$

in the density range $10^3 \leq n_n \leq 10^9$ cm^{-3}. The constant $K \simeq 3 \times 10^{-3}$ cm^{-3} and the exponent $k \simeq 0.5$. Although these "canonical" values of K and k are routinely adopted (e.g., Lizano and Shu 1989; Tomisaka, Ikeuchi, and Nakamura 1990), it should be borne in mind that K can easily differ from this value by a factor of 10 (or more) due to uncertainties in the rate of cosmic-ray ionization and metal depletion onto grains, and k is uncertain by about 20% - 30%. In fact, our recent determination of the charged-species abundances in a fashion consistent with the time evolution of model clouds has shown that no constant k can accurately represent the ion density in the above range of n_n (see § 5). We may nevertheless use equation (25) in a first approach to the problem, to avoid adopting a detailed chemical model, but K and k should then be regarded as free parameters, the former allowed to vary by about a factor of 10 and the latter restricted in the range 0.6 - 0.2. Fiedler and Mouschovias (1992, 1993) and Basu and Mouschovias (1994, 1995a, b) have taken this approach in studying the ambipolar-diffusion-induced protostar formation in axisymmetric geometry in the absence and presence of rotation, respectively, while Ciolek and Mouschovias (1993, 1994, 1995), in their five-fluid simulations, have calculated the ion density through detailed chemical models in a fashion consistent with the time evolution.

The continuity equation (24a) can be used in the left-hand side of the force equation (24b) to write it in the more familiar form $\rho_n (dv_n / dt)$, where $d/dt = \partial / \partial t + v_n \cdot \nabla$ is the (Lagrangian) time derivative comoving with the neutrals.

3.2. Ambipolar Diffusion: General Considerations, and Timescales

With the ion-neutral drift velocity denoted by v_D ($\equiv v_i - v_n$), equation (24c) implies that

$$v_D \simeq \frac{v_A^2 \tau_{ni}}{L} , \tag{26a}$$

$$\simeq 0.1 \left[\frac{v_A}{1 \text{ km s}^{-1}} \right]^2 \left[\frac{\tau_{ni}}{10^4 \text{ yr}} \right] \left[\frac{0.1 \text{ pc}}{L} \right] \quad \text{km s}^{-1} , \tag{26b}$$

where $|\nabla| \sim 1/L$, $v_A = B/(4\pi \rho_n)^{1/2}$ is the Alfvén speed in the neutrals, and in equation (26b) all quantities have been normalized to values typical of the initial stages of core formation in molecular clouds. Equation (26a) suggests a natural ambipolar-diffusion timescale $L/v_D \simeq \tau_A^2 / \tau_{ni}$, so that

$$\tau_{AD} \simeq \frac{\tau_A^2}{\tau_{ni}} , \tag{26c}$$

$$\simeq 10^6 \left[\frac{1 \text{ km s}^{-1}}{v_A} \right]^2 \left[\frac{10^4 \text{ yr}}{\tau_{ni}} \right] \left[\frac{L}{0.1 \text{ pc}} \right]^2 \quad \text{yr} , \tag{26d}$$

where $\tau_A \equiv L/v_A$ is the Alfvén crossing time. The time evolution (due to ion-neutral drift) of the magnetic field threading a predominantly neutral fluid element can be obtained from the plasma flux-freezing equation (24e) by eliminating v_i in favor of v_D and v_n (from the definition of the drift velocity), and using equation (24c) to find

$$\frac{\partial B}{\partial t} - \nabla \times (v_n \times B) = \nabla \times (v_D \times B) , \tag{27a}$$

$$= \nabla \times \left\{ \frac{\tau_{ni}}{\rho_n} \left[\frac{1}{4\pi}(\nabla \times B) \times B \right] \times B \right\}, \tag{27b}$$

$$= \nabla \times \left\{ \frac{\tau_{ni}}{\rho_n} \left[- \nabla \left(\frac{B^2}{8\pi} \right) + \frac{1}{4\pi}(B \cdot \nabla)B \right] \times B \right\}. \tag{27c}$$

This is a nonlinear diffusion equation, with a diffusion coefficient $\mathscr{D} \simeq v_A^2 \tau_{ni}$. Hence, the diffusion timescale is $\tau_{diff} \simeq L^2/\mathscr{D} \simeq \tau_A^2/\tau_{ni}$, which is identical with τ_{AD} (see eq. [26c]).

The above considerations and estimate of τ_{AD} do not account for self-gravity. In order to do so, we consider the early, subcritical phase of ambipolar-diffusion-induced core formation and evolution, during which the acceleration of the neutrals is negligible (i.e., the contraction is quasistatic) and the magnetic forces dominate the thermal-pressure forces in opposing gravity. (During dynamical contraction, magnetic flux tends to get trapped inside the collapsing core; see eq. [31] below, and Fiedler and Mouschovias 1993, Fig. 1b.) Combining the neutral-particle and plasma force equations ([24b] and [24c], respectively), we find that the drift velocity can be expressed as

$$v_D = - \tau_{ni} g_\perp , \tag{28}$$

where $g \equiv - \nabla\psi$, and the subscript $\perp$ denotes a direction perpendicular to the field lines. In the equatorial plane of the forming core, we have that $|g_\perp| \propto G\rho_n r \propto r/\tau_{ff}^2$. Substituting this in equation (28) and using the resulting expression for v_D in the definition of the ambipolar-diffusion timescale in the core, $\tau_{AD} \equiv r/|v_D|$, we find that

$$\tau_{AD} = const \times \frac{\tau_{ff}^2}{\tau_{ni}}, \tag{29a}$$

$$\propto n_i/n_n , \tag{29b}$$

where the definition of τ_{ni} (eq. [11b]) has been used to obtain (29b) from (29a). (This is the origin of eq. [1].) The *const* on the right-hand side of equation (29a) is near 1, independent of geometry (Mouschovias 1987b), and *the expression (29a) for τ_{AD} is seen to be a general property of the equations describing ambipolar diffusion in quasistatically contracting, magnetically supported cores.*

This equation can be used to define a parameter ν_{ff} by

$$\nu_{ff} \equiv \tau_{ff}/\tau_{ni} \simeq \tau_{AD}/\tau_{ff} , \tag{30}$$

which is clearly the factor by which ambipolar diffusion retards the contraction relative to free fall. We therefore refer to it as the *collapse retardation factor.*

After dynamical contraction begins and equation (29a) for τ_{AD} is no longer valid, one may still calculate τ_{AD} analytically from first principles (see Mouschovias 1989, § IIIb). The resulting expression, accounting for the He inertia but ignoring He-ion collisions, is

$$\tau_{AD} = 0.4 \frac{\tau_{ff}^2}{\tau_{ni}} \left[\frac{\Phi_{B,crit}}{\Phi_B} \right]^2 , \qquad \text{for} \quad \Phi_B \le \Phi_{B,crit} ; \tag{31a}$$

$$\simeq 1 \times 10^3 \left[\frac{n_i/n_{H_2}}{10^{-10}} \right] \left[\frac{\Phi_{B,crit}}{\Phi_B} \right]^2 \qquad \text{yr}, \tag{31b}$$

where Φ_B is the actual flux of the core at any stage past the onset of dynamical contraction, and $\Phi_{B,crit}$ is its total flux at the onset of dynamical contraction and is uniquely determined by the mass of the core as $\Phi_{B,crit} = (63G)^{1/2} M$. The normalization of n_i/n_{H_2} in equation (31b) refers to a neutral density $\sim 10^9$ cm^{-3}, above which n_i no longer increases significantly with n_n (Spitzer 1963; Elmegreen 1979; Nakano 1979). Canonical values of K and k have been assumed (see eq. [25]). The free-fall time for a spherical core is

$$\tau_{ff} = (3\pi/32G\rho)^{1/2} . \tag{32}$$

The presence of the factor Φ_B^{-2} on the right-hand side of equation (31) shows clearly the tendency of magnetic flux to get trapped inside a collapsing core, unless the degree of ionization were to decrease very rapidly.

In the presence of grains, if they are well attached to the magnetic field, one can easily show that τ_{ni} in equations (30) and (31a) is replaced by the mean collision time τ_{coll} of neutrals with both ions and charged grains, where $(\tau_{coll})^{-1} = (\tau_{ni})^{-1} + (\tau_{ng^-})^{-1}$. (Physically, collisions of neutrals with ions and grains are processes occurring in parallel, hence the combined timescale is obtained in the same fashion that the net resistance is found in an electrical circuit consisting of a parallel combination of resistances.) A general expression for ν_{ff} is given by Ciolek and Mouschovias (1993) accounting for partial attachment of the charged grains to the field and for the presence of neutral grains. The typical value of ν_{ff} in molecular clouds is near 10, with the range $\simeq 3 - 30$ permissible by observations (see review by Mouschovias 1995).

3.3. Magnetic Braking

The five-fluid equations were presented in §§ 2.1 and 2.2 in vectorial form, and so were the two-fluid equations in § 3.1. Specifically, no component of the velocities or of the magnetic field was assumed to vanish. Rotating model clouds and magnetic braking may therefore be investigated using these equations.

Magnetic braking of quasistatically contracting clouds (or fragments) was studied analytically for both aligned and perpendicular rotators (Mouschovias and Paleologou 1979, 1980), and for magnetically linked, aligned rotators (Mouschovias and Morton 1985a, b). The propagation of the torsional Alfvén waves, including their partial reflection and transmission at cloud surfaces, was followed and interpreted physically using exact analogies with transverse waves on a string of varying density. For typical cloud (or fragment) parameters, magnetic braking was found to be very effective and to wipe out differential rotation on a timescale much shorter than even the free-fall timescale, which is a strict lower limit on the contraction timescale. Hence, if one is interested in the loss of angular momentum by a cloud as a whole or by a protostellar fragment within a cloud, one may ignore (the important for other purposes) transient phenomena involving torsional Alfvén waves and treat the cloud or fragment as a rigid rotator. This approximation, however, would break down during the dynamical phase of contraction of a protostellar core because of the rapid and nonhomologous nature of the contraction. Angular momentum would tend to get trapped inside, and differential rotation would be unavoidable. A simplification, nevertheless, can be introduced even in this case. Two-fluid calculations ignoring rotation have shown that, because of flattening along field lines, balance of forces is rapidly established and maintained *along* field lines even during the dynamical-contraction phase perpendicular to the field lines. Under these conditions magnetic braking can establish isorotation along each magnetic flux tube, while differential rotation can be generated and maintained perpendicular to the field lines. Basu and Mouschovias (1994, 1995a, b) followed this approach to study the combined effects of ambipolar diffusion and magnetic braking on the formation and evolution of protostellar fragments (or cores) in model molecular clouds, up to $n_n \simeq 10^{10}$ cm^{-3}. The dependence of the solution on the free parameters of the problem was also investigated. Some of the results are summarized in § 4.2.

Mouschovias and Paleologou (1986) showed that ambipolar diffusion during the quasistatic phase of contraction lengthens the magnetic-braking timescale by only a few percent. Hence, it is the onset of dynamical contraction, rather than ambipolar diffusion, that brings an end to redistribution of angular momentum by magnetic braking and leads to angular-momentum trapping in a collapsing core (see also Mouschovias 1979a). By the time dynamical contraction sets in ($n_n \gtrsim 10^5$ cm^{-3}), magnetic braking has removed enough angular momentum for binary stars and even solar systems to become dynamically possible; i.e., they can form without interference from centrifugal forces (see § 4.2).

4. Results of Two-Fluid Description of Protostar Formation

We have studied the self-initiated (due to ambipolar diffusion) formation and contraction of protostellar cores in self-gravitating, isothermal, primarily magnetically supported model molecular clouds, by developing and using fully implicit, multifluid, second-order accurate, adaptive-grid codes. A recent review of the main results is found in Mouschovias (1995). The formulation of the problems, the mathematical and numerical techniques, and detailed descriptions of the results obtained from the two-fluid approximation are given in Fiedler and Mouschovias (1992, 1993) for nonrotating clouds, and in Basu and Mouschovias (1994, 1995a, b) for rotating models.

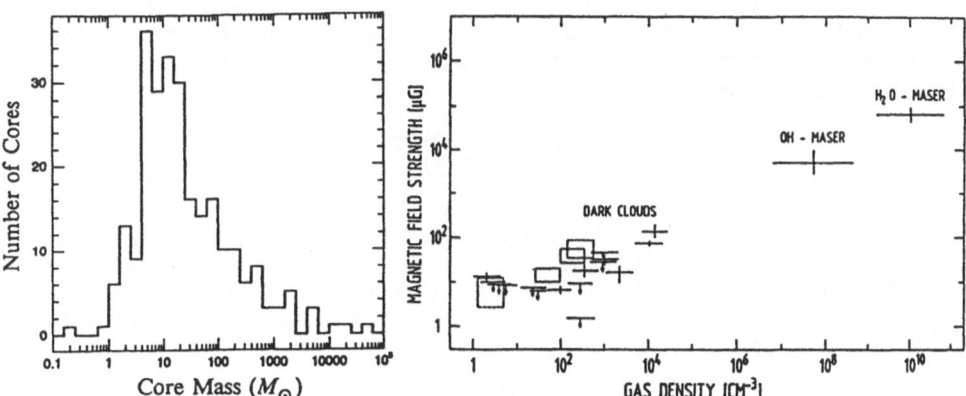

Figure 1. (**a**, *left*) *Number of Cores as a function of their mass, in* $M_{\odot}$ (from Wood *et al.* 1994). (**b**, *right*) *Observed magnetic field strength as a function of gas density* (courtesy of D. Fiebig).

A theory of protostar formation must ultimately explain the observed mass spectrum of protostellar cores (see Fig. 1a) and the observed relation between the magnetic field strength and gas density (see Fig. 1b). It is clear from Figure 1a that the most common core mass is almost 10 $M_\odot$. While it is hardly legitimate to refer to objects with masses greater than about 10^2 $M_\odot$ as cores (they are clouds in their own right), this should not obscure the fact that structure exists in clouds and cloud complexes ranging in mass from about 1 $M_\odot$ to more than 10^3 $M_\odot$. The low-density low-field points in Figure 1b refer to HI Zeeman measurements, as do the boxes, each of which contains several measurements (see Troland and Heiles 1986, Fig. 1). The points labeled as "dark clouds" represent several OH Zeeman measurements. The "error bars" shown for OH and H$_2$O masers are meant to reflect the range of observed densities and fields, not the errors in any one measurement. One may summarize Figure 1b as showing very little, if any, enhancement of the field strength at densities below 10^4 cm^{-3}, and a relation $B \propto n^{1/2}$ above this density. The calculations about to be summarized explain Figure 1b and give as typical protostellar core masses somewhat less than 10 $M_\odot$, which is consistent with the peak in Figure 1a.

4.1. Ambipolar Diffusion in Nonrotating Models

Choosing as units of density, velocity, time, and magnetic field the initial central values of the neutral density $\rho_{n,c0}$, Alfvén speed in the neutrals $v_{A,c0}$, neutral-ion collision time $\tau_{ni,c0}$, and magnetic field strength B_{c0}, it is easily shown that the two-fluid MHD equations (24a) - (25) contain three dimensionless free parameters:

$$\nu_{ff,c0} \equiv \frac{\tau_{ff,c0}}{\pi^{1/2}\tau_{ni,c0}}, \qquad \alpha_{c0} \equiv \frac{B_{c0}^2}{8\pi\rho_{n,c0}C^2}, \qquad k \equiv \frac{d\ln n_i}{d\ln n_n}. \qquad (33a, b, c)$$

The quantity $\nu_{ff,c0}$ is essentially the initial central collapse retardation factor defined in equation (30). The second parameter α_{c0} is the initial value of the ratio of magnetic and thermal pressures at the center; and k is the exponent in the parametrization of the ion density in terms of the neutral density, $n_i \propto n_n^k$ (see eq. [25]). The boundary conditions introduce two free parameters, to which the solution is very insensitive, provided that the model clouds are sufficiently thermally supercritical (as suggested by observations). The initial conditions introduce the central mass-to-flux ratio relative to its critical value for collapse as a fourth free parameter, i.e.,

$$\mu_{c0} \equiv \frac{(dm/d\Phi_B)_{c0}}{(dm/d\Phi_B)_{c,crit}}, \qquad (33d)$$

where the critical value $(dm/d\Phi_B)_{c,crit}$ was calculated from a sequence of exact equilibrium states, and is given by equation (2). [For a uniform (hence, nonequilibrium) thin disk of infinite radius, Nakano and Nakamura (1978) find a smaller critical value by the factor 1.19.] If the initial states are taken to be exact magnetohydrostatic (MHS) equilibrium states with the magnetic field frozen in the matter (i.e., in the absence of ambipolar diffusion), α_{c0} and μ_{c0} are related by

$$(\alpha_{c0}\mu_{c0}^2)_{equil} = const \simeq 1 \qquad (33e)$$

(Mouschovias 1991c). Hence, there are only three significant free parameters in the problem. Observations suggest the following set of typical values: $\nu_{ff,c0} = 8.3$, $\alpha_{c0} = 26$, $\mu_{c0} = 1/4$, and $k = 1/2$. Observations and theoretical considerations limit the range of values to be investigated to $1 \leq \nu_{ff,c0} \leq 10$, $1 < \alpha_{c0} < 100$, $0.1 \leq \mu_{c0} \leq 1.0$, and $1/3 \leq k \leq 1/2$.

In this investigation, model clouds are initially in uniform states, spherical or cylindrical in shape, threaded by a uniform magnetic field aligned with the (z-)axis of symmetry of a cylindrical polar coordinate system (r, ϕ, z) --equilibrium initial states are considered in §§ 4.2 - 5.3. We study first the evolution of a typical model cloud, having $\nu_{ff,c0} = 8.21$, $\mu_{c0} = 0.246$, $\alpha_{c0} = 86.5$, and $k = 1/2$. These values could, for example, represent a region $Z = R = 0.75$ pc in a molecular cloud, of neutral density $n_{n,0} = 300$ cm^{-3}, ion density 5.36×10^{-3} cm^{-3}, temperature $T = 10$ K, mass 45.5 $M_\odot$, threaded by a magnetic field $B_0 = 30$ μG.

Figure 2a exhibits the central density $n_c \equiv n_{n,c}$ (in cm^{-3}) as a function of time (in Myr). *The evolution occurs in three distinct phases.* (1) *The cloud relaxes along field lines to a quasi-equilibrium state*, and the central density increases by a factor $\simeq 8$ in about 6 Myr. (If ambipolar diffusion were not present, an equilibrium state would be reached having a central density greater than its initial value by a factor of 3.4, instead of 8.) (2) *Quasistatic, ambipolar-diffusion-controlled contraction follows.* It takes place on the central flux-loss timescale,

$$\tau_{\Phi,c0} = \left[\frac{\Phi_B}{|d\Phi_B/dt|} \right]_{c0} = \frac{1}{2}\tau_{AD,c0} , \qquad (34a, b)$$

of the quasi-equilibrium state, which is $\simeq 9$ Myr. This phase lasts until $t \simeq 15$ Myr, at which time the central density has increased to about 3.3×10^4 cm^{-3} and the central mass-to-flux ratio has reached the critical value for collapse. Soon thereafter, (3) *dynamic contraction in the radial direction sets in.* Even at the end of the run, however, at a central density greater than 3×10^8 cm^{-3}, the central acceleration is only 30% of the local gravitational field; i.e., the dynamic contraction is by no means a free fall (see Fig. 2b). At this time, the r-components of the central magnetic and thermal-pressure forces are 40% and 30% of the r-component of the gravitational force, respectively. *Along field lines*, once balance of thermal-pressure and gravitational forces is established at the end of the relaxation phase, it is maintained even during the lateral dynamic-contraction phase (see Fig. 2c).

Isodensity contours, field lines, and velocity vectors are displayed in Figure 3a at time $t = 16.135$ Myr, near the end of the run, when the central density is $n_c = 2.9 \times 10^8$ cm^{-3}. The equatorial radius of the supercritical core is $R_{core} = 0.14R = 0.10$ pc, and has been at this value for the last four orders of magnitude enhancement of the central density. During this time, its mass has been $M_{core} \simeq 5.0$ $M_\odot$. Magnetic forces provide very effective support of the envelope beyond $r \gtrsim 0.3R$. *The isodensity contours reveal the formation of a protostellar disk.* The field lines have deformed appreciably only in the innermost $\simeq 15\%$ of the cloud. The innermost 3% is shown in Figure 3b; to better represent the magnetic field, six field lines are displayed for each one shown in Figure 3a. The velocity vectors are normalized to the magnitude of the maximum infall velocity $|v_n|_{max} = 0.49$ km s^{-1}, which occurs on the axis of symmetry at $z \simeq 3 \times 10^{-3}R \simeq 2.3 \times 10^{-3}$ pc ($\simeq 464$ AU). A shock is beginning to form in this region. (The shock evident in Fig. 3a at $z \simeq 0.3R$, near the z-axis, is a result of the early relaxation along field lines; it is not present if equilibrium states are used as initial states.) The mass inside each flux tube visible in Figure 3b (found by integrating from $r = 0$ to the field line and from $-Z$ to $+Z$) is 0.0525, 0.199, 0.419, 0.695, 1.02, and 1.38 $M_\odot$. The corresponding initial masses were 5.61×10^{-3}, 0.0225, 0.0505, 0.0899, 0.140, and 0.202 $M_\odot$. More than 77% of the mass in each of these flux tubes lies inside the isodensity contour labeled "10,000".

Figure 4a shows B_c as a function of n_c for three models (one being that of Figs. 2 - 3) differing only in the initial value of the dimensionless mass-to-flux ratio μ_{c0}, whose values label the three curves. The more subcritical a cloud is, the greater the factor ($\simeq \mu_{c0}^2$; see eq. [5]) by which the central density increases (with hardly any increase in the field strength) before critical conditions are reached and the core begins to contract dynamically. Only then

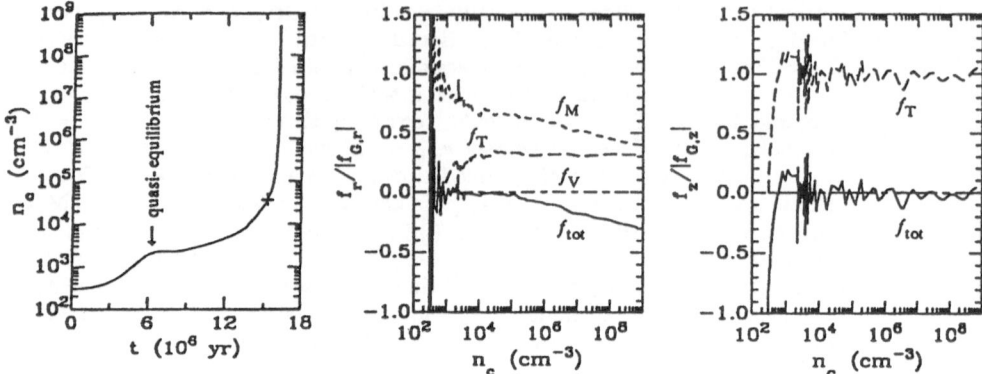

Figure 2. *Evolution of central values.* A + marks the point at which the central mass-to-flux ratio becomes critical. (**a**, *left*) Neutral density as a function of time. (**b**, *middle*) Radial components of forces (per unit volume), normalized to the magnitude of the r-component of the gravitational force $f_{G,r}$, as functions of density. Total force f_{tot}, magnetic force f_M, thermal-pressure force f_T, and artificial-viscosity force f_V. (**c**, *right*) z-components of f_{tot}, f_T, and the other two forces, which are negligible, normalized to the magnitude of the z-component of the gravitational force $f_{G,z}$, as functions of density.

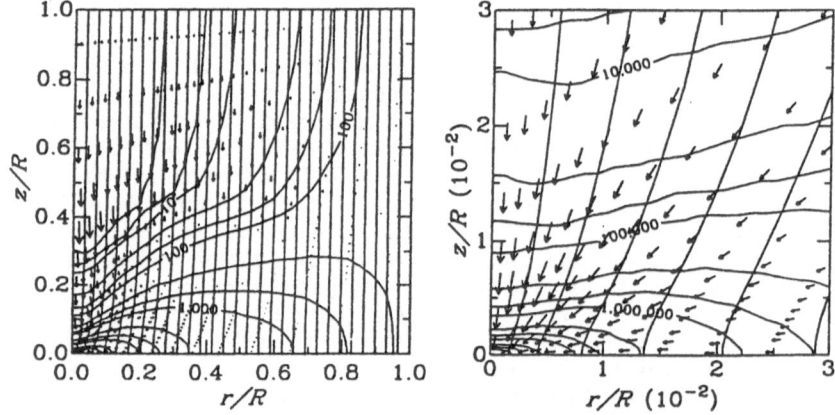

Figure 3. *Isodensity contours, field lines, and velocity vectors*, normalized to the magnitude of the maximum infall velocity (= 0.49 km s⁻¹), near the end of the run (when t = 16.135 Myr, and n_c = 2.9 × 10^8 cm⁻³). The two unlabeled contour levels between labeled ones have density values 3 and 6 times greater than the previous label. The r and z coordinates are normalized to the cloud radius R. (**a**, *left*) The whole cloud. (**b**, *right*) The innermost 3% of the cloud.

does B_c increase (as $n_c{}^\kappa$, with κ = 0.47, a value near the flux–freezing limit of 1/2). Once a supercritical core forms, however, memory of the initial μ_{c0} is lost, and the value of the field is typically in the milligauss range for densities 10^{8-9} cm^{-3}. Herein may lie the explanation for the magnetic-field observations shown in Figure 1b; i.e., for (1) the near constancy of the observed field strength below densities $\simeq 10^4$ cm^{-3}, and (2) the observed B $\simeq$ a few mG - several $\times$ 10 mG in OH and H$_2$O masers, which typically have densities ~ 10^8 cm^{-3} and > 10^9 cm^{-3}, respectively. (This is not to imply, however, that OH and H$_2$O masers are protostars. Simply, the physical conditions of density and associated field strength predicted by our calculations require only a pump for maser action consistent with observations to take place.)

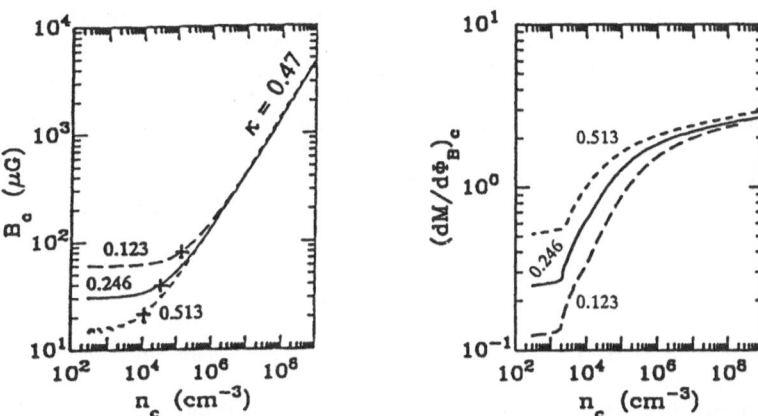

Figure 4. (a, *left*) *Dependence of the B_c-n_c relation and* (b, *right*) *the central mass-to-flux ratio on the free parameter μ_{c0}, whose values label the three curves.*

The central mass-to-flux ratio, normalized to the critical value for collapse, is shown in Figure 4b as a function of central density for the three models of Figure 3a. It increases by a factor $\simeq \mu_{c0}^{-1}$) during the subcritical phase of evolution, and by an additional factor of about 2.6, almost independent of its initial value, during the dynamic phase. The factor by which the mass-to-flux ratio increases during the dynamic phase depends on the values of the other two free parameters, $\nu_{ff,c0}$ and k (see Fiedler and Mouschovias 1993, Figs. 5b and 6b). The smaller these values are, the greater its increase. Physically, the smaller $\nu_{ff,c0}$ is, the smaller the degree of ionization (hence, the more effective the ambipolar diffusion) *at all densities*. The smaller k is, the smaller the degree of ionization (and the more rapid the ambipolar diffusion) becomes *as the neutral density increases*.

4.2. Rotating Models: Ambipolar Diffusion and Magnetic Braking

In order to account for rotation (and, later, other physical phenomena), we take advantage of the result found above, namely, that relatively rapid relaxation along field lines leads to the formation of a thin disk, and establishes and maintains balance of forces along field lines at all times (certainly up to $n_{n,c} \lesssim 10^{10}$ cm^{-3}). We start the calculation from exact MHS equilibrium states at time $t = 0$. Thermal-pressure forces balance gravity along field lines, while magnetic, centrifugal, and thermal-pressure forces do so perpendicular to the field lines. Ambipolar diffusion and magnetic braking are switched on at $t = 0$. The two-fluid MHD equations are integrated over z (i.e., perpendicular to the plane of the disk), thereby reducing the spatial dimensionality of the problem and improving enormously the computational efficiency. For the model cloud described below, the mass-to-flux ratio in the initial equilibrium state is the same as that of a "reference" state threaded by a uniform magnetic field B_{ref}, but with a column density profile $\sigma_n(r) = \sigma_{c,ref}[1 + (r/l_{ref})^2]^{-3/2}$. (In galactic dynamics, this column density distribution is known as "Toomre's model 1"; Toomre 1963.) The quantity $\sigma_{c,ref}$ is the central column density, and l_{ref} is the characteristic length of its distribution; as suggested by observations, l_{ref} must be significantly greater than the critical thermal lengthscale $\lambda_{T,cr} \sim C\tau_{ff,c}$ (see Mouschovias 1991a), where C is the isothermal sound speed and $\tau_{ff,c}$ the central free-fall time for a thin disk. For the initial angular velocity profile, we use that obtained from a linear dependence of the specific angular momentum ($\ell = \Omega r^2$) on mass; this corresponds to the angular momentum of a uniformly rotating disk of uniform column density. The initial equilibrium state is calculated from the corresponding reference state by allowing the model cloud to contract under magnetic-flux and angular-momentum conservation until force balance is achieved. Because the model clouds are magnetically subcritical, the equilibrium central values $n_{n,c0}$, $B_{z,c0}$, and Ω_{c0} are typically greater than their values in the corresponding reference state by only 2% - 4%. Given the column density $\sigma_n(r, t)$ and a fixed external pressure P_{ext}, the mass density $\rho_n(r, t)$ is uniquely determined analytically from the condition of force balance along field lines. The half-thickness is then given by $Z(r, t) = \sigma_n(r, t)/2\rho_n(r, t)$. An important new free parameter is now present in the problem, namely, the ratio of the "external" and central neutral densities in the reference state:

$$\tilde{\rho} = \frac{\rho_{ext}}{\rho_{n,c,ref}} ; \tag{35}$$

its typical value is 0.01. If $\tilde{\rho} = 0$, no magnetic braking can take place, and angular momentum is conserved.

The evolution of a typical model cloud ($n_{n,c,ref} = 3 \times 10^3$ cm^{-3}, $T = 10$ K, $\Omega_{c,ref} = 0.48$ km s^{-1} pc^{-1} = 1.55×10^{-14} rad s^{-1}, $B_{ref} = 30$ μG, degree of ionization $x_{i,c} \equiv n_{i,c}/n_{n,c} = 1.72 \times 10^{-7}$, $l_{ref} = 1.15$ pc, $R_0 = 5l_{ref}$, total mass $M_{cl} = 191$ $M_{\odot}$) is followed to an enhancement of the central density by a factor of 10^6. These dimensional quantities imply that $\mu_{c0} = 0.324$ (using the disk critical mass-to-flux ratio) and $\nu_{ff,c0} = 21.6$; we still take $k = 1/2$. [The greater value of $\nu_{ff,c0}$ in the present model compared to that of the nonrotating model of § 4.1 is almost entirely due to the fact that the central free-fall time of the reference disk exceeds that of a uniform sphere of the same density by the factor 2.4 --the gravitational field of a disk at r is weakened by matter outside the radius r.]

Figure 5a shows the central density and angular velocity as functions of time. [Note: 1 km s^{-1} pc^{-1} = 3.24×10^{-14} rad s^{-1}.] As in the absence of magnetic braking, the density increases slowly during the ambipolar-diffusion-controlled, quasistatic phase, by a factor $\simeq$ 20. Once a supercritical core forms, rapid contraction ensues. *The angular velocity evolves in three distinct phases.* (1) *It decreases exponentially*, to almost its background value, on the magnetic-braking timescale $\tau_J = \tau_{\parallel} \equiv \tau_{\parallel,fan}(R_{cl} = R_0)$ (see eq. [3a]) for an aligned disk rotator with straight-parallel field lines --the quantity $\nu_{A,ext}$ is the Alfvén speed in the external medium near the axis of symmetry. (2) *It remains essentially constant*, because of very effective magnetic braking, during the entire quasistatic phase of contraction. (3) *It increases almost as $n_{n,c}^{1/2}$* during the subsequent collapse phase, implying that angular momentum is nearly conserved in the rapidly contracting core. During this stage of evolution, $\tau_J \simeq \tau_{\Phi} \simeq 10\tau_{\sigma}$ near $r = 0$, and τ_J is well approximated by $\tau_{\parallel fan}$ (see eq. [3a]).

The various timescales evaluated at the center and normalized to the initial central free-fall time of the disk, $\tau_{ff,c0} = 1.48 \times 10^6$ yr, are shown in Figure 5b as functions of the central density enhancement, $n_{n,c}/n_{n,c0}$. The central "dynamical timescale" is operationally defined by

$$\tau_{dyn,c} \equiv \lim_{r \to 0} \left[\frac{r}{|g_r|}\right]^{1/2} , \tag{36}$$

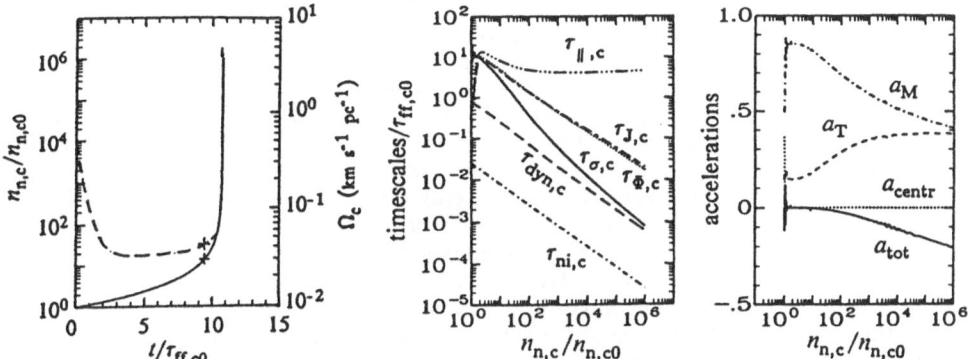

Figure 5. *Evolution in the presence of magnetic braking.* (**a**, *left*) Central neutral density (*solid line*), normalized to its initial value $n_{n,c0} = 3.12 \times 10^3$ cm^{-3}, as a function of time, normalized to the initial central free-fall time $\tau_{ff,c0} = 1.48 \times 10^6$ yr. Also, central angular velocity (in km s^{-1} pc^{-1}; *dashed line*; scale on right side of the frame). Meaning of + is as in Fig. 2a. (**b**, *middle*) Central timescales, normalized to $\tau_{ff,c0}$, as functions of central density enhancement $n_{n,c}/n_{n,c0}$. Magnetic-braking timescale $\tau_{J,c}$ (*long-dash, short-dash line*), magnetic-braking time for rotator with straight-parallel field lines $\tau_{\parallel,c}$ (*dash, triple-dot line*), flux-loss time $\tau_{\Phi,c}$ (*dotted line*), column-density e-folding time $\tau_{\sigma,c}$ (*solid line*), dynamical time $\tau_{dyn,c}$ (*dashed line*), and neutral-ion collision time $\tau_{ni,c}$ (*dash-dot line*). (**c**, *right*) Forces (per gram of neutral matter), normalized to $|g_r|$, in the limit $r \to 0$. Magnetic force a_M (*dash-dot line*), thermal-pressure force a_T (*dashed line*), centrifugal force a_{centr} (*dotted line*), and total force a_{tot} (*solid line*).

and is approximately (but not exactly) equal to the central free-fall time. The timescale $\tau_{J,c}$ (*long-dash short-dash line*) is only slightly smaller than $\tau_{\parallel,c}$ initially, since the field lines are only slightly compressed in the central part of the initial equilibrium state. However, $\tau_{J,c}$ rises quickly in the early evolution as the angular velocity decreases toward the background value; i.e., magnetic braking renders itself ineffective after it achieves corotation with the background. The rise in $\tau_{J,c}$ is halted when it becomes approximately equal to the column-density e-folding time, $\tau_{\sigma,c}$, as a balance is established between the rates of inward lateral advection of angular momentum (perpendicular to the field lines) and its transport away from the core along magnetic field lines. The relation $\tau_{J,c} \simeq \tau_{\sigma,c}$ is maintained during the quasistatic phase of contraction, because magnetic braking can remove angular momentum just as rapidly as it is advected inward. Since the inward advection of mass and angular momentum is governed by ambipolar diffusion during the quasistatic contraction phase, Figure 5b also shows that $\tau_{\Phi,c} \simeq \tau_{J,c} \simeq \tau_{\sigma,c}$. After the onset of dynamic contraction of the core, $\tau_{J,c}$ is found to decrease as $n_{n,c}^{-1/2}$, in accordance with the prediction of equation (3c) --see discussion following equation (4). Although the relation $\tau_{J,c} \simeq \tau_{\Phi,c}$ is maintained even during the dynamic phase, these timescales become significantly greater than $\tau_{\sigma,c}$ as this phase progresses; in fact, $\tau_{\sigma,c}$ approaches $\tau_{dyn,c}$ by the end of the run. The timescale $\tau_{\parallel}$ becomes the longest of all timescales, but is also irrelevant to the process of angular momentum loss by the core, because of the significant deformation of the field lines. Note that $\tau_{\sigma,c}$ is only 4×10^5 yr at a density enhancement of 10^2 (i.e., for typical parameters, at $n_{n,c} = 3 \times 10^5$ cm^{-3}), and only 1.2×10^3 yr at the end of the run (at $n_{n,c} = 3 \times 10^9$ cm^{-3}).

Figure 5c shows the central forces (magnetic, thermal-pressure, centrifugal, and total), normalized to the magnitude of the local gravitational force, as functions of central density enhancement. Magnetic braking is so effective initially that Ω_c decreases very rapidly, which in turn results in a dramatic decrease in rotational support (from 36% to 0.1%) with hardly any increase in central density. However, the magnetic force rises just as rapidly and compensates for the loss of centrifugal support (it increases its share of support against gravity from 50% to 85%) while the share of the thermal-pressure support remains fixed at 14%. This is so because, in a subcritical cloud, only a slight contraction is required for the magnetic-pressure and tension forces to make up for the loss of centrifugal support. After its precipitous initial drop, the centrifugal force never again becomes a significant source of support in the core. It is also clear from Figure 5c that magnetic opposition to gravity decreases and thermal-pressure opposition increases as the evolution progresses into the supercritical phase. However, even at the end of the run, the central magnetic force remains somewhat greater than the thermal-pressure force, and contributes slightly more than 40% to the support of the cloud; it is much greater than the thermal-pressure force everywhere else in the cloud. The magnitude of the total central (infall) acceleration rises to 20% of the gravitational acceleration by the end of the run. This is dynamic contraction, but by no means free fall --even at a central density of 3×10^9 cm^{-3}.

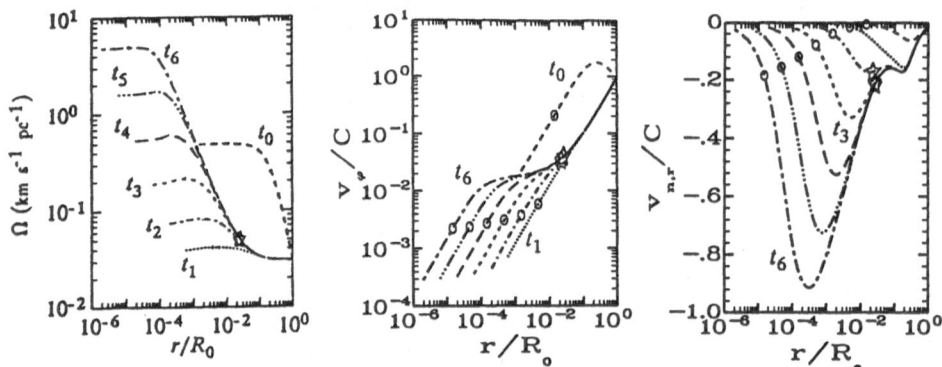

Figure 6. *Angular velocity and components of the linear velocity as functions of radius at seven different times (0, 13.379, 15.560, 15.846, 15.896, 15.907, and 15.910 Myr); R_0 = 5.75 pc. A star on a curve marks the instantaneous radius of the supercritical core. (a, left) Angular velocity. (b, middle) Azimuthal v_ϕ and (c, right) radial v_r components of the velocity, normalized to the isothermal sound speed C (= 0.188 km s^{-1} at T = 10 K).*

The angular velocity as a function of radius is shown in Figure 6a at seven different times t_j (j = 0, 1, ..., 6) chosen such that the central density enhancement is 10^j at time t_j. It is clear that *the initial angular momentum distribution* (curve labeled t_0) *is "forgotten" very quickly because of effective magnetic braking*. Only in the supercritical core does Ω increase with time. Centrifugal forces are rapidly reduced to an insignificant level *everywhere* in the cloud; they remain negligible at subsequent times. Uniform rotation is maintained in the central region and in the envelope, which contains most of the cloud's mass, while most of the supercritical core exhibits differential rotation. In fact, in most of the core at the end of the run, v_ϕ is nearly constant (see Fig. 6b). Significant infall velocities develop only in the supercritical core. Even these, however, remain subsonic (see Fig. 6c) at all times.

The central magnetic field increases by a factor of 1.6 during the subcritical phase of contraction, and by a factor of $\simeq$ 180 by the end of the run. The exponent κ = 0.48 at late times. The maximum value of B_r at the upper surface of the disk ($z = +Z$) is $(B_{r,Z})_{max} = B_z/4$ and occurs at $r \simeq 7 \times 10^{-5} R_0 = 4.0 \times 10^{-4}$ pc $\simeq 83$ AU. Elsewhere, $B_{r,Z} \ll B_z$. It is always the case that $|B_{\phi,Z}| \ll B_z$. Magnetic forces support the envelope very well against gravity. The mass and radius of the supercritical core are $M_{core} \simeq 10\ M_\odot$ (in excellent agreement with the peak in Fig. 1a) and $R_{core} \simeq 0.14$ pc, respectively. Its specific angular momentum has decreased by more than two orders of magnitude, to $\ell = 10^{20}$ cm^2 s^{-1}, which is comparable to that of wide visual binaries. The innermost $1\ M_\odot$ has $\ell = 10^{19}$ cm^2 s^{-1}, which is comparable to that of the protosolar nebula.

As part of an extensive parameter study, the dependence of the physical properties of rotating protostellar cores on the initial mass-to-flux ratio μ_{c0} (in units of its thin-disk critical value) was determined (Basu and Mouschovias 1995b). For the range $0.1 \leq \mu_{c0} \leq 1.0$, suggested by observations, the core mass is found to scale as $M_{core} \propto \mu_{c0}$, while its specific angular momentum scales as $(J/M)_{core} \propto \mu_{c0}^2$. Hence, $(J/M)_{core} \propto M_{core}^2$.

5. Protostar Formation in the Four-Fluid Approximation

As shown in § 2.2, the electrons and the ions are well attached to the magnetic field for densities $< 10^{10}$ cm^{-3}. They are therefore treated as a single fluid (the plasma), leading to a four-fluid (neutrals, plasma, charged and neutral grains) description of protostar formation. The details of the formulation of the problem and the results are found in Ciolek and Mouschovias (1993, 1994, 1995), and the numerical techniques are described in Morton, Mouschovias, and Ciolek (1994). Summaries of the results from different perspectives are given by Mouschovias (1995, 1996), and Ciolek (1996).

5.1. Main Effects of Grains

Microscopically, it is well known that grains play an important role in the chemical reactions that determine the degree of ionization. Macroscopically, the charged grains, if attached to the field, can exert a drag on the neutrals in addition to that exerted by ions, as originally suggested by Baker (1979) and Elmegreen (1979). We use the proper creation/destruction rate equations (7a) - (7c) to calculate in a self-consistent manner the abundances of charged species, i.e., electrons, molecular ions, atomic (metallic) ions, and charged grains, at each time step of the evolution; this includes the conversion of neutral grains to charged grains and *vice versa*. The high-energy ($\gtrsim 100$ MeV) cosmic-ray ionization rate is taken to be $\zeta_{CR} = 5 \times 10^{-17}$ s^{-1} (Spitzer and Tomasko 1968; Payne, Salpeter, and Terzian 1984). Grain-neutral frictional forces are also properly accounted for in the momentum equations (see § 2.1). *The quantities $\nu_{ff,c0}$ and k (see eqs. [33a] and [33c]) are no longer free parameters*, since the ion density is uniquely determined as a function of (the time-dependent) neutral density everywhere in a model cloud. As in the investigations ignoring grains, the initial central mass-to-flux ratio (μ_{c0}) is an important free parameter in the problem. The most significant new free parameters introduced by the physics of grains are (1) the dimensionless grain-neutral collision rate, and (2) the rate of (inelastic) electron attachment onto neutral grains relative to the rate of elastic grain-neutral collisions. They are most sensitive to the value of the grain radius a (see Ciolek and Mouschovias 1993, 1994), which, as is intuitively clear, affects both the rate of chemical reactions involving grains and the grain-neutral drag forces. In what follows we use the "standard" value $a = 3.75 \times 10^{-6}$ cm.

We use a reference state to calculate the mass-to-flux ratio and determine an initial equilibrium state as described in § 4.2. The initial central mass-to-flux ratio, in units of its thin-disk critical value, is now $\mu_{c0} = 0.256$. The initial equilibrium state could represent a

528

cloud with $n_{n,c0} = 2.60 \times 10^3$ cm^{-3}, $B_{z,c0} = 35.3$ μG, $T = 10$ K, $l_{ref} = 0.858$ pc, $R_0 = 5l_{ref} = 4.29$ pc, $M_{cl} = 98.3$ $M_\odot$, and a grain mass fraction $\chi_{g,0} = 0.01$. The main difference between this equilibrium state and that of § 4.2, except for the presence of grains and absence of rotation, is that l_{ref} has now been taken to be smaller than that of the rotating cloud by the factor 1.34. As explained earlier, the results relating to the formation and collapse of protostellar cores are not affected by the choice of l_{ref}, provided that it is significantly greater than the critical thermal lengthscale; in this case, we have that $l_{ref} = 5.5\lambda_{T,cr,ref}$.

Ambipolar diffusion is turned on at time $t = 0$ and the evolution is followed, as usual, to six orders of magnitude enhancement of $n_{n,c}$. Figure 5a describes the evolution of $n_{n,c}$ of the present model cloud as well, except for the fact that, instead of taking $\simeq 10\tau_{ff,c0}$ to achieve supercritical conditions in the central flux tube, it now takes $\simeq 13\tau_{ff,c0}$.

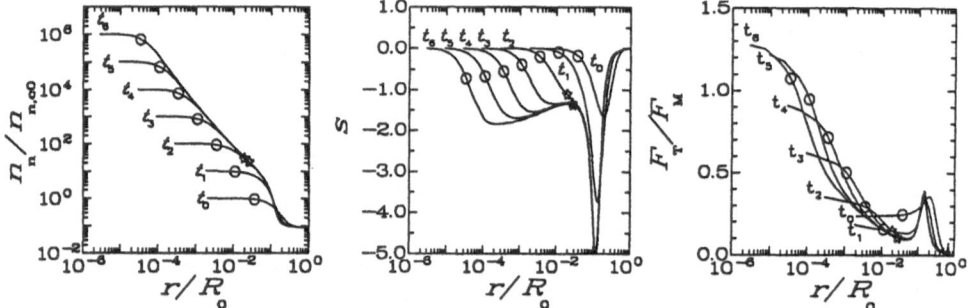

Figure 7. *Evolution in the presence of grain-neutral drag.* Spatial profiles of physical quantities are shown at seven different times ($t = 0$, 10.667, 13.279, 13.614, 13.663, 13.673, and 13.675$\tau_{ff,c0}$; $\tau_{ff,c0} = 1.2437\tau_{dyn,c0} = 1.3295$ Myr). The radial coordinate r is normalized to $R_0 = 4.29$ pc. See text for meaning of *stars* and *open circles*. (**a**, *left*) Neutral density, normalized to its initial central value $n_{n,c0} = 2.60 \times 10^3$ cm^{-3}. (**b**, *middle*) Slope $s \equiv \partial\ln n_n/\partial\ln r$. (**c**, *right*) Ratio of thermal-pressure and magnetic forces.

Figure 7a shows the neutral density n_n as a function of radius r at seven different times t_j ($j = 0, 1, ..., 6$) chosen such that the central density increases by a factor 10^j in the time t_j. A "*star*" on a curve, present only after a supercritical core forms, marks the instantaneous radius of the core. Once such a core forms, it is characterized by a compact, nearly uniform-density central region [whose size is determined by the instantaneous value of the critical thermal lengthscale $\lambda_{T,cr} = C^2/2G\sigma_{n,c}(t)$, marked by an *open circle* on each curve], and a "tail" of infalling matter left behind by the shrinking (both in size and in mass) central region. A near power-law density profile is established in the tail, $n_n \propto r^s$, with $-1.84 \leq s \leq -1.50$ in the region $1.0 \times 10^{-4} \leq r/R_0 \leq 8.0 \times 10^{-3}$ at time t_6 (see Fig. 7b). *A break in the slope of the density profile occurs almost exactly at the boundary of the supercritical core*, reflecting the different physics governing the evolution of the core and that of the subcritical, magnetically well-supported envelope. At the end of the run, the mass and radius of the core are $M_{core} = 8.21 \times 10^{-2}M_{cl} \simeq 8.1$ $M_\odot$ (in excellent agreement, once again, with the observations shown in Fig. 1a), and $R_{core} = 2.67 \times 10^{-2}R_0 \simeq 0.11$ pc (which is also a typically observed core radius). The radius of the uniform-density central region is $\simeq \lambda_{T,cr}(t_6) = 3.38 \times 10^{-5}R_0 \simeq 30$ AU. Although the central density at this time is 2.6×10^9 cm^{-3}, the *mean* density of the supercritical core is only $\langle n_n \rangle_{core} = 39.5 n_{n,c0} = 1.0 \times 10^5$

cm^{-3}, comparable to the density of observed protostellar cores using NH_3 as a tracer molecule (e.g., see Bachiller *et al.* 1990; Crutcher *et al.* 1994). Similarly, the central field is 4 mG, but its mean value in the core is only $\simeq 65$ μG.

Perpendicular to the field lines, the magnetic force F_M dominates the thermal-pressure force F_T in opposing gravity at all times and everywhere in the cloud, including the supercritical core, except in the uniform-density central region (within $r < 10^{-4}R_0$) at the end of the run, where $F_T/F_M \simeq 1.0 - 1.3$ (see Fig. 7c).[5] The maximum infall speed occurs at the end of the run and is only $\simeq 0.6C$, i.e., the infall is subsonic and far from being free fall. Its location is $r \simeq 2 \times 10^{-4}R_0$; it decreases rapidly on either side of this radius, and is negligible farther out in the subcritical envelope. Qualitatively, it behaves as shown in Figure 6c. The magnitude of the maximum infall acceleration occurs at a slightly smaller radius than the infall-speed maximum, and is only $\simeq 16\%$ that of gravity.

The exponent $\kappa \equiv d\ln B_c/d\ln n_{n,c}$ as a function of central density is shown in Figure 7d. Ion depletion onto grains would by itself tend to allow ambipolar diffusion to operate effectively even at high densities, during the collapse phase. However, grain-neutral drag almost compensates for the effect of ion depletion, and κ increases to 0.48 by the end of the run, a value near the flux-freezing limit of 1/2. During the early, quasistatic phase of contraction κ is very small, revealing that the neutrals are contracting through essentially stationary field lines (and plasma).

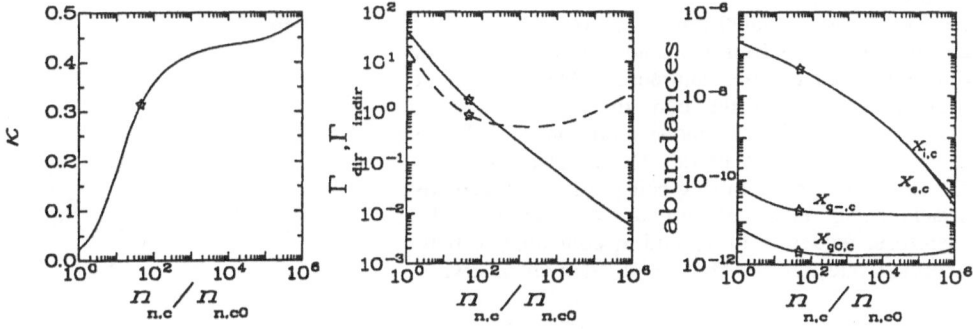

Figure 7, cont'd. *Central values of parameters as functions of central density* $n_{n,c}$, *normalized to its initial value as in Fig. 7a.* (d, *left*) Exponent $\kappa \equiv d\ln B_c/d\ln n_{n,c}$. (e, *middle*) Charged-grain direct (Γ_{dir}; *solid line*) and indirect (Γ_{indir}; *dashed line*) magnetic attachment parameters. (f, *right*) Abundances of ions ($x_{i,c}$), electrons ($x_{e,c}$), charged grains ($x_{g_-,c}$), and neutral grains ($x_{g0,c}$) relative to neutral molecules.

The direct (Γ_{dir}; see eq. [19]) and indirect (Γ_{indir}; see eq. [20]) attachment parameters (of grains to the magnetic field) are shown in Figure 7e as functions of central density. It is clear that, if it were not for the grain-quasiparticle electrostatic attraction, the grains would detach from the field at density enhancements $\gtrsim 10^2$. In reality, they remain partially

---

[5]Hence, calculations that ignore the magnetic field or treat it as a perturbation on the zeroth-order flow of a singular, nonmagnetic, isothermal sphere (e.g., Galli and Shu 1993a, b) are based on a fundamentally incorrect assumption, which is responsible for erroneous conclusions and speculations, such as large magnetic pinching forces and magnetic reconnection during the isothermal phases of star formation.

530

coupled to the field even at the end of the run, at a central density of 2.6×10^9 cm^{-3}.

Customarily, it is thought that the microscopic physical processes that determine the degree of ionization affect the rate at which ambipolar diffusion progresses and, therefore, the evolution of a cloud. Not much attention has been paid to the effect of the evolution on the microscopic physics. Figure 7f shows the central number densities, relative to that of the neutrals, of ions ($x_{i,c}$), electrons ($x_{e,c}$), charged grains ($x_{g^-,c}$), and neutral grains ($x_{g_0,c}$) as functions of neutral density enhancement $n_{n,c}/n_{n,c0}$. It is clear that, during the quasistatic phase of contraction, the dust-to-gas ratio *decreases* in the central flux tubes of a cloud because the grains are well attached to the magnetic field and are thus left behind by the contracting neutrals. This reduces the rate of electron and ion capture by grains. If we represent the relation between the ion and neutral densities in the usual manner, $n_i \propto n_n{}^k$, *no constant k can accurately describe the ion density at all times.* Figure 7g exhibits the central value of the exponent k as a function of central density enhancement. It is seen (1) that $k_c > 1/2$ during the quasistatic phase (i.e., there are more ions than one would deduce from chemistry in static cloud models); and (2) that k_c decreases to $\simeq 0.1$ at the end of the run. *Hence, the magnetic field, through its effects on the evolution, affects significantly the chemistry in contracting cores.*

Figure 7g. *Exponent* $k_c \equiv d\ln n_{i,c}/d\ln n_{n,c}$ *as a function of central density enhancement* $n_{n,c}/n_{n,c0}$. It is greater than 1/2 (the value predicted by chemical models of static clouds) during the quasistatic phase, and becomes significantly smaller than 1/2 during the dynamic phase of core contraction. The decrease of grain abundance in the core during quasistatic contraction (see Fig. 7f) allows more ions and electrons to remain in the gas phase. The constant relative abundance of grains during collapse leads to effective attachment of electrons and ions onto grains and, therefore, reduction of x_e and x_i compared to their expected values in the absence of grain chemistry.

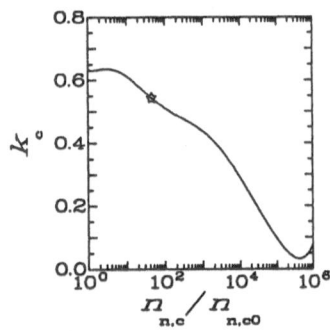

The precise factor by which the dust abundance decreases in a supercritical core relative to its value in the envelope depends on the degree to which grains are attached to the field. For well-attached grains (i.e., $a \lesssim 7 \times 10^{-6}$ cm), this factor is equal to $1/\mu_{c0}$, the initial central mass-to-flux ratio in units of its critical value for collapse. In practice, the grain abundance in a core relative to that of the parent envelope may be easier to determine observationally than the *initial* mass-to-flux ratio of the core, which is essentially that of the envelope in the vicinity of the core. Hence, one may turn the problem around and, *from the observable reduction of the grain abundance in a core, one may straightforwardly infer the initial mass-to-flux ratio relative to its critical value.* Only after a core becomes magnetically and thermally supercritical, and dynamical contraction ensues, does its grain mass fraction become "frozen" in the matter. Since small grains are better attached to the magnetic field lines than more massive grains, the dynamical effect described above implies that the larger grains will tend to be more abundant in high-density cores. To observations this would appear as grain growth with increasing density (see Vrba et al. 1981, 1993), when it may actually be a simple and direct consequence of the fact that the small grains, being better attached to the field lines, are preferentially left behind during the formation and contraction of protostellar cores.

5.2. Comparison with Observations: The Barnard 1 Cloud

The results of these calculations have been used to construct a detailed evolutionary model for the Barnard 1 cloud (Crutcher, Mouschovias, Troland, and Ciolek 1994). We first showed that the profile of the column density (as revealed by $C^{18}O$ and ^{13}CO observations) of the "internal envelope" of the cloud (see Bachiller *et al.* 1990), of mass 600 $M_\odot$, can be well represented by an exact MHS equilibrium model of a disklike cloud, whose axis of symmetry (the z-axis) is inclined by 20° with respect to the plane of the sky so as to explain the observed 3:1 axes ratio. We then turned on ambipolar diffusion. A magnetically supercritical core formed in about $\simeq 10^7$ yr, but, once the core formed, it collapsed in less than 1 Myr. The predicted properties of the protostellar core include a mass = 13.4 $M_\odot$, radius = 0.13 pc, mean density = 1.3×10^5 cm^{-3}, and mean magnetic field strength along the line of sight = 29.1 μG. These are in excellent agreement with the corresponding observed values of the NH_3 core; namely, mass $\simeq$ 13 $M_\odot$, radius $\simeq$ 0.15 pc, mean density > 8×10^4 cm^{-3}, and B_{los} = 30 ± 4 μG. Moreover, the calculated spatial profile of the Alfvén speed compares well with observations of the linewidth as a function of radius along the major axis. The model makes further predictions concerning the structure of the protostellar core that can be tested by higher spatial resolution observations.

5.3. Effect of Ultraviolet Radiation

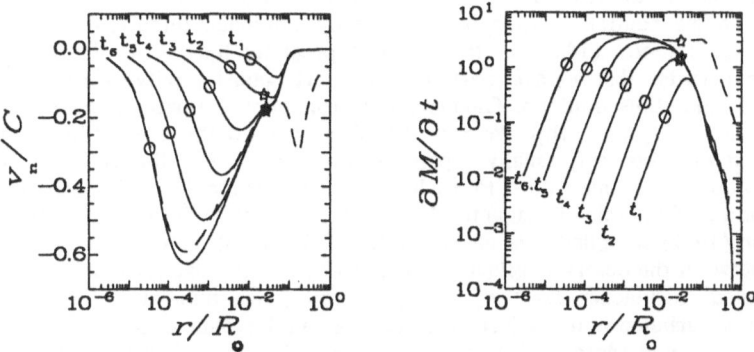

Figure 8. *Effect of UV ionization.* Spatial profiles of (a, *left*) the neutral velocity v_n in the equatorial plane, normalized to the isothermal sound speed C = 0.188 km s^{-1}, and (b, *right*) the mass infall rate (in $M_\odot$/Myr), at six times $t_1, t_2, ..., t_6$ for a model identical to that of Figs. 7a-7g, except that UV ionization is now accounted for. The times are chosen such that the central density at time t_i exceeds its initial value by the factor 10^j; they are equal to 8.614, 10.507, 10.774, 10.818, 10.827, and 10.829$\tau_{ff,c0}$; $\tau_{ff,c0}$ = 1.3295 Myr. The *dashed line* refers to the model of § 5.1, which ignores UV ionization, at the end of the run. The "stars" and open circles have the same meaning as in Fig. 7.

In addition to cosmic-ray ionization and the microscopic and macroscopic effects of grains described in § 5.1, Ciolek and Mouschovias (1995) have included the effect of UV ionization. We take ζ_{uv} = 1.2×10^{-11} s^{-1} at the cloud surface, and we account for the attenuation due to grain absorption and carbon ionization. In addition to the ions HCO$^+$, Mg$^+$, and Na$^+$, other ions (e.g., S$^+$, Si$^+$, and C$^+$) now play a significant role (see Appendix). Figure 8a shows the infall speed of the neutrals as a function of radius at six different times. The dashed line represents the velocity profile at the end of the run in the absence of UV ionization (the model cloud of Figs. 7a - 7g). Clearly, although the infall speed within the core is virtually unaffected by UV ionization, infall is drastically reduced in the envelope. Figure 8b exhibits the mass infall rate dM/dt (in $M_\odot$/Myr) as a function of r/R_0 at six different times. The dashed line represents the infall rate at the end of the run in the absence of UV ionization, as in Figure 8a. Although the formation and evolution of the supercritical protostellar core is unaffected (except for the fact that it now takes 30% *less* time to form[6]), *infall beyond the supercritical core is effectively cut off by the higher degree of ionization in the envelope.* The maximum mass that may accrete onto the forming star is therefore that of the supercritical core. Some of the accreting mass may be subsequently ejected in the form of a wind during the opaque (nonisothermal) phases of star formation.

6. Hydromagnetic Waves, Linewidths, Ambipolar Diffusion, and Star Formation

The presence of HM waves in molecular clouds, and the role they may play in supporting the clouds certainly along field lines (Mouschovias and Morton 1985a, b; Mouschovias 1987a), does not alter in any significant way the ambipolar-diffusion-induced formation and collapse of protostellar fragments (or cores) described above. This is so because, for typical molecular cloud parameters ($n_n \sim 10^3$ cm^{-3}, $T \simeq 10$ K, $B \simeq 30$ μG), the size of the region that can just become gravitationally unstable because of ambipolar diffusion happens to be essentially equal to the Alfvén lengthscale λ_A (Alfvén waves with wavelengths $\lambda \leq \lambda_A$ cannot propagate in the neutrals because of damping by ambipolar diffusion) --see Mouschovias 1987a; or 1991a, eqs. (18a, b). In fact, it is precisely the decay of HM waves due to ambipolar diffusion that removes part of the support against gravity over the critical thermal lengthscale and thus initiates fragmentation (or core formation) in molecular clouds (Mouschovias 1987, § 2.2.6[b]). We may therefore consider the issue of whether HM waves can explain the observed, usually supersonic but subAlfvénic linewidths without concern that attributing the linewidths to such waves might modify substantially our current understanding of the role of magnetic fields and ambipolar diffusion in star formation.

Larson (1981) compiled data on linewidths of 54 clouds, clumps and cores, and found a relation between the observed velocity dispersion σ_v and the size (diameter) L of the objects, the so-called "turbulence law": $\sigma_v \propto L^{0.35}$, which was thought to be the signature of Kolmogorov turbulence in molecular clouds. A similar data analysis was carried out by Leung, Kutner, and Mead (1982) and was extended by Myers (1983; see also Falgarone and Pérault 1987, and Fuller and Myers 1992), who subtracted from each linewidth the thermal contribution and found the "turbulent" (or nonthermal) part to be given by

$$(\Delta v)_{turb} \simeq 1.3 \, (R/1 \text{ pc})^{0.5} \quad \text{km s}^{-1}, \tag{37}$$

where R is the FWHM size (or "radius") of the object. More data have accumulated since then, but they are consistent with the relation (37).

The empirical relation $(\Delta v)_{NT} \propto R^{1/2}$ was explained as a natural consequence of the presence of HM waves in magnetically supported, self-gravitating clouds (Mouschovias

---

[6]The now better magnetic support of the envelope keeps matter farther away from the axis of symmetry of the protostellar *disk*. Hence, there is less dilution of the core's gravitational field by the mass in the magnetically supported envelope. This leads to a faster formation and evolution of a supercritical core. Arguments that UV ionization increases τ_{AD} too much to account for protostar formation are thus seen to be in error.

1987a). For such clouds, the Alfvén speed is always comparable to the free-fall speed, i.e.,

$$v_A \simeq \left[\frac{2GM}{R}\right]^{1/2} = (2\pi G\sigma_m R)^{1/2} \tag{38}$$

(Mouschovias 1978, eq. [16]), where the column density is $\sigma_m = M/\pi R^2$; hence, near-Alfvénic linewidths are naturally expected to be "virialized" in such clouds. The relation $(\Delta v)_{NT} \propto R^{1/2}$ is thus seen to be a consequence of the magnetic support of clouds having comparable column densities σ_m. Also, the issue emphasized by Myers (1985, p. 95) as an unexplained consequence of the so-called "condensation law" $\rho R \simeq const$ of Larson (1981), namely, that cloud column densities seem to vary by less than a factor of about 10, finds a natural explanation in this picture. Simply, self-gravitating clouds are expected to have column densities comparable to $\sigma_{m,crit} = (1/63G)^{1/2}B$, which depends only on the mean field B, which in turn is not expected to vary much from place to place in the interstellar medium under conditions suitable for the formation of self-gravitating clouds. It follows, therefore, that $\sigma_m (= 4\rho R/3) \simeq const$, but only to the extent that the mean magnetic field $B \simeq const$ for different self-gravitating clouds.

We eliminate σ_m from eq. (38) in favor of B as described above, evaluate the constants and find that, if the observed linewidths are due to large-amplitude, long-wavelength ($\lambda \gtrsim 0.1$ pc), hence long-lived, Alfvén waves, they should be related to B and R by

$$(\Delta v)_{wave} \simeq 1.4 \left[\frac{B}{30~\mu G}\right]^{1/2} \left[\frac{R}{1~pc}\right]^{1/2} \quad km~s^{-1} \tag{39}$$

(see, also, Mouschovias 1987a, § 2.2.1 for original derivation). Thus, the "turbulence law" found by Myers is a special case of this more general result.

Equation (39) contains the implicit assumption that the material velocities responsible for the (usually supersonic) linewidths are slightly subAlfvénic or Alfvénic [i.e., $(\Delta v)_{NT} \simeq (\Delta v)_{wave} \lesssim v_A$]. This is certainly the case for the Barnard 1 cloud: the theoretical work described in § 5.2 predicts a spatial profile of the Alfvén speed in excellent agreement with the observed linewidths away from the IRAS source embedded in the NH_3 core of B1 (see Crutcher et al. 1994). Mouschovias and Morton (1985a) have shown that equipartition is established between the magnetic energy in Alfvén waves and the kinetic energy of the material motions associated with the waves, after only a few reflections off fragments (or the cloud surface) and the consequent wave-wave interaction. It follows, therefore, that the relation $(\Delta v)_{wave} \lesssim v_A$ also implies that $(\delta B)_{wave} \lesssim B$. Virialization of linewidths is thus a consequence of two results: (1) the establishment of equipartition between the kinetic energy in the nonthermal motions and the magnetic energy in the Alfvén waves, and (2) the nonlinear nature of the Alfvén waves, as established by observations, which show that linewidths are almost Alfvénic.

Observations to date (e.g., the 14 objects, including masers, studied by Myers and Goodman 1988) for which $(\Delta v)_{NT}$, R (FWHM), and B have been measured are in excellent agreement with equation (39). Mouschovias and Psaltis (1995) used these data to plot the nonthermal linewidth $(\Delta v)_{NT}$ versus size R as in Figure 9a. (The error bars are as estimated by Myers and Goodman; they indicate an uncertainty by a factor of 2.) *There is no obvious correlation between $(\Delta v)_{NT}$ and R.* To test equation (39), they showed as in Figure 9b the same data, but they plotted the quantity $(\Delta v)_{NT}/R^{1/2}$ against B. The solid line is the theoretical prediction, equation (39). The dashed line is a least-squares fit to the data. The difference between the two lines is, for practical purposes, insignificant.

It may appear surprising at first sight that linewidths of OH masers, which are neither massive clouds nor protostellar cores, would obey the relation (39). However, it should be

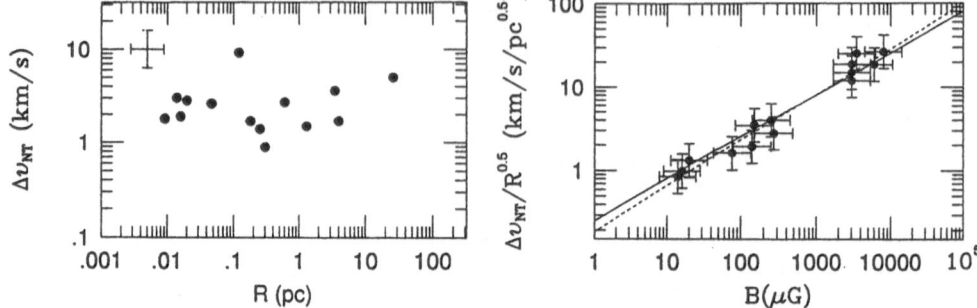

Figure 9. *Hydromagnetic Waves and Linewidths.* (**a**, *left*) Nonthermal linewidth versus (FWHM) size for fourteen objects (data from Myers and Goodman 1988). It is almost a scatter diagram. (**b**, *right*) Same linewidth-size data, but exhibiting the ratio $(\Delta v)_{NT}/R^{1/2}$ as a function of the measured magnetic field strength B. Error bars are as in (*a*). The theoretical prediction (eq. [39]) is shown as a solid line. The dashed line is a least-squares fit to the data.

borne in mind that these objects are found in high-density regions of star forming clouds. For such regions, the detailed numerical calculations described in §§ 4 and 5, which follow the ambipolar-diffusion-initiated protostar formation from densities of a few $\times$ 10^2 cm^{-3} to $\sim 10^{10}$ cm^{-3}, invariably predict milligauss fields at densities $\gtrsim 10^8$ cm^{-3}. In other words, the star formation process creates physical conditions of density and associated field strength that require only a pump for maser action consistent with observations to take place (see Mouschovias 1991a). We expect equation (39) to be valid for H_2O masers as well, which are characterized by even higher densities ($\sim 10^{10}$ cm^{-3}) and stronger fields (almost 10^2 mG).

7. Conclusion

A five-fluid (neutral molecules, ions, electrons, charged and neutral grains) MHD description of protostar formation has been physically motivated and justified. For the early, isothermal phase (neutral density $n_n \lesssim 10^{10}$ cm^{-3}), the electrons and ions behave as a single fluid (the plasma) well coupled to the magnetic field; hence, a four-fluid description becomes appropriate. The charged grains (typical radius $a \simeq 4 \times 10^{-6}$ cm) are directly attached to the field only up to densities $\sim 10^5$ cm^{-3}. Indirect, at least partial attachment, due to electrostatic attraction by well-attached electron-shielded ion quasiparticles, dominates at higher densities. (A similar, indirect attachment of ions to the field is expected during the later, opaque phase of star formation, as n_n increases by a few more orders of magnitude while the electrons remain well attached and the electron density n_e is not negligible compared to the ion density n_i; i.e., provided $n_e \gtrsim 10^{-3} n_i$.)

   Theoretical calculations have predicted and observations have shown that star formation in a typical, self-gravitating molecular cloud is not the result of collapse of the cloud as a whole, despite the fact that its observed mass exceeds the critical thermal ($\simeq$ Jeans) mass by a factor $\sim 10^2$ - 10^5. Clouds are primarily magnetically supported, with hydromagnetic (HM) waves probably providing some support at least along field lines. In such magnetically subcritical clouds, ambipolar diffusion is an unavoidable process. Analytical and numerical two-fluid calculations first, and four-fluid axisymmetric simulations later (accounting for both cosmic-ray and UV ionization), have shown that star formation is a self-initiated process, and have followed the formation and contraction of protostellar fragments (or cores) from typical densities $\sim 10^3$ cm^{-3} to central densities $\sim 10^{10}$ cm^{-3}, by which isothermality begins to break down and the ions are no longer well attached to the field.

Ambipolar diffusion damps the HM waves and allows gravity to initiate quasistatic contraction in the deep interior of a cloud, where the degree of ionization is $x_i \lesssim 10^{-7}$, over a region containing approximately one critical thermal mass (= a few $M_\odot$ at $n_n \sim 10^3$ cm^{-3} and $T \simeq 10$ K), while the field lines and the attached plasma and grains remain essentially stationary. A disklike core begins to form and to contract on a timescale equal to the central magnetic-flux redistribution timescale $\tau_{\Phi,c}$ (= $\tau_{AD,c}/2$) $\lesssim 10^7$ yr, which exceeds the central free-fall timescale $\tau_{ff,c}$ typically by a factor $\simeq 10$. Magnetic braking is very effective during this phase of evolution, and keeps the entire cloud essentially corotating with the background. Centrifugal forces are reduced to a negligible level, never to play a significant role in the evolution during the isothermal phase of star formation ($n_n \lesssim 10^{10}$ cm^{-3}).

The ambipolar-diffusion-induced and controlled redistribution of mass in the central flux tubes eventually increases the central mass-to-flux ratio to the critical value for collapse [=$1.5(63G)^{-1/2}$], and dynamic contraction of the thermally and magnetically supercritical core ensues. The density enhancement at which critical conditions are achieved is essentially equal to μ_{c0}^{-2}, where μ_{c0} (typically 0.1 - 1.0) is the initial central mass-to-flux ratio is units of its critical value for collapse. Even though the lateral contraction (i.e., perpendicular to the field lines) is dynamic, the remarkable result is found that force balance is maintained *along* field lines. Magnetic forces represent the dominant opposition to gravity (in the lateral direction), even in the supercritical core, up to central densities $\simeq 3 \times 10^9$ cm^{-3}. Only in a small central region, of radius approximately equal to the instantaneous value of the critical thermal lengthscale $\lambda_{T,crit}(t) \simeq C\tau_{ff,c}(t)$ ($\simeq 30$ AU; C is the isothermal sound speed), does the thermal-pressure force become comparable to the magnetic force near the end of a typical run, at $n_{n,c} = 3 \times 10^9$ cm^{-3}. Thermal pressure is responsible for maintaining a uniform-density central region of radius $\simeq \lambda_{T,crit}(t)$ during the evolution. This region shrinks in both size and mass, leaving behind a "tail" of matter (all within the supercritical core), in which a near power-law density profile is established $n_n \propto r^s$, with $-1.8 \lesssim s \lesssim -1.5$. Both angular momentum and magnetic flux tend to get trapped inside the core during the collapse phase, so that the magnetic field and the angular velocity vary almost as $B_c \propto n_{n,c}^{1/2}$ and $\Omega_c \propto n_{n,c}^{1/2}$, the magnetic-flux and angular-momentum conservation laws for a disk. The predicted typical physical properties of protostellar cores are in excellent agreement with observations: mass $\simeq 1 - 15$ $M_\odot$; radius $\simeq 0.1 - 0.2$ pc; *mean* field $\langle B \rangle$ greater than that of the parent envelope by less than a factor of 2 (although the *central* field is typically $\simeq$ a few mG at $n_n \simeq 10^8 - 10^9$ cm^{-3}); *mean* density $\langle n_n \rangle \simeq 10^5$ cm^{-3}; specific angular momentum $J/M \simeq 10^{19} - 10^{21}$ cm^2 s^{-1}, which is comparable to that of wide binary systems. The innermost 1 $M_\odot$ has a $J/M \simeq 10^{19}$ cm^2 s^{-1}, which is the inferred value for the protosolar nebula (see Bodenheimer et al. 1993). Hence, magnetic braking resolves the angular momentum problem both for binary and solar-type stars. During the formation and contraction of protostellar cores, the cloud envelopes, which typically contain $\simeq$ 90% of the mass, hardly suffer any evolution; they remain magnetically supported.

The theory also predicts that the presence of large-amplitude, long-wavelength Alfvén waves in self-gravitating clouds implies a linewidth-size relation which is in excellent quantitative agreement with observations, namely, $\Delta v \simeq 1.4 (B/30 \ \mu G)^{0.5}(R/1 \ pc)^{0.5}$ km s^{-1}.

The process of core formation itself affects the relative abundance of grains (hence, the chemistry) in dense cores in a profound way: grains attached to the field are left behind during the quasistatic phase of contraction, thus reducing the dust-to-gas relative abundance by the factor $1/\mu_{c0}$. Aside from its evident implications on the chemistry in dense cores, this result can be used to deduce the initial mass-to-flux ratio of a cloud if the relative abundances of grains in a core and the envelope are observed. Also, since small grains are better attached to the field than larger grains, we predict a stratification of grains according to size during core formation. This effect gives the *impression* of grain growth with optical depth. A number of commonly used assumptions (e.g., a fixed dust-to-gas mass ratio as a function of density, and the parametrization of the ion density in terms of the neutral density as a power law with a single, constant exponent) are found to be incorrect. We have described a modern theory of star formation incorporating these new results.

There is little doubt that the multifluid-MHD description of star formation will uncover many more surprises when the opaque phases of evolution ($n_n > 10^{10}$ cm^{-3}) are explored.

Acknowledgements: This work was supported in part by the National Science Foundation under grant AST 93-20250. Discussions on chemistry issues with Tom Hartquist and Glenn Ciolek are gratefully acknowledged.

Appendix: Relevant Chemical Reactions in the Presence of UV Ionization

Detailed chemical models for molecular clouds have been presented by Solomon and Klemperer (1972), Herbst and Klemperer (1973), Oppenheimer and Dalgarno (1974), Glassgold and Langer (1974, 1976), Black and Dalgarno (1977), Mitchell et al. (1978), Prasad and Huntress (1980), and Graedel et al. (1982). Here we consider only the reactions affecting the equilibrium abundances of the relevant charged species.

Cosmic-ray ionization:
$$H_2 + CR \rightarrow H_2^+ + e . \tag{A1}$$

UV ionization:
$$\mathscr{A} + UV \rightarrow \mathscr{A}^+ + e , \qquad \mathscr{A} = C, Mg, Na, Fe, S, Si. \tag{A2}$$

HCO$^+$ formation:
$$H_2^+ + H_2 \rightarrow H_3^+ + H , \qquad H_3^+ + CO \rightarrow HCO^+ + H_2. \tag{A3a, b}$$

$$C^+ + OH \rightarrow CO^+ + H , \qquad CO^+ + H_2 \rightarrow HCO^+ + H . \tag{A3c, d}$$

$$C^+ + H_2O \rightarrow HCO^+ + H . \tag{A3e}$$

Dissociative recombination:
$$HCO^+ + e \rightarrow H + CO . \tag{A4}$$

Radiative recombination:
$$\mathscr{A}^+ + e \rightarrow \mathscr{A} + hv . \tag{A5}$$

Charge transfer:
$$A + HCO^+ \rightarrow A^+ + HCO , \qquad A = Mg, Na, Fe; \tag{A6a}$$

$$A + M^+ \rightarrow A^+ + M , \qquad M = C, S, Si; \tag{A6b}$$

$$s + C^+ \rightarrow s^+ + C , \qquad s = S, Si; \tag{A6c}$$

$$Si + S^+ \rightarrow Si^+ + S . \tag{A6d}$$

Electron and ion capture by grains:

$$e + g_0 \rightarrow g_- , \qquad HCO^+ + g_- \rightarrow HCO + g_0 , \qquad \mathscr{A}^+ + g_- \rightarrow \mathscr{A} + g_0 . \tag{A7a, b, c}$$

As explained in § 2.1, footnote #3, dissociative recombination of H_3^+, which would compete with reaction (A3b), can be neglected.

References

Bachiller, R., Menten, K. M., and del Rio-Alvarez, S. 1990, *A&A*, **236**, 461
Baker, P. L. 1979, *A&A*, **75**, 54
Basu, S., and Mouschovias, T. Ch. 1994, *ApJ*, **432**, 720

____ . 1995a, *ApJ*, **452**, 386

____ . 1995b, *ApJ*, **453**, 271

Black, J. H., and Dalgarno, A. 1977, *ApJS*, **34**, 405

Bodenheimer, P., Ruzmaikina, T., and Mathieu, R. D. 1993, in *Protostars and Planets III*, ed. E. H. Levy and J. Lunine (Tucson: Univ. of Arizona), 367

Bonnor, W. B. 1956, *MNRAS*, **116**, 351

Ciolek, G. E. 1996, in *The Role of Dust in the Formation of Stars*, ed. H. U. Käufl (Berlin: Springer), *in press*

Ciolek, G. E., and Mouschovias, T. Ch. 1993, *ApJ*, **418**, 774

____ . 1994, *ApJ*, **425**, 142

____ . 1995, *ApJ*, **454**, 194

Crutcher, R. M., Mouschovias, T. Ch., Troland, T., and Ciolek, G. E. 1994, *ApJ*, **427**, 839

Dalgarno, A. 1987, in *Physical Processes in Interstellar Clouds*, ed. G. E. Morfill and M. Scholer (Dordrecht: Reidel), 219

Draine, B. T., and Sutin, B. 1987, *ApJ*, **320**, 803

Ebert, R. 1955, *Zs. Ap.*, **37**, 217

____ . 1957, *Zs. Ap.*, **42**, 263

Elmegreen, B. G. 1979, *ApJ*, **232**, 729

Falgarone, E., and Pérault, M. 1987, in *Physical Processes in Interstellar Clouds*, ed. G. E. Morfill and M. Scholer (Dordrecht: Reidel), 59

Fiedler, R. A., and Mouschovias, T. Ch. 1992, *ApJ*, **391**, 199

____ . 1993, *ApJ*, **415**, 680

Fuller, G. A., and Myers, P. C. 1992, *ApJ*, **384**, 523

Galli, D., and Shu, F. H. 1993a, *ApJ*, **417**, 220

____ . 1993b, *ApJ*, **417**, 243

Glassgold, A. E., and Langer, W. D. 1974, *ApJ*, **193**, 73

Graedel, T. E., Langer, W. D., and Frerking, M. A. 1982, *ApJS*, **48**, 321

Herbst, E., and Klemperer, W. 1973, *ApJ*, **185**, 505

Hollenbach, D., and Salpeter, E. E. 1970, *J. Chem. Phys.*, **53**, 79

Jeans, J. H. 1928, *Astronomy and Cosmogony* (Cambridge: Cambridge Univ.)

Langer, W. D. 1985, in *Protostars and Planets II*, ed. D. C. Black and M. S. Mathews (Tucson: Univ. of Arizona), 650

Larson, R. B. 1981, *MNRAS*, **194**, 809

Larsson, M. et al. 1993, *Phys. Rev. Lett.*, **70**, 430

Leung, C. M., Kutner, M. L., and Mead, K. N. 1982, *ApJ*, **262**, 583

Lizano, S., and Shu, F. H. 1989, *ApJ*, **342**, 834

McDaniel, E. W., and Mason, E. A. 1973, in *The Mobility and Diffusion of Ions and Gases* (New York: Wiley)

Mestel, L., and Spitzer, L., Jr. 1956, *MNRAS*, **116**, 503

Mitchell, G. F., Ginsburg, J. L., and Kuntz, P. J. 1978, *ApJS*, **38**, 39

Morton, S. A., Mouschovias, T. Ch., and Ciolek, G. E. 1994, *ApJ*, **421**, 561

Mott, N. F., and Massey, H. S. W 1971, *The Theory of Atomic Collisions*, 3rd ed. (Oxford: Oxford Univ.)

Mouschovias, T. Ch. 1976a, *ApJ*, **206**, 753

____ . 1976b, *ApJ*, **207**, 141

____ . 1977, *ApJ*, **211**, 147

____ . 1978, in *Protostars & Planets*, ed. T. Gehrels (Tucson: Univ. of Arizona), 209

____ . 1979a, *ApJ*, **228**, 159

____ . 1979b, *ApJ*, **228**, 475

____ . 1983, in *Solar and Stellar Magnetic Fields: Origins and Coronal Effects*, ed. J. O. Stenflo (Dordrecht: Reidel), 479

538

____ . 1985, *A&A*, **142**, 41

____ . 1987a, in *Physical Processes in Interstellar Clouds*, ed. G. E. Morfill and M. Scholer (Dordrecht: Reidel), 453

____ . 1987b, in *Physical Processes in Interstellar Clouds*, ed. G. E. Morfill and M. Scholer (Dordrecht: Reidel), 491

____ . 1989, in *The Physics and Chemistry of Interstellar Molecular Clouds*, ed. G. Winnewisser and J. T. Armstrong (Berlin: Springer), 297

____ . 1991a, *ApJ*, **373**, 169

____ . 1991b, in *The Physics of Star Formation and Early Stellar Evolution*, ed. C. J. Lada and N. D. Kylafis (Dordrecht: Kluwer), 61

____ . 1991c, in *The Physics of Star Formation and Early Stellar Evolution*, ed. C. J. Lada and N. D. Kylafis (Dordrecht: Kluwer), 449

____ . 1995, in *The Physics of the Interstellar Medium and Intergalactic Medium*, ed. A. Ferrara, C. F. McKee, C. Heiles, and P. R. Shapiro (San Francisco: ASP), 80, 184

____ . 1996, in *The Role of Dust in the Formation of Stars*, ed. H. U. Käufl (Berlin: Springer), *in press*

Mouschovias, T. Ch., and Morton, S. A. 1985a, *ApJ*, **298**, 190

____ . 1985b, *ApJ*, **298**, 205

Mouschovias, T. Ch., and Paleologou, E. V. 1979, *ApJ*, **230**, 204

____ . 1980, *ApJ*, **237**, 877

____ . 1986, *ApJ*, **308**, 781

Mouschovias, T. Ch., and Psaltis, D. 1995, *ApJ*, **444**, L105

Mouschovias, T. Ch., and Spitzer, L., Jr. 1976, *ApJ*, **210**, 326

Myers, P. C. 1983, *ApJ*, **270**, 105

____ . 1985, in *Protostars & Planets II*, ed. D. C. Black and M. S. Matthews (Tucson: Univ. of Arizona), 81

Myers, P. C., and Goodman, A. A. 1988, *ApJ*, **326**, L27

Nakano, T. 1979, *PASJ*, **31**, 697

Nakano, T., and Nakamura, T. 1978, *PASJ*, **30**, 671

Nakano, T., and Umebayashi, T. 1986a, *MNRAS*, **218**, 663

____ . 1986b, *MNRAS*, **221**, 319

Nishi, R., Nakano, T., and Umebayashi, T. 1991, *ApJ*, **368**, 181

Oppenheimer, M., and Dalgarno, A. 1974, *ApJ*, **192**, 29

Payne, H. E., Salpeter, E. E., and Terzian, Y. 1984, *AJ*, **89**, 668

Pneuman, G. W., and Mitchell, T. P. 1965, *Icarus*, **4**, 494

Prasad, S. S., and Huntress, W. T., Jr. 1980, *ApJS*, **43**, 1

Solomon, P. M, and Klemperer, W. 1972, *ApJ*, **178**, 389

Spitzer, L., Jr. 1941, *ApJ*, **93**, 369

____ . 1948, *ApJ*, **107**, 6

____ . 1963, in *Origin of the Solar System*, ed. R. Jastrow and A. G. W. Cameron (New York: Academic), 39

____ . 1978, *Physical Processes in the Interstellar Medium* (New York: Wiley-Interscience)

Spitzer, L., Jr., and Tomasko, M. G. 1968, *ApJ*, **152**, 971

Tomisaka, K., Ikeuchi, S., and Nakamura, T. 1990, *ApJ*, **362**, 202

Toomre, A. 1963, *ApJ*, **138**, 385

Troland, T. H., and Heiles, C. 1986, *ApJ*, **301**, 339

Vrba, F. J., Coyne, G. V., and Tapia, S. 1981, *ApJ*, **243**, 489

____ . 1993, *AJ*, **105**, 1010

Watson, W. D. 1976, *Rev. Mod. Phys.*, **48**, 513

Watson, W. D., and Salpeter, E. E. 1972, *ApJ*, **174**, 321

Wood, D. O. S., Myers, P. C., and Daugherty, D. A. 1994, *ApJS*, **95**, 457

JETS ASSOCIATED WITH YOUNG STARS

T.P. RAY
Dublin Institute for Advanced Studies
5 Merrion Square, Dublin 2, Ireland

1. Historical Introduction

1.1. HERBIG-HARO OBJECTS

It is almost half a century since the discovery by Herbig (1951) and Haro (1952) of the enigmatic nebulosities that bear their name. It was clear from very early on that HH objects were associated in some way with star formation and, in fact, it was originally suggested that they were actual sites were stars were made (Herbig, 1970). To this day, popular texts, even such erudite volumes as the Cambridge Atlas of Astronomy, still make statements like "... we also find in this class stars called Herbig-Haro objects, which are stars in the process of being formed"! Of course we now know this idea to be incorrect and as early as 1958 (Osterbrock 1958) it was proposed that HH objects might be the result of an energetic outflow from a young star. Observational support for this idea, however, was lacking until Schwartz (1975) noted the resemblance between the spectra of HH objects and those of knots in supernova remnants. This led him to propose that HH objects were due to radiatively cooled shocks driven by supersonic winds from young stars. Work by Raymond (1976, 1979), Dopita (1978) and others confirmed this suggestion and they found good agreement between theoretical shock models and observed line intensities. Estimated shock velocities, derived from fitting various line ratios, ranged from around 50kms^{-1} up to about 300kms^{-1} depending upon the degree of excitation of the HH object.

The *coup de grâce* that ensured, however, that the outflow model for HH objects was correct was the discovery that many of them had tangential velocities, i.e. proper motions, directed away from young stars (Herbig & Jones 1981; Jones & Herbig 1982). For example in the case of HH 1 and 2, Herbig and Jones (1981) found that these objects were moving in opposite directions at velocities up to 350kms^{-1} away from their source. In

539

K. C. Tsinganos (ed.), Solar and Astrophysical Magnetohydrodynamic Flows, 539–566.
© 1996 *Kluwer Academic Publishers.*

this case it was initially thought that the source was the so-called Cohen-Schwartz star (see Fig. 1) although it is now clear that it is in fact a much more embedded object discovered by the VLA (Pravdo *et al.* 1985) that is located closer to the centre of the HH 1/2 system. Rather interestingly although HH 1 and HH 2 have large tangential velocities, their spectroscopically measured radial velocities are quite low ($V_{rad} \lesssim$ 10-40 kms^{-1}). It follows that the velocity vectors of this outflow must lie close to the plane of the sky. Both HH 1 and 2 are bow-shaped, particularly HH 1, a point to which we will be returning to later. One remarkable and obvious feature of Fig. 1 is the narrow jet emanating from VLA 1 that points towards HH 1. This was amongst the earliest jets to be found (Strom *et al.* 1985) and their discovery marked a watershed in our understanding of star formation.

1.2. THE DISCOVERY OF JETS FROM YOUNG STARS

Early models, for example Königl (1982), proposed that HH objects arose from the interaction of clumps of matter with oppositely directed supersonic flows i.e. jets. It was not suggested at this stage that some HH objects actually represented parts of jets. Thus there were some similarities, for example, between Königl's (1982) hypothesis and the model put forward by Schwartz (1978) a few years earlier, in which a HH object resulted from the interaction of a supersonic stellar wind with dense cloudlets in the surrounding medium.

In his classic review paper, however, Schwartz (1983) noted several "alignments", i.e. strings of HH objects lying on a straight line through an infrared source. These included HH 7-11, HH 24 and a quite striking outflow HH 46/47 (Dopita, Schwartz & Evans 1982). In the latter case, several HH objects, including a "bridge of emission", were found to be flowing with negative radial velocity away from an embedded infrared source, while in the opposite direction one HH object (HH 47C) was found to have a similar but positive radial velocity. This was a clear example of bipolar ejection from a young star. The first time the phrase "jet" was used in the context of observed outflows was by Mundt and Fried (1983) when they announced the discovery of jets from DG Tau B, HH 30 and HL Tau (Fig. 2) systems. Within a few years, many more jets were found (Mundt *et al.* 1984; Strom *et al.* 1985; Reipurth *et al.* 1986; Ray 1987) including what probably remains to this day the archetypal example of its class, the HH 34 system (Mundt 1986; Reipurth *et al.* 1986; Bührke, Mundt & Ray, 1988). With these discoveries came the realisation that several previously discovered HH objects were either parts of jets (e.g. HH 7-11, HH 12, HH 30) or represented the "working surface" where a jet ploughed into its surrounding medium (e.g. HH 1 and HH 34).

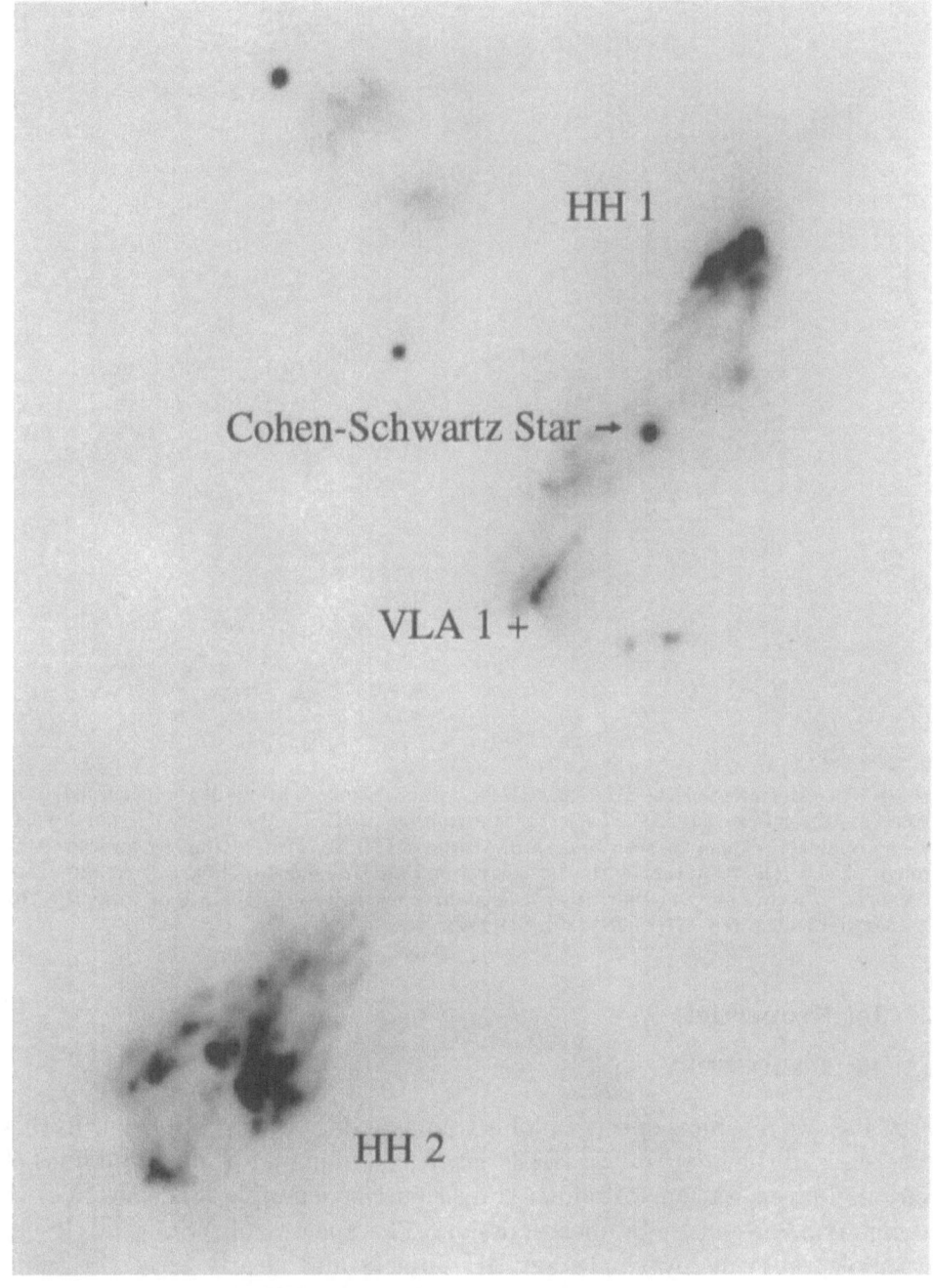

Figure 1. The HH 1/2 region taken in [SII]$\lambda\lambda$6716,6731 light with the New Technology Telescope (NTT) in Chile. Both HH 1 and HH 2 were found to be moving away, roughly in the plane of the sky, from the radio source (VLA 1) located approximately halfway between the two objects. Note the jet from VLA 1 directed towards HH 1 along its symmetry axis. North is up in this and all subsequent images unless specified otherwise.

542

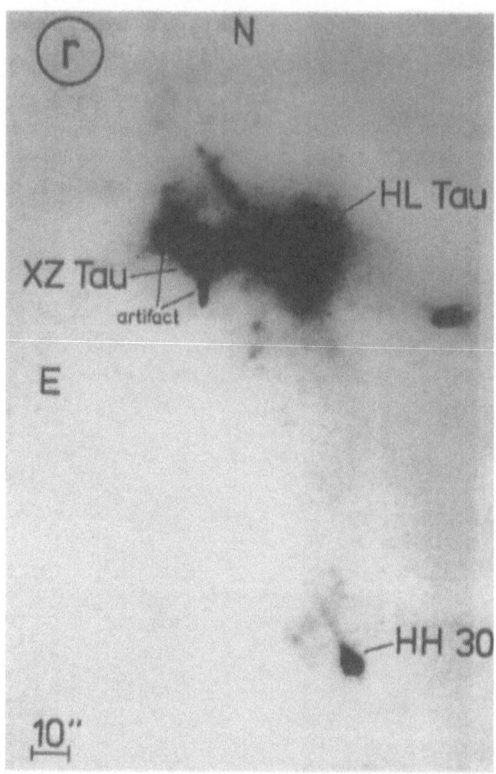

Figure 2. An image of the HL Tau/HH 30 region taken with one of the earliest CCDs from Mundt & Fried (1983). The jet in the vicinity of HL Tau points to the north-east. A much fainter jet can be seen emanating from HH 30. It is interesting to compare this image of the HH 30 system with the corresponding HST image (Fig. 12) to illustrate advances in technology and facilities. The r filter used here included light from the Hα line as well as the red [NII], [OI] and [SII] doublets.

2. Jet Properties

2.1. SPECTROSCOPY

Fig. 3 shows a typical spectrum of a young stellar object (YSO) jet, in this case some of the knots of the HH 34 jet. It is dominated by emission lines of various low excitation forbidden transitions for example [SII]$\lambda\lambda$6716,6731 along with, of course Balmer emission. The spectra of YSO jets closely resemble those of low-excitation HH objects and clearly have the same origin i.e. radiative shocks. In contrast with some HH objects there are no known high excitation jets, for example with strŏng [OIII]λ5007 emission. It should be emphasised however that where high excitation HH objects are found these can usually be identified with bow shocks where a jet is moving into its surroundings and thus driving a powerful shock. Shock models for

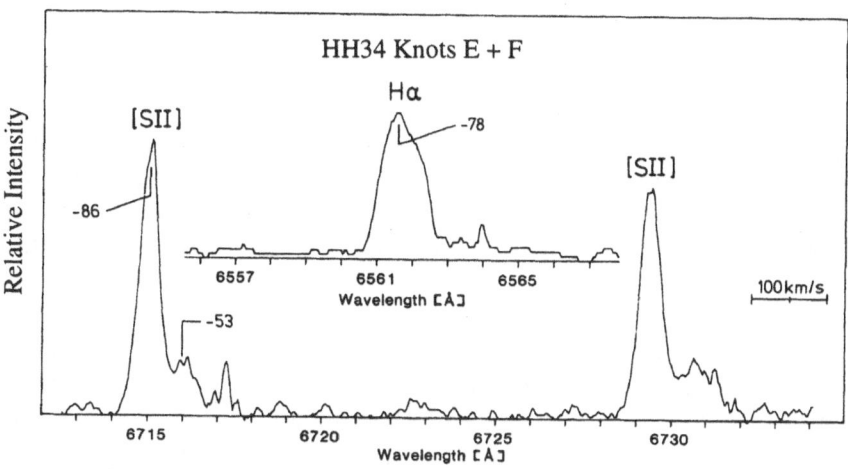

Figure 3. Hα and [SII]λλ6716,6731 spectrum of the HH 34 jet Knots E + F from Bührke, Mundt & Ray (1988). The [SII] emission is stronger than the Hα emission and this is typical of many HH jets.

YSO jets typically imply shock velocities V_{shock} in the range 30-60 kms^{-1}.

Radial velocities measurements show that jets are often blueshifted although bipolar systems (e.g. LkHα 233, Corcoran & Ray 1996) exist. The reason for this are obvious: a jet will only be optically visible if it is not too deeply embedded. YSOs on the other hand form inside molecular clouds, and thus optically we will preferentially pick up systems forming on the edges of clouds with the redshifted jet directed inwards and the blueshifted jet directed outwards i.e. towards us. Typical radial velocities of jets vary and depend to some degree, as one might imagine, on the luminosity of the source. For low mass stars, i.e. $M_* \lesssim 2M_\odot$ $V_{rad} \leq$100-200 kms^{-1} and may be twice as high, or more, for intermediate mass YSOs. This raises an interesting question, why, if the radial velocities of YSO jets are so high (and of course actual jet velocities, V_{jet} even higher), are the jet shock velocities so low? There are two obvious ways around this problem: firstly the internal jet shocks could be oblique or alternatively they might be generated by small variations (e.g. in velocity with $\delta v/v \approx$ 10-20%) in the outflow from the source. As we shall see (§3.1), such temporal changes can give rise to internal working surfaces (smaller scale versions of what occurs at the head of a jet) along its length. We will address the question of which model is more applicable later.

2.2. MORPHOLOGY

YSO jets come in a variety of shapes and sizes. Some representative examples include L1551 IRS 5 (Davis *et al.* 1995), HH 30 (López *et al.* 1995), HH 34 (Bührke, Mundt & Ray 1988) and HH 111 (Reipurth, Raga & Heathcote 1992). All of them contain knots which are often quasi-periodically spaced. The origin of these knots is a matter of discussion (see §3.1) but it seems natural to identify them with the internal shocks mentioned above. Many jets can be resolved transversely, i.e. we can determine their diameters, at least up to a few arcseconds from their source, using even ground based telescopes. Opening angles of YSO jets, defined by the diameter of the jet over the distance to the source, vary enormously (Mundt, Ray & Raga 1991) but at angular distances of 10 or more arcseconds they can be as small as a few degrees (for example HH 30) to as much as 15° (for example the DG Tau B counterjet). Certainly, as we shall see, there are optical outflows which cannot be described as jets, particularly in association with larger luminosity sources (e.g. Cep A, see Corcoran, Ray & Mundt 1993) with much bigger opening angles. Moreover opening angles along individual jets changes with distance from the source (Mundt, Ray & Raga 1991).

In a number of cases, YSO jets are found to be associated with conical reflection nebulae, for example Haro 6-5B (Mundt, Ray & Raga 1991), 1548C27 (Mundt *et al.* 1984) and L 1551 IRS 5 (see Fig. 4) to name but a few. Rather interestingly, in these cases it is always found that the jet axis is along the symmetry axis of the reflection nebula.

2.3. PHYSICAL PARAMETERS

From the type of emission line spectra seen in YSO jets, we can determine that the temperatures in the emitting gas are around 5×10^3–10^4 K. These are, of course, average temperatures in the radiative cooling zones of the shocks, even higher values are reached in the immediate post-shock zone. Once we have a temperature, we can determine a sound speed and, when combined with the velocity of a YSO jet, this fixes its Mach number. Estimates of their Mach numbers $M_j \approx 20$–100 i.e. they are hypersonic. Providing we are not too close to the source where values are higher than critical, electron densities can be estimated for example using the ratio of the [SII]$\lambda\lambda 6716,6731$ lines. Characteristic values $N_e \approx 10^2$–10^4 cm^{-3} are found in most cases. Given the derived shock velocities, as evidenced by the excitation line ratios and using N_e, one can estimate pre-shock densities. These values in turn allow us to calculate jet mass loss rates $\dot{M}_{jet}$ assuming the diameter of the jet is known. Early estimates (for examples, Mundt, Brugel & Bührke 1987) of $\dot{M}_{jet}$ were too low as they underestimated the neutral fraction in the jet (see, for example, Raga, Binette & Cantó 1990). More

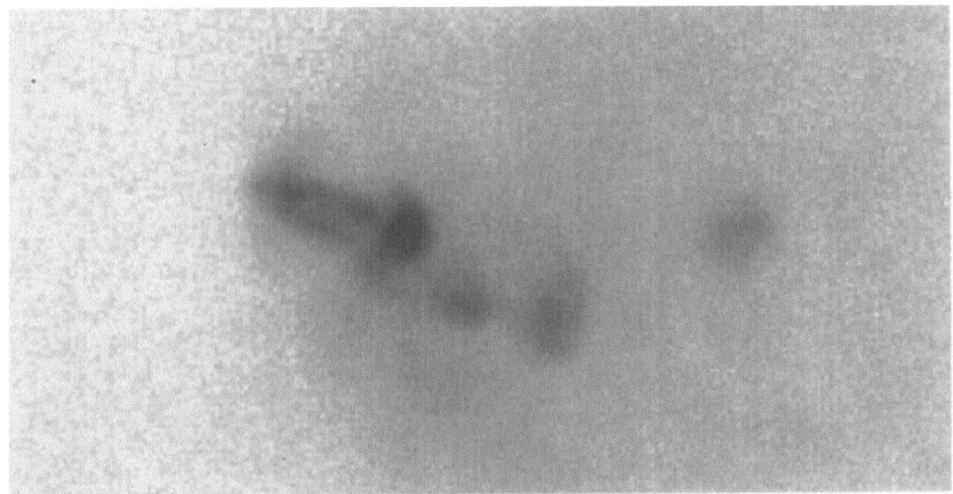

Figure 4. The L 1551 IRS 5 jet in [SII]$\lambda\lambda$6716,6731 light. This jet appears poorly collimated compared to others like the HH 34 and HH 111 systems. Note the faint conical reflection nebula surrounding the jet and that the jet lies on its central axis. In the past the presence of such nebulae have been taken to infer the existence of a neutral wind in addition to the jet component (see §5.1).

recent calculations (Hartigan, Morse & Raymond; Bacciotti, Chiuderi & Oliva 1995) suggest that more than 90% of the jet may be neutral. The extra mass flux in the jet turns out to have important consequences, as we shall see later on, in connection with molecular outflows (see §5.1). Current estimates place $\dot{M}_{jet}$ in the range 10^{-8}–10^{-6} $M_\odot yr^{-1}$ depending largely on the luminosity of the YSO.

Observed YSO jet lengths vary enormously. For example Hirth (1994), Hirth, Mundt & Solf (1994) have shown that the high velocity forbidden line emission observed in many classical T Tauri stars (cTTSs) is due to small-scale ($\lesssim$ 2") jets (see Fig. 5). Such systems have detected jet lengths of at most a few hundred AU. At the other extreme are the so-called "super flows" (Fig. 6) the discontinuous emission of which stretches for 10^5 AU or more. Examples include the HH 34 (Bally & Devine 1994) and HH 1/2 (Ogura 1995) systems. In these cases bow-shaped HH emission is found well beyond known HH objects suggesting the latter are not ploughing into the surrounding interstellar medium but instead are already moving into gas from a previous outflow episode. This idea is supported not only by the existence of super flows but by comparing proper motion and radial velocity data with estimated shock velocities of well-known bow-shaped HH objects (see, for example, Morse *et al.* 1994).

Where bipolar jet systems are observed, both jets are often found to differ considerably in their properties and these differences cannot be ex-

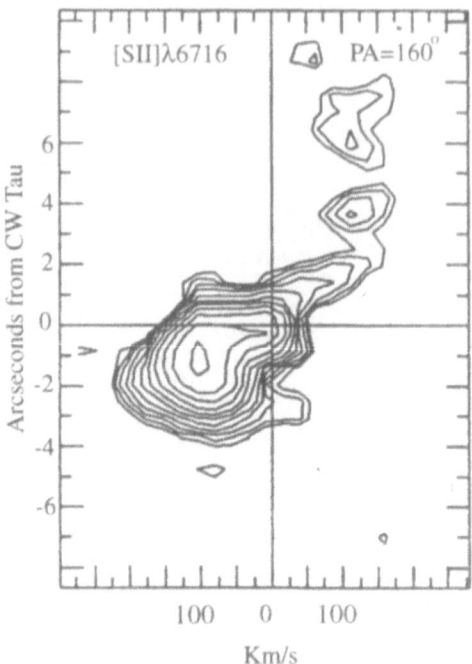

Figure 5. A long slit spectrogram (at a position angle of 160°) of CW Tau in [SII]λ6716 light. Note the spatial extent of the emission at both red and blue-shifted velocities. If one takes spectra through other position angles (see Hirth, Mundt & Solf 1994), the spatial extent of the emission is found to be reduced implying the collimated outflow is in the direction of 160°. This spectroscopic technique is particularly effective when isolating HH jets close to nebulous stars.

plained by extinction effects alone. For example jets and counterjets can differ as regards excitation, opening angles, electron densities (Mundt, Ray & Raga 1993) and even velocity (Hirth *et al.* 1994). Such asymmetries are probably due to differences in environment rather than intrinsic differences in the initial bipolar outflow from the source.

By combining the projected lengths of YSO jet systems with their velocity, one can derive a rough dynamical timescale τ_{dyn}. The usefulness of this timescale however must be questioned as at best it appears to be nothing more than a lower limit to the age of the HH outflow. In part this is because the existence of "super flows" implies that, in at least some cases, what has been considered in the past to be the "working surface" and terminus of the jet has turned out to mark nothing more than a drastic change in the properties of the outflow "τ_{dyn}" ago. It would seem reasonable to suspect that with the increasing use of very large format CCDs, many more such "super flows" will be discovered in the future. Moreover another reason to

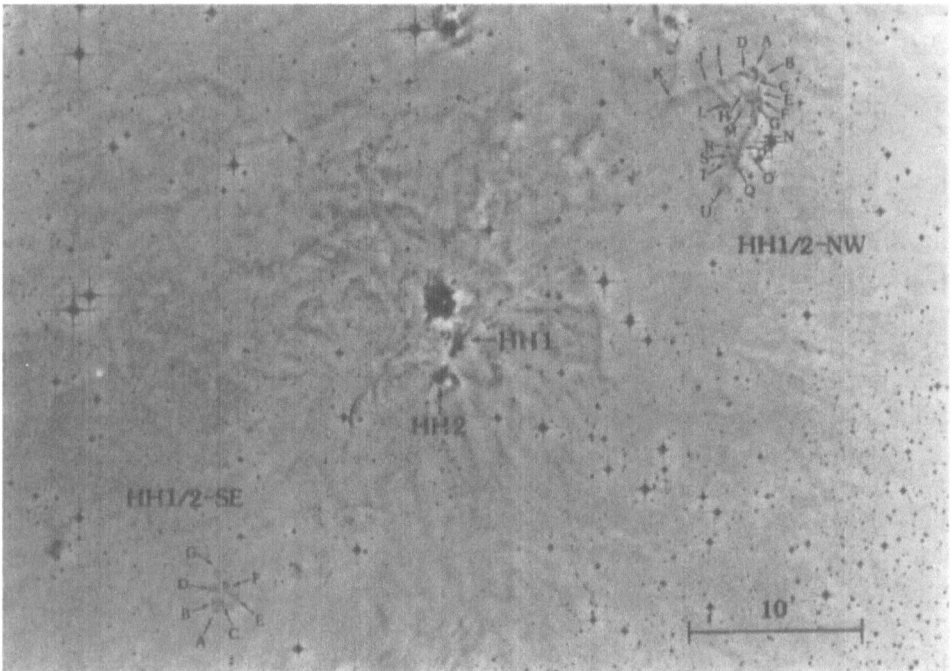

Figure 6. The HH 1/2 super flow from an image kindly supplied by Katsuo Ogura (Ogura 1995). The dynamical timescale of this outflow is at least 3×10^4 yrs although the YSO outflow phenomenon in T Tauri stars is thought to last even longer (see text).

treat τ_{dyn} with suspicion is that in many cases tracing HH outflows can be very difficult once the flow has moved beyond the boundary of the parent cloud. In particular the blueshifted outflow, which is usually on the edge of the ambient cloud anyway, can extend well beyond it and the resultant bow shock could be very dim.

2.4. SOURCES OF YSO JETS

Historically speaking, most of the earliest discovered YSO jets were found to be associated with low mass stars. There are several reasons for this: firstly the type of low mass young star that generates jets i.e. cTTSs, or embedded young stars of similar luminosity, are relatively speaking quite common compared to their high mass counterparts. Secondly lower mass stars evolve less quickly, so presumably they remain in the jet phase for longer. Finally background low excitation HII emission can sometimes make the identification of HH objects/jets more difficult, certainty through imaging alone (see, for example, Poetzel, Mundt & Ray 1992). In the past few years, however, we can come to realise that optical outflows, including jets, occur not only in cTTSs, and stars of a similar ilk, but in their more luminous counterparts the Herbig Ae/Be stars as well. A discussion of the optical outflows seen in the case of the latter is given in Mundt & Ray (1994) and space limitations prevent a review of their properties here.

What appears to be common to all jet/optical outflow sources is that they show evidence for a disk. Moreover it would seem that such disks are actively accreting. Fig. 7 shows the spectral energy distribution of a typical jet source HL Tau. In this case the star is a cTTS i.e. one in which the equivalent width of the Hα emission line is greater than 10Å. The optical part of the spectrum is well-fitted by a Kurucz (i.e. standard) blackbody-like stellar photosphere model, however at infrared/mm wavelengths there is an "excess". It has been shown (see, for example, Edwards, Ray & Mundt 1993) that the degree of excess is in large part due to disk accretion and that the accretion rate determines the mass loss rate through the jet. On a more qualitative note, the connection between disks and outflows is further underpinned by the observation that no known jets have ever been found to be associated with weak-line T Tauri stars (wTTSs) which in most respects resemble the cTTSs with the difference that their spectral energy distributions do not show any evidence for a disk or, at best, considerably less massive structures with central holes (Osterloh & Beckwith 1994; Sargent 1995). If the jet formation process depended solely on the star itself, we would expect to see them in the case of the wTTSs. We shall see in §5.2, that similar arguments can be put forward that outflows from Herbig Ae/Be stars also arise from disk accretion.

Of course the spectral signature of a disk is not quite the same thing as observing it directly. Here we shall not recall the extensive literature on this topic but instead the reader is referred to the excellent review by Sargent (1995). It suffices to say that flattened structures of circumstellar gas with radii of order 1000 AU have been observed around the jet sources HL Tauri (Sargent & Beckwith 1987) and DG Tauri (Sargent & Beckwith 1989) and in both cases the observed jets are perpendicular to these structures. Recent observations (see Sargent 1995) imply, however, that this gas is not rotating around these stars in a quasi-Keplerian fashion but may instead be in free-fall. A smaller scale disk-like structure has been found (with a radius of about 350 AU) around the jet source DO Tau by Koerner & Sargent (1995) at a PA $\approx 160°$). Here the jet has a PA $\approx 70°$ (Hirth et al. 1994) i.e. it is perpendicular to the "disk" but interpretation of the velocity field of this structure is perhaps complicated by a mixture of infall, rotation and outflow (Koerner & Sargent 1995).

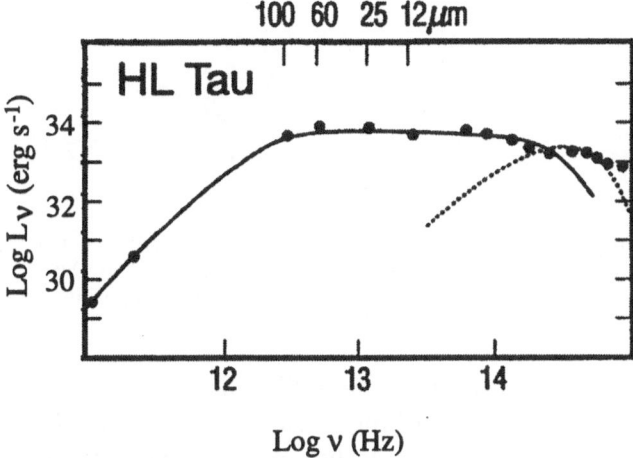

Figure 7. The spectral energy distribution of the classical T Tauri star HL Tau, a known jet source (Mundt *et al.* 1990). While the optical emission can be well fitted by a blackbody-like spectrum, the excess at infrared/mm and uv wavelengths cannot. The excesses are most readily explained as arising from accretion through a disk. The processes of disk accretion and the generation of jets/outflows seem to be intimately linked (see Edwards, Ray & Mundt 1993). Adapted from Sargent (1995).

3. The Propagation of YSO Jets

3.1. WHAT IS THE ORIGIN OF THE KNOTS?

As discussed earlier on, apart from the large scale bow shock-shaped features seen in an number of optical outflows, one obvious morphological characteristic of YSO jets is that they are knotty. Moreover these knots are sometimes quasi-periodically spaced. Originally Falle, Innes & Wilson (1987) proposed that these knots were stationary shocks (reminiscent of those seen in the exhausts of supersonic jet aircraft). However this idea fell out of favour on two accounts: firstly the knot spacing in outflows like the HH 34 jets are typically a few jet diameters apart whereas the stationary shock model would predict a spacing of around $M_j D_j$. More damning still was the discovery that the knots move (see, for example, Eislöffel & Mundt 1994; Eislöffel, 1995) with velocities comparable to the typical velocities of the jets. Obviously this cannot occur if they are stationary shocks!

Borrowing from an earlier idea expressed by Rees (1981) to explain the knots of the M87 jet, Raga & Kofman (1992) proposed that YSO jet knots were due to temporal variations in the outflow from the source. By altering say the jet injection velocity by 10-20%, internal shocks are generated in the jet. Such shocks would, in essence be internal working surfaces (see next section) and in time develop into small-scale bow shocks (see, for example,

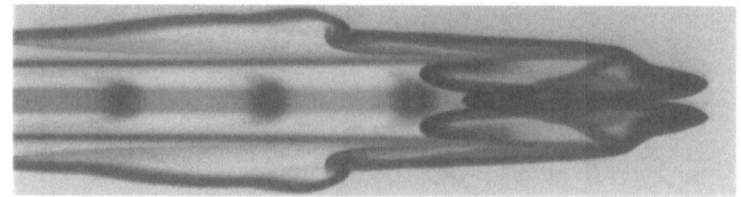

Figure 8. Jet simulations in which the output from the source, i.e. the injection velocity, varies with an amplitude of 10% (from Downes, 1996, see also Downes, Ray & Drury, 1996). Note that as an internal shock develops, its edges steepen into a bow shock.

Fig. 8 and Downes 1996). Of course unlike the stationary shocks mentioned previously, these internal working surfaces are expected to move out at a velocity comparable to that of the jet. Moreover, in the idealised case, i.e. for a steadily varying outflow, they are expected to fade away steeply with distance from the source (see Raga and Kofman 1992). Reality. as always is somewhat more complicated as exemplified by the proper motion data (Eislöffel 1995) although there seems little doubt now that temporal variation of the outflow is responsible for knots in at least some jets. It has been suggested (see, for example, Bodo *et al.* 1994, Bührke, Mundt & Ray 1988, Massaglia *et al.* , 1996, Ray & Mundt 1993) that Kelvin-Helmholtz (KH) pinching instabilities may be a possible cause of the knots. Although this cannot be ruled out, there is still no observational support for this idea. In particular if KH instabilities are to generate the correct knot spacing, one requires either a hot tenuous medium outside the jet (in order to effectively increase the external sound speed while maintaining pressure balance) or the external medium needs to be moving at a velocity comparable to that of the jet (see, for example, the simulations of Gouveia Dal Pino, Birkinshaw & Benz (1996) where KH instabilities occur at the head of the jet at the interface between the jet and the surrounding cocoon).

It should be possible to test these models further in the future by time evolution studies. Incidentally on the subject of temporal variability it is worth pointing out that in a number of sources e.g. L1551 IRS5 (Neckel & Staude 1987), the Serpens Radio Jet (see §4.2 and Curiel 1995) and HH 80/81 (Martí, Rodríguez & Reipurth 1995) knots have been seen to form at the source and move outwards with time.

3.2. BOW SHOCKS

A number of jet systems show large scale bow shocks, for example HH 34 and HH 212 (see Fig. 9). These features differ from jets not only morphologically but kinematically as well. Whereas line profiles for jets are typically narrow (with FWHM $\lesssim$ 30 kms^{-1}), those of bow shocks are much broader

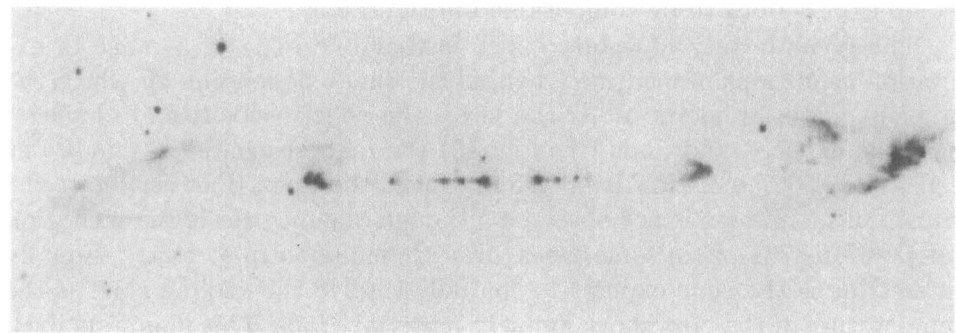

Figure 9. The HH 212 system as observed in shocked molecular hydrogen (2.12μm line) from Zinnecker & McCaughrean (1995). Several bow shocks can be seen on either side of the flow. This is the first YSO jet to be discovered entirely through shocked H_2 emission. Note that the 2.12μm emission extends to the bow shock apexes implying that the bow shock velocities cannot be high in comparison to most optically observed ones. This, however, does not necessarily mean that the underlying jet is slower, instead it could mean that the ambient density is much greater than the jet density.

(FWHM $\lesssim 300\mathrm{kms}^{-1}$) as one would expect from bow shock modelling (Hartigan, Raymond & Hartmann 1987). To be precise there are in fact two shocks in the vicinity of the head of the jet which together mark the boundaries of the so-called "working surface". At the inner jet shock, or Mach disk as it is sometimes called, jet material is decelerated while at the outer bow shock, ambient material is accelerated. In particular, it is at the working surface where one can see a major distinction between extragalactic jets and YSO jets from both an observational and modelling perspective. Here the effect of radiative cooling (insignificant in the case of extragalactic jets) is very apparent. For example while cooled post-shock ambient material is deflected after the bow shock away from the jet axis, the corresponding material from the jet is deflected inwards leading to the development of a cool cap (see, for example, Blondin *et al.* 1990). In time this cap breaks up, partially as a result of Rayleigh-Taylor instabilities, allowing the core of the jet to burst through (Raga, 1995). Whether the jet shock or bow shock is brightest depends on the relative densities of the jet and ambient material. If $\rho_{jet} \gg \rho_{amb}$, the bow shock will of course be very strong and the jet shock weak. Most of the radiation will then come from the former. If $\rho_{amb} \gg \rho_{jet}$, the situation is of course reversed. As shown by Hartigan (1989), however, for a large range of density ratios, the surface brightness of both the jet and bow shocks is comparable. This is because the surface density depends not only on the density of the emitting species, i.e. ρ, but also on the energy per unit mass in the post-shock zone, i.e. V_s^2. Now the ram pressure in the vicinity of the jet shock $\rho_{jet} V_{j,s}^2$ roughly balances the ram pressure $\rho_{amb} V_{a,s}^2$ in the vicinity of the bow shock so we would expect both to be comparable in brightness.

The great beauty of a bow shock is that one expects a range of excitation conditions immediately behind the shock depending on where the ambient material enters it. At the head, the shock velocities are highest, here one expects, and finds (see Fig. 10) the highest excitation conditions e.g. [OIII]λ5007 emission is only found near the apex. Further along the wings, such emission is not observed although one may see lower excitation [SII]$\lambda\lambda$6716,6731 or, in some cases, even H_2 emission in a "skirt". Equally interesting is that one expects to spatially resolve the cooling zone as the temperature of the post shock ambient material drops. This manifests itself as spatial offsets in the peak emission of lines of varying excitation (Bührke, Mundt & Ray 1988).

4. The View from Space and at Non-optical Wavelengths

4.1. HST IMAGING

Prior to the installation of COSTAR and the Wide Field Planetary Camera 2 (WFPC2) only a small number of observations were carried out of HH objects and associated jets with HST. For example, direct imaging by Schwarz *et al.* (1993) of HH 2 showed that a wide range of excitations levels were present within its individual knots. Some features were found to resemble small bow shocks but in the main the sub-structures looked chaotic and appeared to be the result of the HH 2 bow shock breaking-up. Kepner *et al.* (1993) used deconvolution techniques on WFPC images of the DG Tauri jet and found they could trace it to within 0.25" of the star and that it was already collimated at such small angular distances. The corresponding projected distance is only 40 AU and Kepner *et al.* (1993) inferred that the collimation scale is smaller than this. However, it has to be emphasised that this jet is almost pointing directly at us given its high radial velocity (Mundt, Brugel & Bührke 1987) so that the actual collimation scale could be much larger.

Since the refurbishment mission several outflows, including HH 1/2, HH 30, HH 34, HH 46/47 and HL Tau, have been observed in various emission line filters. Figs. 10 and 11 show WFPC2 images, taken through a [SII]$\lambda\lambda$6716,6731 filter of the HH 46/47 and HH 34 jets. These images were reduced from data in the HST Archive. Firstly in the case of HH 46/47, the detailed morphology of the jet and the HH 47A working surface is seen to be extremely complex (Reipurth & Cernicharo 1995). The jet is well-resolved and appears to consist of strands or filaments which spiral from side to side *en route* to HH 47A. The HH 47A Mach disk or jet shock (already observed by Reipurth & Heathcote 1991) from the ground, is very evident in the corresponding Hα image (Fig. 10 inset of HH 47A). Also in

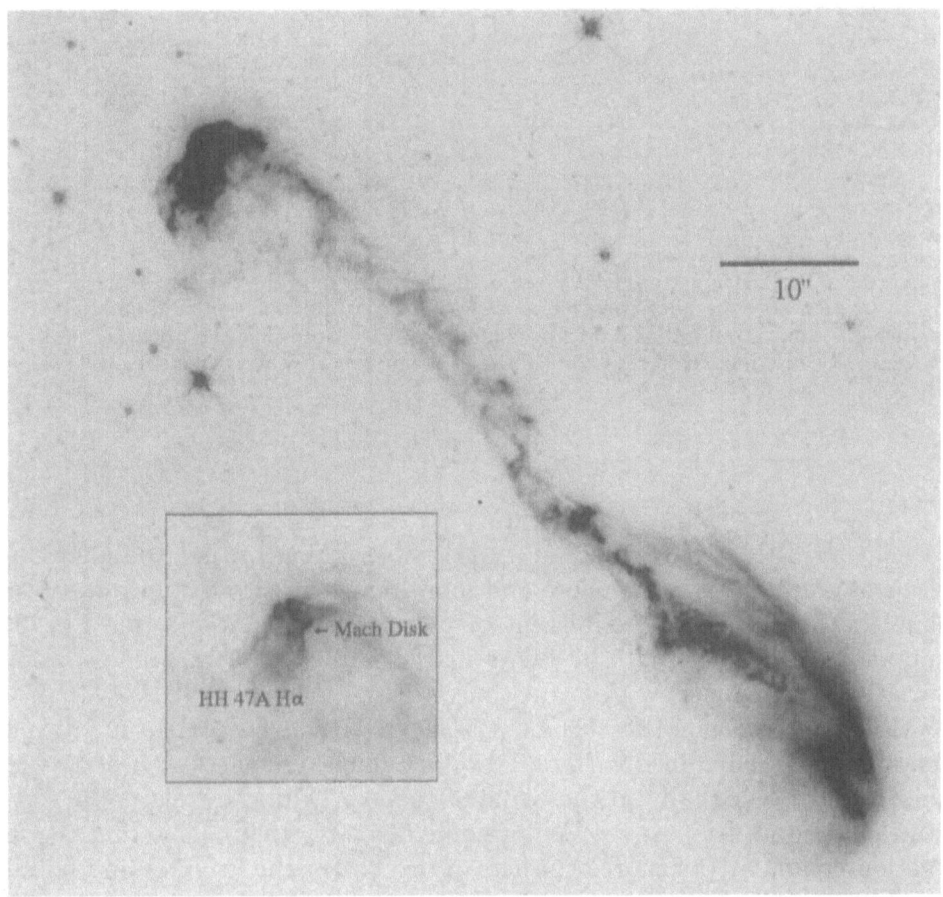

Figure 10. The HH 46/47 jet as seen in [SII]λλ6716,6731 light using the HST. The jet has a braided appearance and is seen to wiggle from side to side as HH 47A (top left) is approached. One section, perpendicular to the jet, of HH 47A is of higher excitation than the main body of the jet (see Hα inset and text). This is thought to be the Mach disk or jet shock (Reipurth & Cernicharo 1995). The cometary-like nebula in the bottom right is largely reflection nebulosity in the vicinity of the source.

the Hα image (not shown here), the jet is less pronounced but one can see high excitation regions, i.e. areas of Hα emission with little corresponding [SII]λλ6716,6731 external to the jet where its bends. This emission was also detected in ground-based images (Reipurth & Heathcote 1991) and has been attributed to sideways shocks in the ambient medium caused by changes in the jet's direction (see Biro 1994 for modelling). The cause of the wiggling motion is uncertain, precession of the jet source appears to be a plausible explanation and such wiggling is known to occur in several other jets (Reipurth 1989; Curiel 1995).

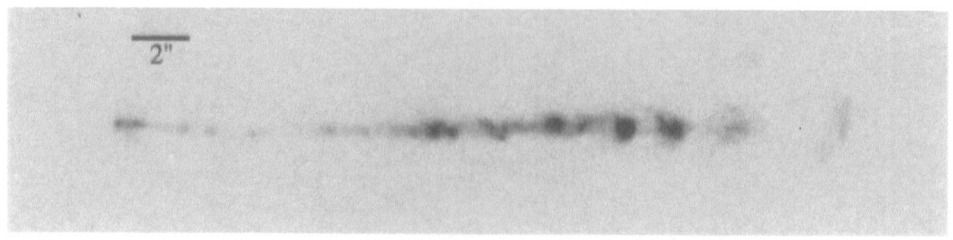

Figure 11. The HH 34 jet in [SII]λλ6716,6731 as observed by the HST with the WFPC2. It can be seen how the knots become more bow-like with distance from the source. This is precisely what a pulsed jet model predicts (see §3.1). Note that the orientation of this image has been adjusted so that the HH 34 jet is horizontal. The source is to the far left.

The WFPC2 HH 34 image [Fig. 11] is intriguing in that the knots are seen to gradually become more and more bow-like with distance from the source. This is highly reminiscent of pulsed jet simulations (see Fig. 8), supporting the idea that the knots in the HH 34 jet arise from episodic variations in the outflow from the source. Fig. 12 shows a [SII]λλ6716,6731 WFPC2 image from Ray *et al.* (1996) of HH 30 near HL Tau. Proper motion studies combined with radial velocity measurements have show that this outflow is virtually in the plane of the sky (Mundt *et al.* 1990). The HST continuum frames (see also Stapelfeldt *et al.* 1996) show that the reflection nebula at the source consists of two cusps, the brighter and fainter one being associated with the blue-shifted and red-shifted part of the flow respectively. This type of cusp structure is exactly what one expects to observe from light scattered in a flared disk seen virtually edge-on. The dark lane passing through the two cusps is presumably the disk itself (see inset to Fig. 12). The jet is seen to emerge perpendicular to the "disk" and, a comparison with the ground based image (see Mundt, Ray & Raga 1991) shows several more knots close to the source than previously detected. If we measure the diameter of the jet as a function of distance from the source, it is found (Ray *et al.* 1996) that it does not change appreciably as one approaches the source, even within 2 arcseconds. This intriguing finding of the HST data (which was suspected on the basis of ground based images, Mundt, Ray & Raga 1991) lends support to models in which the jet is not collimated until its diameter reaches 50 or more AU (see, for example, Fendt, Camenzind & Appl, 1995 for a model that might explain these findings). Finally it is worth pointing out that the HST data on some jets lends support to the idea that the knots are due to internal working surfaces, e.g. part of the HH 34 jet, while in others, like HH 30 or HH 46/47, it does not.

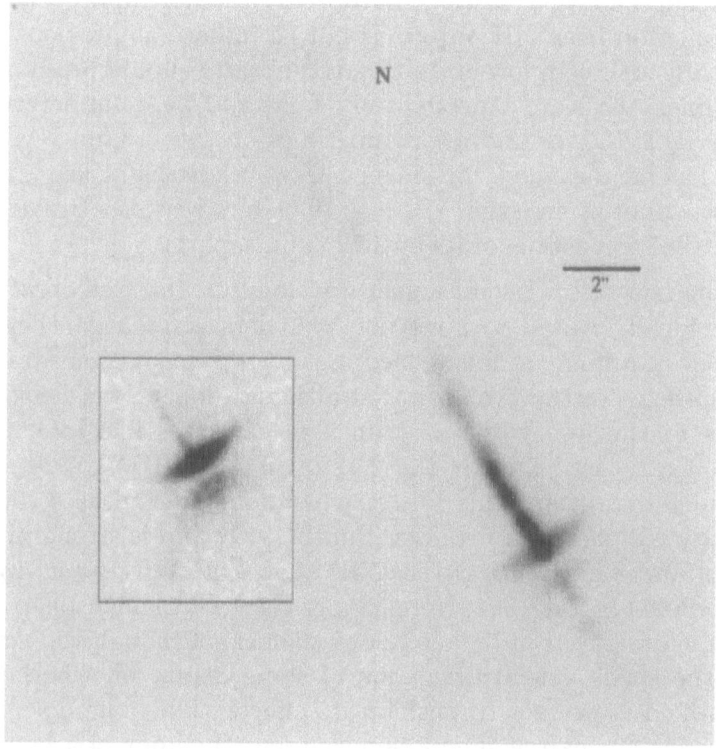

Figure 12. A [SII]$\lambda\lambda$6716,6731 minus continuum WFPC2 image of the HH 30 system from Ray *et al.* 1996. The two cusps of reflected light at the source is what one expects from a flared disk viewed edge on (see inset). The plane of the disk corresponds to the dark lane between the cusps.

4.2. OBSERVING YSO JETS AT NON-OPTICAL WAVELENGTHS.

It was stated earlier on that YSO jets derive their emission from shocks. Depending upon the shock velocity, the immediate post-shock temperatures can range from 3.10^4K – 3.10^6K for V_{shock} = 30-300 kms^{-1}, assuming the pre-shock gas is not ionised. The higher temperatures apply for example at the apexes of "terminal shocks", e.g. HH 1, and the lower values are typical of the knots seen in YSO jets. Correspondingly we expect to observe HH emission at a range of wavelengths from the UV to radio.

For example high excitation objects are expected to produce UV emission from species such as CIV, CIII] and MgII as was found by IUE (Brugel 1989). In fact, it appeared initially that the amount of reddening corrected UV emission from objects like HH 1 and HH 2 was greater than predicted by standard shock models given their optical line intensities. However, as pointed out by Cardelli and Brugel (1988) one cannot apply the galactic extinction curve in these cases. A flatter curve is more appropriate and thus the UV emission from HH objects is not as intense as previously thought. Theoretically although low excitation HH objects should produce some UV emission lines, these are expected to be faint and were not detected, for example, by IUE. IUE did however sometimes observe a faint UV continuum which, it has been argued, for longer UV wavelengths at least, is probably due to two photon emission (Brugel 1989 but see also Brugel, Böhm & Mannery 1981 for details of its optical counterpart).

Turning now to somewhat longer wavelengths, infrared observations are of course ideally suited to studying YSO jets given that they are often enshrouded in a dusty ambient medium. Efforts have concentrated on observing outflows in the near infrared although longer wavelengths studies will come to the fore with the launch of satellites like ISO. One of the most frequently used lines is the 2.12μm $\nu=$ 1-0 S(1) ro-vibrational line of molecular hydrogen. This line is particularly useful as a diagnostic of low velocity ($\lesssim$ 40 kms^{-1}) shocks (Smith 1994) and is strong in many well known outflows e.g. HH 46/47 (Eislöffel et $al.$ 1994 and see Fig. 13.), HH 1/2 (Davis, Eislöffel & Ray 1994; Noreiga-Crespo & Garnavich 1994). Molecular hydrogen is destroyed in higher velocity shocks although the actual upper limit to the shock velocity depends to some extent on whether or not a magnetic field is present to cushion its effects. Thus one does not expect to find molecular hydrogen emission say near the apex of a bow shock (although this can occur to some degree because of the engulfing of ambient material, see Raga 1995). Instead 2.21μm emission is a useful tracer of the lower excitation regions between a collimated jet and its molecular ambient environment. In particular it is expected to be excited in the oblique bow shock wings (see Fig. 13) where the shock velocities are low. Some examples of outflows seen in shocked H$_2$ are shown in Figs. 13 and 14. In Fig. 13 we see the HH 46/47 counterflow in shocked molecular hydrogen. Note the extensive, optically invisible, counter bow shock, calculations (see Eislöffel et $al.$ 1994) would indicate that ambient material, accelerated at this bow shock, is responsible for the redshifted CO lobe associated with HH 46/47. Another very recently discovered example (Davis et $al.$ 1996) is the shocked molecular hydrogen outflow associated with Lynds 1634 (Fig. 14). Here we find multiple bow shocks straddling either side of the central source (IRAS 05173-0555) which is the origin of a known CO outflow (Fukui 1989). In this case no H$_2$ jet is seen, unlike the counterflow in HH 46/47 but presumably this could be because any jet contains very little molecular material. Note that the Lynds 1634 flow contains the RNO 40 optical flow (Bohigas, Persi & Tapia 1993) and that the latter is not a separate flow (as pointed out by Hodapp 1994).

Unlike their extragalactic counterparts, YSO jets have proved to be

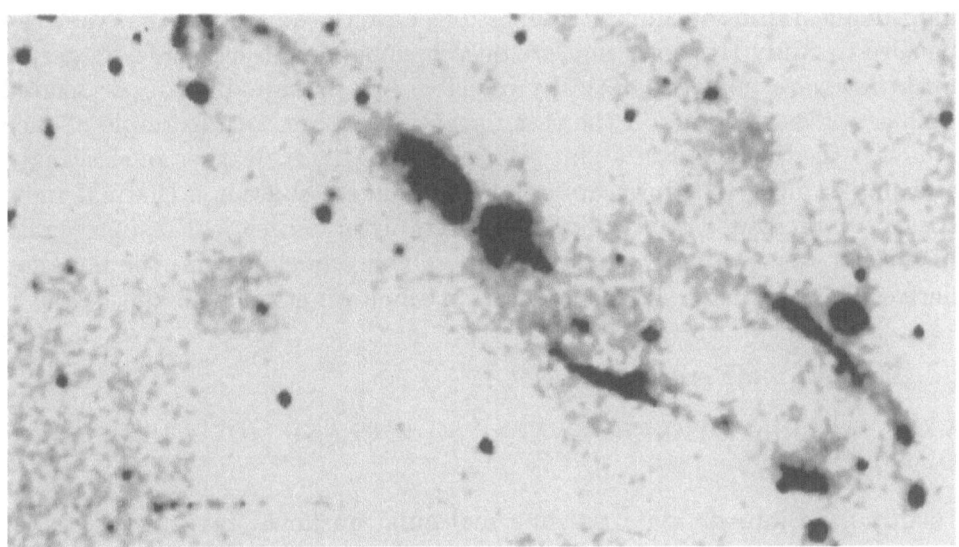

Figure 13.   The HH 46/47 outflow in H$_2$ 2.12μm emission from Eislöffel *et al.* 1994. Optically only HH 47C, the apex of the counterflow to the south-west is seen. In the near-infrared, however, we see the extensive wings of the counter flow bow shock as it ploughs its way through the Bok globule associated with HH 46/47. Note also the counterjet pointing to the south-west, clearly visible at 2.12μm but optically very faint. HH 47A borders the top centre left position.

L 1634

Figure 14.   The L1634 outflow in H$_2$ 2.12μm emission from Davis *et al.* 1996. Note the large number of bow shocks seen either side of the YSO (IRAS 05173-0555) which is slightly to the left of centre and the source of a CO outflow (Fukui, 1989).

difficult, although not impossible, to observe at radio-wavelengths. The ionised component of an outflow produces free-free emission but it is very weak. Despite this difficulty, the number of YSO jet detections has been increasing steadily since the first radio continuum observations of HH 1/2 over a decade ago (Pravdo *et al.* 1985). For a comprehensive review of this topic the reader is referred to Curiel (1995) and it suffices here to say that, in principle, radio continuum observations have the advantage that they can be used to study HH flows that are deeply embedded in molecular material. Such flows are not only optically invisible but presumably are associated with a very early stage in the star formation process. An example of such a flow is the Serpens Radio Jet (Curiel et al. 1993). Rather interestingly, and unlike most HH flows, its outer components show non-thermal radio spectra, indicative of optically thin synchrotron emission. Presumably, as in the hot spots of extragalactic radio galaxies and quasars, the emission derives from diffusive shock acceleration albeit on a much smaller scale.

5. The Wider Perspective

5.1. THE CONNECTION BETWEEN YSO JETS AND MOLECULAR OUTFLOWS

I will not go into details regarding molecular outflows, instead the reader is referred to Henriksen (these proceedings) and, for example the excellent reviews by Bachiller & Gómez-González (1992) and Fukui et al. (1993). It suffices here to say that YSO molecular outflows have been known for many years at mm wavelengths (see, for example, Bally & Lada 1983). In particular large scale (0.1–1 pc) bipolar molecular lobes were discovered using various CO transitions, typically with low velocities of 10-30 kms^{-1}, but carrying significant momentum (Fukui, 1993). The masses of these flows were found to be so high that they had to be accelerated ambient material and they did not originate at the YSO itself. Unlike HH jets these flows were found to be poorly collimated.

In this section we shall focus on the possible connection between HH and molecular flows, the simultaneous study of which has proved difficult because of their conflicting observational requirements. In particular, if sufficient accelerated ambient material is present to constitute a readily observed molecular outflow, this usually implies that the optical extinction towards it is high. This in turn means that it is difficult to optically observe any associated HH flow. Despite this problem, progress has been made in addressing the question whether molecular outflows are somehow driven by HH type jets as we shall see.

It was originally suggested by Mundt, Brugel & Bührke (1987) that YSO jets had insufficient momentum by several orders of magnitude to drive any associated molecular outflow. As there had to be some sort of "prime mover", this led to the idea that the ionised jet was surrounded by an unseen neutral wind and that what we optically observed just represented the core of the outflow from the star (Ray 1993). This idea is now thought to be mistaken for two reasons, the first of which is straightforward and the second more subtle. Early calculations of the mass fluxes in YSO jets based on observed shock velocities and post-shock electron densities (Mundt, Brugel & Bührke, 1987) have proven to be gross underestimates. In fact the neutral fraction is much bigger than previously thought (see, for example, Hartigan, Morse & Raymond 1994) leading to mass fluxes that have been revised upwards by factors of 10 or more. The second argument concerns the use of jet and molecular dynamical timescales as a measure of outflow lifetimes. Generally speaking where HH flows and molecular outflows are seen, the spatial extent of both are similar. Of course if you divide such lengths by the typical velocities of the optical and molecular flows you get dynamical timescales that differ by factors of 10-30, i.e. by the ratio of the HH to molecular velocities. Of course both measures of the outflow lifetime cannot be correct, and in fact it would appear on the basis of statistical arguments (see, for example, Parker, Padman & Scott 1991 for molecular outflows and Mundt *et al.* 1990 for optical jets) that neither are. In particular, the optical jet phenomenon lasts much longer by factors of around 10^2 than one might naïvely expect on the basis of earlier dynamical age estimates. Thus jets have much longer to supply the necessary momentum to drive any associated molecular outflow than was originally thought. Moreover calculations of the molecular outflow age based on the molecular outflow dynamical timescales were also in error: again the tendency was to underestimate the age of the outflow but here the discrepancy is only a factor of 10 or so. The net result is that both the optical outflow phase and the molecular outflow phase probably last for roughly the same length of time.

How can dynamical timescales differ from statistically estimated ones by so much? The simplest explanation is that, at least in the embedded phase, the heads of the YSO jets do not advance as rapidly as the jets themselves. Moreover there may be the additional complication of intermittency (see §3.1). Similar arguments can also be made that molecular outflows are also "quasi-stationary".

Given then that there is enough momentum in a YSO jet to drive any associated molecular outflow, how is it done? Two principle mechanisms have been proposed: gradual entrainment along the jet of ambient material (Stahler 1994) and so-called "prompt entrainment" at the head i.e. the bow shock (see, for example Chernin *et al.* 1994). Numerical simulations of adiabatic jets (De Young 1986) and more recently radiative ones (Chernin *et al.* 1994; Downes, Ray & Drury 1996) show that high Mach number flows like YSO jets, exhibit little entrainment along their length. Raga & Cabrit (1993), however, have found the opposite to be the case with pulsed jets, by virtue of the sideways expansion of their internal working surfaces. Nevertheless, observations, such as those in Fig. 15 of the RNO 43 outflow and others (Davis & Eislöffel 1995), support the idea that prompt entrain-

560

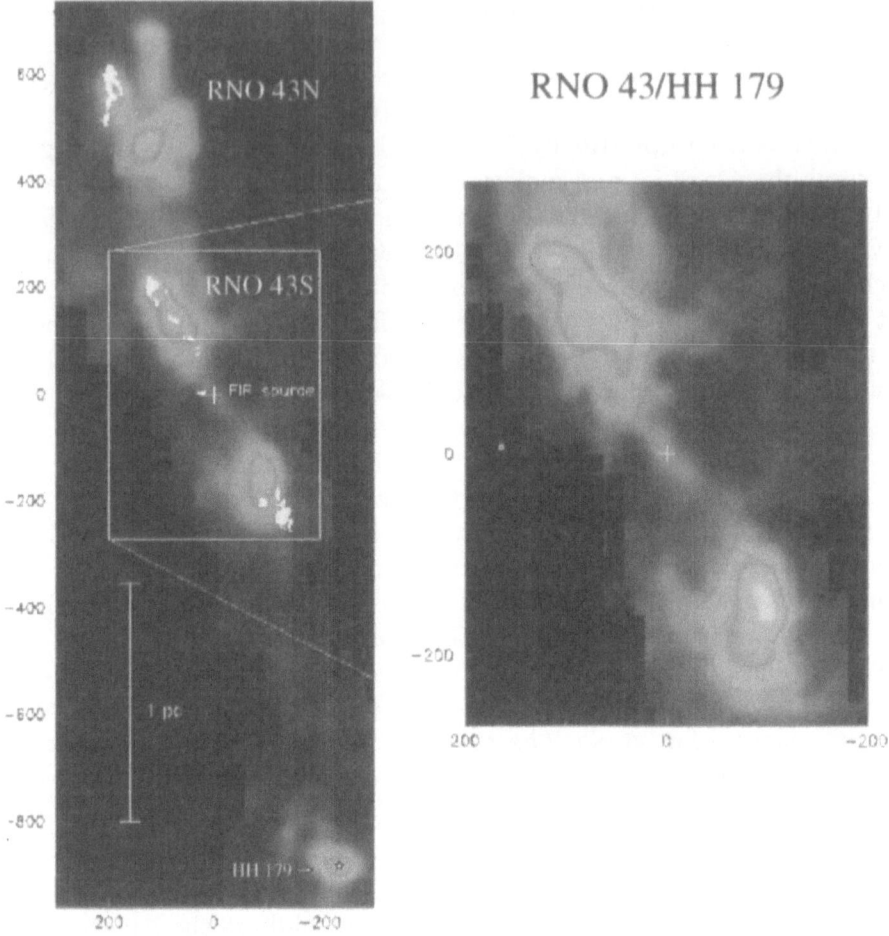

Figure 15. The molecular bipolar flow seen in ^{12}CO $J = 2 \to 1$ transition associated with the RNO 43 optical outflow and adapted from Bence, Richer & Padman (1995). The optical emission (see Mundt, Brugel & Bührke 1987 or Ray 1987) is superimposed in white. The white asterisk is HH 179. Note the similar spatial total extent of the optical and molecular emission and the close association between the molecular emission and the optical bow shocks.

ment is certainly important given the close spatial correlation between CO clumps and optical/near-infrared observed bow shocks.

5.2. THE DISK-JET CONNECTION: A CASE STUDY

In §2.4 it was stated that there is a clear connection between disks and YSO jets in the sense that if we see a YSO jet, or for that matter any HH outflow, there is evidence for the presence of a disk. The converse is, however, not necessarily true. A clear illustration of the disk-jet connection is the previously mentioned dichotomy between cTTSs and wTTSs as regards their outflow properties: for low mass stars, YSO jets are always associated with cTTSs, or what appear to be their embedded counterparts. In comparison, wTTSs show no evidence for the presence of a disk, do not have strong winds and certainly show no signs of extended outflows. This is in spite of the clear similarities between cTTSs and wTTSs as regards basic *stellar, as opposed to circumstellar*, properties.

To show, however, the connection between disks and jets in more detail, it is instructive to look at a recent study (Corcoran & Ray 1996) of Herbig Ae/Be stars. These, as mentioned previously, are the higher mass analogues of the TTSs. I will not review their outflow properties here (see, for example, Mundt & Ray 1994) but merely concentrate on first the evidence for disks around these stars and second the disk-outflow connection.

Spectroscopically Herbig Ae/Be stars, like many cTTSs, are found to show low velocity forbidden line emission and, in a number of cases, high velocity forbidden line emission as well (Corcoran & Ray 1996). Moreover, where high velocity forbidden emission lines are seen they can always be traced to a jet. Equally important is the finding that the low velocity forbidden line emission (which is very common amongst Herbig Ae/Be stars) is almost always blueshifted by values ranging from -10 kms^{-1} to -100 kms^{-1} with respect to the systematic velocity of the star (Corcoran & Ray 1996). As in the case of the cTTSs (see Edwards, Ray & Mundt 1993), this suggests the presence of an obscuring disk, that blocks our view of the redshifted low velocity component.

The above argument for the presence of disks around Herbig Ae/Be stars is a geometrical one. Can we determine a link between say the level of outflow activity and the accretion rate in our putative disks? Details of how this is done, for cTTSs at least, is given in Edwards, Ray & Mundt (1993). Briefly though, the strength of the forbidden line emission of Herbig Ae/Be stars is, of course, a gauge of outflow activity. Or, if we wish to use a relative measure of activity (with respect to the luminosity of the star) then the equivalent width of a forbidden line may be employed. A relative measure of accretion activity is harder to find but it turns out, as in the

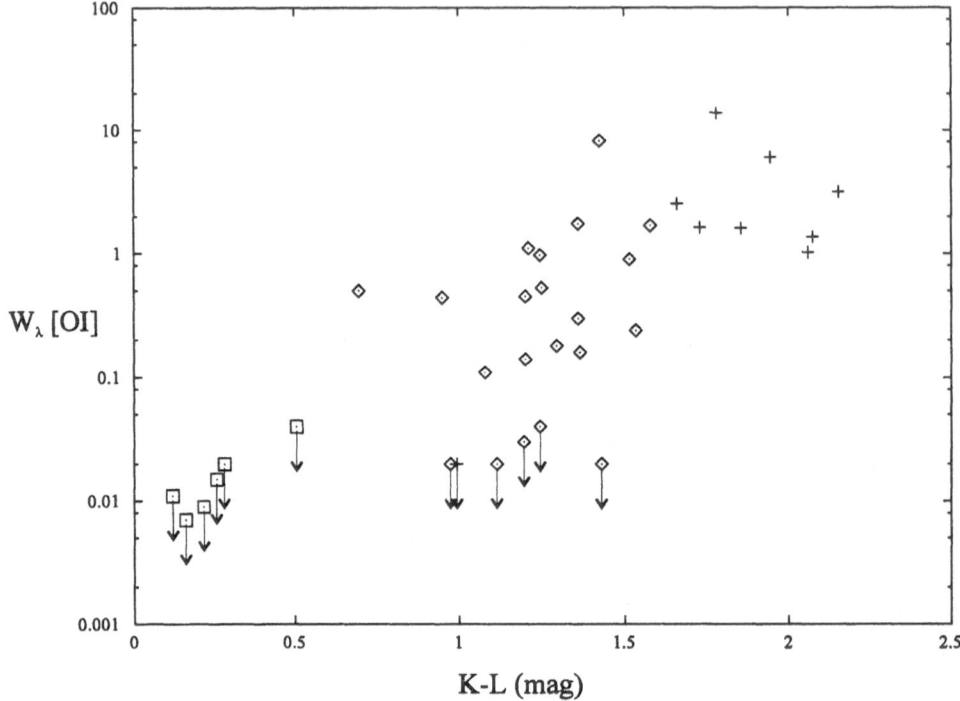

Figure 16. A plot of the equivalent width of the [OI] forbidden line against K-L colour for the Corcoran & Ray (1996) sample of Herbig Ae/Be stars. It is seen that as K-L increases the relative strength of the outflow also increases. Good arguments can be made that the K-L colours for these stars are measures of disk accretion activity (Corcoran & Ray 1996) and thus disk activity and outflows appear to be directly linked in Herbig Ae/Be stars. Downward arrows indicate upper limits.

case of cTTSs (see Edwards, Ray & Mundt 1993), that the star's H-K or K-L colours are appropriate as one might expect on theoretical grounds. If the equivalent width of the forbidden line emission is plotted against, say, K-L colours (Fig. 16) then it is found that the relative strength of the forbidden line emission scales with the colour excess as in the cTTSs (Corcoran & Ray 1996). Moreover there is a continuous smooth transition in the correlation between infrared excess and forbidden line luminosity from T Tauri stars to Herbig Ae/Be stars implying that the same basic physical outflow process must be going in both groups of stars. The clear inference is that the strength of the outflow scales with the rate of accretion.

Finally there is one more interesting point to be made about Fig. 16. It can easily be shown (see Edwards, Ray & Mundt 1993) that a standard optically thick but geometrically thin, flat accretion or reprocessing disk has a K-L colour of 0.8. No forbidden line emission is detected (Fig. 16) in Herbig Ae/Be stars with K-L $\lesssim$ 0.8 even though one might reasonably expect it by extrapolating back the W([OI]) v. K-L relationship from higher K-L values. [OI] emission is only found in stars with redder colours. The fact that the cut-off occurs at the value for a standard disk obviously suggests the optically thick disk hypothesis is the correct one since it is relatively easy to suggest ways *to increase* K-L values above those for a standard disk by, for example, non-standard disk geometries (e.g. flared disks) or temperature distributions although the reverse is not true.

References

Bacciotti, F., Chiuderi, C., Oliva, E., 1995, A&A, 296, 185

Bachiller, R., Gómez-González, J., 1992, A&A Rev., 3, 257

Bally, J., Lada, C.J., 1983, ApJ, 265, 824

Bally, J., Devine, D., 1994, ApJ, 428, L65

Bence, S.J., Richer, J.S., Padman, R., 1995, MNRAS, in press

Biro, S., 1994, PhD Thesis, University of Manchester

Blondin, J.M., Fryxell, B.A., Königl, A., 1990, ApJ, 360, 370

Bodo, G., Massaglia, S., Ferrari, A., Trussoni, E., 1994 A&A, 283, 655

Bohigas, J., Persi, P., Tapia, M., 1993, A&A, 267, 168

Brugel, E.W., 1989, in ESO Workshop on *Low Mass Star Formation and Pre-main Sequence Objects*, ed. B. Reipurth, (ESO, Garching), p311

Brugel, E.W., Böhm, K.H., Mannery, E., 1981, ApJ, 243, 874

Bührke, T., Mundt, R., Ray, T.P., 1988, A&A, 200, 99

Cardelli, J.A., Brugel, E.W., 1988, AJ, 96, 673

Chernin, L., Masson, C., Gouveia Dal Pino, E.M., Benz, W., 1994, ApJ, 426, 204

Corcoran, D., Ray, T.P., Mundt, R., 1993, A&A, 279, 206

Corcoran, M., Ray, T.P., 1996, in preparation

Curiel, S., Rodríguez, L.F., Moran, J.M., Cantó, J., 1993, ApJ, 415, 191

Curiel, S., 1995, Rev. Mex. Astr. Ap. Serie de Conferencias, 1, p59

Davis, C.J., Eislöffel, J., 1995, A&A, 300, 851

Davis, C.J., Eislöffel, J., Ray, T.P., 1994, ApJ, 426, L93

Davis C.J., Mundt R., Eislöffel, J., Ray, T.P., 1995, AJ, 110, 766

Davis C.J., Ray, T.P., Eislöffel, J., Corcoran, D., 1996, in preparation

DeYoung, D.S., 1986, ApJ, 307, 62

Dopita, M.A., 1978, ApJS, 37, 117

Dopita, M.A., Schwartz, R.D., Evans, I., 1982, ApJ, 263, L73

Downes, T., 1996, in *Solar and Astrophysical Magnetohydrodynamic Flows*, Proceedings of the NATO ASI, Heraklion, Crete, Astrophys. Lett. & Comm., ed. K. Tsinganos, in press

Downes, T., Ray, T.P., Drury, L., O'C., 1996, in preparation

564

Edwards, S., Ray, T.P., Mundt, R., 1993, in *Protostars and Planets III*, eds. E. Levy and J. Lunine, (University of Arizona Press), p567

Eislöffel, J. 1995, in *Disks and Outflows Around Young Stars*, Proceedings of the Heidelberg Conference in honor of Prof. H. Elsässer, eds. S.V.W. Beckwith, A. Natta and J. Staude, Lecture Notes in Physics Series, Springer-Verlag, in press

Eislöffel, J., Davis, C.J., Ray, T.P., Mundt, R., 1994, ApJ, 422, L91

Eislöffel, J., Mundt, R., 1994, A&A, 284, 530

Falle, S.A.E.G., Innes, D.E., Wilson, M.J., 1987, MNRAS, 225, 741

Fendt, C., Camenzind, M., Appl, S., 1995, A&A, 300, 791

Fukui, Y., 1989, in *ESO Workshop on Low Mass Star Formation and Pre-main Sequence Objects*, ed. B. Reipurth, (ESO, Garching), p95

Fukui, Y., Iwata, T., Mizuno, A., Bally, J., Lane, A.P., 1993, in *Protostars and Planets III*, eds. E. Levy and J. Lunine, (University of Arizona Press), p603

Gouveia Dal Pino, E.M., Birkinshaw, M., Benz, W., 1996, ApJ, submitted

Haro, G. 1952, ApJ, 115, 572

Hartigan, P., 1989, ApJ, 339, 987

Hartigan, P., Morse, J.A., Raymond, J., 1994, ApJ, 436, 125

Hartigan, P., Raymond, J., Hartmann, L., 1987, ApJ, 316, 323]

Herbig, G.H., 1951, ApJ, 113, 697

Herbig, G.H., 1970, in *Spectroscopic Astrophysics*, ed. G.H. Herbig, (Berkeley, Univ. California Press), p237

Herbig, G.H., Jones, B.F., 1981, AJ, 86, 1232

Hirth, G.A., 1994, PhD Thesis, University of Heidelberg

Hirth, G.A., Mundt, R., Solf, J., Ray, T.P., 1994, ApJ, 427, L99

Hirth, G.A., Mundt, R., Solf, J., 1994, A&A, 285, 929

Hodapp, K.W., 1994, ApJS, 94, 625

Jones, B.F., Herbig, G.H., 1982, AJ, 87, 1223

Kepner, J., Hartigan, P., Yang, C., Strom, S., 1993, ApJ, 415, L119

Koerner D.W., Sargent, A.I., 1995, AJ, 109, 2138

Königl, A., 1982, ApJ, 261, 115

López, R., Raga, A., Riera, A., Anglada, G., Estalella, R., 1995, MNRAS, 274, L19

Martí, J., Rodríguez, L. F., Reipurth, B., 1995, ApJ, 449, 184

Massaglia, S., Bodo, G., Rossi, P., Ferrari, A., 1996, in *Solar and Astrophysical Magnetohydrodynamic Flows*, Proceedings of the NATO ASI, Heraklion, Crete, Astrophys. Lett. & Comm., ed. K. Tsinganos, in press

Morse, J.A., Hartigan, P., Heathcote, S., Raymond, J.C., Cecil, G., 1994, ApJ, 425, 738

Mundt, R., 1986, Can. J. Phys., 64, 407

Mundt, R., Bührke, T., Fried, J.W., Neckel, T., Sarcander, M., Stocke, J., 1984, A&A, 140, 17

Mundt, R., Brugel, E.W., Bührke, T., 1987, ApJ, 319, 275

Mundt, R., Fried, J.W., 1983, ApJ, 274, L83

Mundt, R., Ray, T.P., 1994, in *The Nature and Evolutionary Status of Herbig Ae/Be Stars*, eds. P.S. Thé, M.R. Perez & E.P.J. van den Heuvel, P.A.S.P. Conf. Ser., p237

Mundt, R., Ray, T.P., Bührke, T., Raga, A.C., Solf, J., 1990, A&A, 232, 37

Mundt, R., Ray, T.P., Raga, A.C., 1991, A&A, 252, 740

Neckel, Th., Staude, H.J., 1987, ApJ, 322, L27

Noreiga-Crespo, A., Garnavich, P.M., 1994, AJ, 108, 1432

Ogura, K., 1995, ApJ, 450, L230

Osterbrock, D.E., 1958, PASP, 70, 399

Osterloh, M., Beckwith, S.V.W., 1995, ApJ, 439, 2880

Parker, N.D., Padman, R., Scott, P.F., 1991, MNRAS, 252, 442

Poetzel, R., Mundt, R., Ray, T.P., 1989, A&A, 224, L13

Pravdo, S. H., Rodríguez, L. F., Curiel, S., Cantó, J., Torrelles, J. M., Becker, R.H., Sellgren, K.M., 1985, ApJ, 293, L35

Raga, A.C., 1995, Rev. Mex. Astr. Ap. Serie de Conferencias, 1, p103

Raga, A.C., Binette, L., Cantó, J., 1990, ApJ, 360, 612

Raga, A.C., Cabrit, S., 1993, A&A, 278, 267

Raga, A.C., Kofman, L., 1992, ApJ, 386, 222

Ray, T.P., 1987, A&A, 171, 145

Ray, T.P., Mundt, R., 1993, in *Astrophysical Jets*, eds. M. Fall, C. O'Dea, M. Livio and D. Burgarella, (Cambridge University Press), p145

Ray, T.P., Mundt, R., Dyson, J., Innes, D.E., Falle, S.A.E.G., Raga, A.C., 1996, in preparation

Raymond, J.C., 1976, PhD thesis, Univ. Wis. Madison

Raymond, J.C., 1979, ApJS, 31, 1

Rees, M.J., 1978, MNRAS, 184, L61

Reipurth, B., 1989, in *ESO Workshop on Low Mass Star Formation and Pre-main Sequence Objects*, ed. B. Reipurth, (ESO, Garching), p247

Reipurth, B., Bally, J., Graham, J.A., Lane, A.P., Zealey, W.J., 1986, A&A, 164, 51

Reipurth, B., Cernicharo, J., 1995, Rev. Mex. Astr. Ap. Serie de Conferencias, 1, p43

Reipurth, B., Heathcote, S., 1991, A&A, 246, 511

Reipurth, B., Raga, A.C., Heathcote, S., 1992, ApJ, 392, 145

Sargent, A.I., 1995 in *Disks and Outflows Around Young Stars*, Proceedings of the Heidelberg Conference in honor of Prof. H. Elsässer, eds. S.V.W. Beckwith, A. Natta and J. Staude, Lecture Notes in Physics Series, Springer-Verlag, in press

Sargent, A. I., Beckwith, S.V.W., 1987, ApJ, 323, 294

Sargent, A. I., Beckwith, S.V.W., 1989, in *Structure and Dynamics of the Interstellar Medium: Proc. IAU Coll. 120*, eds. G. Tenorio-Tagle, J. Melnick and M. Moles (Berlin: Springer-Verlag), p215

Schwartz, R.D., 1975, ApJ, 195, 631

Schwartz, R.D., 1978, ApJ, 223, 884

Schwartz, R.D., 1983, ARA&A, 21, 209

Schwartz, R.D., Cohen, M., Jones, B.F., Bohm, K., Raymond, J.C., Hartmann, L.W., Mundt, R., Dopita, M.A., Schultz, A.S.B., 1993, AJ, 106, 740

Smith, M.D., 1994, MNRAS, 266, 238

Stahler, S.W., 1994, ApJ, 422, 616

Stapelfeldt *et al.*, 1996, in preparation

Strom, S.E., Strom, K.M., Grasdalen, G.L., Sellgren, K., Wolff, S., 1985, AJ, 90, 2281

Zinnecker, H., McCaughrean, M.J., 1995, in preparation

MOLECULAR OUTFLOWS FROM PROTOSTARS

Questions, Options and Facts

R.N. HENRIKSEN
Astronomy Group, Dept. of Physics; Queen's University at Kingston
Ontario, Canada K7L 3N6

1. Introduction

In this discussion of molecular outflows I isolate what I regard as the basic theoretical questions, and discuss some of the model options in the light of a few decisive observational facts. In addition several new lines of theoretical argument are introduced regarding the behaviour of the protostar environment. These concern the global energy and momentum balance and the nature of the steady state that is compatible with a non-zero Poynting flux. Finally brief reference is made to a thorough study by (Fiege and Henriksen, 1995) of the global model suggested in (Henriksen and Valls-Gabaud, 1994) and fore-shadowed somewhat in (Tsinganos and Sauty, 1992).

2. Pertinent Theoretical Questions

Although not all of the following questions are without at least tentative answers, they all seem to remain useful for provoking thought about protostellar outflow. For the purposes of this article I suppose that the reader is familiar with the general character of this phenomenon. However in brief, we may refer to the intense, moderately collimated, standard (SV) and high velocity (HV) outflow that is seen in molecular line wings extending to a few tens of kilometres per second; to the rather fainter and better collimated extremely high velocity (EHV) outflow found in the wings of some lines out to 100 $km\ s^{-1}$; and to the optical jets and their attendant shocks that are essentially perfectly collimated and show proper motions up to three or four hundred $km\ s^{-1}$. We may ask:

K. C. Tsinganos (ed.), Solar and Astrophysical Magnetohydrodynamic Flows, 567–583.
© *1996 Kluwer Academic Publishers.*

2.1. WHAT IS THE ENERGY AND MOMENTUM SOURCE FOR THE OUTFLOW?

Early explanations for the impetus behind these outflows depended heavily on stellar 'winds', and their geometrical variant stellar 'jets', interacting with ambient material. Although winds possess the admirable virtue of being known to exist widely in the Hertzsprung-Russell stellar domain (Sauty, 1994), and although winds can be transformed into jets by environmental interactions (Königl, 1982), it must be realized that such explanations only displace the problem for proto-stellar objects. Moreover, prodigious mass loss over a dynamical time (Levreault, 1988) imply that some part of the flow must be a circulation of ambient material.

The correlations between bolometric luminosity and the mechanical luminosity and force(Cabrit and Bertout, 1992) suggest also a source of heat that is related to the outflow itself, especially since the bolometric luminosity can reach extreme values ($\simeq 10^5\ L_\odot$). These hints are incorporated in the global model that is based on magnetically driven winds from accretion discs (Blandford and Payne, 1982),(Pelletier and Pudritz, 1992), (Ferreira and Pelletier, 1994), which deals directly with these questions as do the models of (Fiege and Henriksen, 1995) and of (Shu et al., 1994). For an excellent and thorough review of the magneto-hydrodynamic aspects of such models one should refer to the work by (Sauty, 1994).

2.2. HOW ARE THESE FLOWS COLLIMATED AND WHERE?

This question has been treated in all models and finds various answers. Here I note that observationally the high resolution measurements of (Guilloteau et al., 1992) show direct evidence for collimation on spatial scales $\leq\ 100\ au$. More recently (Chernin, 1995) has studied shock excited water masers in the same source (L1448C) and infers that they are formed at $\simeq\ 31\ au$ along the same position angle as the HV outflow. Clearly these collimated outflows extend essentially to the stellar scale and the transition between the outflow region and the star can be expected to play a vital role.

2.3. HOW MANY 'INDEPENDENT' PARTS OF THE 'MACHINE' ARE THERE?

The observed phenomenon has the SV, HV, EHV, and optical jet components. To these may be added the mm wave observations of large discs, the shock excited masers and the shocked H_2 emission. These are clearly inter-related, but is it because they have the same cause, or because one or more components provoke the others, or because they are simply different aspects of the same underlying global behaviour? Similarly, does the theoretical machine need as distinct elements an accretion disc, a thick large

569

torus, a wind, a jet, or even a proto-stellar core? Or are these again different aspects of the same collapse/accretion phenomenon, reflecting different temporal and spatial scales? The coincidence of the maximum outflow speeds or 'jet' speeds with free fall velocity onto a solar type 'core, is suggestive in this regard. It seems that the most satisfying explanation lies in the latter direction, and (Fiege and Henriksen, 1995) show by making synthetic radio maps of their quadrupolar circulation model that many of the observational components do appear.

2.4. WHAT KINDS OF TIME DEPENDENCE ARE THERE?

Under this heading we may imagine evolutionary development that may be revealed in the observational YSO classes (e.g. (André et al., 1993)), and that may lead to time variable environments and accretion rates. The evolutionary sequence presumably terminates with the T-Tauri stars, while on dynamical time scales there might be 'bang-whimper' or 'whimper-bang' modes as suggested in (Henriksen, 1994). On a shorter time scale (Raga and Cabrit, 1993) have suggested that sequential shocks produced by intermittent source activity may produce the outflows. Clumpy optical jets add support to this view as was already discussed in (Choe and Henriksen, 1986). However some care must be exercized in using this interpretation as (Fiege and Henriksen, 1995) show that some clumping may be due simply to finite 'fields' in both real and velocity space.

We might also look for precession of the outflow axis since this is known to occur in high energy jet phenomena, and there may be some observational evidence for this (André and Bontemps, 1995). However if the evidence consists in seeing different position angles at simultaneously different spatial and velocity scales then (Fiege and Henriksen, 1995) caution that this may simply be a consequence of a rotating, velocity sorted conical outflow.

2.5. WHAT DETERMINES THE CHARACTERISTIC SCALES OF THE OUTFLOWS?

Under 'characteristic scales' we might include a length scale r_o, a 'lifetime' t_o which is probably not independent of the length scale; a bolometric luminosity L_o even if it appears to vary over five or six orders of magnitude and may be time dependent, and the mass flow $\dot{M}$. A characteristic velocity, v_o, is just as moot as the luminosity as the maximum seems to depend on the sensitivity, wave band, and resolution of the observations. Although these do vary considerably from object to object the range is nevertheless characteristic of the protostellar outflow phenomenon. We would like to identify a key parameter or parameter set which determines the others much as the mass fixes many other gross characteristics of a main-sequence star. In proto-stars the basic energy source is probably gravity with perhaps a magneto-rotational-radiative transmission of some fraction of the material to a molecular outflow. One could hope that in the post core collapse phase, $\dot{M}$ itself may be such a key parameter as it is easy to see how this may be linked to ambient conditions and the core mass. Most of this is likely to be in a forced circulatory or 'elevator' mode in which it releases energy on the descent that is then refocussed (by radiative heating or magnetic driving) on an outgoing fraction. Correlations of this quantity with bolometric and mechanical luminosities, with mechanical force and even with mm-band luminosity ((André and Bontemps, 1995);the latter correlation is a kind of tautology from this point of view since the mm emission is sensitive to the quantity of mass in circulation) are thus easy to understand. Other scales are probably more sensitive to the details.

2.6. DOES THE PHENOMENON OCCUR ALSO IN A PRE-PROTOSTAR PHASE?

With the advent of well developed outflow in low luminosity sources with very young cores but extensive outflow, such as VLA 1623 (Bontemps et al., 1995), one suspects that the outflows may extend to self-gravitating phases such as early cloud collapse. The ultimate model would presumably predict the nature of the outflow from conditions in the cloud at the onset of collapse (e.g. (Henriksen, 1994)). This raises difficulties for models in which a thin accretion disc and a central heat source, presumably established by the gravity of a central star, are vital components. It is possible however that only the youngest of hydro-static cores is required.

2.7. HOW CONTINGENT OR UNIVERSAL IS THE PHENOMENON?

This question really poses some of the previous questions in a slightly more abstract form. Is this phenomenon inevitable in the presence of a rotating, self-gravitating, magnetic,dusty, opaque molecular cloud (or some sub-set of these properties) in the same sense that core collapse is inevitable for N-body systems? The analogy may carry some way in that it is the non-homologous gravitational collapse that supplies the energy to a fraction of the material near the rotation axis, much as energy is supplied to the halo by the collapsing core in N-body systems during the gravo-thermal 'catastrophe'. In fact the analogy is quite close for spherical stellar winds, but it is complicated by the magneto-rotational anisotropy and the very massive protostellar 'atmosphere'.

Finally two questions of more specific observational significance:

2.8. WHAT ARE THE 'NEIGHBOURHOOD' MAGNETIC AND VELOCITY FIELDS?

Many models make distinct predictions regarding the global patterns of velocity and magnetic field in the regions adjoining the collimated outflow. It is thus vital to observe wide fields with good dynamic range and velocity resolution. Since dust is almost certainly intimately related to the phenomenon, measures of infra-red polarization from dust emission offer hope of getting the magnetic field direction.

2.9. ARE THERE DIFFERENCES BETWEEN THE MAGNETIC AND ROTATION AXES?

The morphology of the magnetic field is itself a fascinating study. It is usually treated as being dipolar with an axis aligned with the rotation axis, but magnetic dynamo models (Brandenburg et al., 1995) of turbulent accretion discs often suggest quadrupole fields. Moreover recent work by (Aly, 1994), (Lynden-Bell and Boily, 1994) and (Lovelace et al., 1995) suggest that shear instabilities can lead to open field line configurations wherein much of the flux is no longer linked to the central star. Normally (e.g.(Ferreira and Pelletier, 1994)) dissipation and/or ambipolar diffusion is invoked to allow the accreting material to disconnect from the ambient magnetic field.

There must be some such early disconnection phase as otherwise there remains a magnetic flux problem. This is presumably solved at the same time as is solved the angular momentum problem for the central mass, and if it is by magnetic transport then (Pudritz and Norman, 1983) *some field lines must link the 'star' to the outflow during this phase*. This is provided that we include under 'star' regions that are torque coupled to the 'core'. This stage might be identical to the self-gravitating epoch. The implication, in all but the steadily diffusing cases, is that there must be a region of violent transition between the stellar magnetic field and the ambient field. This region plays a key role in some models (Shu et al., 1994) and in others (Fiege and Henriksen, 1995) its scale must set an upper limit to the velocity of the outflowing gas.

Differences in the magnetic and rotational axes might be expected in sub-Alfvénic flow where the field may dictate the geometry. In the self-gravitating phase this could conceivably lead to precession of the source axis.

3. General Considerations

Having posed questions in a provocative but somewhat vague fashion, it is useful to be more precise about the essential elements of any satisfactory

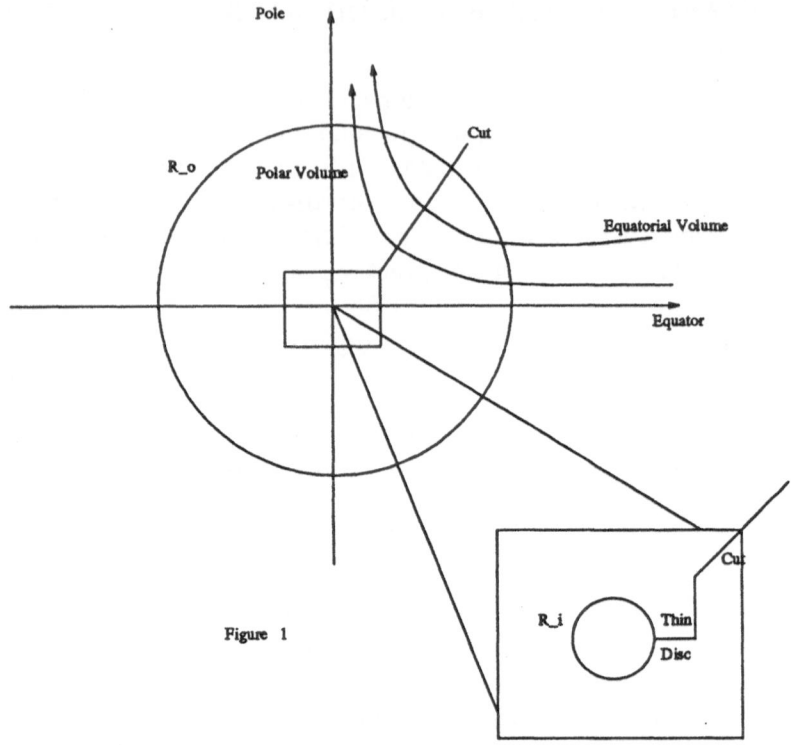

Figure 1

Figure 1. Sketch of a bipolar outflow. The upper right quadrant is to be reflected in the equator and rotated about the polar axis. The sphere indicates the gravitational limit of the protostellar influence and the central 'box' is on a very different spatial scale.

model. We consider first the cartoon in figure 1 which seeks to portray an inflow-outflow circulation with an inner core, core-circulation transition zone, and outer boundary in the most general terms.

We divide the total volume into a 'polar region' and an 'equatorial' region (relative to the rotation axis) by means of a 'cut' as shown. In a steady state a convenient form of the energy equation for a viscous, MHD medium is

$$
\nabla \cdot \left[\vec{S} + \rho \vec{v} \left(\frac{\vec{v}^2}{2} + h + \Phi \right) - \vec{\vec{\sigma}} \cdot \vec{v} \right] =
$$

$$
\rho T \frac{ds}{dt} - \frac{1}{2\eta} \sigma_s^{ij} \sigma_{s\ ij} - \zeta (\nabla \cdot \vec{v})^2. \tag{1}
$$

In this expression Φ is the gravitational potential such that $\Delta \Phi = 4\pi G \rho$, h and s are respectively the gas specific enthalpy and specific entropy and $\vec{\vec{\sigma}}$ is the total viscous stress tensor. The traceless part of the viscous stress tensor $\vec{\sigma_s}$ is given as usual by

$$\vec{\sigma_s}^{ij} \equiv \eta \left(\nabla^j v^i + \nabla^i v^j - \frac{2}{3} \nabla.\vec{v}\, \delta^{ij} \right), \tag{2}$$

where η is the 'viscosity'. In general this latter quantity is neither molecular nor constant of course (not assumed here), being rather an eddy viscosity. The total viscous stress tensor is $\vec{\sigma} = \vec{\sigma_s} + \zeta(\nabla.\vec{v})\vec{1}$, where ζ is the bulk coefficient of 'viscosity' and $\vec{1}$ is the unit tensor.

The vector $\vec{S}$ stands for the Poynting flux vector

$$\vec{S} \equiv \frac{c\vec{E} \times \vec{B}}{4\pi}, \tag{3}$$

and other symbols have their usual meanings.

Strictly equation 1 should include the divergence of the radiation flux on the left, while the enthalpy should be the sum of the gas enthalpy plus the radiation enthalpy $4aT^4/3\rho$. However for a cold, low-velocity medium in thermal equilibrium with the radiation field, the radiative enthalpy is negligible and the divergence of the flux is zero for sufficiently low opacity. The region of strong disequilibrium between the radiation field and the dusty gas close to the star (where say optical photons are degraded to the infra-red) can lead to distributed heating, but this we include in $\rho T ds/dt$.

Equation 1 is thus somewhat grotesque but it allows us at least to do a careful accounting of the sources and sinks of the outflow energy. Thus by integrating over the polar volume while supposing that on the outer boundary R_o the dominant energy flux is the outflow phenomenon itself, and defining the cut and the inner boundary R_i as in figure 1 one finds

$$\int_{polar\ R_o} \rho \vec{v} \frac{\vec{v}^2}{2}.d\vec{A} =$$

$$- \int_{cut+R_i} \vec{S}.d\vec{A} - \int_{cut+R_i} \rho \vec{v} \left(\frac{\vec{v}^2}{2} + h + \Phi \right).d\vec{A} + \int_{polar\ vol} \rho T(\frac{ds}{dt})\,dV$$

$$+ \int_{cut+R_i} \vec{v}.\vec{\sigma}.d\vec{A} - volume\ dissipation. \tag{4}$$

The LHS of this expression is just the energy flux advected by the polar outflow. On the right we find the potential sources and sinks successively as, in order of appearance:

(i) The magnetic acceleration (deceleration) due to the Poynting flux into (out of) the polar volume across the cut and across the stellar surface. Note that as in figure 1 the cut is taken to include a thin circumstellar accretion disc.

(ii) The advection of kinetic energy and enthalpy reduced by the gravitational potential across the cut and the stellar surface. This is positive for flow into the region *provided that the net specific energy is positive.* In general this implies some heating of the equatorial region.

(iii) This term is the external heating supplied to the polar volume. It is through this term that we expect radiative heating from say an accretion shock to play a direct role. Such heating is also probably reponsible for any positive specific energy that is advected from the equatorial volume as in (ii) above.

(iv) This term consists of the work done on material crossing into the polar volume by the viscous stress. The viscous stresses would normally be expected to produce a drag on the material along the outer cut, but over the surface of the star and the thin disc shear stresses might make a positive contribution. This would be so if the rotation speed of the stellar surface (or thin disc) exceeds that of the surrounding gas. However left to itself such an effect would produce an *inward* flow along the rotation axis (Landau and Lifshitz, 1987) for an equatorial distribution of shear stress due to centrifugal acceleration of near disc gas. A more 'funicular' distribution of shear stress, such as may be found in the throat of a thick toroid, might produce a positive contribution (we imagine the cut made coincident with the funnel surface). This condition is similar to the celebrated critical angle of the field lines for acceleration in the original disc driven winds (Blandford and Payne, 1982). The process might not be so efficient however since generally viscosity will dissipate much of the shear into heat. The only exception would be the presence of an inverse turbulent cascade of energy such as one requires from a stochastic dynamo. This would be modelled here as a negative viscosity. The original disc driven winds do not dissipate (Blandford and Payne, 1982), but those that rely on a magnetic 'X' point (Shu et al., 1994) would share this property.

(v) This is the volume viscous dissipation that must be overcome since it is always negative in the absence of a dynamo (or large scale vorticity generator in the absence of a magnetic field).

Thus this rigorous energy accounting allows us to identify Poynting energy flux (term (i)) and radiative heating, be they provided from the equatorial volume (including a possible thick disc), the thin disc or the 'star', as the principle possibilities for driving the polar energy flux. There is moreover the possibility of shear acceleration by viscous stresses (viscous part of term (v)) if the boundary of the polar region has the right form.

This is not a surprising conclusion, but it allows us subsequently to classify various models according to their energy sources.

We can conduct a similar argument for the momentum flux.

In a steady state conservation of momentum can be written in the form

$$\nabla.\vec{\vec{\Pi}} = \Phi\nabla\rho, \tag{5}$$

where the total stress tensor is given by

$$\vec{\vec{\Pi}} \equiv (p + \rho\Phi)\vec{\vec{1}} + \rho\,\vec{v} \otimes \vec{v} - \vec{\vec{\sigma}} - \vec{\vec{M}}, \tag{6}$$

and the Maxwell stress is

$$\vec{\vec{M}} \equiv \vec{B} \otimes \vec{B}/4\pi - (B^2/8\pi)\vec{\vec{1}}. \tag{7}$$

On integrating equation 5 over the polar volume and supposing once again that the outflow momentum flux dominates on R_o one finds

$$\int_{polar\ R_o} \rho\vec{v} \otimes \vec{v}.d\vec{A} = -\int_{cut+R_i} \vec{\vec{\Pi}}.d\vec{A} + \int_{polar\ vol} \Phi\nabla\rho\,dV. \tag{8}$$

Equations 6 and 8 together now provide an accounting of the vector momentum advected in the outflow. This is relevant primarily to the collimation of the outflow. An inspection shows that once again reduced pressure (reduced by the gravitational energy per unit volume), magnetic and viscous stresses and advection whether they act across any or all of the star, thin disc or cut may all play a role. It is interesting to note how conditions near the 'star' may determine the asymptotic collimation as well as may external stresses (pressure, magnetic, viscous) acting across the cut. A delicate transverse balance must be maintained in order to avoid either de-collimation or re-collimation in the manner of (Chan and Henriksen, 1980). Other constraints on asymptotic collimation have been discussed by (Heyvaerts and Norman, 1989) and (Sauty, 1994). The integrated volume term in equation 8 allows for a reduced gravitational energy per unit volume if the flow enters a low density region, but this is not a driving force and is better put on the other side of the equation to be produced along with the advected momentum.

Finally in this section I wish to address the question of steady flow in the presence of magnetic fields. Normally this is achieved by either allowing for a slippage between the gas and the magnetic field due to either ambi-polar diffusion or Ohmic dissipation, or by setting $\vec{v}||\vec{B}$. The latter approach has the virtue of the greatest simplicity, but it is not the only ideal steady state. At this meeting Contopoulos has argued that poloidal field lines may simply be convected into a sink on the axis of the system. But this is at the cost of producing a non-zero azimuthal electric field and hence a time dependent axial sink by Faraday's law. However in axial symmetry the steady flux-freezing condition requires only that

$$\vec{v} \times \vec{B} = \nabla\Psi, \tag{9}$$

where Ψ is the electric potential in the inertial frame. To avoid the axial time dependent sink one needs therefore only insist that ((Chan and Henriksen, 1980) appendix A)

$$\vec{v_p} \| \vec{B_p},$$ (10)

where the subscript p indicates a vector in the poloidal plane. Then the polidal electric field is found from a strictly axi-symmetric potential Ψ where

$$\nabla_p \Psi = \vec{v_p} \times \vec{B_\phi} + \vec{v_\phi} \times \vec{B_p}.$$ (11)

This equation may be solved simulataneously with the other magneto-hydrodynamical equations of the problem for the electric potential that is compatible with non-parallel 3D field and stream lines (despite the projected parallelism in the poloidal plane). The models of the type advocated in (Fiege and Henriksen, 1995) continue to exist, although so far only the simplest case of zero electric field ($\vec{v} \| \vec{B}$) has been studied. However it is not possible to have disc driven solutions where the field lines pass through the disc (Ferreira and Pelletier, 1994) without either diffusion, or the central time dependent sink of Contopoulos.

4. Models; Classification and Critique

4.1. WIND DRIVEN MODELS

To a certain extent a critique of these models serves only to 'flog a dead horse', so that I will be brief. In such models a largely neutral (and therefore conveniently 'dark') wind is weakly collimated by a thin disc and evacuates a cavity that pushes a shell of the ambient molecular material before it. Question 1 above is not answered at all except insofar as stellar winds are widespread (although not neutral). The models imply a weak bipolar collimation by ambient matter, perhaps a thin circumstellar disc, in reponse to question 2. A typical example of such models is that of (Shu et al., 1991) although the idea of a hollow outflow originated earlier with the observers (Snell, 1985). At least the optical jets and the EHV outflow are presumably extra components, in response to question 3. Winds do have the virtue of probably being an inevitable and universal (in the sense of the comments following question 7) consequence of the Hayashi phase in star formation, but the real problem with these models is that they do not explain the observed phenomenon. (Cabrit and Bertout, 1992) showed that the expected position-velocity profiles did not match those observed, and (Masson and Chernin, 1993) made the important observation that most of the swept up matter is at high velocity in these models, which contrasts starkly with the observed low velocity peaks and weak high-velocity wings of the observed line profiles. This is a key observational requirement and can be used to exclude these simple models.

4.2. ENTRAINED OUTFLOW

This model postulates a jet as the prime mover and achieves the collimation and acceleration of the molecular outflow by virtue of 'viscous' entrainment. The viscosity is not of course molecular but may be either turbulent or 'magnetic' (actually an effect of shearing the magnetic field (Henriksen, 1987)). This idea stems from attempts to understand the relation between VLBA and VLA extragalactic jets (see e.g. (Henriksen, 1985),(Henriksen, 1987) and references therein), and there is a rather rigorous theory based on a generalization of Landau-Squires submerged jets. The model has the advantage of explaining the molecular outflow and the collimation in a coherent fashion, but suffers from not caring too much about the jet source. The arguments already encompassed both momentum conserving and energy conserving outflow. They have not been calculated to my knowledge, but the line profiles would probably be about right. Dynamical behaviour suggestive of this kind of model has apparently recently been found in HL Tau (Cabrit et al., 1995) and in at least some cases (Mitchell et al., 1995) optical jets appear to have the expected relation to the surrounding outflow.

However, numerical jet simulations with cooling jets (Dal Pino and Benz, 1994) suggest that entrainment becomes much less effective when the jets cool sufficiently that they are 'ballistic' in the ambient medium. This implies that entrainment may not happen for proto-stellar jets as opposed to extra-galactic jets. It is true of course that close to the star the cooling may be offset by radiative heating, and magnetic shearing can play a part. Perhaps such models are best seen as the 'tail wagging the dog' version of the 'dog wagging the tail' models of (Fiege and Henriksen, 1995). In the latter models it is the large scale circulation that creates the jet rather than vice versa.

4.3. OUTFLOW DRIVEN BY WANDERING AND/OR INTERMITTENT JETS

These models are a variant of the previous one in that rather than assume a steady mean entrainment due to turbulent or magnetic coupling to the surroundings, they imagine a jet which is intermittent, either because of sequential ejection or because of 'precession' of the axis. For proto stellar jets this has been championed recently by (Raga and Cabrit, 1993) and (Masson and Chernin, 1993) but once again similar ideas were proposed in the extragalactic context by P. Scheuer. A relaxation oscillator model for intermittent ejection was already proposed by (Choe and Henriksen, 1986).

But the novelty of the proposal by Raga and Cabrit is that the sideways ejecta from a sequence of shocks can accumulate to produce the wide angle molecular outflow. Masson and Chernin solve this 'decollimation problem' with the wandering jet hypothesis after having found the outflow to be too well collimated if produced by a fixed jet.

As in all such models the questions regarding origin and collimation are by-passed, particularly those concerning precession and ejection intermittency. But if the jet is the prime mover then it must contain all of the momentum seen in the wider outflow. In some cases this appears plausible (with 10% ionization)(Mitchell et al., 1995), but it is far from clear that this is generally true. The presence of shocked H_2 emission (Kitamura et al., 1992) coinciding with 'clumps' in the outflow is taken as support for this picture. However shocks may arise just as readily near the axis in a global circulation picture (Gomez de Castro and Pudritz, 1993), (Fiege and Henriksen, 1995).

These models avoid the suppression of entrainment by cooling mentioned above, but the cooling may present them with other problems. The cooling time is really very short (compared to a dynamical time) in HII regions of densities $10^4 - 10^5 \ cm^{-3}$ and so even the sideways ejecta may become ballistic, which would reduce its inflationary effectiveness. Either, one has to trap the photons in the surrounding medium (not so much a collimating medium as an insulating one)(Choe and Henriksen, 1986) or the jet must dissipate sufficient energy to balance the cooling. This latter option re-poses the 'missing waste heat problem'(Scheuer, 1983) in that one would expect to see i/r knots well removed from the central core. In the low luminosity source VLA 1623 (André et al., 1993) one finds a very well developed molecular outflow, but the 350 μ map shows only a smooth, roughly spherical cloud with a radius of about 2000 au. There is no near i/r source apart from the thermal emission from dust at $T \approx 20K$ and little free- free emission. In short it does not appear to be greatly disturbed by multiple shocks.

Moreover all such models suffer from excessive collimation and from an expectation (Henriksen, 1987) that the higher luminosity sources should be better collimated. Once again VLA 1623 is a notable counter example. In brief my conclusion is that the best verdict for such jet driven models at present is the scottish one of "not proven".

4.4. MAGNETIZED DISC DRIVEN OUTFLOW

This has been the most popular model since the work of Blandford and Payne, and I can not do bibliographic justice here to the many works and authors who have made substantial studies. Fortunately many of the appropriate references are to be found in (Sauty, 1994) or elsewhere in this volume. The most detailed and physical studies of the disc launching and evolution are now probably those of (Ferreira and Pelletier, 1994) and (Wardle and Königl, 1993) at least in the near disc region. Briefly, accretion in a disc is made to continue in the face of the angular momentum barrier by virtue of a magnetic field that subtracts gas and angular momentum from the disc and deposits it in a collimated outflow. Normally the field penetrates the disc in a dipolar mode, and so, much effort has been expended in finding consistent models which both produce a centrifugal wind and allow the disc gas to cross the field lines.

The models have the virtue of dealing directly with many of the questions we raised in our first section. The origin and collimation of the outflow are coherently and elegantly treated and the ultimate energy source is clearly gravitational. The models solve simultaneously in effect the angular momentum and magnetic flux problems of a protostar, as well as allowing the accumulation of its mass, by driving the molecular outflows. This aspect is very satisfying and must probably be maintained in any eventual definitive model.

There remain some difficulties with this picture however, not the least of which is the somewhat artificial distinction between the 'disc' and the 'corona'. A centifugally driven disc wind requires the disc flow to be super-Alfvénic, the coronal flow to be sub-Alfvénic, and the wind to be once again super-Alfvénic. This is just the case for the solar wind (with the photosphere playing the role of the dense disc), but as remarked in (Henriksen, 1993) there is no analogue in these models to the energy flux that is thought to actually drive the solar wind. The theory has shown that the required steady 'free-standing' open magnetic structures are not impossible, but they appear as rather low entropy machines. That is one might expect in practice that instability driven turbulence of various kinds would both erase the distinction between disc and corona and etablish a magnetic field with local sources, perhaps of the type found recently by Brandenburg et al. ((Brandenburg et al., 1995)). That is, the situation might evolve towards the global flow picture advocated in (Henriksen and Valls-Gabaud, 1994), (Narayan and Yi, 1995) and especially (Fiege and Henriksen, 1995).

Another omission from most of these models to date is the inner boundary condition. In (Pelletier and Pudritz, 1992) this is treated asymptotically in a self-similar solution, much as is done in the models of (Henriksen and Valls-Gabaud, 1994) and (Fiege and Henriksen, 1995). However it is virtually certain that the core-disc boundary layer is the seat of energetic activity. It is just this aspect which is emphasized in the next type of model by Shu and his associates (Shu et al., 1994).

4.5. BOUNDARY LAYER MODELS

That the zone of interaction between a 'star' and a surrounding accretion flow should be a crucial boundary layer region is not new (e.g. (Henriksen and Reinhardt, 1974),(Henriksen, 1976),(Henriksen and Reinhardt, 1977), (Ghosh and Lamb, 1979),(Aly, 1980),(Torbett, 1985), (Bertout, 1987)),(Henriksen, 1989)), but the nature of this layer is very different in different models. The novelty of the approach by Shu and his associates lies mainly in a fully 3D treatment of the 'point' (perhaps a region in general) where the radial disc flow makes the transition to the vertical outflow. The formulation is not particularly new (cf e.g. (Henriksen and Reinhardt, 1977), (Chan and Henriksen, 1980)) and many of the qualitative results are more consistently found in the old split monopole models (for example the first mention of the outward transport of angular momentum in magnetic accretion occurs at least as early as 1972 (Henriksen and Chia, 1972) in this context), but the work does treat squarely the difficult dynamical and magnetic problem of the disc-outflow transition. The detailed analytical and numerical work is restricted to the vicinity of the 'X' neutral point so that no global model is yet available, but they do succeed in producing escape velocity material there. This small scale beamed outflow has presumably to entrain the larger scale molecular outflow, or else the outer flow is produced by a magnetized disc as above.

There seem to be a number of outstanding problems however. Perhaps the most difficult is the cross field line flow that is required in their innermost region. Such a region is parameter rich (Ferreira and Pelletier, 1994). Moreover some unspecified torque coupling is assumed to act across this region, which begs some of the important questions. They do avoid the criticism of requiring a 'rootless' super equipartition (sub-Alfvénic) field thanks to the near presence of the star, and so perhaps the low entropy configuration can be maintained. However I doubt that all of the circulating mass will descend to the boundary layer before being reaccelerated. The arguments of (Henriksen and Valls-Gabaud, 1994), (Narayan and Yi, 1995), (Fiege and Henriksen, 1995) show in fact that quadrupolar deflection is possible at large scales. Thus at best this model is likely to be perhaps an optical jet launching mechanism closest to the axis of a more general outflow.

Moreover I do not believe that this model is dynamically complete and therefore unique. In general (Henriksen and Reinhardt, 1974) one expects a magnetic boundary layer to be determined by the slow and fast magneto-sonic points and the intermediate Alfvénic point along each poloidal field-stream line. For a net outward flow of angular momentum the split monopole theory shows that the layer exists between the slow point and the Alfvén point only (referred to as a 'radiative' boundary layer in (Henriksen and Reinhardt, 1974)). In between these points some shearing is normally found in contrast to the uniform rotation assumption of Shu et al, and in fact this provides the necessary torque coupling. The complete boundary layer solution should I think include these points. The situation becomes even more confused if one admits that the stellar magnetic field may be oblique to the rotation axis of the system. In the limit of a perpendicular rotator, it would seem that elements of the split monopole theory might apply directly (Henriksen, 1976). The model has not yet been carried to the point of a direct comparison with radio observations.

4.6. HEATED QUADRUPOLAR ADVECTION MODELS

These models have been introduced recently by (Fiege and Henriksen, 1995) following (Henriksen, 1993), and (Henriksen and Valls-Gabaud, 1994). I have no space here to discuss these models in detail. Basically they solve rigorously for a self-similar gravitationally driven convective circulation in the extended 'atmosphere' (from a fraction of an au to$\geq$ 10,000 au !) of a protostar. The field-stream lines are quadrupolar. An evacuated region exists near the axis of rotation where a high speed outflow is produced. This outflow decreases in speed and increases in mass systematically with angle from the axis. We refer to this structure as a 'nested conical outflow'. Near the equatorial plane a thick rotating extended disc forms naturally when sufficient heating is provided to produce a high-speed axial outflow.

Synthetic radio maps of total intensity and in velocity channels have been calculated based on the dynamical solution, projected on the sky at various angles. The observations of SV, and HV outflows are reproduced in almost every respect. In particular the line profiles, the high velocity-(better) collimation correlations (Guilloteau et al., 1992). (Richer et al., 1992), and 'clumping' are all found in the synthetic maps. Not only the forms of the lines but also their brightness temperatures are reproduced and this in turn leads to predictions for densities, dust temperatures and magnetic field strengths.

In short it would seem that the more massive parts of the outflows are well represented by these rigorous models. Even conical shocks are expected where excited molecules may be created, although no detailed calculations have yet been performed. There are dynamical solutions which tend to the Bondi-Hoyle ((Bondi and Hoyle, 1944)) accretion tail shock limit which would appear as unipolar outflow (confusion effects can also cause such asymmetry).

The difficulty with such models is that the very high velocity optical jets do not appear unless the solution is pushed very close to a central core

582

from which the escape speed is about the jet speed of three or four hundred $km\ s^{-1}$. But here is precisely the region where the inner boundary layer may be dominant and the assumptions of this type of model break down. Moreover, we find that the radiative luminosity really has to be very large ($\geq 10^4\ L_\odot$) in order to drive these outflows if only dust opacity is used. Thus we may have to include Poynting flux as above and/or resonant absorption and possibly self gravity to explain the outflows of low bolometric luminosity. This can readily be done. An alternative view however is to add the boundary layer effects discussed in the previous section as the innermost part of the machine. Such a region need not even be magnetically dominated (Torbett, 1985),(Choe and Henriksen, 1986) in a post core formation phase when the angular momentum and magnetic flux problems have presumably been solved.

5. Acknowledgements

It is a pleasure to thank Kanaris Tsinganos and his organizing committee for their invitation to such a delightful spot, and all the participants for creating a lively meeting using both good science and good manners. This work was supported in part by a grant from the Canadian NSERC.

References

Aly J.J. 1980 *A&A* **86**, 192.
Aly J.J. 1994 *A&A* **288**, 1012.
André P., Ward-Thompson D. and Barsony, M. 1993 *ApJ* **406**, 122.
, private communication.
Bertout C. 1987 *Circumstellar Matter*, eds. Appenzeller I. and Jordan C. , Reidel, Dordrecht; 23.
Blandford R.D. and Payne D.G. 1982 *MNRAS* **199**, 883.
Bondi H. and Hoyle F. 1944 *MNRAS* **104**, 273.
Bontemps S., André P. and Ward-Thompson D. 1995 *A&A* **297**, 98.
Brandenburg A., Nordlund A., Stein R.F. and Torkelsson U. 1995 *ApJ* **446**,741.
Cabrit S. and Bertout C. 1992, *A&A* **261**, 274.
Cabrit S. et al. 1995 *A&A*, in press.
Chan K.L. and Henriksen R.N. 1980 *ApJ* **241**, 534.
Chernin L.M. 1995 *ApJ* **440**, L97.
Choe Seung-Urn and Henriksen R.N. 1986 *ApJ* **305**, 131.
Ferreira J. and Pelletier G. 1994 *A&A* **295**, 807.
Fiege. J. and Henriksen, R.N., 1995, prepublications I and II; sumitted to *MNRAS*.
Ghosh P. and Lamb F.K. 1979 *ApJ* **232**, 259.
Gomez de Castro A. I. and Pudritz R. 1993 *ApJ* **409**, 748.
Gouveia Dal Pino, E.M. and Benz W. 1994 *ApJ* **435**, 261.
Guilloteau S., Bachiller R., Fuente A. and Lucas R. 1992 *A&A* **265**,L49.
Henriksen R.N. 1976 *ApJ Lett* **210**, L19.
Henriksen R.N. 1985 *Can J. Phys* **64**, 403.
Henriksen R.N. 1987 *ApJ* **314**, 33.
Henriksen R.N. 1989 *Accretion discs and Magnetic Fields in Astrophysics*, ed. G. Belvedere, Kluwer; 117.

Henriksen R.N. 1993 *Cosmical Magnetism*, NATO Advanced Study Institute, July 1993, IOA Cambridge,UK.

Henriksen R.N. 1994 *The Cold Universe*eds Montmerle et al., Editions Frontieres, 241.

Henriksen R.N. and Chia T.T. 1972 *Nature* 240, 133.

Henriksen R.N. and Reinhardt M. 1974 *A& A* 31, 195.

Henriksen R.N. and Reinhardt M. 1977 *Ap and Sp Sci.* 49, 3.

Henriksen R.N., Reinhardt M. and Aschenbach B. 1977 *A&A* 28, 47.

Henriksen R.N. and Valls-Gabaud D., 1994,*MNRAS* 266,681.

Heyvaerts J. and Norman C.D. 1989 *ApJ* 347, 697.

Kitamura Y., Kawabe R. and Ishiguro M., 1992 *PASJ* 44, 44.

Königl A., 1982, *ApJ* 261,115.

Landau L.D. and Lifshitz E.M. 1987 *Fluid Mechanics*, Pergamon Press, Oxford.

Levreault R.M. 1988, *ApJ* 330, 897.

Lovelace R.V.E. Romanova M.M. and Bisnovatyi-Kogan G.S. 1995 *MNRAS* 275 , 244.

Lynden-Bell D. and Boily C. 1994 *MNRAS* 267, 146.

Masson C.R. and Chernin L.M. 1993 *ApJ* 414, 230.

Mitchell G.F. et al. 1995 *ApJ Lett*, in press.

Narayan R. and Yi I. 1995 *ApJ* 444,231.

Pudritz R.E. and Norman C.A. 1983 *ApJ* 274, 677.

Pelletier G. and Pudritz R. 1992 *ApJ* 394,117.

Raga A. and Cabrit S. 1993 *A&A* 278,267.

Richer J.S., Hills R.E. and Padman R. 1992 *MNRAS* 254, 525.

Sauty C., 1994, *Ann. Phys. Fr.* 19, 459.

Scheuer P.A.G. 1983 *Highlights Astr.* 6, 735.

Shu F.H., Ruden S.P., Lada C.J. and Lizano S. 1991 *ApJ* 370,L31.

Shu F., Najita J., Ostriker E., Wilkin F., Ruden S. and Lizano S. 1994 *ApJ* 429,781. 49.

Snell R.N. 1985*Can J Phys* 64, 431.

Torbett M. V. 1985 *Can J Phys* 64, 514.

Tsinganos K. and Sauty C., 1992, *A&A* 255,405.

Wardle M. and Königl A. 1993 *ApJ* 410, 218.

JETS AND MHD FLOWS ASSOCIATED WITH SYMBIOTIC STARS

Observational results and overall properties

MENAS KAFATOS
Center for Earth Observing and Space Research
Institute for Computational Sciences and Informatics,
and Department of Physics
George Mason University
Fairfax, VA 22030

Abstract.   Symbiotic stars are evolved systems consisting of a binary containing a late-type star (M-type, red giant or in some cases long-period variable) and a hot subdwarf or white dwarf secondary. These systems are classified as either S-type (showing the presence of a hot stellar continuum) or D-type (dust-type and showing no hot star continuum). D-type symbiotics are associated with long-period variables or Miras. Symbiotics have been studied at a variety of wavelengths, most recently and most successfully at far UV wavelengths using the *International Ultraviolet Explorer* (*IUE*) and even at soft X-rays using *ROSAT*. Observational results and overall properties of the components in symbiotics stars are presented. These systems show evidence of high ionization lines such as He II, C IV and N V as well as a number of semi-forbidden lines such as O IV], O III], N III], C III], and C II] and forbidden lines such as [Ne V] and [O II]. The lines arise in nebular regions surrounding the system, accretion disks and even jets. A number of systems show directed outflow, often in the form of individual jet parcels. Although the exact nature of the outflow process is not determined, it is likely to involve an accretion disk which may be magnetized. In the D-type symbiotics, ejection may be facilitated by radiation pressure onto grains. These jets provide opportunities to study the entire physics of jet phenomena for objects at close distances to the solar system. Some observational results of two jet prototypes are presented, R Aquarii (D-type) and CH Cygni (S-type). R Aqr is the nearest astrophysical jet and affords great opportunities to study these phenomena. *ROSAT* observations of these systems indicate that

K. C. Tsinganos (ed.), Solar and Astrophysical Magnetohydrodynamic Flows, 585–605.
© *1996 Kluwer Academic Publishers.*

the soft X-ray emission is probably arising in "hot spots" which may be MHD-related. A new magnetized accretion model for R Aqr is presented.

1. Introduction

Symbiotic stars as the term indicates, are two disparate stars living together. They are believed to be widely separated interacting binary starts surrounded by a hot ionized gas. These binaries generally consist of a cool M giant, which is often a Mira or late type semi-regular variable, and a hot companion with an effective surface temperature of $50,000K \leq T_{eff} \leq 100,000K$, appropriate to white dwarfs, or most likely central stars of planetary nebulae. Orbital periods in symbiotics can vary from ~1 to several dozen years. This fact alone separates them from other interacting binaries which usually have much shorter periods that range from hours to months. In these widely separated binaries, the stellar components evolve independently as long as both remain on the main sequence, but interact considerably during later stages of evolution. The product of this evolution may be the circumstellar gas observed in all symbiotics.

Often, these systems show extended nebular structure similar to planetary nebulae, and in a few cases, bipolar structure with outflow velocities as high as $\geq 1,000$ km s^{-1} has been observed. The relevant processes observed in symbiotics may include: mass expulsion, accretion disk formation, thermonuclear outbursts, and mass outflows that can lead to the formation of jet-like structures.

A characteristic of the symbiosis is a composite spectrum in the optical and ultraviolet which provide evidence for the existence of the cool red giant and the hot companion. The near and far-UV afford an opportunity to directly probe the high excitation source in symbiotics, because the luminous M star, which dominates the integrated visible and infrared light, makes essentially no contribution at ultraviolet wavelengths, or soft X-rays.

At visual wavelengths, symbiotic stars are characterized by the presence of molecular absorption features and continua appropriate for a late-type M star, which may be a Mira or other long-period variable. Nebular emission lines, on the other hand, include both forbidden and Balmer line emission, and suggest the presence of a high-excitation source (Boyarchuk 1975).

Moreover, the UV allows us to probe the temporal behavior of high-excitation emission lines in symbiotics because the luminous M star, which dominates the integrated visible and infrared light, makes essentially no contribution to UV wavelength lines. Thus, the far-UV provides a means of directly probing the high excitation source in these systems. The UV spectra of symbiotic stars suggest an enormous range of excitation that extends from low-temperature species such as O I $\lambda\lambda$ 1302-1306, C II $\lambda\lambda$ 1334, 1335, Si II $\lambda\lambda$ 1806, 1816, and Mg II $\lambda\lambda$ 2795, 2802 to strong resonance emission lines such as N v $\lambda\lambda$ 1238, 1242, C IV $\lambda\lambda$ 1548, 1550, as well as other high-ionization lines such as He II λ 1640 that are excited by the EUV continuum radiation that corresponds to a source with a T_{eff} of at least 50,000 K to 100,000K. Moreover, the semi-forbidden lines of O IV] $\lambda\lambda$ 1397-1407, N IV] λ 1487, O III] $\lambda\lambda$ 1660, 1666, N III] $\lambda\lambda$ 1747-1753, Si III] λ 1892 and C III] $\lambda\lambda$ 1907, 1909 provide information about the circumstellar nebular conditions of symbiotic stars (Meier *et al* 1994).

Allen (1982) has defined two categories of symbiotic stars based on IR photometry. Type S symbiotic stars have blackbody continuum emission in the K-band, while type D symbiotics exhibit thermal silicate emission from dust. Most known symbiotics ($\approx$110 out of 140) are S types, and about 30 (20%) are classified as type D.

S-type systems contain an M giant and often exhibit a Rayleigh-Jeans stellar blackbody continuum flux distribution in the far-UV. In the 1200-2000 A wavelength range of the *IUE* SWP camera this flux is generally appropriate to a $T_{eff} \geq$ 50,000 K star (Penston & Allen 1985), which is in agreement with the presence of the N V, C IV, and He II emission lines. On the other hand, the presence of strong silicate emission at 10 to 20 μm correlates with the presence of long-period variables, e.g., Miras, in D-type symbiotics (Allen 1982; Kafatos, Michalitsianos, & Feibelman 1982). The D-type symbiotics have a similar UV continuum flux distribution (Penston *et al* 1983) that either rises with increasing wavelength or is flat in the 2000-3200 A wavelength range of the *IUE* LWR/LWP cameras (see Kafatos, Michalitsianos, & Feibelman 1982). Since no hot secondary star has been directly observed in these D-type systems, some secondary stars may be buried in the nebular material. Thermal IR emission, presumably from circumstellar dust, is associated with extensive mass loss from Mira winds.

By observing symbiotic stars we can examine the basic physical processes of: i) mass loss from red giants and the formation of planetary nebulae; ii) accretion onto the compact stars and evolution of nova-like outbursts; iii) photoionization and radiative transfer within gaseous nebulae; iv) formation of extended emission line structures (Kenyon 1986). Symbiotics may represent a transitory phase in the evolution of certain types of binary stars where there is substantial mass transfer from the M giant to the hot companion. More recent theories have even suggested that symbiotic stars may be possible progenitors of type I supernovae (Kenyon *et al* 1993). There is still much to learn about the nature of these systems due to the difficulty of observing the hot secondary stars.

The six known symbiotic stars positively identified with soft X-ray emission are: V 1016 Cyg, RR Tel, HM Sge (Allen 1981), AG Dra, CH Cyg (Anderson *et al* 1981; Leahy and Taylor 1987; Leahy and Volk 1995) and R Aqr (Viotti *et al* 1987; Kafatos *et al* 1996). Four of these are D type (Meier *et al* 1994). Accretion in the D-type systems may occur at rates as high as 10^{-5} or even 10^{-4} $M_\odot yr^{-1}$. These rates are higher than for the two S-type systems (AG Dra and CH Cyg), in which compact regions with T≥10^6 K can explain the observed soft X-ray emission. Although it contains an M giant, the S-type symbiotic CH Cyg is a strong X-ray source. The other S-type, AG Dra, contains a K giant of ambivalent spectral class.

For a review of symbiotic star properties see Boyarchuk (1993); Kenyon (1986); Mikolajewska, Friedjung, Kenyon and Viotti (1988).

2. Symbiotic Star System Components

If indeed symbiotic stars consist of a) a late-type giant star, b) a hot component, and c) an ionized nebula, the combined spectral properties should reflect this fact. The hot star is likely a hot stellar source similar to the central star of a planetary nebula (e.g., Boyarchuk 1975), or an accreting white dwarf or even a main sequence star (e.g., Kenyon and Webbink 1984). Radio and optical observations can provide a rough estimate of the temperature and luminosity of the hot object, although the physical nature of this component is determined by satellite data (e.g. taken with *IUE*, Kenyon 1986).

It is important to realize that physical properties of the cool giant are responsible for the observable properties of an individual symbiotic star. For the 1-3 year orbital periods that are typical of these systems, a normal red giant star with radius, R_g, ~50 $R_\odot$ does not fill its Roche lobe, while a bright asymptotic branch giant or super-giant with $R_g \geq 200R_\odot$ is likely to fill its tidal lobe (Allen 1973; Kenyon 1986). Since mass transfer seems essential for all symbiotic models, the physical size of the giant determines how this material is lost and the manner with which it interacts with the hot component. Lobe-filling giants apparently lose mass at rates >$10^{-6}M_\odot yr^{-1}$ (Webbink 1979; Lipunov 1995), and this is roughly the lower limit required to form a symbiotic system with an accreting main sequence star. Red giants smaller than their tidal lobes can only lose mass in a stellar wind, and observations of single stars suggest wind-driven mass loss rates of ~$10^{-7}M_\odot yr^{-1}$ for red giants and ≥$10^{-5}M_\odot yr^{-1}$ for Mira variables and supergiants (Zuckerman 1979). Calculations described by Tutukov and Yungel'son (1993) and

Lipunov (1995) show that perhaps $\approx 1\%$ of this material could be captured by a hot companion star, implying an accretion rate $\geq 10^{-7} M_{\odot} yr^{-1}$. This can energize a binary containing a white dwarf or hot sub-dwarf, but fails for a main sequence star. Thus, symbiotic stars might be divided into two categories based on the nature of the hot component and the giant component (Kenyon 1986): i) systems containing a giant which has expanded to its Roche limit and a low mass main sequence star, and ii) systems containing a white dwarf or sub-dwarf and a red giant losing mass in a stellar wind.

Whereas the above may be true for most symbiotics, the existence of jets in some (most notably CH Cyg and R Aqr, see below) and possibly the existence of soft X-ray emission may argue for the formation of an accretion disk which likely forms in a Roche-lobe overflow situation by the mass loosing primary. The rather high degree of collimation of the ejected parcels of gas (Kafatos, Michalitsianos and Hollis 1986) favors the presence of an accretion disk; otherwise, it would be difficult to form collimated jets that may be precessing (Hollis and Michalitsianos 1993) if stellar winds were the only means of mass transfer and ejection.

The presence of TiO bands in the visible spectra of symbiotic stars allowed one to classify the giants as M2 or later (Boyarchuk 1993), most likely of luminosity class III; one notable exception being the CI Cyg giant which is likely luminosity class II with $R \sim 300 R_{\odot}$ (Kenyon and Fernandez-Castro 1987; Boyarchuk 1993).

In D-type symbiotics, the cool primary is, instead, a Mira or a late-type long-period variable star. The IR-excess is believed to be due to dust emission with $T \approx 1,000$ K (Boyarchuk 1993). The IR observations of D-type symbiotics and variability (Feast, Robertson and Catchpole 1977) indicated changes characteristic of Miras. The color temperature of these dust envelopes in the IR is ~1,000K although the observed IR emission may be consistent with thermal emission of dust with $T \sim 250$-450K (Boyarchuk 1993). The dust may be concentrated in the circumstellar envelope of the Mira and could account for the weak indication of a cool component (Boyarchuk 1993). The Miras in D-type symbiotics are covered by strongly absorbing dust, silicate, envelopes. There has even been suggestion of eclipses of the Mira primary by a very cool (dust envelope) component. A hypothesis has been proposed that there is a sudden increase of mass loss by the cool component followed by enhanced dust formation (Boyarchuk 1993). Moreover, there seems to be a connection between the increase of radio emission and increased IR emission (Seaquist and Taylor 1987). This would argue in favor of disk formation when enhanced episodes of mass loss such as jets take place (see below), applicable to both S- and D-type symbiotics. The circumstellar envelope is, however, much more massive in D-type symbiotics

($\sim 10^{-5}$-10^{-3} M$_\odot$) compared to S-type symbiotics ($\sim 10^{-6}$M$_\odot$) (see Boyarchuk 1993).

The nature of the hot stars in symbiotic systems is much harder to ascertain. In general, hot star continua can be fitted if $T_{eff} \sim 7 \times 10^4$-$1.4 \times 10^5$K (Boyarchuk 1993). In general, (Kafatos and Michalitsianos, 1984), the hot stars emit as much energy as a normal red giant but in the far UV (say ~ 10-10^3L$_\odot$). Most hot stars in symbiotic would be found in the hot sub-dwarf part of the H-R diagram, the least luminous of them being the hot star in R Aqr (see below).

If, on the other hand, the UV radiation originates in an accretion disk, the accretion rate should be $\geq 10^{-6}$ M$_\odot$ yr^{-1} for a main sequence secondary and $\geq 10^{-4}$M$_\odot$ yr^{-1} for a white dwarf secondary, although the observed colors do not fit such a model (Boyarchuk 1993) except perhaps for the soft X-ray emitting symbiotics including R Aqr and CH Cyg (see below).

The diffuse gas in symbiotic stars emits a rich far UV (and optical) spectrum. Density estimates and electron temperatures, T_e, can be obtained by a variety of methods (cf. Kafatos, Michalitsianos and Fahey 1985; Kafatos, Michalitsianos and Feibelman 1982; Kafatos, Michalitsianos and Hollis 1986, etc.). In general, semi-forbidden line ratios give $n_e \sim 10^9$cm^{-3} for S-type symbiotics and $n_e \sim 10^6$cm^{-3} for D-type symbiotics (see Boyarchuk 1993 and references above). The electron temperature is typically higher than normal H II regions, $15,000 \leq T_e \leq 20,000$K.

3. Overall Properties of Symbiotics

Here we provide some results in tabular form, primarily from Kenyon (1986). UV properties of selected symbiotics obtained with the *IUE* can be found in Meier *et al* (1994).

The above tables are from Kenyon (1986) and help provide an overall picture of the stellar and nebular properties in symbiotic stars. Detailed far UV continuum and line emission properties for the following symbiotics can be found in Meier *et al* (1994): EG And, AX Per, LMC Anon, LMC 563, RX Pup, He 2-38, SY Mus, BI Cru, RW Hya, He 2-106, T CrB, AG Dra, HK Sco, CL Sco, H 1-36 Arae, Y CrA, YY Her, AS 296, AS 295B, AR Pav, V 443 Her, BF Cyg, CH Cyg, HM Sge, CI Cyg, V1016 Cyg, RR Tel, HBV 475, AG Peg, Z And and R Aqr. The reader is referred to that paper for detailed information including distances, spectral type, extinction E_{B-V}, T_e, n_e and T_* (hot star) as well as radial velocities, orbital periods and X-ray and radio (6 cm) fluxes.

Table 1: Binary Periods for Selected Symbiotics

Symbiotic Star	Period (Days)
Z and	756.9
EG And	470
R Aqr	44 years
UV Aur	395.2
TX CVn	70.8
V748 Cen	564.8
T CrB	227.5
BF Cyg	757.3
CH Cyg	15.75 years
CI Cyg	855.25
V1329 Cyg	959
AG Dra	554
RW Hya	372.5
BX Mon	3.8 years
SY Mus	627.0
AR Pav	604.6
AG Peg	827.0
AX Per	681.6
V2601 Sgr	850
V2756 Sgr	243
CL Sco	624.7
BL Tel	778.6

Table 2: Mira Pulsational Periods in D-type Symbiotics

Symbiotic Star	Period (Days)
R Aqr	387
BI Cru	280
V1016 Cyg	470
RX Pup	580
HM Sge	>400
RR Tel	380
He 2-34	370
He 2-38	433
He 2-106	400

Table 3: He II Zanstra Temperatures for Symbiotics

Symbiotic	T_* (K)
Z and	180,000
BF Cyg	75,000
CI Cyg	155,000
V1016 Cyg	160,000
V1329 Cyg	160,000
AG Dra	170,000
YY Her	125,000
V443 Her	110,000
RS Oph	<95,000
AG Peg	180,000
AX Per	135,000
HM Sge	145,000
CL Sco	90,000
AS 289	145,000
BD-21° 3873	135,000
M1-2	80,000

Table 4: Nebular Densities and Temperatures from UV (*IUE*) Spectra

Symbiotic System	n_e (cm^{-3})	T_e(K)
Z And	2×10^{10}	80,000
R Aqr	10^6	15,000
BF Cyg	10^{5-8}	40,000
CI Cyg	10^9	10,000
V1016 Cyg	3×10^6	15,000
V1329 Cyg	3×10^6	12,500
AG Dra	3×10^9	-------
V433 Her	10^9	15,000
RW Hya	10^{8-9}	15,000
SY Mus	10^{10}	15,000
AG Peg	10^{7-11}	15,000
AX Per	2×10^8	12,500
RX Pup	10^{9-11}	15,000
HM Sge	10^{6-7}	15,000
RR Tel	10^{6-8}	15,000
HD 330036	10^6	15,000

Nebular parameters can be obtained from semi-forbidden and forbidden line ratios and absolute intensities (e.g. Kafatos, Michalitsianos and Hollis 1986; Meier and Kafatos 1995). The SiIII] lines, the [OIII]/OIII] ratio and other similar methods yield volume averaged n_e. The CII] ratio can be used to estimate T_e. Absolute intensities of UV lines can yield size estimates, R, once n_e and T_e are known by the above methods.

Properties of the secondary can be obtained once the ionizing radiation has been estimated from R and n_e using the Strömgren relation. For example, for R Aqr, using $n_e \sim 4 \times 10^5 cm^{-3}$, $R \sim 3 \times 10^{14} cm$ and $T_e \sim 20,000 K$, we find $N_i \sim 3 \times 10^{42} s^{-1}$. The He II λ 1640 modified Zanstra method can be used to estimate the effective temperature, T_*, of the hot star. This is not a totally reliable method because the underlying continuum at $\lambda 1640$ may have a nebular contribution (Meier & Kafatos 1995). Once, though, T_* is estimated, the radius of the hot star can be obtained from N_i and T_e.

Alternatively, the photoionizing radiation may arise from an accretion disk and its inner boundary layer (Meier and Kafatos 1995). The relevant relationships are (cf. Kafatos, Michalitsianos and Hollis 1986) as follows: the accretion disk theory of Bath *et al* (1974), Lyden-Bell and Pringle (1974), and Tylenda (1977) can be used. We find the following relationships for the radius R_* of the hot star (assumed equal to the inner radius of the accretion disk); the accretion rate $\dot{M}$ ($M_\odot$ yr^{-1}); the mass of the secondary, M_*; and the disk luminosity L_d ($L_d \sim L_{bl}$), where "bl" refers to the boundary layer:

$$R_* \geq 5.8 \times 10^9 \ (T_{bl}/10^5 \ K)^{-3/2} \ [N_i/(f \times 10^{45} \ s^{-1})]^{1/2} \ cm,$$
$$\dot{M} \ (M_*/M_\odot) \geq 3.2 \times 10^{-8} \ (T_{bl}/10^5 \ K)^{-1/2} \times [N_i/(f \times 10^{45} \ s^{-1})] \ M_\odot \ yr^{-1},$$
$$L_d \sim L_{bl} \geq 5.9 \ (T_{bl}/10^5 \ K) \ [N_i/(f \times 10^{45} \ s^{-1})] \ L_\odot.$$

In the above relations, the "greater than" sign applies if the observed H II region(s) is (are) photoionized by an accretion disk, with a boundary-layer temperature T_{bl}, and the region(s) is (are) density-bound. Otherwise, the equality sign applies. N_i is the number of Lyman continuum photons emitted per second, and the parameter f can be found from blackbody tables: $f = 0.186$, 0.7, and 0.9, for $T_{bl} = 40,000$, 100,000, and 200,000 K, respectively. The ionizing photon flux N_i can be obtained from the Strömgren theory (see above).

The assumption for a single, spherical region to account for the UV emission lines is probably an over-simplification. Nearby symbiotics such as R Aqr and CH Cyg indicate a much more complicated geometry at different densities and even temperatures.

We now confine our discussion to two symbiotics with observed jet structure, R Aqr (D-type) and CH Cyg (S-type). Most of our discussion focuses on R Aqr which has been extensively studied in the far UV and has been known to undergo eruptions in the past. Although the thermo-nuclear outburst model mostly likely applies to many symbiotics (Kenyon 1988; Paczinski and Rudak 1980; Boyarchuk 1993), it is likely that the symbiotics showing evidence of jet formation are powered by accretion (Kafatos, Michalitsianos and Hollis 1986).

4. R Aquarii

Since the discovery of an astrophysical jet, the peculiar symbiotic star R Aquarii (R Aqr) has been the subject of many inquiries. This D-type symbiotic binary star system, which consists of a hot sub-dwarf or possibly white dwarf and a Mira variable, contains a jet which is believed to have formed in the mid 1970's (Herbig 1980) and possibly an accretion disk (Kafatos, Michalitsianos & Hollis 1986). At a distance of 250 pc (Whitelock 1987), R Aqr, along with possibly CH Cyg, is the closest symbiotic star and astrophysical jet, and with a 44 year period (Willson, Garnavich & Mattei 1981), it has the longest binary orbit of any symbiotic. This complex system has been detected at virtually every wavelength from radio to soft X-rays. In addition to the wide range of excitation, R Aqr is the only known symbiotic associated with SiO maser emission (Lepine, Le Squeren & Scalise 1978). A detailed ultraviolet analysis of this extraordinary object could help in understanding the many active objects which eject matter and emit radiation that spans the electromagnetic spectrum. Moreover, this analysis could help in providing clues to the emission mechanisms of astrophysical jets.

R Aqr is believed to consist of a hot sub-dwarf or white dwarf and a 387 day period Mira (M7e+pec) variable (Kenyon & Fernandez-Castro 1987) that is embedded in an H II region. The Mira has a 387 day pulsation period and corresponding ephemeris of Max (V) = JD 2382892 + 386.3*E (Mattei & Allen 1979). This system has been known to undergo erratic outbursts. The most recent outburst occurred between 1928 and 1935 in which the hot companion achieved a visual magnitude m_v of ~8, rivaling the visual light of the Mira (Mattei & Allen 1979).

The existence of extended jet structure is one of the features that makes R Aqr an unique astronomical object (Meier and Kafatos 1995). In 1977, Wallserstein & Greenstein (1980) detected a "jet" of nebular material with objective prism spectroscopy in the optical region extending ~6″ NE off the star while Herbig (1980), confirmed the existence of the jet from optical plates taken in 1980. Observations by Sopka et al (1982) first detected the NE jet in the radio region with the VLA. Further VLA radio frequency observations by Kafatos, Hollis & Michalitsianos (1983) at 6 cm resolved the system into three discrete regions of emission: (1) feature B at 6".5 from R Aqr at a position angle of 29°, (2) feature A at 2".5 and 45°, both related to the NE jet, and (3) the compact H II region. Hollis *et al* (1986), using 2 cm VLA observations, discovered two distinct emitting regions, C1 (the compact H II region) and C2. Recent *Hubble Space Telescope (HST)* observations obtained with the Faint Object Camera (FOC) centered at $\lambda = 1230$ A showed that feature C1 itself consists of four distinct components (Burgarella & Paresce 1992; Paresce & Hack 1994), while using near-UV observations, Mauron *et al* (1985) discovered a counter jet (feature A') in the SW direction. Additional observations at 6 cm by Kafatos *et al* (1989) detected radio continuum emission in the counter-jet, which extends approximately 10" SW of R Aqr at a position angle of 231°. The most recently *HST* discovered emission feature, feature D, was found at 8".4 from the central object C1 with a position angle of 20° (Paresce, Burrows, & Horne 1988).

The position of the *IUE* 10" x 20" aperture relative to the R Aqr radio morphology can be found in Michalitsianos & Kafatos (1988). The large size of the *IUE* aperture allows integrated ultraviolet emission in the HII region nebulosity primarily from features C1, C2, A, and A', whereas the placing of the aperture in the NE jet integrates UV emission primarily from features A, B, and D.

Ultraviolet observations of R Aqr exhibit several emission lines of O I $\lambda\lambda$ 1302-1306, C II $\lambda\lambda$ 1334, 1335, C IV $\lambda\lambda$ 1548, 1550, O III] $\lambda\lambda$ 1660, 1666, N III] $\lambda\lambda$ 1747-1753, Si III] λ 1892, C III] $\lambda\lambda$1907, 1909, C II] $\lambda\lambda$ 2325, 2326, [O II] λ 2470, and Mg II $\lambda\lambda$ 2975, 2803. These emission lines are superposed on a UV continuum which gradually rises with decreasing wavelength in the NE jet (Michalitsianos & Kafatos 1982) and is essentially independent of wavelength for the central star and associated nebulosities (Johnson 1980, 1982; Michalitsianos, Kafatos & Hobbs 1980; Meier *et al* 1994). However, UV spectra of the H II region taken between 1986 October and 1991 December reveal a continuum flux that increases toward shorter wavelengths, similar to the NE jet. The appearance of N V $\lambda\lambda$ 1238, 1240 and He II λ 1640 emission in the NE and SW jets, compared with the weakness of N V and He II in the H II region, gives evidence that higher excitation conditions are present in the two jet structures (Michalitsianos, Hollis, & Kafatos 1986; Kafatos *et al* 1986; Hollis *et al* 1991). The presence of N V emission in the NE jet is also consistent with the detection of soft X-ray emission in the 0.25-1 keV range observed by the *EXOSAT* satellite (Viotti *et al* 1987) and more recently by *ROSAT* (Kafatos *et al* 1996).

By monitoring the temporal variability of R Aqr, insights can be obtained about the physics of astrophysical jets. Moreover, the UV emission lines can help in calculating the physical parameters for the system: electron temperature, line-emitting region, photoionization parameters and possible accretion disk parameters (Kafatos, Meier and Martin 1993; Meier and Kafatos 1995; see also section 3).

The overall UV continuum flux distribution has increased towards both shorter and longer wavelengths over the time period the H II region of R Aqr has been observed with *IUE*. Additionally, the emission line intensities have increased by factors of ~1.5-2 compared to earlier values in 1979 Jan. These spectra indicate that the central Feature A (and to a certain extent the H II region) are increasing in excitation or that the emitting volume is increasing. The emergence of N V and He II around 1986 Oct. argues for higher excitation conditions in the jet, particularly Feature A. Since the 10″ x 20″ *IUE* large aperture receives flux from both regions and N V and He II have been previously detected in the jet (Features A and B), this suggests these high excitation lines originated in Feature A instead of the central H II region. The active phase which is believed to have begun around 1986 Oct. may still be continuing (cf. Meier and Kafatos 1995).

Low resolution UV spectra of the R Aqr jet have revealed the following information: an ultraviolet continuum flux that rises with decreasing wavelength and contains several emission lines. The *IUE* spectrum taken in 1983 May displays emission lines of C IV $\lambda\lambda$ 1548, 1550, C III] $\lambda\lambda$ 1907, 1909, C II] $\lambda\lambda$ 2325, 2326, [O II] λ 2470 and Mg II $\lambda\lambda$ 2795, 2803 that have increased by a factor of ~2, and a continuum flux distribution ~1.5 times stronger in the shorter wavelength region (1200-1900 A) compared to levels in 1982 May. The UV continuum flux and emission lines remained essentially constant throughout 1985 Jan. (cf. Meier *et al* 1995). One interesting feature was the first detection of NV $\lambda\lambda$ 1238, 1242 in the jet. The most radical changes in the UV spectra of the jet took place in 1986 Dec. where the continuum flux distribution increased considerably in the long wavelength region and emission lines of C II, N IV], C IV, Si III], C II], [O II], Mg II, O III were approximately 2 to 4 times greater than levels in 1985 Dec. The overall UV continuum flux distribution of the jet increased by a factor of ~2-3 over the *IUE* observation period and the emission line intensities increased 3-6 times compared to levels in 1982 May and 1983 May. The overall increase in the UV continuum flux and in virtually all emission line intensities provides evidence, as indicated earlier, that the excitation of the jet continues to increase or perhaps the emitting region is expanding. This trend began around 1986 Dec. and may still be continuing. The detection of the high excitation emissions lines of N V $\lambda\lambda$ 1238, 1242 and He II λ 1640 also provides evidence of a high excited region with a temperature greater than 65,000 K (Kafatos, Michalitsianos & Hollis 1986; Michalitsianos, Hollis & Kafatos 1986; Hollis *et al* 1991 and Meier and Kafatos 1995).

Meier and Kafatos (1995) have carried out analysis to obtain the physical parameters for the hot component in R Aqr for different effective temperatures ($T_* = 50,000$ K, 100,000 K) for the central hot star. Their calculations indicate that the secondary star is most likely a white dwarf or even a hot sub-dwarf. Burgarella, Vogel & Paresce (1992) found a radius of $R \geq 0.1 R_\odot$ for the hot star. The modified Zanstra method was employed to determine the effective temperature of the hot component. In this method the He II λ 1640 line intensity is divided by the underlying continuum (assumed stellar) at the same wavelength (Murset *et al* 1991). The resultant effective temperature for R Aqr should be taken as a lower limit since the underlying continuum at λ 1640 may have a nebular component from free-free and free-bound emission. This method depends on the wavelength of the He II Balmer α transition (λ=1640A for the n=3→2 transition), α(n) the recombination coefficients for the nth level, $G_4(T)$ the numerical value of the integral used in the Zanstra method and the electron temperature. Assuming T_e=20,000K for the H II region, they found an effective temperature of $T_* \approx 60,800$K for the hot star (Meier and Kafatos 1995). This estimate is consistent with the number of ionizing photons $\sim 10^{42} s^{-1}$ needed to ionize the central H II region. The value of T_* in this case is higher than the most recent estimate of $T_* \sim 40,000$K (Burgarella, Vogel & Paresce 1992), but is in agreement with previous estimates of $T_* \geq 50,000$K (Michalitsianos *et al* 1980; Kaler 1981; Michalitsianos & Kafatos 1982).

The spectral variations in the UV (Meier and Kafatos 1995) are consistent with a model for the jet in which shock excitation is responsible for the observed emission lines and the increase of emission line flux. This shock may form as the stellar wind from the hot star or a bipolar wind from an accretion disk (Kafatos, Michalitsianos and Hollis 1986) collides with condensations of matter, i.e. Features A, B and D (Solf 1992) (see however, a magnetized model proposed here). *HST* FOC observations by Paresce & Hack (1994) found that the jet collimated within 15 A.U. of the Mira, placing it within the binary orbit. Therefore, the energetic stellar wind most likely originates either from an accretion disk or the secondary star. An accretion disk could have formed in the mid 1970's near periastron as material from the Mira overflowed its Roche lobe. If a bipolar wind emanating from the accretion disk collides with the stationary condensations of matter it could produce emission lines through collisional ionization and recombination as the gas cools. The variations in the UV may be explained by a precessing accretion disk as proposed by Michalitsianos *et al* (1988) and Hollis and Michalitsianos (1993). A precessing disk with a changing direction of collimated flow could be responsible for the increased activity in the jet which began in 1985. *IUE* observations of N V and He II, which have steadily increased since 1986 Oct., suggest an increase in the ionization state of the system for Feature A.

In summary, the ultraviolet spectra obtained during 1979-1991 of the symbiotic star R Aqr confirm that Features A and B of the jet are increasing in excitation, consistent with the prediction of Kafatos, Michalitsianos and Hollis (1986). Shock excitation by a radiative shock may be the primary emission mechanism in the jet. The shock may originate as a bipolar wind emanating from an accretion disk collides with the stationary clumps of matter (Solf 1992). The increase in emission line intensities in the jet around 1983 may be due to a precessing accretion disk (Michalitsianos *et al* 1988) which is directing its collimated flow towards Features A, B and D. In our analysis, a combination of these two models would best explain the ultraviolet observations: We believe a highly collimated bipolar wind is produced by a precessing accretion disk. As the wind interacts with stationary condensations of matter (Features A, B, D & A') a radiative shock is generated. A bipolar wind is needed to explain the NE jet and the presence of the SW counter-jet. Furthermore, a precessing accretion disk may be responsible for the increase in the UV flux distribution and individual emission line intensities if the bipolar wind is highly directional and collimated. On the other hand, the resultant high resolution radio images of R Aqr (Dougherty *et al* 1994) show remarkable small-scale structure reminiscent of AGN jets. This may be indicative of successive ejection of individual jet parcels as favored by Kafatos, Michalitsianos and Hollis (1986) which are photoionized by the EUV flux escaping perpendicular to a thick accretion disk. In either case, the presence of an accretion disk seems necessary.

Unlike CH Cyg where ejection velocities $\sim$1,000 km s^{-1} are indicated, the outflow speeds in R Aqr are low, generally $\leq$100 km s^{-1} and not exceeding 300 km s^{-1} in most extreme cases (Kafatos, Michalitsianos and Hollis 1986). These would be consistent with outflows from the extended envelope of the Mira or the outer regions of a cool, dust-dominated accretion disk.

Hollis *et al* (1995) have presented evidence for sub-arcsec changes in the morphology of the inner 5 arc sec of the R Aqr jet over a two year period (1991-1993). These unsaturated data were taken with *HST* FOC. Images of the R Aqr jet were successfully restored to the original design resolution. Their results suggest that the NE and SW jet/counterjet rotates in projection on the sky in a counter clockwise manner, and moreover, that the jet is expanding in a helical-like structure. They interpret the rotational motion as due to precession (with a period $\sim$600 years) of the accretion disk and the helical structure as due to MHD instabilities (Hollis *et al* 1995).

ROSAT observations of R Aqr were carried out in Sep. 1994. The count rate was $\sim$0.0097cts s^{-1} which translates to a flux $\sim$7.2 x 10^{-13}erg cm^{-2}s^{-1}

(Kafatos *et al* 1996). Our best estimate of the temperature of the X-ray emitting plasma is ~8.4x10^{5}K which for log N$_H$~19.8 appropriate to R Aqr results in a characteristic radius of the emitting region ~10^5 cm. This is obviously too small to be an emitting region on an accretion disk and would be barely consistent with a boundary layer around a ~10^8 cm white dwarf. Such small sizes may be indicative of MHD "hot spots" in an accreting column; alternatively, it may be that we are seeing reflected X-rays as suggested for the *EXOSAT* observations. In that case a larger size can be accommodated. It is important though to emphasize that we have no indication of an extended X-ray emission. Future high resolution X-ray observations with *AXAF* would be needed to unequivocally prove that the source of X-rays is point-like and located on the position of the symbiotic star. Similar small sizes result for CH Cyg (see below).

5. CH Cygni

This S-type symbiotic underwent an eruption in 1984-1985. A radio producing thermal jet was produced, with a two-sided expansion ~1"yr^{-1} corresponding to transverse velocity of ~1,100 km s^{-1} in each direction (Seaquist, Taylor and Button 1984). The formation of the jet is consistent with supercritical accretion onto a white dwarf. The onset of the radio outburst coincided with a remarkable decline in the visual brightness of the star (Taylor, Seaquist and Mattei 1986).

The nebular parameters are $n_e > 2\times10^6$cm^{-3} and a total mass for each jet component ~10^{-6} M$_\odot$. CH Cyg, like R Aqr, is relatively nearby, with estimates as close as ~200 pc (Solf 1995). The two stars would, therefore, provide good information on the physics of astrophysical jets.

Despite their differences, both star systems show soft X-ray emission. Leahy and Taylor (1987) and Leahy & Volk (1995) find a PSPC *ROSAT* detection ~5.8x10^{-12}erg cm^{-2}s^{-1} for the energy range 0.1-2.5 keV. The X-rays may arise in the jet components. They find the need for multiple components including a power law with index ~0.5. *EXOSAT* observations in May 1985 yielded a flux ~1.2x10^{-11} erg cm^{-2}s^{-1} (0.02-2.5 keV). We carried out re-analysis of the CH Cyg data. We find that T~3x10^6 and log(N$_H$)=20.4 give adequate fits if a blackbody is assumed with essentially the same flux used by Leahy and Volk (1995). As was the case with R Aqr, the resultant radius of the emission region is quite small, ~4x10^4 cm, whereas the total luminosity ~10^{32} erg s^{-1} (Kafatos *et al* 1996).

There must be something common in the jet formation/X-ray emission to produce similarly small emission regions in such disparate systems as a D-type symbiotic with slowly expanding jet components ($\sim$100 km s^{-1}) and an S-type symbiotic with a much faster jet ($\sim$1,000 km s^{-1}). It is interesting to note that R Aqr and CH Cyg have the longest binary periods of any symbiotic (see Table 1).

The R Aqr, CH Cyg symbiotic system properties are compared to each other in Table 5.

Table 5: R Aqr, CH Cyg Properties

	Distance	Binary Period (yrs)	M-Primary	Secondary	V_{exp}(jet)	n_e (cm^{-3})
R Aqr	200-250 pc	44	M7e+pec	Sub Dwarf? White Dwarf?	$\sim$100 km s^{-1}	4×10^5 (H II); 6×10^4 (NE jet); 4×10^2 (SW jet)
CH Cyg	200-600 pc	15.75	M6	White Dwarf?	$\sim$1,000 km s^{-1}	$>2 \times 10^6$

	T_e (K)	$\dot{M}_{acc}$ (M$_\odot$ yr^{-1})	F(ROSAT) erg cm^{-2} s^{-1}	T_{BB}(ROSAT) K
R Aqr	19,000 (H II); 18,000-26,000 (NE jet); 33,000 (SW jet)	$\sim$10^{-8}	7×10^{-13}	8.4×10^5
CH Cyg	7,000 K	$\sim$10^{-5}	$\sim$10^{-11}	3×10^6

	F(6cm) mJy	IR Properties	Log (Column Density)
R Aqr	$\sim$10	H$_2$O + CO Bands	19.8
CH Cyg	1.4	CO Absorption Band	20.4

We now turn our attention to some theoretical considerations re. a magnetized white dwarf. We apply these ideas to R Aqr, although the similarity in the smallness of the X-ray emitting regions in both R Aqr and CH Cyg may be indicative that a magnetized white dwarf may also be part of the CH Cyg system.

6. MHD Flows in R Aquarii

Assuming the hot component in R Aqr is a white dwarf/subdwarf with $R_*\sim10^9$cm - 3×10^9cm, one would expect a hot star wind with $V_{esc} \sim$ 2,000-

3,000 km s^{-1}. No such velocities have been observed in R Aqr (typically V~100 km s^{-1} with upper limit ~300 km s^{-1}, Solf 1992) indicating that the jet motions or associated wind originate far from the hot star. Kafatos, Michalitsianos and Hollis (1986) proposed that the mass loss originates in the outer regions of a cool, geometrically thick disk. In these outer regions, r~10^{13} cm , dust grains would form and supercritical accretion would result even at ~10^{-8} M$_\odot$yr^{-1}. Radiation pressure on the grains would drive a strong wind at terminal velocities ~150 km s^{-1}, a reasonable value for R Aqr.

The question remains why the hot star does not drive a fast wind. We propose here a strongly magnetized hot star, with B$_*$~10^7 G at its surface. The magnetic field would completely dominate gas or radiation pressure. In other words, the magnetized hot star does not allow the formation of an inner disk and, therefore, no high velocity material can be ejected. Escaping material is, instead, driven out of the cooler regions of a thick disk. For B$_*$~10^7 G, the co-rotation radius of light cylinder is ~5x10^{11} cm. Outside it, P$_{rad}\geq$P$_B$, radiation pressure dominates magnetic pressure and slow ejection can take place, perhaps by radiation pressure on grains. For r~10^{12}-10^{13} cm, V$_{esc}$~ 150-50 km s^{-1}, respectively, as observed in R Aqr. The disk itself terminates at ~5x10^{11} cm where a turbulent boundary layer forms. Inside the light cylinder, no outflow occur but accretion proceeds along the poles of the dipole field as in AM Her stars. Inside the disk, turbulent B-fields likely exist with $\alpha \leq 1$, where α = P$_B$ / (P$_{rad}$ + P$_{gas}$). Outside the light cylinder mass loss occurs which powers the observed jet.

A summary of the MHD accretion onto a magnetized hot subdwarf and the resultant mass loss is summarized in Table 6.

Table 6: Magnetized Hot Subdwarf & MHD Flow

r (cm)	P$_B$ ***	P$_{gas}$	P$_{rad}$	B (G)	n (cm^{-3})	T(K)	V$_{esc}$ (km s^{-1})	V$_A$ (km s^{-1})
3x10^9 *	4x10^{12}	10^3	10^4	10^7	10^{14}	$\leq$10^5	3,000	$\leq$c
10^{12}	3x10^{-3}	10^{-2}	10^{-1}	0.3	10^{10}	~10^4	~150	70
10^{13} **	3x10^{-9}	10^{-4}	10^{-3}	3x10^{-4}	3x10^8	~3,000	~50	---

*	The stellar radius
**	Gas outflow results from cool outer regions of accretion disk
***	Assuming a dipole field

This model is shown in Figure 1. In this figure, the light cylinder is shown at ~5x10^{11} cm and the direction of the slow wind (V$_{esc}$ ~ 100 km s^{-1}), which forms in the cool regions of an extended disk outside the light cylinder.

Mass flows are shown with thicker arrow lines. The ordered dipole field inside the magnetosphere and turbulent fields inside the thick disk are shown with lighter lines. Finally, the hatched region at the edge of the disk indicates the location of a turbulent boundary layer, where mass inflow originates. Matter follows the B-field lines and would accumulate on the magnetic poles of the hot sub-dwarf.

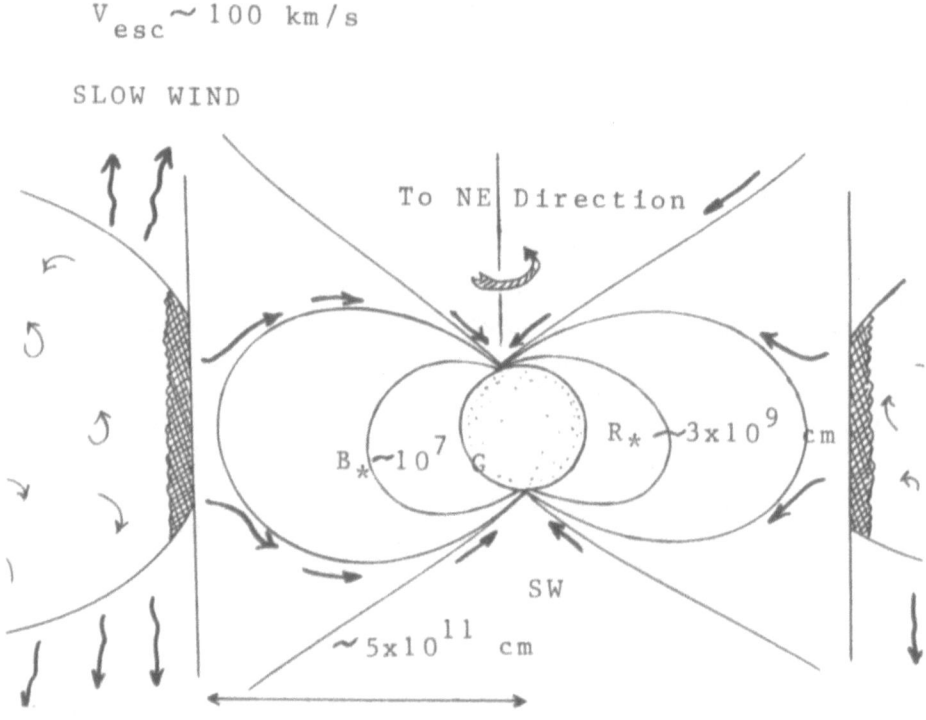

Figure 1: Strongly magnetized subdwarf model for R Aqr

Light cylinder is at $\sim 5 \times 10^{11}$ cm; outside it turbulent B-fields exist in thick, cool accretion disk; inside it high vacuum with accreting gas flowing along the B-field lines onto the poles of the hot subdwarf. Outside the light cylinder radiation pressure drives a "cool" wind.

In this table P_B is the magnetic pressure, P_{gas} the gas pressure and $P_{rad} = L_d/4\pi r^2 c$ the radiation pressure in c.g.s. units; V_A the Alfven speed. The disk luminosity is $\sim 5 L_\odot$ for an accretion rate $\sim 10^{-8} M_\odot yr^{-1}$. The gas pressure is $2nkT$ and the density is found from the accretion rate $\dot{M}$.

In this picture, the difference between R Aqr and CH Cyg is simply the location of the light cylinder and where the disk begins. For CH Cyg, V_{esc} ~ 1,100 km s^{-1}, and the disk begins to form outside ~10^{10} cm. Mass loss occurs because of radiation pressure on electrons, no grains exist at such small distances from the hot star, and accretion rates ~10^{-5} M$_{\odot}$ yr^{-1} are needed for supercritical flows to occur (see Table 5). The hot stars in both systems are, however, fed from Roche lobe overflow from the M-primary.

It may very well be that MHD flows are common among many symbiotics but due to their large distances, individual components cannot be observed. The fact that both nearest symbiotics, one D and one S-type, show jet structures leads us to believe that this may not be a coincidence but a common phenomenon.

7. References

Allen, D.A. (1973), *M.N.R.A.S* **161**, 145.

Allen, D.A. (1981), *M.N.R.A.S* **197**, 739.

Allen, D.A. (1982), in M. Friedjung and R. Viotti (eds.) *IAU Colloq. 70, Nature of Symbiotic Stars*, Reidel, Dordrecht, P. 27.

Anderson, C.M., Cassinelli, J.P., and Sanders, W.T. (1981), *Ap.J.* **247**, L127.

Bath, G.T., Evans, W.D., Papaloizou, J., and Pringle, J.E. (1974), *M.N.R.A.S* **169**, 447.

Boyarchuk, A.A. (1975), in V.E. Sherwood, L. Plaut (eds.) *IAU Symposium No 67, Variable Stars and Stellar Evolution*, Reidel, Dordrecht, p. 377.

Boyarchuk, A.A. (1993), in J. Sahade, G.E. McCluskey and Y. Kondo (eds.) *The Realm of Interacting Binary Stars*, Kluwer Academic Publishers, Dordrecht, pp. 189-207.

Burgarella, D., and Paresce, F. (1992) *Ap. J.* **389**, L29.

Burgarella, D., Vogel, M, and Paresce, F. (1992), *Astr. Astroph.* **262**, 83.

Dougherty, S.M., Bode, M.F., Lloyd, H.M., Davis, R.J., and Eyres, S.P. (1994), *M.N.R.A.S.* Oct. 94.

Feast, M.W., Robertson, B.S.C., and Catchpole, R.M. (1977), *M.N.R.A.S* **179**, 499.

Herbig, G. (1980), *IAU Circ.* **3535**.

Hollis, J.M., and Michalitsianos, A.G. (1993), *Ap. J.* **411**, 235-238.

Hollis, J.M., Lyon, R.G., Dorband, J.E., and Feibelman, W.A. (1995), *B.A.A.S.* **27**, 815.

Hollis, J.M., Michalitsianos, A.G., Kafatos, M., and Welch, W.J. (1986), *Ap. J.* **309**, L53.

604

Hollis, J.M., Oliversen, R.J., Kafatos, M., Michalitsianos, A.G., and Wagner, R.M. (1991), *Ap. J.* **377**, 227.

Johnson, H. (1980), *Ap. J.* **237**, 840.

Kafatos, M., Hollis, J.M. Pedelty, J.A., and Shubow, J.B. (1996), in preparation.

Kafatos, M., Hollis, J.M., Yusef-Zadeh, F., Michalitsianos, A.G., and Elitzur, M. (1989), *Ap. J.* **346**, 991.

Kafatos, M., Meier, S.R., and Martin, J. (1993), *Ap. J. Suppl.* **84**, 201-214.

Kafatos, M., Michalitsianos, A.G. (1984), *Scientific American* **251**, 84-94.

Kafatos, M., Michalitsianos, A.G., and Fahey, R.P. (1985), *Ap. J. Suppl.* **59**, 785.

Kafatos, M., Michalitsianos, A.G., and Feibelman, W.A. (1982), *Ap. J.* **257**, 204.

Kafatos, M., Michalitsianos, A.G., and Hollis, J.M. (1983), *Ap. J. Suppl.* 267, L103.

Kafatos, M., Michalitsianos, A.G., and Hollis, J.M. (1986), *Ap. J. Suppl.* **62**, 853-874.

Kenyon, S.J. (1986) *The Symbiotic Stars*, Cambridge University Press, Cambridge.

Kenyon, S.J. (1988), in J. Mikolajewska, et al (eds.) *The Symbiotic Phenomenon*, Kluwer Academic Publishers, Dordrecht.

Kenyon, S.J., and Fernandez-Castro, T. (1987), *A.J.* **93**, 938.

Kenyon, S.J., and Webbink, R.F. (1984) *Ap. J.* **279**, 252.

Kenyon, S.J., Mikolajewska, J., Mikolajewska, M., Polidan, R.S., and Slovak, M. H. (1993), *A.J.* **106**, 15-73.

Leahy, D.A., and Taylor, A.R. (1987) *Astr. Astrophy.* **176**, 262-266.

Leahy, D.A., and Volk, K. (1995), *Ap. J.* **440**, 847-852.

Lepine, J.R.D., LeSqueren, A.M., and Scalise, E. (1978), *Ap. J.* **225**, 869.

Lipunov, V.M. (1995), in R.A. Sunyaev (ed.) *Soviet Astronomy Review* (in preparation).

Lynden-Bell, D., and Pringle, J.E. (1974), *M.N.R.A.S* **168**, 603.

Mattei, J.A., and Allen, J. (1979), *J.R.A.S.C.* **73**, 173.

Mauro, N., Nieto, J.L. Picat, J.P. Lelievre, G., and Sol, H. (1985), *Astr. Astroph.* **142**, L13.

Meier, S.R., and Kafatos, M. (1995), *Ap. J.* **451**, 359-371.

Meier, S.R., Kafatos, M., Fahey, R.P., and Michalitsianos, A.G. (1994), *Ap. J. Suppl.* **94**, 183-220.

Michalitsianos, A.G., and Kafatos, M. (1982), *Ap. J.* **262**, L 47.

Michalitsianos, A.G., and Kafatos, M. (1988) in J. Mikolajewska, et al (eds.) *The Symbiotic Phenomenon*, Kluwer Academic Publishers, Dordrecht, p. 235.

Michalitsianos, A.G., Hollis, J.M, and Kafatos, M. (1986) *Canadian J. Phys.* **64**, 523.

Michalitsianos, A.G., Kafatos, M., and Hobbs, R.W. (1980), *Ap. J.* **237**, 506.

Michalitsianos, A.G., Kafatos, M., Hobbs, R.W., and Maran, S.P. (1980), *Nature.* **284**, 148.

Michalitsianos, A.G., Oliversen, R.J. Hollis, J.M., Kafatos, M., Crull, H.E., and Miller, R.J. (1988), *A.J.* **95**, 178.

Michalitsianos, A.G., Perez, M., and Kafatos, M. (1994), *Ap. J.* **423**, 441-445.

Mikolajewska, J., Friedjung, M., Kenyon, S.J., and Viotti, R. (eds.) (1988) *The Symbiotic Phenomenon*, Kluwer Academic Publishers, Dordrecht.

Murset, U., Nussbaumer, H. Schmid, H.M., and Vogel, M. (1991), *Astr. Astroph.* **248**, 458.

Paczinski, B., and Rudak, B. (1980), *Astr. Astroph.* **82**, 349.

Paresce, F., and Hack, W. (1994), *Astr. Astroph.* **287**, 154.

Paresce, F., Burrows, C., and Horne, K. (1988), *Ap. J.* **329**, 318.

Penston, M.V. *et al* (1983), *M.N.R.A.S.* **202**, 833.

Penston, M.V., and Allen, D.A. (1985), *M.N.R.A.S.* **212**, 939.

Seaquist, E.R., and Taylor, A.R. (1987), *Ap. J.* **312**, 813.

Seaquist, E.R., Taylor, A.R., and Button, S. (1984), *Ap. J.* **284**, 202-210.

Solf, J. (1992), *Astr. Astroph.* **257**, 228-244.

Solf, J. (1995), *Astr. Astroph.* **257**, in press.

Sopka, R.J., Herbig, G., Kafatos, M., and Michalitsianos, A.G. (1982), *Ap. J.* **258**, L32.

Taylor, A.R, Seaquist, E.R., and Mattei, J.A. (1986), *Nature* **319**, L38.

Tutukov, A.V., Yungel'son, L.R. (1993), *M.N.R.A.S.* **260**, 675.

Tylenda, R. (1977), *Acta Astr.* **27**, 235.

Viotti, R., Piro, L., Friedjung, M., and Cassatella, A. (1987), *Ap. J.* **319**, L7.

Wallenrstein, G., and Greenstein, J.L. (1980), *P.A.S.P.* **92**, 275.

Webbink, R.F. (1979), in H.M. Van Horn, and V. Weidemann (eds.) *White Dwarfs and Variable Degenerate Stars*, Rochester Univ. Press, Rochester, p. 426.

Whitelock, P.A. (1987) *P.A.S.P.* **99**, 573.

Willson, L.A., Garnavich, P., and Mattei, J.A. (1981) *Inf. Bull. Var. Stars,* **1961**.

Zuckerman, B. (1979), *Ap. J.* **230**, 442.

PHENOMENOLOGY AND MODELLING OF LARGE-SCALE JETS

A. FERRARI

Dipartimento di Fisica Generale, Università di Torino

Osservatorio Astronomico di Torino, Pino T.se, Italy

AND

S. MASSAGLIA, G. BODO, P. ROSSI

Osservatorio Astronomico di Torino, Pino T.se, Italy

Abstract.

Supersonic jets ejected from Active Galactic Nuclei were proposed 20 years ago to explain extended radio galaxies. Today that proposal has been verified by many high angular resolution observations at radio and optical frequencies and by indirect data at X and γ rays. The interpretation of these objects requires investigations of nonlinear fluid phenomena to interpret the origin, propagation and termination of collimated outflows and their non-thermal emissivity. On the other hand the observed phenomenology requires to follow the time evolution of the relevant physical processes in physical conditions that cannot be reproduced in laboratory experiments. In these lectures a review of recent investigations on jets is presented, the results are compared with observational data and directions for future studies are indicated.

1. HISTORICAL INTRODUCTION

The first evidence for a common existence of highly collimated, supersonic outflows in astrophysical objects came from the systematic observations of twin lobes in radiogalaxies in the years 1960 - 70. It was soon clear that these sources had enormous dimensions and powers and emitted nonthermal synchrotron radiation: these two facts together excluded the possibility of a singular production event in the nucleus of the parent galaxy.

K. C. Tsinganos (ed.), Solar and Astrophysical Magnetohydrodynamic Flows, 607–642.
© *1996 Kluwer Academic Publishers.*

In fact a single plasmon moving in a constant density atmosphere has a stopping distance $D \sim s/\nu$, where s is the plasmon scale and ν the density ratio atmosphere/plasmon. Unless very dense plasmons are considered, one would get $D \leq s$ in contrast with observations. On the other hand dense plasmons do require uncomfortably large kinetic energies, $\geq 10^{61}$ erg, to be delivered in single events. The same amount of energy delivered continuously on times $\geq 10^7$ years is a much lesser problem.

Concerning the observed luminosity, one must take into account the short synchrotron lifetimes of relativistic electrons: they cannot survive an initial event for more than $\sim 10^6$ years in radio. Instead the copious electron acceleration could be explained more economically in terms of fluid jets continuously transferring energy and momentum from the galactic nucleus into the lobes and along jets maintaining *in situ* particle reacceleration.

In fact, with the increase in sensitivity and angular resolution of radiotelescopes, it became clear in 1970 that one could detect bridges of nonthermal emission connecting galactic nuclei and radiolobes. Although nonthermal continuum did not allow Doppler measurements of motion velocity in these bridges, it was clear that a permanent connection existed that maintained a surprising collimation. Jets velocities were found to be supersonic to avoid rapid opening of the flow.

VLBI observations traced the outflow collimation down to sub-parsec scales and allowed to measure in some cases superluminal proper motions. This fact, together with a statistically significant presence of one-sided jets in strong sources, was considered as an evidence that jet may, at least in some cases, be supersonic.

High velocities were found consistent with the fact that radiogalaxies have extensions up to Mpc scales and their lifetime cannot be longer than 10^8 years.

Later (1970-80) jets were discovered to emit also in the high-frequency bands of optical to X and γ rays, and it was definitely established their connection with very high-energy phenomena originating in the deep cores of active galactic nuclei (AGNs). In this respect two recent observational developments must be mentioned:

1. the Hubble Space Telescope has gathered evidence of a close spatial connection between thermal and nonthermal radiation emissions in the central regions of AGNs; in particular the nonthermal emission corresponds to the initial part of the jet that appears to compress the external interstellar plasma while ploughing its way out (Capetti et al. 1995);

2. the Compton Gamma Ray Observatory has detected strong and highly variable γ-ray emission from blazar-type AGNs, suggesting that these objects owe their enormous brightness to Doppler boosted radiation

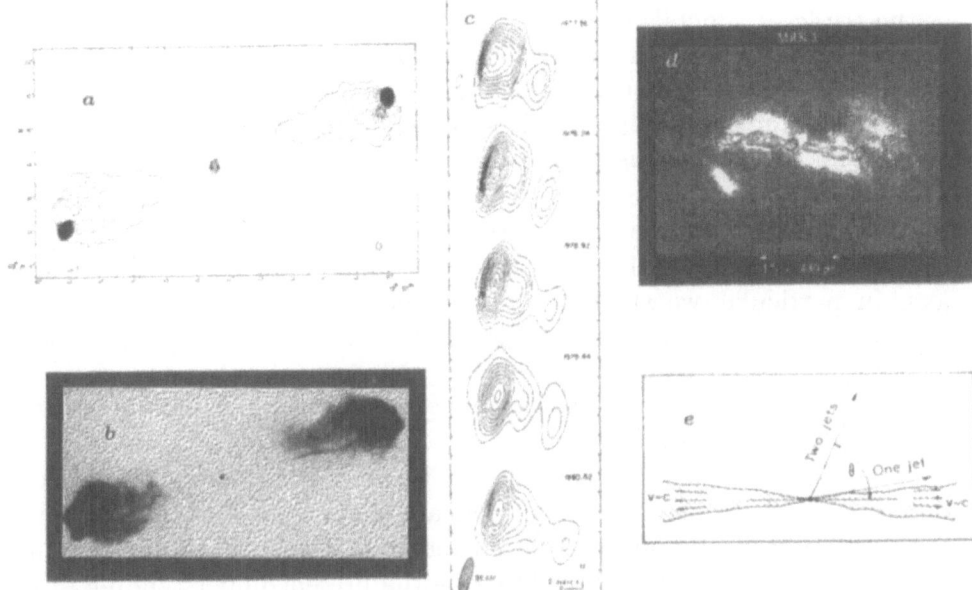

Figure 1. Collimated jets from AGNs - a) Cygnus A in early observations - b) VLA Cygnus A radio map - c) Superluminal motions in 3C 273 - d) HST map of Markarian 3 - e) Geometry of Doppler beaming in one-sided jets and blazars

from jets pointing towards the Earth (Hartman et al. 1993); relativistic jet velocities would then acquire an additional support.

Table 1 summarizes the characteristic physical parameters of AGNs and their jets.

AGN powers	$10^{39} \div 10^{49}$ erg s^{-1}
Variability time scales	hours $\div$ years
Jet extensions	1 pc $\div$ few Mpc
Relativistic Lorentz factors	≤ 10

Table 1

Figure 1 shows instead a collection of sample morphologies of jets from AGNs.

The modelling of supersonic, relativistic, collimated outflows from the AGNs has been one of the most challenging problems in astrophysical hydrodynamics and magnetohydrodynamics in the recent years.

Other contributions in these Proceedings discuss the early development of the study of supersonic HD and MHD outflows in connection with the observations of the solar and stellar winds and plasma motions in solar magnetic loops. Although the global and specific energetics of stellar and galactic phenomena differ by orders of magnitude, most of the dynamical events and the underlying physical processes may not be substantially far apart.

In this paper we shall move from the HD winds and emphasize the new physical elements that have been introduced to model extragalactic jets. In particular we shall show the importance of nonlinear effects in producing the observed morphologies. In fact the important point we want to stress in these lectures is the following: jets are not only a manifestation of relativistic effects, but are shaped by extremely complex nonlinear time-dependent phenomena.

The plan of the lectures is the following. In Section 2 the basic observations of extended radiogalaxies and jets are reviewed, from which we derive the basic scenario and physical parameters in Section 3. The following three Sections 4 to 6 are devoted to illustrate the present understanding of the main physical questions about jets: origin, propagation, termination. The final Section 7 discusses the critical points and future directions of research.

2. OBSERVATIONS OF COLLIMATED OUTFLOWS

Outflows from AGNs show a number of characteristic morphologies that may not be present altogether in a same object, but occur most often. We shall briefly summarize those that are useful to derive physical parameters of jets and define a reasonable phenomenological model.

2.1. RADIO DATA.

We start with radio data collected by high resolution mapping at VLA, MERLIN, VLBI, VLBA (Muxlow & Garrington 1991, Blandford 1991).

1. *Radio cores.* A compact, flat spectrum (perhaps superposition of several peaked components), stationary component, unresolved with angular dimensions ≥ 0.1 arcsec, is observed at the origin of the jet in the central regions of the radiogalaxy. Observations do not allow to decide whether this component is in fact the base of the jet or is related with the emission of the central machine of AGNs.

2. *Jets.* This term indicates the collimated part of the outflow connecting the core to the extended radio components (lobes): a bridge of emission is called jet when the ratio of elongation to transverse size is ≥ 4 (Bridle 1984). As already mentioned, in most cases we observe

two opposite jets, but also one-sided sources are observed. They may show enhanced emissivity toward the core (centre-brightened) or toward the lobes (edge-brightened), can be smooth or knotty, straight or wiggled. Spectral indices of radio emission is ~ 0.6. Another important point is related with the nonthermal emissivity distribution: it is now clear that emissivity is concentrated in knots and filaments. In addition there is evidence that jets may be hollow (at least for the nonthermal component).

3. *Lobes*. These are the extended regions of radio emission, characterized by many types of morphological structures, as tails, bridges, hot spots, haloes, plumes, filaments. The hot spots are compact regions of high emissivity and flat spectral index ($\sim 0.5 \div 1$) where the jet impinges upon the external medium and creates a strong shock, locally accelerating relativistic particles.

Fanaroff and Riley (1974) first noted that radio structures depend on luminosity, an abrupt transition occurring at $P_{178MHz} = 5 \times 10^{25}$ W Hz^{-1}. Sources below this critical luminosity (*FR I objects*) have rather smooth two-sided jets and large scale edge-darkened lobes; a typical example of this class is 3C 449. Sources above the threshold (*FR II objects*) tend to have edge-brightened profiles with strong outer hot-spots; the archetypal of this class is Cygnus A (Fig. 2).

However, although FR II sources are systematically brighter than FR I, the contribution of the main components to total radio luminosity is similar (Leahy 1985).

For modelling jets interesting morphological features appear to be *knots* and *wiggles* showing a remarkable periodicity in some sources, most likely due to intrinsic properties of flow propagation: they are an indirect diagnostics of jet physical parameters.

Additional informations come from the study of polarization (orientation of projected magnetic field) and Faraday rotation and depolarization (density of relativistic electrons). However it must be always kept in mind that a large contribution to the measured values may come in fact from the external interposed medium.

2.2. OPTICAL AND X-RAY DATA.

Recently HST has detected a large number of optical jets showing that nonthermal optical and radio emission follow the same patterns with remarkable accuracy (Macchetto 1995). In particular in strong sources the brightness distribution appears to be consistent with hollow jets wrapped by bright filaments. In weak jets the brightness distribution is more turbulent.

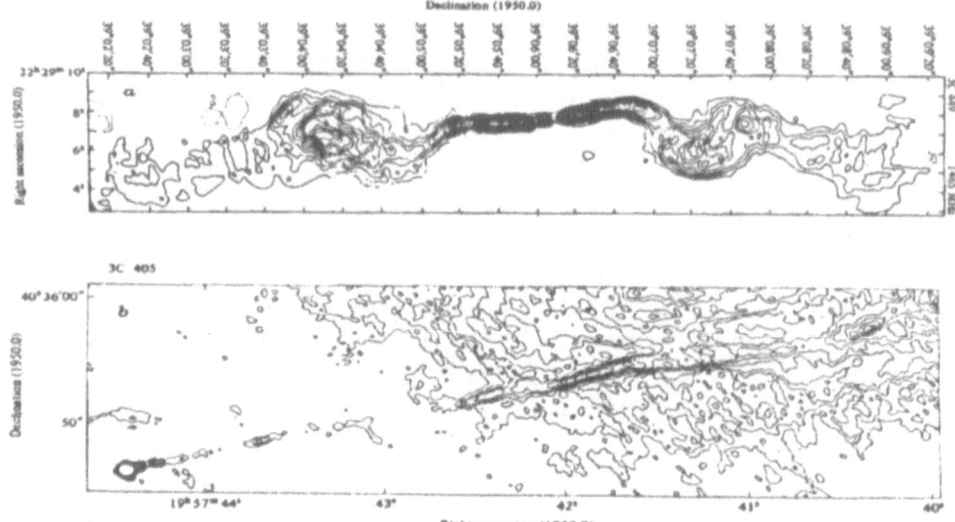

Figure 2. a) VLA map of 3C 449 at 1465 MHz (FR I) - b) VLA map of Cygnus A at 4885 MHz (FR II)

Another important information produced recently by HST (Capetti et al. 1995) is related with line emission from the central (NLR) regions of some Seyfert 2 galaxies (Mrk 3, Mrk 78, Mrk 573, Mrk 348) that spatially coincides with the nonthermal radio and optical components. This indicates a likely physical link between the supersonic plasma in jets and a surrounding thermal plasma: the line-emitting gas is somewhat heated by the passage of the supersonic flow.

Concerning X-ray emission, within the angular resolution limits of X-ray telescopes, profiles also correspond remarkably (Feigelson 1980, Brinkmann 1995). The overall spectra show a nonthermal continuum with a steepening around optical frequencies: in M 87 the radio power index is ~ 0.7 increasing to ~ 1.7 in optical.

2.3. PHYSICAL PARAMETERS.

In the last 20 years the main discussion about radiogalaxies has been concentrated on deriving physical parameters from the above observations. Given the fact that emission from jets and lobes is nonthermal and polarized, the phenomenological model assumed is the *optically-thin synchrotron source*. In particular, to minimize the energy budget, estimates are made in the limit of *minimum energy requirement* corresponding to equipartition between relativistic electrons (and protons) and magnetic fields. For cores the optically-thin approximation breaks down and other models are used, which are not relevant to this presentation.

Using various diagnostics, one can compile the list of typical average physical parameters of extended radiogalaxies (Table 2).

			Core	Jet		Hot-spot	Lobe
D (size)		kpc	$\leq 10^{-3}$	$2 \div 10^3$		5	$50 - 10^3$
B_{eq}		gauss		$\leq 10^{-3}$		10^{-3}	10^{-5}
$n_{e,rel}$		cm^{-3}					
$n_{e,th}$		cm^{-3}		$10^{-2} \div 10^{-5}$		$\leq 10^{-2}$	$\leq 10^{-4}$
polarization		%	≤ 2	$0 \div 60$		15	$0 \div 60$
spectral index		α	0.0	0.6		0.6	0.9
v_{flow}/c			$\rightarrow 1$	10^{-1}		10^{-3}	10^{-3}

Table 2

In addition, since the evolution of physical parameters along the jet does not follow a simple adiabatic expansion law (Scheuer 1974), pressure equilibrium with an external medium and/or magnetic field is requested by the high degree of collimation (see a recent discussion by Fabian and Rees 1995): the thermal plasma characteristics for confinement are listed in Table 3.

n_{th}	cm^{-3}	4×10^{-3}
T	K	10^8
B	nT	0.1

Table 3

However we mention that the estimates of the external medium pressure, coming from X-ray surveys, when compared with the estimates of the jet pressure from synchrotron model, do not confirm the idea of equilibrium between internal and external pressures. Strong jets appear overpressured (Feigelson 1981), weak jets underpressured (Feretti et al. 1995): either the observed features are transients and other plasma components must be considered, or our model dependent estimates require some rethinking.

614

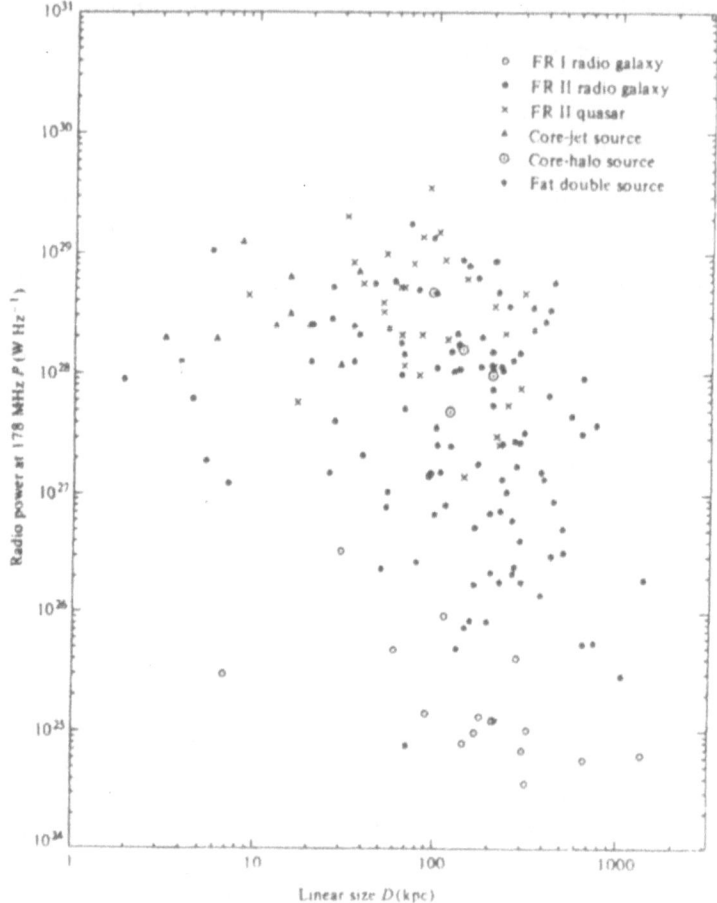

Figure 3. Luminosity -Linear Size diagram for a 3C sample (Laing et al.1983)

2.4. LUMINOSITY - LINEAR SIZE DIAGRAM.

A sort of Hertzsprung-Russell diagram for radio galaxies is represented by a plot $P - D$ (see Fig. 3). Many observational selection effects do not allow yet to develop a real evolutionary theory from this diagram, but a general trend for classifying various types of morphologies is apparent (in these notes there is no space to elaborate on morphologies). This challenges to work on unified models for radio galaxies and for AGNs in general.

2.5. UNIFIED MODELS.

Meanwhile jets appear to be an important element in many radiogalaxies and quasars, not all AGNs show directly the presence of outflows. On the other hand there seems to be a strong correlation between outflows and all

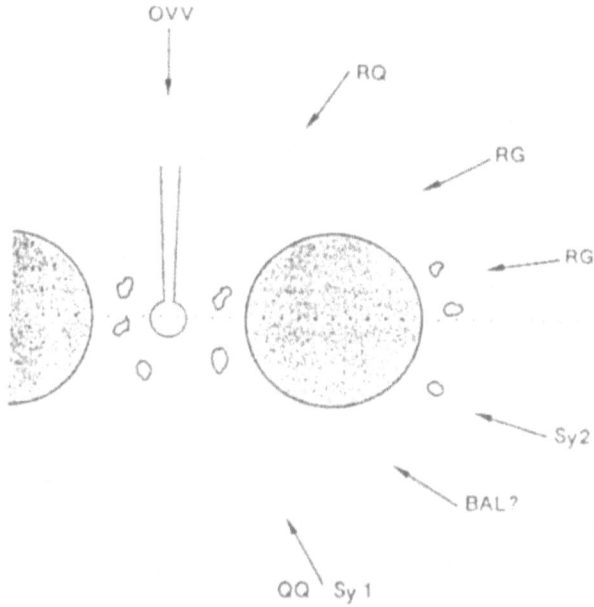

Figure 4. A unified model of AGNs

other types of nuclear activity in galaxies.

Therefore phenomenological scenarios in which all types of AGNs, from Seyfert galaxies to blazars and quasars, are explained as different appearances of a similar galactic core, but with different jet orientation with respect to the observer's line-of-sight, have become very popular.

In Fig. 4 we illustrate graphically a typical interpretation of AGNs based on such assumption. However a real understanding of unified models awaits to be tested on theoretical grounds (Blandford 1991).

3. PHYSICS OF COLLIMATED OUTFLOWS

The paradigm for AGNs was originally proposed by Scheuer (1974) and Blandford and Rees (1974) and is schematically represented in Fig. 5.

Jets are produced in the inner core of AGNs by some still unclear mechanisms based on general relativistic electrodynamics in the throat of accretion disks, thrusting continuously supersonic and/or superalfvenic magnetized plasma along the angular momentum axis. The twin jets plough their way through the ambient intergalactic gas transferring energy and momentum far away from the parent core. Jets are structurally affected by the interaction with the external medium originating shocks, filaments and wiggles. Local electron acceleration to relativistic energies supports synchrotron emission. The *head* where the jet pushes against the external

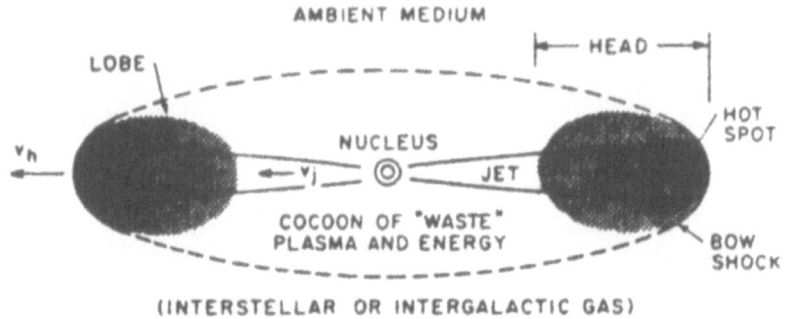

AMBIENT MEDIUM

LOBE

HEAD

NUCLEUS

HOT SPOT

v_h

v_j — JET

COCOON OF "WASTE" PLASMA AND ENERGY

BOW SHOCK

(INTERSTELLAR OR INTERGALACTIC GAS)

Figure 5.   Schematic diagram of a strong radio source

medium is a turbulent working surface producing a bow shock and a cocoon around the entire source.

Modelling this scenario is difficult because of the high nonlinearity of the physics involved. Therefore the various building blocks of the model have been attacked separately by different groups of theorists. In particular we can select three classes of problems related with jets: (1) origin, (2) propagation, (3) termination. They are treated separately in the following Sections.

We dedicate the present Section to discuss some of the common basic physical assumptions.

3.1. RELATIVISTIC PLASMAS AND MAGNETIC FIELDS.

The synchrotron plasma model suggests that radio-emitting regions must contain a fully ionised, collisionless, relativistically hot and magnetized gas, although this could coexist with a cold component (protons and nuclei) that would dominate the mass content. Some authors have discussed the possibility that outflows are made of electron-positron pairs to interpret the fact that energy is not deposited along jets and to reduce the kinetic energy content in radio lobes.

When describing the structure and dynamics of outflows, a fluid approximation is used, assuming that magnetic fields provide a collective behaviour although particle collisional mean-free-paths are very large ($\lambda_{gyr} \ll D \ll \lambda_{coll}$). It is not clear what is the field filling factor in the various regions and how important is the turbulent with respect to the ordered component. However they are an essential ingredient for both the radiation and dynamics, and any realistic model must include them: in particular fluid models must use magnetohydrodynamics.

Two considerations are worth mentioning in this respect. The first has to do with the observed morphologies of the magnetic fields, as derived from polarized emissivity measurements. One assumes the minimum (equipartition) energy condition when applying the synchrotron model: on the other hand the emission comes from structures (knots and filaments) that appear not to be in pressure equilibrium, possibly transients.

Another consideration is about the often assumed adiabatic expansion of emitting structures (blobs) along the jet. Actually emissivity is very strongly dependent on the source projected cross-section($\propto R^{-(5+4\alpha)}$), which does not seem to be verified by observations, notwithstanding uncertainties in emissivity estimates. This may indicate that turbulent amplification of magnetic fields is occurring everywhere, or transient non-equilibrium formation like shocks or filaments dominate the dynamics.

3.2. AGE, EXPANSION SPEED AND SPECTRAL BREAK.

The obvious lower limit to the age of radio galaxies is the light-travel time from the nucleus to the projected distance of the lobes $\tau \geq \beta_{\max} cD/2$. If the source is actually expanding relativistically, its morphology must be asymmetrical because of Doppler boosting/dimming of the approaching/receding component. For symmetric sources $\beta_{\max} \leq 0.2$. With this approach $\tau \geq$ few 10^6 years.

An upper limit comes instead from the analysis of the spectrum in the weakest/oldest parts of the source, i.e. faint parts of the lobes. In these regions we may assume that no reacceleration occurred in recent times, and therefore the spectrum must have a steep cut-off above a certain frequency. This argument yields $\tau \leq$ few 10^7 years, and indirectly $\beta_{\max} \geq 0.01$ (Alexander and Leahy 1987).

However these must be considered average values. In strong radio sources, superluminal motions have been detected that indicate velocities very close to c in nuclear regions. At the same time we must remember the existence of one-sided jets that are interpreted through the above mentioned relativistic Doppler effect.

Still in connection with speed, jets are usually considered supersonic flows (or superalfvenic). This requirements comes from considerations on collimation, presence of shocks, low thermal emissivity, which all point to small sound speeds ($v_s \ll v_j$).

3.3. THE MINIMUM ENERGY HYPOTHESIS.

A long standing issue is the validity of the assumption of minimum energy for estimating the physical parameters (Burbidge 1956). Emitting regions would be close to equipartition between relativistic electron and magnetic energies (or pressures). Several circumstantial facts appear to agree with this idea, that is also confirmed by laboratory plasma experiments predicting instabilities when such limit is violated by large factors.

On the other hand shocks and filaments do not seem to be in equipartition, with a prevalence of electron pressure (Bodo et al. 1992). One must also remember that a hidden thermal component may largely exceed the synchrotron plasma.

4. PARSEC-SCALE JETS

In other chapters of these Proceedings acceleration mechanisms and stationary outflow configuration models are presented. They are obtained balancing gaseous, magnetic and rotational forces to form accretion disks and considering ejection of supersonic/relativistic gas along rotation funnels by radiation pressure or electrodynamic effects.

Until recently the behaviour of jets in the cores of AGNs was known only through estimates coming from radiation models. Now, owing to radio (VLBI) and optical (HST) high-resolution imaging, we have direct informations on the appearance of outflows close to their origin. These data are consistent with accretion disk models around a compact, massive object, and therefore support the black hole hypothesis as the most efficient (and evolutionarily inescapable) energy source (Rees 1984).

In this Section we discuss the initial propagation phases of supersonic outflows from accretion disks using elementary conservation laws.

4.1. ACCRETION DISKS.

First of all we recall that accretion disks can provide a typical luminosity (e.g. Pringle 1981):

$$L = \frac{\dot{M}m}{2r_{\min}} \leq 1.3 \times 10^{38} \frac{m}{m_\odot} \text{ergs}^{-1} \tag{1}$$

where $\dot{M}$ is the accretion rate, m the black hole mass, $r_{\min}$ the internal radius of the disks; gravitational units are used. This luminosity corresponds to the energy released by viscous stresses while gas moves down to the hole.

The basic structure of disks depends on the efficiency by which the energy released is actually radiated away (thin disks) or stored in the plasma (thick disks) or advected into the hole (proton tori). In any case a fraction of L is available under the form of photons or plasma waves or electrodynamic flux for accelerating a flow along the rotation axis.

4.2. DYNAMICS OF EQUILIBRIUM JETS.

An essential role in the propagation of jets is played by the boundaries of the channels. Collimation can be produced either by real "walls", as in the case of thick accretion disks or tori, or by magnetic fields anchored in the disk and wrapped by spin to form a confining shield around the rotation axis region (Camenzind, these Proceedings). Farther away from the disk gas or magnetic pressure of the surrounding interstellar medium can be the origin of collimation. However this matter is still controversial, due to the paucity of informations. Nevertheless collimation appears to start very deep in the core of AGNs, in the same regions where acceleration occurs.

If a laminar jet is confined by external agents we can study its propagation in equilibrium conditions. For instance we can employ the hydrodynamic conservation laws of Parker's solar wind theory (1963). In the newtonian, nonrelativistic limit and for a polytrophic gas $P = \rho^\delta$, mass conservation and Bernoulli equations are:

$$\rho v A = \dot{M} \Rightarrow P^{(\delta+1)/2\delta} \mathcal{M} A = \text{const} \tag{2}$$

$$P^{(\delta-1)/\delta} \left[\frac{\mathcal{M}^2}{2} + \frac{\delta}{\delta-1} + \frac{\phi}{c_s^2} \right] = \text{const}, \tag{3}$$

where $\mathcal{M}$ is the flow Mach number and ϕ the gravitational potential. The power and thrust of the flow can be written as:

$$
\begin{aligned}
L &= \int \left[\left(u_{tot} + \rho\phi + \rho v^2/2 \right) \mathbf{v} + \mathbf{\Psi}_{tot}.\mathbf{v} + \mathbf{q}_{tot} \right] \cdot \mathbf{dS} = \\
&= \frac{1}{2} \dot{M} v^2 \left(1 + \frac{\delta}{\delta-1} \frac{2}{\mathcal{M}^2} - \frac{v_{esc}^2}{v^2} \right),
\end{aligned}
\tag{4}
$$

$$
\begin{aligned}
\mathbf{T} &= \int \left(\rho \mathbf{v}\mathbf{v} + \mathbf{\Psi}_{tot} \right) \cdot \mathbf{dS} = \\
&= \dot{M} v \left(1 + \frac{\delta}{\delta-1} \frac{1}{\mathcal{M}^2} \right).
\end{aligned}
\tag{5}
$$

To calculate the integrals we assumed the total stress tensor $\mathbf{\Psi}_{tot}$ isotropic and including stresses due to tangled magnetic fields and radiation.

We can obtain the solutions of Eqs. 2 and 3 in the limit of given power and discharge $\dot{M}$ (Blandford and Rees 1974). The cross-sectional area shows typically a minimum value for $P_{min} \sim (1/2)P_{base}$, where the jet speed becomes transonic. Therefore, an adiabatic jet accelerates to supersonic speed passing through a *converging de Laval nozzle*. In the subsonic part of the flow, pressure and density are substantially constant and $A \propto v^{-1}$. The difficulty of the Blandford and Rees solution comes from the fact that the transonic point is rather far away from the nucleus, contradicting the observations that show the flow supersonic already well below the parsec scale.

4.3. RELATIVISTIC JETS.

Ferrari et al. (1986) have discussed general solutions of the problem of relativistic equilibrium flows for given profiles of the propagation channels. They obtain an equation for the flow Mach number:

$$\gamma^2 \left(1 - \frac{\beta_s^2}{\beta^2}\right) \frac{d\beta^2}{dZ} = -\frac{B}{Z^2} + \frac{2\beta_s^2}{S} \frac{dS}{dZ} + \frac{2D(Z)z_0}{\gamma^2 \rho c^2} \tag{6}$$

where:

$$Z = \frac{z}{z_0}, \quad B = \frac{2GM}{z_0 c}, \quad S(Z) = \frac{A(Z)}{A(z_0)}, \quad \beta_s = \frac{v_s}{c}.$$

The first term on the right-hand side is the gravitational term. The second one is very interesting: it can be easily derived that when the channel $S(Z)$ expands more (less) rapidly than spherically ($\propto Z^2$), its walls will deposit (subtract) momentum. The last term indicates momentum addition by external forces; for the case of radiation in the limit of optically thin plasma:

$$D(Z) = \frac{GM\rho\gamma^3 L}{z_0^2} \left[H\left(1 + \beta^2\right) - \beta\left(J + K\right)\right]$$

where H, J, K are the momenta of the radiation field.

As discussed by Ferrari et al. (1986) the solution of relativistic flows from accretion funnels can be characterized by many critical points, as compared to the single critical point (de Laval nozzle) of the wind solution. In particular the sudden expansion at the exit of the funnel brings the first critical point close inside the nucleus.

We notice that in these solutions the thrust is not constant, decreases in the subsonic regime and increases in the supersonic regime. In fact the surface of the jet is not parallel to the flow and this corresponds to an external pressure force acting consistently on the field: where the channel narrows (widens), the flow is slowed down (accelerated).

The asymptotic velocity of flows accelerated by radiation pressure or plasma waves reaches at most $\beta_{asym} \leq 0.28$ due to Compton drag in the wave field.

4.4. COLLIMATION.

For the hypersonic part of the flow, $\mathcal{M} \gg 1$, Eqs. 2 and 3 give $\mathcal{M} \propto P^{(1-\delta)/2\delta}$ and $A \propto P^{-1/\delta}$, implying $v = $ const; for an adiabatic flow $\delta = 5/3$, the scaling is $\mathcal{M} \propto P^{-4/5}A^{-1}$, $A \propto P^{-3/5}$. Pressure distribution in galaxies is found to be $P_{ext} \propto r^{-2}$ and one calculates that the angle subtended by the jet width as seen from the nucleus decreases as $A^{1/2}r^{-1} \propto r^{-2/5}$. Therefore equilibrium jets can be collimated even though their area expands.

On the other hand the estimated equipartition synchrotron pressure inside a jet scales with cross sectional area as $P \propto A^{-2/7}$, i.e. falls much less rapidly than the equilibrium solution above $P \propto A^{-5/3}$. This means that some form of internal dissipation must operate. This point will be discussed later when treating large scale propagation.

4.5. MINI-JETS AS CHOCKED JETS.

The flow pattern depends at some extent from the profile of the gravitational potential in the nucleus. In particular it has been proven for adiabatic jets that a mass distribution $M_{gal} \propto r^{-s}$, with $s \leq \delta$, does not allow the flow to reach a transonic point: the flow chocks inside the galaxy (Ferrari et al. 1986). This result may be related to mini-jets in the cores of Seyferts of spiral galaxies in general where the effective gravitational potential decreases less rapidly because of the rotational centrifugal force.

4.6. FREE JETS.

An alternative to confined jets are finite pressure jets expanding into a steep negative density gradient. As shown in laboratory experiments (Thompson 1972), after the nozzle the flow, with adiabatic index Γ, expands with a maximum bending of streamlines fixed by the local Mach number $\mathcal{M}_0$; the opening angle ξ increases as:

$$\xi - \xi_0 \approx \frac{2}{\Gamma - 1}\frac{1}{\mathcal{M}_0} \tag{7}$$

which shows that in order to maintain confinement the jet must be highly supersonic. Additional effects due to cooling of the expanding gas and recollimation shock formation maintain the opening angle below this value.

4.7. MAGNETOHYDRODYNAMIC AND CENTRIFUGAL MODELS.

Acceleration models and stationary flows based on magnetic and centrifugal effects are studied elsewhere in these Proceedings. The real advantages of these models come in terms of collimation conditions and extraction of energy from the central black hole. From the point of view of propagation however the scalings remain typically those discussed above.

4.8. DIFFICULTIES AND IMPROVEMENTS.

In conclusion our understanding of the initial phases of propagation of jets is in relatively good agreement with fluid models predicting transonic points inside the deep cores of AGNs. However acceleration models have not yet provided a clear explanation of bulk velocities with Lorentz factors γ up to 10 without strong braking by photon fields, although several interesting ideas have been put forward (e.g. Ghisellini et al. 1990).

5. LARGE-SCALE JETS, MORPHOLOGIES, RADIATION

The same elementary dynamics presented in the previous section applies obviously to large scale jets. In these regions we are definitely in the hypersonic regime were thrust and velocity remain substantially constant. Also in terms of wind-type hydrodynamics we are in the asymptotic part of the solution, at least in first approximation.However one has now to deal with the morphologies that modulate the observed flow pattern.

Shocks, filaments, wiggles are structures common to all sources and do not appear to strongly modify the basic flow. Actually this is consistent with the fact that the power carried along in the propagation is much larger than that emitted radiatively, $L_{kin} \gg L_{rad}$.

In laboratory plasma flow we have several examples of beautiful patterns excited by instabilities resembling those in radiogalaxies (Van Dyke 1982). However we must be very careful in making comparison: in laboratory fluids one maps the thermal component, the bulk flow; in astrophysical jets we map the nonthermal component responsible for emissivity and we can only guess that a direct link exists with the thermal component carrying the bulk flow. Nevertheless, also in laboratory experiments, morphologies do not necessarily destroy the flow pattern: the parent instabilities must reach some sort of saturation.

5.1. FLUID INSTABILITIES: LINEAR ANALYSIS.

After the initial proposal by Blandford and Rees that extended radiogalaxies were the result of continuous outflows from galactic nuclei, it was soon realized that jets, according to experiments, can maintain their directionality for relatively short distances due to onset of fluid instabilities, typically 10 times their diameter. Astrophysical jets appear instead much longer lived. Therefore several calculations have, since then, been addressed to the question of jet stability. In most cases the fluid approach has been adopted, consistently with considerations in previous sections.

The instability analysis of fluids, in the limit of small departures from equilibrium, is done by linearizing the perturbed equations and Fourier developing perturbations in wave modes $\propto \exp\left[-i\left(\omega t - \mathbf{k} \cdot \mathbf{r}\right)\right]$. The resulting (algebraic) dispersion relation $D\left(\omega, \mathbf{k}\right) = 0$ is then studied to find modes with $\mathcal{I}m\ \omega \leq 0$ (locally growing modes) or $\mathcal{I}m\ k \geq 0$ (spatially growing modes. When such modes exist, the equilibrium of the fluid can be destroyed, unless saturation effects stop the growth of the perturbations in the nonlinear regime.

5.1.1. Kelvin-Helmholtz instability.

In the case of fluid beams in relative motion with respect to an external medium the typical instability is the *Kelvin-Helmholtz instability*. Its origin can be derived simply from Bernoulli equation: if a ripple develops at the interface between the two fluids, the flow over the ripple has to be faster to pass over it, and therefore it will exert less pressure and allow the ripple to grow further. This will cause mixing of the two fluids and transfer of momentum across the boundary with a progressive slowing down of the beam.

This instability has been studied in much detail for supersonic, compressible, astrophysical jets in cylindrical, slab or conical geometries, both for axisymmetric and non axisymmetric perturbations, with and without magnetic field and radiation; for a review see Birkinshaw (1991).

Two types of modes are found: ordinary surface modes, with amplitude steeply decreasing away from the interface, and reflected body modes, which affect the whole plasma in the jet; body modes are typical of supersonic jets.

The fastest timescales and the most unstable wavelengths are:

$$\tau_{KH} \sim \frac{1}{\mathcal{I}m\omega} \sim \text{few times} \tau_c\ \frac{\lambda_{KH}}{R_j} \tag{8}$$

$$\lambda_{KH} \sim 2\pi R_j \mathcal{M}_j \tag{9}$$

$\tau_c = 2R_j/c_s$ is the beam crossing time at the speed of sound and R_j is the beam radius; for magnetized beams c_s must be replaced by c_A. High density contrasts between jet and external medium and strong magnetic fields reduce the effect of instability.

Time scales are rather short with respect to propagation and therefore these modes can affect (destroy) the beam soon after the exit from the nozzle.

The linear analysis does not include interaction between modes, but this must be present and, as discussed by Benford et al. (1980), produces a turbulent spectrum in the beam. Long wavelength modes, that correspond to maximum instability, modulate the morphology of the beam, while short wavelengths coming from the turbulent cascade can lead to thermal dissipation and suprathermal particle acceleration, as we shall discuss below.

5.1.2. *Filamentation instability.*

Another type of instability that appears to be relevant in jet structuring is the synchrotron "thermal" instability (Bodo et al. 1990). If we assume that the pressure in jets is mainly due to the relativistic electron component, synchrotron losses in regions compressed by fluid instability will start the runaway effect of thermal instabilities (Field 1965): the compressed gas will radiate more and reduce the gas pressure leading to further compression. One predicts a stationary situation in which condensation modes with $k_\parallel \ll k_\perp$ modulate the magnetic field in filaments. These filaments may be overpressured with respect to the external plasma and should appear brighter than the surrounding medium because of enhanced emission, provided a suitable input of fresh particles is guaranteed (Rossi et al. 1993).

These filaments may be connected with the observations of radio emission in jets and, recently, also optical emission in some jets as in M 87 and 3C 66B. Growth rates of radiative modes are consistent with a full development of filaments away from the core (Rossi et al. 1995).

5.1.3. *Resistive instabilities, etc.*

Important instabilities of magnetically confined jets should be reconnection modes. They develop in magnetic neutral sheets or shears where a component of the field is inverted. For a beam with an elicoidal field wrapped around cylindrical surfaces, the equilibrium conditions predict a field pitch-angle decreasing away from the axis. When a non-axisymmetric perturbation is excited along the beam, its pitch-angle will match the equilibrium pitch-angle at some well defined radius $R_{crit} \propto B_\theta/B_\parallel$. That cylindrical surface becomes a neutral sheet for the component of the field vector perpendicular to the perturbation wavevector. The end result is field annihilation and local heating and/or particle acceleration. The growth rate of these modes is relatively long, but indications of an explosive nonlinear phase have been obtained in laboratory experiments.

5.2. NONLINEAR EVOLUTION OF HYDRODYNAMIC INSTABILITIES.

Since extragalactic jets last much longer than predicted by the linear stability theory, it is clear that the nonlinear evolution must be studied to investigate possible saturation effects. Two approaches have been followed: (i) looking for dissipation of the energy released by instabilities and (ii) investigating numerically the long term nonlinear evolution of unstable modes.

In this paragraph we concentrate on the second approach. However we must start with a warning. Numerical simulations have become very sophisticated in recent years owing to the availability of supercomputers. Nevertheless the solution of Navier-Stokes or Euler equation is necessarily limited by the discretization by finite difference schemes to relatively low Reynolds numbers (high viscosity) $\mathcal{R}e \leq 1000$; in fact the size of the mesh points is fixed by the finite dimensions of the grid over which the integration domain is represented. This enhances energy dissipation away from the flow and smooths down sharp discontinuities that instead may be very relevant to fluid problems.

Recently a definite improvement to this limitation has been introduced by the adoption of hybrid algorithms that use more physical schemes in the regions of sharp discontinuities. The most successful method as of today is the upwind differentiation along characteristics: inside the single integration cells one solves, in Godunov schemes, a Riemann problem to calculate the flux transported by nonlinear waves at the interfaces between adjacent cells (Van Leer 1974). The accuracy level at interfaces fixes the accuracy order of the method: the best codes use today a parabolic interpolation, Piecewise Parabolic Method (PPM).

One of the difficulties of the method comes from the number of characteristics, i.e. the number of nonlinear waves to match at cell interfaces. For a hydro problem we have 3 characteristics, for an MHD problem we have 7 characteristics (Zachary et al. 1993).

For the hydrodynamic case the long term evolution of unstable modes in an infinite adiabatic jet in pressure equilibrium with the external homogeneous medium (no gravity) is studied by solving the following set of Euler equations with the PPM code:

$$\frac{\partial \rho}{\partial t} + \nabla \cdot (\rho \mathbf{v}) = 0 \tag{10}$$

$$\frac{\partial \mathbf{v}}{\partial t} + (\mathbf{v} \cdot \nabla)\,\mathbf{v} = -\nabla P/\rho \tag{11}$$

$$\frac{\partial P}{\partial t} + (\mathbf{v} \cdot \nabla)\,P - \Gamma \frac{P}{\rho} \left[\frac{\partial \rho}{\partial t} + (\mathbf{v} \cdot \nabla)\,\rho \right] = 0 \tag{12}$$

Bodo et al. (1994, 1995) have followed instabilities of infinite (periodic boundary conditions) cylindrical jets and slabs up to the onset of a quasi-stationary state after $t \sim 40 R_j/c_s$. For details about computations, integration domain, grid cell size, boundary conditions refer to the quoted papers; we only point out here that the whole set of computations has been classified and is representative of the astrophysical configurations. Bodo et al. indicate the existence of four evolutive phases.

1. *Linear phase.* Modes excited by an initial perturbation evolve according to the linear theory. At the end of this phase internal shocks form and make the jet expanding. Little energy and momentum exchange between jet and ambient is observed.
2. *Acoustic phase.* The growth of deformations of the jet produces piston-like fronts that drive shocks into the external medium, leading to conspicuous energy and momentum transfer from the jet to the ambient. Light jets transfer more momentum than heavy jets.
3. *Mixing phase.* At this stage light jets break up because of matter entrainment from external medium and become very turbulent; heavy jets mix vigorously with the external material.
4. *Quasi-stationary phase.* In this final phase jets reach a quasi-stationary configuration that depends on the density ratio. Heavy jets maintain a coherent directionality, light jets appear completely mixed and diffused.

In conclusion the persistence of jets depends principally on the density contrast with the ambient and the Mach number. Examples of the evolution is illustrated in Fig. 6.

Coming to radio galaxies, the application of the above results predicts that, if jets have similar initial Mach number, FR I jets, that are turbulent and strongly decelerated, should correspond to dense environments (light jets); and FR II to lighter environment: the morphology of jets would depend on the ambient density, in agreement with a recent suggestion by De Young (1993).

However the Mach number may also have an influence. In particular one finds that, for a same density ratio, highly supersonic jets tend to be less turbulent (FR II sources), and mildly supersonic jets more turbulent (FR I sources).

Such ideas have been discussed qualitatively by many authors; numerical simulations define quantitatively the trend of solutions, although one still needs informations on the ambient medium to decide which is the global picture in terms of density contrasts and Mach numbers.

5.3. SHOCKS.

Finally we discuss the evolution of *internal shocks*, already predicted in linear studies, but now properly followed by numerical methods. These shocks arise in the form of conical structures inside a cylindrical jet forming a typical angle related to the flow Mach number (Norman et al 1982). Later, due to mass entrainment and momentum diffusion, shocks tend to extend into the external medium perpendicularly to the flow and become substantially transverse structures.

a)

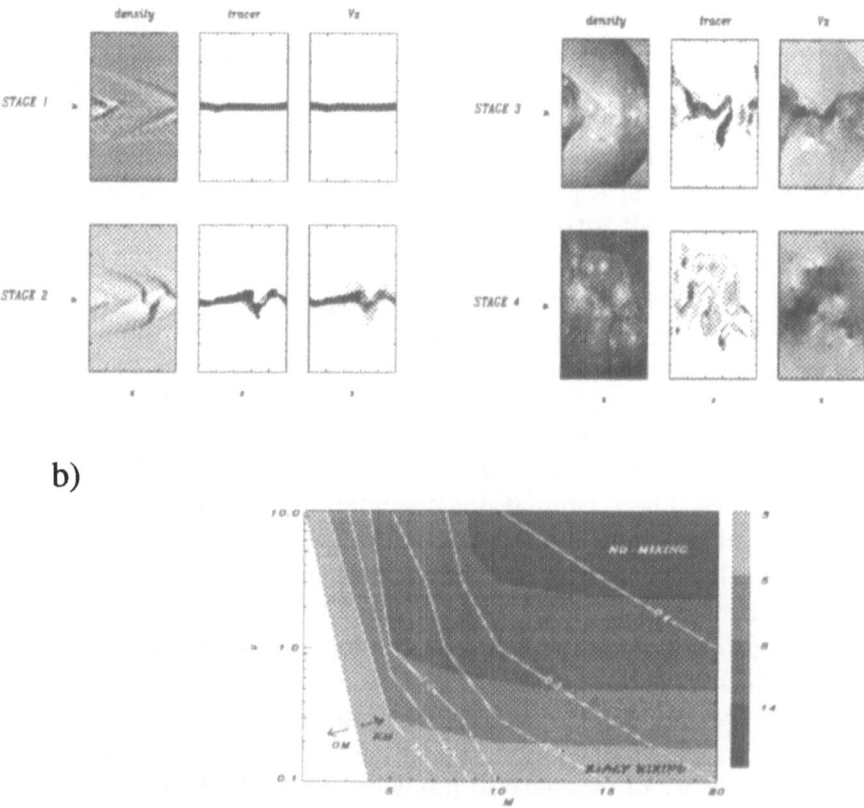

Figure 6. Long term evolution of Kelvin-Helmholtz instabilities: (a) the four phases of evolution; (b) phenomena in physical parameter space M, ν

Another numerical experiment to localize the formation of shocks in jets propagating out of a potential well can be done using the wind equation 6. As we have discussed above, the geometry of the flow channel, when different from purely spherical expansion, corresponds to momentum deposition or subtraction. If we superimpose a Kelvin-Helmholtz-like perturbation on the channel, the wind equation acquires new critical points. The general topology, as discussed in Ferrari et al. (1985), allows wind solutions that become successively supersonic and again subsonic through shocks at compressions in the channel. An application to M 87 provides rather interesting results.

5.4. NONLINEAR MHD INSTABILITIES.

Magnetic fields are an important component in radio galaxies as shown by their synchrotron emission. As discussed in the previous Section, they may also be at the root of the physical mechanisms leading to jet acceleration and collimation.

On large scales their importance may however be less crucial for the outflow dynamics, especially because magnetic fields cannot exceed the equipartition limit: otherwise pinching instability would rapidly break up the directional motion.

The analysis of linear MHD instabilities has been already presented above. Here we only comment on the difficulties of numerical nonlinear studies. Most experiments performed so far have used the standard finite difference scheme with finite cells and, correspondingly, a rather large numerical viscosity. Shibata and Uchida (1985) have used a Lax-Wendroff scheme, Stone & Norman (1992a, 1992b) have developed a 3-D MHD code named Zeus. These simulations are unable to represent properly discontinuities and therefore tend to smooth out strong instabilities.

Recently Malagoli et al. (1995) have finally succeeded in producing a 2-D high order Godunov MHD code. This code has been tested on the standard problem of the Kelvin-Helmholtz instability of a magnetic shear layer; the formation of cat's eyes has been followed and the subsequent series of reconnection events yielding asymptotically to a stationary turbulent thick layer. We expect therefore to have the possibility to investigate the phenomenology of magnetized jets with more physical insights.

Benford (1979) proposed a different picture stressing the importance of magnetic fields in regions of jets that appear to be overpressured. He supposed that a jet can carry an electric current so that consistently an azimuthal field is created decreasing with distance from the axis $B_\phi \propto r^{-1}$. For consistency a return current must be present far away. In between a sheath will form around the jet where magnetic stresses are small (no current) but magnetic pressure can confine pressure 10 - 100 times larger. Instabilities in this configuration may be relevant, especially of kinetic type. A complete study does not exist, and also numerical simulations are lacking.

5.5. INSTABILITIES AND RADIATION.

Nonthermal radiation emission from jets needs to be supported by *in situ* particle acceleration, as the lifetime of synchrotron electrons is much shorter than the travelling time from the parent nucleus. We have discussed the observational and theoretical evidences that shocks may be regions where electrons are reaccelerated. But we also have emission from the weak diffuse regions in the flow.

So far we have discussed instabilities from a negative point of view, i.e. as processes that can destroy jets: in fact they appear to grow fast enough to impair collimated flow propagation downstream of the transonic nozzle up to the distances observed in radio galaxies. However in reality instabilities may also have a positive aspect.

Saturation effects, to avoid the growth of instability to such large values that would damage the dynamics, must lead to dissipation of the energy stored in perturbations. Simple dissipation to the thermal level would create problems overpressuring jets. Coupling to radiative modes would instead be more attractive.

Benford et al. (1980) have proposed a scenario in which long wavelength fluid modes start a nonlinear cascade towards short wavelength modes supporting a Fermi-like acceleration by interaction with MHD random waves. The electromagnetic component escapes from the system and *de facto* saturates the dynamical effectiveness of the instability. We report in Fig. 7 a general scheme of the various processes that are likely to complicate the real situation.

In such a scheme Ferrari et al. (1979), Eilek et al. (1979), Lacombe (1977) have calculated the timescales of nonlinear coupling of modes, showing how they can be efficient enough to guarantee a constant input of energy towards particle acceleration and radiation.

Of course particle acceleration in shocks is more efficient and may be dominant in FR II type sources, but it cannot explain sources of FR I type that require a diffuse, turbulent distribution of scattering centers.

6. JET TERMINATION

Jets terminate in extended radio lobes. These structures represent the "wastebasket" where all the energy transported by the continuous flow is finally released through complex interactions with the external medium.

In Fig. 8 we show the western radio lobe of Cygnus A (Perley et al 1984) as a typical example. One can recognize a working surface with a *hot spot* at the very head of the beam; the whole regions is threaded by *filaments* and polarization indicates magnetic fields structures consistent with a naive interpretation in terms of a high velocity fluid propagating in a stationary fluid of similar density. The diffuse part of the lobes has the shape of a bow shock; recently this impression is being supported by observations of stellar jets by HST (Ray these proceedings).

Two main difficulties arise with radio lobes. One is the energetics: if one assumes that the plasma contains protons and electrons with identical number density consistent with the synchrotron model, the estimate of the total kinetic energy in lobes L_j largely exceed the radiative luminosity

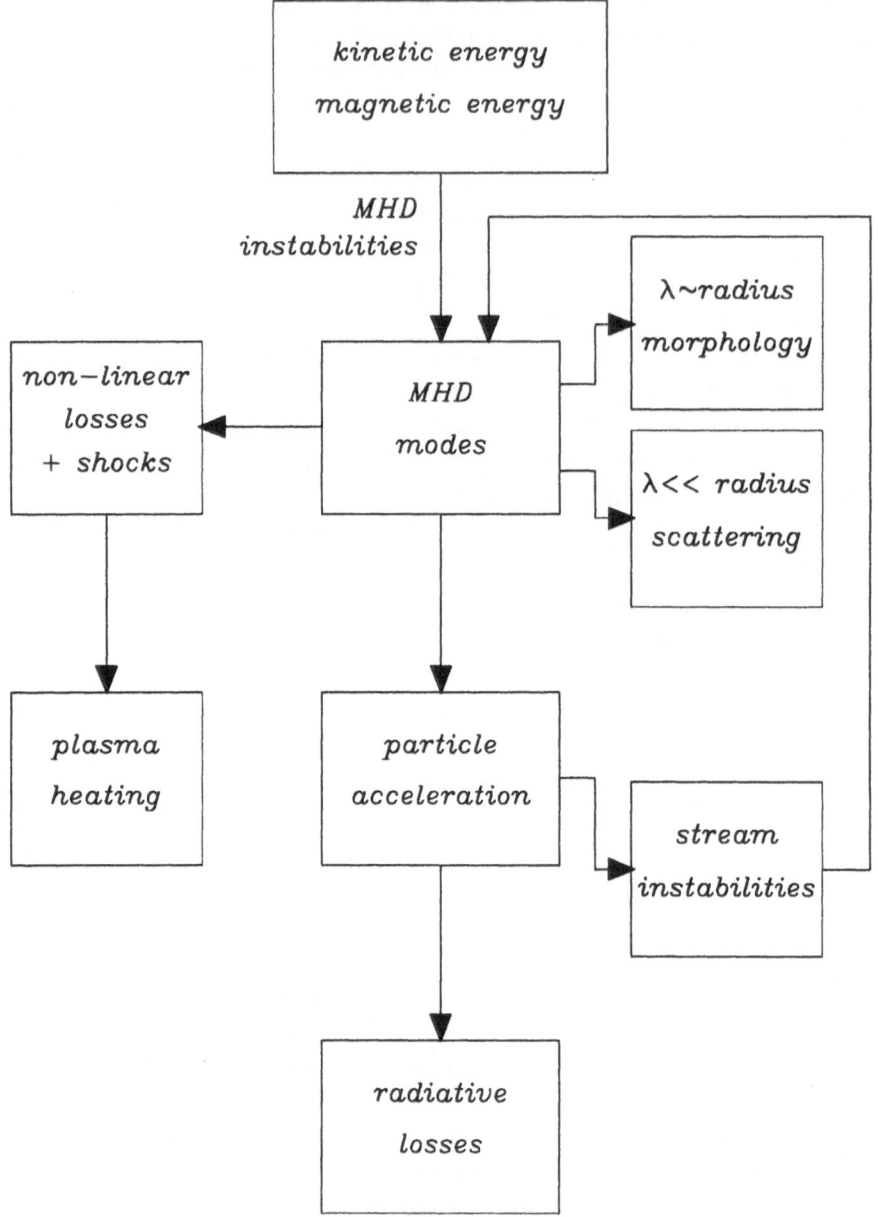

Figure 7. Physical processes in jets

Figure 8. Detailed radio map of the western lobe of Cygnus A

integrated over the source lifetime, and its value can be embarrassingly as high as $10^3 L_{rad}$ (Scheuer 1974). A way out could be to adopt an electron - positron pair model (Kundt 1987), but still one would require relativistic motions to reduce the annihilation rate and this would imply large inertial masses anyway.

Another relevant fact is that the dimensions of radio lobes are much larger the jet cross section, and also larger than expected from an expansion of the overpressured gas in the head or hot spots over the source lifetime. Scheuer (1982) has proposed that this is due to an oscillation in the direction of the flow ejection related to some short periodicity in the AGN. In this case the jet's head would drive its thrust onto different areas at different times, injecting relativistic particles and magnetic flux over extended regions. On the other hand it looks very strange to find the signature of these oscillations in the outer lobes and not at all in the jets that are so narrow and well collimated.

In the framework of radiation emission, again as in jets, local particle acceleration for synchrotron emission must operate following the observed patterns: hot spots and filaments must be the site of efficient acceleration, but diffusion brings particles all over the lobe.

6.1. HOT SPOTS.

Parker in his solar wind theory (1963) suggested that the supersonic out-flow links to the subsonic dynamics of the undisturbed interstellar medium through a *termination shock* matching the asymptotic wind solution to a subsonic breeze with $P_\infty \neq 0$. In the case of jets this shock should be lo-calized at the head of the collimated flow: the best evidence that this is what happens are the hot spots, compact overpressured regions character-istic of strong jets (FR II), absent in weak sources (FR I) because turbulent dissipation has reduced the activity of the flow.

The condition for the formation of hot spots comes from balancing the jet's thrust and external medium ram pressure force:

$$\Pi_j \sim \frac{L_j}{(v_j - v_h)} \sim \frac{\rho_j (v_j - v_h)^3 A_j}{(v_j - v_h)} = \Pi_{ram} \sim \rho_e v_h^2 A_j \qquad (13)$$

where L_j is the jet's kinetic luminosity, v_j the flow velocity, v_h the head's advancing velocity, ρ_j and ρ_e the jet's and ambient's velocity respectively. Then:

$$v_h \sim \frac{v_j}{1 + \sqrt{\nu}} \qquad (14)$$

(ν the ratio of external to internal density) and the distance travelled de-pends on the source lifetime; in particular it can be much larger than any stopping distance of a single emitted plasmon (see Introduction).

Blandford (1978) calculated the particle acceleration by a first-order Fermi-mechanisms in shocks, obtaining for rather simple conditions that power-law particle spectra do actually form with slope $\sim 2 \div 3$, in perfect agreement with the observed radiation spectra $\sim 0.5 \div 1$ in the synchrotron model.

6.2. COCOONS.

The advancing termination shock creates an extended perturbation in the intergalactic medium (IGM) through a bow shock enveloping the jet chan-nel. The bow-shocked plasma forms a *cocoon*, where the internal jet pressure and the external IGM pressure adjusts to equilibrium through appropriate gradients in the macroscopic quantities.

The cocoon's length is determined by the balance between the jet's thrust and the ram pressure of the external medium, its width essentially by the internal pressure. This situation has been modelled in a sketchy way by Begelman and Cioffi (1989) and Cioffi and Blondin (1992) using the same arguments of Eq. 14; from those papers we report the cartoon of Fig. 9.

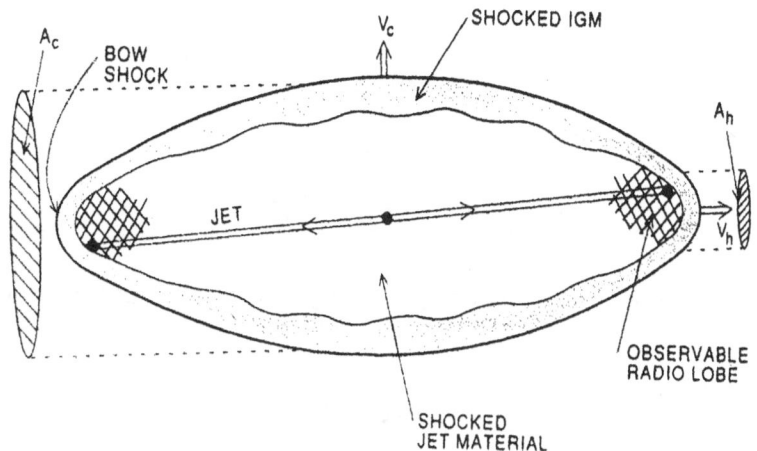

Figure 9. Schematic diagram of overpressured cocoons around jets

The head of the bow shock has now a cross-section $A_h \geq A_j$, in accordance with the proposal by Scheuer (1982) that the jet direction fluctuates on short timescale; it advances at velocity v_h determined by the corrected balance equation 14; for $v_j \gg v_h$:

$$\frac{L_j}{v_j} \sim \rho_e v_h^2 A_h$$

$$v_h \sim \left(\frac{L_j}{\rho_e v_j^3 A_h}\right)^{1/2} v_j \sim \left(\frac{\rho_j A_j}{\rho_e A_h}\right)^{1/2} v_j \qquad (15)$$

The cocoon pressure P_c drives a sideways shock into the IGM at a speed v_c that is fixed by the balance of P_c and ram pressure $\rho_e v_c^2$. P_c is estimated from the total energy deposited over the source lifetime in the observed volume V_c; at first approximation:

$$P_c \sim \frac{1}{V_c} \times 2 \int L_j \, dt \sim \frac{2L_j t}{2A_c v_h t} \sim \rho_e v_c^2$$

where A_c is typically much larger than the jet radius or hot spots; assuming $A_c \sim v_c^2 t^2$:

$$v_h \sim \frac{L_j t^2}{\rho_e A_c^2} \qquad (16)$$

Eqs. 15 and 16 can be combined to give the lateral cocoon expansion:

$$A_c \sim \left(\frac{L_j v_j A_h}{\rho_e}\right)^{1/2} t \, , \quad v_c \sim \left(\frac{L_j v_j A_h}{\rho_e}\right)^{1/4} t^{-1/2} \qquad (17)$$

634

When the expansion velocity of the cocoon decreases below the speed of sound c_s in the IGM, the cocoon is no longer overpressured. In fact we can write the condition for overpressure in the form:

$$P_c \sim \rho_e v_c^2 \geq P_e \sim \rho_e c_s^2$$

$$\rho_e v_c^2 \sim \frac{\rho_e A_c}{t^2} \sim \frac{(L_j \rho_e v_j A_h)^{1/2}}{A_c} \geq \rho_e c_s^2$$

At the same time we must require for jets that the cocoon is elongated, $v_h \geq v_c$:

$$L_j \geq \rho_e v_j^3 A_h \left(\frac{A_h}{A_c}\right)^2$$

When this condition is not satisfied, the outflow assumes a spheroidal shape.

We stress two interesting results originating from the presence of the cocoon. First of all, as the whole body of radio galaxies is embedded in an overpressured region, jet collimation is made easier. Due to limitations in angular resolution, X-ray measurements of gas pressure around radio galaxies may not be relevant to beam collimation as it may not come from the regions of the cocoon at the interface of the flow. Secondly the dimension of the lobes are related to the bow-shock expansion, which largely depends on the jet kinetic luminosity, and are made wide through wave propagation and not simple diffusion.

6.3. FILAMENTS.

Another characteristic structure observed in lobes are filaments. We have already mentioned in the study of jet propagation that a synchrotron "thermal" instability excite compressional modes with wavevector transverse to the local magnetic field and giving rise to filamentary structures elongated parallel to the field. These structures are non-stationary and evolve rapidly with the timescale of synchrotron emission. Numerical results by Bodo et al. (1992) have shown the nonlinear development of the instability to make the filaments brighter than the background, provided a continuous supply of relativistic electrons is guaranteed. In terms of timescales of particle diffusion, the origin of these particles could be the hot spots themselves, when present, or weak shocks diffused in a turbulent medium.

An application to Cygnus A shows that the thickness of the filaments, obtained as a result of the instability evolution, is of the order of 2 kpc, which can be compared with the observed value of about 1 arcsec that correspond to ~ 1 kpc. Also the emissivity contrasts of the order of 10 are compatible with the observational data.

6.4. NUMERICAL SIMULATIONS OF JET'S HEAD EVOLUTION.

The physical processes described above have been tested in numerical simulations to understand their nonlinear time evolution. Several calculations have been published since the pioneering work by Norman et al. (1982), devoted to the analysis of the propagation of a supersonic jet shot into an ambient medium.

The basic picture that has emerged can be summarized as follows:

1. the deceleration of the jet flow at its head is accomplished through the formation of a strong shock (Mach disk) which thermalizes the jet bulk kinetic energy;
2. the overpressured shocked jet material forms a backflow along the sides of the jet and inflates a cocoon whose size increases decreasing the density ratio between jet and ambient material;

Finally, a bow shock is driven into the external medium. This basic picture appears to fit with the structures seen in radio maps of extragalactic jets and includes substantially the main themes discussed above. Recently several new elements have been introduced in the numerical model (for a recent review see e.g. Burns, Norman & Clarke 1991): they include variability of the injection properties of the jet (Clarke & Burns 1991), variation of the physical parameters of the ambient medium along the jet propagation path (Norman, Burns & Sulkanen 1988), nonadiabaticity of the flow (Blondin, Fryxell & Königl 1990), fully 3-D geometry (Norman, Stone & Clarke 1991, Hardee & Clarke 1992, Hardee, Clarke & Howell 1995) and MHD effects (Clarke, Norman & Burns 1986; Lind et al. 1989).

More recently, numerical simulations of relativistic jets have been carried out by Martí, Müller & Ibáñez (1994) and Duncan & Hughes (1994), for low Mach number jets, and by Martí et al. (1995) for high Mach number jets.

In spite of these efforts many aspects of the problem are still not well understood, because of the complexity of the jet-cocoon structure: in fact, the cocoon excites perturbations in the jet flow, which in turn can be amplified by the Kelvin–Helmholtz mechanism and induce a strong activity of the jet's head that affects the cocoon structure.

Finally, when trying a more direct comparison of the results with observations, one must remember that what is observed is an outcome of the distribution of energetic particles and magnetic field and not the bulk of the flowing plasma, and therefore direct comparisons could be misleading.

In a recent analysis based on a sophisticated 2-D hydrocode of the PPM type, Massaglia et al. (1995) have calculated in much detail the dynamics of the interaction with an extensive exploration of the parameter space especially towards high Mach numbers, following the jet propagation up to very long times through a specific algorithm that allowed to keep all parts of the cocoon inside the computational domain.

The aim of the research was the dynamics of the interaction of supersonic, underdense, cylindrical fluid jets with a homogeneous, undisturbed medium. The integration was performed over the full set of adiabatic, inviscid fluid (Euler) equations 10, 11, and 12. In order to follow the mixing and entrainment effects between jet and IGM, one solves an additional advection equation for a scalar field f:

$$\frac{\partial f}{\partial t} + (\mathbf{v} \cdot \nabla) f = 0 \qquad (18)$$

where f is initially set 1 inside the jet and 0 outside. A collimated flow is injected from the left boundary into a medium at rest and in pressure equilibrium. The velocity and density profiles transverse to the jet are constant, rapidly changing to external values across the interface.

The phases of evolution correspond to the unfolding of the discontinuity between jet and external material into: (i) a reverse shock propagating in the jet against the flow, (ii) a contact discontinuity, separating the jet material from the external medium, and (iii) a bow-shock propagating in the external material.

Five different regions appear: (1) the jet; (2) the shocked jet material still flowing in the forward direction; (3) the shocked jet material reflected backwards at the contact discontinuity and flowing back at the jet side; (4) the shocked external material; (5) the unshocked external material. The shocked jet and external material form the cocoon.The high pressure cocoon squeezes the jet and drives shock waves into it, which reflecting on the axis assume the characteristic biconical shape.

The aspect of the interaction depends essentially on $\mathcal{M}$: a stronger interaction between biconical shocks and jet head is obtained for high Mach number jets (Fig. 10). The dependence on ν is much weaker.

The jet thrust is modulated by the biconical shocks impinging on its head and this can produce a periodic increase in the advance velocity of the head, leading to a strong change in the cocoon morphology.

6.4.1. Dynamics of the jet's head.

The jet's head advances following on the average the Begelman and Cioffi solution. The typical evolution timescale is:

$$\tau = t \frac{\mathcal{M}}{1 + \sqrt{\nu}} \qquad (19)$$

Two classes of dynamical evolution exist:

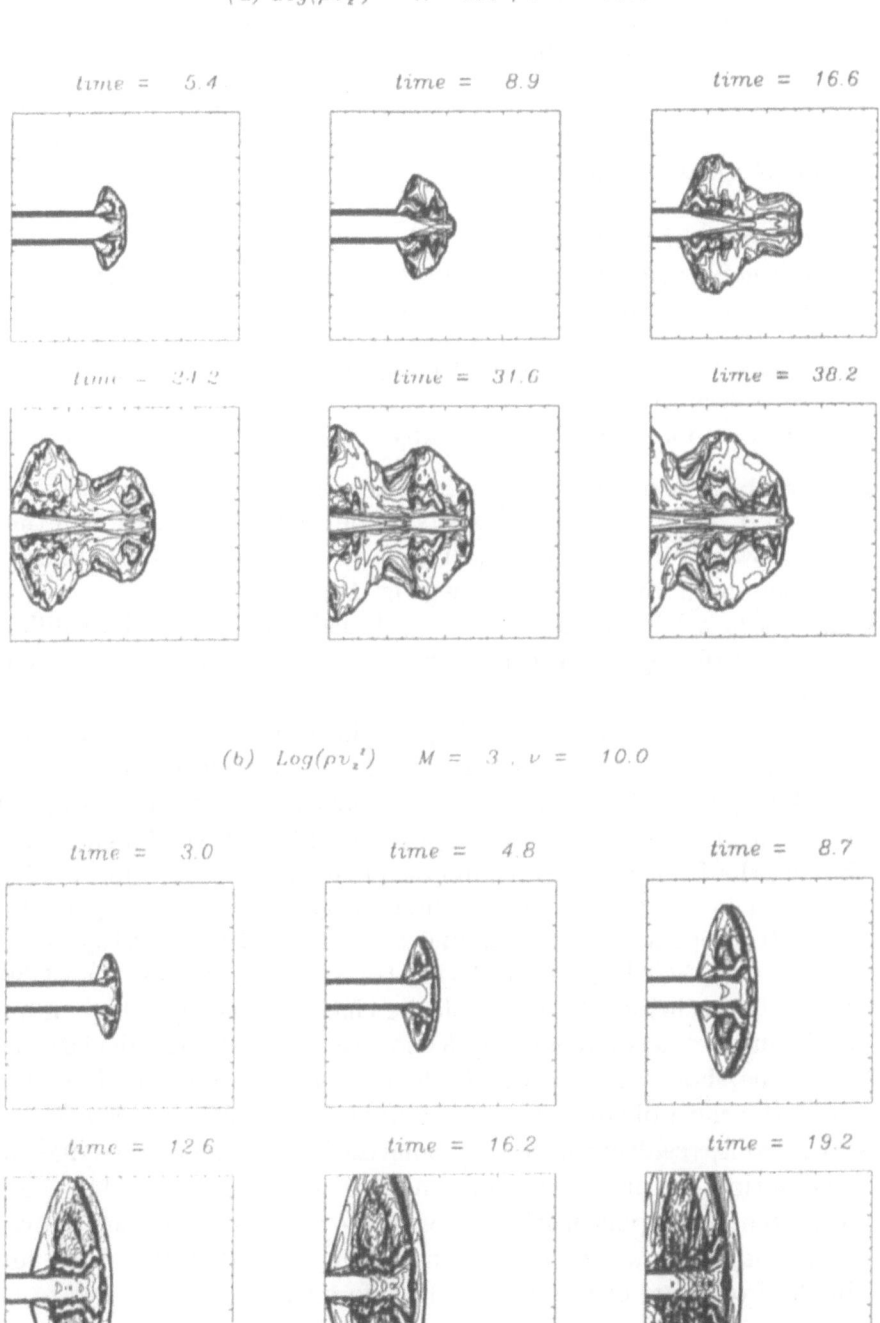

Figure 10. Evolution of jet's head: (a) high Mach number jet, (b) low Mach number jet

1. jets with high $\mathcal{M}$ and ν have faster v_h and show recurrent acceleration phases as due to strong thrust by the biconical shocks (Fig. 11);
2. jets with low $\mathcal{M}$ and ν have slower v_h as the shock thrust is weak.

The critical parameter is the inclination angle of the biconical shocks that determines the thrust behind the head: shocks must have a small inclination angle on the axis.

6.4.2. *Dynamics of the cocoon.*

Obviously the cocoon's evolution follows the head, but the influence of the density contrast is stronger. One gets two classes:

1. spear-headed cocoons, bearing the sign of the recurrent acceleration of jets with high $\mathcal{M}$ (Fig. 11);
2. fat cocoons, more extended laterally.

One can better understand the general trend looking at the behaviour of the tracer f, that provides snapshots of the spatial distribution of the jet particles. Slow jets and fast jets with high density ratios behave differently from fast jets with low density ratios (Fig. 12).

Can this fact be, at least in a very broad sense, related to different radio lobe morphologies? In other words, is the cocoon representative of radio lobes?

For high Mach numbers, $\mathcal{M} = 100$, the shock that surrounds the cocoon involves matter of the external ambient. Is this shocked region site of particle acceleration? If the answer is positive we can say that the form of the lobe resembles the density distribution of Fig. 12, with an elongated structure having the front part protruding from the lobe. Similar morphologies can be found in the sample of high luminosity radio sources by Leahy & Perley (1991); representative examples can be: 3C42 $((X_M/L)_{\mathrm{obs}} \sim 0.75)$, 3C184.1 (~ 0.8), 3C223 (~ 0.7), 3C441 (~ 0.7), 3C349 (~ 0.77), 3C390.3 (~ 0.6). In this scheme, the jet would be characterized by a high value of the Mach number accompanied by a moderate value of the density ratio. Moreover, the shock that surrounds the cocoon, according to simulations, must have the effect of enhancing the component of the magnetic field along the shock front, resulting in a high polarization at the source edge, with the polarization vector directed normally to the edge itself. This effect is clearly visible in the polarization maps of the sources mentioned above.

In the case of slow jet, we note from Fig. 10a) that the shock forms only in the front part of the cocoon, therefore the actual lobe has to have a morphology similar to that given by the tracer in Fig. 12b. Examples of this second kind o f morphology can be found in the sample of Leahy & Perley (1991): 3C296 $((X_M/L)_{\mathrm{obs}} \sim 0.34)$, 3C296 (~ 0.33) and 3C173.1 (~ 0.32) are good examples of this class, while 3C382 (~ 0.4) and 3C457

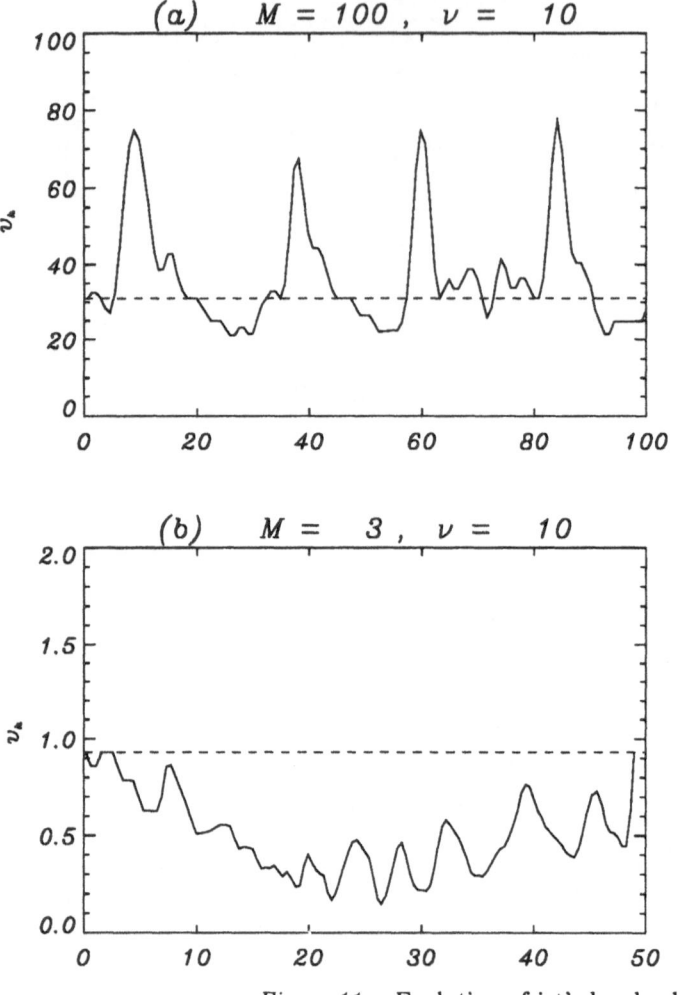

Figure 11. Evolution of jet's head velocity

(~ 0.25) are less so; the remaining sources of the sample are more irregular and many appear to bear some characteristics of both kind.

Acknowledgements. This work has been supported by ASI and MURST grants.

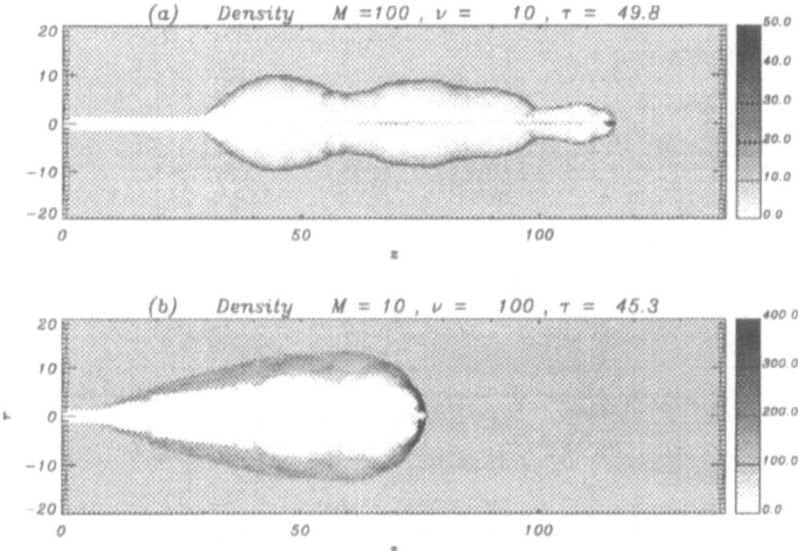

Figure 12. Cocoon morphology: (a) high Mach number, high density contrast, (b) high Mach numbers, low density contrast

7. References

Alexander P., Leahy J.P., 1987, *MNRAS*, **255**, 27

Begelman M. C., Cioffi D. F., 1989, *Ap. J.*, **345**, L21.

Benford G., 1979, *MNRAS*, **183**, 29.

Benford G., Ferrari A., Trussoni E., 1980, *Ap. J.*, **241**, 98

Birkinshaw M., 1991, in: Hughes P.A. (ed.) *Beams and Jets in Astrophysics* (Chap. 6). Cambridge Univ. Press, Cambridge.

Blandford R. D., Rees M.J., 1974, *MNRAS*, **169**, 395

Blandford R.D. & Ostriker J.P., 1978, *Ap. J.*, **221**, L29

Blandford R.D., 1991, in *Active Galactic Nuclei*, Berlin Springer Verlag

Blondin J.M., Fryxell B.A., Königl A., 1990, *Ap. J.*, **360**, 370.

Bodo, G., Ferrari, A., Massaglia, S., and Trussoni, E., 1990, *MNRAS*, **244**, 530.

Bodo G., Ferrari A., Massaglia S., Rossi P., Uchida Y., Shibata K., 1992, *A&A*, **256**, 689.

Bodo G., Massaglia S., Ferrari A., Trussoni E., 1994, *A&A*, **283**, 655.

Bodo G., Massaglia S., Rossi P., Rosner R., Malagoli A., Ferrari A., 1995, *A&A*, in press.

Bridle A.H. & Perley R.A., 1984, *Ann. Rev. Astron. Astrophys.*, **22**, 319.

Brinkmann W., 1995 in IAU 175 *Extragalactic Radio Sources*, Reidel Dordrecht, in press.

Burbidge G., 1956, *Ap. J.*, **124**, 416.

Burns J. O., Norman M. L., Clarke D. A., 1991, *Science*, **253**, 522.

Capetti A., Axon D.J., Macchetto F., Sparks W.B. and Boksenberg A., 1995, *Ap. J.*, **446**, 155.

Cioffi D. F., Blondin J. M., 1992, *Ap. J.*, **392**, 458.

Clarke D.A., Norman M.L., Burns J.O., 1986, *Ap. J.*, **311**, L63.

Clarke D.A. and Burns J.O., 1991, *Ap. J.*, **369**, 308.

De Young D.S., 1993, *Ap. J.* **405**, L13.

Duncan G.C., Hughes P.A., 1994, *Ap. J.*, **436**, L119.

Eilek J.A., 1979, *Ap. J.*, **230**, 373.

Fanaroff B.L., Riley J.M., 1974, *MNRAS*, **167**, 31.

Feigelson E.D., 1980, PhD Thesis Harvard University.

Feigelson E.D., Schreir E.J., Delvaille J.P., Giacconi R., Grindlay J.E., Lightman A.P., 1981, *Ap. J.*, **251**, 31.

Feretti L., Fanti R., Parma P., Massaglia S., Trussoni E., Brinkmann W., 1995, *A&A*, **298**, 699.

Ferrari A., Trussoni E., Zaninetti L., 1979, *A&A*, **79**, 190.

Ferrari A., Trussoni E., Rosner R., Tsinganos K., 1985, *Ap. J.*, **294**, 397.

Ferrari A., Trussoni E., Rosner R., Tsinganos K., 1986, *Ap. J.*, **300**, 577.

Field G., 1965, *Ap. J.*, **162**, 531.

Ghisellini G., Bodo G., Trussoni E., Rees M.J., 1990, *Ap. J.*, L362.

Hardee P.E., Clarke D.A., 1992, *Ap. J.*, **400**, 9.

Hardee P.E., Clarke D.A., Howell D.A., 1995, *Ap. J.*, **441**, 644.

Hartman R.C. et al. , 1993, *Ap. J.*, **407**, L41.

Kundt W., 1987 in Kundt W. (ed): *Astrophysical Jets and their Engines*, (Reidel: Dordrecht).

Lacombe C., 1977, *A&A*, **54**, 1.

Laing R.A., Riley J.M., Longair M.S., 1983, *MNRAS*, **204**, 151.

Leahy J.P. & Perlay R.A., 1991, *A. J.*, **102**, 537.

Lind K.R., Payne D.G. Meyer D.L., Blandford R.D., 1989, *Ap. J.*, **344**, 89.

Malagoli A., Bodo G., Rosner R., 1995, *Ap. J.*, in press.

Macchetto F., 1995 in IAU 175 *Extragalactic Radio Sources*, Reidel Dordrecht, in press.

Martí J.Mª, Müller E., Ibáñez J. Mª, 1994, *A&A*, **281**, L9.

642

Martí J.Mª, Müller E., Font J.A., Ibáñez J. Mª, 1995, *Ap. J.*, **448**, L105.

Massaglia S., Bodo G., Ferrari A., 1995, *A&A*, in press.

Muxlow T.W.B. and Garrington S.T., 1991, in: Hughes P.A. (ed.) Beams and Jets in Astrophysics (Chap. 2). Cambridge Univ. Press, Cambridge.

Norman M.L., Smarr L., Winkler K.H.A., M. D. Smith M.D., 1982, *A&A*, **113**, 285.

Norman M.L., Burns J.O., Sulkanen M.E., 1988, *Nature*, **335**, 146.

Norman M.L., Stone J.M., Clarke D.A., 1991, AIAA, Aerospace Sciences Meeting, 29th, Reno, NV, Jan. 7-10, p. 12.

Parker E.N., 1963, *Interplanetary dynamical processes*, (Interscience: New York).

Perley R.A., Dreher J.W., Cowan J.J., 1984, *Ap. J.*, **285**, L35.

Pringle J.E., 1981, *Ann. Rev. Astron. Astrophys.*, **19**, 137.

Rees M.J., 1984, *Ann. Rev. Astron. Astrophys.*, **22**, 471.

Rossi, P., Bodo, G., Massaglia, S., & Ferrari, A., 1993, *Ap. J.*, **414**, 112.

Rossi, P., Bodo, G., Massaglia, S., & Ferrari, A., in preparation.

Scheur, P.A.G., 1974, *MNRAS*, **166**, 513.

Scheur, P.A.G., 1982, in IAU 97, Eds. Heeschen D.S. & Wade C.M., Reidel Dordrecht, 163.

Shibata K. & Uchida Y., 1985, *PASJ*, **37**,31.

Stone J.M. & Norman M.L., 1992a, *Ap. J. Suppl.*, **80**, 759.

Stone J.M. & Norman M.L., 1992b, *Ap. J. Suppl.*, **80**, 791.

Thompson P.A., 1972, *Compressible Fluid Dynamics*, (McGraw-Hill: New York).

Van Dyke, 1982, *An album of fluid motion*, The parabolic Press, Standford California.

van Leer B. 1974, *Journal of Computational Physics*, **14**, 361.

Zachary A.L., Malagoli A., Colella P., 1994, SIAM *J.Sci.Comput.*, **15**, 263.

MHD ACCRETION-EJECTION FLOWS
IN ACTIVE GALACTIC NUCLEI

G. PELLETIER
Laboratoire d'Astrophysique
Observatoire de Grenoble. France

J. FERREIRA
Observatoire de Grenoble, France
and Landessternwarte, Heidelberg, Germany

AND

G. HENRI AND A. MARCOWITH
Observatoire de Grenoble, France

1. Introduction

Is it possible to launch jets from a Keplerian accretion disk? Is the magnetic field able to transfer the angular momentum through the outflow? If yes, what are the conditions? And how deeply the structure of the accretion disk is modified? These issues are relevant for Young Stellar Objects (YSO), Active Galactic Nuclei (AGN) and so-called "microquasars" (compact objects of the Milky Way displaying similar features). We will present an overview of these topics by considering first the MHD flow.

Then the last part of the presentation will be more specifically devoted to the role of MHD in the AGN phenomenon, especially in the case of the radio-loud class of AGNs that displays hard X-rays, γ-rays, synchrotron sources in relativistic motions. We will describe the main lines of a "triptych" for AGNs where the magnetic field is the link between the accretion disk, the jets and the high energy source.

2. The MHD engine

For many years, astrophysicists have invoked the accretion power to account for the most luminous phenomena of Nature. However the way to liberate

K. C. Tsinganos (ed.), Solar and Astrophysical Magnetohydrodynamic Flows, 643–657.
© *1996 Kluwer Academic Publishers.*

the mechanical energy (gravitation and rotation) remains a very opened issue. Purely hydrodynamical models face to the viscosity problem: the luminosity is a direct measure of the required turbulent viscosity. No theory so far provide with a turbulent viscosity that would have the expected value. It is interesting for the development of MHD astrophysics to note that the most promising instabilities that would create the favorable turbulence are due to the magnetic field. In this section, we will focuse mostly on the laminar aspect of the MHD flow, where opened field lines threading an accretion disk are expected to play a major role in the energetics of the engine.

The growth of the toroidal component of the magnetic field by differential rotation creates a situation where the magnetic field can achieve the three expected functions, namely: first, the angular momentum transfer by the magnetic braking $d\Gamma_m \propto -B_\phi B_p r dr$, second, the propulsion of jets by the poloidal Poynting flux $\propto E \times B_\phi$ and third, the collimation of the jets by the tension effect due to B_ϕ also. An important concept is the Alfvén surface where the flow has a poloidal velocity that reaches the local Alfvén velocity V_A at a distance from axis called the Alfvén radius r_A that governs the transfer of angular momentum. When the flow crosses smoothly this critical surface, the acceleration is more efficient in the subAlfvénic region when the Alfvén is large. Moreover the angular momentum transfer is also more efficient and requires a weaker mass flux. So one of the goal of the investigation of this issue is to find the conditions that insure the largest possible Alfvén radius.

In this presentation, we consider a Keplerian accretion disk and define a small aspect ratio $\varepsilon \equiv h(r)/r$, where $h(r)$ is the half width of the accretion disk at radius r. We assume a bipolar magnetic field (a quadrupolar configuration is also possible, especially if dynamo action is at work. However it seems less efficient at extracting the angular momentum). The theory can be used to modelize young stellar object environment or Schwarzshild black hole environment as proposed by many authors (Blandford and Payne 1982, Pudritz and Norman 1983, Lovelace et al. 1987, Königl 1989, Sakurai 1985, Uchida and Shibata 1985, Pelletier and Pudritz 1992 etc.). In principle accretion is enough to obtain a magnetic field having a significant intensity such that it plays a dynamical role. Indeed as is well known, the compression of the magnetic field by infalling matter is an obstacle to create stars.

2.1. ANGULAR MOMENTUM TRANSFER

Let us compare the size of the viscous torque with that of the magnetic torque by defining the parameter

$$\gamma \equiv \frac{viscous\ torque}{magn.\ torque} \sim \alpha_v \frac{h}{r} \frac{p_{gas}}{p_{mag}} \frac{B_0}{B_\phi^+} , \tag{1}$$

where α_v is the dimensionless coefficient that measures the size of the turbulent viscosity in the Shakura and Sunyaev prescription; $B_0 \equiv B_z(r,0)$ and B_ϕ^+ is the value of the toroidal field on the disk surface that will be estimated later on. The condition for angular momentum extraction mostly by the magnetic field is easily achieved a priori, since $\gamma \ll 1$ is equivalent to

$$\frac{B_\phi^+}{B_0} \gg \alpha_v \frac{h}{r} \frac{p_{gas}}{p_{mag}} . \tag{2}$$

The radial current that insures the angular momentum transfer by the magnetic field is such that

$$J_0 B_0 = \frac{1}{2} \rho_0 u_0 \Omega_0 \tag{3}$$

(where $J_0(r) \equiv J_r(r,0)$, $\rho_0(r) \equiv \rho(r,0)$ and $u_0(r) = -u_r(r,0)$ is the velocity of the accretion flow on the midplane, and Ω_0 the rotation velocity on the midplane).

2.2. DIFFUSION

The poloidal magnetic flux is decribed by a scalar $a(r,z) \equiv r A_\phi(r,z) = \frac{flux}{2\pi}$ and is governed by a diffusion equation involving the magnetic diffusivity ν_m:

$$\frac{\partial}{\partial t} a + \vec{u}_p . \nabla a = \nu_m \Delta' a , \tag{4}$$

where $\vec{u}_p$ is the poloidal velocity field. It is then useful to define a magnetic Reynolds number associated to the accretion flow:

$$\mathcal{R}_m \equiv \frac{r u_0}{\nu_0} , \tag{5}$$

where $\nu_0 \equiv \nu_m(r,0)$. The higher the Reynolds number the more bent the field lines, because they are more frozen in the accretion flow. This is simply estimated by introducing the curvature radius of the field lines on the midplane $l(r)$ such that

$$\frac{\partial^2 a}{\partial z^2}(r,0) = -\frac{a(r,0)}{l^2} . \tag{6}$$

Then for a distribution of flux on the midplane proportional to r^β, we obtain

$$\mathcal{R}_m = \frac{r^2}{\beta l^2} \, . \tag{7}$$

If $\mathcal{R}_m \sim \frac{r^2}{h^2}$ or more, the curvature is so strong that the accretion disk would be highly unstable.

If $\mathcal{R}_m \sim 1$ or less, the field lines are so straight that no jet can be launched, unless maybe by a corona.

If $\mathcal{R}_m \sim 1/\varepsilon$, the curvature radius is at an intermediate scale between h and r: $l \sim \sqrt{hr}$. Then the bending is reasonable. This is the best regime to launch jets that will also carry the angular momentum (Ferreira and Pelletier 1995). However this good regime cannot allow any simplification of the MHD equations, the flow is intrinsically two-dimensional.

2.3. CURRENT PROFILE

The electromotive force $\Omega B_z r dr$ maintains the radial current responsible for the magnetic braking of the unipolar inductor (Barlow wheel). However in a plasma, the differential rotation $r\frac{\partial \Omega}{\partial r} B_r$ modifies the vertical profile of J_r. Precisely the induction equation implies (Ferreira and Pelletier 1995):

$$\frac{1}{2} \frac{h^2}{\nu_m J_r} \frac{\partial^2}{\partial z^2}(\nu_m J_r)\,|_{z=0} = 3 \frac{V_{A0}^2 h^2}{\nu_0^2} \, , \tag{8}$$

where V_{A0} is the Alfvén velovity on the midplane. Thus a smooth current profile requires some magnetic diffusivity that scales like

$$\nu_0 = \alpha_m V_{A0} h \tag{9}$$

with, similarly to Shakura Sunyaev prescription, a dimensionless coefficient $\alpha_m \sim 1$. In fact as it will turn out that such engine works only if there is a rough equipartition between gas pressure and magnetic pressure, such amount of magnetic diffusivity is achieved by a turbulence that leads to an effective magnetic Prandl number of order unity (Pouquet et al. 1976). It is interesting to remark that it corresponds to a plasma that has a diffusion time τ_D across the disk comparable to the Alfvén time τ_A, since $\tau_D = h^2/\nu_0$ and $\tau_A = h/V_{A0}$. This condition is close to the condition of saturation of resistive instabilities.

We can now estimate B_ϕ^+ with this diffusivity scaling:

$$\frac{B_\phi^+}{B_0} \sim \alpha_m \mathcal{R}_m \varepsilon \frac{\Omega_0 h}{V_{A0}} \, . \tag{10}$$

We introduce an important parameter which is roughly the ratio of the magnetic pressure over the gas pressure:

$$\mu \equiv \frac{V_{A0}^2}{\Omega_0^2 h^2} \ . \tag{11}$$

Thus the torque parameter

$$\gamma = \frac{\alpha_v}{\alpha_m} \mu^{-1/2} \mathcal{R}_m^{-1} \ . \tag{12}$$

Therefore in the good regime, where $\mu \sim 1$ and $\mathcal{R}_m \sim 1/\varepsilon$, the torque parameter is of order ε. Thus the magnetic torque dominates largely the viscous torque.

2.4. CROSSING THE SLOW MAGNETOSONIC SURFACE

The requirement that the flow must cross smoothly the slow magnetosonic surface puts constraints over three parameters, namely the pressure ratio μ, the magnetic Reynolds number $\mathcal{R}_m$ and the ejection index ξ defined in

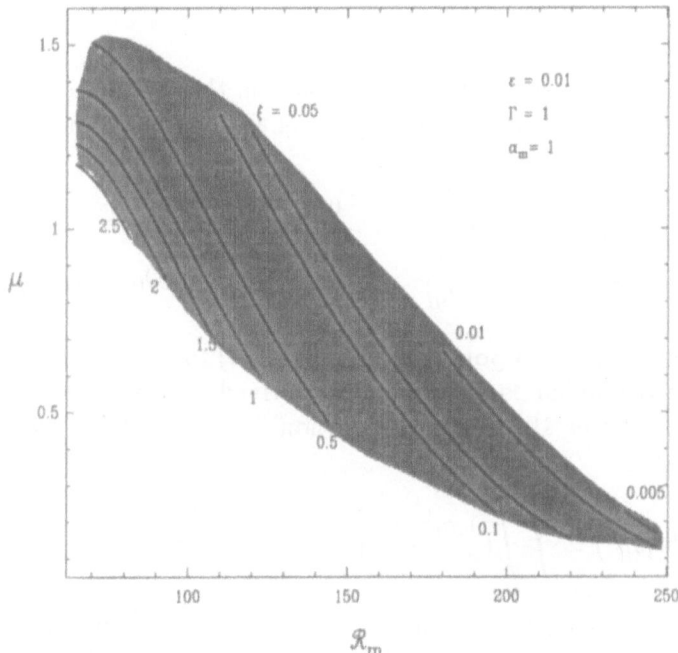

Figure 1. Parameter space for a fixed value of the aspect ratio $\varepsilon = 0.01$.

term of the accretion rate $\dot{M}_a$ by

$$\xi = \frac{dLn \ \dot{M}_a}{dLn \ r} \ .$$
(13)

The ejection index is a measure of the amount of accreted matter rerouted in the outflow; the accretion rate would be a constant in absence of outflow. The conservation of mass relates simply the accretion rate up to some radius r to the mass flux in the outflow between the internal radius and the radius r. The figure 1, extracted from Ferreira and Pelletier (1995) shows the conditions and reveals that a rough equipartition must be fulfilled between gas pressure and magnetic pressure, and that the magnetic Reynolds number must be between few tens and few hundreds; roughly of order ε^{-1}.

3. Motion and confinement coupled

The jet flow can be correctly described by ideal MHD. Moreover, stationarity and axisymmetry reduce the problem to two coupled equations involving the two scalar functions: the flux function $a(r, z)$ and the square of the Alfvén number which is defined with the poloidal fields by

$$\vec{u}_p \equiv m\vec{V}_{Ap} \ .$$
(14)

These two equation are the Bernoulli equation that governs the motion and the Grad-Shafranov equation that governs the confinement:

$$\frac{1}{2}\frac{(\nabla a)^2}{r^2} + \frac{\partial U}{\partial m^2} = 0 \qquad (Bernoulli)$$
(15)

$$\nabla.(\frac{m^2 - 1}{r^2}\nabla a) - \frac{\partial U}{\partial a} = 0 \qquad (Grad - Shafranov) \ .$$
(16)

We introduced a pseudo potential $U(a, m^2; r, z)$ for formal convenience. Indeed these two coupled equations can be derived from a variational principle (Rosso and Pelletier 1994) with the action functional $S(a, \nabla a, m^2)$ defined by:

$$S(a, \nabla a, m^2) = \int \int (\frac{1}{2}\frac{m^2 - 1}{r^2}(\nabla a)^2 + U(a, m^2; r, z))rdrdz \ .$$
(17)

Jet acceleration and jet confinement are tightly linked, because the widening of the magnetic surfaces allow the acceleration. And when the jet crosses the Alfvén surface, the self collimation becomes possible because of the increase of the toroidal field due to differential rotation. This coupling together with the existence of several unknown critical surfaces and successive changes of EDP type (elliptical and hyperbolic) make the numerical resolution difficult. Only recent analytical results brought some new understanding (Tsynganos and Sauty 1992). A convenient method to calculate the jet dynamics and self-collimation consists in extremalizing the functional under the constraints of boundary conditions and of smooth crossing the critical surfaces (Rosso and Pelletier 1994). Several classes of solutions were found: sub-A, super-A but sub-FM, and super-A and super-FM flows. Our calculation confirmed that a flow that crosses the fast magnetosonic surface undergoes cylindrical self-collimation (Heyvaerts and Norman 1989).

3.1. CROSSING THE ALFVÉN SURFACE

The main goal of the calculation is to exhibit the hidden law that controls the Alfvén surface crossing. Our numerical results indicate that, although $\Omega_* r_A$ (where Ω_* is the angular velocity of the field lines) must be larger than the Alfvén velocity $V_{A|A}$ at the Alfvén surface as shown analytically (Pelletier and Pudritz 1992), nevertheless it must not be much larger. The precise law could probably be accurately calculated as suggested by Jean Heyvaerts (this meeting). Anyway our estimate is enough to get informations about the location of the surface and the size of the "lever arm" r_A.

The Alfvén radius controls ejection and thus accretion through a simple relation derived from angular momentum transfer:

$$\xi = \frac{r_0^2}{2r_A^2} \, . \tag{18}$$

So taking account of this relation together with the regularity constrain, we get a new relation that gives the ejection index in terms of the main accretion disk parameters:

$$\xi \simeq \frac{1}{8\mu} (\alpha_m \varepsilon \mathcal{R}_m)^2 \, . \tag{19}$$

This is in fact an upper bound (see Ferreira, this proceeding).

3.2. GLOBAL ENERGY BUDGET

We define the fraction χ of liberated mechanical power that is converted into radiation by matter falling in the gravitational well of the black hole of mass M_* down to some internal radius r_i:

$$L = \chi \frac{1}{2} G \frac{\dot{M}_a M_*}{r_i} \, . \tag{20}$$

The power injected in each jet is therefore

$$P_j = \frac{1-\chi}{2}\frac{1}{2}G\frac{\dot{M}_a M_*}{r_i} . \tag{21}$$

We can derive the maximum bulk Lorentz factor of the flow by assuming that most of the injected power goes into kinetic energy:

$$P_j \simeq P_{kin} \equiv \int \frac{d\dot{M}_j}{da}(\gamma_\infty - 1)c^2 da \equiv \dot{M}_j(\bar{\gamma}_\infty - 1)c^2 , \tag{22}$$

where $\dot{M}_j$ is the mass flux in each jet. The last relation defines an average bulk Lorentz factor $\bar{\gamma}_\infty$. It is also suitable to define an average ejection index $\bar{\xi}$ such that

$$\dot{M}_j = \frac{1}{2}\int_{r_i}^{r_e} \xi \dot{M}_a(< r)\frac{dr}{r} \equiv \frac{1}{2}\bar{\xi}\dot{M}_a . \tag{23}$$

The asymptotic Lorentz factor is thus simply related to χ and $\bar{\xi}$:

$$\bar{\gamma}_\infty = 1 + \frac{1-\chi}{2\bar{\xi}}\frac{V_K^2(r_i)}{c^2} , \tag{24}$$

where $V_K(r)$ is the Keplerian velocity at radius r.

3.3. TURBULENT HEATING

The fraction χ of the liberated mechanical energy converted into radiation has two contributions. One comes from the turbulent Joule effect and the other from turbulent viscosity. The Joule contribution gives:

$$\chi_{Joule} = \alpha_m^2 \mathcal{R}_m \varepsilon^2 \sim \mathcal{R}_m^{-1} . \tag{25}$$

And the viscosity contribution gives:

$$\chi_{visc.} = \frac{\gamma}{1+\gamma} \sim \mathcal{R}_m^{-1} . \tag{26}$$

Since the magnetic Reynolds number is rather high, most of the liberated power is injected in the jets! The range of the accretion disk where the jets are launched is not bright. This should be considered to understand the soft X-ray part of AGN spectra, whose presumed accretion disk radiates mostly in the UV band.

In the case of AGNs and microquasars, the jets could be relativistic with Lorentz factors of order 10 provided that the ejection index is as low as $\bar{\xi} \sim 10^{-2}$, since

$$\bar{\gamma}_\infty = 1 + \frac{1-\chi}{12\bar{\xi}} . \tag{27}$$

Subrelativistic flows depend on the Keplerian motion around the internal radius:

$$\bar{u}_\infty = (\frac{1 - \chi}{\bar{\xi}})^{1/2} V_K(r_i) . \tag{28}$$

Our estimate of the ejection index from (19) is more likely $\bar{\xi} \simeq 0.1$. Which leads to subrelativistic velocities in the case of Schwartzschild black holes: $\bar{u}_\infty$ between 0.1 and 0.4 light velocity. For young stellar objects, we find the expected velocities of few $100 km/s$.

4. The issue of the turbulent transport

The standard theory of accretion disk needs a turbulent viscosity (Shakhura and Sunyaev 1972). A promising clue to solve this problem is to take into account the magnetic field (Balbus and Hawley 1992) that changes the Rayleigh criterium in such a way that Keplerian motions become unstable (Velikhov 1959, Chandrasekhar 1960). From recent investigation (Curry et al. 1995, Foglizzo and Tagger 1995) we can infer that this shearing instability is quenched when the vertical magnetic field is greater than a critical value B_c such that $V_{A0} \sim \Omega_0 h$, whatever the toroidal component. In our model, the viscosity is no more important, but we need a magnetic diffusivity. The development of MHD turbulence would lead to a magnetic Prandl number of order unity (magnetic diffusivity and turbulent viscosity of the same order of magnitude). As we mentioned previously, such a situation eventually achieves our diffusivity requirement provided that a rough equipartition of magnetic and gas pressures is met. Which means that MHD accretion-ejection structures work for $B_0 \sim B_c$. In order to get advantage of the shearing instability the structure should work a little below the critical condition.

It is also interesting to note that under those conditions (magnetic Prandl number of order unity and $B_0 \sim B_c$) resistive instability are close to saturation since $\tau_A \sim \tau_D$. However these anomalous dissipation processes could lead to an anisotropic magnetic diffusivity.

We mention also that the detailed analysis of the vertical structure of the accretion flow revealed the existence of convection layers. Thus a theory of the turbulent heat flux under these conditions (strong differential rotation and magnetic field) remains to be done.

5. A Magnetic Cauldron for AGNs and Microquasars

The high energy physics of AGNs and microquasars manifests as hard X-ray, γ-ray emission, probably as electron-positron pair production, relativistic ejection of high energy particles clouds. The possible role of MHD in this issue is, first, to confine this suprathermal plasma and, second, to maintain, through wave turbulence, scattering and stochastic acceleration of these particles. A "magnetic cauldron" was proposed by Henri and Pelletier in 1991, mostly to confine, power and launch pair plasma clouds or beams that likely explain the high energy physics. The inverse Compton process involving UV photons, X-rays and relativistic electrons generates γ-rays that are immediately converted into pairs in a compact source, optically thick to γ-rays. These new relativistic pairs are accelerated by MHD turbulence, so that they participate also to Compton process and thus new pair creation. The annihilation is not enough to prevent an unavoidable pair creation catastrophe provided that the turbulence level be reasonably high.

5.1. WHY AN ELECTRON POSITRON COMPONENT?

There are observational suggestions in favour of a pair plasma. First the so called "compactness" is high. Second, the pair model explains the γ-ray spectrum of "blazars" quite satisfactorily (Marcowith, Henri, Pelletier 1995). About 45 blazars has been observed up to now by GRO experiments (Comptel and EGRET) (see Von Montigny et al. 1995, for instance). Their intense hard X-ray and γ-ray emission was interpreted as due to the relativistic beaming of the Compton radiation (Dermer and Schlickeiser 1993). And it has been proposed that this relativistic flow was composed of pairs that becomes optically thin to γ-rays (Henri, Pelletier and Roland 1991, 1993) at about few hundred Schwartzschild radii of the central black hole. The radiation is mostly composed of a Compton emission with a small annihilation bump around a few MeV. The main challenge is to account for the spectral break, with a variation of the index greater than 0.5, at few MeV correctly. Detailed calculations taking into account anisotropy, inhomogeneity and opacity effects for $\gamma\gamma$-absorption has been performed (Marcowith, Henri, Pelletier 1995) and they lead to a remarkable fit of the data for the quasars 3C273 and 3C279, with one main free parameter: the black hole mass. An exemple is given on figure 2, extracted from Marcowith et al. (1995). The assumption of a pair plasma is crucial to obtain these fits. In its source at about 10 Schwartzschild radii, the pair plasma has a particle density n_* of about $10^9 cm^{-3}$, with a population of highest energy characterized by a Lorentz factor γ_{max} between 10^3 and 10^4 (in fact, these maximum Lorentz factors are obtained along the magnetic field; the distribution is very anisotropic even in the pair plasma restframe, because of the anisotropic cooling), and a bulk Lorentz factor γ_b between 4 and 5. How are these pairs accelerated and heated?

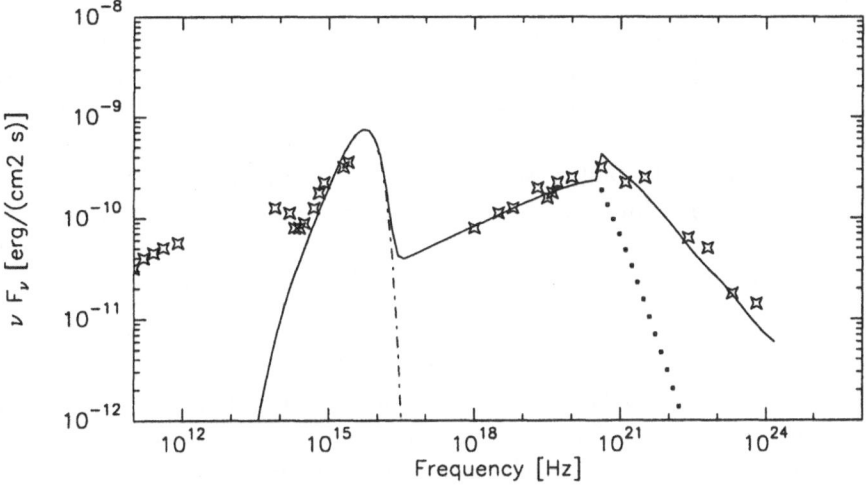

Figure 2. Fit of 3C273 spectrum

5.2. HEATING AND BOOSTING THE PAIR PLASMA

The dynamics of the relativistic plasma must be analyzed according to its distance to the central source of radiation, namely the inner part of the accretion disk. The Compton radiation field is very anisotropic along the axis. Primary run away electrons are efficiently boosted, because Coulomb collisions are negligible. So a two stream instability is trigged in the ambient plasma. Therefore a transient stage of electronic microturbulence, together with pair creation, leads rapidly to high energy pair plasma with parameters having values comparable to those just mentioned, taking account of the Compton cooling. A Fokker-Planck equation has been solved for the anisotropic distribution function taking into account also of pitch angle scattering (Marcowith et al. 1995b). This relativistic leptonic plasma likely becomes heavier than the ambient protonic component. But then the electronic two stream instability is quenched.

A new stage raises where the main plasma is the pair flow and the perturbing beam is the back stream of ambient protons with respect of the pair plasma restframe. We will come to this point later on. We just point out that usual radiation hydrodynamics does not describe these kinetic effects, that turn out to be crucial for this issue.

What can further boost the pair plasma? We envisage three possibilities: i) the Compton force on hot pairs, ii) the internal pressure force in widening magnetic surfaces and iii) the Poynting flux of a fastly rotating black hole. Fastly rotating black holes are interesting candidates to launch relativistic beams provided that the Compton drag be overcome (M. Camenzind 1986, V. Beskin et al. 1991). The two other possibilities were considered by Henri and Pelletier (1991). Somehow the Compton drag can be more easily overcome with a proton plasma. About the first possibility, S. Phinney in 1987 showed that cold electrons in a Compton field in a region of compactness l reach bulk Lorentz factor $\gamma_b \simeq l^{1/7} \simeq 2-3$. Electrons having a smaller Lorentz factor are efficiently accelerated, whereas electrons having a larger Lorentz factor suffer an efficient drag. However Henri and Pelletier (1991) showed that this result is significantly modified for hot electrons of energy $\bar{\gamma} m_e c^2$:

$$\gamma_b \simeq (l\bar{\gamma}^2)^{1/7} \sim 10 . \tag{29}$$

For the anisotropic distribution we consider, the $\bar{\gamma}$ must be considered as γ_{max}. Therefore the main problem is to heat the pair plasma to compensate Compton cooling. Then the anisotropy of the Compton radiation field accelerates the relativistic pair clouds quite sufficiently to account for the observed motions.

5.3. THE PROTON BACKSTREAM INSTABILITY

The situation where ambient protons are considered as a beam that pervades a relativistic pair plasma is quite unusual and deserve a special investigation. The relativistic electrons and positrons are supposed to be on one hand the heaviest component and on other hand confined by the magnetic field. Their pressure p_* is of the same order of magnitude than the magnetic pressure p_{mag} (at about $100 r_G$ from the black hole, we expect a magnetic field of the order of $100 Gauss$ and we have to bear in mind that $\bar{\gamma}$ is the typical energy of the most energetic pair in the restframe of the pair plasma; and the distribution is very anisotropic even in this rest frame, as mentioned previously). Under these circumstancies, the Alfvén waves are modified by the displacement current and relativistic effects. In the restframe of the pair plasma the dispersion relation of the waves is $\omega = k_\parallel V_*$ and the phase velocity

$$V_* = \frac{c}{\sqrt{1 + 2p_*/p_{mag}}} . \tag{30}$$

This velocity is always close to light velocity. These modes are absorbed at Landau-synchrotron resonnance with the high energy pairs; which occurs at the wave number $k_a = \omega_{ce}/\bar{\gamma}c$, corresponding to a wavelength comparable with the Larmor radius of the high energy pairs. The absorption rate is of order $\omega_{ce}/\bar{\gamma}$. These waves do not exist for wave numbers larger than k_a.

Ambient protons constitute a back stream that triggers an instability of these waves at Landau-synchrotron resonnance, as long as the bulk velocity

$v_b > V_*$. The maximum of instability occurs for the mode of wave number

$$k_0 = -\frac{\omega_{ci}}{\gamma_b}\frac{1}{v_b - V_*} , \qquad (31)$$

that propagates backwards, the maximum growth rate being

$$\Gamma = \frac{\omega_{pi}}{\gamma_b}(\frac{v_b - V_*}{2V_*})^{1/2} \qquad (32)$$

(ω_{pi} and ω_{ci} are respectively the plasma and cyclotron pulsations of protons). However this unstable mode exists only if $|k_0| < k_a$, which gives an interesting condition for the development of the resonnant interaction:

$$\bar{\gamma} < \gamma_b \frac{m_p}{m_e} . \qquad (33)$$

This interaction heats also the pair plasma and entrains the ambient protons that are then coupled to the relativistic flow. However the latter condition indicate that this coupling ceases when the pair plasma has been so heated that the inequality is reversed.

5.4. TURBULENT CASCADE IN THE RELATIVISTIC PLASMA

As long as the relativistic pair plasma is the heaviest component, any turbulent cascade of Alfvén waves toward shorter wavelength will transfer energy mostly to the pair plasma, because the resonance with protons occurs at a wavelength shorter than the resonnance with high energy pairs. Any long Alfvén wave produced by the accretion disk or the MHD jet will be directly transmitted by pressure to the high energy source as long as its wavelength is larger than its radius a (of few $10r_G$). There is no refraction effect, and coupling with magnetosonic waves is not necessary for this transmission; moreover magnetosonic waves are often rapidly damped. So we can expect that the large instabilities of the accretion disk and of the MHD jet power the high energy source through Alfvén waves.

Any cascade of Alfvén waves maintains a stochastic acceleration of the pair plasma. It is worth to realise that the stochastic acceleration is efficient because V_* is close to c. The usual expansion of the Fokker-Planck operator in power of V_A/c is odd. In this unusual situation where the phase velocity is close to c, a "first order" Fermi process at a shock would not be more efficient than the "second order" acceleration process. So no shock is needed to explain the high energy source! Moreover a shock would accelerate protons also, which is not at all interesting. It is much more interesting and efficient to have an acceleration process that favors immediately the radiating particles!

For an inertial spectrum $F(k) \propto k^{-p}$, with a ratio $\eta_t \equiv < \delta B^2 > /B_0^2$, we can estimate the luminosity of the high energy source of length b:

$$L \sim \eta_t \omega \left(\frac{\omega_{ce}}{\overline{\gamma}\omega}\right)^{2-p} \frac{B_0^2}{2\mu_0} \pi a^2 b \ . \tag{34}$$

Even for $p = 2$ and $\eta_t \sim 10^{-2}$, the high energy source can be explained by a turbulent cascade powered by a long Alfvén wave (of frequency ω), bearing in mind the Doppler beaming effect.

6. Synthesis

An open magnetic field in rough equipartition with the internal pressure can extract the angular momentum of a Keplerian disk and control the accretion on a Schwartzschild black hole. The angular momentum is carried out by an outflow provided that the magnetic Reynolds number be of order 100. The innermost region of the disk where the jets are launched is much less bright than the other part that radiates the black body spectrum. In AGNs and microquasars such MHD accretion-ejection flows are very powerful and reach terminal velocity of a sizable fraction of the velocity of light.

The MHD jet is suitable to confine the compact source of high energy radiation. The large instabilities of the structure can drive long powerful Alfvén waves that supply energy to the high energy source. The MHD jet as energy and momentum carrier for the compact source is very suitable, because it has much power and is free from Compton drag because its velocity remains mildly relativistic or even subrelativistic.

In a short transient region governed by electronic microturbulence, the anisotropic Compton radiation field generates a high energy pair plasma that is likely more massive than the proton component. Then ambient protons forms merely a back stream that drive an instability of the dominant pair plasma. Consequently they are entrained and they contribute to heat the pair plasma.

As long as the pair plasma is heated to $\bar{\gamma} \sim 10^3$, which seems to be achieved by turbulent waves, the anisotropy of the Compton force, that is helped also by the widening of the flux tubes, propels pair clouds to expected bulk Lorentz factors. A Fokker-Planck equation was approximately solved to find the distribution function; this work is in course of publication.

The description of the pair creation catastrophe needs still some developments. The decrease of the Alfvén velocity when the pair density grows makes the stochastic acceleration to slow down; which stops the creation process. However the sudden widening of the flux tubes due to the catastrophe is also an important effect that we expect to calculate in the near future.

These are the main aspects of the physics of the "magnetic cauldron". This paradigm opens some plausible understanding of AGNs and new developments of the kinetics of relativistic plasmas together with unusual magnetohydrodynamics.

7. References

Balbus, S.A., Hawley, J.F. (1992), *ApJ.*, **Vol. 392**, pp. 662-666

Beskin V.S., Istomin Y.A., Pariev V.I. (1991), *Extragalactic Radio Sources*, proc. 7th IAP meeting, Paris.

Blandford, R. D. and Payne, D. G. (1982), *MNRAS*, **Vol. 199**, pp. 883

Camenzind, M. (1986), *Astron. & Astrophys.*, **Vol. 156**, pp. 137

Chandrasekhar, S. (1960), *Proc. N.A.S.*, **Vol. 46**, pp. 56

Curry, C., Pudritz, R.E., Sutherland, P. (1995), *ApJ.*, **Vol.** , pp. in press

Dermer, C.D., Schlickeiser, R. (1993), *ApJ.*, **Vol. 416**, pp. 458

Ferreira, J., Pelletier, G. (1995), *Astron. & Astrophys.*, **Vol.295**, pp. 807-832

Foglizzo, T., Tagger, M. (1995), *Astron. & Astrophys.*, **Vol.** , pp. in press

Henri, G., Pelletier, G. (1991), *ApJ. Lett.*, **Vol. 383**, pp. L7

Henri, G., Pelletier, G., Roland, J. (1993), *ApJ. Lett.*, **Vol. 404**, pp. L41

Heyvaerts, J., Norman, C. Z. (1989), *ApJ.*, **Vol. 347**, pp. 1055

Konigl, A. (1989), *ApJ.*, **Vol. 342**, pp. 208

Lovelace, R.V.E., Wang, J.C.L., Sulkanen, M.E. (1987), *ApJ*, **Vol. 315**, pp. 504

Marcowith, A., Henri, G., Pelletier, G. (1995), *MNRAS*, **Vol. 277**, pp. xx

Marcowith, A., Henri, G., Pelletier, G. (1995b), *Astron. & Astrophys.*, submitted

Pelletier, G., Henri, G., Roland, J. (1991), *The Nature of the Compact Object in AGNs*, Cambridge University Press, pp. 368

Pelletier, G., Pudritz, R. E. (1992), *ApJ.*, **Vol. 394**, pp. 117

Pouquet, A., Frisch, U., Léorat, J. (1976), *J. Fluid Mech.*, **Vol. 77**, pp. 321

Pudritz, R.E., Norman, C. A. (1983), *ApJ.*, **Vol. 274**, pp. 677

Pudritz, R.E. (1985), *ApJ*, **Vol. 293**, pp. 216

Sakurai, T. (1985), *Astron. & Astrophys.*, **Vol. 152**, pp. 121

Shakura, N.I., Sunyaev, R.A. (1973), *Astron. & Astrophys.*, **Vol. 24**, pp. 337

Tsinganos, K., Sauty, C. (1992), *Astron. & Astrophys.*, **Vol. 257**, pp. 790

Uchida, Y., Shibata, K. (1985), *PASJ*, **Vol. 37**, pp. 515

Velikhov, E. (1959), *Sov. Phys. JETP*, **Vol. 36**, pp. 1398

Von Montigny, C., et al. (1995), *ApJ.*, **Vol. 440**, pp. 525-553

INTERACTION OF TURBULENT ACCRETION DISKS WITH EMBEDDED MAGNETIC FIELDS

JEAN F. HEYVAERTS AND ANN BARDOU
Observatoire de Strasbourg, Université Louis Pasteur
11, Rue de l'Université, 67000 Strasbourg, FRANCE
heyvaert@cdsxb6.u-strasbg.fr

AND

ERIC R. PRIEST
Solar Physics Group, Institute of Applied Mathematics
University of St Andrews
North Haugh, St Andrews, KY169SS, Scotland, UK
eric@dcs.st-and.ac.uk

Abstract. We show that the level of turbulence in accretion disks can be derived from a self-consistency requirement that the associated effective viscosity should match the instantaneous accretion rate. When turbulence originates in the magnetic shearing instability, the effective kinematic viscosity coefficient is shown to be describable by a Shakura-Sunyaev law with $\alpha \approx 0.04$. It is shown that thin disks supported by any turbulence with injection scale of order of the disk thickness, are very low magnetic Reynolds number systems. Turbulent viscosity-driven solutions with negligible field dragging and no emission of cold winds or jets are natural consequences of such regimes. Such disks are shown to expell the magnetic field of the accreting object from their kleplerian regions radially outwards, resulting in a flux distribution in the disk which differs very much from a dipolar one.

1. Turbulence and accretion

Matter which is to accrete on a star must lose most of its specific angular momentum. Several processes which may achieve this have been discussed in the literature. One of them consists in transfering the excess angular momentum to some other part of the accretion disk by viscous transport.

K. C. Tsinganos (ed.), Solar and Astrophysical Magnetohydrodynamic Flows, 659–672.
© *1996 Kluwer Academic Publishers.*

Another way of getting rid of this excess angular momentum is to shed it to infinity, in a magnetized MHD wind. Following Blandford and Payne (1982), this wind is often conceived as centrifugally driven, although this certainly is not the only possibility. Angular momentum loss by MHD winds also requires some degree of turbulence in the disk, for otherwise no consistent disk-wind connection can be built (Ferreira and Pelletier 1995).

In this communication we discuss in some more detail viscous-turbulent angular momentum transport and associated effects, in particular magnetic field dragging by the accretion flow.

The viscous transport has to be of a turbulent nature, for otherwise accretion time scales would be unreallistically long. This idea then calls for an identification of the instability in which turbulence originates, for a quantitative estimate of its saturated state and associated effective viscosity, as well as of other transport effects produced by the turbulence, such as heat conduction, magnetic field diffusion and dynamo effect.

We develop here the idea that the level of turbulence in an accretion disk is regulated by the macro-scale parameters of the accreting system and is predictable from rather general self-consistency requirements in terms of them.

A natural consequence is that viscous-turbulent accretion disks are, by nature, very low magnetic Reynolds number systems. We discuss the implications for field dragging and for the interaction of disks with the magnetic field of the accreting star. We find that the latter is severely modified by the interaction and almost completely expelled from any region of the disk which is not adjacent to the magnetopause boundary, if any.

2. What determines the level of turbulence?

The level of turbulence developed in an accretion disk is not independent of the global dynamics of the accreting system. If the turbulence level is too large, the mass transfer becomes too effective, and the disk locally empties, while if it is too small, accreted matter piles up. The disk thickness would then change. Since the turbulence development depends on it, were it only because of limitations imposed to the vertical size of turbulent eddies, a feedback effect should result. In a stationary state, the turbulence level should then adjust in such a way to allow precisely that rate of mass transfer that it is imposed to the system by global processes taking place in it and in the donor star. This feedbaack process is summarized in the flow-chart shown in Figure 1.

2.1. SELF-CONSISTENT TURBULENCE IN GENERAL

Let us begin to follow the feedback loop in Figure 1 by assuming a certain turbulence level, with associated turbulent viscosity. The cascading process drives to dissipation a certain power per gram, ϵ, which depends on this turbulence level. This heating rate enters as an element of local thermal balance. If $\dot{M}$, or the local mass surface density, Σ, is known, the disk thickness results from it and from local vertical force balance. This geometrical constraint controls some aspects of the local dynamics of turbulent perturbations. For example, it may determine the so-called "injection scale" of the turbulence. Some understanding of the sources and development of the turbulence are necessary to take these geometrical effects into account. When the injection scale and the rate of energy transfer ϵ in the turbulent cascade are both known, turbulence physics would give the associated effective viscosity. This effective viscosity should be the one initially assumed. So, it should be possible to write a self-consistency equation to determine the actual turbulence level at radius r in terms of the imposed mass accretion rate, or, equivalently, of the mass surface density.

2.2. SELF-CONSISTENT KOLMOGOROFF TURBULENCE

Let us illustrate these ideas by first adopting the naïve view that the turbulence in the disk is, near any given radius, developed into an isotropic Kolmogoroff spectrum, with injection scale equal to a fraction $1/f$ of the local disk thickness.

We start the feedback loop in Figure 1 on the left hand side, assuming we know the effective viscosity $\nu_\star$. The heating rate per unit mass, assuming keplerian rotation, is

$$\epsilon = \frac{9}{4}\nu_\star \Omega^2. \tag{1}$$

Balancing viscous heating with optically thick radiative cooling, we obtain the disk temperature at radius r :

$$\sigma_B T^4 = \frac{9}{8}\nu_\star \Sigma \frac{GM_\star}{r^3}. \tag{2}$$

In this equation, $M_\star$ is the mass of the accreting star, G the gravitational constant and σ_B the Stefan-Boltzmann constant. Vertical force balance gives the thickness of the disk if the surfacic mass density is assumed to be known, which is equivalent to assume $\dot{M}$ to be known. We consider the disk as vertically isothermal both for simplicity and because we expect the turbulence to be a very efficient heat conducting agent. Then,

$$h^2 = \frac{2k_B T/m}{GM_\star/r^3}. \tag{3}$$

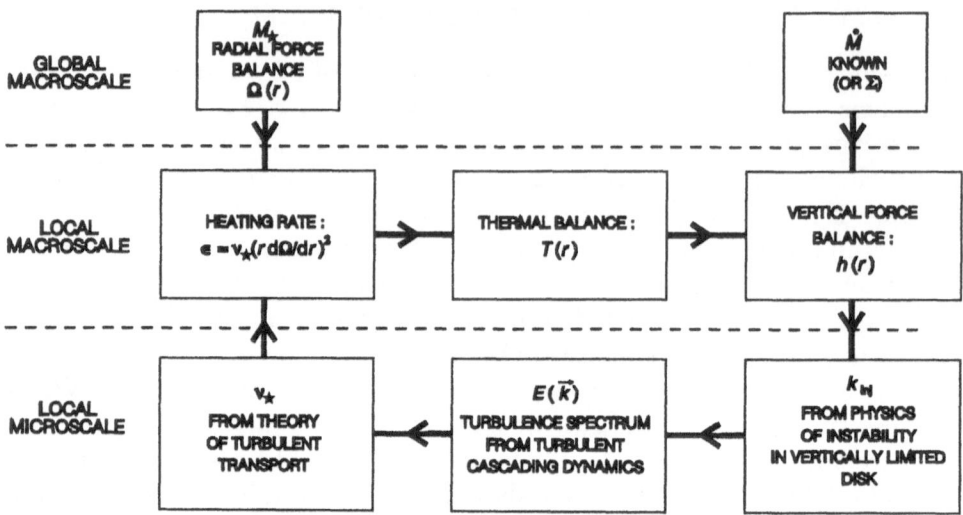

Figure 1. The turbulence level in a thin accretion disk is established as a result of a self-consistency process controlled by the mass of the accreting star, $M_\star$, and the accretion rate, $\dot{M}$. There is a feed-back from the effective viscosity, $\nu_\star$, onto the local disk thickness, h, which partly controls the size of turbulent eddies, k_{inj}^{-1}, and so $\nu_\star$ itself . The figure illustrates this feed-back with arrows indicating the sense of the causal relations between various physical effects. The self-consistency process puts turbulent phenomena at the local micro-scale in relation to phenomena at the local or global macro-scale.

Define a reference length R_0 and a reference mass M_0 and dimensionless variables

$$x = \frac{r}{R_0} \qquad \nu = \frac{\nu_\star}{R_0^2 \sqrt{GM_\star/R_0^3}} \qquad \sigma = \frac{\Sigma R_0^2}{M_0}$$

$$T_0 = \left(\frac{9}{8\sigma_B} M_0 \left(\frac{GM_\star}{R_0^3} \right)^{3/2} \right)^{1/4}. \tag{4}$$

Using equation(2) in equation (3), we obtain

$$h^2 = R_0^2 \left(\frac{2k_B T_0/m}{GM_\star/R_0} \right) x^{9/4} \nu^{1/4} \sigma^{1/4}. \tag{5}$$

We now turn to the lower part of Figure 1, and describe how our assumed understanding of the turbulence provides a "bootstrap" equation for the effective viscosity, $\nu_\star$. Let us adopt a simple representation of effective viscosity in terms of the parameters describing the spectrum, the injection scale, $2\pi/k_{inj}$, and the r.m.s velocity fluctuation, $< v^2 >$:

$$\nu_\star = \frac{1}{3}k_{inj}^{-1}\sqrt{< v^2 >}. \qquad (6)$$

More sophisticated expressions for effective viscosity in terms of the characteristics of the turbulent spectrum could, and should, be used instead, but this simple one is enough to illustrate the self-consistency idea. The root mean square velocity fluctuation is given in terms of the energy spectrum $E(k)$ by:

$$< v^2 > = 2 \int_{k_{inj}}^{\infty} E(k)dk. \qquad (7)$$

A Kolmogoroff spectrum with a rate of energy transfer in the cascade ϵ, can be written as:

$$E(k) = C\epsilon^{2/3}k^{-5/3}, \qquad (8)$$

where C is the Kolmogoroff constant. Using equation (8) to calculate the effective viscosity according to equations (6) and (7), we find:

$$\nu_\star = \sqrt{\frac{2C}{27}}\epsilon^{1/3}k_{inj}^{-4/3} = \lambda\epsilon^{1/3}k_{inj}^{-4/3}. \qquad (9)$$

Had we used a more sophisticated relation than equation (1), a slightly different value of λ would have resulted. If we now take into account the fact that the injection scale equals a fraction $1/f$ of the scale height, so that

$$k_{inj} = \frac{\pi f}{h}, \qquad (10)$$

we obtain, using equation (1) (the quantity ϵ is the same there as it is in equation (8))

$$\nu_\star = \frac{3\lambda^{3/2}}{2\pi^2 f^2}h^2\sqrt{\frac{GM_\star}{r^3}}. \qquad (11)$$

Note that this is a Shakura-Sunyaev type of formula. Since the scale height h is half the disk thickness H, the α factor is

$$\alpha_{ss} = \frac{3\lambda^{3/2}}{8\pi^2 f^2}. \qquad (12)$$

Finally substituting expression (5) of the scale height h, and using the variables defined in equation (4), the relation (11) gives the required self-consistency equation:

$$\nu = \frac{3\lambda^{3/2}}{2\pi^2 f^2}\left(\frac{2k_B T_0/m}{GM_\star/R_0}\right) x^{3/4}\nu^{1/4}\sigma^{1/4} \qquad (13)$$

which can be solved for the effective viscosity in terms mass surface density.

3. Turbulence from the magnetic shearing instability

One particular possible source of turbulence in an accretion disk is the magnetic shearing instability (Balbus and Hawley, 1991). For instability, the magnetic tension should not exceed the destabilizing excess centrifugal force. For a vertical seed field B_0, and axisymmetric perturbations, instability occurs only if the vertical wave number is less than $k_\star = (3\Omega^2/v_{A0}^2)^{1/2}$. Since this wavelength has to fit in the disk thickness this requires that $k_\star < \pi/h$, which can be translated into $v_{A0} < c_S$. If the plasma β in the disk becomes smaller than unity, the instability is quenched.

3.1. FULLY DEVELOPED MAGNETIC SHEARING INSTABILITY

To evaluate the effective viscosity actually present in the disk, it is useful to precise which instability feeds the turbulence, in as much as this may control some of the parameters of the turbulent spectrum which is to develop from it. Numerical studies of the three-dimensional development of this instability have been conducted by Gammie, Hawley and Balbus (1995), and by Brandenburg et al.(1995). The results obtained by the latter authors are also reported at this meeting. Both calculations were carried in a numerical simulation box of finite size, and the boundary conditions, which are different in each of them, have been slightly idealized as compared to the real situation.

Nevertheless, some general features can be recognized in these simulations. On the one hand, the turbulence is observed to develop in the meridional plane up to a scale of the order of the thickness of the simulation box. Because of the shear in the initial flow, the turbulence spectrum develops a strong anisotropy, being less extended in wave-vector space in the azimuthal direction. On each individual axis of the wave-vector space (radial, r, vertical, z and azimuthal, θ) the spectrum is Kolmogoroff-like, i.e with a slope $-11/3$.

3.2. ACCOUNTING FOR TURBULENCE ANISOTROPY

To incorporate these features of the simulation spectrum in our modeling, we have chosen to represent the turbulence spectrum as a squashed spectrum, in which the spectral energy density in wave vector space is of the form:

$$W(\vec{k}) = W(K), \tag{14}$$

where the scalar quantity K is related to the actual wave vector $\vec{k}$ by

$$K = (k_r^2 + k_z^2 + \frac{k_\theta^2}{q^2})^{1/2}, \tag{15}$$

and q is an anisotropy parameter which measures the extent to which the spectrum is compressed in the azimuthal direction in wave vector space. Surfaces of constant spectral energy density in $\vec{k}$-space are then ellipsoids. ¿From simulations, the parameter q appears to be close to the ratio of the poloidal size of the computation box to its toroidal size. In nature, we expect this ratio to become approximately h/r. Within the framework of this modelization, we can repeat Kolmogoroff's classical dimensional argument. In the inertial range, the spectrum depends only on a quantity K having the dimension of a wavenumber, on a rate of energy transfer in the direct cascade, ϵ, and on the dimensionless anisotropy parameter q, but not on the injection wave number. It can then be shown that this spectrum must necessarily be of the form:

$$W(\vec{k}) = C(q)\epsilon^{2/3}K^{-11/3}. \tag{16}$$

The associated effective viscosity is

$$\nu_\star = \frac{1}{3}l_{corr\perp} < \delta v_\perp^2 >^{1/2}. \tag{17}$$

Because the transport of interest is in the radial direction, the correlation length in the poloidal plane is the relevant one. The characteristic transport velocity is the poloidal velocity. The model-spectrum has a poloidal correlation length equal to

$$l_{corr\perp} = \frac{2\pi^2}{5k_{inj}}. \tag{18}$$

and the rms velocity fluctuation can similarly be calculated from equation (16). We obtain

$$\nu_\star = \frac{2\pi^2}{15}\sqrt{6\pi q C(q)}\epsilon^{1/3}k_{inj}^{-4/3}. \tag{19}$$

The number $C(q)$ is not precisely known, except when $q = 1$, in which case it reduces to the Kolmogoroff constant. Here its value for very small q is needed. It can be estimated by the argument that the nonlinear energy transfer time be, for eddies of any size, of order of their turnover time. It has been checked that no dynamical time scale bearing on eddy evolution could, in the present situation, be much shorter than the turnover time. This leads to the conclusion that

$$4\pi q C(q) \approx 1. \tag{20}$$

Adopting as a value of the parameter f in equation (10) $f = 2$, we obtain

$$\nu_\star = \frac{2\pi^2}{15}\sqrt{\frac{3}{2}}\epsilon^{1/3}\frac{h}{2\pi}^{4/3}, \qquad (21)$$

which can be translated into a value of the Shakura-Sunyaev parameter

$$\alpha_{ss} = \frac{3}{16\pi^2}\left(\frac{2\pi^2}{15}\sqrt{\frac{3}{2}}\right)^{3/2} = 0.04. \qquad (22)$$

This value is still subject to uncertainties, both in the actual value of λ and in that of $C(0)$. Note that this relation is of the same form as the one calculated for the case of isotropic Kolmogoroff turbulence, but for the value of the parameter λ which is different in both cases.

3.3. INHIBITION OF THE MAGNETIC SHEARING INSTABILITY AND CONSEQUENCES

When the vertical component of the seed-field becomes so large that $v_{A0}^2 \geq c_S^2$, the magnetic shearing instability is quenched. This happens at the point where the β of the plasma calculated with the seed-field becomes of order unity. The magnetopause of an accreting star, if any, is located where the magnetic energy density becomes comparable to the keplerian kinetic energy density. At the magnetopause, then, $v_A^2 \approx \Omega^2 r^2$. However, for a thin disk, the Mach number of the keplerian velocity is large. This implies that at the magnetopause, the β parameter of the plasma should not be much larger unity, or else the thin disk approximation would have become invalid before this boundary. Therefore a region should exist in the disk between the radius at which $\beta = 1$ and the magnetopause where the Balbus-Hawley instability cannot develop in the linear approximation. The size of this region is unknown, because the flux distribution in the disk could differ from the dipolar one, as discussed in section 5.

4. Magnetic field dragging by turbulent accretion disks

We have used the expression derived above for the turbulent viscosity to calculate the spreading of mass injected in the accretion disk at a given radius and at a constant rate (Bardou, Heyvaerts and Priest, 1996). The inwards flowing matter eventually reaches a steady state when less and less matter is diverted away from the injection point. We also studied magnetic field dragging by this flow. Let A be the magnetic flux function, which, for passive field dragging, obeys the advection-diffusion equation:

$$\frac{\partial A}{\partial t} + \vec{v}_{pol}.\vec{\nabla}A = \eta_\star\left(\frac{\partial^2 A}{\partial z^2} + r\frac{\partial}{\partial r}\frac{1}{r}\frac{\partial A}{\partial r}\right). \qquad (23)$$

Since the disk is assumed to be thin, it is desirable to transform it into one that describes flux distribution in the plane of the disk only. This can be done by Taylor-expanding the magnetic flux function in the vicinity of the equator. The term which contains derivatives with respect to z in equation (23) is in this approximation expressed in terms of the disk thickness and of the partial derivative $\partial A/\partial z$ at the upper disk surface. The latter can be expressed in terms of the flux distribution in the equatorial plane by solving a potential problem, assuming the field external to the disk to be potential. This assumption would be acceptable for open fields when the medium outside of the disk is very tenuous. If the field lines that span the disk were also anchored in the accreting star, the assumption of a potential external field would, on the contrary, be unacceptable. We come back on this point in section (5). Let us separate the flux function into an homogeneous field part and a part produced by electric currents flowing in the disk:

$$A(r,z) = \frac{B_0 r^2}{2} + a(r,z). \tag{24}$$

For a potential outer field, we obtain, after some maths, the following integro-differential equation for flux distribution in the disk's plane:

$$\frac{\partial a}{\partial t} + r v_r B_0 + v_r \frac{\partial a}{\partial r} = \frac{\eta_\star}{h(r)} P \int_0^\infty dx \frac{a(x,0) - a(r,0)}{\pi(r-x)^2} - \frac{\eta_\star a(r,0)}{\pi r h(r)}. \tag{24}$$

The integral on the r.h.s. has to be understood as a principal part distribution, P. An order of magnitude estimate of the ratio R_m of the advection term $v_r \partial a/\partial r$ to the largest diffusive term, i.e. the one which is represented by the principal part, gives

$$R_m \approx \frac{\nu_\star}{\eta_\star} \frac{h}{r} \ll 1. \tag{25}$$

Indeed, if $\nu_\star$ and $\eta_\star$ are both due to turbulence, we expect their ratio, the magnetic Prandtl number, to be of order unity. We conclude that a self-consistently turbulent disk which accretes by viscous-turbulent transport is a very low magnetic Reynolds number system. A consequence is that the external uniform field is barely perturbed, and that the matter penetrates through the field rather easily. Also, the resulting magnetic structure is unfavourable to the emission of cold centrifugally-driven winds, since the Blandford-Payne criterium is far from being satisfied.

Let us stress again that these conclusions rest on the assumption that turbulence has been able to develop eddies with a size comparable to the local disk thickness. This seems to be possible when the magnetic shearing instability is able to operate, but may be hindered if the seed magnetic field is too large. The thin disk approximation has been made. Note that our conclusion that centrifugally-driven cold winds cannot be emitted from such disks does not imply that thermally-driven winds cannot be. However such winds would still have only little lever-arm, because the uniform field seems to suffer only little deformation, and the torque exerted by such a wind may be comparatively weak. Such an idea also requires examination of the processes by which an open disk's corona could be heated.

5. Interaction between a thin turbulent accretion disk and the magnetic field of the accreting star

5.1. FIELD REJECTION BY THE DISK: PHYSICAL MECHANISM

The magnetic Reynolds number of turbulent accretion disks is small when estimated from the radial accretion velocity, but it is large when estimated from the azimuthal quasi-keplerian one. One may then wonder what happens when a magnetic field anchored in the accreting star interacts with a turbulent disk, for it seems that it should be frozen into azimuthal motions but unfrozen into radial ones.

If the magnetic Reynolds number of radial motions were also large, the answer to this question would be known from earlier work on the evolution of force free fields driven by boundary motions, which hqs been carried out mainly for understanding solar flares (Aly(1984, 1991, 1994), Heyvaerts et al.(1982), Lynden-Bell and Boily (1994)). When the shear between magnetically conjugated foot points of the magnetic field lines becomes large, the magnetic configuration asymptotically opens. When close to opening, it develops current sheets which become subject to reconnection instabilities, developing into a flare-like process. Persistent shearing of magnetic foot points in the disk by the keplerian motion should then trigger an intermittent sequence of field opening and flaring reconnection events (Van Ballegooijen(1994), Lovelace et al. (1995)). The expansion of the force-free structure develops tension forces in the boundary, where magnetic foot points are anchored. If instead of being perfectly conducting the boundary is resistive, these tensions are free to relax. This leads to an outwards expansion of the structure, because Lyndell-Bell and Boily (1994) have shown that in this geometry these forces pull in the outward radial direction. The magnetic structure is then expected to expand outwards if field diffusivity is taken into account. This expansion will ultimately be limited by diffusivity itself and by the finite disk extent.

5.2. FIELD INFLATION: A MORE FORMAL DERIVATION

The magnetic structure of the accretion disk corona, the region threaded by the magnetic field connecting the disk to the star, can be modeled as a force-free field. Axial symmetry is assumed. The poloidal field can be written as

$$\vec{B}_P = \vec{\nabla} \times \left(\frac{a(r,\theta)}{r\sin\theta} \ \vec{e}_\phi \right) \tag{26}$$

The flux function a depends on the poloidal spherical coordinates r and θ. Surfaces of constant a are magnetic surfaces. Force-free fields are such that the electric current flows on magnetic surfaces, so from Ampère's law it follows that the quantity $I = r\sin\theta \, B_\phi$ is a function of a. The flux function a obeys the Grad-Shafranoff equation

$$\frac{\partial}{\partial r}\left(\frac{1}{\sin\theta}\frac{\partial a}{\partial r} \right) + \frac{\partial}{\partial \theta}\left(\frac{1}{r^2 \sin\theta}\frac{\partial a}{\partial \theta} \right) = -\frac{II'}{\sin\theta} \tag{27}$$

The quantity I is proportional to the electric current enclosed between the axis and the magnetic surface a. It can be calculated from an analysis of the electric circuit between the star and the disk. This analysis consists in using Ohm's law in the disk, while assuming the star and the coronal medium to be perfectly conducting. The star's rotation induces a potential drop between magnetic surfaces which images at their foot-point in the disk when the plasma in the disk's corona is perfectly conducting. Ohm's law then gives the radial electric current surface density on the disk as a function of the thickness-integrated conductibility Σ_D and of the flux distribution on the disk. Altogether, the current surface density, $\vec{i}$, is found to depend on the distance R from the axis as:

$$i_R = \Sigma_D \frac{da}{dR}\left(\Omega_\star - \Omega(R) \right) \tag{28}$$

This surface current is supported by a jump of the azimuthal component of the coronal magnetic field between upper and lower disk's surfaces. So, equation (28) gives the value of B_ϕ at the disk's surface, and $I(a)$ is finally found to be given by

$$I(a) = \frac{\mu_0}{2}\Sigma_D \frac{(\Omega_\star - \Omega(R))}{dlog(R(a))/da} \tag{29}$$

where $R(a)$ is the distance to the axis of the intersection of magnetic surface a with the disk. If Σ_D were to vanish, no electric current would flow in the system, and the coronal field would be potential. Our self-consistent turbulent model indicates that $\mu_0\Sigma_D R\Omega(R)$ actually scales as $R^{-1/8}$ in the disk and of course vanishes outside of it. It then scales as $R^{-1/8}f(R)$, where $f(R)$ is a function which decreases from unity to zero near the outer edge of the disk, when R increases from distances interior to distances exterior to it. To sum up, $I^2(a)$ is a function of $R(a)$:

$$I^2(a) = C \quad i^2(R(a)) \tag{30}$$

where C is a constant which can be determined from equation (29). It is interesting to discuss the shape assumed by the coronal magnetic structure for different disk conductivities, assuming equation (30) to remain valid, C being replaced by a control parameter λ, varying between zero (potential field) and C (actual field). Since the constant C is numerically quite large, the limit of large control parameters can be considered. Using this modified form of equation (30) in equation (27), multiplying it by $\partial a/\partial r$, and integrating over r and θ on that part of the poloidal plane which is exterior to the magnetosphere and to the star, one obtains an equation that can be turned into the following inequality (more details can be found in Bardou and Heyvaerts (1995)):

$$\lambda \int_0^\pi \frac{d\theta_0}{\sin\theta_0} \left(1 - \frac{\Omega_\star}{\Omega(R)}\right)^2 \left(\frac{da}{dR}\right)^2 \frac{f^2(R)}{R^{1/4}} \leq \int_0^\pi d\theta_0 \, r_m^2 B_r^2(r_m, \theta_0) \tag{31}$$

In this formula, θ_0 is the colatitude of a point on the star's magnetosphere surface, defined by r_m. In the polar regions, we consider the star's surface to be the natural extension of the magnetosphere boundary. Because flux is frozen in the star, there is a definite value of the flux function a associated with each θ_0. As earlier, $R(a)$ is the distance from disk's center to where the magnetic surface a meets the plane of the disk. So, indirectly, R is a function of θ_0. The function $R(a)$ describes the flux distribution in the disk that eventually emerges from this dissipative interaction between the field and the keplerian differential rotation. $R(a)$ actually is the unknown that we are trying to solve for. The right hand side of inequality (31) is known from flux distribution on the star's surface. Call it J. Changing from variable θ_0 to a in the integral on the left hand side of inequality (31) we find that the function $R(a)$ is constrained by the inequality:

$$\int_0^{a_m} \psi(a)da \left(\frac{\Omega(R(a)) - \Omega_\star}{\Omega(R(a))}\right)^2 \frac{1}{R^{1/4}(a)R'^2(a)} \leq \frac{J}{\lambda} \tag{32}$$

where $\psi(a)$ is a known function. When the control parameter grows larger and lsarger, the right-hand-side of this inequality approaches zero. This can be shown to imply that dR/da approaches infinity when λ grows, or, stated anotherway, that the z-component of the magnetic field becomes smaller and smaller at any given place of the disk's surface as λ grows larger. Bardou and Heyvaerts (1995) numerically checked this behaviour of the magnetic structure with increasing disk conductibility. They also provide an approximate analytic solution for large λ in the manner of Lynden-Bell and Boily (1994), looking for a self-similar solution of the form $a = r^{-p}g(\cos\theta)$. Boundary conditions at the disk have been idealized by assuming that, because of magnetic diffusivity, the disk can sustain no jump in B_r between its lower and upper faces. This solution also shows that the magnetic field anchored on the star tends to be repelled out of the disk by turbulent diffusion. The flux remaining in the disk in stationary state is but a very small fraction of the flux that would thread it if the magnetic structure were potential and dipolar. Our conclusion is then that non-dynamically-significant coronal magnetic fields anchored in the star interacting with a keplerian turbulent accretion disk are largely expelled radially outwards of it. The consequences of this process for rotational coupling between star and disk may be very significant. The expelled fields should form a potential field structure wrapping around the accretion disk, possibly interacting with the field of the donor star. This change in the vertical component of the disk's magnetic field would react on the development of the magnetic shearing instability, which should be more easily triggered in the innermost regions of the disk than if the field were dipolarly distributed. What happens if some central part of the disk does not develop turbulence remains to be studied, as is the structure of the intermediate region between the magnetopause and the more remote regions of the disk where the magnetic field is not dynamically significant.

References

1. Aly, J.J., (1984), On some properties of force-free magnetic fields in infinite regions of space , *Astrophys.J.*, **283**, pp. 349–362
2. Aly, J.J., (1991), How much energy can be stored in three-dimensional force-free magnetic fields? , *Astrophys. J.*, **375**, pp. L61–L64
3. Aly, J.J., (1994), Asymptotic formation of a current sheet in an indefinitely sheared force-free field: an analytical example , *Astron. Astrophys.*, **288**, pp. 1012–1020
4. Balbus, S.A. and Hawley, J.F. (1991), A powerful local shear instability in weakly magnetized disks, *Astrophys.J.*, **376** pp. 214–222
5. Bardou, A., and Heyvaerts, J., (1995), Interaction of a stellar magnetic field with a turbulent accretion disk , *Astron. Astrophys.*, in press
6. Bardou, A., Heyvaerts, J. and Priest, E.R., (1996), Magnetic field diffusion in self-consistently turbulent accretion disks, submitted to *Astrophys. J.*
7. Blandford, R. D., and Payne, D.G., (1982), Hydromagnetic flows from accretion discs and the production of radio jets, *Month. Not. Roy. Astr. Soc.*, **199**,pp. 883–903
8. Brandenburg, A., Nordlund, A., Stein, R.F. and Torkelsson, U., 1995, Dynamo-generated turbulence and large-scale magnetic fields in keplerian shear flow, *Astrophys.J.*, **446**, pp. 741–754
9. Ferreira, J., and Pelletier, G., 1995, Magnetized accretion-ejection structures III,

Astron. Astrophys., **295**, pp. 807–832

10. Hawley, J.F., Gammie, C.F. and Balbus, S.A., (1995) *Astrophys.J.*, Local three dimensional magnetohydrodynamic simulation of accretion disks **440**, pp. 742–763

11. Heyvaerts, J., Lasry, J.M., Schatzman, M. and Witomsky, P., 1982 Blowing up of two-dimensional magnetohydrostatic equilibria by an increase of electric current or pressure, *Astron. Astrophys.*, **111**, pp. 104–112

12. Heyvaerts, J. and Priest, E.R. (1992) A self-consistent turbulent model for solar coronal heating , *Astrophys.J.*, **390**, pp. 297–308

13. Lovelace, R.V.E. , Romanova, M.M., and Bisnovatyi-Kogan, G.S., (1995), Spin up/ spin down of magnetized stars with accretion discs and outflows, *Month. Not. Roy. Astr. Soc.*, **275**, pp. 244–254

14. Lubow, S.H., Papaloizou, J.C.B., and Pringle, J.E., (1994), Magnetic field dragging in accretion disks, *Month. Not. Roy. Astr. Soc.*, **267**, pp. 235–240

15. Van Ballegooijen, A.A., (1994), *Space Sci.Rev.*, Energy release in stellar magnetospheres,*Space Sci.Rev.*, **68**, pp. 29–38

MAGNETIC RECONNECTION IN ACCRETION DISK CORONAE

It's not only working on the Sun!

H. LESCH
*Institut für Astronomie und Astrophysik der Universität München,
Scheinerstraße 1, 81679 München, Germany*

1. Introduction

There is now general consensus that active galactic nuclei (AGN) consist of an accreting supermassive black hole (BH) (up to $10^{10} M_\odot$) surrounded by an accretion disk. The disk consists of thermal plasma which is heated up via viscous accretion to temperatures of about $5 \cdot 10^4 K$. The existence of thermal plasma in this temperature range has been established beyond any doubt by optical observations which exhibit broad line width up to $3 \cdot 10^4 \, kms^{-1}$ and UV observations which clearly show spectra compatible with the thermal spectrum of an accretion disk.

AGN also show extremely energetic outflows extending even to scales beyond the outer edge of the galaxy in the form of strongly collimated radio jets (e.g. Perley 1989). The emerging components clearly exhibit show superluminal motion, i.e. they move with relativistic speed, with bulk Lorentz factors $\Gamma \leq 10$ (e.g. Blandford 1990). There is substantial evidence that magnetic forces are involved in the jet driving mechanism (Blandford 1990; Camenzind; Heyvaerts; Pelletier, this volume) and that the magnetic fields will also provide the collimation of the flow, since huge currents are involved in the jet flows (Benford 1978; Lesch et al. 1989; Appl, this volume).

We consider as a possible scenario a rotating black hole surrounded by a magnetized accretion disk (Camenzind, this volume). As the disk's plasma accretes in the potential of the central mass, magnetic field lines are convected inwards, amplified and finally deposited on the horizon of the black hole. A dynamo in the disk may be responsible for the maintenance and amplification of the magnetic field in the disk (for the details of dynamo action in accretion disks see Khanna, this volume). The interplay of differential rotation and convective turbulence (ascending (descending) turbulent cells) amplifies the magnetic field to a strength whereas processes like magnetic buoyancy limit further amplification (Stella & Rosner 1984).

K. C. Tsinganos (ed.), Solar and Astrophysical Magnetohydrodynamic Flows, 673–682.
© 1996 *Kluwer Academic Publishers.*

674

ACCRETION DISK MHD-PROCESSES

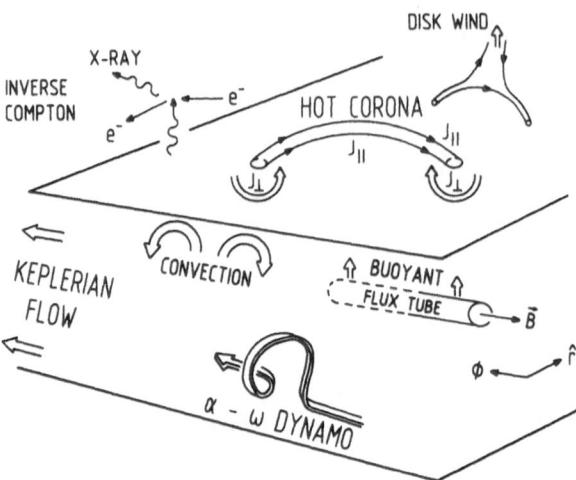

Figure 1. Hydromagnetic phenomena in thin accretion disks and the buildup of a corona. Whenever the magnetic field in the disk (generated by a dynamo) is in equipartition with the amplifying forces (differential shear and convective motions), magnetic flux tubes, convective turbulent cells, etc. will ascend and feed magnetic energy into the corona.

An inevitable consequence of convective turbulence and magnetic buoyancy will be to form a hot, magnetically active corona by transferring magnetic energy into the coronal plasma (see Figure 1).

Whenever the magnetic field in the disk (generated by a dynamo) is in equipartition with the amplifying forces (differential shear and turbulent motions) magnetic flux tubes, convective turbulent cells, etc. will ascend, which by driving parallel currents feed magnetic energy into the corona. The closure of the current circuit will be provided by flux tubes bending back to the disk surface. Moreover, in order for a current to flow ($\nabla \cdot j = 0$, i.e $\nabla \cdot j_{\parallel} = -\nabla \cdot j_{\perp}$), a resistor (load) within the current circuit is necessary, which may be provided by a perpendicular current. This leads to localized dissipative regions (current sheets) where magnetic energy is transferred into particle acceleration via magnetic reconnection. Similar phenomena are known from the sun, where particle acceleration in the first phase of a flare (the injection phase) is associated with strong magnetic dissipation in small current sheets (Tsuneta, this volume).

The necessity of very rapid and efficient particles acceleration is evident from recent observations in the radio and optical (see Wagner and Witzel 1995 and references therein) which reveal variability in flat spectrum sources on time scales shorter than one day. For example the BL Lac object 0954+65 exhibits an appearant luminosity fluctuation of $10^{46} ergs^{-2}$ during the steepest increase within 6 hours (assuming isotropic emission).

This implies brightness temperatures in the GHz radio range of the order of 10^{18}K, exceeding the Compton limit for an incoherent synchrotron source of 10^{12}K by six orders of magnitude. Since in a relativistically moving plasma the apparent brightness temperature can be amplified by factor Γ^3, Begelmann et al. (1994) proposed that the involved bulk Lorentz factors have to be of the order of 30-100, in order to keep the hypothesis of incoherently radiating sources.

However, observations do not reveal such high bulk Lorentz factors, therefore we propose that coherent emission of anisotropic relativistic electron beams (REB) is responsible for the very intense radiation. Coherent radiation can provide the small rising time for the variable emission, its enormous brightness temperatures and its high luminosities in the radio range without invoking relativistic beaming effects (Lesch 1991; Weatherhall and Benford 1991; Lesch and Pohl 1992). We propose that the REB are produced via **magnetic reconnection** in the immediate surrounding of an accreting black hole, i.e. in the coronae of accretion disks or in the magnetized jets.

We note, that isotropic beam-like distributions reveal themselves via an exceptional optically thin synchrotron spectrum, $S_\nu \propto \nu^{0.3}$. Such a spectrum has been observed in the centre of the Milky Way and in M81 (Duschl and Lesch 1994; Reuter and Lesch 1995).

2. Magnetic reconnection and particle acceleration

Reconnection is an intrinsic property of a magnetized plasma with shearing motions. The encounter of magnetic field components with different polarity correspond to parallel electric currents, which attract each other. In that sense not only shearing motions, but any kind of plasma motion can trigger reconnection (Priest, this volume).

If the plasma pressure between the opposite fields $\pm B$ is insufficient to keep the fields apart (e.g. by pushing the different flux systems together), the plasma squeezes from between and the two fields approach each other. The field gradient steepens and eventually the current density $\frac{c}{4\pi}\nabla \times B$ becomes so large that there is strong dissipation. The problem of reconnection is to know how the dissipation of the currents with the density $j = en_e v_D$ is provided (e is the charge, n_e is the electron number density in cm^{-3} and v_D denotes the drift velocity of the electrons). We start with the fact that reconnection corresponds to the dissipation of a current. The dissipation rate Q is

$$Q = \frac{j^2}{\sigma}. \tag{1}$$

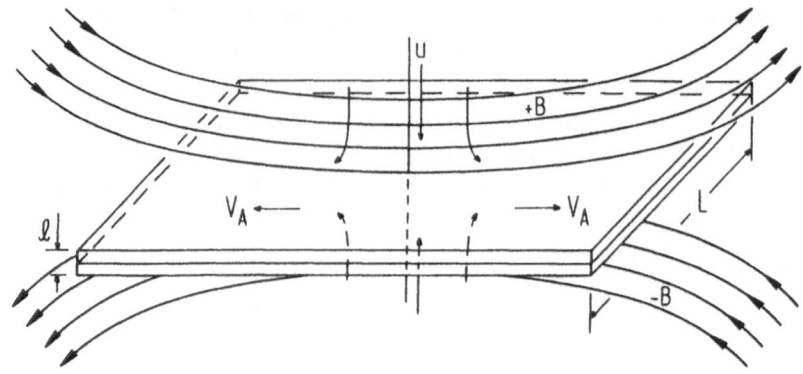

Figure 2. A perspective view of two attracting filaments, which could have their origin in unbalanced forces in a magneto-active accretion disk. When the magnetic field gradient is strong enough, rapid dissipation sets in and an electric field E is induced which accelerates the particles. L denotes the characteristic length of the reconnection zone.

If the current density is constant, enhanced dissipation is equivalent to the reduction of the classical electrical conductivity σ provided by Coulomb collisions $\sigma = 6.3 \cdot 10^{10}\, T^{3/2}$ which means that dissipation is very weak. This can be understood as follows: when the conductivity is high, the magnetic field lines are "frozen into" the plasma. Field lines embedded within a volume element of plasma are carried along by the moving plasma itself. Any two plasma elements that are threaded by the same lines of force will remain threaded in this way, i.e. two plasma elements that are threaded by different lines of force can never be threaded together. Thus a high conductivity prevents the field lines from merging. If reconnection occurs, it can happen only in those plasma regions where the electrical conductivity is drastically reduced below its classical Coulomb value.

The reduction of the electrical conductivity is provided by plasma instabilities. If the plasma is locally unstable, microscopic instabilities will excite waves which enhance the collision frequency by wave-particle interactions; they lead to *anomalous conductivity*. The effective σ becomes anomalously low, greatly enhancing the dissipation and reconnection of the lines of force. Then the concept of "frozen in" field lines is no longer valid and the plasma moves relative to the field lines. When this occurs, strong electric fields along the reconnection length L are induced. It is the dissipation of the electric field E which transforms the current energy into particle energy.

Schindler et al. (1988, 1991) have shown that in three dimensions the deviations from an ideal plasma state (infinitely high conductivity) in the reconnection zone, i.e. from the frozen-in concept, correspond to a magnetic field aligned electric potential $U = -\int E_{\parallel} ds$. This results was used to consider the acceleration of particles in cosmical plasmas and it was concluded that parallel electric fields associated with nonideal plasma flows might play an important role in cosmic acceleration (Schindler et al. 1991).

The actual value of $E_{\parallel}$ depends on the details of the microscopic instability, which is responsible for the deviations of the plasma from the ideal high conductivity state. It has been shown that in the case of the slow reconnection mode first proposed by Parker (e.g. Parker 1979) it is possible to get an estimate of the maximum Lorentz factor that particles can attain via parallel electric potentials (Lesch and Pohl 1992).

Starting from the assumption of a stationary Ohmic dissipation in a three-dimensional reconnection sheet with an area of side length L and a thickness l. The dissipation across the sheet lj^2/σ is just to devour the influx of magnetic energy $uB_0^2/8\pi$. u is the approaching velocity of the field lines. Conservation of fluid requires that the net magnetic field inflow balances the outflow $uL = v_A l$.

In terms of the magnetic Reynolds number $R_M = \frac{2Lv_A}{\eta}$, where $\eta = \frac{c^2}{4\pi\sigma}$ denotes the magnetic diffusivity, one reaches

$$l = \frac{2L}{\sqrt{R_M}}$$

and

$$u = \frac{2v_A}{\sqrt{R_M}}$$

The thickness of a reconnection sheet is also defined by Maxwell's equation $\nabla \times \mathbf{B} = \frac{4\pi}{c}\mathbf{j}$, which means $l \simeq \frac{cB}{4\pi j}$. The accelerating electric field is given by $E \simeq \frac{u}{c}B$. Now the whole problem is shifted to the microscopic level of description. The reduction of σ is provided by current driven plasma instabilities, whose excited plasma wave frequency depend on the drift velocity between electrons and ions (e.g. Mikhailovski 1974). We choose in the following the wave mode which has the lowest instability threshold where the drift velocity v_D is equal to the ion thermal velocity $v_{thi} = \sqrt{\frac{k_B T_i}{m_i}}$ the *lower hybrid (LH) wave* $\omega_{LH} \simeq 4.2 \cdot 10^5 B$. LH waves depend on the magnetic field strength, which means they are also suitable for 3D-reconnection, where the magnetic field is not zero in the reconnection zone. A fully developed LH-instability results in an effective collision frequency of the order of ω_{LH} (Sotnikov et al. 1978). Inserting u into the electric field and l into L, using $\nu_{coll} \simeq \omega_{LH}$ and with

$$\gamma m_e c^2 \simeq eEL$$

one gets the maximum Lorentz factor electrons can achieve in a three-dimensional reconnection zone, where the conductivity is determined via lower-hybrid waves (Lesch 1991; Lesch and Pohl 1992)

$$\gamma_{LH} \simeq 6 \cdot 10^4 \frac{B^2}{n_e} \frac{1}{\sqrt{\beta}},$$

where $\beta = \frac{n_e k_B T}{\frac{B^2}{8\pi}}$ denotes the ratio of thermal pressure to magnetic pressure. X-ray observations of AGN's indicate (e.g. Blandford 1990) thermal densities $n_e \simeq 10^6 cm^{-3}$ and temperatures $T \simeq 10^8 K$. Magnetic field strength in the global MHD models give (Heyvaerts, this volume) $B \simeq 10^2 G$. With an effiency of the reconnection process of 10% one gets

$$\gamma_{LH} \simeq 10^5 \left[\frac{B}{100\,G}\right]^3 \left[\frac{n_e}{10^6\,cm^{-3}}\right]^{-3/2} \left[\frac{T}{10^8\,K}\right]^{-1/2}$$

Since magnetic reconnection is a very effective acceleration mechanism the particles will flow along the field lines and the distribution function of the electrons will be quasi-monoenergetic and anisotropic in pitch angle (de Jager 1986). The achievable energy depends on the length of the reconnection zone, i.e. it is a linear accelerator. Since the final energy distribution will show an accumulation of particles either at the maximum energy or at the energy where radiation losses and acceleration mechanism exactly cancel, we obtain an energy distribution which exhibits pronounced low and high energy cutoffs, i.e. a relativistic electron beam (REB) (Lesch and Schlickeiser 1987).

During their propagation through the background plasma the electrons will be isotropized by magnetohydrodynamical turbulence (e.g. Schlickeiser 1984). Depending on the ratio of background electron density to beam density, such energy distributions reveal themselves in two ways: first, if they are aniostropic beams as very effective coherent radio emitters (Weatherhall and Benford 1991; Lesch and Pohl 1992), and second as isotropized distributions, they exhibit a very special incoherent synchrotron spectrum, where the radio flux S_ν increases with increasing frequency ν with the spectral index 0.33 (Duschl and Lesch 1994; Reuter and Lesch 1995).

3. Observational consequences of unstable relativistic electron beams

It is well known that a distribution function which exhibits a positive momentum gradient $\frac{\partial f}{\partial p}$ is unstable (e.g. Mikhailovski 1974 and references therein). In particular electron distribution functions with $\frac{\partial f}{\partial p} > 0$ excite the high frequency electrostatic Langmuir wave at frequencies almost equal to the electron plasma frequency $\omega_{\text{pe}} = \sqrt{\frac{4\pi e^2 n_e}{m_e}} \simeq 5.6 \cdot 10^4 \left[\frac{n_e}{1\,\text{cm}^{-3}}\right]^{1/2}$. A plasma in which a relativistic electron beam propagates exhibits a positive momentum gradient and is unstable (Kaplan & Tsytovich 1973). The instability responsible for wave excitation is called the *beam-plasma instability* and operates on the time scale (Lesch & Schlickeiser 1987) (Γ_{lin} denotes the linear growth rate of the instability)

$$t_{\text{lin}} \simeq \Gamma_{\text{lin}}^{-1} \simeq \frac{n_e}{n_b} \frac{\gamma}{\omega_{\text{pe}}}, \tag{2}$$

where n_b is the number density of the relativistic beam electrons.

The excited Langmuir waves slow down the beam through resonant wave particle interactions towards the plateau distribution function with $\frac{\partial f}{\partial p} \simeq 0$ in the resonant momentum intervall. For our purposes it is only necessary to estimate the field energy of the waves being generated during the relaxation process of the beam towards the plateau distribution function. A rough estimate is provided by the difference between the energy density of the relativistic electrons before and after the relaxation process (Lesch & Schlickeiser, 1987)

$$\Delta W \simeq \frac{1}{4} n_b \gamma m_e c^2. \tag{3}$$

In the final state of the relaxation process a large fraction ($\simeq 25\%$) of the energy of the relativistic electrons is transferred into plasma waves. ΔW presents the energy reservoir for any kind of coherent radiation process which transfers beam energy into radiative energy.

4. Coherent emission of relativistic electron beams

In the presence of anisotropic electron streams the naturally occuring beam instability excites plasma waves. The excited fluctuations saturate in a turbulent state of intense localized regions of electrostatic field $E_0^2/8\pi \simeq \Delta W$. Energetic electrons accelerate while transversing the soliton-like electric field structures. Their radiation is extremely broad band in frequency, a bremsstrahlung type spectrum in which the minimum impact parameter is the scale size of the soliton $D \simeq c/\omega_{\text{pe}}$. This wavelength will be considered as the typical length scale at which the electrostatic field associated with a Langmuir wave changes spatially. As the beam passes through the Langmuir turbulence, it encounters a spatially varying electrostatic field, causing it to oscillate and therefore radiate electromagnetic waves. Radiation extends many orders of magnitude higher in frequency than the plasma frequency $\omega_{\text{pe}} < \omega < \gamma^2 c/D$ and is relativistically beamed.

The underlying physical mechanism is scattering of the beam particles on intense localized electrostatic fields presented by strong Langmuir waves. This process arises only in the case of strong turbulence, where the soliton structures are build up. Since the intensity of the plasma waves depend on the beam density, the *coherent collisionless bremsstrahlung by electron beams* is only present in sources where the beam density is about 10^{-2} times the background density (Benford 1991). The coherence is due to a resonance feedback well known from free electron lasers - the trapping of particles in the potential field of a plasma wave. The number of particles which can be in coherence is $N = n_b D^3 \simeq n_b c^3/\omega_{pe}^3 = 10^{15}/\sqrt{n_e}$.

An upper limit to the electromagnetic wave power emitted by the Langmuir turbulence induced process is given by (Baker et al. 1988)

$$P_{coh} = \frac{\sqrt{\pi m_e e^2 c^4 n_e^3}}{650} \left(\gamma \frac{n_b}{n_e}\right)^4 f V_{plasma}$$

or

$$P_{coh} \simeq 3.6 \cdot 10^{-2} \, erg \, s^{-1} \left(\frac{n_e}{10^6 \, cm^{-3}}\right)^{3/2} \left(\gamma \frac{n_b}{n_e}\right)^4 f V_{plasma}.$$

P_{coh} is an estimated maximum power from a plasma volume $V_{plasma} = 4\pi/3 R^3$ whose fraction f is filled with coherent clumps of size $\propto D^3$.

For a galactic nucleus we obtain the luminosity

$$L_{coh} = 2 \cdot 10^{48} ergs^{-1} \left[\frac{n_b/n_e}{10^{-2}}\right]^4 \left[\frac{n_e}{10^6 \, cm^{-3}}\right]^{3/2} \left[\frac{\gamma}{10^4}\right]^4 \left[\frac{R}{10^{15} \, cm}\right]^3 \left[\frac{f}{10^{-5}}\right].$$

The emitted frequency is between 10 Mhz and $10^{16} Hz$. We expect to see intraday variability between radio and optical. However a different origin for the optical and higher frequencies will be discussed in the last Section.

We emphasize that the filling factor of the volume is only 10^{-5} and that even without relativistic boosting effects $L_{obs} \propto \Gamma^3 L_{coh}$ by the global flow, one easily reaches the observed high luminosities via coherent plasma emission. The brightness temperature of a coherent source is $T_{bc} = N T_{bic}$. Since maximum incoherent brightness temperature is $T_{bic} = \gamma m_e c^2/k_B \simeq 10^{12}$K we easily reach the observed values of 10^{18}K.

5. Discussion

We presented arguments that in the immediate surrounding of accreting black holes magnetic reconnection produces relativistic electron beams. This anisotropic energy distributions reveal themselves as efficient coherent emitters. This may explain the radio observations of intraday variability with extremely high brightness temperatures far beyond the Compton limit. The optical intraday variability exhibits much lower brightness tmeperatures which direct towards an incoherent process, for example synchrotron radiation. The typical emitted frequency is then $\nu \simeq \gamma^2 eB/(2\pi m_e c)$, which gives

$$B \simeq 1G \left[\frac{\gamma}{10^4}\right]^{-2} \left[\frac{\nu}{10^{15}Hz}\right]$$

The corresponding energy loss time is

$$t_{\text{syn}} = \frac{5 \cdot 10^8}{\gamma B^2} s \leq 1day \left[\frac{\gamma}{10^4}\right]^{-1} \left[\frac{B}{1G}\right]^{-2},$$

which is in accordance with the observed time scales.

The magnetic field strength of about 1 Gauß is in accordance with energy equipartition between magnetic fields and relativistic electrons $B \geq \sqrt{\gamma n_b m_e c^2}$. Since the beams are supposed to be produced by magnetic reconnection their energy cannot exceed the magnetic energy but should be slightly less.

The electrons emit only synchrotron radiation when they gyrate around the magnetic field, i.e. they posses an isotropic pitch angle distribution. This is the major difference to the coherent process, where the electrons move along the field lines and have an anisotropic pitch angle distribution.

A quasi-monoenergetic isotropic electron distribution produces an exceptional optically thin synchrotron spectrum, where the radiation flux increases with increasing frequency up to a critical frequency and then cuts off exponentially, i.e. $S_\nu \sim \nu^{1/3} \exp\left(-\frac{\nu}{\nu_c}\right)$ with $\nu_c \simeq \gamma^2 eB/(2\pi m_e c)$ Such a spectrum has been observed in the centres of our Milky Way (Sgr A*) and of M81 (Lesch and Reich 1992; Duschl and Lesch 1994; Reuter and Lesch 1995) (Fig. 3). These spectra support the idea that magnetic reconnection in galactic nuclei is present and produces distribution functions which can be clearly distinguished from the power-law electron distributions produced by shock wave acceleration or stochastic acceleration.

I would like to thank K. Schindler for many stimulating discussions and suggestions. This work was supported by the Bennigsen-Foerder Award (1994) of the government of Nordrhein-Westfalen. I express my sincere thanks to the organizers for this enlightning conference

References

Baker, D.N., Borovsky J.E., Benford, G. and Eilek J.A. (1988) *Astrophys. J.* **326**, 110

Begelman, M.C., Rees, M.J. and Sikora, M. (1994) *Astrophys. J*, **429**, L57

Benford, G. (1978) *Monthly Notices of the Royal Astronomical Society*, **183**, 29

Benford, G. (1991) in *Jets and beams in Astrophysics*, Eds. H. Sol, J. Roland, G. Pelletier, Cambridge University Press, Cambridge, p. 85

682

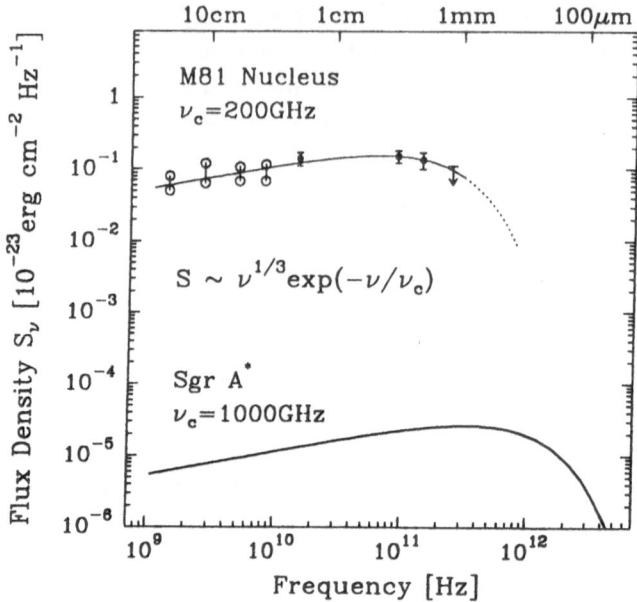

Figure 3. The radio spectra of the nucleus of M81 and of Sgr A* in the m-mm range. Connected open circles mark the time variability range of the M81 nucleus. The thin solid line with the dotted continuation corresponds to a $S_\nu \sim \nu^{1/3} \exp\left(-\frac{\nu}{\nu_c}\right)$ spectrum scaled with $\nu_c = 200 GHz$. The fat line in the lower part of the figure is the well measured spectrum of Sgr A* scaled to the distance of M81 (3.25 Mpc). It shows the same spectral behaviour with $\nu_c = 10^3$ GHz (Reuter and Lesch 1995).

Blandford, R.D. (1990) in *Active Galactic Nuclei*, Saas-Fee Advanced Course 20, Eds. T.J.L. Courvosier and M. Mayor, Springer, Berlin, p. 161

de Jager, C. (1986) *Space Sci. Rev.* **44**, 43

Duschl, W. and Lesch, H. (1994) *Astron. Astrophys.* **286**, 431

Kaplan, S.A. and Tsytovich, V.N. (1973) *Plasma Astrophysics*, Pergamon Press, Oxford

Lesch, H. and Schlickeiser, R. (1987) *Astron. Astrophys.* **179**, 93

Lesch, H., Appl, S. and Camenzind, M. (1989) *Astron. Astrophys.* **225**, 341

Lesch, H. (1991) *Astron. Astrophys.* **245**, 48

Lesch, H. and Pohl, M. (1992) *Astron. Astrophys.* **254**, 29

Lesch, H. and Reich, W. (1992) *Astron. Astrophys.* **264**, 493

Mikhailovskii, A.B. (1974) *Theory of Plasma Instabilities*, Plenum Press New York

Parker, E. (1979) *Cosmical Magnetic Fields*, Clarendon Press, Oxford

Perley, R. (1989) in *Hot Spots in Extragalactic Radio Sources*, eds. K. Meisenheimer and H.J. Röser, Lecture Notes in Physics, Springer (Berlin), 1 SP-50, p. 425

Reuter, H.P. and Lesch, H. (1995) *Nature* (submitted)

Schindler, K., Birn, J. and Hesse, M. (1988) *J. Geophys. Res* **93**, 5547

Schindler, K., Birn, J. and Hesse, M. (1991) *Astrophys. J.* **380**, 293

Schlickeiser, R. (1984) *Astron. Astrophys.* **136**, 227

Sotnikov, V.I., Shapiro, V.D. and Shevchenko, V.I. (1978) *Sov. J. Plasma Phys* **4**, 252

Stella, L. and Rosner, R. (1984) *Astrophys. J.* **277**, 312

Wagner, S. and Witzel, A. (1995) *Ann. Rev. Astron. Astrophys.* (in press)

Weatherhall, J.C. and Benford, G. (1991) *Astrophys. J.* **378**, 543

RELATIVISTIC OUTFLOWS IN THE GALAXY

I.F. MIRABEL
Service d'astrophysique/CEA/DSM/DAPNIA
Centre d'Etudes de Saclay. 91911 Gif/Yvette, France

AND

L.F. RODRIGUEZ
Instituto de Astronomía/UNAM
Apdo. Postal 70-264, México, DF 04510, México

1. Introduction

The improved positions of compact high energy sources provided by the new X-ray and gamma-ray space observatories allowed the discovery at radio wavelengths of several sources of relativistic outflows in the Galaxy. In addition to the classic stellar jet source SS433 (Margon, 1984), two new jet sources that exhibit apparent superluminal motions were identified. Although it is commonly believed that in these accretion-power X-ray sources the acceleration and collimation of the ejecta is magnetohydrodynamic, a number of observational and theoretical questions on the nature of these "microquasars" remain open. Here we will review the observational properties of the superluminal sources GRS 1915+105 (Mirabel & Rodríguez, 1994) and GRO J1655-40 (Hjellming & Rupen, 1995; Tingay et al. 1995) and will compare them with other known sources of relativistic jets in the Galaxy. The aspects of these phenomena that are specifically relevant for relativistic astrophysics were previously discussed by Mirabel & Rodríguez (1995).

2. GRS 1915+105

GRS 1915+105 (also known as the hard X-ray transient in the constellation of Aquila) was discovered by the satellite GRANAT (Castro-Tirado et al. 1994) on August 15, 1992. Since then, the source was observed as one of the brighter sources in the sky at energies $\geq$ 20 keV (Harmon et al.

K. C. Tsinganos (ed.), Solar and Astrophysical Magnetohydrodynamic Flows, 683–698.
© *1996 Kluwer Academic Publishers.*

1994) during repeated periods that lasted several months. Unlike typical X-ray novae, this transient exhibits a slow rise to maximum X-ray luminosity and frequent recurrent activity that lasts several months. The fairly hard spectrum observed by SIGMA (Finoguenov et al. 1994) and BATSE (Harmon et al. 1994), with emission up to 220 keV and changing photon index between -2.5 and -3.0, are consistent with it being a collapsed object in a binary system. The primary is likely to be a black hole because: 1) as most black hole candidates the X-ray spectrum shows a hard X-ray tail at $\geq$ 150 keV (Harmon, A. et al. 1994; Finoguenov et al. 1994), and 2) it has often been observed with integrated X-ray luminosities $L_x \geq 4\ 10^{38}$ erg s^{-1}, which correspond to Eddington luminosities of objects with mass ≥ 3 M$_\odot$. We note that during the last 3 years GRS 1915+105 has frequently become the most luminous hard X-ray source in the Galaxy.

Due to the large position errors (of the order of a few degrees) of the gamma-ray telescopes, the association of this recurrent hard X-ray transient with one of the only three known repeaters of soft gamma-ray bursts is still uncertain (Kouveliotou et al. 1993; Grindlay, 1994; Mirabel et al. 1994). Fortunately, the position of the X-ray source could be first determined with an accuracy of 3' by SIGMA (Finoguenov et al. 1994), and later within an error circle of 10'' diameter with ROSAT (Greiner 1993). A time-variable radio source was found with the VLA inside that error circle (Mirabel et al. 1994). GRS 1915+105 is close to the galactic plane (l = 45.37°, b = -0.22°) and no optical counterpart to a limiting magnitude of 21 in the R band had been found (Castro-Tirado et al. 1994; Mirabel et al. 1994). On the other hand, the observations in the H(1.65 μm) and K(2.2μm) bands revealed the presence of a time variable infrared counterpart (Mirabel et al. 1994), with K magnitudes between 13 and 14. Later, Böer, Greiner, & Motch (1995) and Mirabel et al. (1995) have detected the source in the I band with a magnitudes in the range of 23-24. The available X-ray, optical, infrared, and radio observations suggest that the source is beyond a visual extinction of A_v = 27±1 magnitudes (Greiner et al. 1994; Mirabel et al. 1994; Chaty et al. 1995; Böer et al. 1995).

Based on the infrared resemblance in absolute magnitude, color, and time variability of GRS 1915+105 with SS 433, Chaty et al. (1995) suggest that GRS 1915+105 is likely to be a black hole in a high mass binary.

2.1. SUPERLUMINAL MOTIONS IN GRS 1915+105

This candidate black hole draw our attention since 1992 because among the GRANAT hard X-ray sources of our VLA monitoring program it was the one that had exhibited the most striking changes in the position of compact radio components. On December 1993 GRS 1915+105 produced a strong radio outburst, and by February 1994 we already realized the existence of jets that appeared to change in a few days time. At a distance of $\geq$ 10 kpc the observed displacements appeared superluminal. To understand the nature of such displacements, we remained on the look-out for another strong radio outburst. Such outburst occured in March-April 1994, when we were already prepared to follow the evolution of the source by regular observations with the VLA in the configuration that provides the highest angular resolution.

Figure 1 shows a pair of bright radio condensations emerging in opposite directions from a compact, variable core (Mirabel & Rodríguez 1994). We find that both before and after the remarkable ejection event shown in Figure 1, GRS 1915+105 ejected other pairs of condensations but with flux densities one to two orders of magnitude weaker. One of these weaker pairs can be seen in the first four maps of Figure 1, where one can see a fainter pair of condensations moving ahead of the bright ones at about the same speed and direction. So far we have followed the motions of four pairs of plasma clouds in aproximately the same direction on the sky with similar proper motions. The similar proper motions of the different ejection events can be appreciated when we plot the angular displacements of the respective condensations with respect to the stationary core (Figure 2). The time separation between ejections suggests a quasiperiodicity with intervals in the range of 20 to 30 days.

The proper motions from the stationary core of the bright pair ejected on March 19, 1994 could be determined very accurately. The angular displacements are consistent with ballistic (that is, unaccelerated) proper motions over an interval of 75 days while they remained detectable. At a kinematic distance of 12.5 kpc (Rodríguez et al. 1995) the proper motions of the approaching ($\mu_a = 17.6 \pm 0.4$ mas d^{-1}) and receding ($\mu_r = 9.0 \pm 0.1$ mas d^{-1}) condensations imply apparent velocities on the plane of the sky of $v_a = 1.25c$ and $v_r = 0.65c$, respectively. Since the apparent velocities in the sky are given by:

$$v_{a,r} = \frac{v \, sin\theta}{(1 \mp (v/c) \, cos\theta)}, \tag{1}$$

we find that the ejecta move with a true speed of $v = 0.92c$ at an angle $\theta = 70°$ to the line of sight.

The relativistic interpretation receives strong support from the fact that, as expected (Mirabel & Rodríguez 1994; 1995), the flux density ratios of the approaching and receding condensations at a given angular displacement give values of 8-10, consistent with Doppler boosting of radiation from a twin pair moving apart with the derived true speed and angle with respect to the line of sight.

686

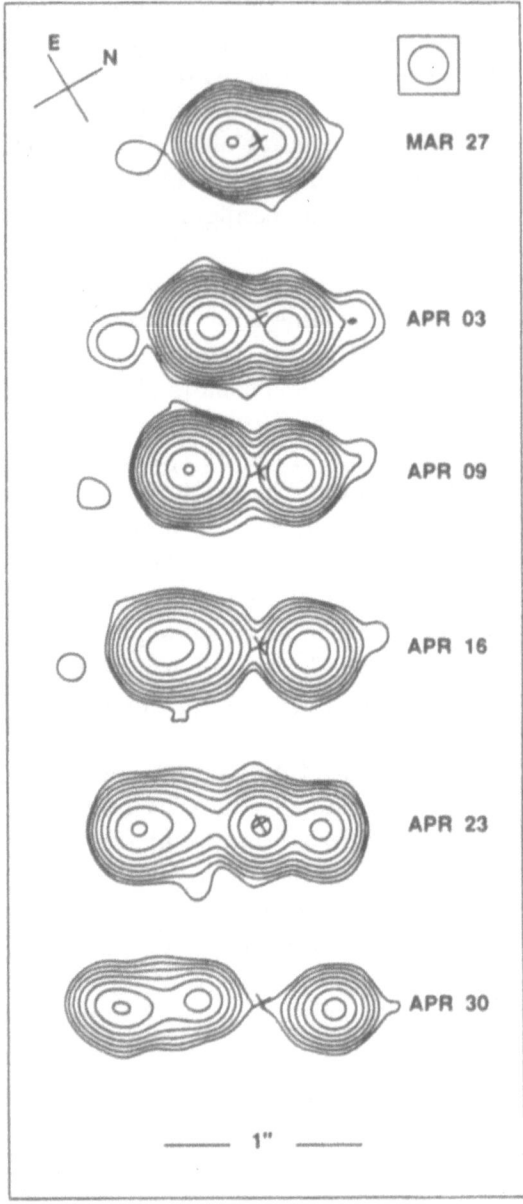

Figure 1. Pair of radio condensations moving away from the hard X-ray source GRS 1915+105. These uniform-weight VLA maps were made at λ3.5-cm for the 1994 epochs on the right side of the figure. Contours are 1,2,4,8,32,64,128,256 and 512 times 0.2 mJy/beam for all epochs except for March 27 where the contour levels are in units of 0.6 mJy/beam. The half power beam width of the observations, 0.2 arc sec, is shown in the top left corner. The position of the stationary core is indicated with a small cross. The maps have been rotated 60° clockwise for easier display and comparison. Note in the first four epochs the presence of a fainter pair of condensations moving ahead the bright ones and in the last epoch the presence of a new southern component.

GRS1915+105

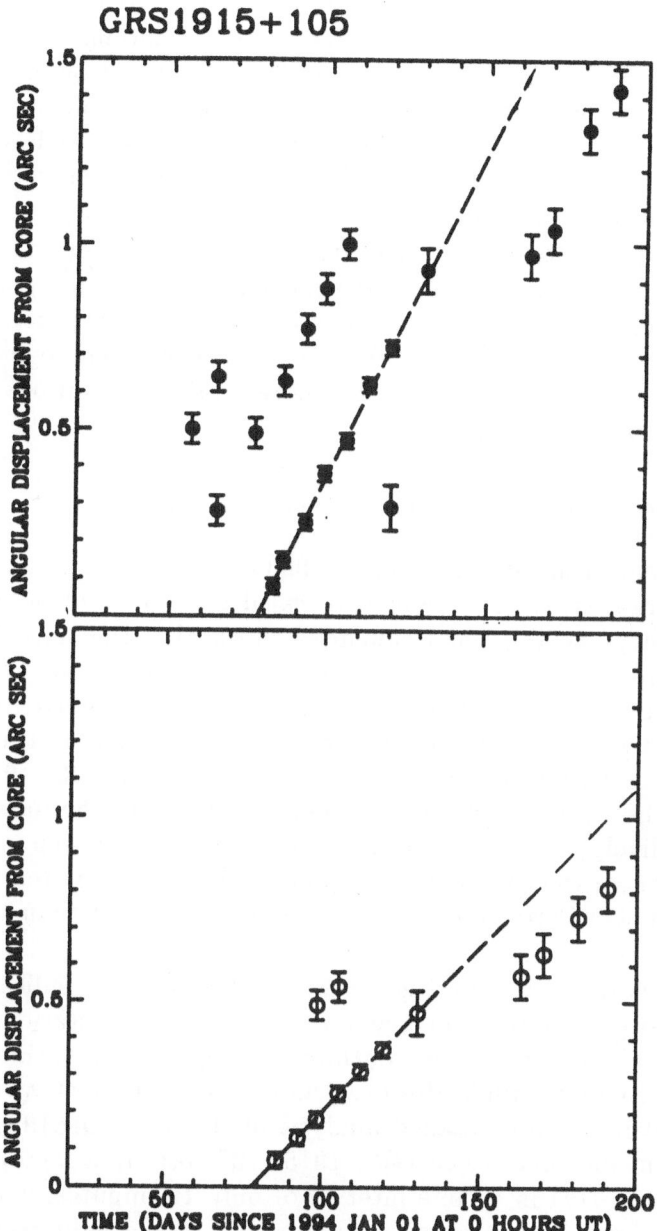

Figure 2.   Upper: Angular displacements as a function of time for four blueshifted condensations corresponding to ejections that took place on (from left to right) 1994 January 29, February 19, March 19, and April 21). Lower: Angular displacements as a function of time for three redshifted condensations corresponding to ejections that took place on (from left to right) 1994 February 19, March 19, and April 21). The 1994 January 29 ejection was relatively weak and we could not detect the redshifted component in our data. The dashed lines are the least squares fit to the angular displacements of the 1994 March 19 event, the brighter and better studied.

GRS 1915+105 is the first superluminal source where the proper motions of both the approaching and receding condensations can be detected and measured. In the case of quasars, because of strong Doppler favoritism, only the proper motions of the approaching condensation has been measured. The consistency of the relative proper motions and relative flux densities in GRS 1915+105 gives a firm basis to the interpretation of superluminal motions in terms of true relativistic ejections.

The ballistic motions, flux density ratios, and smooth fading of the clouds as a function of time (see Mirabel & Rodríguez, 1995) suggest that we observe actual bulk motions of massive plasma clouds rather than the propagation of shocks. In the later case, due to the inhomogeneity of the interstellar medium one would expect more erratic fluctuations in the flux and proper motions.

2.2. TIME VARIABILITY IN THE RADIO AND X-RAYS

The radio monitoring since November 1993 by (Rodríguez et al. 1995) from the Very Large Array and Nançay revealed several radio outbursts. The radio-light curve at different frequencies shows that the time evolution of the fluxes and spectral indeces change from burst to burst. We point out that the outburst of March-April 1994, which corresponded to the major ejection event shown in Figure 1 was extraordinarily long in time. It lasted more than one month, whereas most radio outburst in GRS 1915+105 appear to last between 2 and 10 days. This difference in the duration of the outbursts is likely to be related to the differences we have seen in the brightness evolution of the individual radio emitting clouds ejected at different times. It appears that fainter clouds fade more rapidly than the brighter ones.

The observations by (Rodríguez et al. 1995) already indicated the existence of large variations in the flux density of the stationary core in timescales of hours. We have continued seeing similar variations in our VLA monitoring over 1995. However, an exceptionally fast variation event was detected in a more detailed analysis of the data. On 1994 November 16 the 3.5-cm flux density of GRS 1915+105 rose by a factor of 5 (from about 8 to 40 mJy) in a time interval of only 10 minutes, with variation of a factor of two in $\sim$ 4 minutes (Figure 3). Using light travel time arguments, this result suggests that the radio continuum emission from the core probably originates from a region with dimensions $\leq 10^{13}$ cm. If this estimate is correct, the brightness temperature during this event reached $T_B \simeq 5 \times 10^{15}$ K, implying that the brightness is inverse-Compton limited.

GRS 1915+105 also exhibits fast variability in the X-rays. Photon count variations of $\sim$ 20% in few minutes has been measured with ASCA (Kotani,

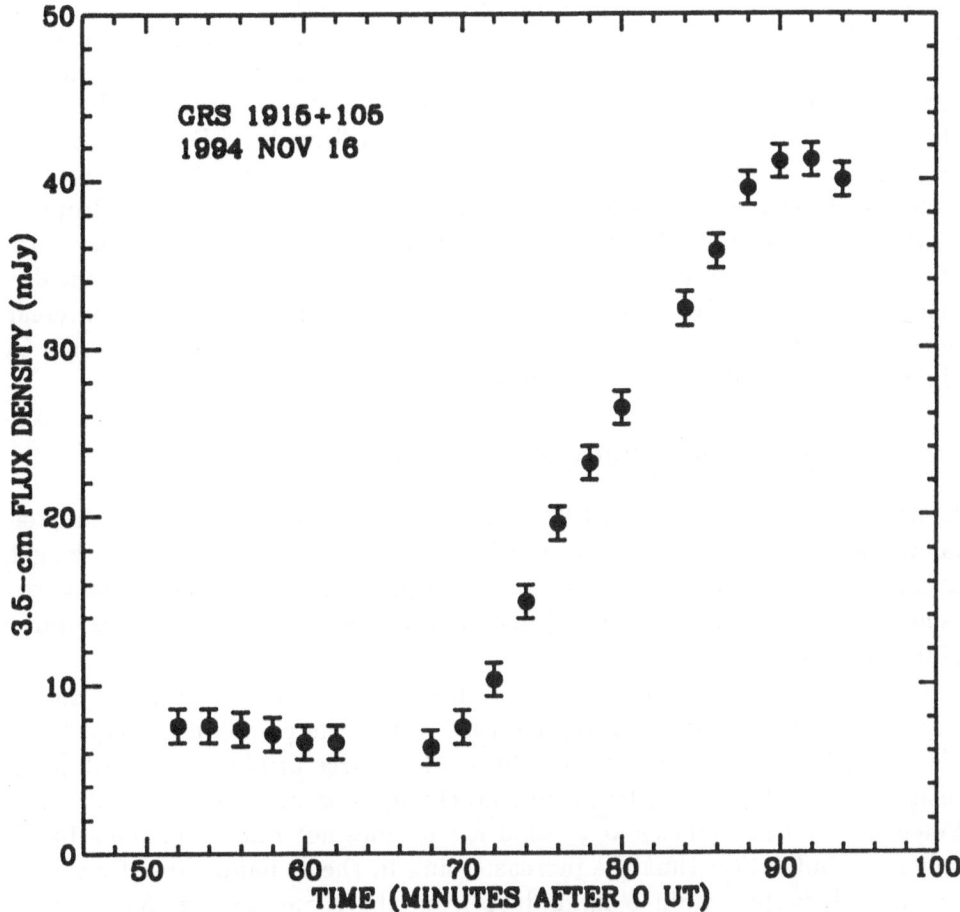

Figure 3. Flux densities of GRS 1915+105 at 3.5-cm during the fast 1994 November 16 event as a function of UT time. This rapid variability suggests a scale of $\sim 10^{13}$ cm for the radio-emitting core and that the brightness temperature during the event was inverse-Compton limited.

et al. 1995). Furthermore, Paciesas et al. (1995) have reported the detection with BATSE of an unprecedented extreme degree of variability in the 20-100 keV range. In 1992, during the initial stages of a long period of actvity three strong flares were detected in an interval of about 1 hour.

The BATSE (Harmon et al. 1994) and GRANAT/WATCH (Sazonov et al. 1994) X-ray light curves, together with the radio light curves of GRS 1915+105 indicate that strong activity in the X-rays is associated to the activity at radio wavelengths. However, the detailed time correlation between the X-rays and the radio during outbursts is still unclear. During the 1994 March outburst the X-ray peak in the 8-20 keV energy band preceeded the radio outburst and the X-ray photon counts (Sazonov et al. 1994) dropped as the radio flux rose. However, during an outburst on 1995 August 9-10, the photon counts in the 20-100 keV band appeared to have peaked one day after the intense radio flare (Ghigo et al. 1995; Zhang et al. 1995), whereas no significant variations in intensity were detected in the source's flux in the 8-20 keV band (Sazanov & Sunyaev 1995).

2.3. EVIDENCE FOR A CIRCUMSTELLAR COCOON

In August 1995, GRS 1915+105 appeared to be embedded in a circumstellar cocoon. The observations in the infrared, X-rays, and radio wavelength around the outbusts of 1995 August 9-10 provided three independent evidences for the presence of interstellar dust and gas very close to the source (Mirabel et al. 1995).

First, the infrared observations 5 days before and 5 days after the X-ray/radio outburst of 9-10/August 1995 reveal the emission of very warm circumstellar dust, since after the burst the source brightened by 1.2 magnitudes in the $K(2.2\mu m)$ band, but no change was observed at $J(1.25\mu m)$. Since the dust sublimates at $\geq$ 2000 K and does not radiate in the J band, this is an indication that the increased flux in the K band came from very hot dust located within 5 light days from the X-ray source. Assuming a maximuum dust temperature of T $\leq$ 2000 K we derive a minimuum mass of dust within a radius of 5 light days of

$$M_d \geq 4.15 \ 10^{-10} \ 10^{-m_K/2.5} \ (e^{65424/T}-1) \ D^2_{kpc} = 5.4 \ 10^{-7} \ M_\odot \ , \ (2)$$

which implies a minumuum gas mass of 5.4 10^{-5} $M_\odot$, assuming a conservative gas-to-dust mass ratio of 100.

Second, the 1995 August 9-10 X-ray outburst was rather soft when compared with previous outbursts, since in the 8-20 keV the source's flux was $\sim$ 700 mCrab with no significant variation during several days (Sazonov & Sunyaev, 1995), whereas in the 20 to 100 keV it showed an average flux of 150 keV (Harmon et al. 1995) with a peak of 220 mCrab (Zhang et al. 1995). This is an indication that the hard X-rays produced during this outburst were Compton-diffused by a high density circumstellar gaseus cocoon.

The third independent indication of a high density interstellar environment came from the short life-time of the radio outburst of August 9-10, which lasted $\leq$ 2 days. This rises the possibility that the kinetic energy of ejected plasma clouds was rapidly dissipated by shocks with the surrounding gas, and that -as in SN1987A- after certain time, most of the energy is being radiated in the infrared, which may be used as a calorimeter.

2.4. ENERGY AND POWER CONSIDERATIONS

Mirabel and Rodríguez (1995) estimate the mass of a typical individual ejecta pair in GRS 1915+105 to be of the order of 10^{24} grams. Assuming that we have an ejection event every 25 days, an average mass loss rate of $\sim 10^{-8}$ $M_\odot$ yr^{-1} is derived for the system. At a true speed of 0.92c, the average mechanical luminosity injected into the interstellar medium is $\sim 10^5$ $L_\odot$. The similarity between the average mechanical luminosity and the X-ray luminosity from the accretion disk supports the notion that the gravitational energy of the accreting gas ends roughly divided in equal parts: one half being radiated in the X-rays and the other as mechanical outflow.

This observational result has been predicted by Shakura & Sunyaev (1973) in their classic theory on accretion disks in black holes. They have demonstrated that the radiated flux -mostly in the X-rays- from the accretion disk in a black hole is $L_R = GM/2R_o$ dM/dt, which corresponds to half the gravitational energy of the matter infalling toward the black hole. The remaining energy is being transformed into mechanic outflow energy in the inner regions of the accretion disk.

It is interesting to note that, most likely, the acceleration of the ejecta takes place in timescales similar to the shortest ones observed in the radio continuum (a few minutes, see section 2.2). In this case, the peak mechanical power involved could be as large as $\sim 10^{10}$ $L_\odot$ (comparable to the radiation energy of an AGN) for the most massive ejecta (10^{25} grams), such as those of the 1994 March 19 ejection event.

2.5. SEARCH FOR EXTENDED LOBES IN GRS 1915+105

To search for extended jets and lobes associated with GRS 1915+105, we did sensitive 20-cm continuum observations of the region with the VLA in the C configuration during 1994 November and December. A map made with these data is shown in Figure 4.

No evidence for extended jets was found in this map, but our interest was aroused by the existence of two radio sources located symmetrically about 17' from GRS 1915+105 and at a position angle of 157° ± 1°, consistent with the position angle of ejection determined by Mirabel & Rodríguez (1994). However, these two radio objects are bright sources in the IRAS catalog (IRAS 19124+1106 and IRAS 19132+1035, respectively), and their

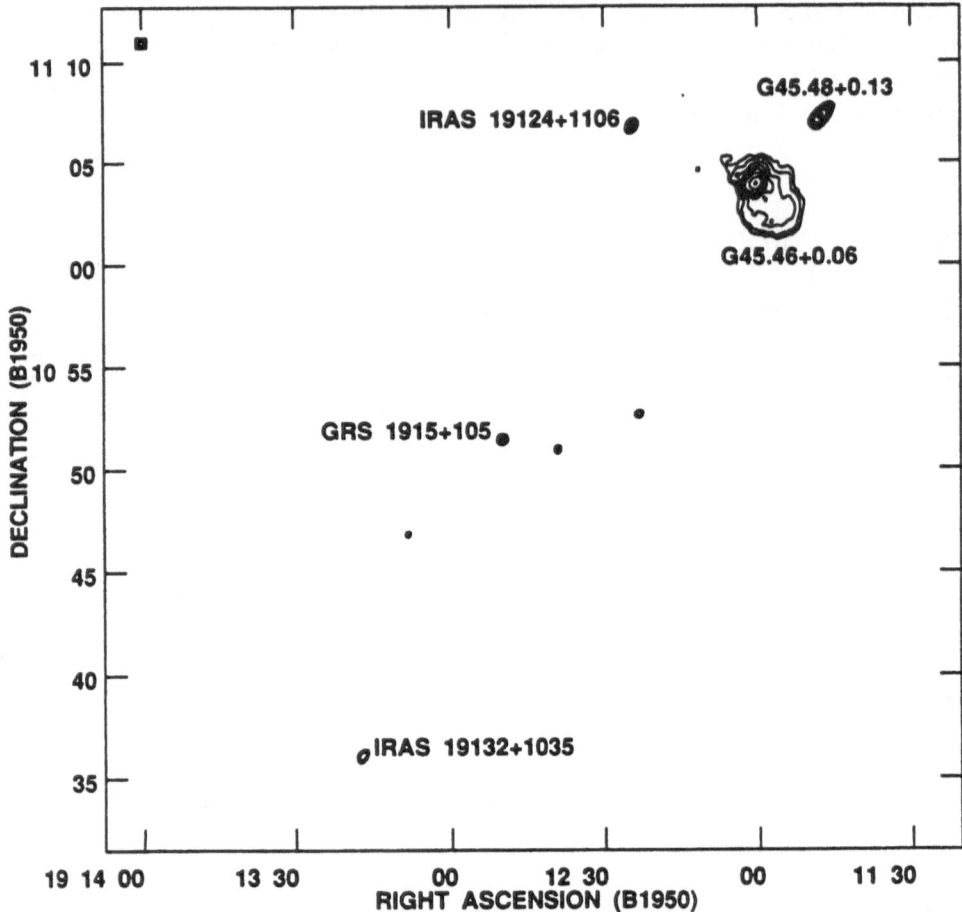

Figure 4. VLA map at 20-cm of the GRS 1915+105 region made in the C configuration. Contours are -4, 4, 8, 16, 32, 64, 128, 256, and 512 times 0.5 mJy beam^{-1}, the rms noise of the map. The half power contour of the beam is shown in the top left corner. The sources G45.48+0.13 and G45.46+0.06 are well-known H II regions. We detected radio continuum emission asssociated with the bright IRAS sources IRAS 19124+1106 and IRAS 19132+1035, that appear symmetrically located with respect to GRS 1915+105. Most probably, these IRAS sources are H II regions powered by late O-type stars. The map is not corrected for primary beam response and sources away from the phase center appear weaker than they are. The sources away from the phase center also show bandwidth (radial) smearing.

20-cm radio continuum flux densities (80 and 40 mJy, respectively) and 100-μm infrared flux densities (580 and 490 Jy, respectively) are consistent with them being H II regions located at $\sim$10 kpc and powered by late O-type stars. Nevertheless, given the remarkable symmetry with respect to GRS 1915+105 and the coincidence in position angle with the compact ejecta, we have started a program to study these IRAS sources in detail. It is intriguing that one of these sources has a non-thermal radio spectrum.

The ballistic motion of the plasma clouds (Figure 2) indicate that the density of the medium through which they move up to distances of a few light months must be very low, otherwise some deceleration should have been observed. On the other hand, for the relativistic outflows would not reach as far as 17 arcmin from the source (60 pc) unless there is a large-scale cavity around GRS 1915+105. Assuming masses $\leq 2\ 10^{25}$ gr for the relativistic clouds from the conservation of linear momentum we derive for the interstellar density an upper limit of 0.05 protons cm^{-3}. If the radio sources in Figure 4 are extended lobes from GRS 1915+105, the later must be inside a cavity. This is the case of analog source SS 433 which is at the center of the remanant W50.

3. The Two Superluminal Sources and SS 433

The discovery of superluminal motions in GRS 1915+105 estimulated a search for similar relativistic ejecta in other transient hard X-ray sources. Soon after the superluminal motions in GRS 1915+105 had been reported (Mirabel & Rodríguez 1994), similar phenomena were searched and immediately found by two different groups in the first hard X-ray nova (GRO J1655−40; X-ray nova Scorpii) that appeared during 1994 August-September (Tingay et al. 1995, Hjellming & Rupen 1995).

In GRO J1655−40 a hard X-ray tail was detected up to 600 keV, and from the X-ray properties it was proposed that it is a black hole in a binary system (Harmon et al. 1995). This has been confirmed by Bailyn et al. (1995a, 1995b) who found in the optical that GRO J1655−40 is an eclipsing and spectroscopic low mass binary where the primary star is a black hole with a mass between 4 and 5.2 solar masses. The fast rise of the X-ray flux suggests that GRO J1655−40 is an X-ray nova powered by accretion. In contrast, GRS 1915+105 cannot be classified as a X-ray nova since in the X-rays it had a rise time of a few months, it has been active from 1992 until present, usually undergoing long duration flares interspersed with quiescent periods (Harmon et al. 1994). GRO J1655−40 is at $\sim$ 3.2 kpc from the Sun and has an optical counterpart similar to other accreting-black-hole candidates in low mass binaries. The optical spectrum during outburst showed a blue spectrum with prominent broad hydrogen emission lines (Bailyn et al. 1995a), but during its quiescent state six months later it showed a redder spectrum with narrower emission lines (Mirabel et al. 1995). In contrast, GRS 1915+105 is obscured by intervening dust and no prominent emission lines have been detected in the J, H, and K band spectra of the variable infrared counterpart.

The relativistic ejections observed in the radio in GRO J1655–40 have striking similarities as well as differences with those in GRS 1915+105. Bright components moving apart with proper motions in the range of 40 to 65 mas d^{-1} were independently observed by the Southern Hemisphere VLBI Experiment array (Tingay et al. 1995), the VLA and VLBA (Hjellming & Rupen 1995). At a distance of 3.2 kpc the motions of the ejecta have been fit by Hjellming & Rupen (1995) using a kinematic model with a velocity of 0.92, and a jet axis inclined 85° to the line of sight at a position angle of 47°, about which the jets rotate every three days at an angle of 2°.

In contrast to repeated ejections in GRS 1915+105, the flux ratios of the blobs on either side of GRO J1655–40 cannot be ascribed to relativistic Doppler boosting. In GRO J1655–40 the asymmetry in brightness flips from side to side (Hjellming & Rupen 1995). Not only the jets appear to be intrinsically asymmetric, but also the sense of that asymmetry changes from event to event. Therefore, although similar spatial velocities of 0.92c are found in both superluminal sources, due to intrinsic asymmetries the ultimate physical interpretation of the superluminal expansions in GRO J1655-40 remains ambiguous.

In Table 1 we summarize the parameters for the two superluminal sources together with the parameters of the classic galactic source of relativistic jets SS 433. The kinetic power of the ejecta in SS 433 was derived from the optical emission lines (Margon, 1984), whereas the kinetic power in GRS 1915+105 and GRO J1655-40 are derived from the radio continum emission. If the radio sources 17' away from GRS 1915+105 are actually radio lobes associated to GRS 1915+105, they would be at about the same distance as are the "ears" of the remnant W50 from SS 433 (Elston & Baum, 1987). We note that in SS 433 the ejecta are symmetric as in GRS 1915+105 (Spenser, 1984; Vermeulen et al, 1993) and the brightnesses consistent with Doppler boosting. Light travel time effects on the radio brightness of the ejecta have been observed in GRS 1915+105 (Mirabel & Rodríguez, 1995) and SS 433 (Spencer, 1984). Contrary to GRS 1915+105, the X-ray outbursts in GRO J1655-40 are sometimes associated to radio outbursts, but sometimes they are not.

TABLE 1. Galactic Superluminal Sources and SS433.

Property	GRS 1915+105	GRO J1655–40	SS 433
X-ray Activity	≥3 Years	~4 Months	Persistent
Distance	12.5 kpc	3.2 kpc	5.1 kpc
Nature of Binary	BH+High Mass	BH+Low Mass	NS?+High Mass
Velocity of Ejecta	0.92c	0.92c	0.26c
Kinetic Power	$3 \ 10^{38}$ erg/s	$2 \ 10^{37}$ erg/s	$5 \ 10^{39}$ erg/s
Angle of Ejection	70°	85°	80°
Precession Angle	5° ?	~2°	20°
Precession Period	60 days ?	3 days	164 days
Ejecta Symmetry	Symmetric	Asymmetric	Symmetric
Doppler Boosting	Yes	No	Yes
X-rays/Radio association	Yes	Yes/No	Yes
Large-scale lobes	60 pc ?	?	50 pc

4. Search for "moving" emission lines

A method to determine the distance to relativistic jet sources using the proper motions and Doppler factors of the ejecta has been proposed by Mirabel & Rodríguez (1994, 1995). However the searches for moving emission lines -as seen in SS433- in the two superluminal sources GRS 1915+105 and GRO J1655-40 had negative results. Because there are about 27 mag of optical absorption along the line of sight, the searches in GRS 1915+105 have been conducted in the near infrared and X-rays. In GRO J1655-40 one can, in addition, obtain optical spectra. The ASCA spectra of both sources show an absorption edge at ~ 7.1 keV (Kotani et al. 1995), but no emission lines as in SS433. In the 1.0-2.5 μm band the observed spectra of GRS 1915+105 show no lines. In GRO J1655-40 were detected emission lines with no significant Doppler shifts, but no moving lines from jets. These negative results could be due to Doppler smearing of the emission lines from plasma clouds with intrinsic relativistic expansion. From the size evolution of the plasma clouds in the jets of GRS 1915+105 (Mirabel & Rodríguez, 1994), we derive intrinsic expansions of 0.7c and 0.1c in the directions parallel and perpendicular to the bulk motions. The emission lines from clouds expanding at relativistic velocities would be "washed-out" and perhaps show up as extremely broad bumps which would be difficult to identify.

5. Other Sources of Relativistic Jets in the Galaxy

In Table 2 are listed the known sources of relativistic jets in the Galaxy. Persistent and transient X-ray black hole candidates appear to produce two different types of jets. Persistent sources of hard X-rays (e.g. 1E1740.7-2942, GRS 1758-258) are usually associated to faint, double-sided radio structures that have sizes of several arcmin (parsec scales). The core of these *micro-FR II* type of galactic radio sources are rather faint (few mJy). Although variable, the cores of these persistent sources do not exhibit dramatic variability, neither in the X-rays nor in the radio. The radio properties of these synchrotron double-sided jets indicate that black holes that are persistent X-ray sources also are persistent sources of relativistic outflows. On the contrary, violently variable hard X-ray transients (e.g. GRS 1915+105 and GRO J1655-40) exhibit associated variations in the X-rays and radio fluxes of several orders of magnitude in short intervals of time. Because these black-hole X-ray transients produce rather sporadic ejections of discrete plasma clouds, their proper motions can be measured.

Conservation of angular momentum in accretion disks indicates that probably all hard X-ray sources that accrete at super-Eddington rates must produce relativistic jets. However, the observational study of these jets presents several difficulties in the practice. In persistent hard X-ray sources like Cygnus X-1 one may find that they are surrounded by faint non-thermal radio features extending several arcmin (Martí et al. 1995, 1996a), but unless they are perfectly aligned with the variable core radio counterpart -as in 1E1740.7-2942- it is difficult to prove that they are associated with the X-ray source. On the other hand, in transient black hole binaries one may observe transient sub-arcsec jets, but unless the interferometric observations are properly scheduled, the evolution is too rapid and it may not be possible to follow up the proper motions of discrete clouds. This may have been the case in the radio observations X-ray Nova Ophi 93 (Martí, Rodríguez & Mirabel 1996b).

In Table 2 we summarize the velocities of the ejecta that have been measured so far. Proper motions of relativistic ejecta have been determined with accuracy in the superluminal sources GRS 1915+105 and GRO J1655-40, and in SS 433. Besides these three sources, proper motions were also measured -but with less accuracy- for moving features in Cygnus X-3 (Schalinski et al. 1995), and for the expanding H_α filaments in the Crab nebula (Oort & Walraven, 1956). It is interesting that the ejecta from neutron star candidates have velocities in the range of 0.2c-0.3c, whereas the bulk motion of ejecta from black hole candidates have larger velocities of about 0.9c. These jet velocities are comparable to the Keplerian rotational velocities expected at the base of the jets, close to neutron stars (0.3c) and black holes (0.9c)

TABLE 2. Sources of Relativisticc Jets in the Galaxy.

Source	X-ray emission	Candidate object	V_{ejecta}	Reference
Circinus X-1	Persistent/soft	Neutron Star		Stewart et al. (1993)
Cygnus X-3	Persistent/soft	Neutron Star	0.3c	Schalinski et al. (1995)
SS 433	Persistent/soft	Neutron Star	0.26c	Margon (1988)
Crab Nebula	Persistent/hard	Neutron Star	0.2c	Oort & Walraven (1956)
1E 1740.7-2942	Persisten/hard tail	Black Hole		Mirabel et al. (1992)
GRS 1758-258	Persistent/hard tail	Black Hole		Rodríguez et al. (1994)
Cygnus X-1	Persistent/hard tail	Black Hole		Martí et al. (1995, 1996)
GRS 1915+105	Transient/hard tail	Black Hole	0.92c	Mirabel et al. (1994)
GRO J1655-40	"X-Nova"/hard tail	Black Hole	0.92c	Hjellming & Ruppen (1995)

respectivelly. Kudoh & Shibata (1995) have shown that in magnetically driven jets the velocities of the mass outflows are comparable to the inner circular velocities of the accretion disks. These jet properties could be used in the future to discriminate between neutron stars and black holes.

IFM thanks Marc Sauvage for his help with the interpretation of the infrared observations, and Philppe Laurent and Jacques Paul for their comments on the X-ray aspects of this paper. LFR is grateful to CONACyT, México and DGAPA, UNAM for their continued support.

References

1. Bailyn, C. D. et al. (1995a) *Nature*, **374**, 701
2. Bailyn, C. D. et al. (1995b) *IAUC* 6173
3. Boër, M., Greiner, J., & Motch, C. (1995), *Astr. Ap.* in press
4. Briggs, M.S., Pendleton, G.N., & Brock, M.N. (1993) *Nature* **362**, 728
5. Castro-Tirado, A. et al. (1994) *Astrophys. J. Supp.* **92**, 469
6. Chaty, S., Mirabel, I.F., Duc, P. A., Wink, J. E. & Rodríguez, L.F. (1995) *Astr. Ap.* in press
7. Della Valle, M., Mirabel, I. F., & Rodríguez, L. F. (1994) *Astr. Ap.*, **290**, 803
8. Elston, R. & Baum, S. (1987) *Astron. J.* **94**, 1633
9. Finoguenov, F. et al. (1994) *Astrophys. J*, **424**, 940
10. Gilmore, W. S., & Seaquist, E. R. (1980) *Astron. J.* **85**, 1486
11. Ghigo, F.D. et al. (1995) *IAUC* 6204
12. Greiner, J. (1993) *IAUC* 5786
13. Greiner, J. et al. (1994) *AIP Conference Proceedings* **304**, eds. C. E. Fichtel, N. Gehrels, & J. P. Norris, New York, p 260
14. Grindlay, J. E., (1994) *Astrophys. J. Supp.* **92**, 465
15. Harmon, A. et al. (1994) *AIP Conference Proceedings* **304**, eds. C. E. Fichtel N. Gehrels, & J. P. Norris, New York, p 210
16. Harmon, A. et al. (1995) *Nature* **374**, 703
17. Harmon, A. et al. (1995) *IAUC* 6204

698

18. Hjellming, R. M., & Johnston, K. J. (1985) *Astrophys. J.* **328**, 600
19. Hjellming, R. M., & Rupen, M. P (1995) *Nature* **375**, 464
20. Kotani, T. (1995) private communication
21. Kouveliotou, C. et al. (1993) *Nature* **362**, 728
22. Kudoh, T. & Shibata, K. (1995) *Astrophys. J.* **452**, L41
23. Margon, B.A. (1984) *Rev. Astr. Astrophys.*, **22**, 507
24. Martí, J., Rodríguez, L.F., Mirabel, I.F., Paredes, J.M. (1995) *Astr. Ap*, in press
25. Martí, J., Rodríguez, L.F. & Mirabel, I.F. (1996a) in preparation
26. Martí, J., Rodríguez, L.F. & Mirabel, I.F. (1996b) in preparation
27. Mirabel, I.F., Rodríguez, L.F. Cordier, B., Paul, J., & Lebrun, F. (1992) *Nature* **358**, 215
28. Mirabel, I.F., Duc, P.A., Rodríguez, L. F., Teyssier, R., Paul, J., Claret, A., Auriere, M., Golombek, D., & Martí, J. (1994) *Astr. Ap.* **282**, 17
29. Mirabel, I.F. & Rodríguez, L.F. (1994) *Nature* **371**, 46
30. Mirabel, I.F. & Rodríguez, L.F. (1995) *Proceedings of the 17th Texas Symp. on Relativistic Astrophysics, Annals of the New York Academy of Sciences*, in press
31. Mirabel, I.F. et al. (1995) in preparation
32. Oort, J.H. & Walraven, Th. (1956) *Bull. Astron. Inst. Netherlands*, **462**, 285
33. Paciesas, W.S. et al. (1995) *Proceedings of the Third Compton Symp.* in press
34. Rodríguez, L.F., Mirabel, I.F., & Martí, J. (1994) *Astrophys. J.* **401**, L15
35. Rodríguez, L.F., Gerard, E., Mirabel, I.F., Gómez, Y. & Velázquez, A. (1995) *Astrophys. J.* in press
36. Sazonov, S.Yu. et al. (1995) *Astron. Letters* **20**, 787
37. Sazonov, S.Yu. & Sunyaev, R.A. (1995) *IAUC* 6209
38. Schalinski, C.J. et al. (1995) *Astrophys. J.*, **447**, 752
39. Seaquist, E. R. (1993) *Rep.Prog.Phys.* **556**, 1145
40. Shakura, N.I. & Sunyaev, R.A. (1973) *Astron. Astrophys.* **24**, 337
41. Spencer, R.E. (1984) *MNRAS*, **209**, 869
42. Stewart R.T., Caswell, J.L., Haynes, R.F. & Nelson, G.J. (1993) *MNRAS* **261**, 593
43. Tingay, S.J. et al. (1995) *Nature* **374**, 141
44. Vermeulen, R. C., McAdam, W. D., Trushkin, S. A., Facondi, S. R., Fiedler, R. L., Hjellming, R. M., Johnston, K. J., & Corbin, J. (1993) *Astron. Ap.* **270**, 188
45. Zhang et al. (1995) *IAUC* 6209

STATIONARY RELATIVISTIC MHD FLOWS

Accretion to Rotating Black Holes and Formation of Relativistic Jets

MAX CAMENZIND
Landessternwarte Königstuhl
D-69117 HEIDELBERG, Germany

Abstract. It is widely accepted that the formation of jets in active galactic nuclei and in young stellar sources is ultimately related to the existence of gaseous disks around some central object. Rotating black holes are still thought to be the prime–mover behind the activity detected in centers of galaxies, while, in the case of protostellar jets, rapidly rotating stars and disks are responsible for the ejection of bipolar outflows. In both cases, magnetic fields are invoked for the acceleration and the collimation of these outflows.

MHD flows near rapidly rotating compact objects must include the general relativistic effects of the underlying metric. Relativistic jet flows on the parsec–scale in active galactic nuclei have to be based on special relativistic elements which go beyond the traditional Newtonian MHD description. We give a comprehensive introduction into the theory of relativistic MHD for rapidly rotating compact objects with special emphasis on axisymmetric flows. In addition, we show that accreting black holes dispose of two different energy channels – the accretion power as well as dissipation of rotational energy of a black hole by means of magnetic processes. The angular momentum can only be tapped from black holes by interaction with rotating magnetospheres that are built up e.g. by the inner accretion disk.

The gravitational field of rotating black holes is more complex than that of Newtonian objects. In addition to the ordinary gravitational force, the rapid rotation of compact objects also generates the *gravitomagnetic force* which couples with electromagnetic fields over Maxwell's equations. This effect has interesting consequences e.g. for the time–evolution of magnetic fields advected from the interstellar matter towards the black hole. The shearing of the absolute space around rapidly rotating black holes acts as a gravitomagnetic dynamo effect which amplifies any seed field near a rotating object. This process will provide the dipolar magnetic structures that are behind the bipolar outflows seen as relativistic jets in elliptical galaxies. The magnetic fields also influence the accretion towards the rotating

K. C. Tsinganos (ed.), Solar and Astrophysical Magnetohydrodynamic Flows, 699–725.

black hole. For sufficiently rapidly rotating holes, the accretion can carry negative angular momentum inwards, spinning down in this way the black hole. For extremely fast rotating holes, accretion could even occur with total negative energy.

The presence of magnetic fields near rotating black holes has other consequences. The rapid rotation of the disk–magnetosphere initiates outflows which are collimated on the scale of a few hundred Schwarzschild radii into cylindrical relativistic jets. Solutions of the force–balance equation are discussed which demonstrate the mechanism. The structure of the magnetic fields dragged along into the parsec–scale jets is essential for the understanding of the emission mechanisms of flat–spectrum quasars and BL Lac objects. Self–collimation also works in the protostellar case, since typical jet radii derived from observations are of the order of a few light cylinder radii, which makes Newtonian MHD obsolete on this scale.

1. Introduction

In 1963 Maarten Schmidt identified the mysterious emission lines that have been found in the object 3C 273 from a list of unknown radio sources. 3C 273 is still the brightest known quasar. Its host galaxy is one of the biggest elliptical galaxies (Röser & Meisenheimer 1991), and the quasar itself belongs to that class of radio–loud objects which have been found by the COMPTON satellite to be strong gamma–ray emitters (Fichtel et al. 1994). Recent observations by the Hubble space telescope (Bahcall et al. 1995) have revealed the very faint helical structure of the optical jet of this source. Giant elliptical galaxies are now thought to host the most massive centers in the form of collapsed rotating black holes. Disks on the parsec–scale should turn out as the ultimate evidence for the presence of these monsters. This has now been shown by HST for various radio galaxies and, in particular, for M 87 (Ford et al. 1994; Harms et al. 1994).

Matter on the parsec–scale, which finally accretes onto the central black hole, also brings in magnetic fields. The equipartition field strength on the parsec–scale is of the order of a few milli–Gauss. These fields are advected inwards by the accretion process, restructured and amplified by strong shear and turbulence in the Keplerian disk. Once these fields accrete onto a rapidly rotating black hole, new forces appear which are due to the spin of the black hole. The gravitomagnetic field of a rapidly rotating black hole couples into Maxwell's equations and amplifies the disk field to local equipartition in a few revolution times. The structure of the resulting rotating magnetosphere is most probably behind the various phenomena such as coronal activity and relativistic bipolar outflows.

Investigations of the role of magnetic fields in the physics of active galactic nuclei is a fairly young discipline of research compared e.g. with similar activity going into the structure of the solar corona. It is, however, nowadays commonly assumed that the generation of relativistic jets can only be achieved over magnetic effects (Blandford et al. 1993; Camenzind 1993). It is still not yet common knowledge that also the emission processes behind the optical, X–ray and gamma–activity seen in quasars are ultimately due to the magnetic structure of the jets on the parsec–scale. Magnetic models predict a definite structure which can be tested in the future. Magnetic effects are, however, also important for the accretion process in Seyfert galaxies and QSOs. Standard accretion disk models can definitely not account for the observed variations and the hard X–ray emission. We have to wait for the next generation of disk models based on relativistic MHD.

2. The Disk–Jet Connection in AGN and Young Stellar Objects

It is now widely accepted that the formation of jets is ultimately related to the existence of gaseous disks around some central object. In the case of elliptical galaxies that harbour the most efficient collimated outflows on the extragalactic scale, this central object is most probably a spinning black hole. Matter accumulating towards the center of a galaxy brings in specific angular momentum and can therefore not diretly accrete onto the central object – accretion disks must be formed in this case (Fig. 1). The parsec–scale extensions of these disks can now be traced by HST observations.

NGC 4261 is the radio galaxy 3C 270 which consists of a nuclear point source, two oppositely directed jets and two radio–lobes. The HST image shows a smooth absorbing dusty disk with sharp edges with a diameter of $\simeq 120$ parsecs, essentially perpendicular to the axis of the radio jets. Since the rotation period in this inner part of the galaxy is a few million years, the life–time of the jet phenomenon must be considerably longer than this. These facts combined with the alignment of the disk angular momentum and the radio axis lead to the conclusion that this parsec–scale disk is indeed the fuel supply for the central engine. Recent observations by HST on the inner part of M 87 have revealed a similar structure in this archetypical radio galaxy. Here, a disk is rotating at a speed of $\simeq 550$ km/s on a scale of 14 parsecs, also seen perpendicular to the jet axis. The high rotation velocity signals the presence of more than $2 \times 10^9 \, M_\odot$ inside a radius of 14 parsecs. Since the central stellar density in large elliptical galaxies does not exceed 1000 $M_\odot$ pc^{-3}, this huge mass can only be accounted for by a collapsed central compact object. These are the structures which are essential for the understanding of the central activity in radio galaxies,

Dusty Torus

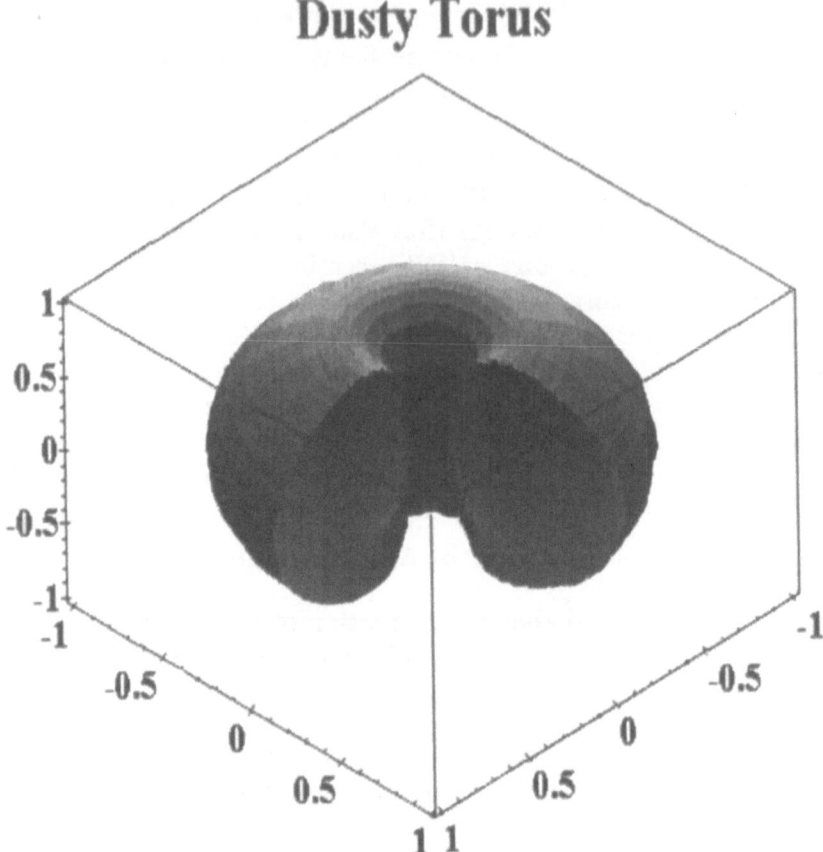

Figure 1. The disk–jet connection in Quasars and radio galaxies. The high gas and dust mass assembled in the core of a giant elliptical galaxy forms a toroidal structure which has a funnel–like opening along the axis. Coordinates are given in units of the core radius (typically about 100 pc). This gas is the fuel for the central accretion disk surrounding a supermassive black hole with mass $M_H \simeq 10^9 \ M_\odot$. Disk winds collimated into jets escape unimpeded along the funnels of the torus.

quasars and BL Lac objects (Camenzind 1995, Fig. 1).

AGN jets are best resolved at radio and millimeter wavelengths using the VLBI–technique (Krichbaum et al. 1993; 1994). Since disks are not visible at these wavelengths, the disk–jet connection cannot yet be directly made visible. VLBI–maps show however highly collimated outflows on the parsec–scale with intriguing small–scale structure. These jets also show quite often superluminal motion as a sign of a highly relativistic flow in the jet channel.

In the case of jets from young stellar objects, there is still very little direct observational data in the optical/IR on the connection between disks and outflows. One exception is the recent imaging of the central source of the Herbig–Haro object HH 30 by HST (see Fig. 2 from the HST image archive). Here, one can directly see the disk photosphere edge–on in light from the protostar scattered by gas and dust in the accretion disk. The jet splashes away from the disk exactly along its axis. Such disks have an extension of a few hundred AU, and the jets are probably already collimated on a scale of 10 to 100 AU, corresponding to 1000 – 10000 stellar radii. In distinction to the AGN case, one knows here the central mass (about one solar mass) and its radius (about two solar radii). The outflows are non–relativistic and show generally a strong shock excitation in forbidden emission lines (Hirth et al. 1994). Using HST, one will be able to resolve this innermost structure of jets in young stellar sources in the near future.

3. MHD of Rapidly Rotating Objects

Relativistic magnetohydrodynamics (RMHD) has evolved over the last 20 years. The basics of the theory are generally known (Anile 1989), but applications are not well known in the astrophysical community. The theory has important consequences for rotating compact objects, such as neutron stars and black holes.

3.1. THE GRAVITATIONAL FIELD OF COMPACT OBJECTS

The gravitational field of rotating compact objects is not just a generalization of the Newtonian potential. There are additional effects which are generated by the spin J_H of a rotating object. These effects are essential for an understanding of the physics occurring within a few Schwarzschild radii. Despite a general scepticism of many astrophysicists to use general relativistic formulations of the laws of physics, in the case of compact objects there is no ground for this. All laws can be formulated in the 3+1–split of rotating space–time which represents a kind of special relativistic formulation of the laws of physics (Thorne et al. 1986). The only thing which

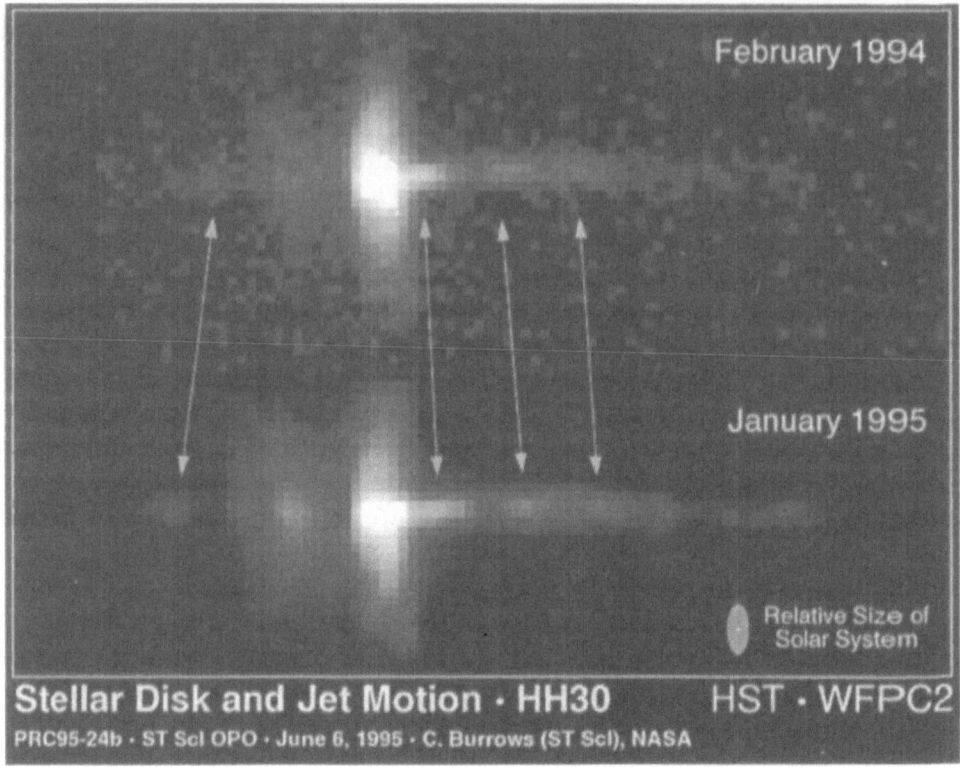

Figure 2. The disk–jet connection for the young stellar outflow in the object HH 30 in the constellation of Taurus (HST image on the WWW). The optical light from the central star is scattered in the outskirts of the surrounding accretion disk. The star itself is heavily obscured by the midplane of the accretion disk which extends to a few hundred A Us (edge–on view of the system). The line–emission from the jet matter is visible having more or less a cylindrical shape. Exposures taken one year apart show the motion of blobs of gas in the jet ejected from the center at a speed of about 200 km/s. The width of the optical jet is about 50 AU corresponding to about 3000 stellar radii.

has to be considered is the presence of the gravitomagnetic field due to the spinning object.

The metric of rapidly rotating compact objects is in general given as

$$ds^2 = -\alpha^2 c^2\,dt^2 + h_{ik}\,(dx^i + \beta^i\,dt)(dx^k + \beta^k\,dt) \tag{1}$$

with spatial indices $i, k = 1, 2, 3$. In this representation, the structure function α represents the redshift between local observers and infinity, while h is the metric of the local 3–space (Thorne et al. 1986). β^i are the components of a vector potential that represents the so–called gravitomagnetic forces which are not present in a pseudo–Newtonian description. The most important examples for the use of this metric in astrophysics are axisymmetric gravitational fields of rotating neutron stars and black holes.

3.1.1. *Rotating Neutron Stars*
Rapidly rotating neutron stars do exist. The neutron star behind the radio pulsar PSR 1937+214 has a rotation period $P = 1.558$ milliseconds, which is about a factor three above the critical rotation period for a neutron star with a realistic equation of state. Non–Newtonian effects are, however, important for even moderate rotations. Unfortunately, in the case of neutron stars, Einstein's equations for the metric functions in equation (1) cannot be solved analytically. Early calculations (Friedman et al. 1986) show the form of these structure functions numerically. Numerical solutions have been given recently using a very elegant method based on the method of Finite Elements for the form of these structure functions (Herold et al. 1996).

3.1.2. *Rotating Black Holes*
The metric of rotating black holes is stationary and axisymmetric and usually given in a parametrized form for the redshift factor (in spherical coordinates (r, θ, ϕ), for further details see e.g. Chandrasekhar (1990))

$$\alpha = \rho\,\sqrt{\Delta/A} \tag{2}$$

and for the axisymmetric gravitomagnetic potential

$$\beta^r = 0 = \beta^\theta \quad , \quad \beta^\phi = -\omega = -\frac{2a_H\,M_H r}{A} , \tag{3}$$

where the following abbreviations have been introduced (given in dimensionless form, $G = 1 = c$)

$$\Delta = r^2 + a_H^2 - 2M_H r , \tag{4}$$
$$\rho^2 = r^2 + a_H^2 \cos^2\theta , \tag{5}$$
$$A = (r^2 + a_H^2)^2 - a_H^2\,\Delta\,\sin^2\theta , \tag{6}$$
$$R = \frac{\sqrt{A}}{\rho}\,\sin\theta . \tag{7}$$

R is the cylindrical radius in Kerr geometry which plays a fundamental role in the following. The total mass M is only used to scale the coordinates in terms of the Schwarzschild radius $M_H = GM/c^2$. The other essential parameter which determines this Kerr metric is the total angular momentum J_H of the rotating black hole, expressed in these units, $a_H = J_H/M$.

The two potentials α and $\vec{\beta}$ represent the local gravitational potentials, $\vec{g} = -\nabla \ln \alpha$ is a measure for the standard gravitational force, while $\vec{H} = \frac{1}{\alpha} \nabla \wedge \vec{\beta}$ is a kind of Lorentz force generated by the angular momentum of the compact object (Thorne et al. 1986). All the metric coefficients of 3–space are then also known functions $(i, k = r, \theta, \phi)$

$$h_{rr} = \frac{\rho^2}{\Delta} \quad , \quad h_{\theta\theta} = \rho^2 \quad , \quad h_{\phi\phi} = R^2 . \tag{8}$$

As in Newtonian physics, the gravitational force is produced in lowest order by the total mass of the object. The gravitomagnetic potential β^ϕ, on the other hand, is proportional to the total angular momentum, and the corresponding force field forces all matter into rotation near compact objects (this is known as the frame–dragging effect). This induced rotation is so strong near a black hole that all matter and, in particular also observers, must corotate with the horizon. Even strong magnetic fields in accreting plasmas cannot hinder this corotation with the 3–space of the black hole. The angular velocity Ω of any object moving on the Kerr background is limited by $\Omega_-(r) \le \Omega(r) \le \Omega_+(r)$ with

$$\Omega_-(r) = \omega(r) - \frac{c\alpha}{R} \quad , \quad \Omega_+(r) = \omega + \frac{c\alpha}{R} . \tag{9}$$

There is therefore a particular observer field, called ZAMO, which just corotates with the absolute space, $\Omega \equiv \omega$. At the horizon, where $\alpha(r_H) = 0$, they also corotate with the black hole, $\Omega(r_H) = \Omega_H = \omega(r_H) = M_H a_H/(2r_H)$. In the case of a non–rotating black hole, $a_H = 0$, this also means that all matter is forced to non–rotation near the horizon, $\Omega_H = 0$, independent of the initial angular momentum. This fact is a fundamental difference between stellar objects and black holes. Matter accreting towards a non–rotating black hole must loose angular velocity until it reaches the horizon with $\Omega = 0$.

For this reason, a kind of extended boundary layer is built up in the inner region around a black hole (see Fig. 3), in distinction to very narrow boundary layers found around stellar objects. The rotation of accreting plasma probably deviates from standard Keplerian motion already at distances far away from the horizon (Fig. 3). These effects have not yet been studied in details and are crucial for understanding e.g. the emission from the inner part of accretion disks around rotating black holes.

3.2. ROTATIONAL ENERGY OF COMPACT OBJECTS

It is now well–known that the rotational energy of neutron stars

$$E_{\rm rot} = \frac{1}{2} I_* \Omega_*^2 \tag{10}$$

drives the pair winds of radio pulsars, I_* is the moment of inertia of the star and Ω_* its angular velocity. Most of the dissipated energy is transformed into kinetic energy of the pair wind with kinetic luminosity

$$L_w \leq \dot{E}_{\rm rot} = 4\pi^2 I_* \frac{-\dot{P}}{P^3} = \frac{E_{\rm rot}}{t_{1/2}}, \tag{11}$$

where $t_{1/2} = P/(-2\dot{P})$ is a measure for the life–time of a pulsar. This mechanism is a consequence of special relativistic MHD (Camenzind 1989b).

It is by far less well–known that a rotating black hole also has a free energy that can be liberated in magnetic processes. This is a consequence of the angular momentum J_H of a black hole. As usual in relativity, energies add quadratically

$$M_H = \sqrt{M_{\rm irr}^2 + \left(\frac{J_H}{2GM_{\rm irr}/c}\right)^2}. \tag{12}$$

The irreducible mass $M_{\rm irr}$ is a kind of rest mass for a rotating black hole that cannot be further reduced by any physical processes, except Hawking radiation – which is however unimportant for macroscopic black holes. Therefore, each black hole contains a rotational energy

$$E_{\rm rot} = (M_H - M_{\rm irr})c^2 = M_H c^2 \left(1 - \sqrt{\frac{1}{2}(1 + \sqrt{1 - (a_H/M_H)^2})}\right), \tag{13}$$

which can be extracted by means of some electrodynamic processes. Since $E_{\rm rot} \leq 0.29\, M_H c^2$, the rotational energy is a considerable amount of energy

$$E_{\rm rot} < 5 \times 10^{55}\ {\rm Watt\ s}\ \frac{M_H}{10^9\, M_\odot}, \tag{14}$$

and no other object can approach this upper limit for the rotational energy. As we have seen, the mass M_H of black holes residing in giant elliptical galaxies can easily exceed $10^9\, M_\odot$. If this rotational energy could be dissipated in a kind of pulsar process, this would represent a considerable luminosity

$$L_{\rm rot} \simeq \frac{E_{\rm rot}}{t_{\rm diss}} \simeq 1.6 \times 10^{40}\ {\rm Watt}\ \frac{M_H}{10^9\, M_\odot} \frac{10^9\, {\rm yr}}{t_{\rm diss}}, \tag{15}$$

essentially comparable to the mean luminosity of radio quasars at redshift of 2 – provided the system is able to dissipate the energy on a time–scale of a few billion years. This process could be behind the energetization of bright quasars, such as 3C 273 and 3C 345. In fact, Rees et et al. have already in 1982 suggested that the non–thermal power of radio–loud objects (quasars and radio galaxies) could be accounted for by this rotational energy. The way this energy can be extracted from the black hole is, however, still unknown. The dissipation of rotational energy would provide besides the classical accretion power a very efficient second energy channel.

The above energy loss rate is an interesting number which can be compared to the bulk kinetic power Q_j in jets estimated from the by–products of the jets, the large–scale lobes (Rawlings & Saunders 1991)

$$Q_j = \frac{k\,U}{\tau_j}\,. \tag{16}$$

Here U is the energy stored in the lobes, taken from equipartition energy U_{eq}, $k \simeq 2$ allows for $P\,dV$ work expended by the jet on pushing back and warming up the extended medium. τ_j is the age of the jet, estimated e.g. from the spectral age of the radio lobes. This amounts to a maximal jet power $\simeq 10^{40}$ Watt for bright radio quasars and NLRGs for $0.5 < z < 1.0$. On the other hand, material radiatively excited by an AGN cools by line emission. Radio galaxies are usually only narrow–line emitters and one can therefore easily estimate the total narrow–line luminosity L_{NLR} in all narrow–lines. Rawlings & Saunders (1991) obtained for an unbiased sample of FR II radio galaxies and low–power FR Is a correlation between the jet power and L_{NLR}

- $Q_j \propto L_{NLR}$ for FR I and FR II radio galaxies, as well as for radio quasars;
- $Q_j \simeq 100\,L_{NLR} \simeq 10^{36} - 10^{40}$ Watt;
- Radio–quiet quasars do not lie on the $Q_j - L_{NLR}$ correlation.

The last point strongly indicates that the jet power is some extra power provided e.g. by the rotational energy of the central source. But also radio–loud objects do have photoionizers which by virtue of $Q_j \propto L_{NLR}$ are controlled by the jet driving mechanism. It is also interesting that jet sources with given Q_j have a higher low–frequency luminosity when the sources are in a dense cluster environment. Since the above correlation extends over more than 4 orders of magnitude, this could just reflect the scaling of the central mass from $\simeq 10^6\,M_\odot$ in faint ellipticals to $\simeq 10^{10}\,M_\odot$ in the most giant ellipticals. If jet power were only related to rotational energy of the black hole, then its essentially the mass that dictates the jet power.

3.3. STATIONARY MHD FLOWS

The basic equations describing relativistic flows on a given background metric follow from the conservation of the energy–momentum tensor **T**, $T^{\alpha\beta}_{;\beta} = 0$,

$$T = (\rho + P)\, u \otimes u + P\, g + u \otimes q + q \otimes u + t \,. \qquad (17)$$

u is the 4–velocity, ρ the total energy density, P the pressure, q the heat and photon flux and t the viscous stress tensor (Straumann 1991). The symmetries of the Kerr background imply the existence of two conserved currents, the energy–current

$$P^{\alpha} = (\mu u_t)\, N^{\alpha} + P\, k^{\alpha} + t^{\alpha}_t + u^{\alpha} q_t + q^{\alpha} u_t \quad , \quad \nabla \cdot P = 0 \,, \qquad (18)$$

as well as the angular momentum–current

$$J^{\alpha} = (\mu u_{\phi})\, N^{\alpha} + P\, m^{\alpha} + t^{\alpha}_{\phi} + u_{\phi} q^{\alpha} \quad , \quad \nabla \cdot J = 0 \,. \qquad (19)$$

$N = nu$ is the conserved particle number current (mass conservation), $\nabla \cdot N = 0$, and $\mu = (\rho + P)/n$ the relativistic enthalpy, $\vec{k}$ and $\vec{m}$ are the Killing vectors of the Kerr geometry. As an application of these equations, we show disk accretion profiles for rapidly rotating black holes in Fig. 3 (Peitz 1994).

For a plasma in the one–fluid approximation, the energy–momentum tensor has to be supplemented by the electromagnetic energy–momentum tensor $\mathbf{T}^{\mathrm{EM}}$ given in terms of electric fields $\vec{E}$ and magnetic fields $\vec{B}$ as measured by ZAMOs. This tensor has the following energy density ϵ, Poynting flux $\vec{S}$ and stress–tensor t

$$\epsilon^{\mathrm{EM}} = \frac{1}{8\pi} \left(\vec{E}^2 + \vec{B}^2 \right), \qquad (20)$$

$$\vec{S}^{\mathrm{EM}} = \frac{1}{4\pi} \left(\vec{E} \times \vec{B} \right), \qquad (21)$$

$$t^{\mathrm{EM}} = \frac{1}{4\pi} \left(-\vec{E} \otimes \vec{E} - \vec{B} \otimes \vec{B} + \frac{1}{2} h \left(\vec{E}^2 + \vec{B}^2 \right) \right). \qquad (22)$$

Electric and magnetic fields are measured in Minkowski space with respect to static observers. This notion is useless in curved space, unless we specify some preferred frame of reference that has a global existence. The ZAMOs are the natural frames defined globally on Kerr space–time. One measures therefore electromagnetic fields with respect to ZAMOs. In the 3+1–split, Maxwell's equations assume the same form as in Minkowski space (Thorne et al. 1986)

$$\nabla \cdot \vec{E} = 4\pi \rho_e \,, \qquad (23)$$

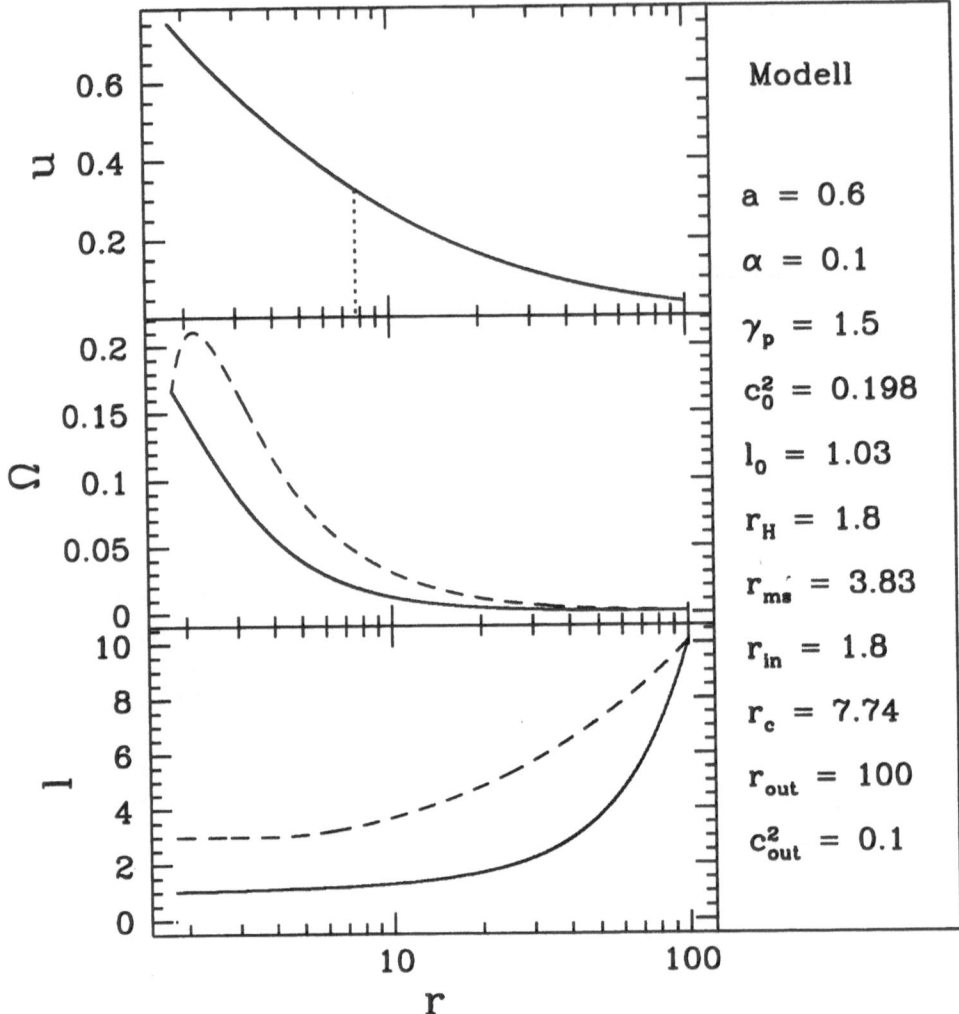

Figure 3. A transonic solution for the relativistic polytropic accretion towards a rapidly rotating black hole with Kerr parameter $a_H = 0.6$, viscosity parameter $\alpha = 0.1$ and $\Gamma = 1.5$ (Peitz 1994). Depending on viscosity, the rotation can become sub–Keplerian in an appreciable fraction of the disk (dashed curve: standard accretion disk model). The accretion velocity u is in units of the speed of light, the plasma angular velocity Ω in units of c^3/GM and the specific angular momentum l in units of GM/c. Hydrodynamic accretion still brings in a finite angular momentum, $l_H = GM/c$. In distinction to slow rotation, the maximum in the rotation law moves towards the horizon with increasing Kerr parameter a_H. The sonic transisition occurs at $7.54 GM_H/c^2$.

$$\nabla \cdot \vec{B} = 0, \tag{24}$$

$$\nabla \times (\alpha \vec{E}) = -\frac{1}{c}\frac{\partial \vec{B}}{\partial t} + \frac{1}{c}(\vec{B}_p \cdot \nabla \omega) R \, \vec{e}_\phi, \tag{25}$$

$$\nabla \times (\alpha \vec{B}) = \frac{1}{c}\frac{\partial \vec{E}}{\partial t} - \frac{1}{c}(\vec{E} \cdot \nabla \omega) R \, \vec{e}_\phi + \frac{4\pi}{c}\vec{j}, \tag{26}$$

except for two additional terms resulting from the interaction with the gravitomagnetic field. Faraday's equation has an additional induction term produced by shearing the poloidal magnetic field, and Ampère's law has a similar expression resulting from the shearing of electric fields. The redshift factor α just accounts for the transformation between local time and global time.

In addition to particle number conservation and angular momentum conservation, the conservation of total energy (required by stationarity of the problem) determines the poloidal velocity u_p along a given magnetic flux surface (Camenzind 1986a, b; 1987; 1991; 1994; Takahashi et al. 1990). In particular, in the case of high conductivity, plasma is guided by the structure of the axisymmetric flux surfaces $\Psi = const$

$$\vec{u} = \frac{\eta(\Psi)}{\alpha n}\vec{B} + \gamma(\Omega^F - \omega)\frac{R}{\alpha}\vec{e}_\phi, \tag{27}$$

and, as a consequnce of Faraday's induction law, the rotation of the field lines remains constant at $\Omega^F(\Psi)$ (Camenzind 1986a, b). Similar to the non-magnetic case, one can find a conserved energy–flow and conserved angular momentum flow vector field

$$P^\alpha = E(\Psi)\,N^\alpha \quad, \quad J^\alpha = L(\Psi)\,N^\alpha. \tag{28}$$

For stationary and axisymmetric plasma flows, the total energy E remains constant along a given magnetic flux surface $\Psi = const$

$$E(\Psi) = \mu\,(\alpha\gamma + \omega u_\phi) + \frac{\Omega^F I}{2\pi\eta} \tag{29}$$

and similarly, the total angular momentum also stays constant

$$L(\Psi) = \mu\,u_\phi + \frac{I}{2\pi\eta}. \tag{30}$$

$I = -\alpha c R B_\phi/2$ represents the total electric current flowing inside a region $\Psi = const$, $\mu = (\rho + P)/n$ is the relativistic enthalpy with $\mu \simeq mc^2$ for cold plasmas. The energy equation shows the coupling between the plasma angular momentum u_ϕ and the gravitomagnetic potential ω, α accounts for the redshift of the plasma energy. As in the Newtonian case, such plasma flows allow in general for the exchange of energy and angular momentum between the plasma and the magnetic fields.

In addition to these relations, an adequate equation of state is needed. For $n = n(P, s)$ and $T = T(P, s)$ the first law of thermodynamics implies

$$d\mu = \frac{1}{n} dP + T \, ds. \tag{31}$$

In particular, for $P = K_0(s) \, n^\Gamma$ with a polytropic index Γ, one finds

$$\mu = mc^2 + \frac{\Gamma}{\Gamma - 1} K_0(s) \, n^{\Gamma - 1}. \tag{32}$$

For given integrals of motion $E(\Psi)$, $L(\Psi)$, $\Omega^F(\Psi)$, $\eta(\Psi)$ and $K(\Psi)$ the conservation laws for stationary and axisymmetric plasma flows can be reduced to give relations for the Lorentz factor, the angular momentum and the poloidal current flux function $I(\Psi)$

$$\gamma = \gamma(\Psi, R) \quad = \quad \frac{E}{\alpha \mu \eta} \frac{\alpha^2(1 - x_A^2) - M^2(1 - \omega L/E)}{\alpha^2 - (\Omega^F - \omega)^2 R^2 - M^2}, \tag{33}$$

$$u_\phi = u_\phi(\Psi, R) \quad = \quad \frac{E}{\mu \eta} \frac{(1 - x_A^2)(\Omega^F - \omega)R^2 - M^2 L/E}{\alpha^2 - (\Omega^F - \omega)^2 R^2 - M^2}, \tag{34}$$

$$I = I(\Psi, R) \quad = \quad 2\pi \frac{\alpha^2 L - (\Omega^F - \omega)R^2(E - \omega L)}{\alpha^2 - (\Omega^F - \omega)^2 R^2 - M^2}, \tag{35}$$

where the quantity M, defined as

$$M^2 = \frac{4\pi \mu \eta^2}{n} = \frac{\alpha^2 u_p^2}{u_A^2}, \tag{36}$$

represents the Alfvén Mach number. The redshift factor α corrects for the singular behaviour of the poloidal velocity at the horizon of the black hole. The Alfvén Mach number M is a quantity well-behaved at the horizon. The quantity x_A defined by

$$x_A^2 = \frac{\Omega^F L}{E} \tag{37}$$

is a measure for the amount of energy carried by the electromagnetic fields. In special relativity, $x_A^2 \leq 1$, and $x_A^2 \equiv R_A^2/R_L^2$ is a direct measure for the position of the Alfvén surface in terms of the light cylinder radius R_L. For relativistic flows, $R_A \to R_L$. In the curved background of a black hole, this quantity is only a parameter that, however, still determines the position of the Alfvén surfaces (positions, where the denominators in the above equations vanish)

$$\alpha^2(R_A) - (\Omega^F - \omega(R_A))^2 R_A^2 = M_A^2 = 1 - x_A^2 . \tag{38}$$

This equation has in general, for given rotation Ω^F, two solutions, the outer Alfvén surface corresponding to the special relativistic one and an inner Alfvén surface near the horizon of a black hole. The light cylinder surfaces are special solutions of this equation for $x_A^2 = 1$. A rotating magnetosphere of a black hole has two light surfaces, the outer one slightly deformed by the underlying metric and an inner one due to the existence of the frame–dragging effect. For slow rotation, the outer light surface moves to infinity, and the inner one towards the static limit. In the special case, $\Omega^F = \Omega_H$, the inner light surface coincides with the horizon. Near a black hole, the quantity x_A^2 can exceed unity and even become negative, depending on the signs of L and E.

The regularity condition at the Alfvén point also determines the total angular momentum L as a function of the position of the Alfvén point

$$\frac{L}{E} = \frac{R_A^2(\Omega^F - \omega(R_A))(1 - x_A^2)}{\alpha^2(R_A)} \tag{39}$$

which simply reads in the Newtonian case as $L = R_A^2 \Omega^F$. At the same time, the Mach number has to satisfy

$$M_A^2 = 1 - x_A^2 . \tag{40}$$

The last two equations are well–known in Newtonian MHD ($\alpha = 1$, $\omega = 0$), where they form the basis for a treatment of axisymmetric MHD winds in stellar systems (for this see e.g. Paatz & Camenzind 1995)

$$j = R^2 \Omega = \frac{R^2 \Omega^F - M^2 L}{1 - M^2} , \tag{41}$$

$$RB_\phi = -4\pi\eta \frac{L - R^2 \Omega^F}{1 - M^2} . \tag{42}$$

In Newtonian MHD, the Mach number is simply $M^2 = v_p^2/v_A^2$. The first equation indicates that plasma is corotating with the field lines for low Mach numbers, $M^2 \ll 1$, and that the specific angular momentum j is equal to the total angular momentum L for high Mach numbers, $M^2 \gg 1$. These equations tell us that Newtonian MHD is only valid inside the light cylinder $R_L = c/\Omega^F$. This approximation is therefore not justified for rapidly rotating magnetic surfaces gnerated by accretion disks around black holes. It is even not justified for magnetic surfaces generated by rapidly rotating young stars (Fendt et al. 1995).

3.4. POLYTROPIC WIND EQUATION

In ideal MHD, plasma flows along the magnetic surfaces $\Psi = const$. Since the Lorentz factor, the angular momentum (or angular velocity Ω) and the poloidal current function I are essentially only functions of the radius along the flux surface and the Mach number, the normalisation of the plasma 4–velocity, $u_\alpha u^\alpha = -1$, leads to an algebraic equation either for the Mach number $M = M(R)$, or for the poloidal velocity $u_p(R)$

$$\frac{K}{R^2 D^2} = \frac{1}{16\pi^2} M^4 \frac{B_p^2}{\eta^2} + \alpha^2 \left(\frac{\mu}{E}\right)^2 , \tag{43}$$

where

$$D = \alpha^2 - R^2(\Omega^F - \omega)^2 - M^2 \tag{44}$$

and

$$\begin{aligned} K &= \alpha^2 R^2 (1 - x_A^2)^2 \left(\alpha^2 - R^2(\Omega^F - \omega)^2 - 2M^2\right) \\ &+ M^4 \left(R^2(1 - x_A^2)^2 - \alpha^2 (L/E)^2\right) \end{aligned} \tag{45}$$

For cold plasmas, $\mu = mc^2$, this is a polynomial of 4th order in M^2. Their solution branches have been discussed in Camenzind (1986) for Special Relativity, and by Takahashi et al. (1990) and Camenzind & Englmaier (1996) for accretion branches of rotating Kerr black holes. For a general polytropic equation with the specific enthalpy μ given by equation (32), this equation can still be reduced to a polynomial of a certain degree. This procedure has the advantage that all solution branches are known.

The wind equation (43) can be written in dimensionless form, when we introduce the *magnetization parameter* σ, defined as follows

$$\sigma(\Psi) = \frac{R_*^2 B_{p,*} c}{4\pi \eta(\Psi) R_L^2(\Psi)} , \tag{46}$$

where we introduced the light cylinder radius $R_L(\Psi)$ for a given flux surface

$$R_L(\Psi) = \frac{c}{\Omega^F(\Psi)} . \tag{47}$$

$\Psi_* = R_*^2 B_{p,*}$ is a measure for the magnetic flux included in the flux surface, for a given footpoint R_* of the flux surface. We normalize all radii in units of R_L, $x = R/R_L$, and obtain in this way the normalized wind equation

$$x^2 \frac{\bar{K}}{D^2} = \sigma^2 \Phi^2 M^4 + \alpha^2 x^4 \left(\frac{\mu}{E}\right)^2 , \tag{48}$$

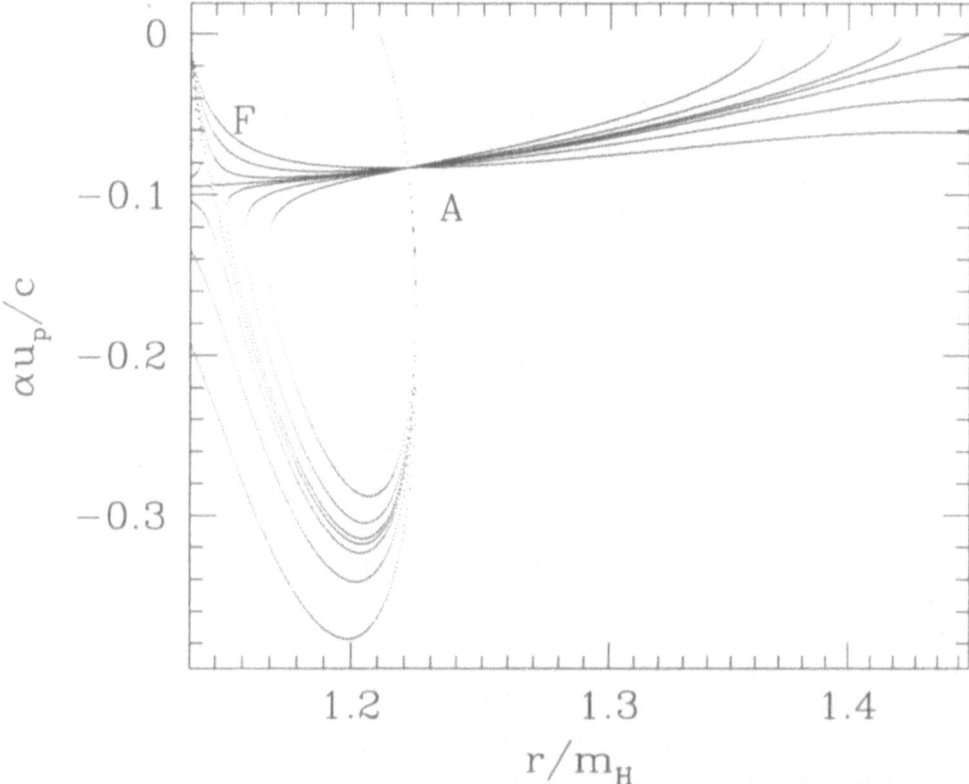

Figure 4. The accretion branches of the cold wind approximation on a rotating Kerr background (Camenzind & Englmaier 1996). For extremely rapidly rotating black holes, the Alfvén point (A) moves into the ergosphere ($r_A < 2M_H$) and the fast magnetosonic point (F) occurs immediately before the plasma falls through the horizon. In this case, the accretion brings in a total negative angular momentum.

where we used $\bar{K} = K/R_L^2$. This equation for the Mach number contains three essential parameters: the total energy E/μ, the angular momentum parameter $x_A^2 = \Omega^F L/E$ and the magnetisation paramater σ. In addition, the topology of the magnetic flux function enters into this equation over the flux–tube function Φ, defined as

$$\Phi = \frac{R^2 B_p}{R_*^2 B_{p,*}} = \Phi(\Psi, R)\,, \qquad (49)$$

which is constant for a monopole–type geometry. When the pressure of the plasma is not negligible, the quantity $K(\Psi)$ is a fourth parameter of the problem together with the polytropic index Γ. The above form of the wind equation shows that the Mach number is a well–behaved quantity on Kerr space. Since the last term vanishes at the horizon, $\alpha(r_h) = 0$, we get some constraint on the Mach number at the horizon itself, independent of the total energy E. Only the angular momentum parameter x_A^2 is important for this constraint.

One can easily study this dimensionless form of the wind equation, once the flux–tube function Φ is known. This is a nice exercise which can be done on any computer using some polynomial solver. Not all four parameters in the wind equation are free parameters due to the existence of critical points. The general relativistic wind equation has three critical points exactly as in the case of the Newtonian counterpart. There is no new element that appears for magnetic accretion onto rotating black holes. As a consequnce, these three critical points fix three of the four parameters in the general case. x_A^2 is fixed by the position of the Alfvén point, the total energy E by the position of the fast magnetosonic point and the plasma load parameter σ by the position of the slow magnetosonic point. Therefore, only the entropy parameter K is a free parameter that is usually fixed up by density n_* and temperature T_* of the injection region.

In the case of cold plasma flows, we have only three parameters, but only two critical points (the slow magnetosonic point coincides in this case with the injection radius). Cold plasma flows are in general very good approximations for relativistic ion flows, but certainly not for hot electron–positron winds ejected e.g. from rotating neutron stars. In this cold approximation, we can use the magnetisation as a free parameter of the problem. Its value depends on the strength of the magnetic field in the injection region, the rotation of the magnetosphere and the plasma load, or wind loss rate $\dot{M}$

$$\sigma \simeq \frac{\Psi_*^2}{2\dot{M}R_L^2} \, . \tag{50}$$

Under astrophysical conditions, the value of σ varies from extremely small values, $\sigma \simeq 10^{-8}$ for protostellar winds (Fendt et al. 1995) to $\sigma \simeq 1$ for radio galaxies and $\sigma \simeq 20$ for outflows from Quasars (Camenzind 1990, 1993).

As an example we consider here the accretion of a cold plasma towards a rotating black hole along some magnetic flux surface $\Psi = const$. As expected from the general discussion, this problem has three critical points. For plasma accretion from regions near the marginal stable orbit, the magnetisation is probably of order unity, since the magnetic field is anchored in the inner part of the disk. The plasma flows through the slow magnetosonic

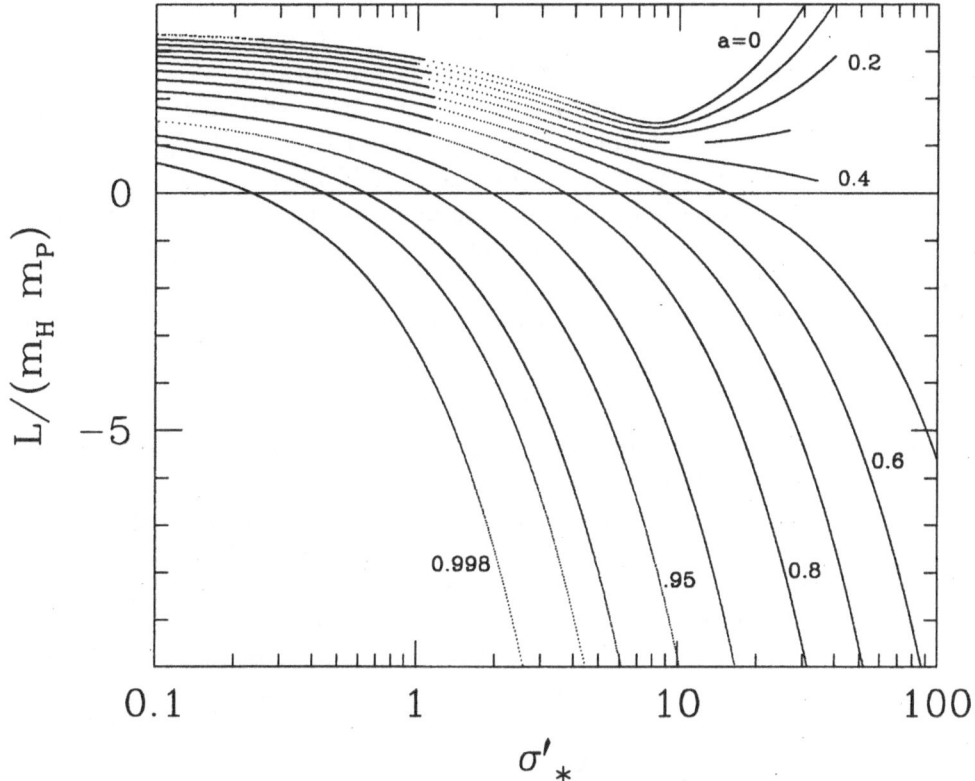

Figure 5. The total angular momentum in adiabatic magnetic accretion as a function of the magnetisation in the disk (Camenzind & Englmaier 1996). The Kerr parameter is indicated for the individual curves. For slowly rotating black holes, the total angular is always positive, and only for extremely rapidly rotating holes accretion with negative angular momentum can occur for a realistic magnetisation of order unity.

point (not shown here), the Alfvén point, and crosses the fast magnetosonic point just before it enters into the horizon (Fig. 4). For causal reasons, all these crtitical points must occur, but relativistic MHD adjusts itself to this behaviour.

The position of the Alfvén point is a question of the magnetisation σ and the spin of the black hole. For rapidly rotating black holes, the Alfvén point can move into the ergosphere, and the total angular momentum L of the accretion flow can become negative (Camenzind & Englmaier 1996), indicating the possibility of extracting angular momentum from the hole (Fig. 5). This is a general relativistic effect which has no analog in stellar accretion. In particular, when the fields are rotating slower than the horizon, $0 < \Omega^F < \Omega_H$, solutions can occur with either negative angular momentum, or negative total energy, or both (for $a_H > 0.95M_H$). Since the black hole rotates faster than the magnetic field lines, it can transfer via the gravitomagnetic coupling angular momentum to the plasma and from there to the fields. This process would deposit angular momentum in the inner disk which must be carried off by some disk wind. This is probably the injection mechanism for the strong jet sources.

3.5. AXISYMMETRIC MAGNETOSPHERES

In the case of stars or neutron stars, the magnetic flux is anchored in the stellar source itself. This is different in the case of black holes. The source of the magnetic flux is external to the black hole, most probably in the innermost part of the accretion disk itself (Khanna & Camenzind 1994; 1995). This means that we have a flux distribution $\Psi_d(R)$ along the disk surface. External to the disk, the flux comes into force–equilibrium with the boundary condition $\Psi = 0$ along the rotational axis. This force equilibrium can be derived from the equations of motion

$$\nabla_k T_a^k + \frac{1}{\alpha} S_\phi \nabla_a \omega + (\epsilon \delta_a^k + T_a^k) \frac{1}{\alpha} \nabla_k \alpha = 0 \tag{51}$$

The index $a = r, \theta$ is the poloidal index. This force–equilibrium has to be projected into the direction of the normal $\vec{n} \propto \nabla \Psi$ of the flux surfaces. After somewhat lengthy calculations, one can derive in this way the general relativistic Grad–Schlüter–Shafranov (GSS) equation for the form of the magnetic surfaces (Nitta et al. 1991; Beskin & Pariev 1994)

$$\frac{R}{\alpha} \nabla \cdot \left(\frac{D}{\alpha R^2} \nabla \Psi \right) + \frac{\Omega^F - \omega}{\alpha^2} |\nabla \Psi|^2 \frac{d\Omega^F}{d\Psi} +$$
$$\frac{32\pi^4}{\alpha^2 R^2} \frac{E^2}{M^2} \frac{\partial}{\partial \Psi} \left(\frac{G}{D} \right) - 16\pi^3 \mu n \frac{1}{\eta} \frac{d\eta}{d\Psi} - 16\pi^3 nT \frac{ds}{d\Psi} = 0. \tag{52}$$

Here

$$G = \alpha^2 R^2 (1 - x_A^2)^2 + \alpha^2 M^2 (L/E)^2 - M^2 R^2 (1 - \omega L/E)^2. \tag{53}$$

This equation is regular at the Alfvén surface $D = 0$, since there also $G = 0$ and all the other terms behave well at this surface.

This equation is the general relativistic extension of the GSS equation derived by Camenzind (1987) for special relativistic flows. The form of the equation does not change, when the 3+1–split is used. The gravitomagnetic forces only enter into the plasma motion. An interesting limit of this equation is obtained in the force–free limit, $M^2 \ll 1$ globally (Okamoto 1992; Haehnelt 1991)

$$\nabla \cdot \left(\frac{D}{\alpha R^2} \nabla \Psi\right) + \frac{\Omega^F - \omega}{\alpha} |\nabla \Psi|^2 \frac{d\Omega^F}{d\Psi} + \frac{16\pi^2}{\alpha R^2} I \frac{dI}{d\Psi} = 0. \qquad (54)$$

This treatment is well–justified for the description of coronal magnetic fields. Since $I = I(\Psi)$ this represents a highly non–linear equation which can be solved with a suitable FE–solver e.g. (Camenzind 1987; Haehnelt 1991; Fendt 1994). In general, only flux outside the inner light cylinder surface can escape towards the outer light cylinder surface, and carry, therefore, angular momentum away by means of some disk winds.

3.6. PROSPECTS FOR TIME–DEPENDENT MHD

Even in the axisymmetric case, the time–dependent Maxwell's equations (23–26) still represent a complicated system for the electromagnetic fields. For recent solutions of the gravitomagnetic dynamo equations see Khanna & Camenzind (1994; 1995).

4. Relativistic Jets as Self–Collimated Plasma Flows Driven by Rapid Rotation

The appearance of the light cylinder in rapidly rotating magnetospheres has a great effect on the collimation of magnetized winds. This can easily be shown by looking at the plasma confinement problem.

4.1. THE JET–PLASMA CONFINEMENT

Instead of using the magnetic flux function, the force–balance perpendicular to the magnetic surfaces can be formulated in terms of fields (Li et al. 1992; Appl & Camenzind 1993a; Li 1993)

$$\kappa \frac{B_p^2}{4\pi} \left(1 - M^2 - \frac{R^2}{R_L^2}\right) = \left((1 - \frac{R^2}{R_L^2})\nabla_\perp \frac{B_p^2}{8\pi} + \nabla_\perp \frac{B_\phi^2}{8\pi} + \nabla_\perp P\right)$$
$$- \frac{B_p^2 \Omega^F}{4\pi c^2} \nabla_\perp (R^2 \Omega^F) + \left(\frac{\mu n u_\phi^2}{R} - \frac{B_\phi^2}{4\pi R}\right)(-\vec{n} \cdot \nabla R). \qquad (55)$$

$\kappa = \vec{n} \cdot (\vec{B} \cdot \nabla)\vec{B}/B_p^2$ is the curvature of the poloidal field lines, M the Alfvénic Mach number, and $R_L = c/\Omega^F$ the light cylinder radius of the

720

rotating field lines, which is related to the mass M_H of a rapidly rotating black hole

$$R_L = c/\Omega^F \simeq 10\,GM_H/c^2 \simeq 0.01\,\mathrm{lyr}\,(M_H/10^{10}\,M_\odot)\,. \qquad (56)$$

When all radii are within the light cylinder, $R \ll R_L$, this equilibrium condition reduces to the classical plasma confinement equation: centrifugal forces, curvature forces, magnetic and plasma pressure gradients are balanced by the pinch force exerted by the toroidal field B_ϕ. The rapid rotation of the field lines brings in a new element: the electric field $E_p = (R/R_L)B_p$ can no longer be neglected. For this reason, the light cylinder appears as a critical point in the magnetic pressure. This has a profound consequence for the question of the existence of self–collimated solutions.

The existence of self–collimated solutions, where the pinch force balances all competing pressure forces, can easily be investigated for conical outflows (Appl & Camenzind 1993a,b). For this purpose, we return to the original Grad–Shafranov equation in the force–free limit with rigid rotation. In the asymptotic domain, $z \gg R_L$, the above equation reduces to the one–dimensional pulsar equation (rigidly rotating)

$$(1-x^2)\frac{d^2\Psi}{dx^2} - \frac{1+x^2}{x}\frac{d\Psi}{dx} + g\,\frac{dI^2(\Psi)}{d\Psi} = 0\,. \qquad (57)$$

The nature of the solutions only depends on the form of the dimensionless current distribution $I(\Psi)$ and the dimensionless coupling constant g (Appl & Camenzind 1993b). Due to the existence of the singular point at the light cylinder surface, $x = 1$, one can prescribe only one boundary condition, which is naturally chosen as $\Psi(x = 0) = 0$. The second condition follows from the singular point in this differential equation. This is the main reason for the existence of self–collimated solutions of the relativistic Grad–Shafranov equation. Li (1993) has constructed solutions of Equ. (55) by assembling together magnetic surfaces. For technical reasons, the boundary conditions at the disk surface had to be chosen in a somewhat unrealistic way. But these solutions clearly demonstrate the existence of self–collimation in the high σ domain. The 2D solution shown in Fig. 6 based on the GSS equation indeed converges towards the asymptotic solution of Equ. (57).

The pulsar equation shows the interesting property that for a too weak current, i.e. for $g < g_\infty$ no self–collimated solutions are possible. Approaching g from above, the jet radius moves to infinity. *Self–collimation requires some amount of current flowing in the jet.*

4.2. SELF–COLLIMATED SOLUTIONS OF THE GSS EQUATION

The boundary conditions along the disk surface are naturally given in terms of a flux distribution $\Psi_d(R)$. For this reason, it is more natural to solve directly the Grad–Schlüter–Shafranov equation (52) together with the wind equation (Camenzind 1987). This still represents a highly non–linear problem, and no analytical solutions are known, except for some linaer cases (Michel 1973; Lovelace et al. 1987). The non–linearity is extremely important for this problem.

Differential rotation of a magnetosphere produces an uncomfortable term in the GSS–equation. For numerical reasons, we constrained ourselves upto now to rigidly rotating magnetospheres $\Omega^F(\Psi) = \Omega_*$ for all flux surfaces. Also the general relativistic effects have been neglected, using $\alpha = 1$ and $\omega = 0$. Despite this restriction, the equation is still of some complexity (Fendt et al. 1995)

$$x\nabla \cdot \left(\frac{D}{x^2} \nabla \Psi \right) = -\frac{1}{\sigma^2} \frac{\gamma x}{M^2} (E' - \Omega L')$$
$$-\frac{1}{x} \left(I^2 + M^2 |\nabla \Psi|^2 \right) (\ln \sigma)' . \tag{58}$$

Its force–free version is somewhat simpler

$$x\nabla \cdot \left(\frac{D}{x^2} \nabla \Psi \right) = -g_{\mathrm{I}} \frac{1}{x} I I' . \tag{59}$$

A prime denotes the derivative with repsect to Ψ. $I(\Psi)$ is the total current enclosed by a flux surface. The wind equation along a flux surface has a simple polynomial form

$$\sum_{m=0}^{4} A_m(x; E, x_A^2, \sigma; \Phi) u_{\mathrm{p}}^m = 0 , \tag{60}$$

where the coefficients A_m depend explicitly on the magnetization σ, the flux tube function $\Phi = x |\nabla \Psi|$, and the flow parameters energy and angular momentum. The solutions of this equation along each flux surface provides us with the quantities $E(\Psi)$ and $L(\Psi)$, as well as the functions $M^2(\Psi, R)$ and $I(\Psi, R)$. A self–collimated 2D solution of the force–free GSS equation is shown in Fig. 6 (Fendt et al. 1995). The outermost flux surface is a free boundary and has to be found in such a way that the magnetic field structure does not show any kinks at the light cylinder (which is the Alfvén surface in the force–free limit).

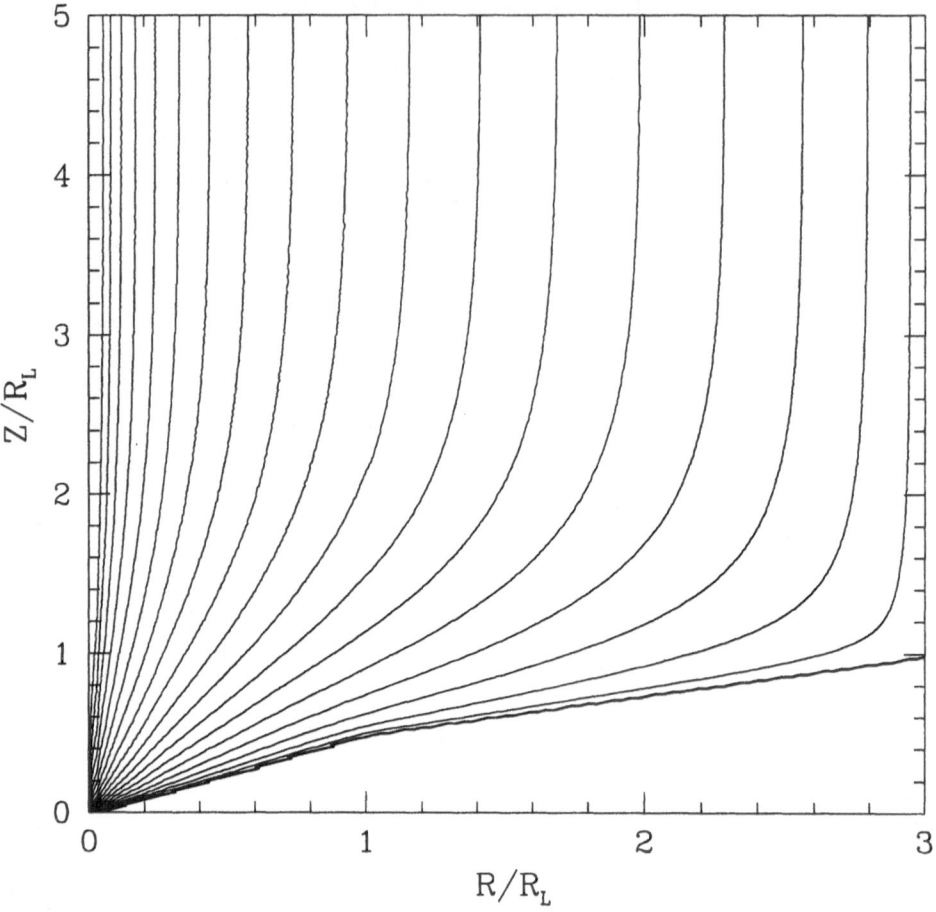

Figure 6.   The poloidal magnetic flux surfaces in a self-collimated disk outflow (Fendt et al. 1995). The last magnetic flux surface represents the outer boundary which is a free boundary surface essentially determined by the regularity of the field lines at the Alfvén surface (which coincides with the light cylinder in the force-free limit). This outer surface cannot be prescribed, but is determined by the self-collimation property of the solution.

4.3. PLASMA DIAGNOSTICS FOR PARSEC–SCALE JETS

Since the jet radii in relativistic outflows are only a few light cylinder radii, the jet plasma has still a high rotation velocity, typically about 10 percent of the speed of light. Plasma moves therefore along helical structures with low pitch angles. Such structures have been observed in the jets of bright quasars on the milliarcsecond scale (Krichbaum 1993; 1994). Emission from blobs moving along the parsec–scale jets would show then some quasi–

periodic modulation (Camenzind & Krockenberger 1992). In the last years, various observations have confirmed this prediction (Schramm et al. 1993; Wagner et al. 1995). It is fascinating that relativistic effects reduce the time–scales in such a way that one can observe the motion of plasma from a few light years to a few hundred parsecs within a few years.

5. Conclusions

Jets cannot be formed without magnetic fields. Though relativistic MHD is a very young field of research, compared e.g. with the activities in solar physics, it is my impression that we already made some progress in the last ten years. I hope that I have convinced everybody to use in the future relativistic MHD instead of the Newtonian one which must fail when outflows are driven by rapidly rotating objects. In this case, even general relativistic aspects are important as shown by the gravitomagnetic dynamo of rapidly rotating black holes.

Jets driven by accretion onto rotating black holes have radii which just scale with the Schwarzschild radius. In this sense, jets are found having radii of a few astronomical units in the Galactic center upto to a few light years in the brightest quasars. These structures are now resolvable using VLBI in the millimeter regime.

On the theoretical side, one would be happy to have some analytical solutions of the non–linear Grad–Schlüter–Shafranov equation. In the near future, it will be possible to show the existence of self–collimated outflows for some broader class of outflows including differential rotation, centrifugal forces and plasma pressure. I think it is also a question of time to include differential rotation into the 2D solution procedures.

We are, however, far away from having a time–dependent solver for the propagation of relativistic MHD jets. Only with such a scheme, we will be able to simulate the helical motion found in the jet of 3C 273, as well as the self–colllimated cocoon structure of this most extreme jet outflow in the Universe. The plasma processes that accelerate the electrons upto energies of 10^{12} eV must be related to these instabilities. The Universe provides us with an extraordinary jet lab in form of the radio jets of quasars and radio galaxies.

Acknowledgements: Many of the results discussed in this review grew out of numerous discussions with my collaborators in Heidelberg. Special thanks go to Stefan Appl, Ramon Khanna, Jonathan Ferreira, Martin Haehnelt, Martin Krockenberger, Peter Englmaier, Christian Fendt, Oliver Dreissigacker, Gernot Paatz and Jochen Peitz. Projects on jets and AGN were funded by our Sonderforschungsbereich 328 in Heidelberg.

References

Anile, A.M.: 1989, *Relativistic Fluids and Magneto–Fluids*, Cambridge Univ. Press (Cambridge)

Appl, S., Camenzind, M.: 1992, A&A **256**, 354

Appl, S., Camenzind, M.: 1993a, A&A **270**, 71

Appl, S., Camenzind, M.: 1993b, A&A **274**, 699

Bahcall, J.N. et al. (1995), Princeton University preprint

Beskin, V.S., Pariev, V.I.: 1993, Phys. Uspekhi, June

Blandford, R.D., Znajek, R.L.: 1977, MNRAS **179**, 433

Blandford, R.D.: 1992, in *Theory of Accretion Disks – 1*, eds. F. Meyer et al., Kluwer (Dordrecht), p. 35

Blandford, R.D., Netzer, H., Woltjer, L.: 1993, *Physics of Active Galactic Nuclei*, Springer–Verlag (Heidelberg)

Blandford, R.D., Payne, D.G.: 1982, MNRAS **199**, 883

Camenzind, M.: 1986a, A&A **156**, 137

Camenzind, M.: 1986b, A&A **162**, 32

Camenzind, M.: 1987, A&A **184**, 341

Camenzind, M.: 1989a, in *Accretion Disks and Magnetic Fields in Astrophysics*, ed. G. Belvedere, Kluwer (Dordrecht), p. 129

Camenzind, M.: 1989b, in *Neutron Stars and their Birth Events*, ed. W. Kundt, Kluwer (Dordrecht), p. 139

Camenzind, M.: 1990, in *Reviews of Modern Astronomy* 3, ed. G. Klare, Springer–Verlag (Heidelberg), p. 234

Camenzind, M.: 1991, in *Texas/ESO–CERN Symp. Rel. Astrophys., Cosmology and Fundamental Physics*, Ann. N.Y. Acad. Sci. **647**, 610

Camenzind, M.: 1993, in *The Jets of Radio Galaxies*, eds. H.J. Röser & K. Meisenheimer, Lecture Notes in Physics **421**, Springer-Verlag (Heidelberg), p. 109

Camenzind, M.: 1994, in *Theory of Accretion Disks – 2*, eds. W.J. Duschl et al., Kluwer (Dordrecht), p. 313

Camenzind, M.: 1995, in *Reviews of Modern Astronomy* 8, ed. G. Klare, Astron. Gesellsch. (Hamburg)

Camenzind, M., Krockenberger, M.: 1992, A&A **255**, 59

Camenzind, M., Englmaier, P.: 1996, A&A, submitted

Chandrasekhar, (1990), *The Mathematical theory of Black Holes*, Clarendon Press

Englmaier, P.: 1993, *Magnetische Akkretion auf rotierende Schwarze Löcher*, Diploma thesis, University of Heidelberg

Contopoulos, J., Lovelace, R.V.E.: 1994, ApJ **429**, 139

Fendt, Chr.: 1994, Ph.D. thesis, University of Heidelberg

Fendt, Chr., Camenzind, M., Appl, S.: 1995, A&A **300**, 791

Fichtel, C. et al.: 1994, ApJ Suppl. **94**, 551

Ford, H.C. (1994), *ApJ Lett.* **435**, L27

Friedman, J.L., Ipser, J.R., Parker, L. (1986), *ApJ* **304**, 115

Haehnelt, M.: 1991, diploma thesis, University of Heidelberg

Harms, R.J.: 1994, ApJ Lett. **435**, L35

Herold, H., Müther, H., Riffert, H., Ruder, H.: 1996, in *Springer Tracts in Modern Physics*, Springer–Verlag (Heidelberg), in press

Hirth, G.A., Mundt, R., Solf, J. (1994), *A&A* **285**, 929

Khanna, R.: 1993, Ph.d. thesis, University of Heidelberg

Khanna, R., Camenzind, M.: 1992, A&A **263**, 401

Khanna, R., Camenzind, M.: 1994, ApJ Lett. **435**, L129

Khanna, R., Camenzind, M.: 1995, A&A, in press

Krichbaum, T.P., Witzel, A., Graham, D.A. (1993), *A&A* **275**, 375

Krichbaum, T.P., Witzel, A., Standke, K.J., et al. (1994), in *Compact Extragalactic Radio Sources*, eds. J.A. Zensus & K.I. Kellermann, NRAO–workshop, Socorro, p. 39

Li, Zh.-Yu.: 1993, ApJ **415**, 118

Li, Zh.-Yu., Chen, T., Begelman, M.C.: 1992, ApJ **394**, 459

Lovelace, R.V.E., Wang, J.C.L., Sulkanen, M.E.: 1987, ApJ **315**, 504

Nitta, S., Takahashi, M., Tomimatsu, A.: 1991, Phys. Rev. **D 44**, 2295

Okamoto, I.: 1992, MNRAS **254**, 192

Peitz, J.: 1994, *Relativistische Akkretion auf rotierende Schwarze Löcher*, diploma thesis, University of Heidelberg

Pelletier, G., Pudritz, R.E.: 1992, ApJ **394**, 117

Rawlings, S., Saunders, R.: 1991, Nature **349**, 138

Röser, H.-J., Meisenheimer, K.: 1991, A&A **252**, 458

Schramm, K.-J., Borgeest, U., Camenzind, M. et al.: 1993, A&A **278**, 391

Straumann, N.: 1991, *General Relativity and Relativistic Astrophysics*, Springer–Verlag (Heidelberg), Chapter 9

Takahashi, M., Nitta, S., Tatematsu, Y., Tomimatsu, A.: 1990, ApJ **363**, 206

Thorne, Kip S., Price, R.H., MacDonald, D.M: 1986, *Black Holes: The Membrane Paradigm*, Yale University Press (New Haven)

Wagner, S.J., Camenzind, M. et al.: 1995, A&A, in press

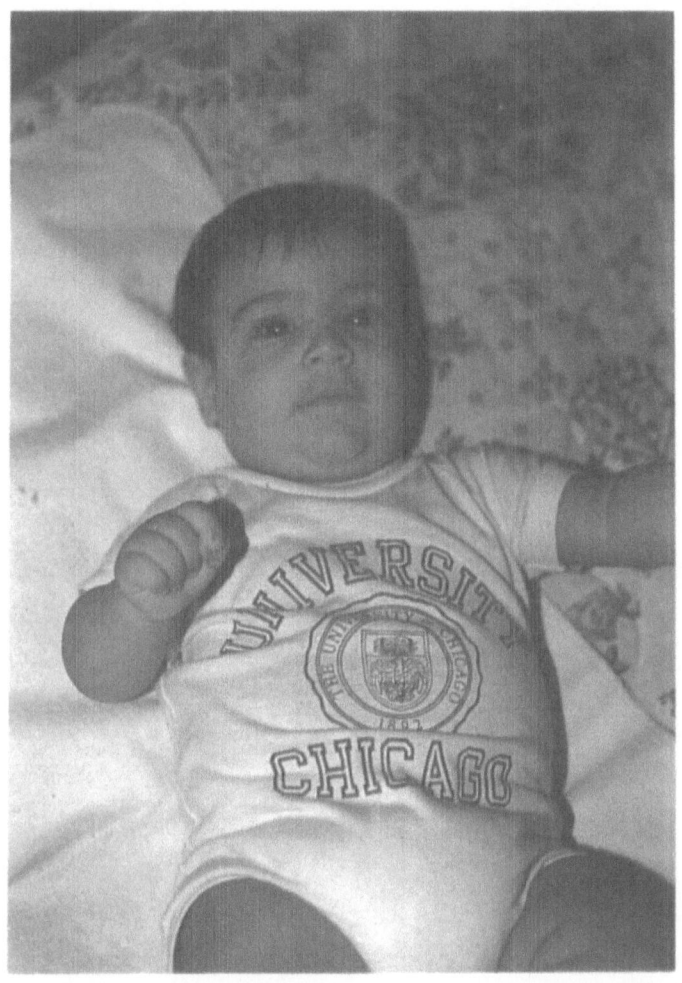

Stars are born like humans are. This is the picture of a baby who was actually born during the very same day at which starbirth was discussed in the Advanced Study Institute on "Solar and Astrophysical MHD Flows", June 19th, 1995.

INDEX

abundances
    electron, 529-530
    grain, 529-530
    ion, 529-530
acceleration
    centrifugal, 421
    gravitational, 328
    magnetic, 333
    magnetic-centrifugal, 317
    radial, 328
    radiative, 332
accretion disk, 363, 561, 643, 659,
       673, 702
    anomalous conductivity, 676
    instability, 664
    Keplerian flow, 674
    magnetic fields, 664, 676
    magnetized, 245
    relativistic e^-–beams, 679
    surface, 231
    thick, 230
    thin, 230
    turbulence, 659, 664, 674
    viscosity in, 240
acoustic
    oscillations, 64
    power, 64
    waves, 64
active
    region, 88, 95
        corona, 95
adiabatic process, 477
aerodynamic drag force, 21
AGN, 220, 643
    magnetic cauldron, 651
    powers, 609

radio-loud, 643
Alfvén
    disturbance, 334
    Mach number, 49, 326
    radius, 310
    speed, 310, 342
    wave-driven winds, 324
    waves, 324, 342
        resonant, 64, 67
        torsional, 242
Alfvén density, 463
Alfvén lengthscale, 532
Alfvén lever arm, 463
Alfvén Mach number, 387, 463
Alfvén radius, 463
Alfvén regularity condition, 388,
       467
Alfvén singularity, 388, 433
Alfvén surface
    spherical, 389
Alfvén waves, 475, 483, 492
    damping, 532
    torsional, 493, 507, 518
ambipolar diffusion, 476, 484, 491,
       502, 505-538
    effect on hydromagnetic waves,
       532, 535
    in star formation, 505, 538
angular momentum, 493, 507, 508,
       518-519, 526, 535
    conservation, 493, 518, 519, 524,
       535
    problem, 508, 535
        resolution, 508, 519, 523-527
    scaling with mass, 527
    specific, 524, 527, 535

732